ENCYCLOPEDIA OF TELECOMMUNICATIONS

ENCYCLOPEDIA OF
TELECOMMUNICATIONS

ROBERT A. MEYERS, Editor   TRW, Inc.

Academic Press, Inc.
Harcourt Brace Jovanovich, Publishers
San Diego New York Berkeley Boston
London Sydney Tokyo Toronto

ACADEMIC PRESS, INC.
San Diego, California 92101

United Kingdom Edition published by
ACADEMIC PRESS LIMITED
24-28 Oval Road, London NW1 7DX

Library of Congress Cataloging-in-Publication Data

Encyclopedia of telecommunications.

   1. Telecommunication—Dictionaries.   I. Meyers,
Robert A. (Robert Allen), Date.
TK5102.E645   1988   621.38   88-22343
ISBN   0-12-226691-9   (alk. paper)

PRINTED IN THE UNITED STATES OF AMERICA
88 89 90 91   9 8 7 6 5 4 3 2 1

EDITORIAL ADVISORY BOARD

CONTENTS

CONTRIBUTORS

AHAMED, SYED V. *City University of New York*. Intelligent Networks.

ALBER, ANTONE F. *Bradley University*. Videotex and Teletext.

ANDERSON, KENNETH D. *Naval Ocean Systems Center*. Radio Propagation (in part).

BAUMGARTNER, JR., GERALD B. *Naval Ocean Systems Center*. Radio Propagation (in part).

BERGER, U. S. *Bell Telephone Laboratories* (retired). Microwave Communications.

CAMPANELLA, S. J. *COMSAT Laboratories*. Satellite Communications.

CHERIN, A. H. *AT&T Bell Laboratories*. Optical Fiber Communications (in part).

DORDICK, HERBERT S. *Temple University*. Telecommunications.

FARVARDIN, N. *University of Maryland*. Source Coding, Theory and Applications.

FLANAGAN, J. L. *AT&T Bell Laboratories*. Digital Speech Processing (in part).

GIBBS, H. M. *University of Arizona*. Optical Circuitry (in part).

GIBSON, U. J. *University of Arizona*. Optical Circuitry (in part).

HAYKIN, SIMON. *McMaster University*. Communication Systems, Civilian.

HERSHEY, JOHN E. *BDM Corporation*. Signal Processing, General (in part).

HITNEY, HERBERT V. *Naval Ocean Systems Center*. Radio Propagation (in part).

JENKINS, W. K. *University of Illinois*. Signal Processing, Analog.

KING, RONOLD W. P. *Harvard University*. Antennas.

KREAGER, PAUL S. *Washington State University*. Data Communication Networks.

LIENTZ, BENNET P. *University of California, Los Angeles*. Computer Networks.

PAPPERT, RICHARD A. *Naval Ocean Systems Center*. Radio Propagation (in part).

PEYGHAMBARIAN, N. *University of Arizona*. Optical Circuitry (in part).

RABINER, L. R. *AT&T Bell Laboratories*. Digital Speech Processing (in part).

RICHTER, JUERGEN H. *Naval Ocean Systems Center*. Radio Propagation (in part).

RUSCH, W. V. T. *University of Southern California*. Antennas, Reflector.

SARID, D. *University of Arizona*. Optical Circuitry (in part).

SCHENKER, LEO. *AT&T Bell Laboratories*. Telephone Signaling Systems, Touch-Tone.

SEATON, C. T. *University of Arizona*. Optical Circuitry (in part).

SOBOLEWSKI, JOHN S. *University of Washington*. Data Transmission Media.

STEGEMAN, G. I. *University of Arizona*. Optical Circuitry (in part).

STUBBLEFIELD, ROBERT W. *AT&T Bell Labs*. Packet Switching.

SULLIVAN, EDMUND J. *Naval Underwater Systems Center*. Acoustic Signal Processing.

TARIYAL, B. K. *AT&T Technologies, Inc*. Optical Fiber Communications (in part).

TAYLOR, F. F. *AT&T Bell Laboratories*. Telecommunication Switching.

TAYLOR, FRED J. *University of Florida*. Signal Processing, Digital.

WARREN, M. *University of Arizona*. Optical Circuitry (in part).

WEIMER, PAUL K. *RCA Laboratories*. Television Image Sensors.

WEINSTEIN, STEPHEN B. *Bell Communications Research*. Voiceband Data Communications.

WEISZ, WILLIAM J. *Motorola, Inc*. Radio Spectrum Utilization.

YARLAGADDA, RAO. *Oklahoma State University*. Signal Processing, General (in part).

ZIEMER, RODGER. *University of Colorado at Colorado Springs*. Modulation; Multiplexing.

PREFACE

Telecommunications is the means by which information-bearing signals are transmitted from a source to a user by electromagnetic phenomena. It is clearly the basis for the emerging global information society.

The *Encyclopedia of Telecommunications* has been prepared in response to a well-defined need for a comprehensive, in-depth and yet concise treatment of the latest in telecommunications theory and applications for the intelligent layperson, advanced high school student, bachelor's degree candidate, graduate student, and practicing professional. Users of this *Encyclopedia* will include office managers in all fields of government and business (for informed selection among LAN, PBX, Videotex, Teletext, facsimile, Telemail, and other office communications systems); virtually all scientists and engineers (for better use of existing telecommunications systems used to transmit their scientific data and to choose and specify improved systems to advance communications with their colleagues and remote equipment); and telecommunications specialists (as a ready reference at an appropriate scientific level). The encyclopedia format was chosen to allow two modes of retrieving information: a major subject entry and the usual detailed subject index at the end of the volume.

A review of telecommunications books currently in print reveals that they either lack the depth of this publication or are so detailed as to be of use only to those already expert in the field.

This *Encyclopedia* treats each telecommunications discipline in a separate article, prepared by a recognized expert, that includes basic theory, derivations of mathematical relationships, hardware, history, status, and a forecast of future directions. Often this information is presented in a graphical format with photographs of devices presented as well as relational diagrams and electrical circuitry. The approximately 570-page volume contains more than 400 such illustrations, 50 tables, 200 bibliographic entries, and 300 glossary entries. Authors are affiliated with leading telecommunications firms such as AT&T Bell Laboratories, Bell Communications Research, RCA Laboratories, Motorola, and COMSAT Laboratories as well as a number of universities and government laboratories.

The articles average 20 pages in length and contain a glossary of terms specific to the subject covered. The glossaries will facilitate understanding by readers not familiar with the subject matter covered. Further, each article begins with an outline and an introductory definition of the subject, which together with the glossary allow the reader to understand the specific portions of the article containing the desired information. This approach is not found in any other book on telecommunications.

The articles cover, in an interlocking manner, the three major components of telecommunications: (1) the transmitter, (2) the communications channel, and (3) the receiver. Complete communications systems are presented in articles titled "Communication Systems, Civilian"; "Computer Networks"; "Data Communication Networks"; "Intelligent Networks"; "Microwave Communications"; "Satellite Communications"; "Telecommunications"; "Telephone Signaling Systems, Touch-Tone"; "Videotex and Teletext"; and "Voiceband Data Communications." Transmission and reception are covered in the articles "Antennas"; "Antennas, Reflector"; and "Digital Speech Processing." Transmission is discussed in the articles "Acoustic Signal Processing"; "Modulation"; "Multiplexing"; "Packet Switching"; "Signal Processing, Analog"; "Signal Processing, Digital"; "Signal Processing, General"; "Source Coding, Theory and Applications"; "Telecommunication Switching"; and "Television Image Sensors." The communications channel is presented in articles titled "Data Transmission Media"; "Optical Circuitry"; "Optical Fiber Communications"; "Radio Propagation"; and "Radio Spectrum Utilization."

Robert A. Meyers

ACOUSTIC SIGNAL PROCESSING

Edmund J. Sullivan   *Naval Underwater Systems Center*

GLOSSARY

A/D Converter: Device that samples an analog (A) quantity at some (usually) periodic rate and converts it to a digital (D) quantity of a specified number of bits.

Biologics: Biological sources of acoustic noise and scattering (usage in underwater acoustics and sonar).

Convolution: The convolution of two functions is the integral of the product of two functions where one of the functions has its independent variable reversed in sign. The convolution of the two functions $f(x)$ and $g(x)$ is given by

$$c(y) = \int_0^y f(x)g(y-x)\,dx$$

Decibel notation: Logarithmic means of representing a quantity as compared with a reference quantity. For example, if $10\log(A/B) = +25\ (-25)$, then A is said to be 25 decibels (dB) "above" ("below") B.

Deconvolution: Inverse process of convolution. If $c(y)$ is the convolution of $f(x)$ and $g(x)$, then the deconvolution would be the process that recovers $g(x)$ $(f(x))$ given $f(x)$ $(g(x))$ and $c(y)$.

Dirac delta function: Usually designated $\delta(x)$, this function (more properly called a distribution) has the properties that it is zero everywhere except at $x = 0$, and its integral over the range $-a$ to $+a$, where a is any number greater than zero, is equal to unity.

Doppler shift: Shift in the frequency of a source, as perceived by the listener, that is caused by any relative motion along the line between the source and the listener.

Gaussian process: Any statistical process governed by the Gaussian probability density function. The Gaussian probability density function for a random variable x is given by

$$\frac{1}{\sigma\sqrt{2\pi}}\,e^{-x^2/2\sigma^2}$$

Here, σ is called the standard deviation of the process.

Gyre: Large-scale closed circulatory system (oceanography usage).

Nyquist rate: Lowest rate at which a finite-bandwidth signal can be periodically sampled in order to reproduce the signal completely and faithfully. The Nyquist rate is equal to twice the bandwidth of the signal.

Platform: Physical location of some relevant component of a sonar system, such as an acoustic array. It commonly designates a ship or submarine.

Seismics: Study of the structure of the earth by the use of sound.

Tomography: Process whereby a three-dimensional object is sectioned into plane sections for viewing. It is usually accomplished by performing an inverse scattering process using electromagnetic, acoustic, or X radiation.

White noise: Noise whose spectrum is "flat," that is, constant as a function of frequency. It is characterized by an autocorrelation function that is a Dirac delta function.

Acoustics is the science of sound, including its production, transmission, and reception. Sound itself is the propagation of mechanical energy in a material medium. As such, it can carry information. Acoustic signal processing is the discipline concerned with the separation of desired acoustic information from undesired information, or "noise."

I. Basics

The transfer of acoustic information can be conveniently represented by a model with three parts: the source, the medium, and the receiver. The source transmits acoustic energy of a given form through the medium, and some of this energy, along with unwanted energy, or "noise," is picked up by the receiver. In practice, such systems can be found in either of two modes: the active or the passive. Examples of these two modes are realized in shipborne sonar (*sound navigation and ranging*) systems. An active sonar system can be found in two forms. These are referred to as monostatic and bistatic and are depicted in Fig. 1A,B. The monostatic system has the source and receiver located on the same platform, and indeed both may employ the same acoustic transducer. The key point is that, for active systems, the source is under the control

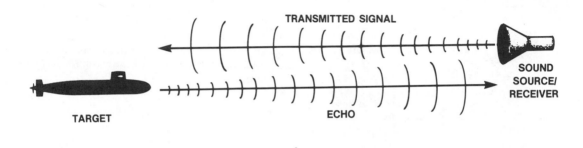

A

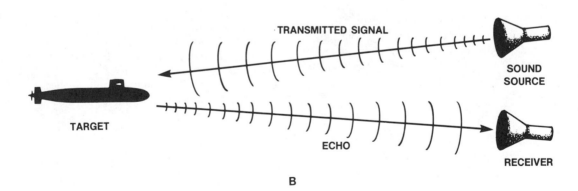

B

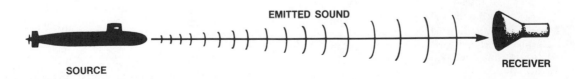

C

FIG. 1.   Sonar system configurations. A, Active monostatic system; B, active bistatic system; C, passive system.

of the listener. In a passive sonar system, only the receiver is under the control of the listener, as in the case of a surveillance system that listens for sounds that may emanate from a submarine. This is shown in Fig. 1c.

These ideas can be made clearer by comparing sonar systems with radar (*radio detection and ranging*) systems, which are always active. A radar system transmits electromagnetic energy and, as in sonar systems, receives scattered or reflected energy both monostatically and bistatically. Generally, there is no passive radar, however, since the "listener" must always provide the "illuminating" energy.

The comparison with radar systems can also be useful when discussing the medium. Both sonar and radar systems can suffer degraded performance due to scattering from matter in the medium and scattering from boundaries. The problem is somewhat more severe in the sonar system, however, since the boundaries are usually in closer proximity, and particulate matter and biologics in the medium are a major source of reverberation.

Another major difference between the two systems arises from the fact that sonar system performance is more profoundly affected than that of radar systems by the rather strong dependence of the speed of sound both on temperature and on pressure. Consequently, unlike the situation with radar systems, straightline propagation is essentially impossible in most cases in sonar systems. (In some acoustic applications other than sonar, such as seismics and tomography, the importance of the medium remains a major concern.)

Once the acoustic energy reaches the receiver, the signal-processing task begins. Acoustic signal processing is concerned with both detection and estimation. Detection is defined as the determination of the existence or nonexistence of a signal at the receiver. Estimation, which is a somewhat broader term than detection, is the quantification of certain parameters or descriptors of the signal, the medium, or the contents of the medium. [*See* SIGNAL PROCESSING, DIGITAL; SIGNAL PROCESSING, GENERAL.]

An example of detection is the determination of the presence of a submarine in a volume, termed the detection volume, by an active sonar system. That is, the only concern is the existence or nonexistence of an echo arising from the detection volume and due to a submarine.

On the other hand, the determination of the existence of flaws in a piece of metal by the use of high-frequency sound would require more information from the signal than simply its existence. At the very minimum, an estimate of the time delay of the reflected acoustic pulse would be necessary.

The question of estimation leads to a class of acoustic signal-processing problems called inverse problems. This is not a well-defined concept and is best described in terms of its relation to the "forward" problem. The forward problem can be stated as follows. If the source transmits acoustic energy of a given form through the medium, what will the receiver receive? The inverse problem is concerned with the question: If the receiver receives a signal of a given form, what does this tell us about the medium, its contents, or the source?

Examples of the forward problem are sonar, audio systems, and communication systems. The inverse problem is concerned with acoustic tomography, ocean bottom profiling, seismic deconvolution, and sonar target classification. This perspective can be represented mathematically by Eq. (1).

$$f(\mathbf{x}) = \int K(\mathbf{x}, \mathbf{x}')g(\mathbf{x}')\,d^3\mathbf{x}' \qquad (1)$$

If Eq. (1) represented an acoustic scattering problem, then $K(\mathbf{x}, \mathbf{x}')$ would represent the acoustic energy scattered from a differential volume $d\mathbf{x}'$ or "source" at the point denoted by the vector coordinate $\mathbf{x}'$, and $g(\mathbf{x})$ would represent the distribution of sources and therefore the distribution of the scattering medium. The scattered field would be given by the function $f(\mathbf{x})$. The forward problem, then, would be concerned with determining $f(\mathbf{x})$ given some information about $g(\mathbf{x})$ and $K(\mathbf{x}, \mathbf{x}')$. An example of an inverse acoustic scattering problem would be the determination of $g(\mathbf{x})$ given some information about $K(\mathbf{x}, \mathbf{x}')$ and measured values of $f(\mathbf{x})$. Sometimes $K(\mathbf{x}, \mathbf{x}')$ is called the Green's function and here would describe the acoustic field scattered from a point "source."

Because the inverse problem is concerned with extracting a great deal more information from the signal than simply knowledge about its existence or nonexistence or the estimation of a few parameters, it can require a very computationally intensive process. In most cases the degree of computation is so great that only with the advent of modern computer technology has such processing become feasible.

Acoustic signal processing can be two-dimensional in the sense that it involves both spatial and temporal aspects. This is in contrast to certain communications problems, for example, that are concerned only with the time and frequency domains. The temporal aspect of the problem concerns itself with the behavior in time of the signal or its spectral characteristics. The spatial aspect concerns itself mainly with the directional properties of the signal. This leads to the study of arrays or antennas whose spatial extent and directional sensitivities are of interest. [See ANTENNAS.]

Signal-processing engineers are, in many cases, concerned with the design of optimal or suboptimal systems. This basically leads to the design of optimal or suboptimal filters, that is, devices that efficiently extract the desired information in the time or frequency domain. The receiving array or antenna can be thought of as a spatial filter. A functional example of a spatial filter is the two-element array formed by the human ears, which enables us to determine the direction of incoming sound.

Acoustic signal processing is a methodology leading to many applications. These include audio systems, communications, acoustic holography, acoustic pattern recognition, acoustic tomography, seismics, ocean bottom profiling, nondestructive testing, speech recognition and synthesis, and sonar target classification. In spite of the broad range of applications, acoustic signal processing lends itself to a rather well defined formal exposition, which is the subject of the next section.

II. Signal Processing in the Time Domain

A. TOTALLY KNOWN SIGNALS

In the case of acoustic signal processing one is interested in at least two kinds of signals. These are deterministic signals that are corrupted by noise and random signals that are corrupted by noise. Examples of deterministic signals corrupted by noise are active sonar and nondestructive testing with high-frequency sound. In both cases, a known deterministic signal is transmitted and subsequently received after being acted on by the medium and its inclusions and corrupted by noise. An example of the second kind of signal, a random signal corrupted by noise, is the case of a passive sonar listening to the acoustic emissions from a submarine.

When considering such signals, there are two levels of knowledge that one may seek. These are commonly referred to as detection and estimation. In detection, as discussed in the introduction, one simply desires to know whether the signal is present at the receiver. In the case of estimation, one seeks to estimate one or more parameters of the signal. In the case of active sonar detection, one can determine the likelihood of the presence of an echo. Estimation then might be the estimation of the amplitude, frequency, or time of arrival of the echo. More complicated cases of estimation fall under the category of inverse problems.

We begin with the problem of detection of deterministic signals corrupted by noise. Formally, detection is cast as a problem in hypothesis testing. Basically, one forms two hypotheses:

$$H_0 \equiv \text{hypothesis that signal is not present (null hypothesis)} \quad (2a)$$

$$H_1 \equiv \text{hypothesis that signal is present} \quad (2b)$$

In terms of the received signal x, the actual signal s, and the noise n, these two hypotheses can be expressed as

$$H_0; x = n \quad (3a)$$

$$H_1; x = s + n \quad (3b)$$

Expressing the probability of occurrence of x under the hypothesis that H_1 is true as P_1, and the probability of occurrence of x under the hypothesis that H_0 is true as P_0, a decision is made based on the likelihood ratio Λ. That is, if

$$\Lambda = P_1/P_0 > \eta \quad (4)$$

where η is a number to be determined, it is assumed that the signal is present. Clearly, this decision is strongly influenced by η. Decreasing η will increase the probability that the decision "signal present" will be made. However, the cost of doing so will increase the probability that such a decision is wrong. Such wrong decisions are referred to as "false alarms." The strategy is to choose η as low as possible without exceeding an acceptable false-alarm rate (FAR). The procedure, then, is to (1) determine acceptable FAR, (2) determine η based on FAR, (3) determine Λ, and (4) compare Λ with η.

The determination of Λ is the task of the processor. Such a processor, that is, one whose output is the likelihood ratio Λ, is referred to as an optimum processor. As a simple example of these ideas, we choose a case where the signal is a constant equal to unity and the noise is a zero-mean Gaussian process with variance $= \sigma^2 = 2$. Thus,

$$p_0 = \frac{1}{\sqrt{4\pi}} \exp\left[-\frac{x^2}{4}\right] \qquad (5)$$

$$p_1 = \frac{1}{\sqrt{4\pi}} \exp\left[-\frac{(x-1)^2}{4}\right] \qquad (6)$$

These probability densities are shown in Fig. 2.

The probability that the signal exceeds the threshold x_T is equal to the area labeled region 2. Thus, the probability of false alarm is given by

$$P_{FA} = \int_{x_T}^{\infty} \frac{1}{\sqrt{4\pi}} \exp\left[-\frac{x^2}{4}\right] dx \qquad (7)$$

Similarly, the probability of detection, given that the signal is present, is given by

$$P_D = \int_{x_T}^{\infty} \frac{1}{\sqrt{4\pi}} \exp\left[-\frac{(x-1)^2}{4}\right] dx \qquad (8)$$

Choosing a false-alarm probability of 0.1, we can solve Eq. (7) for x_T. The result is $x_T = 1.8$. Since the likelihood ratio can also be written in terms of the probability density functions, η is given as

$$\eta = \frac{\exp[-(1.8-1)^2/4]}{\exp[-(1.8)^2/4]} = 1.9 \qquad (9)$$

Thus, if η is picked to be 1.9, the probability of false alarm will be 0.1 and the probability of detection will be

$$P_D = \frac{1}{\sqrt{4\pi}} \int_{1.8}^{\infty} \exp\left[-\frac{(x-1)^2}{4}\right] dx = .285 \qquad (10)$$

These results are valid for a single observation of x.

Although this is a simple example, the procedure to be followed in more complicated cases is, in principle, the same. The mathematics, however, becomes more complex.

On examination of Fig. 2, it can be seen that the greater the separation of the mean for signal present from that for signal absent, the higher the probability of detection can be for a given false-alarm rate. This separation is conveniently represented by the detection parameter d. For the example just given, d is defined as

$$d = (m_1 - m_0)^2/\sigma^2 \qquad (11)$$

where m_1 is the mean with signal present, m_0 the mean with signal absent, and σ the standard deviation of noise. The quantities P_D, P_{FA}, and d are commonly combined and represented by a family of curves called receiver operating characteristics, or ROC curves, an example of which is shown in Fig. 3.

The preceding example is a binary decision problem; that is, there are only two hypotheses to be considered. The example is based on a single sample. In the case of time-varying signals, however, there are potentially an infinite number of samples available. It is possible, however, to reduce this case to a binary decision problem. An example of this is shown in the sequel.

Suppose one were to make N observations (samples) of a continuous time-varying signal. If these observations were independent samples of

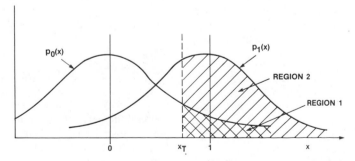

FIG. 2. Probability densities for the two corresponding hypotheses. Here x_T is the threshold value. The area designated region 1 is equal to the false-alarm probability, and the area designated region 2 is equal to the probability of detection.

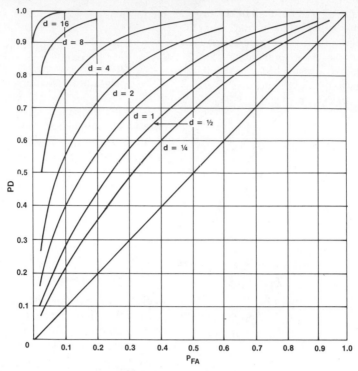

FIG. 3. Receiver operating characteristics. The probability of detection is plotted as a function of the probability of false alarm for several values of the detection parameter d. [From Urick, R. J. (1967). "Principles of Underwater Sound for Engineers," McGraw-Hill, New York.]

a Gaussian process, the joint probability densities for the two hypotheses H_0 and H_1 would be given by the products of the densities for each sample. That is,

$$p_0 = \prod_{i=1}^{N} \frac{1}{\sigma\sqrt{2\pi}} \exp\left[-\frac{x_i^2}{2\sigma^2}\right] \qquad (12)$$

$$p_1 = \prod_{i=1}^{N} \frac{1}{\sigma\sqrt{2\pi}} \exp\left[-\frac{(x_i - s_i)^2}{2\sigma^2}\right] \qquad (13)$$

where x_i and s_i are the ith measurement and signal samples, respectively, and $\prod_{i=1}^{N}$ denotes multiplication of all N densities. The likelihood ratio is then given by

$$\Lambda = \frac{\displaystyle\prod_{i=1}^{N} \exp\left[-\frac{(x_i - s_i)^2}{2\sigma^2}\right]}{\displaystyle\prod_{i=1}^{N} \exp\left[-\frac{x_i^2}{2\sigma^2}\right]} > \eta \qquad (14)$$

where the $1/\sigma\sqrt{2\pi}$ term has been canceled. This relation can be conveniently expressed in logarithmic form. Equation (14) then can be expressed as

$$\frac{1}{\sigma^2}\sum_{i=1}^{N} x_i s_i > \frac{1}{2\sigma^2}\sum_{i=1}^{N} s_i^2 + \ln\eta \qquad (15)$$

Making the assumption that the samples are taken at a frequency equal to twice the bandwidth of the noise (Nyquist sample frequency) and further assuming that the noise is white, that is, its power spectral density is equal to a constant, $N_0/2$, for all frequencies, the sums in Eq. (15) can be well approximated by integrals for the case of large bandwidth. One then finds

$$\frac{2}{N_0}\int_0^T x(t)s(t)\,dt > \frac{E}{N_0} + \ln\eta \qquad (16)$$

Here,

$$E = \int_0^T s^2(t)\,dt \qquad (17)$$

is the energy of the signal. Here $N_0/2$, the power spectral density of the noise, is equal to the noise power in a 1-Hz band. In arriving at Eq. (16) we have used the fact that, for the case of Nyquist sampling, $\sigma^2 = (N_0/2) \times$ sample frequency.

The integral of the product of $x(t)$ and $s(t)$ is simply the correlation of the received signal with the known signal. This is to be compared with a threshold that is determined by η (which in turn is determined by P_{FA}, the false-alarm probability) and the signal-to-noise ratio. The point here is that, in spite of the fact that the observation vector can be very large, the decision statistic is a single number, the sufficient statistic. The detection parameter d now becomes

$$d = 2E/N_0 \qquad (18)$$

The quantity $2E/N_0$ is the signal-to-noise ratio. Here, it is to be noted that the decision is based on only the sum (an integral) of the observations, weighted by a replica of the known signal. This is the sufficient statistic. Also, the ROC curves, shown in Fig. 3, now hold for this case with d as given above, where d depends only on the signal energy and the noise power spectral density.

The correlation process in Eq. (16) when directly implemented is referred to as a replica correlator. Its realization as a filtering process is expressed by setting

$$s(t) = h(T - t) \qquad (19)$$

where $h(t)$ is the time domain representation of the matched filter. That is, the optimium (matched) filter is simply the time-reversed version of the replica and the process can be seen to be equivalent to a convolution in the time domain. The integration is carried out to time $t = T$, and the comparison for decision is then made. This is the optimum processor for totally known signals.

B. Not Totally Known Signals

The matched filter also plays an important role in the case of not totally known signals. In the case of unknown phase, the optimum detector turns out to be a square-law detector. This is usually realized in terms of a quadrature receiver, shown in Fig. 4. Here, it can be seen that the output is simply the sum of the squares of the outputs of two matched filters, where the signals being matched are $\sin \omega t$ and $\cos \omega t$, where $\omega = 2\pi f$ and f is the frequency of the signal. Clearly, it can be seen that the phase information in the signal is discarded, as one would intuitively expect. This can be viewed as an "incoherent matched filter."

Further cases, although more complicated, all involve either the coherent or the incoherent matched filter in their implementation. For the case of completely unknown signals, Gaussian noise, low values of the signal-to-noise ratio, and large time–bandwidth products, the detection parameter is

$$d = BT(S/N)^2 \qquad (20)$$

where B is bandwidth, T integration time, S signal power, and N noise power. The term BT is usually referred to as the time–bandwidth product and occurs frequently in signal-processing formalism.

C. Estimation

So far, only the case of pure detection has been considered. When one is concerned with a determination of one or more of the signal pa-

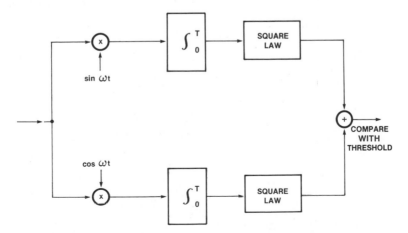

FIG. 4. Quadrature detector. This constitutes the optimum linear processor for the case of a signal of unknown phase. [From Whalen, A. D. (1971). "Detection of Signals in Noise," Academic Press, New York.]

rameters, the situation is conventionally referred to as a problem in estimation. In some cases of detection, an estimation of one or more of the signal parameters is obtained as a fringe benefit, but the original goal was purely that of detection. Here we are concerned with the direct estimation of one or more of the signal parameters as the actual goal. The subject of estimation covers a large area. For the purposes of the subject discussed here, only the so-called maximum likelihood estimate will be discussed, since it is the one most pertinent to the subject. The maximum likelihood estimate is based on finding the maximum of the likelihood function. This is a conditional probability density function that can be expressed as $p(x/a)$, where $p(x/a) \, dx$ is the probability that x lies in the interval dx given the parameter a.

The acoustic maximum likelihood problem is posed as follows. Given a received signal

$$x(t) = S(t, \alpha) + n(t) \qquad (21)$$

it is desired to estimate the value of the parameter α.

Assuming that the noise is Gaussian and white, the likelihood function can be written

$$p(x/\alpha) = A \exp \left\{ -\frac{1}{N_0} \int_0^T [x(t) - S(t, \alpha)]^2 \, dt \right\} \qquad (22)$$

This is simply a continuous-time representation of Eq. (13). The quantity A is a normalization factor and is of no concern to us here. Upon maximizing this function with respect to the parameter α, the result is

$$\int_0^T [x(t) - S(t, \alpha)] \frac{\partial S(t, \alpha)}{\partial \alpha} \, dt = 0 \qquad (23)$$

The estimation of the amplitude offers a simple example of how Eq. (23) is used. If the signal is written as

$$S(t, \alpha) = \alpha S_0(t) \qquad (24)$$

it quickly follows that the estimate is the solution of

$$\int_0^T [x(t) - \alpha' S_0(t)] S_0(t) \, dt = 0 \qquad (25)$$

or

$$\alpha' = \int_0^T x(t) S_0(t) \, dt \Big/ \int_0^T S_0^2(t) \, dt \qquad (26)$$

Note that, for a given integration time T, the denominator can be considered to be a constant,

and thus the amplitude estimator is equivalent to a matched filter. If $S_0(t)$ is picked such that

$$\int_0^T S_0^2(t) = 1 \qquad (27)$$

then the mean of the estimate is given as

$$E\{\alpha'\} = E \left\{ \int_0^T [\alpha S_0(t) + n(t)] S_0(t) \, dt \right\} = \alpha \qquad (28)$$

since the mean of $n(t)$ is zero. Here E indicates the expected value operation. When the expected value of the estimated parameter is equal to the true value, the estimator is said to be unbiased.

The preceding example was concerned with the estimate of a single parameter. The process can be generalized to the estimation of multiple parameters.

A further increase in the complexity of the estimation problem leads to a class of problems generally referred to as inverse problems. Examples of acoustic inverse problems are tomography, or three-dimensional viewing via "sectioning" of an object; acoustic imaging; ocean bottom profiling; and seismics.

There exists no clear definition of when a given problem is an inverse problem. As mentioned in the introduction, an inverse problem is generally deemed to be inverse to a forward problem. It usually is more computationally intensive than the forward problem. The general linear acoustic inverse problem is conveniently discussed in terms of Eq. (1), which is reproduced below in two forms.

$$f(\mathbf{x}) = \int K(\mathbf{x}, \mathbf{x}') g(\mathbf{x}') \, d\mathbf{x}' \qquad (1a)$$

$$f_i = \sum_i K_{ij} g_j \qquad (1b)$$

For convenience, we assume that Eq. (1) represents the general linear acoustic scattering problem. Here $g(\mathbf{x})$ represents the "source" or scatterer distribution, $K(\mathbf{x}, \mathbf{x}')$ represents the scattering from a differential volume, $f(\mathbf{x})$ represents the scattered field, and $\mathbf{x}, \mathbf{x}'$ are position vectors.

The acoustic scattering problem leads to an equation of the form of Eq. (1a) as follows. Starting from the inhomogeneous Hemholtz equation, we have

$$(\nabla^2 + k^2)\phi(\mathbf{x}) = s(\mathbf{x}) \qquad (29)$$

Here, $\nabla^2 = \partial^2/\partial x^2 + \partial^2/\partial y^2 + \partial^2/\partial z^2$ in Cartesian coordinates. This operation is known as Laplacian; k, as before, is the wave number and is given by $k = 2\pi/\lambda$, where λ is the wavelength. An integral equation solution for the scattered field $\phi_s(x)$ can be found as

$$\phi_s(\mathbf{x}) = \int s(\mathbf{x}') \frac{e^{ik|\mathbf{x}-\mathbf{x}'|}}{|\mathbf{x} - \mathbf{x}'|} \, d\mathbf{x}' \qquad (30)$$

where $\phi(x)$ in Eq. (29) is the total acoustic field and is given by

$$\phi(\mathbf{x}) = \phi_i(\mathbf{x}) + \phi_s(\mathbf{x}) \qquad (31)$$

and $\phi_i(\mathbf{x})$ is the incident field. Equation (30) simply states that the scattered field at the point denoted by the vector $\mathbf{x}$ is that component scattered from a differential volume $d\mathbf{x}'$ located at $\mathbf{x}'$ summed over the volume distribution $s(\mathbf{x})$. The problem, then, is to invert the system to obtain $s(\mathbf{x})$, which is a description of the distribution of the scatterers.

Because of its generality, the inverse problem does not lend itself to a formal structure in the same sense as do detection and estimation. It is probably more realistic to consider it a collection of techniques. There are, however, some insights that can be gleaned from examination of Eq. (1). First, it is clear that it is generally an integral equation problem. Except for rare special cases, solution in closed form will not be possible, so that a numerical solution will be called for. If the data are in discrete form, it becomes a problem in linear equations. Also, there can arise severe mathematical problems in the solution of such problems. One of these is that the problem can be what mathematicians refer to as ill posed. This is a problem that can be due to noisy data, not enough data, or a combination of the two. A problem is well posed when a unique solution exists that is stable to changes in the data. The problem of nonuniqueness can, in many cases, be dealt with by introducing a priori information.

The stability problem arises when, given the existence of a solution, the solution is extremely sensitive to small perturbations in the data. As an example, consider a measurement of $f(x)$ denoted by $f(x_1)$. Let this measurement incur an error Δf, and let f_0 be the true value. Then,

$$f(x_1) = f_0 + \Delta f$$
$$= \int K(x_1, x)[g(x) + \Delta g] \, dx \qquad (32a)$$
$$f_0 = \int K(x_1, x)g(x) \, dx \qquad (32b)$$

Subtracting Eq. (32b) from (32a) yields

$$\Delta f = \int K(x_1, x) \, \Delta g(x) \, dx \qquad (33)$$

It can be seen from Eq. (33) that Δf can be thought of as a weighted average of $\Delta g(x)$, where $K(x_1, x)$ is the weight. Thus, one is free to select a function $\Delta g(x)$ whose weighted average is as close to zero as desired, but can produce large errors on $g(x)$.

A heuristic explanation of how the scattering problem can be ill posed is to note that the scattered field, which is the source of the observed data, can depend rather weakly on rather large changes in the scattering object or objects. Thus, a small amount of noise or simply the addition of a new datum can represent a very large fictitious change in the scattering object, which is the object of interest.

A second problem that can arise is that an inordinate computational load may ensue. It is, as previously mentioned, this very problem that prevented many inverse problems from being solved until the advent of modern computational capabilities. As a result, much of the effort expended in the inverse problem is focused on particular formulations and algorithms.

III. Spatial Processing and Antenna Arrays

A. ACOUSTIC TRANSDUCERS

As previously mentioned, acoustic signal processing is a two-dimensional problem in that it involves both time domain processing and spatial processing. The spatial-processing aspect of the problem concerns itself with the directional aspects of the transmission and reception of the sound. This is accomplished by means of transducers. The term *transducer* is used in many ways. It sometimes means a single device; other times it is used to describe an array of devices. Historically, the transmitting transducer was called a projector and the receiving transducer or sensor was called a hydrophone. Here, the term *transducer* means a single device that is either a transmitter or a receiver, or sometimes both. A collection of transducers is called an array.

Most receiving transducers are usually configured such that their output is a voltage that is proportional to the acoustic pressure at the transducer. For this reason, the pressure is usually the variable of interest. When discussing the flow of acoustic energy, it is convenient to define the acoustic intensity. This is the average rate of flow of energy across a unit area, where the direction of flow is normal to the area. For the case of a fluid medium, the pressure and intensity of a plane wave are related by

$$I = \overline{\rho^2}/\rho c \qquad (34)$$

where $\overline{\rho^2}$ is the time average of squared pressure, ρ the density of the fluid medium, and c the speed of sound in the medium.

Excluding audio microphones, there are generally four types of transducers: electromagnetic, piezoelectric, magnetostrictive, and optical. As suggested by its name, the electromagnetic transducer operates as a projector by utilizing the magnetic force produced by an electromagnet in whose coil an appropriately modulated electric current exists. An electromagnetic receiving transducer uses the principle of magnetic induction. The propagating acoustic pressure wave causes a motion of a permanent magnet in a coil. An electric current is then produced in the coil. A magnetostrictive transducer uses the fact that nickel and some nickel alloys change their dimensions under the action of a changing magnetic field. This effect, in similarity to the electromagnetic effect, is reciprocal (i.e., straining the magnetostrictive material changes the magnetic field and thereby produces an electric current); hence, there are magnetostrictive receiving transducers. Piezoelectric transducers, which are also reciprocal, make use of the phenomenon that certain materials, such as quartz crystals, Rochelle salt, barium titanate, and some polymers (e.g., polyvinylidene fluoride) change their dimensions under the action of an electric field. And since it is a reciprocal effect, straining these materials produces an electric field. Optical transducers are not reciprocal. They operate only as receivers of sound. They generally fall into two categories: coherent and incoherent. Coherent optical transducers detect sound by using a change in the phase of very narrowband light energy produced by a laser. Although this effect can be produced in many ways, one way would be to use the acoustically induced motion of a mirror to change the optical path length of light compared with a reference beam of light whose path length is not affected by the sound. When the two beams are combined, the sound is detected as a change in the optical interference pattern. Incoherent optical transducers depend on a change of light intensity caused by the sound for their operation. Again, this can be accomplished in many ways, one of which is to use a cladding on an optical fiber whose index of refraction is affected by the sound pressure. The cladding is a material surrounding the fiber that has an index of refraction less than that of the fiber and is therefore responsible for the capacity of the fiber to contain the light with little leakage. This type of transducer modulates the light in the fiber by acoustically controlling the amount of light that leaks out of the fiber.

B. ACOUSTIC ARRAYS

By combining acoustic transducers in an array, one can control the directional properties of acoustic transmission and reception. For example, in many cases it is desirable to transmit sound only in a particular direction. Similarly, it may be important to receive sound only from a particular direction, to separate an acoustic signal from noise, or to resolve received sound into its various sources. In the case of reception (which is all that is considered here) such operations on the received sound waves are referred to as spatial processing and are accomplished by a combination of an array of transducers and a beam former.

Consider Fig. 5, where plane waves of sound pressure, assumed to be at a single frequency, are arriving with a direction of propagation given by θ. The N point receiving transducers are equally spaced with spacing d in a line whose length is given by $L = d(N - 1)$. The quantity L is commonly referred to as the aperture of the array. The output of each transducer is assumed to be a voltage whose value is proportional to the instantaneous acoustic signal, where the acoustic signal at the hydrophone labeled $n = 0$ is taken to be the real part of

$$A(t) = A_0 e^{i\omega t} = A_0[\cos(\omega t) + i \sin(\omega t)] \quad (35)$$

where $\omega = 2\pi f$ and f is the frequency of the sound. The nth transducer, then, has output voltage given by

$$V_n(t) = V_n \exp[i\omega(t + \tau_n)] \qquad (36)$$

where it is easily shown that

$$\tau_n = nd \sin \theta/c \qquad (37)$$

Here, $c = f\lambda$ is the speed of sound and λ the wavelength.

Substitution of Eq. (37) into Eq. (36) yields

$$V_n(t) = V_n \exp(i\omega t + inkd \sin \theta) \qquad (38)$$

where the relation $k = 2\pi/\lambda$ has been used. The quantity k is commonly called the wave number and can be thought of as a spatial frequency.

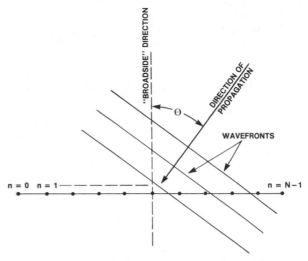

FIG. 5. Line array of uniformly spaced point elements. The plane waves are impinging on the array at an angle θ from the 0°, or "broadside," direction.

Since we are interested only in the spatial aspects of the problem, we shall set the time t equal to zero to simplify matters. The outputs of the N transducers are now added together, resulting in

$$V(\theta) = \sum_{n=0}^{N-1} V_n \exp(inkd \sin \theta) \qquad (39)$$

Equation (39) describes the spatial (angular) response function of a linear array of equally spaced transducer elements to a plane-wave acoustic signal impinging on it at angle θ. The term $nkd \sin \theta$ is the phase of the nth output.

The absolute magnitude of Eq. (39) is plotted in Fig. 6 for the case $d = \lambda/2$ and all V_n equal. This function is approximately equal to $|\sin(\xi)/\xi|$. As the spacing d approaches zero and the number of elements approaches infinity, the approximation approaches an identity.

A great deal can be learned about the properties of linear arrays by inspection of Eq. (39) and Fig. 6. First, the array has a region of maximum response to signals centered at $\theta = 0°$ (broadside). This region of maximum response is usually referred to as the main lobe, and the other maxima are called sidelobes. Second, the response drops off rapidly as the signal is moved away from broadside. The power response is down to about one-half of the maximum when

$$\frac{N}{2} kd \sin \theta' = \frac{\pi}{2} \qquad (40)$$

$\Delta\theta = 2\theta'$ is referred to as the half-power beamwidth.

Solving for this beamwidth from Eq. (40) results in

$$\Delta\theta \simeq \lambda/Nd \qquad (41)$$

where the approximation $\sin \theta \simeq \theta$ for small θ has been used. Since $Nd \simeq L$, for large N, it follows from Eq. (41) that

$$\Delta\theta L' \simeq 1 \qquad (42)$$

where $L' = L/\lambda$ is the aperture in units of wavelength. This is sometimes referred to as the beamwidth–aperture product and shows that, to

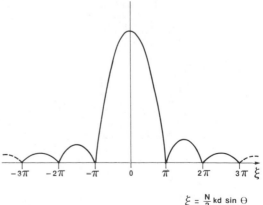

$$\xi = \frac{N}{2} kd \sin \Theta$$

FIG. 6. Response function of a line array of equally spaced point elements.

achieve a given beamwidth, the nondimensional aperture must be approximately equal to the reciprocal of the desired beamwidth in radians. Finally, by inspection of Eq. (39), it is seen that $V(\theta)$ is a periodic function that repeats at values of θ that satisfy

$$\sin \theta = \pm q \frac{\lambda}{d} \qquad (43)$$

where q is an integer. Thus, if d/λ is equal to unity, then the $V(\theta)$ has other maxima at $\theta = \pm 90°$ (end fire). For larger values of d/λ, more maxima occur. This phenomenon is called aliasing and is a consequence of the aperture being undersampled. In practice, it would cause an ambiguity in the determination of the arrival angle of a signal. (These extra main lobes are sometimes called grating lobes.) For this reason, array elements are usually spaced at $d = \lambda/2$. This is the spatial version of Nyquist sampling.

On further inspection of Eq. (39) it is seen that the maximum response occurs where the relative phases between the element outputs is zero. Thus, the main lobe can be "steered" to any desired angle θ_s by introducing appropriate phase shifts δ_n, where

$$\delta_n = nkd \sin \theta_s \qquad (44)$$

Thus, the expression for a line array of equally spaced elements, steered to angle θ_s, is

$$V(\theta, \theta_s) = \sum_{n=0}^{N-1} V_n \exp(inkd \sin \theta - i \delta_n) \qquad (45)$$

or

$$V(\theta, \theta_s) = \sum_{n=0}^{N-1} V_n \exp[inkd(\sin \theta - \sin \theta_s) \qquad (46)$$

This is a simple example of beam forming and is illustrated in Fig. 7.

Rather than phase shifts, which have meaning only for the single-frequency case as treated here, the beam steering can be more generally effected by time delays, where

$$\tau_n = \frac{nkd \sin \theta_s}{2\pi f} = \frac{nd \sin \theta_s}{c} \qquad (47)$$

Equation (39) can be generally represented as an integral by defining an aperture function $V(x)$, namely,

$$V(\theta) = \int_0^L V(x) \exp(ikx \sin \theta) \, dx \qquad (48)$$

For the example of the uniformly spaced line array, the aperture function is

$$V(x) = \sum_{n=0}^{N-1} V_n \delta(x - nd) \qquad (49)$$

where $\delta(x)$ is the Dirac delta function. On observation of Eq. (48), it is seen that $V(x)$ and $V(\theta)$ are Fourier transform pairs with transform variables $\kappa = k \sin \theta$ and x. That is,

$$V(\kappa) = \int_0^L V(x) \exp(i\kappa x) \, dx \qquad (50)$$

It is in this context that the acoustic array can be

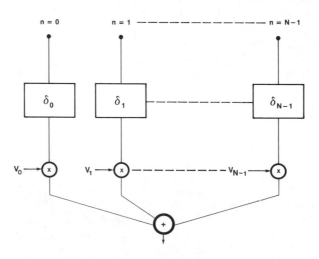

FIG. 7. Phase-shift beam former. After the appropriate phase shifts are applied, the element outputs are multiplied by the weights, or "shading coefficients," before being summed.

seen as a spatial filter, and the beamwidth aperture product naturally follows. Extending this most useful and illuminating analogy, we rewrite Eq. (46) as

$$V(\theta, \theta_s) = \sum_{n=0}^{N-1} V'_n \exp(-inkd \sin \theta_s) \quad (51)$$

where V'_n represents the output of the nth element; that is,

$$V'_n = V_n \exp(inkd \sin \theta) \quad (52)$$

If we make the identification

$$\sin \theta_s = s\lambda/Nd \quad (53)$$

where s is an integer, Eq. (51) becomes

$$V_s = \sum_{n=0}^{N-1} V'_n \exp(-i2\pi ns/N) \quad (54)$$

which is seen to be identical to a discrete Fourier transform of the element outputs, where the spectral lines are the beam outputs. For the case of broadband acoustic signals, Eq. (54) becomes

$$V_{sj} = \sum_{n=0}^{N-1} V'_{nj} \exp(-i2\pi ns/N) \quad (55)$$

Here, the index j labels the jth spectral output of a discrete spectrum analysis. This approach is referred to as $k-\omega$ beam forming, a possible real-ization of which is depicted in Fig. 8. Here, the element outputs undergo a discrete Fourier transform in the time domain to yield a set of narrowband signals, each of which is then Fourier transformed in the spatial domain to yield the beam set.

C. ARRAY OPTIMIZATION

Drawing again on the concept of the array as a spatial filter, the V_n in Eq. (39), sometimes referred to as shading coefficients, can be looked on as spatial filter coefficients. The beam-forming problem can now be stated in a more general form as the selection of the sets $\{V_n\}$ and $\{\delta_n\}$ to achieve an optimum performance, based on some set of criteria, against a given acoustic situation. There are two different approaches to this problem: pattern optimization and gain optimization. Pattern optimization proceeds first by the selection of a desired array response pattern that is deemed efficient for the given purpose and then, via some analytical or numerical technique, the selection of values of V_n to best achieve this goal. A commonly desired pattern is one with a narrow main lobe and low sidelobes. This will give good performance against noise that is uniformly distributed in angle for an array that seeks to detect a signal arriving from a particular direction. A common example of this is

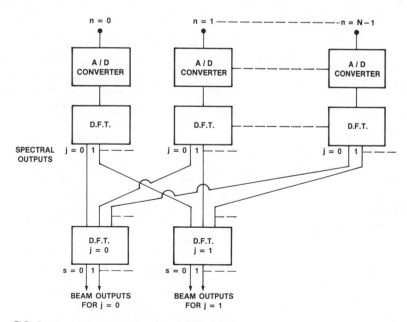

FIG. 8. Frequency domain or $k-\omega$ beam former. The number of outputs is equal to the product of the number of beams and the number of spectral outputs (D.F.T., discrete Fourier transform).

the so-called Chebyshev shading, which gives the lowest possible sidelobe height for a given beamwidth, given that all the sidelobes are constrained to be of the same height. Chebyshev shading is applicable only to a line array of uniformly spaced elements. When the array configuration becomes more complex, the behavior of the array response pattern is also more complex, such as in the case of arrays that are mounted on the curved surfaces of hulls or other platforms (usually called conformal arrays).

The gain optimization approach can be stated as follows. Given a particular noise field and signal, maximize the gain of the array, where the gain is defined as the signal-to-noise ratio of the array divided by the signal-to-noise ratio of a single element.

Suppose the noise amplitude is described by $\eta(\theta)$. Then, from Eq. (39), the total response of the array to this noise field is found by integrating over all directions, namely,

$$R_{\text{noise}} = \sum_{n=0}^{N-1} V_n \int d\Omega \ \eta(\theta) \exp(inkd \sin \theta) \quad (56)$$

where $d\Omega$ represents the differential solid angle. The average noise power output is given by

$$P_{\text{noise}} = \sum_{n=0}^{N-1} V_n \sum_{m=0}^{N-1} V_m M_{n,m} \quad (57)$$

where

$$M_{n,m} = \int d\Omega \ \overline{|\eta(\theta)|^2}$$
$$\times \exp[i(n-m)kd \sin \theta] \quad (58)$$

The overbar denotes the average.

Equation (58) is derived from Eq. (57) under the assumption that the noise field can be represented by spatially uncorrelated point sources. The optimization requires finding those values for the V_n that maximize the gain G, where

$$G = \frac{\text{signal power}}{\text{noise power}} = \frac{\displaystyle\sum_{n=0}^{N-1}\sum_{n=0}^{N-1} V_n V_m S_{n,m}}{\displaystyle\sum_{n=0}^{N-1}\sum_{n=0}^{N-1} V_n V_m M_{n,m}} \quad (59)$$

Here, $S_{n,m}$ is the signal matrix and, for the case under consideration, is given by

$$S_{n,m} = \exp[i(n-m)kd \sin \theta_s]$$

Here, θ_s is the angle of the direction of the incoming plane-wave signal from broadside. Although the gain of an array is defined as the

signal-to-noise ratio of the array divided by the signal-to-noise ratio of a single element, here, for simplicity, we assume that the signal-to-noise ratio of a single element is equal to unity so that G is also equal to the signal-to-noise ratio of the array as well as the gain.

In Eq. (59), $M_{n,m}$ is the covariance matrix of the noise, whose elements are proportional to the correlation of the noise outputs between the nth and mth elements. The conclusion to be drawn from Eq. (59) is that it is desirable to have uncorrelated noise and highly correlated signals. In the case where the noise is completely uncorrelated and the signal is perfectly correlated, the gain is

$$G = N \quad (60)$$

That is, the gain is equal to the number of elements in the array. This can be intuitively understood by noting that the signal amplitudes add coherently so that the signal power output is proportional to N^2, whereas the noise amplitudes add incoherently, giving a noise power output proportional to N. Equation (60) is usually written in decibel notation, that is,

$$G_{\text{dB}} = 10 \log N \quad (61)$$

When the noise is spatially uniform, the gain is called the directivity factor and, in decibels, it is called the directivity index.

D. ADAPTIVE ARRAYS

Adaptive processing, which plays a role in most areas of signal processing, also finds application in array processing. Generally, adaptive processing uses information about the noise to maintain the optimality or near optimality of the system. As can be seen in Eq. (58), the covariance matrix is determined by $\eta(\theta)$, which is a description of the noise field. Since the optimal gain is determined by the covariance matrix, any changes in $\eta(\theta)$ would require a recomputation of the V_n. A sampling of the noise field is thereby required. In practice, this sampling could be effected by a set of auxiliary beams, either from the array itself or an auxiliary array. In some cases, a "suboptimal" approach is sufficient. In the case of a source of noise that is spatially localized, that is, one emanating from a single direction different than that of the signal, the output of the auxiliary beam that is receiving the noise can be subtracted from the output of the beam former. As intuitively expected, this would effectively produce a notch or "null" in the array response function. For this reason, adaptive processing against localized noise sources is sometimes called null steering.

IV. Noise and the Medium

A. BACKGROUND

The medium can affect an acoustic-signal-processing problem in two ways. First, the propagation properties of the medium may vary a great deal from place to place, or the medium may be strongly absorptive. Second, the medium may be a source of noise since contained matter or its boundaries can scatter the sound, or there may be sources of unwanted sound in the medium, such as biologics, shipping, wind, and rain. Also, in the high-frequency end of the spectrum, thermal noise, which is due to the thermal motion of the molecules, can be an important source of oceanic noise. Although the medium can be a problem in most acoustic-signal-processing problems, such as absorption of high-frequency sound by human flesh in the case of medical acoustic imaging, the case of the impact of the ocean on sonar problems will be emphasized here, since the ocean is a difficult—if not the most difficult—medium. As such, it has been extensively studied and can provide an excellent example of the problems encountered.

Figure 9 is a plot of the ambient noise in the ocean due to three sources: shipping, wind, and thermal sources. Here it is seen that which of the three sources is of importance depends on the frequency band of interest. Thus, in the case of most sonar problems, shipping and wind are important sources, whereas the thermal source becomes important only at relatively high frequencies. Two other sources of ambient noise in the ocean are rain and biologics.

A further source of noise, one that is important in the case of active sonar, is reverberation. As discussed, reverberation can occur due to scattering of sound from particulate matter and biologics in the volume of water. This is, appropriately enough, called volume reverberation. Reverberation from both the surface and bottom is called boundary reverberation. Reverberation can be particularly problematical since its level is proportional to the level of the transmitted signal. Thus, increasing the energy of the signal, which can give an increase in gain over ambient noise, is of no use in the case of reverberation. The level of the noise is given in terms of its spectrum level. This is the level in a 1-Hz band of the noise, which is considered to be broadband. A spectrum level of 10 dB would be expressed as

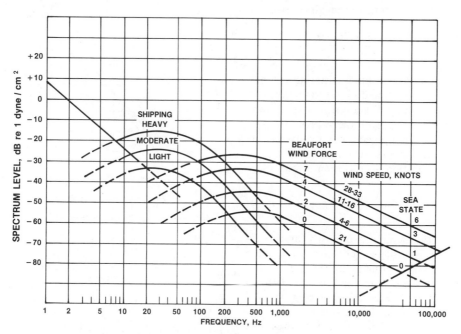

FIG. 9.   Ambient noise spectrum in the ocean. [From Urick, R. J. (1967). "Principles of Underwater Sound for Engineers," McGraw-Hill, New York.]

spectrum level = 10 dB re 1μPa/Hz   (62)

This states that the level is 10 dB "above" a reference level of 1 μPa (micropascal) in a 1-Hz band. The micropascal is a metric unit of pressure equal to 10^{-6} N/m^2 (newton/meter^2). In the case of wind-related noise, there is a family of curves labeled sea state. This is a qualitative and somewhat subjective measure of the level of agitation of the sea surface. Sea state zero indicates a relatively calm sea, and at sea state 6, the ocean is quite agitated.

Next to its spectrum level there are two other properties of ambient noise that are important from a signal-processing standpoint. These are directionality and correlation. For the case of distant shipping, the noise is usually strongest within about ±5° of the horizontal, whereas the surface-generated noise would appear to be radiating downward. For the case of thermally generated noise, the directionality is generally isotropic. The spectrum of the noise is important both for the time-domain- and spatial-domain-processing problem, as can be understood from the preceding section. The spatial correlation and directionality play a strong role in the spatial processing tasks. As can be seen from Eq. (58), the directionality $\eta(\theta)$ strongly influences the covariance matrix, which in turn plays a major role in the determination of the "optimal" array.

The description of reverberation makes use of a quantity called scattering strength, where

$$s_{v,s} = \frac{I_r/\text{unit volume}}{I_i}   (63)$$

Here, I_i is the intensity at the scattering volume, where the subscripts v and s refer to volume and surface, respectively. Theoretical expressions can be derived for I_r given the values of s_v and s_s, which can be computed theoretically for certain cases and, of course, can be measured. The volume reverberation level is then given by

$$RL = 10 \log I_s = \frac{s_v}{r^4} I_0 A   (64)$$

where I_s is the scattered intensity at the receiver, I_0 the intensity of the source, and r the distance to the scattering region; A is a number determined by the transmit and receive patterns of the sonar, the size of the scattering volume, and the characteristics of the sonar signal. This expression is not always useful, since the range to the reverberating volume may not be well defined, or the volume may, in fact, encompass the

sonar system itself. A similar expression can be derived for surface reverberation.

Absorption can be a major problem in areas of acoustic signal processing other than sonar. In particular, acoustic imaging and nondestructive testing may involve media that produce excessive absorption. In the case of sonar, absorption is caused by certain salts contained in the ocean and is a function of the temperature, depth, and frequency. The major cause of this absorption is the acoustically driven disassociation of magnesium sulfate ($MgSO_4$). The effect of this phenomenon is shown in Fig. 10.

Other sources of noise can enter into an acoustic-signal-processing problem. Electrical noise, that is, noise arising from the electronics involved in the processing devices, can under certain circumstances cause problems. Also, a phenomenon commonly called flow noise can cause severe problems for receiving transducers and arrays on moving platforms. This noise arises from the fact that turbulent flow over the hydrophones can cause pressure fluctuations that appear as sound to the transducers. This

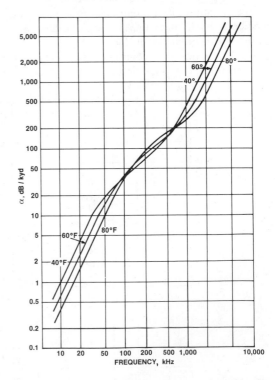

FIG. 10.   Sound absorption in the ocean as a function of frequency and temperature for a salinity of 35 parts per thousand. [From Urick, R. J. (1967). "Principles of Underwater Sound for Engineers," McGraw-Hill, New York.]

can arise in the case of a sonar receiver mounted on the hull of a moving ship or submarine. Also, propulsion machinery can cause mechanical vibrations that can propagate through the structure of the vessel and introduce noise into the array.

B. Reverberation Processing

Reverberation poses a major problem to sonar systems, and the means of processing against it are limited. One technique is to take advantage of any Doppler shifts in the returned signal. If the sonar system is operating against a moving target, the echo will be Doppler-shifted by an amount

$$\Delta f \simeq 2 \frac{v}{c} f_0 \qquad (65)$$

where v is the speed of the target in the direction of the receiver and f_0 the frequency of the transmitted signal. Thus, if $2 \Delta f$ is greater than the bandwidth of the reverberation, the spectrum of the echo will lie outside the spectrum of the reverberation and can easily be detected. This is illustrated in Fig. 11.

Note that the reverberation spectrum is broader than the echo spectrum. This is the case for volume reverberation and certain cases of surface reverberation. In surface reverberation, the spread is due mainly to surface motion caused by wind, whereas in volume reverberation the spread arises due to motion of the scattering particles themselves. Also, when the sonar source is moving, the Doppler shift of the reverberation depends on the direction from which it arrives at the receiver, thus contributing to the broadening of the spectrum.

Another means of processing against reverberation is to spread its spectral width somehow. If the spectrum of the reverberation can be spread, its power spectral density will be reduced since its power is proportional to the

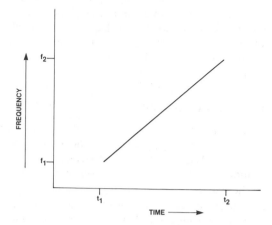

FIG. 12.   Linear FM slide.

power spectral density times the bandwidth. One means of accomplishing this is with an FM slide or a pulse-modulated signal, both of which have increased signal bandwidth. An FM slide signal is a pulse whose frequency changes with time. This process is illustrated in Fig. 12.

If the frequency slide is not too rapid, the bandwidth of the pulse is approximately given by

$$\Delta f \approx f_2 - f_1 \qquad (66)$$

In the case of a fixed-frequency pulse, the bandwidth of the pulse would be given approximately by

$$\Delta f \approx 1/T = 1/(t_2 - t_1) \qquad (67)$$

Thus, an FM slide pulse can be designed whose bandwidth is greater than that of the fixed-frequency pulse of the same duration. When this pulse is processed by a matched filter (assuming that the target does not distort the pulse, an assumption that is not generally true), the detection parameter, and hence the signal-to-noise ratio, is increased over that for a fixed-frequency pulse of the same energy, since the power spec-

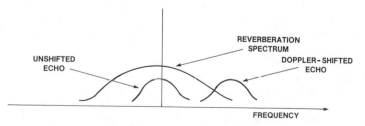

FIG. 11.   Spectra of Doppler-shifted and unshifted echo returns in reverberation.

tral density of the reverberation is reduced. This can be seen by referring to Eq. (18). Although the target does distort the echo somewhat, in many cases considerable gain against reverberation can still be obtained.

C. EFFECTS OF THE MEDIUM

The other problem addressed in this section is the impact of the medium on the processing problem. Again, the ocean provides a rather graphic example of how severe this can be.

The two major factors that have an impact on oceanic propagation are absorption and variations in the sound speed. The variation in sound speed is due mainly to pressure and temperature effects, as noted earlier. The speed of sound increases with increasing pressure and temperature. Thus, if the ocean were of a constant temperature, the sound speed would increase with depth. However, owing to solar and weather effects, the ocean is heated from its surface. This thermal layer, although subject to change with time and geographic location, has a thickness of the order of 1 or 2 km in deep ocean environments. It is subdivided into three sections. The surface layer is the part that is most strongly affected by short-term changes in the weather. It can undergo major changes in its temperature in a single day and can become quite mixed by the wind. This layer is of the order of 100 m in thickness. Beneath the surface layer lies the seasonal thermocline. This is a layer whose temperature gradient is stable for many days and undergoes major changes seasonally. This layer is also of the order of ~100 m. The next layer, which can extend to a depth of a few thousand meters, is called the main thermocline and possesses a temperature gradient that is stable over time scales of several years. Finally, below the main thermocline is the deep isothermal layer, where there is essentially no temperature gradient and the sound velocity dependence is determined primarily by the increase in pressure with depth. Figure 13 depicts this situation for a particular case. Here the sound speed is plotted against depth. An increase in sound speed with depth is called a positive gradient and will cause the direction of sound propagation to deflect upward. This can be intuitively seen by visualizing plane waves of sound propagating in a horizontal direction in the deep isothermal layer. Since the sound speed increases with depth, the lower regions of the plane waves travel faster than the

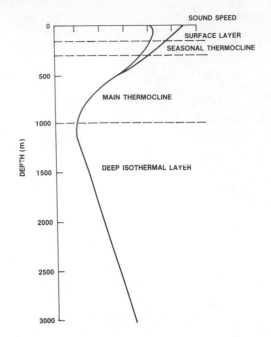

FIG. 13.   Typical oceanic speed-of-sound profile as a function of depth. [From Burdic, W. S. (1984). "Underwater Acoustic System Analysis," Prentice-Hall, Englewood Cliffs, New Jersey.]

upper parts, thus tilting them upward. This is depicted in Fig. 14. Thus, the ray (i.e., the line depicting the direction of propagation) curves upward. By the same argument, sound rays in a negative gradient will be deflected downward. This phenomenon can trap sound at the depth where the sound speed gradient is zero, that is, at the depth where the sound speed is minimum. This creates a duct called the deep sound channel, where sound can propagate for thousands of kilometers in deep ocean environments.

A second phenomenon is known as a surface duct. This occurs when the surface layer becomes mixed by the wind to the extent that there is no significant temperature gradient. The sound speed is then determined solely by pressure, and the layer then has a positive gradient. Since a positive gradient deflects (or refracts) the sound ray upward, the effect depicted in Fig. 15 ensues.

A third phenomenon is known as convergence zone propagation. This is also depicted in Fig. 15. The consequence of this effect is the existence of regions of increased intensity of the received sound at distances from the source that are multiples of ~50 km in deep ocean environ-

mated by the reciprocal of the aperture length in wavelengths, as discussed in Section III. Usually the aperture size of the array is constrained by the size of the platform. This maximum aperture size thus determines the lowest frequency that can be used. Of course, one could simply increase the frequency, thus achieving any desired narrowness of the receiving array beam, but as discussed in Section IV, the higher the frequency, the greater is the absorption of the acoustic energy by the seawater.

In the above example, no consideration was given to reverberation. If this were a problem, the single-frequency pulse could be replaced by an FM slide, with the appropriate changes made in the matched filter.

B. Oceanic Tomography

An example of acoustic signal processing as applied to an inverse problem is that of oceanic tomography. Large-scale experiments have been performed that have used the time delay incurred in transmissions between pairs of transducers. This method recommends itself, since the number of data is equal to the product of the number of transmitters and the number of receivers. In practice, the problem is quite difficult, since as seen in Section IV, the long-range propagation can be very complex due in part to the existence of multiple paths (sometimes called multipaths). The time delays along these multiple paths are then considered as individual data and reflect in part the speed-of-sound profile. The method has been used to determine the large-scale speed-of-sound map of the ocean. This in turn yields information on temperature-driven gyres. Thus, such an inverse problem that directly measures time delay gives indirect information on temperature distribution and thus gyre structure. The effect of the currents themselves does not strongly influence the time delay measurements, since the temperature is the main determinant of the sound speed, at least above the deep sound channel.

In order to reduce the complexity of the example it will be assumed that the temperature is a constant and that the process is intended to measure the current speed itself via time delay measurement. Although a much simpler problem than that discussed above, it clearly demonstrates the inverse nature of the process.

Consider the region being subjected to measurement to be surrounded by N transducers, which are both transmitters and receivers. Thus,

there are $(N^2 - N)/2$ paths. In practice, all of these paths may not be useful due to constraints on the positioning of the transducers, but here we will assume that they are. Let the region in which the flow is to be determined be divided into subregions. Labeling the path by i and the subregion by j, we give the sound speed along path i in region j by

$$c_{ij} = c_0 \pm (U_{xj} \cos \delta_i + U_{yj} \sin \delta_i) \quad (68)$$

where U_{xj} is the x component of U_j, U_{yj} the y component of U_j, U_j the current speed in subregion j (assumed constant over the subregion), δ_i the angle of the path from the x axis, and c_0 the speed of sound (assumed to be the same in all subregions). The travel time of the acoustic pulse along the ith path is then given by summing over the regions transversed by path i:

$$\tau_{i\pm} = \sum_j \frac{L_{ij}}{C_0 \pm (U_{xj} \cos \delta_i + U_{yj} \sin \delta_i)} \quad (69)$$

Here, L_{ij} is the length of path i in subregion j; the plus and minus subscripts indicate sound propagation with or against the flow. Assuming c_0 to be much greater than the current speed, Eq. (69) can be approximated by

$$\tau_{i\pm} \simeq \sum_j \frac{L_{ij}}{C_0^2} [C_0 \mp (U_{xj} \cos \delta_i + U_{yj} \sin \delta_i)]$$

$$(70)$$

The difference between the backward and forward travel times is defined as

$$T_i = \tau_{i-} - \tau_{i+}$$

$$= \sum_j \frac{2L_{ij}}{C_0^2} [U_{xj} \cos \delta_i + U_{yj} \sin \delta_i] \quad (71)$$

If there are Q subregions, there are $2Q$ current velocity components to be determined. Hence, $2Q$ acoustical paths are required. Clearly, Q cannot be greater than $(N^2 - N)/4$. Since the geometry of the situation is taken to be known, the values of the δ_i are known along with the L_{ij}. Defining

$$A_{ij} = \frac{2L_{ij}}{C_0^2} \cos \delta_i \quad (72)$$

$$B_{ij} = \frac{2L_{ij}}{C_0^2} \sin \delta_i \quad (73)$$

we can rewrite Eq. (71):

$$T_i = \sum_i [A_{ij} U_{xj} + B_{ij} U_{yj}] \quad (74)$$

The index i ranges from 1 to $2Q$, and the index j ranges from 1 to Q. Equation (74) can be rewritten in matrix form as

$$T_i = K_{ij}U_j \qquad (75)$$

From Eqs. (72) to (75) we see that

$$K_{ij} = (A_{ij}, B_{ij}) \qquad (76)$$

$$U_j = (U_{xj}, U_{yj}) \qquad (77)$$

Equation (75) has the same form as Eq. (1) in Section I. And as is usually the case in inverse problems, the matrix **K** must be inverted to yield the solution **U**. That is, the solution vector of flow velocities is given by

$$\mathbf{U} = \mathbf{K}^{-1}\mathbf{T} \qquad (78)$$

The process by which one inverts Eq. (74) [or Eq. (75)] could be any one of several schemes, the selection of which can depend on the physical situation itself, the noise content of the data, the computational load required by the inversion process, and whether the matrix is well conditioned. Whatever technique is chosen, the solution will allow a two-dimensional map of the flow field of the region of measurement. The accuracy of the solution will depend on the performance of the algorithm chosen for the solution, how well posed the problem is, the noise content of the data, and the fineness of the mesh (i.e., how many subregions the area of interest was broken into).

An example of how this problem could become ill posed is given by the case where the currents are very small. In this case, since the measurement vector **T** has elements that are differences between measured values of travel time and these travel times will differ by very little, large errors could occur in the solution vector **U**, given sufficient noise or insufficient measurement accuracy. Also, depending on the values of δ_i and L_{ij} in Eqs. (72) and (73), the matrix may or may not emphasize or "amplify" these errors. Thus, great care must be taken in designing the algorithm given the accuracy of the measurements. And it may, in fact, be the case that no algorithm will suffice if the measurements are not accurate enough.

The implementation of such a scheme, as is the case with just about every other inverse problem, is software intensive. A flow chart of how Eq. (78) would be solved for the vector **U** is shown in Fig. 18.

Finally, it should be emphasized that every inverse problem necessarily contains a forward problem. That is, the inputs to an inverse prob-

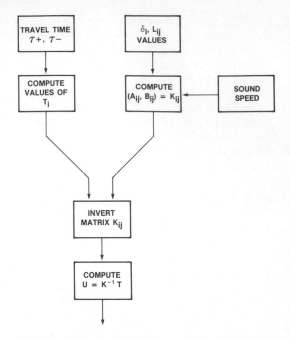

FIG. 18.   Flow chart for inverse-problem algorithm.

lem contain measured data. These data are determined by a detection and estimation process. In the foregoing problem, the travel time data would most likely be determined by transmitting pulses whose spectral and temporal characteristics would be specified by the forward problem of detecting the pulses in the given noise environment and estimating the time delays between their transmission and reception.

BIBLIOGRAPHY

Behringer, D., Birdsall, T., Brown, M., Cornuelle, B., Heinmiller, R., Knox, R., Metzger, K., Munk, W., Spresherger, J., Spindel, R., Webb, D., Worcester, P., and Wunsch, C. (1982). *Nature (London)* **299,** 121.

Burdic, W. S. (1984). "Underwater Acoustic System Analysis." Prentice-Hall, Englewood Cliffs, New Jersey.

Cox, H. (1973). *J. Acoust. Soc. Am.* **54,** 1289.

Hassab, J. C. (1983). *In* "Issues in Acoustic Signal/Image Processing and Recognition" (C. H. Chen, ed.), p. 1. Springer-Verlag, Berlin and New York.

Hudson, J. E. (1981). "Adaptive Array Principles." Peter Peregrinus, Ltd., London.

Metherell, A. F., ed. (1980). "Acoustical Imaging," Vol. 8. Plenum, New York.

Oppenheim, A. V., ed. (1978). "Applications of Digital Signal Processing." Prentice-Hall, Englewood Cliffs, New Jersey.

Sarkar, T. K., Weiner, D. D., and Jain, V. K. (1981).

IEEE Trans. Antennas Propag. **AP-29,** No. 2, 373.

Sato, T., and Shiraki, M. (1984). *J. Acoust. Soc. Am.* **76** (5), 1427.

Sullivan, E., and Kemp, K. (1981). *J. Acoust. Soc. Am.* **70** (2), 631.

Urkowitz, H. (1983). "Signal Theory and Random Processes." Artech House, Inc., Dedham, Massachusetts.

Wegman, E. J., and Smith, J. G., eds. (1984). "Statistical Signal Processing." Dekker, New York.

Whalen, A. D. (1971). "Detection of Signals in Noise." Academic Press, New York.

ANTENNAS

Ronold W. P. King *Harvard University*

GLOSSARY

Antenna array: Configuration of antennas arranged to produce a directional effect.

Complex wave number: Quantity $k = \beta + i\alpha$ in a traveling wave in a dissipative medium and represented by Re $\exp[-i(\omega t - kr)] = \exp(-\alpha r) \cos(\omega t - \beta r)$.

Dipole, dipole antenna: Thin metal cylinder or wire excited by an alternating current generator at its center so that the ends are oppositely charged.

Electrical length: Actual length multiplied by 2π and divided by the wavelength.

Field pattern: Polar diagram of the magnitude of the radiation field.

Monopole, monopole antenna: Thin metal cylinder or wire erected vertically over a conducting plane and excited by an alternating current generator connected between the base of the cylinder and the conducting plane.

Periodic time dependence: $\exp(j\omega t)$ or $\exp(-i\omega t)$. In some applications the first form is used; in others the second. In the technical literature the symbols j and i are used interchangeably with either sign. In this article j is always used with the plus sign, i with the minus sign so that $i = -j = \pm\sqrt{-1}$.

Probe or sensor: Antenna especially designed to measure electric or magnetic fields.

Radiation: Emission and propagation of electromagnetic waves.

Radiation field: Electromagnetic field at all distances r from the antenna sufficiently great so that the magnitude of the field decreases as $1/r$.

Radiation resistance: Quotient of the power radiated by the antenna and the square of the effective current in the antenna at a specified point.

Standing wave: Wave that has the form $\cos \omega t \cos kr$.

Traveling wave: Progressive wave that has the form $\cos(\omega t - kr)$.

Wavelength: Distance (measured in the direction of propagation of a wave) between two successive points that are in the same phase in the oscillation.

Antennas are well known to the nonspecialist as the "rabbit ears" connected to the "ANT" terminals of a television receiver or as the "whip" standing on the fender of an automobile with a hidden cable leading to the car radio. Also familiar are the more elaborate antennas for television and FM radio mounted on roof tops. All of these are made of metal to provide conducting paths for the electric currents induced in them by passing electromagnetic waves that carry television and radio signals from more or less distant transmitters. Antennas for transmission are similar but often much larger and higher metal structures in which large oscillating electric currents are maintained by a generator. These currents produce and radiate the electromagnetic waves that travel outward in space to receiving antennas. The capacity of an alternating current to radiate significant outward-traveling electromagnetic waves depends on the shape and size of its geometric contour and on the amplitude and frequency of its oscillation. [*See* ANTENNAS, REFLECTOR; RADIO PROPAGATION.]

I. The Dipole Antenna in Air

The functions of the sending and receiving antennas in the transmission of information by means of electromagnetic waves can be learned from a study of one of the simplest and most effective antennas: the dipole and the closely related monopole on a large conducting surface like a metal plane or, more approximately, on the surface of the earth or ocean. These are illustrated, respectively, in Fig. 1a,b.

A. CURRENT AND ADMITTANCE

The dipole consists of a metal rod of length $2h$ and radius a ($a \ll h$) center-driven from a two-wire transmission line, which supplies power from a more or less distant generator. The monopole consists of a similar rod of length h base-driven by a coaxial line that penetrates the ground plane. The alternating voltage $V(t) = V_0^e \cos \omega t = \mathrm{Re}\, V_0^e \exp(j\omega t)$ at the end of the transmission line maintains an oscillating current $I_z(z, t) = \mathrm{Re}\, I_z(z) \exp(j\omega t)$ in the antenna. $I_z(z)$ is determined by the following integral equation:

$$\int_{-h}^{h} I_z(z')K(z, z')dz' = -\frac{j4\pi}{\zeta_0} (C \cos k_0 z$$

$$+ \frac{1}{2} V_0^e \sin k_0|z|) \quad (1)$$

$$K(z, z') = \exp(-jk_0 R)/R$$
$$R = \sqrt{(z - z')^2 + a^2} \quad (2)$$

$\zeta_0 = 120\pi$ ohms is the wave impedance of air, $k_0 = \omega\sqrt{\mu_0 \varepsilon_0} = 2\pi/\lambda_0$ is the wave number, λ_0 the wavelength in air, and C is a constant to be determined from $I_z(h) = 0$. A simple approximate solution of Eq. (1) for $k_0 h \leq 5\pi/4$ is

$$I_z(z) = \frac{j2\pi V_0^e}{\zeta_0 \Psi \cos k_0 h} [\sin k_0(h - |z|)$$

$$+ T(\cos k_0 z - \cos k_0 h)] \quad (3)$$

When $k_0 h$ is near $\pi/2$, a useful alternative form is

$$I_z(z) = -\frac{j2\pi V_0^e}{\zeta_0 \Psi} [\sin k_0|z| - \sin k_0 h$$

$$+ T'(\cos k_0 z - \cos k_0 h)] \quad (4)$$

The driving-point admittance for the dipole is $Y_0 = I_z(0)/V_0^e$ and twice this value for the monopole. In Eqs. (3) and (4), Ψ is a real constant, T and $T' = -(T + \sin k_0 h)/\cos k_0 h$ are complex constants. An accurate graph of the admittance $Y_0 = G_0 + jB_0$ is shown in Fig. 2 for a typical antenna with $k_0 a = 0.044$. For this, $\Psi = 6.22$, $T' = 2.65 + j3.79$ when $k_0 h = \pi/2$; $\Psi = 5.74$, $T = -0.17 + j0.18$ when $k_0 h = \pi$. The electrical length near $k_0 h = 1.45$ at which $B_0 = 0$ and G_0 has a maximum is known as *resonance* with $I_z(0)$ near a maximum. The electrical length near $k_0 h = 2.45$ where $B_0 = 0$ and G_0 is near its minimum is known as *antiresonance*. The current

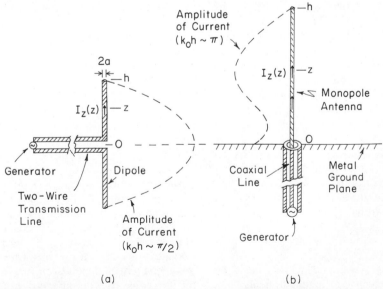

(a)   (b)

FIG. 1.   (a) Dipole in air; (b) monopole in air over metal plane, with current distributions.

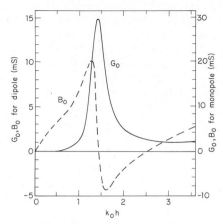

FIG. 2. Admittance $Y_0 = G_0 + jB_0$ of dipole antenna with half-length h and radius a; $k_0 = \omega/c$ ($k_0 a = 0.044$).

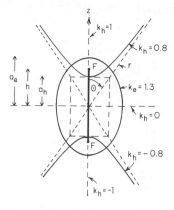

FIG. 3. Spheroidal coordinates k_e and k_h. Antenna of length $2h$ between the two foci F of spheroids and hyperboloids. Spherical coordinates r, Θ.

$I_z(0)$ is near a minimum, but the current $I_z(h - \lambda_0/4)$ is large.

Dipoles are most often designed for operation near resonance. A convenient electrical half-length is $k_0 h = \pi/2$, so that the leading term in Eq. (4) is

$$I_z(z) = V_0^e Y_0 \cos k_0 z$$
$$I_z(z, t) = V_0^e |Y_0| \cos k_0 z \cos(\omega t + \theta_0) \tag{5}$$

where $\theta_0 = \tan^{-1}(B_0/G_0)$. The associated charge per unit length on the surface of the dipole near resonance is

$$q(z) = \frac{j}{\omega} \frac{dI_z(z)}{dz} = -\frac{j}{c} V_0^e Y_0 \sin k_0 z$$
$$q(z, t) = \frac{V_0^e}{c} |Y_0| \sin k_0 z \sin(\omega t + \theta_0) \tag{6}$$

where $c = 3 \times 10^8$ m/sec. It is seen that the charge is in quadrature with the current.

B. COMPLETE ELECTROMAGNETIC FIELD

1. Formulas

The alternating current [Eq. (5)] on the antenna and the associated oscillating charges [Eq. (6)] generate outward-traveling electromagnetic waves known in their entirety as an *electromagnetic field;* it includes both electric and magnetic components known as the electric and magnetic fields and represented by the vectors **E** and **B** (or **H** = **B**/μ_0). For the current and charge distributions in Eqs. (5) and (6) the electromagnetic field is conveniently expressed in the spheroidal coordinates k_e and k_h defined by

$$k_e = a_e/h, \qquad 1 \le k_e \le \infty$$
$$k_h = a_h/h, \qquad -1 \le k_h \le 1 \tag{7}$$

where a_e is the semimajor axis of a spheroid with foci at the ends of the dipole, that is, at $z = \pm h$, and a_h is the semimajor axis of the hyperboloids of revolution with foci also at $z = \pm h$. These are shown in Fig. 3. The electric field has two components in each plane (ρ, z). They are E_e tangent to the spheroids, E_h tangent to the hyperboloids. The magnetic field has only one component B_Φ tangent to circles around the antenna and, therefore, perpendicular to the electric field. For $k_0 h = n\pi/2$, $n = 1, 3, 5, \ldots$, they are given by

$$B_\Phi(t) = -\frac{\mu_0 I_z(0)}{2\pi h} \frac{\cos(n\pi k_h/2)}{\sqrt{(k_e^2 - 1)(1 - k_h^2)}}$$
$$\times \sin(\omega t - n\pi k_e/2) \tag{8}$$

$$E_e(t) = -\frac{\zeta_0 I_z(0)}{2\pi h} \frac{\cos(n\pi k_h/2)}{\sqrt{(k_e^2 - k_h^2)(1 - k_h^2)}}$$
$$\times \sin(\omega t - n\pi k_e/2) \tag{9}$$

$$E_h(t) = \frac{\zeta_0 I_z(0)}{2\pi h} \frac{\sin(n\pi k_h/2)}{\sqrt{(k_e^2 - 1)(k_e^2 - k_h^2)}}$$
$$\times \cos(\omega t - n\pi k_e/2) \tag{10}$$

At large distances from the antenna, such that $r^2 = \rho^2 + z^2 \gg h^2$, $k_e \to r/h$, $k_h \to z/r = \cos \Theta$, where r, Θ, Φ are spherical coordinates with origin at the center of the dipole and the azimuthal angle Θ is measured from the positive z axis. It follows that the radiation field is given by

$$B_\Phi(t) = -\frac{\mu_0 I_z(0)}{2\pi r} \frac{\cos[(n\pi/2) \cos \Theta]}{\sin \Theta}$$
$$\times \sin(\omega t - k_0 r) \tag{11}$$

$$E_e(t) = E_\Theta(t) = -\frac{\zeta_0 I_z(0)}{2\pi r} \frac{\cos[(n\pi/2)\cos\Theta]}{\sin\Theta}$$

$$\times \sin(\omega t - k_0 r) \qquad (12)$$

$$E_h(t) = E_r(t) = \frac{\zeta_0 h I_z(0)}{2\pi r^2} \sin[(n\pi/2)\cos\Theta]$$

$$\times \cos(\omega t - k_0 r) \qquad (13)$$

These formulas completely characterize the electromagnetic waves generated by the current and charges in the antenna.

2. Electromagnetic Waves

The principal properties of the electromagnetic waves are best revealed from the simpler formulas (11)–(13) for points at distances r from the dipole that are large compared with the half-length h of the antenna, $r^2 \gg h^2$. Here the components E_Θ and $B_\Phi = E_\Theta/c$ ($c = 1/\sqrt{\mu_0\varepsilon_0} = 3 \times 10^8$ m/sec) are alone important since they decrease slowly as $1/r$, whereas E_r decreases rapidly as $1/r^2$. The radial dependence of $E_\Theta(t)$ and $B_\Phi(t) = E_\Theta(t)/c$ is contained in the factor $r^{-1} \sin(\omega t - k_0 r)$. At any instant $t = t_1$, this represents a spherical surface of radius r with the phase $\psi = \omega t_1 - k_0 r$. Since the sine is multivalued, spherical surfaces with radii $r_n = (\omega t_1 - \psi + 2n\pi)/k_0$, $n = 0, 1, 2, \ldots$, are all in the same phase. The radial distance between adjacent spherical surfaces that are in the same phase is

$$r_{n+1} - r_n = 2\pi/k_0 \equiv \lambda_0 \qquad (14)$$

where λ_0 is the *wavelength*.

Each spherical surface with a given phase $\psi = \omega t - k_0 r = $ const or $d\psi/dt = \omega - k_0 \, dr/dt = 0$ has the radial velocity $dr/dt = \omega/k_0 = c = 3 \times 10^8$ m/sec. Thus, the electromagnetic waves generated by the antenna consist—at some distance from the antenna—of spherical surfaces of constant phase and constant amplitude for $E_\Theta(t)$ and $B_\Phi(t)$ that expand with the radial velocity c and in the process decrease in amplitude as $1/r$.

Near the antenna the structure of the electromagnetic waves is more complicated but is readily understood from formulas (8)–(10). These show that the surfaces of constant phase or wave fronts are actually spheroidal with foci at the ends of the antenna and not spherical with origin at the center of the antenna. They are defined by $k_e = $ const. On any such surface, $E_e(t)$ and $B_\Phi(t)$ are in phase and in phase with the current in the antenna when $k_e = 1$. $E_h(t)$, on the other hand, is 90° out of phase with $E_e(t)$ and $B_\Phi(t)$ and in phase with the charge on the antenna when $k_e = 1$. Note that $E_h(t)$ is very large near the ends of the antenna, where $q(z)$ has a

maximum amplitude. The expanding spheroidal wave fronts begin with very great eccentricity at the antenna and gradually become spherical as they travel outward. Their velocity is always $c = 3 \times 10^8$ m/sec along the vertical axis, but initially much greater in the equatorial plane. Since both $E_e(t)$ and $E_h(t)$ are present and have a 90° phase difference, the electric vector $\mathbf{E}(t)$ at every point (k_e, k_h) is elliptically polarized; that is, it rotates and changes its length so that its end traces an ellipse. Examples of these ellipses are shown in Fig. 4. They range from straight lines on the equatorial plane ($k_h = 0$, where $E_h = 0$), and along the vertical axis ($k_h = \pm 1$, where $E_e = 0$) to circles between these extremes.

3. Electric Field

The actual loci of the electric vector are defined by the equation $dz/dr = [E_z(t)/E_r(t)]$. This leads to the relation

$$\cos(n\pi k_h/2)\cos(\omega t - n\pi k_e/2) = C \qquad (15)$$

which yields the contours shown in Fig. 5 when the constant of integration C is assigned values between -1 and $+1$. Also shown is the dipole with the charge $q(z, t)$ in Fig. 5a,c and the current $I_z(z, t)$ in Fig. 5b,d. The lines of the electric field initially end on the charges on the antenna, as illustrated in Fig. 5a at the instant t for which $\omega t + \theta_0 = \pi/2$. The upper half of the dipole is instantaneously charged $+$, the lower half $-$,

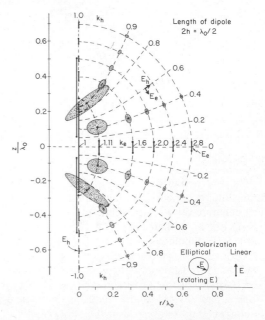

FIG. 4. Elliptically polarized electric field near dipole with $k_0 h = \pi/2$; confocal coordinates k_e, k_h.

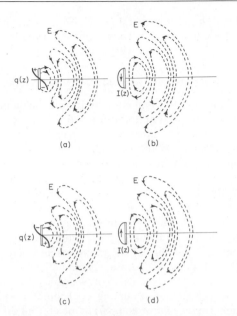

FIG. 5.   Charges $(+, -)$, currents $(\uparrow, \downarrow)$, and electric field $(--- \rightarrow ---)$ of dipole antenna with $k_0 h = \pi/2$ at instants (a) $\omega t + \theta_0 = \pi/2$; (b) $\omega t + \theta_0 = \pi$; (c) $\omega t + \theta_0 = 3\pi/2$; (d) $\omega t + \theta_0 = 2\pi$.

and there is no current. The electric field forms contours from the $+$ charge at the top to the $-$ charge at the bottom (the two dashed contours nearest to the antenna; the other contours shown originated at earlier times). As time passes, a downward current of $+$ charge or upward current of $-$ charge develops which increases to the time t for which $\omega t + \theta_0 = \pi$. At this instant, as shown in Fig. 5b, there is a maximum downward current in the antenna; all charges in moving toward the center of the antenna have been neutralized. In the process they carry the ends of the electric field lines with them so that these have become closed loops detached from the antenna (the two closed dashed contours nearest the antenna shown in Fig. 5b), while simultaneously moving radially outward from the antenna. As time passes, the downward current charges the lower end of the antenna $+$, and leaves the upper end $-$, as shown in Fig. 5c. In the process new electric lines emerge from the antenna ending on the charges and directed from $+$ to $-$ (the two dashed contours closest to the antenna). The instant of maximum charge and zero current is shown in Fig. 5c. The electric lines ending on the antenna in Fig. 5a, after becoming closed as shown in Fig. 5b, have now moved out and begun to curve around to assume a more spherical

shape. The next quarter-period leads to a maximum upward current (Fig. 5d) with no charge on the antenna and a new set of closed electric field lines that move outward. Note that the original field lines that ended on the antenna in Fig. 5a have moved out farther from the antenna. The process is repeated and the next quarter-period is again shown by Fig. 5a. Note that the electric lines originally attached to the antenna (Fig. 5a) have now moved out to be the outermost two closed dashed contours. The outward-traveling electric contours and the associated magnetic ones (which are perpendicular to the plane shown in Fig. 5) form the outward-traveling electromagnetic waves. The associated surfaces of constant phase are those shown in Fig. 4.

C. RADIATED POWER

The power radiated by the dipole and carried outward by the expanding, first spheroidal, then spherical wave fronts is equal to the power supplied to the antenna, $P_{in} = (V_0^e)^2 G_0$, minus any losses due to heating of the never perfectly conducting antenna. The latter are usually negligible. The time-dependent radiated power $T_e(t)$ can also be determined directly by integration of the outwardly directed Poynting vector $S_h(t) = \mu_0^{-1} E_e(t) B_\Phi(t)$ over any wave front $k_e = $ const to which it is perpendicular. The result for $k_0 h = \pi/2$ is

$$T_e(t) = \frac{\zeta_0 I_z^2(0)}{4\pi} \sin^2(\omega t - \pi k_e/2) \, \text{Cin} \, 2\pi \quad (16)$$

where $\text{Cin} \, 2\pi = \int_0^{2\pi} [(1 - \cos x)/x] \, dx = 2.438$. The time-average value is independent of k_e and given by

$$\bar{T}_e = \frac{\zeta_0 I_z^2(0)}{8\pi} \, \text{Cin} \, 2\pi = \frac{1}{2} I_z^2(0) R_0^e \quad (17)$$

where the *radiation resistance* is $R_0^e = 30 \, \text{Cin} \, 2\pi = 73.1$ ohms. This value is correct for an extremely thin antenna. For thicker antennas and with $k_0 h = \pi/2$, R_0^e is somewhat larger.

D. RADIATION (FAR) FIELD; FIELD PATTERNS

Of primary interest in radio transmission is the electromagnetic field at large distances from the dipole. This is given by Eqs. (11) and (12). These formulas contain a trigonometric factor that defines the directional properties when $k_0 h = n\pi/2$. A more general approximate form is obtained from the leading term in Eq. (3) and can be referred to the maximum current at $k_0 z = k_0 h - \pi/2$ in $F_m(\Theta)$ or the current at $z = 0$ in

$F_0(\Theta)$. Thus,

$$F_m(\Theta) = \frac{\cos(k_0 h \cos \Theta) - \cos k_0 h}{\sin \Theta}$$

$$= F_0(\Theta) \sin k_0 h \qquad (18)$$

This formula reduces to the trigonometric factor in Eqs. (11) and (12) when $k_0 h = n\pi/2$. Graphs of $F_m(\Theta)$ for $k_0 h = \pi/2$, π, and $3\pi/2$ and the associated generating currents are shown in Fig. 6. They are known as *field patterns*. When the dipole is electrically short, the field factor is

$$F_0(\Theta) = \frac{k_0 h}{2} \sin \Theta, \qquad k_0^2 h^2 \ll 1 \qquad (19)$$

and the field pattern is a circle.

The electromagnetic field of the single dipole antenna is omnidirectional in its equatorial plane. The vertical field pattern depends on the electrical length, as shown in Fig. 6. The slightly flattened circle for $k_0 h = \pi/2$ becomes increasingly flatter and longer as the length of the antenna is increased to $k_0 h = \pi$, when an additional pair of lobes begins to develop and the main lobe at $\Theta = \pi/2$ begins to decrease. The new lobes grow to a large size when $k_0 h = 3\pi/2$ with maxima at $\Theta = 42°$ and $138°$. The lobe structure with nulls is a consequence of the oppositely directed currents in different parts of the antenna. If the currents are all made codirectional, as in Fig. 6c, a very different field pattern results, as in Fig. 6d. This is quite directive in any vertical plane but is still omnidirec-

tional in the horizontal plane. Additional collinear elements with codirectional currents will further improve the vertical directivity.

E. DIPOLE ARRAYS WITH *N* DRIVEN ELEMENTS

For point-to-point communication the omnidirectional horizontal pattern is inadequate. Significant *horizontal* directivity can be achieved by arranging *N* dipoles side by side in a "curtain" array (Fig. 7) with each element center-driven and at a distance *b* from the adjacent elements. The currents in the individual elements can be determined in terms of the driving voltages from *N* simultaneous integral equations like Eq. (1) but each with $N - 1$ additional integrals that take account of the interelement coupling. The solutions are like Eq. (3) but with the coefficient *T* replaced by a sum of *N* such coefficients. The interelement coupling contributes only to the term $(\cos k_0 z - \cos k_0 h)$ in the current distribution and not to the term $\sin k_0(h - |z|)$.

In important applications, the driving voltages are chosen so that the amplitudes of all currents are the same (uniform arrays). When the currents are also all in phase (broadside array), the horizontal field pattern has maxima in the two directions perpendicular to the line of the array, as shown in Fig. 7a with $b/\lambda_0 = 0.5$. When the phases of the currents increase uniformly from one end of the curtain to the other, the maximum is in the direction of lagging phase, as shown in Fig. 7b with $b/\lambda_0 = 0.25$ and a progressive phase shift per element of $\pi/2$. The beamwidth of the broadside array can be reduced with more elements. Its bidirectional field pattern can be made unidirectional with a second row of elements, parallel to the first, $\lambda_0/4$ from it, and driven with currents of equal magnitude but lagging or leading in phase by $\pi/2$.

The field of a uniform array of *N* elements is the field of one element multiplied by the follow-

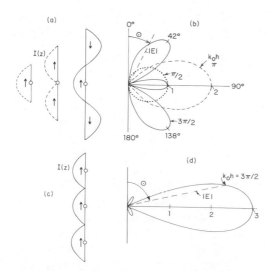

FIG. 6.   (a) Currents in and (b) vertical field patterns of dipoles with $k_0 h = \pi/2$, π, and $3\pi/2$; (c) currents in and (d) vertical field patterns of three-element collinear array.

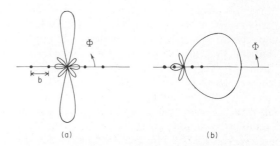

FIG. 7.   Horizontal field patterns of five-element uniform arrays. (a) Broadside, $b/\lambda_0 = 0.5$; (b) endfire, $b/\lambda_0 = 0.25$ (lagging currents on right).

ing array factor,

$$A(\Theta, \Phi) = \frac{\sin[(N\pi/2) \sin \Theta \cos \Phi]}{\sin[(\pi/2) \sin \Theta \cos \Phi]} \quad (20)$$

for the broadside array with $b/\lambda_0 = 0.5$, and by

$$A(\Theta, \Phi) = \frac{\sin[(N\pi/4)(1 - \sin \Theta \cos \Phi)]}{\sin[(\pi/4)(1 - \sin \Theta \cos \Phi)]} \quad (21)$$

for the unilateral endfire array with $b/\lambda_0 = 0.25$. In more complicated arrays designed for special purposes, the amplitudes of the currents in the elements need not be the same but may, for example, be tapered to reduce the sidelobe level. Alternatively, the phases of the currents may be controlled in time to produce a beam that scans a sector in a selected direction.

F. Parasitic Arrays; the Yagi–Uda Antenna

Instead of driving all elements in an array, it is often more convenient to drive only one and then adjust the lengths and distances between the $N - 1$ parasitic elements to obtain the desired field pattern and admittance. A preferred parasitic array is the Yagi–Uda antenna illustrated in Fig. 8 with six elements. It consists of the driven element (2), a single reflector (1), and four directors (3–6). Lengths and spacings for a maximum front-to-back ratio of the electric field are given on the figure. The horizontal field pattern is also in Fig. 8. It is broadly unidirectional.

G. Dipoles for Reception

The electromagnetic waves radiated by a center-driven dipole and traveling outward in space consist of mutually perpendicular electric and magnetic vector fields that propagate with the velocity of light c. As they pass, they induce oscillating currents in any metallic structures to which the incident electric field is not perpendicular. If such a structure is a single dipole or an array of dipoles designed as for transmission but with a receiver (radio or television) connected to it instead of a transmitter, the currents induced by the passing electromagnetic waves maintain an oscillating voltage across the terminals of the receiver. The magnitude of this voltage is proportional to the magnitude of the incident electric field and a quantity called the effective length of the antenna. This incorporates the directional properties of the antenna, which are the same as the field pattern of the antenna when used for transmission. Thus, the Yagi–Uda antenna shown with its field pattern in Fig. 8 is an effective receiving antenna if the load—e.g., a transmission line leading to a television receiver—is connected across the driving terminals of element 2. The directional properties for reception are exactly the same as those for transmission shown in Fig. 8b.

H. Electrically Short Dipole or Monopole as a Probe or Sensor

An important application of the bare metal dipole or vertical monopole over a metal surface is as a sensor or probe for measuring the electric field parallel to its axis. For this purpose, it is usually advantageous to have the antenna electrically short so that $k_0 h \leq 1$. Since the induced current (or the linearly related voltage across a terminating load) is proportional to the incident electric field, the short antenna can be used as a dipole to measure the electric field in air and also as a monopole to measure distributions of charge per unit length along antennas or transmission lines (Fig. 9). The normal component of the electric field on a metal surface is proportional to the charge density.

Actually the response of an electrically short

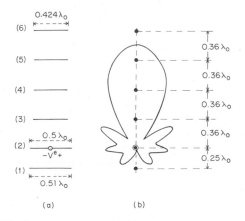

(a)   (b)

FIG. 8.   (a) Yagi–Uda array with four directors; (b) field pattern. Designed for maximum front-to-back ratio; $a/\lambda_0 = 0.00337$.

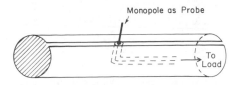

FIG. 9.   Coaxial-line-loaded monopole over slotted tubular conductor (antenna, transmission line).

antenna is proportional not to the electric field, but to its time derivative. At any given frequency, the time derivative $dE/dt \sim \omega E$, so that when the frequency is known, E is determined. When the electric field to be measured is not a continuous wave but a transient or pulse, the electrically short dipole measures dE/dt and E must then be determined by Fourier transformation.

II. Antennas Used in and near Material Media

Many antennas are located on or near the surface of the earth or sea. Others, designed for subsurface communication, remote sensing, or localized heating, may be embedded in the earth, the sea, or even in a living organism. The properties of such antennas are quite different from those of the theoretically much simpler antennas in air far from material media.

A. VERTICAL MONOPOLE ON THE SURFACE OF THE EARTH OR SEA

1. Direct and Reflected Electric Field

The electromagnetic field in the radiation zone of a vertical monopole on the surface of the earth or sea (region 1, Fig. 10a) is given by Eqs. (11) and (12) when the earth or sea is treated as a perfect conductor. Actually, the conductivity of the sea is near 4 S/m, that of the earth between 0.01 and 0.001 S/m, depending on moisture content. These values are far from infinity. An improved formula for the complex amplitude of E_Θ

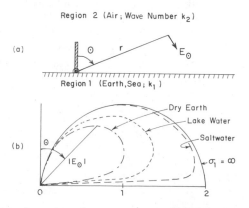

FIG. 10.   (a) Vertical monopole on surface of earth or sea. (b) Field pattern of vertical monopole over perfect conductor, sea, lake, earth; $|[1 + f(\Theta)] \sin \Theta|$.

for an electrically short monopole with $hI_z(0) = 1\ \text{A} \cdot \text{m}$ is

$$E_\Theta = \frac{i\omega\mu_0}{4\pi} \frac{\exp(ik_2 r)}{r} [1 + f(\Theta)] \sin \Theta \quad (22)$$

where $f(\Theta)$ is the *plane-wave reflection coefficient* given by

$$f(\Theta) = \frac{N^2 \cos \Theta - \sqrt{N^2 - \sin^2 \Theta}}{N^2 \cos \Theta + \sqrt{N^2 - \sin^2 \Theta}} \quad (23)$$

Here $N = k_1/k_2$ is the *index of refraction* of the earth or sea with the complex wave number $k_1 = \beta_1 + i\alpha_1 = \omega\sqrt{\mu_0(\varepsilon_1 + i\sigma_1/\omega)}$; $k_2 = k_0 = \omega\sqrt{\mu_0\varepsilon_0}$ is the wave number of the air (region 2). The physical picture is that of electromagnetic waves traveling outward from the antenna to the point of observation at (r, Θ) in two parts: one that travels directly, another that is reflected from the surface of the earth or sea. It is evident from Eq. (22) with Eq. (23) that when $\sigma_1 \sim \infty$, $N \sim \infty$ and $f(\Theta) \sim 1$. In this case, Eq. (22) agrees with Eq. (12) [with $F_m(\Theta) = \sin \Theta$ for the short antenna instead of $\cos[(\pi/2) \cos \Theta]/\sin \Theta$ for the quarter-wave monopole], and the field pattern is given by $[1 + f(\Theta)] \sin \Theta = 2 \sin \Theta$ since $f(\Theta) = 1$, as shown in Fig. 10b for $\sigma_1 = \infty$. When σ_1 is finite, the field pattern is again given by $[1 + f(\Theta)] \sin \Theta$ but now with $f(\pi/2) = -1$, so that the field vanishes when $\Theta = \pi/2$. Patterns for saltwater, lake water, and dry earth are also shown in Fig. 10b.

2. Surface Wave Field

Actually the zero value along the surface resulting from the cancellation of the direct and reflected fields is incomplete. In addition to the direct and reflected waves contained in Eq. (22), the vertical monopole also generates a *surface wave*, which must be added to Eq. (22) in order to obtain the complete electromagnetic field. The surface wave (also called a *lateral wave*) is not a transverse electromagnetic wave [like that defined by Eq. (22) with the associated $B_\Phi = E_\Theta/c$] since it includes a radial component. In the cylindrical coordinates, ρ, ϕ, z, the surface wave consists of the component $E_{2\rho}$, E_{2z}, and $B_{2\phi}$ in the air (region 2) and the components $E_{1\rho}$, E_{1z}, and $B_{1\phi}$ in the earth or sea (region 1). At a depth z in the earth or sea and along the surface in the air ($z = 0$), they are

$$E_{1z}(z) = \frac{\omega\mu_0 k_2}{2\pi k_1^2} g(\rho) \exp(ik_2\rho)$$
$$\times \exp(ik_1 z), \qquad z \geq 0 \quad (24)$$

$$E_{1\rho}(z) = -\frac{\omega\mu_0}{2\pi k_1} f(\rho) \exp(ik_2\rho) \exp(ik_1 z)$$

$$= \frac{\omega}{k_1} B_{1\phi}(z), \qquad z \geq 0 \qquad (25)$$

$$E_{2z}(0) = (k_1^2/k_2^2)E_{1z}(0), \qquad z = 0 \qquad (26)$$

$$\left.\begin{array}{l} E_{2\rho}(0) = E_{1\rho}(0), \\[2mm] B_{2\phi}(0) = B_{1\phi}(0), \end{array}\right\} \qquad z = 0 \qquad (27)$$

where, in general,

$$g(\rho) = f(\rho) - \frac{i}{k_2\rho^3} = \frac{ik_2}{\rho} - \frac{1}{\rho^2} - \frac{i}{k_2\rho^3}$$

$$- \frac{k_2^3}{k_1} \sqrt{\frac{\pi}{k_2\rho}} \exp(-ip)\mathcal{F}(p) \qquad (28)$$

and, at large distances,

$$g(\rho) \sim f(\rho) \sim -k_1^2/k_2^2\rho^2, \qquad p \geq 4 \quad (29)$$

The so-called *numerical distance* p and the Fresnel integral $\mathcal{F}(p)$ are defined by

$$p \equiv k_2^3\rho/2k_1^2; \qquad \mathcal{F}(p) = \int_p^\infty \frac{\exp(it)}{\sqrt{2\pi t}} \, dt \quad (30)$$

The name *surface wave* is readily understood from Eqs. (24) and (25). The exponential factors $\exp(ik_2\rho) \exp(ik_1 z)$ indicate a wave that travels radially along the surface a distance ρ in region 2 (air) and then vertically downward in region 1 (earth, sea) a distance z to the point of observation or location of a receiver.

Radio communication in air between vertical transmitting and receiving antennas erected on the surface of the earth or sea is explained by Eq. (26), which gives the vertical component of the surface wave. With Eqs. (26) and (24), it is given by

$$E_{2z}(0) = \frac{\omega\mu_0}{2\pi k_2} g(\rho) \exp(ik_2\rho) \qquad (31)$$

Radio communication with submarines can also be achieved with a vertical monopole on the surface of the earth or sea. However, in seawater, the radial component of the electric field is an order of magnitude greater than the vertical component. $E_{1\rho}(z)$ is given by Eq. (25). Also, submerged submarines make use of horizontal trailing-wire antennas. Graphs of $|E_{1\rho}(0)|$ due to a vertical monopole (Fig. 11a) with unit electric moment are shown in Fig. 11b as a function of the radial distance with the frequency as the parameter. $|E_{1\rho}(z)|$ as obtained from Eq. (25) is shown in Fig. 11c as a function of the frequency

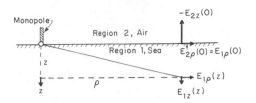

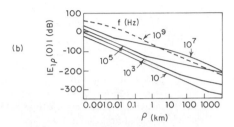

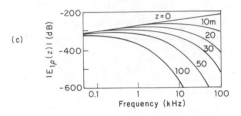

FIG. 11.   (a) Vertical monopole on surface of sea; field vectors at surface and at depth z. (b) $|E_{1\rho}(0)|$ as function of radial distance along surface. (c) $|E_{1\rho}(z)|$ at $\rho =$ 5000 km as function of frequency.

with the depth of the receiver as the parameter. It is seen that the signal strength at the surface $z = 0$ at any fixed radial distance increases with frequency but that it decays exponentially with the depth of the receiver.

B. Insulated Antenna in the Earth or Sea

1. Current and Admittance

A bare metal dipole is generally not useful when embedded in a medium like the earth or sea because these are sufficiently conducting to permit large radial leakage currents to leave the antenna along its entire length. This can be prevented by enclosing the antenna in an insulating sheath, as shown in Fig. 12a for the completely insulated, center-driven antenna and in Fig. 12b for the terminated traveling-wave antenna. The insulation greatly changes the properties of the antenna. If a good dielectric with low permittivity is chosen, its wave number $k_d = \omega\sqrt{\mu_0\varepsilon_d}$ is real and small compared with the wave number k_1 of the ambient medium (earth, water). When

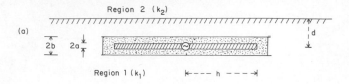

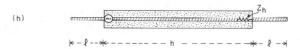

FIG. 12.   Insulated antennas. (a) Completely insulated ($Z_0 = \infty$, $Z_h = \infty$), center-driven conductor (radius a) in insulating sheath (radius b) in region 1 at distance d from region 2. (b) End-driven traveling-wave antenna terminated in bare sections of length l and lumped impedance Z_h at load end.

this is true, the insulated antenna has the properties of a generalized coaxial line with the current distribution

$$I_x(x) = -\frac{iV_0^e}{2Z_c}\frac{\sin k_L(h - |x|)}{\cos k_L h} \quad (32)$$

and the driving-point impedance

$$Z_{in} = \frac{V_0^e}{I_x(0)} = R_{in} - iX_{in} = i2Z_c \cot k_L h \quad (33)$$

for the circuit in Fig. 12a. For Fig. 12b,

$$I_x(x) = \frac{V_0^e}{Z_0 + Z_c}\exp(ik_L x)$$
$$Z_{in} = Z_0 + Z_c \quad (34)$$

The wave number k_L and characteristic impedance Z_c are given by

$$k_L = \beta_L + i\alpha_L \equiv \sqrt{-y_L z_L}$$
$$Z_c = \sqrt{z_L/y_L} \quad (35)$$

where the series impedance per unit length, z_L, and shunt admittance y_L are

$$z_L = z^i - i\omega l + z^e; \quad y_L = g - i\omega c \quad (36)$$

The formulas for the internal impedance per unit length, z^i, the inductance per unit length, l, the leakage conductance per unit length, g, and the capacitance per unit length, c, are the same as for a coaxial line. The external impedance per unit length, z^e, is entirely different from that for the outer conductor of the coaxial line. It takes account of radiation per unit length from the insulated antenna; a coaxial line does not radiate. The radiation impedance per unit length is large when the insulation is thin or b/a is near 1 (see

Fig. 12a), quite small when it is thick with $b/a \gg 1$. The line constants are

$$l = \frac{\mu_0}{2\pi}\ln\left(\frac{b}{a}\right); \quad g = \frac{2\pi\sigma_d}{\ln(b/a)}$$
$$c = \frac{2\pi\varepsilon_d}{\ln(b/a)} \quad (37)$$

$$z^e = -\frac{i\omega\mu_0}{2\pi}\left[\frac{H_0^{(1)}(k_1 b)}{k_1 b H_1^{(1)}(k_1 b)}\right] \quad (38)$$

The internal impedance per unit length, $z^i = r^i - ix^i$, is negligible compared with the external (radiation) impedance per unit length, $z^e = r^e - ix^e$, with good conductors of adequate cross-sectional size when the operating frequency is not too low ($f \geq 1$ kHz). At very low frequencies ($f \leq 100$ Hz), $|z^e| \ll |z^i|$ and $z^i = r_0$, the direct-current (dc) resistance. When the insulated antenna is at a distance d from the boundary with a region 2 (Fig. 12a) for which $|k_2|^2 \ll |k_1|^2$, the mutual impedance per unit length, z_{12}, given by

$$z_{12} = -\frac{i\omega\mu_0}{2\pi}\left[\frac{H_0^{(1)}(2k_1 d)}{k_1 b H_1^{(1)}(k_1 b)}\right] \quad (39)$$

must be added to z_L in Eq. (36).

2. Electric Field

The electromagnetic field of the insulated antenna in region 1 (earth or sea) near its boundary with region 2 (air), as shown in Fig. 12a, is obtained from that of the horizontal infinitesimal electric dipole located at the distance d from the boundary. This is quite complicated and consists of three components of the electric field and three of the magnetic field that each include direct, reflected, and surface wave components. However, except very close to the antenna, the direct and reflected fields are negligible in dissipative media like the earth, seawater, and even lake water, so that only the surface wave must be considered. The largest component at a distance z from the surface in region 1 is

$$E_{1\rho}(\rho, \phi, z) = -\frac{\omega\mu_0 k_2}{2\pi k_1^2} \cos \phi \, g(\rho)$$
$$\times \exp(ik_2\rho) \exp[ik_1(z + d)]$$
(40)

where $g(\rho)$ is given by Eq. (28). It is very similar to the surface wave generated by the vertical monopole and shown in Fig. 11b,c. The form in Eq. (40) indicates that it travels vertically from the dipole at $z = d$ in region 1 to the bounding surface at $z = 0$, then radially outward a distance ρ in region 2, and finally vertically to the receiver at z in region 1.

The field of the traveling-wave antenna shown in Fig. 12b is given by Eq. (40) when multiplied by the electric moment of the antenna. This must take account of the phase difference between the surface wave in region 2 and the current in the antenna. Thus,

$$|[E_{1\rho}(\rho_0, \phi_0, z)]_{ant}|$$
$$= |I_x(0)h_e(\phi_0)E_{1\rho}(\rho_0, \phi_0, z)| \quad (41)$$

where ρ_0, ϕ_0 refer to the origin at the point on the surface above the generator and

$$h_e(\phi_0) = \frac{i\{1 - \exp[i(k_L - k_2 \cos \phi_0)h]\}}{k_L - k_2 \cos \phi_0} \quad (42)$$

The magnitude of Eq. (42) is maximized when $h = \pi/(\beta_L - \beta_2)$.

3. Subsurface Communication

Long insulated traveling-wave antennas and arrays of such antennas can be designed to operate at low frequencies when submerged a small depth d in the ocean for communicating with submarines at great depths. At higher frequencies and buried in the earth, they provide a means of communicating with underground installations via surface waves. At extremely low frequencies, they are used on the sea floor in geophysical studies of the electrical properties of the oceanic crust. In this case the surface wave travels from the antenna down into the oceanic crust, then proceeds in it close to the boundary, and finally vertically up to the receiver in the seawater.

4. Subsurface Heating

Insulated antennas are also used in a vertical position with respect to the boundary surface between regions 1 and 2. In one type of application, it is necessary to provide heating in a localized volume at a significant distance from the surface. An insulated antenna is embedded in that volume and driven coaxially in a manner that limits significant radiation to the antenna proper. This requires a design that suppresses radiating currents on the outside surface of the feedline. A specific application of an embedded insulated antenna is to the hyperthermia treatment of malignant tumors. Extremely thin antennas operated at very high frequencies are inserted into the tumor, where localized heating is due to the radiated direct field. On an enormously larger scale, at correspondingly lower frequencies, insulated antennas are lowered into deep boreholes for the localized heating of oil shale to separate out the oil for subsequent pumping to the surface.

C. HORIZONTAL-WIRE ANTENNA OVER THE EARTH OR SEA; WAVE ANTENNA

1. Properties

Insulated antennas do not have to be rotationally symmetric. The conductor with radius a can be located eccentrically in the insulating sheath and, with it fixed at a distance d from the boundary between the insulator and the ambient region 1, the radius b of the insulation can be allowed to become infinite. If the insulator is air, this leaves the conductor in air at a distance d over a plane region 1, which can be earth or sea. Evidently, the horizontal-wire antenna is simply an eccentrically insulated antenna lying on the earth or sea. It follows that the same formulas (32) and (34) for the current and (33) for the impedance apply, as well as (35), for k_L and Z_c. However, in formula (36) for z_L and y_L,

$$l = \frac{\omega\mu_0}{2\pi} \ln\left(\frac{2d}{a}\right); \qquad g = 0; \qquad c = \frac{2\pi\varepsilon_0}{\ln(2d/a)}$$
(43)

$$z^e = -\frac{i\omega\mu_0}{\pi}\left\{\frac{1}{(2k_1d)^2} - \frac{K_1(2k_1d)}{2k_1d} + \frac{i\pi I_1(2k_1d)}{4k_1d}\right.$$
$$\left. - i\left[\frac{2k_1d}{3} - \frac{(2k_1d)^3}{45} + \frac{(2k_1d)^5}{1575} + \cdots\right]\right\} \quad (44)$$

where K_1 and I_1 are the modified Bessel functions. In practice the horizontal-wire antenna is especially useful for communication and radar when terminated to support a traveling wave of current. In this form it is known as the wave antenna.

The wave antenna can be terminated in its characteristic impedance in two ways (Fig. 13). In Fig. 13a, the terminations are simply resonant extensions of the horizontal wire in series with suitable lumped impedances Z_L so that the effective terminations are equal to Z_c. In Fig. 13b, the terminations are lumped impedances Z_L in series with vertical conductors grounded in the earth. This second form is known as the Beverage antenna after its inventor. It is usually more convenient, but only when an adequate ground connection can be provided. The form in Fig. 13a is independent of the conductivity of region 1 so that it can be used over lake water and dry earth as well as over seawater. The wave antenna can be a long wire quite close to the earth.

2. Electromagnetic Field and Applications to Radar

The electromagnetic field generated by the horizontal-wire antenna along the air–earth boundary is the same surface wave generated by the horizontal insulated antenna in the earth or sea. The contributions to the field by the terminations—whether horizontal or grounded vertical—are negligible compared with the field generated by the long horizontal part with its traveling wave of current. In the air above the surface, the vertical component of the electric field is greatest; in the earth or sea, the radial component is largest. They are

$$[E_{2z}(\rho_0, \phi_0, 0)]_{\text{ant}} = I_x(0)h_e(\phi_0)E_{2z}(\rho_0, \phi_0, 0) \quad (45)$$

where

$$E_{2z}(\rho_0, \phi_0, 0) = \frac{\omega\mu_0}{2\pi k_1} f(\rho_0) \cos \phi_0 \quad (46)$$

$$[E_{1\rho}(\rho_0, \phi_0, z)]_{\text{ant}} = I_x(0)h_e(\phi_0)E_{1\rho}(\rho_0, \phi_0, z) \quad (47)$$

where

$$E_{1\rho}(\rho_0, \phi_0, z) = -\frac{\omega\mu_0 k_2}{2\pi k_1^2} g(\rho_0)$$
$$\times \cos \phi_0 \exp(ik_1 z) \quad (48)$$

The factors $f(\rho_0)$ and $g(\rho_0)$ are defined in Eq. (28); the effective length $h_e(\phi_0)$ is defined in Eq. (42).

The wave antenna is the basic element in arrays of horizontal-wire antennas used for so-called over-the-horizon radar. A signal in the form of an electromagnetic wave of finite duration radiated by the array travels along the surface of the earth or sea, and when its principal component $E_{2z}(\rho, \phi, 0)$ impinges on an obstacle (e.g., a rocket or missile at the horizon), it induces a current that in turn radiates or reflects a signal. This travels back to the array of wave antennas, where it induces currents that travel to the receiver load—which is switched in, the generator out, for the receiving mode. The induced currents are quite complicated since they include components that propagate in the conductors with the wave number k_L and other components that propagate with the wave number $k_2 = k_0$ for air.

3. Application to Remote Sensing

A center-driven horizontal-wire dipole located on, but insulated from, the surface of the earth can be used to generate a surface wave field that travels along the surface but continuously also sends a wave down into the earth. If this encounters an object or a large-scale discontinuity in the properties of an otherwise homogeneous earth, a *scattered field* is sent out (Fig. 14). At the surface the scattered field combines with the incident surface wave field from the transmitter to produce an interference pattern. This can be detected and analyzed with a receiver moved over the surface, and in this way

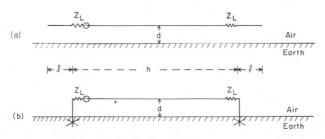

FIG. 13.   Terminated horizontal-wire or wave antennas. (a) Terminated in Z_L in series with open-ended length l; (b) Beverage antenna terminated in Z_L and grounded.

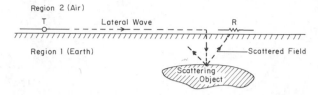

FIG. 14. Fixed horizontal dipole transmitter T and movable receiver R to locate subsurface object by scattering of incident surface wave field.

the location of the boundaries of a buried object or of a region of discontinuity (ore deposit, oil, etc.) can be determined.

D. INSULATED ANTENNA AS A PROBE FOR MEASURING THE ELECTRIC FIELD

If an electrically short, bare metal dipole or monopole is used to measure the electric field in the interior of a material medium like water or biological tissue, its response is linear in the permittivity of the ambient medium. If this is known, the electric field can be determined from the observation. If the permittivity is not known, the electric field in the medium cannot be obtained from measurements with a bare metal probe. Because of the insulation, the insulated probe is relatively insensitive to the properties of the ambient medium and its response to the electric field is almost independent of the permittivity of the ambient medium. This means that the insulated electrically short probe can be used to measure the electric field in a dielectric medium even if the permittivity varies from point to point.

III. The Loop Antenna

The dipole antenna is a form of *open circuit* in which the current oscillates between definite ends of the conductor, where it must vanish as it leaves a concentration of electric charges. A different type of antenna is the *closed loop,* in which there are no ends and the current circulates around or oscillates within a closed metallic contour. A circular transmitting loop is shown in Fig. 15 when driven at diametrically opposite points by equal voltages (a) in phase opposition to generate a circulating current of constant amplitude when the loop is not too large and (b) in phase to generate currents in the dipole model. When the two pairs of generators are superimposed, they provide a single generator at one point and a cancelled pair at the dia-

metrically opposite point. This leaves the loop driven by one generator with both types of current excited. Their relative magnitudes are determined by the impedance of the loop to each mode. When this is electrically small ($kb \ll 1$), the impedance to the circulating mode is small, the impedance to the dipole mode very large. It follows that in the superposition leading to a single generator the dipole mode currents are small compared with the circulating current.

A. ELECTRICALLY SMALL CIRCULAR LOOP WITH A SINGLE GENERATOR

The current in an electrically small, single-turn circular loop driven at $\Phi = 0$ is predominantly in the circulating mode. Its amplitude is sensibly constant around the loop and given by

$$I_\Phi = V_0^e/Z_0; \qquad Z_0 = R_0 + j\omega L_0 \quad (49)$$

where the impedance Z_0 includes the resistance

$$R_0 = R_0^i + R_0^e; \qquad R_0^i = \frac{s}{2\pi a}\sqrt{\frac{\omega\mu_c}{2\sigma_c}};$$

$$R_0^e = 20k^4S^2 \quad \text{ohms} \tag{50}$$

where σ_c is the conductivity, μ_c the permeability, and a the radius of the conductor; $s = 2\pi b$ is the circumference and $S = \pi b^2$ the area enclosed by the wire. The inductance is

$$L_0 = \mu_0 b[\ln(8b/a) - 2] \tag{51}$$

The far field generated by the circulating current I_Φ in the loop includes the two spherical

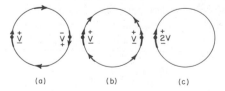

(a)   (b)   (c)

FIG. 15. Circular loop driven at diametrically opposite points by equal voltages (a) in phase opposition (circulating mode); (b) in phase (dipole mode); (c) superposition of a and b.

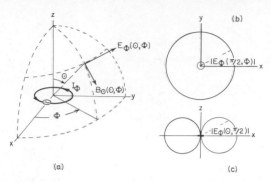

(a)   (b)   (c)

FIG. 16. (a) Circular loop antenna in the xy plane, driven at $\Phi = 0$; uniform current I_Φ. (b) Horizontal field pattern in xy plane. (c) Vertical field pattern in any plane containing the z axis.

components E_Φ and B_Θ (Fig. 16a). These have the values

$$E_\Phi(\Theta, \Phi) = \frac{V_0^e \zeta_0}{4Z_0} \frac{\exp(ikr_0)}{r_0} k^2 b^2 \sin \Theta$$

$$B_\Theta(\Theta, \Phi) = -\frac{k}{\omega} E_\Phi(\Theta, \Phi)$$

(52)

The field patterns of $E_\Phi(\Theta, \Phi)$ are shown in Fig. 16b in the xy plane and in Fig. 16c in the xz plane. Note the sharp nulls in the directions of the $\pm z$ axis. When the loop is not electrically small, more complicated distributions of current, impedances, and field patterns obtain. In general, Eqs. (50)–(52) are good approximations when the circumference or perimeter of a single-turn loop is less than a half-wavelength. For other than the circular shape, Eq. (50) is valid with s the perimeter and S the enclosed area of the loop.

B. Electrically Small Receiving Loop

The most important application of the loop antenna is reception. Many television receivers use a small circular loop for UHF channels; it is a standard tool for measuring magnetic fields; and in a shielded, multiturn form it is the usual antenna for direction finders. In its use for reception the generator (Figs. 15c and 16a) is replaced by the terminals of a transmission line leading to a radio or television receiver or other detector. The incident electromagnetic field induces a current in the loop, which maintains a voltage across the terminals of the load. The mechanism by which this is accomplished is of interest and importance in understanding the operation of the loop antenna.

Consider the circular loop shown in perspec-

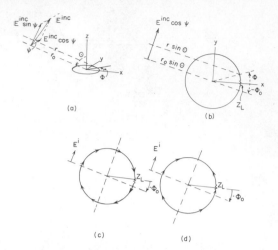

(a)   (b)

(c)   (d)

FIG. 17. Receiving loop in incident field. (a) Perspective view; (b) broadside view; (c) circulating current; (d) dipole-mode current.

tive with the incident electric field in Fig. 17a and lying in the xy plane in Fig. 17b. The incident electric field is propagating in a direction that subtends the angle Θ with the z axis and the angle Φ_0 with the x axis. The current induced in the loop includes both the circulating mode (Fig. 17c) and the dipole mode (Fig. 17d), the latter oriented to have zero currents at the ends of the diameter that lies in the incident plane wave front. For the electrically small loop, the dipole mode currents are negligible only when $kb \ll 1$. When kb is as large as 0.2, the dipole mode currents may be very significant, depending on the location of the load relative to the incident electric field.

The current in the load at $\Phi = 0$ is given by

$$I_\Phi(0) = -\frac{E_0^i h_e(\Theta, \psi)}{Z_0 + Z_1 + Z_L}$$

where E_0^i is the magnitude of the electric field at the center of the loop and the effective length is defined as follows:

$$h_e(\Theta, \psi) = 2\pi b \frac{Y_0 f_0 + Y_1 f_1}{Y_0 + Y_1}$$

where

$$f_0 = \tfrac{1}{2} jkb \cos \psi \sin \Theta$$

$$f_1 = \tfrac{1}{2}(\cos \psi \cos \Phi_0 + \sin \psi \sin \Phi_0 \cos \Theta)$$

For maximum reception, $\psi = 0$ and

$$f_0 = \tfrac{1}{2} jkb \sin \Theta; \qquad f_1 = \tfrac{1}{2} \cos \Phi_0$$

If the load is rotated so that $\Phi_0 = \pi/2$, the load is located at a point where the dipole mode current is zero. Hence, $f_1 = 0$ and the dipole mode contributes nothing to the current in the load. This condition is essential if the loop is used either to measure the magnetic field or as a direction finder.

C. SHIELDED LOOP

When the incident field is a linearly polarized plane wave, it is possible to locate the load at $\Phi = \pm \pi/2$ so that the incident electric vector is perpendicular to the loop at the load. With this orientation, the current in the load is due only to the circulating mode and, hence, proportional to the magnetic field that links with the loop.

When the load consists of a coaxial line from the output terminals of the loop to a receiver, the orientation of the loop into the desired position is not always convenient. Since an electromagnetic wave propagating over the surface of the earth has its electric field perpendicular, the magnetic field parallel to that surface, the load is conveniently located at the top of the loop. This can be accomplished in the manner shown in Fig. 18. The coaxial line is run to the top of the loop inside its conducting surface and is connected to the outside surface of the loop by way of a gap. So long as the incident electric field is linearly polarized and vertical, the presence of a dipole mode on the loop has no effect on the current in the load.

The receiving loop has the same directional characteristics as the driven loop, which are shown in Fig. 16b,c. The sharp nulls in the pattern in Fig. 16c are used for direction finding.

D. LOOP ANTENNA AS A PROBE OR SENSOR

The electrically small, shielded loop antenna is especially useful as a probe for measuring the magnetic field tangent to a metal surface in the arrangement shown in Fig. 19 for a tubular cylinder. Since the transverse magnetic field, which links with the loop, is proportional to the axial

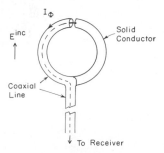

FIG. 18.   Shielded loop.

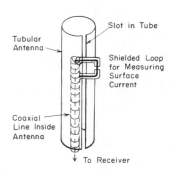

FIG. 19.   Shielded loop current probe.

surface current density on the metal surface, this can also be measured. When the conductor is cylindrical as in Fig. 19, the magnetic field is proportional to the total axial current. Thus, a small shielded loop protruding from a narrow axial slot can be moved along the conductor to measure the current distribution.

IV. The Slot Antenna

In the dipole antenna, periodically varying electric currents and charges are confined to the surface of a metal cylinder, where they are distributed in a manner suitable for the generation of electromagnetic waves. In the half-wave dipole, the current distribution is approximately $I_z(z) = I_z(0) \cos k_0 z$, the charge per unit length $q(z) = q(h) \sin k_0 z$. It is possible, in effect, to interchange these distributions by exciting the currents on the surface of a large plane of metal into which a slot of length $2h$ has been cut (Fig. 20). The inner conductor of a coaxial feedline maintains a driving current across the center of the slot. This spreads out along the edges of the slot, as shown on the right in Fig. 20, travels around the ends of the slot and back along the other edges to enter the outer conductor of the coaxial line. When $2h = \lambda_0/2$, the driving-point admittance is very high and the driving-point current $I_z(0)$ is very small. The axial distribution of current is approximately $I_z(z) = I_z(h) \sin k_0 z$. A quarter-period later the current has ceased, leaving distributions of charge along the edges of the slot. Their axial distribution is approximately $q(z) = q(0) \cos k_0 z$, as indicated on the left in Fig. 20. The associated electric- and magnetic-field lines are shown in Fig. 20 in a transverse section for the electric field lines that end on the conducting plane across the slot and for a longitudinal section for the magnetic field lines that form closed loops around the central conductor through the slot.

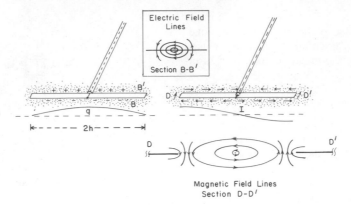

FIG. 20. Current and charge distributions along edges of slot antenna and electric and magnetic fields at two instants of time, one-quarter period apart.

The electromagnetic waves generated by the currents circulating around the edges of a slot in a sufficiently large—ideally infinite—metal plane are readily determined from those of the dipole antenna shown in Figs. 4–6. The first step is to replace the tubular metal dipole by a flat strip of width $w \sim 2a$ that is highly (perfectly) conducting and very thin. Except within distances comparable with the width, the field of the strip dipole is the same as that of the tubular dipole. The surface current on the strip is given by the boundary condition to be $I_z = wH_x = wB_x/\mu_0$.

It is seen in the inset of Fig. 20 that the electric field is transverse across the slot. If the slot is replaced by a fictitious flat strip of magnetically conducting material carrying an axial "magnetic current" given by $I_{mz}(z) = wE_x(z) = q(z) = q(0) \cos k_0 z$ (when $k_0 h = \pi/2$) and this is used to calculate the complete electromagnetic field, the result is precisely that for the dipole but with *electric and magnetic fields interchanged*. This means that the diagrams in Figs. 4–6 for the electric field of the dipole apply to the magnetic field of a slot antenna. The magnetic field lines that end on the fictitious magnetic strip (that replaces the slot) end on fictitious "magnetic charges" with opposite signs on the two sides of the strip. The complementarity between the slot and the strip dipole is a convenient artifice, but should not conceal the fact that the electromagnetic field of the slot is actually generated by the electric currents and charges on the metal plane around the slot and not by the fictitious "magnetic currents and charges" in the slot. The driving-point impedance of a center-fed slot open on both sides is given by the formula $Z_{\text{dipole}} Z_{\text{slot}} = \zeta_0^2/4$, where $\zeta_0 = 120\pi$ ohms.

Slot antennas are often used in conjunction with waveguides. When suitably cut in the walls so that a transverse electric field is maintained across each of them, they provide useful antennas excited by waveguides. Since the metal surfaces forming the outside of a waveguide in which the slots are cut are not large plane surfaces, the electromagnetic field generated by a waveguide slot is necessarily different from that of a slot in a large metal plane.

V. Horn Antennas

A. WAVEGUIDE HORNS

When a metal waveguide is terminated in an open end, the electromagnetic waves propagating in its interior according to the well-known modal distributions of waveguide theory are partly reflected back into the guide, partly transmitted out into the surrounding space. The open end of the waveguide thus behaves like an antenna. The directional properties of the radiated field as well as the fraction of power transmitted and not reflected can be modified by flaring the end of the waveguide in various ways and over a wide range of lengths. The resulting flared structure is known as an electromagnetic horn. Shapes have been designed with different far-field patterns in both the horizontal and vertical planes. A few types of electromagnetic horn are shown in Fig. 21. The directivity is always increased in the plane with the flare. The aperture of the horn acts something like a broadside array of dipoles.

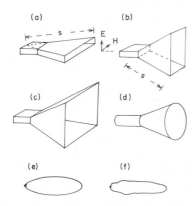

FIG. 21. Horn antennas. (a) H-plane sectoral; (b) E-plane sectoral; (c) pyramidal; (d) conical; (e) horizontal field pattern of H-plane sectoral horn with $s = 6\lambda_0$, flare angle 30°; (f) vertical field pattern of E-plane sectoral horn with $s = 6\lambda_0$, flare angle 35°.

B. BICONICAL HORN AND ANTENNA

An important hornlike antenna that is driven from a transmission line and not a waveguide is the biconical structure shown in Fig. 22a,b and the conical antenna over a ground plane. A complete theory is available for the cone with spherical end cap shown in Fig. 22a,c,d. The wide-angle form in Fig. 22c has broadband properties; that is, its impedance properties do not vary greatly with frequency over a wide range near resonance. The thin conical antenna shown in Fig. 22d (and its biconical counterpart) have

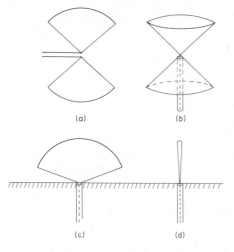

FIG. 22. Biconical and conical antennas. (a) Center-driven from two-wire line; (b) center-driven from coaxial feedline; (c) wide-angle conical antenna over ground plane; (d) thin conical monopole over ground plane.

properties closely resembling those of the cylindrical dipole.

VI. The Dielectric-Rod Antenna

All of the antennas described so far consist of metal surfaces that provide conducting paths for oscillating electric currents and charges. It is possible to have oscillating electric charges and the associated high-frequency alternating currents in a dielectric, and these can generate electromagnetic waves. An example of this type of antenna is shown in Fig. 23a. It consists of a thick dielectric cylinder filling the interior and extending out beyond the end of a circular waveguide excited in the dominant H_{11} mode with the electric field transverse (Fig. 23b). When exposed to this field, the dielectric rod is polarized so that a transverse polarization current $I = j\omega P$ is generated proportional to the electric field. This propagates in the dielectric into the long section of length L not enclosed in metal walls. Here each transverse section of the rod is equivalent to a continuous row of dipoles all oscillating in phase. The successive cross sections have progressively different phases, so that the entire dielectric rod radiates like an endfire combination of broadside arrays of oscillating dipoles. A typical far-field pattern is shown in Fig. 23c when the length of the rod is two wavelengths in air, that is, $L = 2\lambda_0$. The minor lobes can be reduced by tapering the extension of the rod in air.

VII. Pulses and Nonsinusoidal Waves

It is advantageous for some purposes (high-resolution radar, target-signal analysis, short-range communication links) to use a time depen-

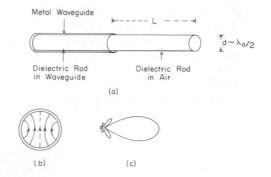

FIG. 23. (a) Dielectric-rod antenna; (b) electric field in dielectric-filled waveguide; (c) far-field pattern.

dence other than sinusoidal for the excitation of an antenna. What is the nature of the electromagnetic field radiated by the current in the dipole shown in Fig. 1a when, for example, the generator applies a single voltage pulse (or a regular succession of such pulses) across the input terminals of the feeding transmission line? If the upper conductor is charged $+$, the lower $-$, during the short duration of the voltage pulse, positive and negative pulses of current travel toward the dipole, respectively along the upper and lower conductors. When these reach the dipole, they radiate an outward-traveling electromagnetic pulse until the entire current has left the transmission line and is traveling outward on the two arms of the antenna. If the pulse is short enough so that this occurs before the forward edges of the positive and negative current pulses reach, respectively, the upper and lower ends of the antenna, radiation virtually ceases until reflection at these ends begins and radiation of an oppositely directed electromagnetic pulse occurs during the time when the outward current is reflected to become an inward moving current (with somewhat reduced amplitude due to energy loss in radiation). There can be a sequence of such reflections alternately at the ends and center of the dipole as the pulses travel back and forth with continuously decreasing amplitude. The associated outward-traveling electromagnetic field consists of a succession of alternating positive and negative pulses, as shown in Fig. 24a, which reproduces a measured pulse sequence. Thus, a single voltage pulse produces an electromagnetic field that consists of a sequence of pulses with decreasing amplitude.

In order to generate only a single outward-traveling electromagnetic-field pulse, it is necessary to suppress the reflections at the ends of the antenna. This is accomplished with a traveling-wave antenna terminated to minimize reflections at the ends at all frequencies. Examples are shown in Fig. 12b and Fig. 13. With such an antenna the current pulse that travels along the antenna is absorbed in, rather than reflected from, the termination. When this is true, radiation is limited to the electromagnetic pulse radiated during the short interval in which the current pulse is transferred from the transmission line to the antenna. It follows that only a single electromagnetic-field pulse travels out from the antenna. A lateral electromagnetic pulse actually generated by an antenna T of the type in Fig. 12b is shown in Fig. 24b at a succession of distances ρ from the end of the antenna.

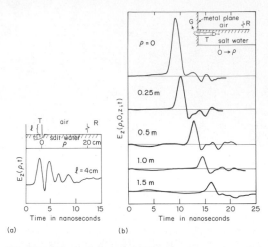

(a)   (b)

FIG. 24.   Vertical electric field over saltwater at radial distance ρ due to a single voltage pulse applied to (a) vertical monopole of length l in air, $\rho = 20$ cm; (b) horizontal terminated insulated traveling-wave antenna in saltwater near the air surface. Pulse width is 2 nsec.

The simplest antenna for receiving pulses and other nonsinusoidal waves is a dipole that is electrically short at all component frequencies that constitute the spectrum of the pulse to be received.

VIII. Conclusion

Antennas are the eyes and ears of the modern world. They transmit and receive all that is seen and heard over television and radio and much of what is said over the telephone. They are essential parts of radar, navigation, radio astronomy, and remote control. They have many shapes and locations on earth, on satellites, on orbiters, and on rockets traveling into space.

BIBLIOGRAPHY

Collin, R. E., and Zucker, F. J., eds. (1969). ''Antenna Theory,'' Parts 1 and 2. McGraw-Hill, New York.
Elliott, R. S. (1981). ''Antenna Theory and Design.'' Prentice-Hall, Englewood Cliffs, New Jersey.
Flügge, S., ed. (1958). ''Encyclopedia of Physics,'' Vol. XVI. Springer-Verlag, Berlin and New York.
Johnson, R. C., and Jasik, H., eds. (1984). ''Antenna Engineering Handbook,'' 2nd ed. McGraw-Hill, New York.
King, R. W. P., and Prasad, S. (1986). ''Fundamental Electromagnetic Theory and Applications.'' Prentice-Hall, Englewood Cliffs, New Jersey.
King, R. W. P., and Smith, G. S. (1981). ''Antennas in Matter.'' MIT Press, Cambridge, Massachusetts (translated into Russian, 1984).

ANTENNAS, REFLECTOR

W. V. T. Rusch   *University of Southern California*

GLOSSARY

Antenna: That part of a transmitting system that converts electrical currents to electromagnetic waves; conversely, that part of a receiving system that converts electromagnetic waves to electrical currents in the receiver.

Aperture blockage (blocking): Shadowing condition that results from objects lying in the path of rays arriving at or departing from the aperture of an antenna.

Aperture illumination efficiency: Ratio of the maximum directivity of an antenna to its standard directivity.

Bandwidth: Range of frequencies within which the performance of an antenna, with respect to some characteristic, conforms to a specified standard.

Beamwidth: In a planar cut through a radiation pattern containing the direction of the maximum of a lobe, the angle between the two directions in which the radiation intensity is one-half the maximum value, one-tenth the maximum value, or zero is defined to be, respectively, the half-power (3-dB), tenth-power (10-dB), or null beamwidth.

Decibel: Twenty times the base-10 logarithm of the ratio of two field magnitude values or 10 times the base-10 logarithm of the ratio of two radiation intensity values.

Directivity: Ratio of the radiation intensity in a given direction from the antenna to the radiation intensity averaged over all directions.

Feed: In a transmitting system, the feed is a small antenna serving as a primary radiator of energy to illuminate the reflector. In a receiving system, the feed accepts the energy that has been transformed by the reflector into a converging wave, which can then be converted to electrical signals in the receiver.

Gain: Ratio of the radiation intensity in a given direction to the radiation intensity that would be obtained if the power accepted by the antenna were radiated isotropically.

Polarization: Spatial orientation of the electric field radiated by an antenna.

Radiation intensity: In a given direction, the power radiated. from an antenna per unit solid angle.

Sidelobe level: Maximum relative directivity of a secondary lobe with respect to the maximum directivity of an antenna, usually expressed in decibels.

Standard directivity: Maximum directivity from a planar aperture of area A when excited with a uniform-amplitude equiphase distribution. For planar apertures the area A of which is much larger than the square of the wavelength λ, and from which the radiation is confined to a half-space, the maximum directivity is $4\pi A/\lambda^2$.

A reflector antenna is a device consisting of one or more reflecting surfaces and a radiating or absorbing feed that converts electrical currents to electromagnetic waves in a transmitting system or converts electromagnetic waves to electrical currents in a receiving system. Specific reflector antennas often carry the name of the reflector shape, such as the paraboloidal reflector antenna. Reflector antennas are widely used in the transmission or reception of microwaves over long distances, such as in radio and radar astronomy, telecommunications, and remote sensing.

I. Prime-Focus Paraboloidal Reflectors

A. HISTORICAL OVERVIEW

A paraboloidal reflector (mirror) has the unique property of transforming rays emerging radially from a point source at its focus into a bundle of parallel rays. Similarly, a parabolic cylinder reflector transforms rays from a line source on its focal line into a bundle of parallel rays. Thus energy initially emitted diffusely over a wide angular range can be entirely redirected in a particular direction. Hertz applied this principle to radio waves in 1888 when he used a 2-m-high, 1.2-m-wide, wood-frame-supported, sheet-zinc parabolic cylinder to focus electromagnetic radiation of 66-cm wavelength. Marconi also used a deep parabolic cylinder for his 25-cm experiments. One of the first large-aperture, doubly curved paraboloidal reflectors was constructed by Reber in 1937. This 31.5-ft-diameter paraboloid, with an F/D ratio of 0.6, consisted of thin aluminum plates on a framework of 72 radial wooden rafters (Fig. 1). [See ANTENNAS.]

During World War II the military value of radar devices stimulated vigorous development of microwave reflectors, and during the postwar decades the stringent demands of radio astronomy and telecommunications, together with the simultaneous development of low-noise receivers, solid-state circuits and devices, and integrated electronics, have stimulated reflector antennas to a highly advanced state.

B. DEFINITION OF RADIO-FREQUENCY PERFORMANCE

An axially symmetric, prime-focus paraboloidal reflector antenna in a transmitting system generally consists of the reflector and a point source of electromagnetic radiation located at its focus (Fig. 2). The laws of geometric optics dictate that the energy from the source be reflected into space as a bundle of rays parallel to the axis of the reflector. However, diffraction effects due to the reflector truncation cause energy to be radiated in all directions of space. A predominance of the radiation is confined to a narrow beam surrounding the reflector axis, but significant levels are radiated at wide angles and even behind the reflector.

Characterization of the complete radiation diagram or directivity of the antenna is a function of the properties of the focal-point source or "feed" and the ratio of the reflector's focal length F to the diameter D of its circular aperture. Simple projections from geometric optics, however, permit the feed properties to be translated into a distribution of the reflected electric field in the reflector aperture. This aperture–field distribution can be used to predict the directivity characteristics.

A traditional approximation for many practical aperture–field distributions is

$$E(r) = E_0[B + (1 - B)(1 - r^2/a^2)^p] \quad (1)$$

where E_0 is a constant, r the radial coordinate of a point in the circular aperture (Figs. 3 and 4), a the maximum radius of the circular aperture ($a = D/2$), and B a constant that specifies the "edge taper" of this axially symmetric distribution. One can convert the parameter B, a dimensionless quantity, to decibels (dB) by taking the logarithm and multiplying by 20.0.

The aperture illumination efficiency of a paraboloidal reflector for which Eq. (1) describes the aperture–field distribution function is plotted in Fig. 5 for $p = 0$, 1, and 2, and B varying from 1 (0.0 dB) to 0.032 (-30.0 dB). The beamwidths of the radiation intensity pattern radiated by this aperture function are plotted in Fig. 6, where the ordinate value is the beamwidth in degrees multiplied by the diameter D divided by the wavelength λ. The levels of the first sidelobe adjacent to the main lobe relative to the peak of the main lobe are plotted in Fig. 7. When $p = 0$ or $B = 1$ (0.0 dB), the radiation intensity is the typical Airy pattern and the first sidelobe is -17.6 dB.

C. ABERRATIONS

1. Effects of Random Surface Distortions

Deviations from a pure paraboloidal surface cause degradation of its directivity pattern. If a zero-mean Gaussian phase error of root mean square (rms) value δ radians is superimposed on its aperture–field distribution [Eq. (1)], the resulting directivity for $p = 0$ is

$$G(\theta, \phi) = D_0(\theta, \phi)e^{-\delta^2}$$
$$+ \left(\frac{\pi D}{\lambda}\right)^2 e^{-\delta^2} \left(\frac{2c}{D}\right)^2 \sum_{n=1}^{\infty} \frac{(\delta^2)^n}{n!n}$$
$$\times \exp\left[\frac{-(\pi c \sin \theta/\lambda)^2}{n}\right] \quad (2)$$

where $D_0(\theta, \phi)$ is the no-error directivity, (θ, ϕ) the standard polar angles, and c a "correlation interval." The complete directivity is thus the

FIG. 1.   Original Reber paraboloid on display at the National Radio Astronomy Observatory (NRAO). (Courtesy of W. E. Howard III, NRAO.)

power sum of a reduced-amplitude, no-error directivity plus a more diffuse pattern containing the "lost" power. The maximum directivity at $\theta = 0$ is thus

$$D(0, 0) = \eta \left(\frac{\pi D}{\lambda}\right)^2 e^{-\delta^2} \left[1 + \frac{1}{\eta} \left(\frac{2c}{D}\right)^2 \sum_{n=1}^{\infty} \frac{(\delta^2)^n}{n!n}\right]$$

(3)

where η is the aperture illumination efficiency. For a shallow reflector with an rms surface error of ε,

$$\delta \cong 2(2\pi\varepsilon/\lambda) \quad \text{radians} \tag{4}$$

For a small correlation interval, the second term in Eq. (3) can usually be neglected, leaving

$$D(0, 0) \cong \eta \left(\frac{\pi D}{\lambda}\right)^2 \exp\left[-\left(\frac{4\pi\varepsilon}{\lambda}\right)^2\right] \tag{5}$$

2. Transverse Feed Defocusing

Displacement of the antenna feed from its position at the focus can cause limited movement of the primary beam of a paraboloidal antenna without moving the reflector. Feed displacement in a direction transverse to the reflector axis will cause the beam to "scan" on the opposite side of the axis. However, operation of the antenna under these defocused conditions will cause some degradation of the directivity pattern. Figure 8 shows the effects on the main antenna beam when it is scanned in this manner: (1) The peak directivity decreases; (2) the beamwidth increases; and (3) the sidelobe on the axis side of the beam rises (coma lobe) while the sidelobe on the other side merges with the main beam and second sidelobe.

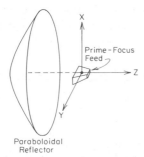

FIG. 2.   Geometry of paraboloidal reflector antenna with prime-focus feed.

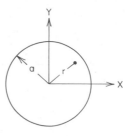

FIG. 3.   Geometry of circular aperture of paraboloidal antenna.

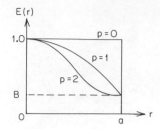

FIG. 4. Aperture–field distribution [Eq. (1)] versus radial coordinate of circular aperture.

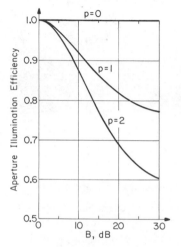

FIG. 5. Aperture illumination efficiency of paraboloidal reflector antenna versus edge taper B. [From Rusch, W. V. T. (1984). *IEEE Trans. Antennas Propag.* **AP-32**, 312–329. Copyright 1984 IEEE.]

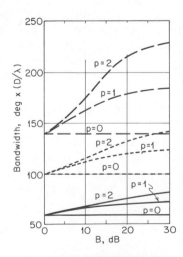

FIG. 6. Beamwidth of paraboloidal reflector antenna versus edge taper B (–––, null beamwidth; - - -, 10-dB beamwidth; ——, 3-dB beamwidth). [From Rusch, W. V. T. (1984). *IEEE Trans. Antennas Propag.* **AP-32**, 312–329. Copyright 1984 IEEE.]

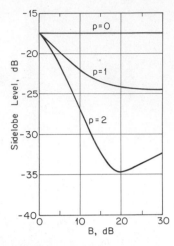

FIG. 7. Sidelobe level of paraboloidal reflector antenna versus edge taper B. [From Rusch, W. V. T. (1984). *IEEE Trans. Antennas Propag.* **AP-32**, 312–329. Copyright 1984 IEEE.]

3. Aperture Blocking

The presence of objects in front of the aperture of a reflector antenna causes significant changes in its directivity pattern. In general, these objects can be classified as (1) large, centrally located objects such as a feed, subreflector, or instrument package; and (2) long, thin cylindrical structures (support struts) used for mechanical support. Ray tracing reveals that the aperture-blocking shadows cast by a point source at the focus of a paraboloid are those shown in Fig. 9. The circular shadow and the rectangular shadows are cast by the parallel rays reflected from the reflector. The triangularlike shadows are cast by the spherical wave from the focus striking the oblique struts.

In general, the peak directivity of the antenna will decrease by an amount proportional to a weighted sum of the total shadow. Effects on the sidelobe characteristics are considerably more difficult to calculate from a diffraction theoretical viewpoint because of the complexity of the problem.

II. Other Reflector Shapes

A. TRADITIONAL DUAL-REFLECTOR ANTENNA SYSTEMS

The basic properties of the Cassegrain dual-reflector antenna (Fig. 10) are derived from the principles of ray optics. A small, convex hyper-

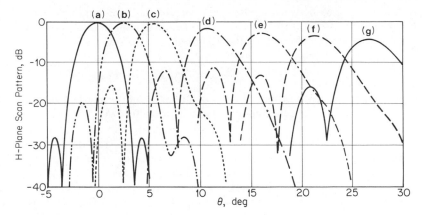

FIG. 8. Radiation patterns of 25-wavelength-diameter paraboloid illuminated by laterally defocused feed. (a) Focused conditions; (b) 1-BW scan; (c) 2-BW scan; (d) 4-BW scan; (e) 6-BW scan; (f) 8-BW scan; (g) 10-BW scan.

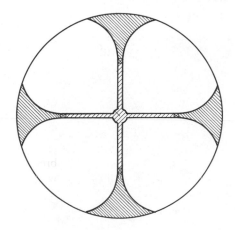

FIG. 9. Aperture-blocking shadows. [From Rusch, W. V. T. (1984). *IEEE Trans. Antennas Propag.* **AP-32,** 312–329. Copyright 1984 IEEE.]

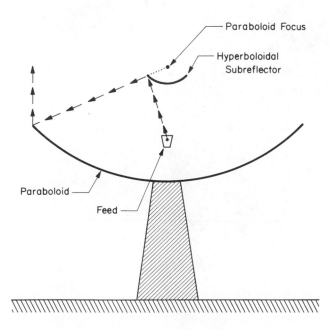

FIG. 10. Geometry of Cassegrain dual-reflector antenna. [From Rusch, W. V. T. (1984). *IEEE Trans. Antennas Propag.* **AP-32,** 312–329. Copyright 1984 IEEE.]

boloidal subreflector is placed between a point-source feed and the prime focus of the paraboloid. Rays from the feed are transformed by the subreflector into rays appearing to emerge from the paraboloid focus. These rays reflect from the paraboloid parallel to its axis. The NASA/JPL 64-m Cassegrain antenna is shown in Fig. 11.

The Gregorian dual-reflector antenna employs a concave ellipsoidal subreflector placed beyond the paraboloid focus. Rays emerge radially from the feed, reflect from the subreflector, converge toward the focus, and then diverge toward the paraboloid as if they were coming from its focus. Thus the ray behavior of the Gregorian system is similar to that of the Cassegrain except for longer ray paths and "inversion" of the ray system. The above descriptions have assumed the antennas to be transmitting. When operating as receiving antennas, the ray paths in these dual-reflector systems are the same, although the ray directions are reversed.

B. OFFSET-FED REFLECTORS

Blocking of the aperture of a reflector antenna by its feed, feed supports, or subreflector generally degrades its directivity pattern. Consequently, many high-performance microwave systems employ offset-fed antennas to eliminate the effects of aperture blocking. This configuration, illustrated by the *Westar* communications satellite antenna in Fig. 12, consists of an offset section of a "parent" paraboloid illuminated by a tilted feed located at the focus. The relatively complicated mechanical geometry of this configuration causes serious fabrication and assembly

FIG. 11.   NASA/JPL 64-m Cassegrain antenna. (Courtesy of D. Bathker, Jet Propulsion Laboratory.)

FIG. 12. *Westar* communications antenna. [Courtesy of D. Nakatani, Hughes Aircraft Company. From Rusch, W. V. T. (1984). *IEEE Trans. Antennas Propag.* **AP-32,** 312–329. Copyright 1984 IEEE.]

problems. Nevertheless, many high-performance requirement specifications can be met only by the offset geometry (Rudge, 1978).

A quasi-offset dual-reflector "open Cassegrain" antenna consisting of an offset section of a paraboloid and an offset hyperboloidal subreflector is shown in Fig. 13a. However, the feed protrudes through the main reflector, and, consequently, aperture blocking is not entirely eliminated. Alternative configurations consist of sec-

tions of conventional coaxial Cassegrain and Gregorian systems (Fig. 13b). A 7-m-diameter, millimeter-wave offset Cassegrain antenna is shown in Fig. 14. System performance of an offset Cassegrain dual-reflector antenna can be significantly improved by depressing the axis of the parent subreflector (Fig. 13c).

Another version of the offset reflector antenna is the horn–reflector antenna (Fig. 15), a combination of electromagnetic horn and a reflecting section of a paraboloid. The apex of the horn coincides with the prime focus of the parent paraboloid. The horn walls produce sufficient shielding to yield ultralow sidelobes and backlobes in most directions. This antenna is widely used in microwave relay systems.

C. Shaped Reflectors

The paraboloid and its dual-reflector variations considered in the preceding section radiate narrow "pencil" beams of circular or elliptical cross section. These cross sections are determined by the laws of diffraction and consequently are termed *diffraction-limited*. Many applications, however, require that the reflector

(a)
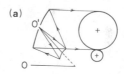

(b)
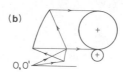

(c)
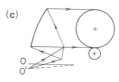

FIG. 13. Offset dual-reflector configurations with paraboloid vertex at O and feed at O'. (a) Open Cassegrain; (b) double offset; (c) optimized double offset. [After Rudge, A. W. and Adatia, N. A. (1978). *Proc. IEEE* **66,** 1592–1618. Copyright 1978 IEEE.]

FIG. 14.   Crawford Hill 7-m offset Cassegrain millimeter-wave antenna. [From Chu, T. S., Wilson, R. W., England, R. W., Gray, D. A., and Legg, W. E. (1978). *Bell Syst. Tech. J.* **57**, 1257–1288.]

contour be modified in order to achieve a substantially different directivity pattern. For example, ray optical procedures have been used to modify a horn–reflector antenna so that the resulting pattern conforms to the islands of Japan.

The techniques of ray optics can also be used to modify the shape of axially symmetric dual reflectors in order to achieve arbitrary aperture distribution functions. Enhancing the central curvature of a Cassegrain subreflector, for example, causes the feed energy to be distributed more uniformly over the paraboloid aperture. Thus it is possible to achieve aperture illumination efficiencies approaching 100%.

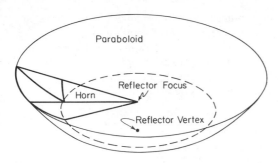

FIG. 15. Horn–reflector antenna. [From Rusch, W. V. T. (1984). *IEEE Trans. Antennas Propag.* **AP-32,** 313–329. Copyright 1984 IEEE.]

The 1977 *Voyager* spacecraft used such a shaped dual-reflector system for high-data-rate telemetry and optimum efficiency at 8.4 GHz (Fig. 16). The dual-frequency subreflector was virtually transparent to radiation from an S-band (2.1-GHz) feed nestled behind it. At S band, the shaped main reflector differed only slightly from a paraboloid, and the S-band feed was placed at the focus of the resultant best-fit paraboloid.

D. SCANNING REFLECTORS

The geometry of a paraboloid is not naturally amenable to single-feed beam scanning without the penalty of scan loss and coma lobes. Reflectors that are symmetric about the scan axis are more suitable for extremely wide angle scanning. For example, a spherical reflector is completely symmetric in any scan plane. A bundle of parallel rays incident on the sphere generates reflected rays, all of which cross a line that passes through the reflector vertex and that parallels the incident ray bundle. A suitable feed can then be placed so as to intercept most of these rays. Phase-correcting subreflectors can also be incorporated to eliminate spherical aberration, although at the expense of significant ap-

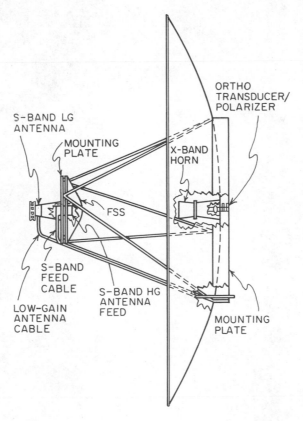

FIG. 16. *Voyager* spacecraft shaped dual-reflector system (FSS, frequency-selective surface; LG, low-gain; HG, high-gain). [Courtesy of A. G. Brejcha, Jet Propulsion Laboratory, From Rusch, W. V. T. (1984). *IEEE Trans. Antennas Propag.* **AP-32,** 312–329. Copyright 1984 IEEE.]

erture blocking. Massive interest in spherical reflectors was generated in 1960 through 1963 by the construction of the 1000-ft-diameter spherical radiotelescope at the National Astronomy and Ionosphere Center in Arecibo, Puerto Rico (Fig. 17).

The spherical geometry is a special case of the toroidal geometry (Fig. 18). The reflector is symmetric about the y axis. In the xz plane the reflector is a circle of radius R_0. Any plane containing the y axis intersects the reflector in its generating function,

$$x^2 + y^2 = f(y, R_0) \qquad (6)$$

Thus, to the extent that the truncation of the torus can be neglected, a feed rotated about the y axis will always encounter the same reflector environment and thus generate a beam shape that is independent of rotation angle.

The parabolic torus represents a compromise between the focusing properties of the paraboloid and the scanning properties of the torus. In this case the generating function is

$$f(y, R_0) = (y^2/2R_0) - R_0 \qquad (7)$$

which, in an attempt to reduce aperture phase errors, has placed the focus on the parabola at the distance $R_0/2$ from the origin. Movement of the feed in the xz plane on a circle of radius $R_0/2$ thus yields essentially identical beams over a wide range of azimuth. Nevertheless, the system suffers from relatively high sidelobes and less than desirable aperture illumination efficiency.

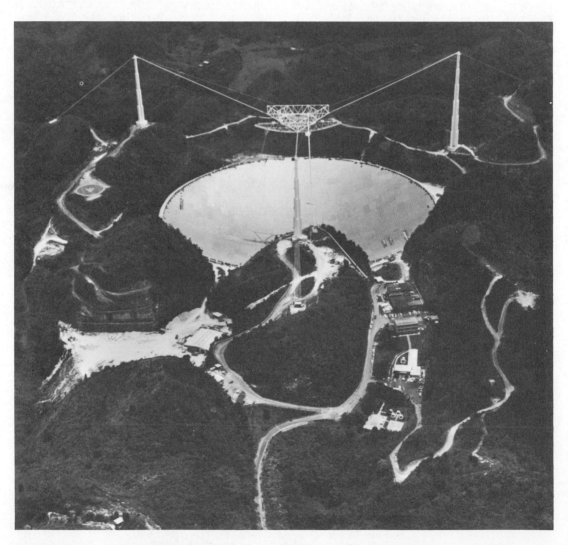

FIG. 17.   Arecibo 1000-ft-diameter fixed spherical reflector. (Courtesy of H. D. Craft, NAIC.)

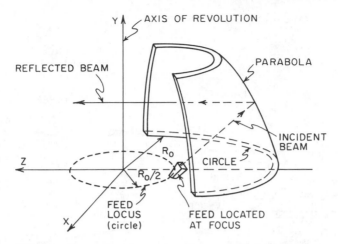

FIG. 18. Geometry of toroidal reflector. [Courtesy of F. A. Young, Hughes Aircraft Company. From Rusch, W. V. T. (1984). *IEEE Trans. Antennas Propag.* **AP-32,** 312–329. Copyright 1984 IEEE.]

III. Feeds for Reflector Antennas

Every reflector antenna has an associated radiating or receiving feed system. Antenna feed design is a separate subject area of considerable sophistication. Feeds can be used to provide high aperture illumination efficiency, low sidelobes, or a special beam contour. For the purposes of a minimal degree of completeness, a single high-performance feed type, used for many prime-focus paraboloidal reflectors, the corrugated horn, will be described here.

The corrugated horn has achieved wide acceptance as a prime-focus illuminator of reflector antennas because of its large bandwidth, symmetric pattern, and relatively low sidelobes. The name is derived from the deep grooves in the inside wall of the horn. The geometry of a millimeter-wave corrugated horn used for the offset Cassegrain radiotelescope shown in Fig. 14 is shown in Fig. 19a. The measured 28.5-GHz patterns in Fig. 19b show symmetry to levels below −25 dB.

IV. Reflectors with Contoured Beams

An antenna designed to have a prescribed pattern shape, generally in the form of a geographic target, but differing substantially from a circle or ellipse, is said to be a contoured-beam antenna. Communication satellite antennas require contoured beams to (1) increase the minimum radiation density over the coverage area, (2) reduce

radiation interference into contiguous areas, and (3) conserve the available radio-frequency spectrum by spatial or polarization isolation. The principal parameters in the design of a contoured-beam antenna are coverage, ripple over the coverage area, sidelobes, polarization, and bandwidth. Successful systems generally employ offset reflectors (to eliminate aperture blockage) and an array of feed elements numbering from a few to more than 100.

The world's first domestic communications satellite in synchronous orbit was the *Telstar* satellite launched in 1972. Its antenna system produced the first contoured beam from synchronous orbit. *Westar,* launched in 1975, was a modified version of the *Telstar* satellite, with beam coverage of the contiguous United States, Alaska, and Hawaii. The *Westar* reflector (Fig. 12) consisted of two parts: a skeleton structure onto which a metallic mesh material was stretched, and a three-rib structure supporting the skeleton. Both the skeleton and the support ribs were fashioned from an aluminum honeycomb graphite-fiber composite. The feed provided maximum directivity at 4 and 6 GHz throughout the three coverage zones.

The *Intelsat IV-A* antenna system consisted of three offset-fed parabolas, each fed by a multiple-horn array (Fig. 20). Each of the three beams was isolated from the other two, and the sidelobes of each directivity pattern were controlled to minimize interference with adjacent beams operating at the same frequencies. The first of

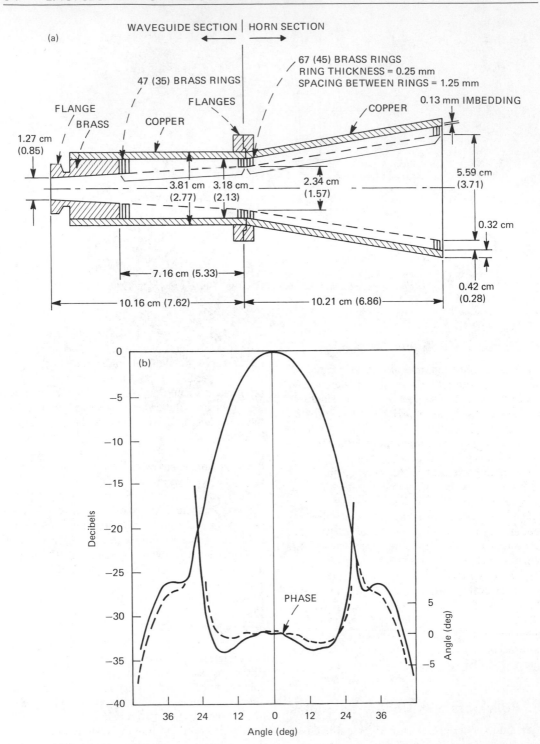

FIG. 19.   Bell Laboratories millimeter-wave corrugated horn. (a) Conical corrugated horn and its hybrid launcher; (b) measured radiation patterns at 28.5 GHz (——, *E* plane; - - -, *H* plane). [From Chu, T. S., Wilson, R. W., England, R. W., Gray, D. A., and Legg, W. E. (1978). *Bell Syst. Tech. J.* **57,** 1257–1288.]

FIG. 20. *Intelsat IV-A* communications antenna. [Courtesy of D. Nakatani, Hughes Aircraft Company. From Rusch, W. V. T. (1984). *IEEE
Trans. Antennas Propag.* **AP-32,** 313–329. Copyright 1984 IEEE.]

the *Intelsat IV-A* series was launched in September 1975.

The *Comstar I* communications satellite (launched in 1976) used a pair of high-directivity reflectors to provide service within the contiguous United States, Alaska, Hawaii, and Puerto Rico (Fig. 21). Isolation between the two antennas was accomplished by means of orthogonal polarization screens in the aperture plane of each of the two offset parabolas.

The *Intelsat VI* spacecraft provides communications between 4 and 6 GHz and between 11 and 14 GHz. Figure 22 shows the 4-GHz transmit array, a planar array of 146 conical horns, arranged so that the resulting secondary pattern from a large offset-fed parabola is contoured in the required manner.

V. Reflector Structures and Materials

A. SATELLITE GROUND STATIONS

The requirements for a cost-effective, high-performance, earth-based satellite tracking antenna have led to three basic structural types: (1) the kingpost structure, which employs a kingpost arrangement for the primary azimuth

FIG. 21. *Comstar I* communications antenna. [Courtesy of Dr. Nakatani, Hughes Aircraft Company. From Rusch, W. V. T. (1984). *IEEE Trans. Antennas Propag.* **AP-32,** 313–329. Copyright 1984 IEEE.]

rotating mechanism; (2) the wheel-and-track structure, with a large azimuth track capable of supporting structures greater than the kingpost; and (3) the beam waveguide structure, which utilizes a mirror system to guide the microwave signals from the reflector to the receiver room, thus eliminating rotary joints and long waveguide runs. Selection of the appropriate structure depends on cost, desired mechanical properties, and mission of the station.

B. HIGH-PRECISION MILLIMETER-WAVE REFLECTORS

Because of an abundance of millimeter-wave and submillimeter-wave spectral-line radio sources, interest has grown in large-aperture, high-precision, radio-astronomical reflectors capable of operating at these very short wavelengths. A 10.4-m Cassegrain antenna was constructed by the California Institute of

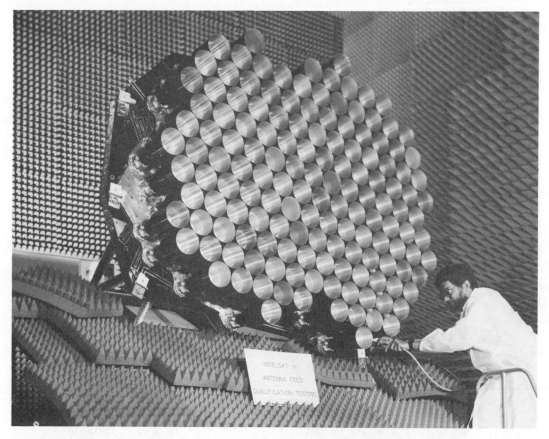

FIG. 22.   *Intelsat VI* 4-GHz transmit feed array. [Courtesy of M. Caulfield, Hughes Aircraft Company. From Rusch, W. V. T. (1984). *IEEE Trans. Antennas Propag.* **AP-32,** 313–329. Copyright 1984 IEEE.]

Technology and operates at 230 GHz with an illumination efficiency of greater than 50%. The surface consists of 84 panels, each consisting of sheet aluminum cemented to accurately machined honeycomb panels, which are supported on a disassemblable tubular steel framework.

The Bell Telephone Laboratories 7-m-diameter offset Cassegrain antenna (Fig. 14) is another state-of-the-art millimeter-wave reflector. The primary mirror consists of 27 cast-aluminum surface panels, and the overall surface tolerance permits operation up to 300 GHz.

C. Ultralarge Ground Reflectors

The largest reflector in the world is the 1000-ft-diameter Arecibo spherical reflector (Fig. 17). Beam scanning for this antenna is provided by the earth's rotation and limited movement of the feed. The largest fully steerable reflector is the Max-Planck-Institut für Radioastronomie 100-m-diameter Gregorian antenna in Germany (Fig. 23). Controlled elastic deformations of the steel supports permit the surface to convert to a continuous family of approximately paraboloidal shapes as the focal point is tracked by computer-controlled movement of the feed. The inner 80 m of the surface consists of solid panels; the remainder is perforated plate and mesh. The antenna has been operational since 1972 at frequencies as high as 36 GHz.

D. Spacecraft Reflectors

Reflectors are inherently broadband, relatively easy to fabricate and test, and lightweight (1–5 kg/m^2). They constitute the largest class of spacecraft-borne high-directivity antennas.

1. Solid Spacecraft Reflectors

Solid prime-focus reflectors have been successfully employed on numerous missions in near and deep space, for example, the *Viking*

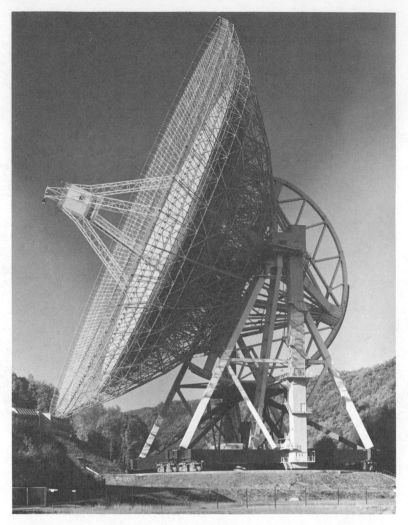

FIG. 23. Max-Planck-Institut für Radioastronomie (MPIfR) 100-m-diameter Gregorian reflector for radio astronomy. [Courtesy of R. Wielebinski, MPIfR. From Hochenberg, D., Grahl, B. H., and Wielbinski, R. (1973). *Proc. IEEE* **61**, 1288–1295. Copyright 1973 IEEE.]

Orbiter (1975) dual-frequency, 1.5-m-diameter paraboloid. This reflector was of sandwich construction, with a 1.3-cm-thick honeycomb core and 0.25-mm graphite epoxy face skins. Offset-fed single reflectors have been used to eliminate blocking aberrations (e.g., the *Westar, Comstar,* and *Intelsat* reflectors described previously). A 3.7-m-diameter solid Cassegrain antenna was used on the *Voyager* mission (Fig. 16).

2. Structural Considerations for Solid-Surface Spacecraft Reflectors

The harsh demands of the launch and space environments have established unusual structural and material requirements: (1) high strength to survive the loads during launch, (2) high stiffness, (3) low mass to permit larger payloads, (4) low thermal expansion coefficient, (5) high contour accuracy, (6) long-term (5–10 yr) stability, and (7) structural compatibility with postfabrication adjustments. Advanced composite materials with fibers of graphite, Kevlar, or boron embedded in an epoxy adhesive matrix have been found to satisfy most of these requirements. Structural configurations that have proved most compatible with these advanced materials are (Fig. 24) (1) thick sandwich unstiffened shell (e.g., the *Viking Orbiter* antenna), (2)

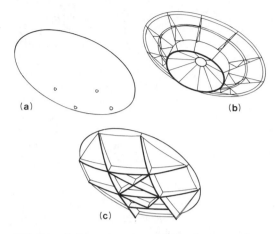

FIG. 24.   Solid-surface spacecraft reflector configurations. (a) Thick sandwich unstiffened shell; (b) truss-supported thin membrane shell; (c) rib-stiffened thin sandwich shell. [From Archer, J. S., (1978). High-performance parabolic antenna reflectors, *7th AIAA Comm. Satellite Sys. Conf., San Diego.*]

truss- or rib-stiffened thin membrane shell, and (3) rib-stiffened thin sandwich shell (e.g., the *Voyager* antenna).

3. Deployable Spacecraft Reflectors

The maximum diameter of a spacecraft-borne solid-surface reflector is limited by the size of the launch vehicle. Larger apertures can be achieved only by deploying, erecting, or manufacturing the antenna in orbit. One of the first large-aperture deployable reflectors was the ATS F & G 30-ft-diameter gored parabola (Fig. 25). This flexible radial wrap rib configuration consisted of a toroidal hub to which were attached 48 flat radial ribs; between the ribs was stretched a lightweight reflecting mesh. To furl the reflector, the ribs were wrapped around the hub with the mesh folded between them. Deployment was accomplished by releasing the stored elastic strain energy of the wrapped ribs.

FIG. 25.   Thirty-foot-diameter 48-gore ATS F reflector. [NASA photo. From Rusch, W. V. T. (1984). *IEEE Trans. Antennas Propag.* **AP-32**, 313–329. Copyright 1984 IEEE.]

FIG. 26.   Dichroic subreflector for *Voyager* antenna system. [Courtesy of A. G. Brejcha, Jet Propulsion Laboratory. From Rusch, W. V. T. (1984). *IEEE Trans. Antennas Propag.* **AP-32,** 313–329. Copyright 1984 IEEE.]

The outer diameter of the stowed configuration was 6.5 ft.

E. DICHROIC REFLECTORS

Dichroic or frequency-selective surfaces are designed to exhibit different ratio-frequency properties, usually of a complementary nature, at two or more separated frequencies. In its simplest form, a dichroic surface is virtually perfectly reflecting at one frequency and virtually transparent at another.

The most common application of dichroic reflectors has been as subreflectors in Cassegrain antenna systems. The subreflectors consist of an array of crosses etched on a shaped dielectric substrate. Near the resonant frequency of the crosses, the field from a feed located on the convex side of the subreflector is almost completely reflected, whereas at frequencies significantly below the resonant frequency the subreflector is virtually transparent, thus permitting a wave from a feed on the concave side of the subreflector to pass through it unattenuated.

A dichroic subreflector was used for the 2.1/8.4-GHz shaped Cassegrain antenna on the *Voyager* spacecraft (Fig. 16). The subreflector was fabricated from 8.4-GHz resonant aluminum crosses, etched on a Mylar sheet, which was bonded to a Kevlar epoxy skin (Fig. 26). A complementary dichroic surface, with resonant slots to transmit high frequencies while reflecting lower frequencies, was used for the ground antenna for the *Voyager* project.

BIBLIOGRAPHY

Galindo, V. (1964). *IEEE Trans. Antennas Propag.* **AP-12,** 403–408.

Hachenberg, O., Grahl, B. H., and Wielebinski, R. (1973). *Proc. IEEE* **61,** 1288–1295.

Hertz, H. (1893). "Electric Waves." Macmillan & Co., Ltd., London.

Marconi, G. (1896). British Patent 12039.

Rudge, A. W., and Adatia, N. A. (1978). *Proc. IEEE* **66,** 1592–1618.

Rudge, A. W., Milne, K., Olver, A. D., and Knight, P., eds. (1982). "The Handbook of Antenna Design," Chapter 4. Peter Peregrinus, Ltd., London.

Rusch, W. V. T. (1984). *IEEE Trans. Antennas Propag.* **AP-32,** 313–329.

Ruze, J. (1965). *IEEE Trans. Antennas Propag.* **AP-13,** 660–665.

Ruze, J. (1966). *Proc. IEEE* **54,** 633–640.

Silver, S., ed. (1949). "Microwave Antenna: Theory and Design," Chapter 6. McGraw-Hill, New York.

Takagi, F., and Takeichi, Y. (1975). *IEEE Trans. Antennas Propag.* **AP-23,** 757–763.

COMMUNICATION SYSTEMS, CIVILIAN

Simon Haykin   *McMaster University*

GLOSSARY

Amplitude modulation (AM): Process in which the amplitude of a carrier is varied about a mean value, linearly with a baseband signal.

Average probability of symbol error: Probability that the symbol recreated at the receiver output of a digital communication system differs from that transmitted, on the average.

Baseband: Band of frequencies representing an original signal as delivered by a source of information.

Bit: Acronym for *bi*nary digi*t*.

Double-sideband suppressed-carrier (DSBSC) modulation: Modulated wave resulting from multiplying a baseband signal by carrier wave.

Frequency modulation (FM): Form of angle modulation in which the instantaneous frequency is varied linearly with the baseband signal.

Frequency-shift keying (FSK): Digital modulation in which the frequency of a carrier takes on a discrete value chosen from a preselected set in accordance with the input data.

Phase modulation (PM): Form of angle modulation in which the angle of a carrier is varied linearly with the baseband signal.

Phase-shift keying (PSK): Digital modulation in which the phase of a carrier takes on a discrete value chosen from a preselected set in accordance with the input data.

Prediction: Filtering operation used to make an inference about the future value of a baseband signal, given knowledge of past behavior of the signal up to a certain point in time.

Pulse-analog modulation: Modulation process in which some characteristic feature (e.g., amplitude, duration, or position) of each pulse in a periodic pulse train (used as carrier) is varied in accordance with the baseband signal.

Pulse-code modulation (PCM): Process that involves the uniform sampling of a baseband signal, quantization of the amplitude of each sample to the nearest one of a finite set of allowable values, and representation of each quantized sample by a code.

Signal-to-noise ratio: Ratio of the average power of a message signal to the average power of additive noise, both being measured at a point of interest in the system (e.g., receiver output).

Single-sideband (SSB) modulation: Form of amplitude modulation in which only the upper sideband or lower sideband is transmitted.

Vestigial sideband (VSB) modulation: Form of amplitude modulation in which one sideband is passed almost completely, whereas just a vestige, or trace, of the other sideband is retained.

In this article, the word *communication* refers to the process of conveying information-bearing signals from one point to another that is physically separate. The information-bearing signal, or baseband signal, may be in analog form as in the case of voice and video signals or in digital form as in the case of computer data. In any event, the purpose of a communication system is to transmit information-bearing signals from a source, located at one point in space, to a user destination, located at another point. Typically, the message produced by the source is not electrical in nature. Accordingly, an input transducer is used to convert the message generated by the source into a time-varying electrical signal called the message signal. By means of another transducer at the receiver, the original message is recreated at the user destination. In this article we discuss various issues pertaining to analog and digital communication systems for civilian applications. [*See* SIGNAL PROCESSING, ANALOG; SIGNAL PROCESSING, DIGITAL; SIGNAL PROCESSING, GENERAL.]

I. Model of a Communication System

Figure 1 shows the block diagram of a communication system. The system consists of three major parts: (1) transmitter, (2) communication channel, and (3) receiver. The main purpose of the transmitter is to modify the message signal into a form suitable for transmission over the channel. This modification is achieved by means of a process known as modulation. [*See* DATA COMMUNICATION NETWORKS.]

The communication channel may be a transmission line (as in telephony and telegraphy), an optical fiber (as in optical communications), or merely free space in which the signal is radiated as an electromagnetic wave (as in radio and television broadcasting). In propagating through the channel, the transmitted signal is distorted due to nonlinearities or imperfections in the frequency response of the channel. Other sources of degradation are noise and interference picked up by the signal during the course of transmission through the channel. Noise and distortion constitute two basic problems in the design of communication systems. Usually, the transmitter and receiver are carefully designed so as to minimize the effects of noise and distortion on

the quality of reception. [*See* DATA TRANSMISSION MEDIA.]

The main purpose of the receiver is to recreate the original message signal from the degraded version of the transmitted signal after propagation through the channel. This recreation is accomplished by a process known as demodulation, which is the reverse of the modulation process in the transmitter. However, owing to the unavoidable presence of noise and distortion in the received signal, the receiver cannot recreate the original message signal exactly. The resulting degradation in overall system performance is influenced by the type of modulation scheme used. Some modulation schemes are less sensitive to the effects of noise and distortion than others.

In any communication system, there are two primary communication resources namely, transmitted power and channel bandwidth. A general system design objective is to use these two resources as efficiently as possible. In most communication channels, one resource may be considered more important than the other. We therefore classify communication channels as power-limited or band-limited. For example, a telephone circuit is a typical band-limited channel, whereas a space communication link or a satellite channel is typically power-limited.

When the spectrum of a message signal extends down to zero or low frequencies, we define the bandwidth of the signal as that upper frequency above which the spectral content of the signal is negligible and, therefore, unnecessary for transmitting the pertinent information. For example, the average voice spectrum extends well beyond 10 kHz, though most of the energy is concentrated in the range of 100 to 600 Hz, and a band from 300 to 3400 Hz gives good articulation. Accordingly, telephone circuits that respond well to the latter range of frequencies give quite satisfactory commercial tele-

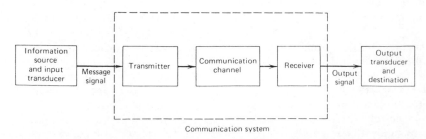

Communication system

FIG. 1. Model of an electrical communication system. [From Haykin, S. (1983). "Communication Systems," 2nd ed. Wiley, New York. © John Wiley & Sons, Inc.]

phone service. [*See* RADIO SPECTRUM UTILI-ZATION.]

II. Analog Modulation

Modulation is formally defined as the process by which some characteristic of a carrier is varied in accordance with a modulating wave. The baseband signal is referred to as the modulating wave, and the result of the modulation process is referred to as the modulated wave.

In analog or continuous-wave (cw) modulation, the modulating wave consists of an analog signal (e.g., voice signal, video signal), and the carrier consists of a sine wave. Basically, there are two types of analog modulation: amplitude modulation and angle modulation. In amplitude modulation the amplitude of the sinusoidal carrier wave is varied in accordance with the baseband signal. In angle modulation, on the other hand, the angle of the sinusoidal carrier wave is varied in accordance with the baseband signal. [*See* ACOUSTIC SIGNAL PROCESSING.]

A. AMPLITUDE MODULATION

Consider a sinusoidal carrier wave $c(t)$ defined by

$$c(t) = A_c \cos(2\pi f_c t) \qquad (1)$$

where A_c is the carrier amplitude and f_c the carrier frequency. For convenience, we have assumed that the phase of the carrier wave is zero in Eq. (1). Let $m(t)$ denote the baseband signal that carries the specification of the message. The carrier wave $c(t)$ is independent of $m(t)$. An amplitude-modulated (AM) wave is described by

$$s(t) = A_c[1 + k_a m(t)] \cos(2\pi f_c t) \qquad (2)$$

where k_a is a constant called the amplitude sensitivity of the modulator.

The envelope of the AM wave $s(t)$ equals

$$a(t) = A_c|1 + k_a m(t)| \qquad (3)$$

where $|\cdot|$ denotes the absolute value of the enclosed quantity. The envelope of $s(t)$ has the same shape as the baseband signal $M(t)$, provided that two requirements are satisfied:

1. The amplitude of $k_a m(t)$ is always less than unity, that is,

$$|k_a m(t)| < 1 \qquad \text{for all } t$$

The absolute maximum value of $k_a m(t)$ multiplied by 100 is referred to as the percentage modulation.

2. The carrier frequency f_c is much greater than the highest frequency component of the baseband signal $m(t)$.

The spectrum of the AM wave $s(t)$ consists of a carrier, upper sideband, and lower sideband. Let W denote the highest frequency component of the baseband signal $m(t)$. For positive frequencies, the carrier is located at f_c, the upper sideband extends from f_c to $f_c + W$, and the lower sideband extends from $f_c - W$ to f_c. Hence, the transmission bandwidth of the AM wave equals $2W$, that is, exactly twice the message bandwidth. For negative frequencies, the spectrum of the AM wave is the mirror image of that for positive frequencies.

Figure 2 illustrates the amplitude modulation process. Figure 2a shows a modulating wave $m(t)$ that consists of a single tone or frequency component, Fig. 2b the sinusoidal carrier wave, and Fig. 2c the corresponding AM wave. The amplitude spectra of the modulating wave, the carrier, and the AM wave are shown on the right-hand side of the figure.

1. Double-Sideband Suppressed-Carrier Modulation

The carrier wave $c(t)$ is completely independent of the information-carrying signal or baseband signal $m(t)$, which means that the transmission of the carrier wave represents a waste of power. This points to a shortcoming of amplitude modulation, namely, that only a fraction of the total transmitted power is affected by $m(t)$. To overcome this shortcoming, we may suppress the carrier component from the modulated wave, resulting in double-sideband suppressed-carrier (DSBSC) modulation. Thus, by suppressing the carrier, we obtain a modulated wave that is proportional to the product of the carrier wave and the baseband signal.

To describe a DSBSC-modulated wave as a function of time, we write

$$s(t) = c(t)m(t) = A_c \cos(2\pi f_c t)m(t) \qquad (4)$$

This modulated wave undergoes a phase reversal whenever the baseband signal $m(t)$ crosses zero. Accordingly, unlike amplitude modulation, the envelope of a DSBSC-modulated wave is different from the baseband signal. For obvious reasons, a device that performs the operation described in Eq. (4) is called a product modulator.

The transmission bandwidth required by DSBSC modulation is the same as that for am-

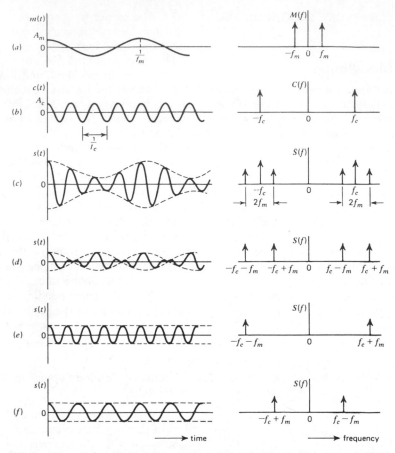

FIG. 2.   Time-domain (left) and frequency-domain (right) characteristics of different modulated waves produced by a single tone. (a) Modulating wave. (b) Carrier wave. (c) AM wave. (d) DSBSC wave. (e) SSB wave with the upper-side frequency transmitted. (f) SSB wave with the lower-side frequency transmitted. [From Haykin, S. (1983). "Communication Systems," 2nd ed. Wiley, New York. © John Wiley & Sons, Inc.]

plitude modulation, namely, $2W$. Figure 2d illustrates the DSBSC-modulated wave for single-tone modulation and the corresponding amplitude spectrum.

2. Quadrature-Carrier Multiplexing

The quadrature-carrier multiplexing or quadrature-amplitude modulation (QAM) scheme enables two DSBSC-modulated waves (resulting from the application of two independent message signals) to occupy the same transmission bandwidth, and yet it allows for the separation of the two message signals at the receiver output. It is therefore a bandwidth-conservation scheme. The transmitter part of the system involves the use of two separate product modulators that are supplied with two carrier waves of the same frequency but differing in phase by 90°.

The transmitted signal $s(t)$ consists of the sum of these two product modulator outputs, as shown by

$$s(t) = A_c m_1(t) \cos(2\pi f_c t) + A_c m_2(t) \sin(2\pi f_c t)$$

(5)

where $m_1(t)$ and $m_2(t)$ denote the two different message signals applied to the product modulators. Thus, $s(t)$ occupies a transmission bandwidth of $2W$, centered at the carrier frequency f_c, where W is the message bandwidth of $m(t)$ or $m_2(t)$.

Quadrature-carrier multiplexing is used in color television. In this system, synchronizing pulses are transmitted in order to maintain the local oscillator in the receiver at the correct frequency and phase with respect to the carrier used in the transmitter.

3. Single-Sideband Modulation

Amplitude modulation and DSBSC modulation are wasteful of bandwidth because they both require a transmission bandwidth equal to twice the message bandwidth. In either case, one-half of the transmission bandwidth is occupied by the upper sideband of the modulated wave, whereas the other half is occupied by the lower sideband. However, the upper and lower sidebands are uniquely related to each other by virtue of their symmetry about the carrier frequency; that is, given the amplitude and phase spectra of either sideband, we can uniquely determine the other. This means that insofar as the transmission of information is concerned, only one sideband is necessary, and if the carrier and the other sidebands are suppressed at the transmitter, no information is lost. Thus, the communication channel has to provide only the same bandwidth as the baseband signal, a conclusion that is intuitively satisfying. When only one sideband is transmitted, the modulation system is referred to as a single-sideband (SSB) system.

The precise frequency-domain description of an SSB wave depends on which sideband is transmitted. In any event, the essential function of an SSB modulation system is to translate the spectrum of the modulating wave, either with or without inversion, to a new location in the frequency domain; the transmission bandwidth requirement of the system is one-half that of an amplitude or DSBSC modulation system. The benefit of using SSB modulation is therefore derived principally from the reduced bandwidth requirement and the elimination of the high-power carrier wave. The principal disadvantage of the SSB modulation system is its cost and complexity.

The time-domain description of the SSB wave is much more complicated than that of the DSBSC wave. Specifically, we write

$$s(t) = \tfrac{1}{2}A_c m(t) \cos(2\pi f_c t) \pm \tfrac{1}{2}A_c \hat{m}(t) \sin(2\pi f_c t) \tag{6}$$

where $m(t)$ is the Hilbert transform of the baseband signal $m(t)$. A Hilbert transformer consists of a two-port device that leaves the amplitudes of all frequency components of the input signal unchanged, but it produces a phase shift of $-90°$ for all positive frequency components of the input signal and a phase shift of $+90°$ for all negative frequency components. The minus sign on the right-hand side of Eq. (6) refers to a SSB-modulated wave that contains only the upper sideband, whereas the plus sign refers to an SSB-modulated wave that contains only the lower sideband.

Figure 2e shows the SSB wave (with upper-side frequency) resulting from the use of single-tone modulation. Figure 2f shows the corresponding SSB wave with lower-side frequency.

4. Vestigial Sideband Modulation

Single-sideband modulation is rather well suited for the transmission of voice because of the energy gap in the spectrum of voice signals between zero and a few hundred hertz. When the baseband signal contains significant components at extremely low frequencies (as in the case of television and telegraph signals and computer data), however, the upper and lower sidebands meet at the carrier frequency. This means that the use of SSB modulation is inappropriate for the transmission of such baseband signals due to the difficulty of isolating one sideband. This difficulty suggests another scheme known as vestigial sideband (VSB) modulation, which is a compromise between SSB and DSBSC modulation. In this modulation scheme, one sideband is passed almost completely, whereas just a trace, or vestige, of the other sideband is retained. The transmission bandwidth required by a VSB system is therefore

$$B_T = W + f_v \tag{7}$$

where W is the message bandwidth and f_v the width of the vestigial sideband.

Vestigial sideband modulation has the virtue of conserving bandwidth almost as efficiently as SSB modulation, while retaining the excellent low-frequency baseband characteristics of double-sideband modulation. Thus, VSB modulation has become standard for the transmission of television and similar signals where good phase characteristics and transmission of low-frequency components are important but the bandwidth required for double-sideband transmission is unavailable or uneconomical.

It is of interest, however, that in commercial television broadcasting the transmitted signal is not quite VSB-modulated, because the shape of the transition region is not rigidly controlled at the transmitter. Instead, a VSB filter is inserted in each receiver. The overall performance is the same as VSB modulation except for some wasted power and bandwidth. Figure 3 shows the idealized frequency response for a VSB filter designed for television receivers.

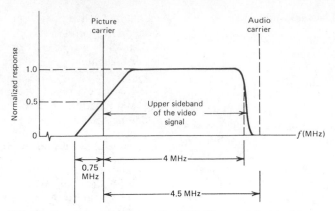

FIG. 3.   Frequency response of a VSB filter used in television receivers. [From Haykin, S. (1983). "Communication Systems," 2nd ed. Wiley, New York. © John Wiley & Sons, Inc.]

5. Discussion

The characteristics and practical merits of the different forms of amplitude modulation can be summarized as follows:

1. In ordinary amplitude modulation systems, the sidebands are transmitted in full, accompanied by the carrier. Accordingly, demodulation is accomplished rather simply by the use of an envelope detector or square-law detector. On the other hand, in suppressed-carrier systems the receiver is more complex because additional circuitry must be provided for the purpose of carrier recovery. For this reason, in commercial AM radio broadcast systems, which involve one transmitter and numerous receivers, full amplitude modulation is the preferred method.

2. Suppressed-carrier modulation systems have an advantage over full amplitude modulation systems in that they require much less power to transmit the same amount of information, which makes the transmitters for such systems less expensive than those required for full amplitude modulation. Suppressed-carrier systems are therefore well suited for point-to-point communication involving one transmitter and one receiver, which would justify the use of increased receiver complexity.

3. Single-sideband modulation requires the minimum transmitter power and minimum transmission bandwidth possible for conveying a message signal from one point to another. Thus, SSB modulation is the preferred method of modulation for long-distance transmission of voice signals over metallic circuits, because it permits longer spacing between the repeaters, which is a more important consideration here than simple terminal equipment. A repeater is simply a wideband amplifier that is used at intermediate points along the transmission path so as to make up for the attenuation incurred during the course of transmission.

4. Vestigial-sideband modulation requires a transmission bandwidth that is intermediate between that required for SSB and DSBSC systems, and the saving can be significant if modulating waves with large bandwidths are being handled, as in the case of television signals and wide-band data.

B. ANGLE MODULATION

In angle modulation the angle of the carrier wave is varied according to the baseband signal. In this method of modulation the amplitude of the carrier wave is maintained constant. An important feature of angle modulation is that it can provide better discrimination against noise and interference than amplitude modulation. However, this improvement in performance is achieved at the expense of increased transmission bandwidth; that is, angle modulation provides a practical means of exchanging transmission bandwidth for improved noise performance. Such a trade-off is not possible with amplitude modulation.

Two forms of angle modulation are distinguished: phase modulation and frequency modulation. These two methods of modulation are closely related in that the properties of one can be derived from those of the other. It is customary to define these two forms of angle modulation as follows:

1. Phase modulation (PM) is that form of angle modulation in which the angle $\theta_i(t)$ is varied linearly with the baseband signal $m(t)$, as shown by

$$\theta_i(t) = 2\pi f_c t + k_p m(t) \qquad (8)$$

The term $2\pi f_c t$ represents the angle of the unmodulated carrier, and the constant k_p represents the phase sensitivity of the modulator. For convenience, we have assumed in Eq. (8) that the angle of the unmodulated carrier is zero at $t = 0$. The PM wave $s(t)$ is thus described in the time domain by

$$s(t) = A_c \cos[2\pi f_c t + k_p m(t)] \qquad (9)$$

2. Frequency modulation (FM) is that form of angle modulation in which the instantaneous frequency $f_i(t)$ is varied linearly with the baseband signal $m(t)$, as shown by

$$f_i(t) = \frac{1}{2\pi} \frac{d_i(t)}{dt} = f_c + k_f m(t) \qquad (10)$$

The term f_c represents the frequency of the unmodulated carrier, and the constant k_f repre-

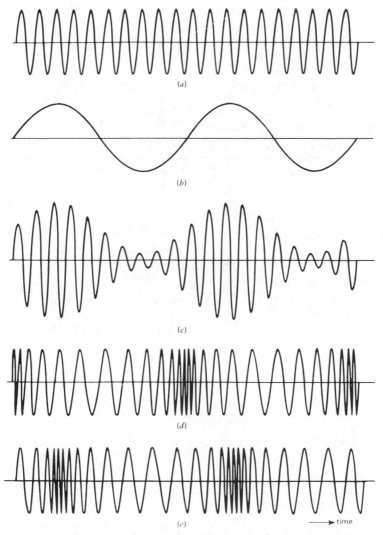

FIG. 4. AM, PM, and FM waves produced by a single tone. (a) Carrier wave. (b) Sinusoidal modulating wave. (c) AM wave. (d) PM wave. (e) FM wave. [From Haykin, S. (1983). "Communication Systems," 2nd ed. Wiley, New York. © John Wiley & Sons, Inc.]

sents the frequency sensitivity of the modulator. Integrating Eq. (10) with respect to time and multiplying the result by 2π, we get

$$\theta_i(t) = 2\pi f_c t + 2\pi k_f \int_0^t m(t)\, dt \qquad (11)$$

where, for convenience, we have assumed that the angle of the unmodulated carrier wave is zero at $t = 0$. The FM is therefore described in the time domain by

$$s(t) = A_c \cos\left[2\pi f_c t + 2\pi k_f \int_0^t m(t)\, dt\right] \qquad (12)$$

A consequence of allowing the angle $\theta_i(t)$ to become dependent on the message signal $m(t)$ as in Eq. (8) or on its integral as in Eq. (11) is that the zero crossings of PM wave or FM wave no longer have a perfect regularity in their spacing; zero crossings refer to the instants of time at which a waveform changes from negative to positive value or vice versa.

The differences between amplitude-modulated and angle-modulated waves are illustrated in Fig. 4 for the case of single-tone modulation. Figures 4a and b refer to the sinusoidal carrier and modulating waves, respectively; Figs. 4c, d, and e show the corresponding AM, PM, and FM waves, respectively.

Figure 5a shows that an FM wave can be gen-

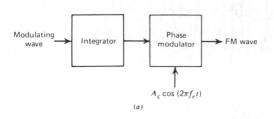

(a)

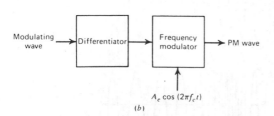

(b)

FIG. 5.   Relationship between frequency modulation and phase modulation. (a) Scheme for generating an FM wave by using a phase modulator. (b) Scheme for generating a PM wave by using a frequency modulator. [From Haykin, S. (1983). "Communication Systems," 2nd ed. Wiley, New York. © John Wiley & Sons, Inc.]

erated by first integrating $m(t)$ and then using the result as the input to a phase modulator. Conversely, a PM wave can be generated by first differentiating $m(t)$ and then using the result as the input to a frequency modulator, as in Fig. 5b.

Consider an FM wave produced by using a single-tone modulating wave of frequency f_m. Let Δf denote the frequency deviation defined as the maximum departure of the instantaneous frequency of the FM wave from the carrier frequency f_c. The ratio $\Delta f/f_m$ is called the modulation index, commonly denoted by β. When β is less than 0.5, the modulated wave is referred to as narrow-band FM. When β is large, it is referred to as wide-band FM.

Transmission Bandwidth of FM Waves

In theory, an FM wave contains an infinite number of side frequencies so that the bandwidth required to transmit such a signal is similarly infinite in extent. In practice, however, the FM wave is effectively limited to a finite number of significant side frequencies compatible with a specified amount of distortion. We can therefore specify an effective bandwidth required for the transmission of an FM wave.

For the case of single-tone modulation, we calculate the transmission bandwidth of the FM wave by using Carson's rule:

$$B_T \simeq 2\,\Delta f + 2f_m = 2\,\Delta f(1 + 1/\beta) \qquad (13)$$

When β is small compared with 1 rad, as in the case of narrow-band FM, the transmission bandwidth B_T is approximately $2f_m$. On the other hand, in the case of wide-band FM for which β is large compared with 1 rad, the transmission bandwidth B_T is approximately $2\,\Delta f$.

C. Effects of Noise

One of the basic issues in the study of modulation systems is the analysis of the effects of noise on the performance of the receiver and the use of the results of this analysis in system design. Another matter of interest is the comparison of the noise performances of different modulation–demodulation schemes. In order to carry out this analysis, we obviously need a criterion that describes in a meaningful manner the noise performance of the modulation system under study. In the case of cw modulation systems, the customary practice is to use the output signal-to-noise ratio as an intuitive measure for describing the fidelity with which the demodulation process in the receiver recovers the original message from the received modulated signal in the presence of noise. The output signal-to-noise ratio is defined as the ratio of the average power of the message signal to the average power of the noise, both measured at the receiver output. Such a criterion is perfectly well defined as long as the message signal and noise appear additively at the receiver output.

In performing noise analysis, it is customary to model the noise as white, Gaussian, and stationary. Under this condition, we find that for the same average transmitted (or modulated message) signal power and the same average noise power in the message bandwidth, an SSB receiver will have exactly the same output signal-to-noise ratio as a DSBSC receiver, when both receivers use coherent detection for the recovery of the message signal. Furthermore, in both cases, the noise performance of the receiver is the same as that obtained by simply transmitting the message signal itself in the presence of the same noise. The only effect of the modulation process is to translate the message signal to a different frequency band.

Coherent detection assumes the availability of a local carrier in the receiver with the same frequency and phase as that of the carrier in the transmitter. The noise performance of an AM receiver is always inferior to that of a DSBSC or SSB receiver. This is due to the wastage of transmitter power that results from transmitting the carrier as a component of the AM wave. Also, an AM receiver suffers from a threshold effect that arises when the carrier-to-noise ratio is small compared with unity. By threshold we mean a value of the carrier-to-noise ratio below which the noise performance of a detector deteriorates much more rapidly than proportionately to the carrier-to-noise ratio. A detailed analysis of the threshold effect in envelope detectors is complicated, however.

Since frequency modulation is a nonlinear modulation process, the noise analysis of an FM receiver is also quite complicated. Nevertheless, we can state that when the carrier-to-noise ratio is high, an increase in the transmission bandwidth B_T of the FM system provides a corresponding quadratic increase in the output signal-to-noise ratio of the system. However, this result is valid only if the carrier-to-noise ratio is high compared with unity. It is found experi-

mentally that as the input noise is increased so that the carrier-to-noise ratio is decreased, the FM receiver breaks. At first, individual clicks are heard in the receiver output, and as the carrier-to-noise ratio decreases still further, the clicks rapidly merge into a crackling or sputtering sound. Under this condition, the FM receiver is said to suffer from the threshold effect.

In certain applications, such as space communications, there is particular interest in reducing the noise threshold in an FM receiver so as to operate the receiver satisfactorily with the minimum signal power possible. Threshold reduction in FM receivers can be achieved by means of an FM demodulator with negative feedback (commonly referred to as an FMFB demodulator) or a phase-locked loop demodulator.

Preemphasis and Deemphasis

The noise performance of an FM receiver can be improved significantly by the use of preemphasis in the transmitter and deemphasis in the receiver. With this method, we artificially emphasize the high-frequency components of the message signal before modulation in the transmitter and therefore before the noise is introduced in the receiver. In effect, the low-frequency and high-frequency portions of the power spectral density of the message are equalized in such a way that the message fully occupies the frequency band allotted to it. Then, at the discriminator output in the receiver, we perform the inverse operation by deemphasizing the high-frequency components, so as to restore the original signal-power distribution of the message. In such a process the high-frequency components of the noise at the discriminator output are also reduced, thereby effectively increasing the output signal-to-noise ratio of the system.

The use of the simple linear preemphasis and deemphasis filters is an example of how the performance of an FM system can be improved by utilizing the differences between characteristics of signals and noise in the system. These simple filters also find application in audio tape recording. In recent years nonlinear preemphasis and deemphasis techniques have been applied successfully to tape recording. These techniques (known as Dolby-A, Dolby-B, and DBX system) use a combination of filtering and dynamic range compression to reduce the effects of noise, particularly when the signal level is low.

III. Pulse and Digital Modulation

In cw modulation, some parameter of a sinusoidal carrier wave is varied continuously in accordance with the message. This is in direct contrast to pulse modulation, in which some parameter of a pulse train is varied in accordance with the message. There are two basic types of pulse modulation: pulse-analog modulation and pulse-code modulation. In pulse-analog modulation systems, a periodic pulse train is used as the carrier wave, and some characteristic feature of each pulse (e.g., amplitude, duration, or position) is varied in a continuous manner in accordance with the pertinent sample value of the message. On the other hand, in pulse-code modulation, a discrete-time, discrete-amplitude representation is used for the signal, and as such it has no cw counterpart.

A. Sampling Theorem

An operation that is basic to the design of all PM systems is the sampling process whereby an analog signal is converted to a corresponding sequence of numbers that are usually uniformly spaced in time. For such a procedure to have practical utility, it is necessary that we choose the sampling rate properly, so that this sequence of numbers uniquely defines the original analog signal. This is the essence of the sampling theorem, which can be stated in two equivalent ways:

1. A band-limited signal of finite energy, which has no frequency components higher than W hertz is completely described by specifying the values of the signal at instants of time separated by $1/2W$ seconds.
2. A band-limited signal of finite energy, which has no frequency components higher than W hertz, can be completely recovered from a knowledge of its samples taken at the rate of $2W$ per second.

The sampling rate of $2W$ samples per second, for a signal bandwidth of W hertz, is often called the Nyquist rate; its reciprocal $1/2W$ is called the Nyquist interval. The sampling theorem serves as the basis for the interchangeability of analog signals and digital sequences, which is so valuable in digital communication systems.

The derivation of the sampling theorem, as described above, is based on the assumption that the signal $g(t)$ is strictly band-limited. Such a requirement, however, can be satisfied only if $g(t)$ has infinite duration. In other words, a strictly band-limited signal cannot be simultaneously time-limited, and vice versa. Nevertheless, we can consider a time-limited signal to be essentially bandlimited in the sense that its frequency components outside some band of interest have negligible effects. We can then justify the practical application of the sampling theorem.

Aliasing

When the sampling rate $1/T_s$ exceeds the Nyquist rate $2W$, there is no problem in recovering the original signal $g(t)$ from its sampled version $g_\delta(t)$. When the sampling rate $1/T_s$ is less than $2W$, however, the aliasing effect arises. That is, a high-frequency component in the spectrum of the original signal seemingly takes on the identity of a lower frequency component in the spectrum of its sampled version. It is evident that if the sampling rate $1/T_s$ is less than the Nyquist rate $2W$, the original signal cannot be recovered exactly from its sampled version, and information is thereby lost in the sampling process.

Another factor that contributes to aliasing is the fact that a signal cannot be finite in both time and frequency, as mentioned previously. Since this violates the strict bandlimited requirement of the sampling theorem, we find that whenever a time-limited signal is sampled, there will always be some aliasing produced by the sampling process. Accordingly, in order to combat the effects of aliasing in practice, we use two corrective measures:

1. Before sampling, a low-pass pre-alias filter is used to attenuate those high-frequency components of the signal that lie outside the band of interest.
2. The filtered signal is sampled at a rate slightly higher than the Nyquist rate.

B. Pulse-Analog Modulation

In pulse-amplitude modulation (PAM), the amplitudes of regularly spaced rectangular pulses vary with the instantaneous sample values of a continuous message signal in a one-to-one fashion. This method of modulation is illustrated in Figs. 6a, b, and c, which represent a message signal, the pulse carrier, and the corresponding PAM wave, respectively.

The PAM wave $s(t)$ is easily demodulated by a low-pass filter with a cutoff frequency just large

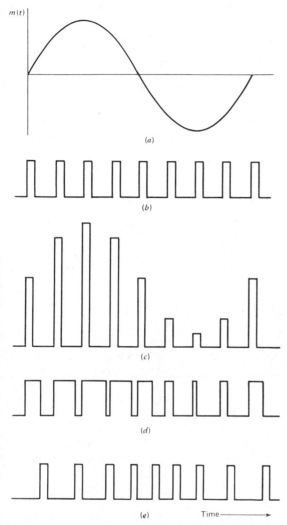

FIG. 6.   Different forms of pulse-analog modulation for the case of a sinusoidal modulating wave. (a) Modulating wave. (b) Pulse carrier. (c) PAM wave. (d) PDM wave. (e) PPM wave. [From Haykin, S. (1983). "Communication Systems," 2nd ed. Wiley, New York. © John Wiley & Sons, Inc.]

enough to accommodate the highest frequency component of the message signal $m(t)$. However, the reconstructed signal exhibits a slight amplitude distortion caused by the aperture effect due to lengthening of the samples.

In pulse-duration modulation (PDM), the samples of the message signal are used to vary the duration of the individual pulses. This form of modulation is also referred to as pulse-width modulation or pulse-length modulation. The modulating wave may vary with the time of occurrence of the leading edge, the trailing edge,

or both edges of the pulse. In Fig. 6d the trailing edge of each pulse is varied in accordance with the message signal.

In PDM, long pulses expend considerable power during the pulse while bearing no additional information. If this unused power is subtracted from PDM, so that only time transitions are preserved, we obtain a more efficient type of pulse modulation known as pulse-position modulation (PPM). In PPM the position of a pulse relative to its unmodulated time of occurrence is varied in accordance with the message signal (Fig. 6e).

C. Pulse-Code Modulation

In PAM, PDM, and PPM, only time is expressed in discrete form, whereas the respective modulation parameters (namely, pulse amplitude, duration, and position) are varied in a continuous manner in accordance with the message. Thus, in these modulation systems, information transmission is accomplished in analog form at discrete times. On the other hand, in pulse-code modulation (PCM), the message signal is sampled and the amplitude of each sample is rounded off to the nearest one of a finite set of allowable values, so that both time and amplitude are in discrete form. This allows the message to be transmitted by means of coded electrical signals, thereby distinguishing PCM from all other methods of modulation.

The use of digital representation of analog signals (e.g., voice, video) offers the following advantages: (1) ruggedness to transmission noise and interference, (2) efficient regeneration of the coded signal along the transmission path, and (3) the possibility of a uniform format for different kinds of baseband signals. These advantages, however, are attained at the cost of increased transmission bandwidth requirement and increased system complexity. With the increasing availability of wide-band communication channels, coupled with the emergence of the requisite device technology, the use of PCM has become a practical reality.

1. Elements of PCM

Pulse-code modulation systems are considerably more complex than PAM, PDM, and PPM systems, in that the message signal is subjected to a greater number of operations. The essential operations in the transmitter of a PCM system are sampling, quantizing, and encoding (Fig. 7). The quantizing and encoding operations are usu-

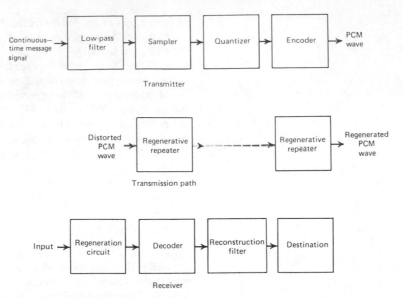

FIG. 7.   Basic elements of a PCM system. [From Haykin, S. (1983). "Communication Systems," 2nd ed. Wiley, New York. © John Wiley & Sons, Inc.]

ally performed in the same circuit, which is called an analog-to-digital converter. The essential operations in the receiver are regeneration of impaired signals, decoding, and demodulation of the train of quantized samples. Regeneration usually occurs at intermediate points along the transmission route as necessary.

2. Sampling

The incoming message wave is sampled with a train of narrow rectangular pulses so as to approximate closely the instantaneous sampling process. In order to ensure perfect reconstruction of the message at the receiver, the sampling rate must be greater than twice the highest frequency component W of the message wave in accordance with the sampling theorem. In practice, a low-pass filter is used at the front end of the sampler in order to exclude frequencies greater than W before sampling. Thus, the application of sampling permits the reduction of the continuously varying message wave to a limited number of discrete values per second.

3. Quantizing

A continuous signal, such as voice, has a continuous range of amplitudes, and therefore its samples have a continuous amplitude range. In other words, within the finite amplitude range of the signal we find an infinite number of amplitude levels. It is not necessary in fact to transmit the exact amplitudes of the samples. Any human sense (the ear or the eye), as ultimate receiver, can detect only finite intensity differences. This means that the original continuous signal can be approximated by a signal constructed of discrete amplitudes selected on a minimum-error basis from an available set. The existence of a finite number of discrete amplitude levels is a basic condition of PCM. Clearly, if we assign the discrete amplitude levels with sufficiently close spacing, we can make the approximated signal practically indistinguishable from the original continuous signal.

The conversion of an analog (continuous) sample of the signal to a digital (discrete) form is called the quantizing process.

4. Encoding

In combining the processes of sampling and quantizing, the specification of a continuous baseband signal becomes limited to a discrete set of values, but not in the form best suited to transmission over a line or radio path. To exploit the advantages of sampling and quantizing, we require the use of an encoding process to translate the discrete set of sample values to a more appropriate form of signal. Any plan for representing each of this discrete set of values as a particular arrangement of discrete events is called a code. One of the discrete events in a code is called a code element or symbol. For example, the presence or absence of a pulse is a symbol. A particular arrangement of symbols used in a code to represent a single value of the discrete set is called a code word or character.

In a binary code, each symbol can be either of two distinct values or kinds, such as the presence or absence of a pulse. The two symbols of a binary code are customarily denoted 0 and 1. In a ternary code, each symbol can be one of three distinct values or kinds, and so on for other codes. However, the maximum advantage over the effects of noise in a transmission medium is obtained by using a binary code, because a binary symbol withstands a relatively high level of noise and is easy to regenerate.

Suppose that in a binary code, each code word consists of n bits (*bit* is an acronym for *bi*nary digi*t*). Then, using such a code, we can represent a total of 2^n distinct numbers. For example, a sample quantized into one of 128 levels can be represented by a seven-bit code word.

5. Regeneration

The most important feature of PCM systems is the ability to control the effects of distortion and noise produced by transmitting a PCM wave through a channel. This capability is accomplished by reconstructing the PCM wave by means of a chain of regenerative repeaters located at sufficiently close spacing along the transmission route.

6. Decoding

The first operation in the receiver is to regenerate (i.e., reshape and clean up) the received pulses. These clean pulses are then regrouped into code words and decoded (i.e., mapped back) into a quantized PAM signal. The decoding process involves generating a pulse the amplitude of which is the linear sum of all the pulses in the code word, with each pulse weighted by its place value (2^0, 2^1, 2^2, 2^3, ...) in the code.

7. Filtering

The final operation in the receiver is to recover the signal wave by passing the decoder output through a low-pass reconstruction filter whose cutoff frequency is equal to the message bandwidth W. Assuming that the transmission path is error free, the recovered signal includes no noise, with the exception of the initial distortion introduced by the quantization process.

8. Noise in PCM Systems

The performance of a PCM system is influenced by two major sources of noise:

1. Transmission noise, which may be introduced anywhere between the transmitter output and the receiver input. The effect of transmission noise is to introduce bit errors into the received PCM wave, with the result that in the case of a binary system, a symbol 1 occasionally is mistaken for a symbol 0, or vice versa. Clearly, the more frequently such errors occur, the more dissimilar to the original message signal is the receiver output. The fidelity of information transmission by PCM in the presence of transmission noise is conveniently measured in terms of the error rate, or probability of error, defined as the probability that the symbol at the receiver output differs from that transmitted.

2. Quantizing noise, which is introduced in the transmitter and is carred along to the receiver output. Typically, each bit in the code word of a PCM system contributes 6 dB to the signal-to-noise ratio at the receiver output.

D. DIFFERENTIAL PULSE-CODE MODULATION

When a voice or video signal is sampled at a rate higher than the Nyquist rate, the resulting sampled signal exhibits high correlation between adjacent samples. The meaning of this high correlation is that in an average sense, the signal does not change rapidly from one sample to the next, with the result that the difference between adjacent samples has a variance that is smaller than the variance of the signal itself. When these highly correlated samples are encoded, as in a standard PCM system, the resulting encoded signal contains redundant information. This means that symbols that are not absolutely essential to the transmission of information are generated as a result of the encoding process. By removing this redundancy before encoding, we obtain a more efficient coded signal.

Now, if we know a sufficient part of a redundant signal, we can infer the rest, or at least make the most probable estimate. In particular, if we know the past behavior of a signal up to a certain point in time, it is possible to make some inference about its future values; such a process is commonly called prediction. The fact that it is possible to predict future values of the signal $m(t)$ provides motivation for the differential quantization scheme shown in Fig. 8. The input to the quantizer is a signal that consists of the

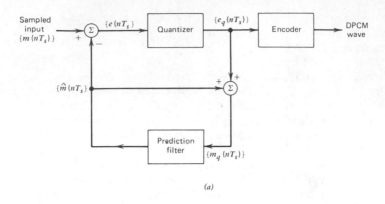

(a)

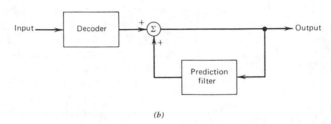

(b)

FIG. 8.   DPCM system. (a) Transmitter. (b) Receiver. [From Haykin, S. (1983). "Communication Systems," 2nd ed. Wiley, New York. © John Wiley & Sons, Inc.]

difference between the unquantized input symbol $m(nT_s)$. This predicted value is produced by using a prediction filter. The difference signal $e(nT_s)$ is called a prediction error. By encoding the quantizer output, as in Fig. 8a, we obtain a variation of PCM known as differential pulse-code modulation (DPCM). This encoded signal is used for transmission.

The receiver for reconstructing the quantized version of the input is shown in Fig. 8b. It consists of a decoder to reconstruct the quantized error signal. The quantized version of the original input is reconstructed from the decoder output using the same design of prediction filter in the transmitter.

E. DELTA MODULATION

The exploitation of signal correlations in DPCM suggests the further possibility of oversampling a baseband signal (i.e., at a rate much higher than the Nyquist rate) purposely to increase the correlation between adjacent samples of the signal, so as to permit the use of a simple quantizing strategy for constructing the encoded signal. Delta modulation (DM), which is the one-bit (or two-level) version of DPCM, is precisely such a scheme.

In its simple form, DM provides a staircase approximation to the oversampled version of an input baseband signal (Fig. 9). The difference between the input and the approximation is quantized into only two levels, namely, $\pm\delta$, corresponding to positive and negative differences, respectively. Thus, if the approximation falls below the signal at any sampling epoch, it is increased by δ. If, on the other hand, the approximation lies above the signal, it is diminished by δ. Provided that the signal does not change too rapidly from sample to sample, the staircase approximation remains within $\pm\delta$ of the input signal.

The principal virtue of DM is its simplicity. It can be generated by applying the sampled version of the incoming baseband signal to a modulator that involves a summer, quantizer, and accumulator interconnected (Fig. 10).

In comparing the DPCM and DM networks of Figs. 8 and 10, respectively, we note that they are basically similar, except for two important differences, namely, the use of a one-bit (two-level) quantizer in the delta modulator and the

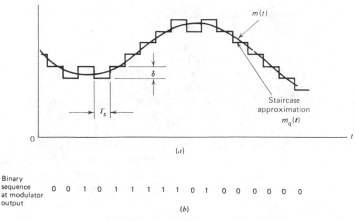

FIG. 9. Delta modulation. [From Haykin, S. (1983). "Communication Systems," 2nd ed. Wiley, New York. © John Wiley & Sons, Inc.]

replacement of the prediction filter by a single delay element.

IV. Data Transmission

In this final section we consider some of the issues involved in transmitting digital data (of whatever origin) over a communication channel. The transmission may be in baseband or passband form. For baseband transmission, we use discrete pulse modulation, in which the amplitude, duration, or position of the transmitted pulses is varied in a discrete manner in accordance with the digital data. However, it is cus-

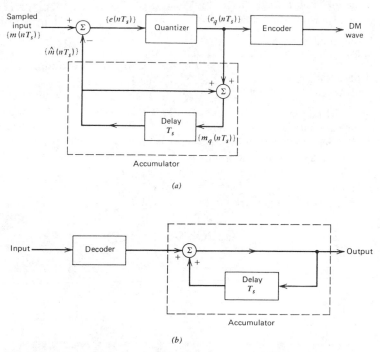

FIG. 10. DM system. (a) Transmitter. (b) Receiver. [From Haykin, S. (1983). "Communication Systems," 2nd ed. Wiley, New York. © John Wiley & Sons, Inc.]

tomary to use discrete PAM because it is the most efficient in terms of power and bandwidth utilization.

To transmit digital data over bandpass channels (e.g., satellite channels) we require the use of digital modulation techniques whereby the amplitude, phase, or frequency of a sinusoidal carrier is varied in a discrete manner in accordance with the digital data. In practice, the variation of phase or frequency of the carrier is the preferred method.

A. Baseband Data Transmission

The basic elements of a baseband binary PAM system are shown in Fig. 11. The signal applied to the input of the system consists of a binary data sequence $[b_k]$ with a bit duration of T_b seconds; b_k is in the form of 1 or 0. This signal is applied to a pulse generator, producing the pulse waveform

$$x(t) = \sum_{k=-\infty}^{\infty} a_k g(t - kT_b) \qquad (14)$$

where $g(t)$ denotes the shaping pulse that is normalized, so that we write

$$g(0) = 1 \qquad (15)$$

The amplitude a depends on the identity of the input bit b; specifically, we assume that

$$a_k = \begin{cases} +a & \text{if the input bit } b_k \text{ is represented by symbol 1} \\ -a & \text{if the input bit } b_k \text{ is represented by symbol 0} \end{cases} \qquad (16)$$

The PAM signal $x(t)$ passes through a transmitting filter of transfer function $H_T(f)$. The resulting filter output defines the transmitted signal, which is modified in a deterministic fashion as a result of transmission through the channel of transfer function $H_C(f)$. In addition, the channel adds random noise to the signal at the receiver input. Then the noisy signal is passed through a receiving filter of transfer function $H_R(f)$. This filter output is sampled synchronously with the transmitter, the sampling instants being determined by a clock or timing signal that is usually extracted from the receiving filter output. Finally, the sequence of samples thus obtained is used to reconstruct the original data sequence by means of a decision device. The amplitude of each sample is compared with a threshold. If the threshold is exceeded, a decision is made in favor of symbol 1 (say). If the threshold is not exceeded, a decision is made in favor of symbol 0. If the sample amplitude equals the threshold exactly, the flip of a fair coin will determine which symbol was transmitted.

1. Baseband Shaping

Typically, the transfer function of the channel and the pulse shape are specified, and the problem is to determine the transfer functions of the transmitting and receiving filters so as to enable the receiver to recognize the sequence of values in the received signal wave.

In solving this problem, we have to overcome the intersymbol interference (ISI) caused by the overlapping tails of other pulses adding to the particular pulse $A_i p(t - iT_b)$, which is examined at the sampling time iT_b. If this form of interference is too strong, it may result in erroneous decisions in the receiver. Clearly, control of ISI in the system is achieved in the time domain by controlling the function $p(t)$.

The receiving filter output $y(t)$ is sampled at time $t = iT_b$ (with i taking on integer values), yielding

$$y(t_i) = \sum_{k=-\infty}^{\infty} A_k p[(i - k)T_b)] + n(t_i)$$

$$= A_i + \sum_{\substack{k=-\infty \\ k \neq i}}^{\infty} A_k p[(i - k)T_b)] + n(t_i) \qquad (17)$$

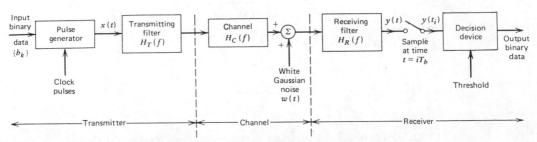

FIG. 11.   Baseband binary data transmission system. [From Haykin, S. (1983). "Communication Systems," 2nd ed. Wiley, New York. © John Wiley & Sons, Inc.]

where $A_k p(t)$ is the response of the cascade connection of the transmitting filter, the channel,
and the receiving filter, which is produced by the
pulse $a_k g(t)$ applied to the input of the cascade.
In Eq. (17), the first term A_i represents the ith
transmitted bit. The second term represents the
residual effect of all other transmitted bits on the
decoding of the ith bit; this residual effect is
called the intersymbol interference. The last
term $n(t_i)$ represents the noise sample at time t_i.

In the absence of ISI and noise, we observe
from Eq. (17) that

$$y(t_i) = A_i$$

which shows that under these conditions, the ith
transmitted bit can be decoded correctly. The
unavoidable presence of ISI and noise in the
system, however, introduces errors in the decision device at the receiver output. Therefore, in
the design of the transmitting and receiving filters, the objective is to minimize the effects of
noise and ISI and thereby deliver the digital data
to their destination in an error-free manner as far
as possible.

One signal waveform that produces zero ISI
in a realizable fashion is defined by

$$p(t) = \text{sinc}(2B_T t) \frac{\cos(2\pi\rho B_T t)}{1 - 16\rho^2 B_T^2 t^2} \quad (18)$$

where $\text{sinc}(2B_T t)$ is the sinc function

$$\text{sinc}(2B_T t) = \frac{\sin(2\pi B_T t)}{2\pi B_T t} \quad (19)$$

and the transmission bandwidth B_T is related to
the bit duration T_b as

$$B_T = 1/2T_b$$

The parameter ρ is called the rolloff factor; its
value lies within the range $0 \le \rho \le 1$.

The function $p(t)$ of Eq. (18) consists of the
product of two factors: the factor $\text{sinc}(2B_T t)$ associated with an ideal low-pass filter and a sec-
ond factor that decreases as $1/|t|^2$ for large $|t|$.
The first factor ensures zero crossings of $p(t)$ at
the desired sampling instants of time $t = iT$, with
i an integer (positive and negative). The second
factor reduces the tails of the pulse considerably
below that obtained from the ideal low-pass filter, so that the transmission of binary waves using such pulses is relatively insensitive to sampling time errors. In fact, the amount of ISI
resulting from this timing error decreases as the
rolloff factor ρ is increased from zero to unity.

The time response $p(t)$ is plotted in Fig. 12 for
$\rho = 0$, 0.5 and 1. For the special case of $\rho = 1$,
the function $p(t)$ simplifies as

$$p(t) = \frac{\text{sinc}(4B_T t)}{1 - 16B_T^2 t^2} \quad (20)$$

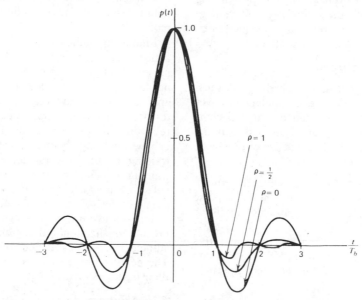

FIG. 12. Time response $p(t)$ for varying roll off factor ρ. [From Haykin, S.
(1983). "Communication Systems," 2nd ed. Wiley, New York. © John
Wiley & Sons, Inc.]

This time response exhibits two interesting properties:

1. At $t = \pm T_b/2 = \pm 1/4B_T$, we have $p(t) = 0.5$; that is, the pulse width measured at half-amplitude is exactly equal to the bit duration T_b.
2. There are zero crossings at $t = \pm 3T_b/2$, $\pm 5T_b/2$, ... in addition to the usual zero crossings at the sampling times $t = \pm T_b, \pm 2T_b, ...$

These two properties are particularly useful in generating a timing signal from the received signal for the purpose of synchronization. However, this requires the use of a transmission bandwidth double that required for the ideal case corresponding to $\rho = 0$.

2. Correlative Coding

Thus far we have treated ISI as an undesirable phenomenon that produces a degradation in system performance. Indeed, its very name connotes a nuisance effect. Nevertheless, by adding ISI to the transmitted signal in a controlled manner, it is possible to achieve a signaling rate of $2B_T$ symbols per second in a channel of bandwidth of B_T hertz. Such schemes are called correlative coding or partial-response signaling schemes. The design of these schemes is based on the premise that since ISI introduced into the transmitted signal is known, its effect can be interpreted at the receiver. Thus, correlative coding can be regarded as a practical means of achieving the theoretical maximum signaling rate of $2B_T$ symbols per second in a bandwidth of B hertz, as postulated by Nyquist, using realizable and perturbation-tolerant filters.

3. Adaptive Equalization

An efficient approach to the high-speed transmission of digital data (e.g., computer data) over a voice-grade telephone channel (which is characterized by a limited bandwidth and high signal-to-noise ratio) involves the use of two basic signal processing operations:

1. Discrete PAM by encoding the amplitudes of successive pulses in a periodic pulse train with a discrete set of possible amplitude levels.
2. A linear modulation scheme that offers bandwidth conservation (e.g., VSB) to transmit the encoded pulse train over the telephone channel.

At the receiving end of the system, the received wave is demodulated and then synchronously sampled and quantized. As a result of dispersion of the pulse shape by the channel, the number of detectable amplitude levels is often limited by ISI rather than by additive noise. In principle, if the channel is known precisely, it is virtually always possible to make the ISI (at the sampling instants) arbitrarily small by using a suitable pair of transmitting and receiving filters, so as to control the overall pulse shape in the manner described above. The transmitting filter is placed directly before the modulator, whereas the receiving filter is placed directly after the demodulator. Thus, in so far as ISI is concerned, we can consider the data transmission to be essentially baseband.

However, in a switched telephone network, two factors contribute to the distribution of pulse distortion on different link connections: (1) differences in the transmission characteristics of the individual links that can be switched together and (2) differences in the number of links in a connection. The result is that the telephone channel is random in the sense of being one of an ensemble of possible channels. Consequently, the use of a fixed pair of transmitting and receiving filters designed on the basis of average channel characteristics may not adequately reduce ISI. To realize the full transmission capability of a telephone channel, there is need for adaptive equalization. By *equalization* we mean the process of correcting channel-induced distortion. This process is said to be adaptive when it adjusts itself continuously during data transmission by operating on the input signal.

Among the philosophies for adaptive equalization of data transmission systems are prechannel equalization at the transmitter and postchannel equalization at the receiver. Because the first approach requires a feedback channel, it is customary to use only adaptive equalization at the receiving end of the system. This equalization can be achieved, before data transmission, by training the filter with the guidance of a suitable training sequence transmitted through the channel so as to adjust the filter parameters to optimum values. The typical telephone channel changes little during an average data call, so that precall equalization with a training sequence is sufficient in most cases encountered in practice. The equalizer is positioned after the receiving filter in the receiver.

4. Eye Pattern

One way to study ISI in a PCM or data transmission system experimentally is to apply the

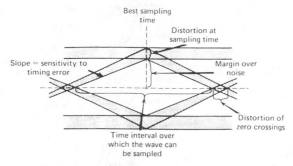

FIG. 13.   Interpretation of the eye pattern. [From Haykin, S. (1983). "Communication Systems," 2nd ed. Wiley, New York. © John Wiley & Sons, Inc.]

received wave to the vertical deflection plates of an oscilloscope and to apply a sawtooth wave at the transmitted symbol rate $1/T$ to the horizontal deflection plates. The waveforms in successive symbol intervals are thereby translated into one interval on the oscilloscope display. The resulting display is called an eye pattern because, for binary waves, its appearance resembles the human eye. The interior region of the eye pattern is called the eye opening.

An eye pattern provides a great deal of information about the performance of the pertinent system (Fig. 13):

1.   The width of the eye opening defines the time interval over which the received wave can be sampled without error from ISI. It is apparent that the preferred time for sampling is the instant of time at which the eye is open widest.

2.   The sensitivity of the system to timing error is determined by the rate of closure of the eye as the sampling time is varied.

3.   The height of the eye opening, at a specified sampling time, defines the margin over noise.

When the effect of ISI is severe, traces from the upper portion of the eye pattern cross traces from the lower portion, with the result that the eye is completely closed. In such a situation, it is impossible to avoid errors due to the combined presence of ISI and noise in the system.

B. BANDPASS DATA TRANSMISSION

When digital data are to be transmitted over a bandpass channel, it is necessary to modulate the incoming data onto a carrier wave (usually sinusoidal) with fixed frequency limits imposed by the channel. The data may represent digital computer outputs or PCM waves generated by

digitizing voice or video signals, and so on. The channel may be a microwave radio link or satellite channel, and so on. In any event, the modulation process involves switching or keying the amplitude, frequency, or phase of the carrier in accordance with the incoming data. Thus, there are three basic signaling techniques: amplitude-shift keying (ASK), frequency-shift keying (FSK), and phase-shift keying (PSK), which can be viewed as special cases of amplitude modulation, frequency modulation, and phase modulation, respectively. Figure 14 illustrates these three basic signaling techniques for binary data input.

Ideally, FSK and PSK signals have a constant envelope. This feature makes them impervious to amplitude nonlinearities, as encountered in microwave radio links and satellite channels. Accordingly, in practice, FSK and PSK signals are much more widely used than ASK signals.

The system is said to be coherent when the locally generated carrier in the receiver is synchronized to that in the transmitter in both frequency and phase. It is said to be noncoherent when the phase of the incoming signal is destroyed in the receiver.

1. Coherent Binary PSK

In a coherent binary PSK system, a pair of signals, $s_1(t)$ and $s_2(t)$ are used to represent binary symbols 1 and 0, respectively; they are defined by

$$s_1(t) = \sqrt{2E_b/T_b} \cos(2\pi f_c t) \qquad (21a)$$

$$s_2(t) = \sqrt{2E_b/T_b} \cos(2\pi f_c t + \pi)$$

$$= -\sqrt{2E_b/T_b} \cos(2\pi f_c t) \qquad (21b)$$

where $0 \le t \le T_b$ and E_b is the transmitted signal energy per bit. In order to ensure that each

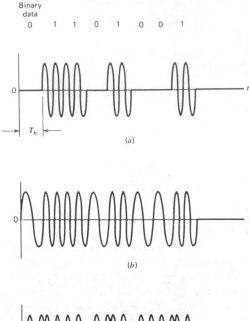

Binary data
0 1 1 0 1 0 0 1

FIG. 14. Three basic forms of signaling binary information. (a) Amplitude-shift keying. (b) Frequency-shift keying. (c) Phase-shift keying. [From Haykin, S. (1983). "Communication Systems," 2nd ed. Wiley, New York. © John Wiley & Sons, Inc.]

transmitted bit contains an integral number of cycles of the carrier wave, the carrier frequency f_c is chosen equal to n_c/T_b for some fixed integer n_c. A pair of sinusoidal waves differing only in phase by 180°, as defined above, are referred to as antipodal signals.

2. Coherent Binary FSK

In a binary FSK system, the symbols 1 and 0 are distinguished from one another by the transmission of one of two sinusoidal waves that differ in frequency by a fixed amount. A typical pair of sinusoidal waves is described by

$$s_i(t) = \begin{cases} \sqrt{2E_b/T_b}\,\cos(2\pi f_i t), & 0 \le t \le T_b \\ 0, & \text{elsewhere} \end{cases}$$

(22)

where the carrier frequency f_i equals one of two possible values: f_1 and f_2. The transmission of frequency f_1 represents symbol 1, and the transmission of frequency f_2 represents symbol 0.

3. Coherent Quadrature Signaling Techniques

Channel bandwidth and transmitted power constitute two primary "communication resources," the efficient utilization of which provides the motivation for the search for spectrally efficient modulation schemes. The primary objective of spectrally efficient modulation is to maximize the bandwidth efficiency, defined as the ratio of data rate to channel bandwidth (in units of bits per second per hertz) for a specified probability of symbol error. A secondary objective is to achieve this bandwidth efficiency at a minimum practical expenditure of average signal power or, equivalently, in a channel perturbed by additive white Gaussian noise, a minimum practical expenditure of average signal-to-noise ratio.

Examples of the quadrature-carrier multiplexing system include the following signaling techniques:

1. Quadriphase-shift keying (QPSK), which is an extension of the binary PSK. In this technique, the phase of the carrier takes on one of four possible values. Specifically, the binary pairs 10, 00, 01, 11 are assigned the phase values $\pi/4$, $3\pi/4$, $5\pi/4$, and $7\pi/4$, respectively. This is illustrated in Fig. 15.

2. Minimum shift keying (MSK), which is a special form of continuous-phase frequency-shift keying, with the detection in the receiver being performed in two successive bit intervals. The nominal carrier frequency f_c is equal to the arithmetic mean of the two frequencies f_1 and f_2 used to represent symbols 1 and 0, respectively. Moreover, the carrier frequency f_c is chosen equal to an integral multiple of one-fourth of the bit rate in order to make the phase of the transmitted signal continuous at the bit transition instants. One other requirement is to make the deviation ratio

$$h = T_b(f_1 - f_2)$$

equal to one-half. Figure 16 illustrates the waveform of the MSK signal.

4. M-Ary Signaling Techniques

In an M-ary signaling scheme, we send any one of M possible signals, $s_1(t)$, $s_2(t)$, ..., $s_M(t)$, during each signaling interval of duration T. For almost all applications, the number of possible signals M is 2^n, where n is an integer. The sym-

Input binary sequence   0   1   1   0   1   0   0   0

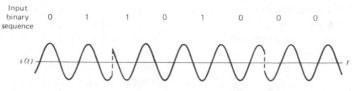

FIG. 15.   QPSK wave $s(t)$. [From Haykin, S. (1983). "Communication Systems," 2nd ed. Wiley, New York. © John Wiley & Sons, Inc.]

bol duration T is equal to nT_b, where T_b is the bit duration. These signals are generated by changing the aplitude, phase, or frequency of a carrier in M discrete steps. Thus, we have M-ary ASK, M-ary PSK, and M-ary FSK digital modulation schemes. The QPSK system considered above is an example of M-ary PSK with $M = 4$.

M-Ary signaling schemes are preferred over binary signaling schemes for transmitting digital information over bandpass channels when the requirement is to conserve bandwidth at the expense of increasing power. In practice, a communication channel rarely has the exact bandwidth required for transmitting the output of an information source by means of binary signaling schemes. Thus, when the bandwidth of the channel is less than the required value, we use M-ary signaling schemes so as to utilize the channel efficiently.

5. Detection of Signals with Random Phase in the Presence of Noise

Up to this point, we have assumed that the information-bearing signal is completely known at the receiver. In practice, however, in addition to the uncertainty due to the additive noise of a receiver, there is often an additional uncertainty due to the randomness of certain signal parameters. The usual cause of this uncertainty is distortion in the transmission medium. Perhaps the most common random signal parameter is the phase, which is especially true for narrow-band

signals. For example, transmission over a multiplicity of paths of different and variable lengths or rapidly varying delays in the propagating medium from transmitter to receiver may cause the phase of the received signal to change in a way that the receiver cannot follow. Synchronization with the phase of the transmitted carrier may then be too costly, and the designer may simply choose to disregard the phase information in the received signal at the expense of some degradation in the noise performance of the system.

For example, in the noncoherent detection of binary FSK signals, one can simplify receiver complexity by using a pair of frequency discriminators tuned to the frequencies f_1 and f_2, which represent binary symbols 0 and 1, respectively.

Another example is that of differential phase-shift keying (DPSK), which can be viewed as the noncoherent version of binary PSK. It eliminates the need for a coherent reference signal at the receiver by combining two basic operations at the transmitter: (1) differential encoding of the input binary wave and (2) phase-shift keying— hence the name differential phase-shift keying. In effect, to send symbol 0 we phase-advance the current signal waveform by 180°, and to send symbol 1 we leave the phase of the current signal waveform unchanged. The receiver is equipped with a storage capability, so that it can measure the relative phase difference between the waveforms received during two successive bit intervals. Provided that the unknown phase contained in the received wave varies slowly

Input binary sequence
0   1   1   0   1   0   0   0

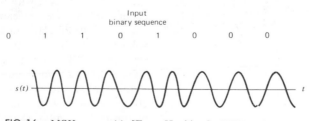

FIG. 16.   MSK wave $s(t)$. [From Haykin, S. (1983). "Communication Systems," 2nd ed. Wiley, New York. © John Wiley & Sons, Inc.]

(i.e., slow enough to be considered essentially constant over two bit intervals), the phase difference between waveforms received in two successive bit intervals will be independent of θ.

6. Discussion

It is customary to use the average probability of symbol error to assess the noise performance of digital modulation schemes. It should be realized, however, that even if two systems yield the same average probability of symbol error, their performances, from the user's viewpoint, may be quite different. In particular, the greater the number of bits per symbol, the more the bit errors will cluster together. For example, if the average probability of symbol error is 10^{-3}, the expected number of symbols occurring between any two erroneous symbols is 1000. If each symbol represents one bit of information (as in a binary PSK or binary FSK system), the expected number of bits separating two erroneous bits is 1000. If, on the other hand, there are two bits per symbol (as in a QPSK system), the expected separation is 2000 bits. Of course, a symbol error generally creates more bit errors in the second case, so that the percentage of bit errors tends to be the same. Nevertheless, this clustering effect may make one system more attractive than another, even at the same symbol error rate. In the final analysis, which system is preferable will depend on the situation.

Two systems having an unequal number of symbols can be compared in a meaningful way only if they use the same amount of energy to transmit each bit of information. It is the total amount of energy needed to transmit the complete message, not the amount of energy needed to transmit a particular symbol satisfactorily, that represents the cost of the transmission. Accordingly, in comparing the different data transmission systems considered, we shall use as the basis of our comparison the average probability of symbol error expressed as a function of the bit energy-to-noise density ratio E_b/N_0.

In Table I, we have summarized the expressions for the average probability of symbol error P_e for the coherent PSK, conventional coherent FSK with one-bit decoding, DPSK, noncoherent FSK, QPSK, and MSK, when operating over an AWGN channel. In Fig. 17 we have used these expressions to plot P_e as a function of E_b/N_0.

In summary, we can state the following:

1. The error rates for all the systems decrease monotonically with increasing values of E_b/N_0.

2. For any value of E_b/N_0, the coherent PSK produces a smaller error rate than any of the other systems. Indeed, in the case of systems restricted to one-bit decoding perturbed by additive white Gaussian noise, the coherent PSK system is the optimum system for transmitting binary data in the sense that it achieves the minimum probability of symbol error for a given value of E_b/N_0.

3. The coherent PSK and the DPSK require an E_b/N_0 that is 3 dB less than the corresponding values for the conventional coherent FSK and the noncoherent FSK, respectively, to realize the same error rate.

4. At high values of E_b/N_0, the DPSK and the noncoherent FSK perform almost as well (to within ~1 dB) as the coherent PSK and the conventional coherent FSK, respectively, for the same bit rate and signal energy per bit.

5. In QPSK two orthogonal carriers $\sqrt{2/T} \cos(2\pi f_c t)$ and $\sqrt{2/T} \sin(2\pi f_c t)$ are used, where the carrier frequency f_c is an integral multiple of the symbol rate $1/T$, with the result that two independent bit streams can be transmitted and subsequently detected in the receiver. At high values of E_b/N_0, coherently detected binary PSK and QPSK have the same error rate performances for the same value of E_b/N_0.

6. In MSK the two orthogonal carriers $2\sqrt{T_b} \cos(2\pi f_c t)$ and $2\sqrt{T_b} \sin(2\pi f_c t)$ are modulated by antipodal symbol shaping pulses $\cos(\pi t/2T_b)$ and $\sin(\pi t/2T_b)$, respectively, over $2T_b$ intervals, where T_b is the bit duration. Correspondingly,

TABLE I. Summary of Formulas for the Symbol Error Probability for Different Data Transmission Systems

System	Error probability P_e
Coherent binary signaling	
Coherent PSK	$\frac{1}{2} \mathrm{erfc}(\sqrt{E_b/N_0})$
Coherent FSK	$\frac{1}{2} \mathrm{erfc}(\sqrt{E_b/2N_0})$
Noncoherent binary signaling	
DPSK	$\frac{1}{2} \exp(-E_b/N_0)$
Noncoherent FSK	$\frac{1}{2} \exp(-E_b/2N_0)$
Coherent quadrature signaling	
QPSK	$\mathrm{erfc}(\sqrt{E_b/N_0})$
MSK	$\quad - \frac{1}{4} \mathrm{erfc}^2(\sqrt{E_b/N_0})$

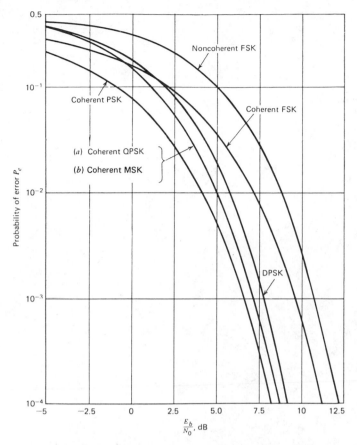

FIG. 17.   Comparison of the noise performances of different PSK and FSK systems. [From Haykin, S. (1983). "Communication Systems," 2nd ed. Wiley, New York. © John Wiley & Sons, Inc.]

the receiver uses a coherent phase decoding process over two successive bit intervals to recover the original bit stream. Thus, MSK has exactly the same error rate performance as QPSK.

ACKNOWLEDGMENTS

The material presented in this article is summarized from the author's book, "Communication Systems" (see Bibliography). The author is indebted to John Wiley and Sons, publisher of this book, for permission to write this article.

BIBLIOGRAPHY

Haykin, S. (1983). "Communication Systems," 2nd ed. Wiley, New York.

Jayant, N. S., and Noll, P. (1984). "Digital Coding of Waveforms." Prentice-Hall, Englewood Cliffs, New Jersey.

Proakis, J. G. (1983). "Digital Communication." McGraw-Hill, New York.

Schwartz, M. (1980). "Information Transmission, Modulation and Noise," 3rd ed. McGraw-Hill, New York.

Shanmugam, K. S. (1979). "Digital and Analog Communication Systems." Wiley, New York.

Stark, H., and Tuteur, F. B. (1979). "Modern Electrical Communications: Theory and Systems." Prentice-Hall, Englewood Cliffs, New Jersey.

Stremler, F. G. (1982). "Introduction to Communication Systems." Addison-Wesley, Reading, Massachusetts.

Viterbi, A. J., and Omura, J. K. (1979). "Principles of Digital Communication and Coding." McGraw-Hill, New York.

Ziemer, R. E., and Peterson, R. L. (1985). "Digital Communications and Spread Spectrum Systems." Macmillan, New York.

COMPUTER NETWORKS

Bennet P. Lientz   *University of California, Los Angeles*

GLOSSARY

Architecture: Structured method for connecting computers and communications components and protocols in such a way that the computer network is practical in terms of performance, cost, and flexibility.

ARPA network: One of the first computer networks developed by the Advanced Research Projects Agency.

Computer network: Computer and communications network linking multiple computers together to share information and resources.

Protocol: Method of communicating messages between computer devices and computers.

T-1 carrier: High speed, high capacity communications link to support volume traffic between computers.

A computer network is a category of computer systems wherein multiple computers are intertied by high speed data communications links. The purpose of a computer network is to share the resources of all computers among various users that are connected to the network. The resources to be shared include data bases, software, and computing resources not available at the computer site normally used by an organization.

I. Definitions

The Advanced Research Projects Agency network (ARPANET) was one of the first networks that supported the interconnection of different computer makes and models so that communication was possible. The ARPANET demonstrated feasibility of technology.

Computers require high speed, high capacity transmission facilities. Physically, computers can be linked by microwave, satellite, fiber optics, or standard cable. Many computers in networks today are often tied together by communications links with speed of over 1 million bps, called T-1 carriers. [*See* COMMUNICATION SYSTEMS, CIVILIAN; TELECOMMUNICATIONS.]

In some computer networks the communications transmission links do not connect directly to the computer, but instead are attached to an interface minicomputer. This minicomputer acts as a front end processor for the data processing computer into the network. This minicomputer performs functions such as:

1. Message formatting and transmission
2. Error analysis and logging of problems
3. Notification of problems to the computers in the network
4. Support of communications protocols
5. Encryption of data
6. Routing of traffic in the network

The size, features, and cost of these minicomputers depend on the characteristics of the specific network. Obviously, the more incompatible the data processing computers are, the more functions and hence the more expensive the minicomputers will have to be.

In order to communicate, computers must share a common method of "handshaking." This means that the network has a common protocol supported by software and hardware that allow the computers to share information.

A protocol packages a message to be sent into a frame with header and trailer information involving addressing, error detection support, and other information. After packaging, the message is transmitted. At the receiving computer, the message is extracted and the header and trailer are removed. As a message moves through the network, the network computers search the header information for the address of the next computer to which to route the message.

Two popular protocols are X.25 and SDLC (Synchronous Data Link Control). X.25 is probably the more popular protocol, since it has several features that are more suitable for networking and is compatible with a broader range of equipment. SDLC is the protocol of System Network Architecture (SNA) developed by the IBM Corporation.

Current protocols in use function at the lower levels of the open systems interconnection (OSI) architecture. The OSI model is composed of seven layers. From lowest to highest these layers are:

1. Physical link—physical transmission of information bits across the network
2. Data link—transmission of frames across network links
3. Network—supports connection of multiple network links;
4. Transport—transfer of data along a complete network path from an origin to a destination;
5. Session—support of communications between application systems;
6. Presentation—provides for application systems to be independent of the form and representation of the data;
7. Application—network access to application systems.

Most protocols work in the first three layers or levels. A network normally operates with one protocol for efficiency reasons. If, for example, a computer network used X.25 as the standard for the network, any non-X.25 computer would require its messages or frames to be repackaged into X.25 for transmission in the network at the origin. At the destination, the message is reestablished in the format of the originating protocol.

A computer network usually spans the organization rather than merely one department. Thus, a computer network can be contrasted with a network to support a limited number of users in a contiguous geographic area, or a local area network. There are other differences, however, beyond geography. Most local area networks deliver data and software to the user workstation or microcomputer. Data and software are stored in a file server that serves the network. An interface unit supports communications between external networks and the local area network. In a computer network there are multiple mainframe and minicomputers interconnected. User devices can be simple terminals.

II. Components of Computer Networks

A computer network is composed of system components that support transmission and communications, and applications function oriented components that provide for end user applications. The following are major components of a computer network.

1. Data processing computers and software are the traditional data processing computers that support applications software systems. For most computer networks, these are computers that are in use at the time the network is established.
2. Network support hardware includes interface hardware between the traditional computers and the data communications links. The ARPANET uses minicomputers called interface message processors for the interface to the network.
3. Data communications facilities are comprised of transmission facilities such as dedicated high speed lines, switching centers, and satellite facilities.
4. Network control software is the software system that handles the transmission and routing of communications traffic in the network. This includes the software to support the communications protocol in use in the network.
5. Terminals, printers, and microcomputers are the hardware components available to end users of the computer network. A network normally supports a wide variety of these devices. One of the more difficult devices to support is the printer because different printers often have noncompatible printer control codes and characters.
6. Terminal control units, multiplexors, and modems are the hardware that interfaces between the terminals and printers and the network transmission facilities.

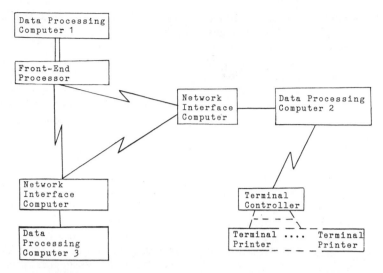

FIG. 1. Computer network components. Computer 1 is a mainframe computer with its own front-end processor. The other two computers are minicomputers and require network interface computers. The terminal cluster is just one of many such in the organization. The cluster connects to the network at the nearest center.

Figure 1 gives an example of the hardware and communications components of the computer network linked together. This can be contrasted with the communications network in Fig. 2 that is not a computer network, but a central set of computers that support on-line transactions.

How these components work together can be seen by observing a general user transaction in the computer network. A user who is seated at a desk with a terminal signs on to the computer network by entering a user identifier and a password. The network may contain software to validate the identifier and the password. If it does not, then the network routes the information to the computer to which the user is assigned. The network support software handles the routing of the information through the network.

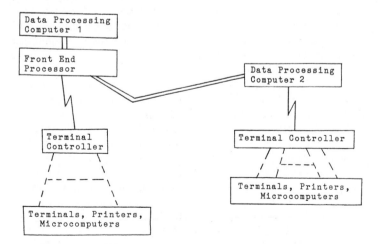

FIG. 2.   Example of on-line computer systems that are not computer networks. Each computer pictured has its own terminal network. The two computers communicate via a remote job entry link for transfer of files. There is no on-line link allowing a terminal in one system to address on-line applications in the other system.

After establishing valid access, the user terminal continues to use the network to communicate with one or more of the computers in the network. If the protocol of the network is not supported by the computer in use, then the network will reformat the messages from the user terminal for the computer at the computer end. The same process occurs at the user end if the terminal is not compatible with the network protocol.

In most computer networks once a user has established connection with the selected computer, then all transactions in that session are carried out with this computer because it has the data and software that the user requires on a continuing basis. However, the computer network supports the connection and shift of the user to other computers.

The communications support components provide a base for the applications systems and electronic mail-type functions. Electronic mail is an example of a generalized software capability that spans different machines to support the interchange and sharing of data, text files, messages, and other information between users. It is an example of a utility function. Electronic mail includes electronic filing, messaging, and may support the interface of electronic messages with an on-line applications system.

An applications system is typically programmed in a higher level language, such as CO-BOL, a language such as C, or in a data base management system language. Early applications systems that employed data communications relied on assembly language and were heavily dependent on the specific hardware employed. Each time there was a significant hardware or system software change, there had to be changes to the applications system programs. Since the early 1980s, the applications systems have been isolated from the physical computer and communications network. Devices could be added, changed, or deleted and there would be no changes required for the applications systems. The network management layer of software was added to the networks to remove the applications software from direct linkage to the physical network.

Computer networks have evolved rapidly with improvements in hardware, software, and communications. These improvements included the following.

1. More sophisticated system and network software that supported applications systems being written in higher level languages opened computer networks to standard business and government organizations.

2. Improvements in the reliability, price-performance, and compatibility of hardware made networking economical and practical.

3. Greater capability in system software reduced the extent of programming necessary for applications systems.

4. Improved data communications services and facilities.

These improvements made computer network economically attractive and technically feasible. These factors are on the supply side of the technology. On the demand side, there has been a growth in the transmission of large volumes of text and images to support teleconferencing and other means of electronic communications.

III. Use of Computer Networks

There are several categories of computer networks. One type is the computer utility. Examples are public timesharing systems offered by General Electric, Computer Science Corporation, and more than 100 other suppliers. These firms offer value added services for sharing of information, program development, and access to data bases. In the early years of data communications, these services were a valuable tool in developing systems since many firms did not have internal timesharing systems available for programmers. As on-line computing became widespread, the use of such networks changed to the value-added, unique services that were uneconomical to offer internally. Access to economic, financial, and other business data bases has been one area of use of growing interest. Use of the network to input, share, and distribute information is also important. These networks span state and national boundaries and offer the benefits of rapid access without a long development or implementation cycle. International networking has been hampered somewhat by regulation and restriction on the flow and content of information that can cross national boundaries. This set of issues is referred to as Transborder Data Flow. A number of countries restrict the flow of detailed transactions across borders. Other countries mandate that firms use local data processing facilities.

A second category of computer network is the interorganization networking supported by the government or by a consortium. An example of a defense network is the military communications network for data transmission. The ARPANET is an example of a network connecting a number of different government agencies, universities, and companies.

The fastest growing computer network category is that within specific, individual organizations. This is a natural trend since communications within organizations far outnumber and exceed the communications between organizations. A computer network in an organization links together separate business units to share information and provide for faster, more efficient processing of information. Here, three brief examples of computer networks in different industries are described. (A detailed example appears in Sec. VIII.)

1. Automotive manufacturing. The individual manufacturing plants are connected to distributors, to headquarters, and to each other via a computer network. Attached to the network are terminals and computers of suppliers as well (see Fig. 3). The network does not stop there. Within the factory, a massive effort is underway to link the various robots and automated functions into a system. While individual automation efforts result in some improvements, large scale benefits accrue only when the different computers are tied together in a network. The network can then be controlled by other computers that perform load balancing, scheduling, sequencing, and other functions.

2. Defense and engineering. Defense contractors are tied together in project networks so that the government contract managers can assess the status of projects and identify problems early. Within the contractor facility it is possible to link in a network the following departments: design, engineering, project management, documentation and quality control/assurance, and manufacturing (see Fig. 4).

3. Financial networks. Funds transfer, auto-

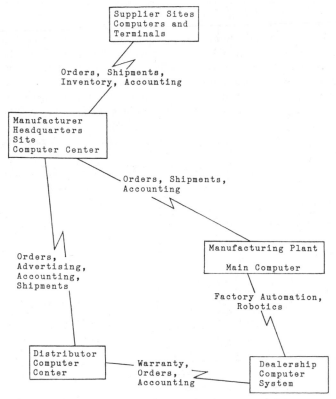

FIG. 3. Example of network for automotive manufacturing. There are five entities shown in this simplified diagram of an automotive manufacturing network. Some of the major pieces of information are shown in the information flow.

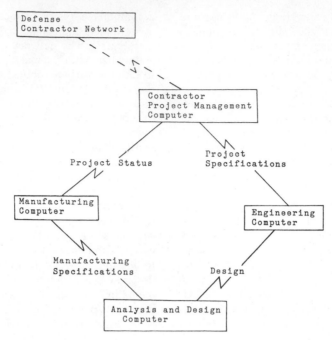

FIG. 4. Example of contractor/engineering computer network.

mated clearing house for checks, and automated teller machine networks are three commonly encountered financial networks (see Fig. 5). These link the computers in different banks and financial institutions with the computers of the Federal Reserve Board and other agencies. Another financial network is the emerging networks linking exchanges together.

There are also networks to support industry or specialty functions. There are networks for agriculture, medicine, library and bibliographic systems, postal services and facsimile, transportation and reservation systems, to mention a few examples. In fact, networks have been created for almost every segment of established industries where there is a need for communications.

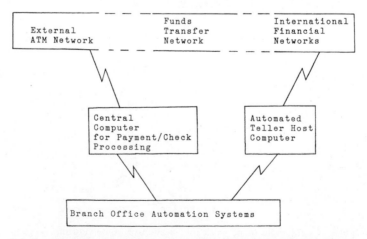

FIG. 5. Example of financial network. Shown are the interfaces between the central computers of a financial institution and the external networks. The branch offices of the organization are tied on-line to the central computers.

IV. Benefits and Risks

Computer networks have yielded a number of benefits. Some of these are general across many networks. Others are specific to the applications in the network. Several general benefits are as follows:

1. Faster flow of information. With different, incompatible systems there is more time and effort involved in transferring information from one system to another. Transfer in the computer network can occur in a matter of seconds.

2. Low variable cost. Once the network is in place, the majority of the costs have been paid. Transmission of information by dial-up telephone line, courier service, or other means is expensive and represents a variable cost.

3. Improved control. Through a network, managers and employees can obtain information more readily, can organize and summarize information effectively, and so can take actions more quickly. In some networks the staff and management can see the effects of their decisions almost immediately.

4. Greater economies of scale. By having access to a number of different computers in the network, one can allocate the location of data bases and applications systems so that a bottleneck does not occur. This results in greater economies of scale.

5. Sharing of information and software. Through the network and the proper password and identification, users can share files and data and can use the same software. A computer network makes an electronic mail capability feasible.

6. Less redundant equipment and software. By having a computer network, an organization can ensure that users need only one terminal or personal computer, instead of one per system. A computer network supports multiple applications systems through the same communications lines. Without a network there would likely be separate, redundant lines into various offices. Redundant lines mean additional communications hardware and terminals, as well as additional software.

For some organizations there are more local and significant benefits to a computer network. One is that without the network a user would normally be able to access only one on-line system. With the network a user could access multiple systems. For example, in a bank branch one could use the same terminal and network to determine a customer's balance for check approval, perform nonmonetary changes to an account, input a loan application, set up a new account, and handle deposits. The automated teller machines in the lobby or outside of the office could share the same network facilities.

There are risks in a computer network due to access availability to a wide variety of possible users. The stories of computer crime and misuse of network systems point to the vulnerability of many networks. However, while the risks cannot be removed totally, there are reasonable steps that one can take to minimize the risk. One is to restrict or eliminate the use of dial-up telephone connections. Many computer crimes involved the access of the computer by dial-up communications ports. Another method is to install monitoring software to detect penetration. Encryption of data transactions and files can also be employed. These methods, however, are not going to be effective against a person with computer expertise. Further steps involve logging transactions, backing up files, and taking other precautions so that, in the event of problems, the systems and data can be restored.

A more serious problem and one that has more likelihood of occurrence is computer network failure. In order to understand failure, two aspects deserve attention. One is the recovery from the failure; the other, the potential source or cause of failure. The steps that need to be taken when a computer network has failed either partially or totally begin with determining the cause and effects of the problem or failure. It might be easy to determine what has failed, but it may be difficult to determine what led to failure.

After this initial step, the systems analysts must determine the best and most expeditious approach to restore the system. It may mean shutting down the entire system and then bringing the system back to life in stages. Before this can be done, the data must be restored to the point at which failure occurred; otherwise, the data in the files may be contaminated. The time to restore the hardware and communications might be measured in minutes; the time to restore the data files might be in terms of hours. This restoration procedure is common in financial networks involving automated teller machines.

After restoration, testing is necessary. Moreover, during the initial operational period there is substantial overhead communications traffic and file setup that can lead to system instability.

A system does not need to fail catastrophically to cause problems for the organization. Poor response time is another problem. In many computer networks there are periods of peak activity. Often, these are in the late morning and the early afternoon. At other times the load on the system in terms of the number of transactions and active users is lighter. While it is possible to have some additional capacity to accommodate peak loads, it is usually uneconomic to have massive spare or excess hardware. Some improvements can be carried out by system tuning, work scheduling, and other methods.

A variety of steps can be taken to improve the reliability of a computer network. One is to have multiple communications paths between any two computers in the network. Thus, if one link in a network becomes overloaded or fails, an alternate route exists. Another step is to use fault-tolerant or redundant hardware, which provide for more availability and reliability.

V. Costs

The costs of a computer network include both one-time and recurring costs. Setup costs must cover communications hardware such as modems, multiplexors, controllers, concentrators; additional hardware for the main computers in the network, including additional ports or a larger front-end processor and memory for support of new users; and hardware to support the network functions of message protocol handling, routing and switching, security, transaction logging, and features such as electronic mail. Network software to support these network functions must also be purchased initially, as well as communications lines such as leased and backup lines and possibly the use of value-added carrier services. Installation of cable, junctions, and interfaces, in addition to staff and support costs for installation must also be considered.

After these initial costs, computer networks require support. This translates into dedicated staff who work on enhancements to the network, network problems, and installation of new users. After all a network is a living system that is sensitive to the user population. If a network is at all successful, then use can expand rapidly, creating a sizable demand for network services. The extent of required support depends on the activities in the network.

The cost of a network obviously depends on the scope and nature of the applications and purpose. Some small networks may cost less than $50,000 to establish and $5,000 per month to operate. Other, larger networks can cost several million dollars to establish and more than $1,000,000 annually to support.

The costs of the network can be offset, however, by more than 50% by the savings accrued when duplicate communications lines and hardware are removed. A well-designed computer network will also allow new users to be attached easily. Without the network, the cost and time to attach new users can be substantial.

VI. Implementation Approach

A preliminary assessment of the current state of communications and on-line systems in the organization is necessary. Facilities and costs are identified. Next, a long-range information systems plan should be available or be developed. This plan identifies the high-priority targets of automation as well as objectives and strategies for the organization. Based on the current systems and future demand, the applications systems that would be available through a computer network can be defined.

If the benefits are not substantial, then the analysis should stop here. There are a number of potential barriers to establishing computer networks. One barrier is that the current computer systems are extremely incompatible. This means that after the network has been setup, massive programming and development to overcome the incompatibilities must still occur.

A second barrier is that the staff has partitioned its use of computers in such a way that there is little need for the network. Each user department works with the software and data at one computer site and does not require or benefit from access to other computers.

Technical barriers are related to the architecture of the network. In order for a computer network to be feasible, there must be a network architecture that is technically possible in terms of implementation. This relates to the compatibility of protocols and standards between machines.

Assuming that there are sufficient benefits in terms of cost savings and increased service, the cost and schedule to implement a computer network through the development of the architecture should now be assessed. Because there are many different methods and approaches to the network, it is necessary to develop a computer-communications architecture. The architecture specifies the method of connecting the computer equipment together and determines the supporting software and hardware.

From the architecture, the components and software development can be estimated in terms of costs and schedule. The architecture can be evaluated in terms of cost, robustness, and compatibility to the organization. Robustness here means how flexible the network will be to accommodate growth and change.

These activities conclude the analysis part of the implementation. The next steps involve the actual installation and testing of equipment and software. Services, data, and software should be prioritized during installation. The initial installation will link the major computers in the network to form a network backbone. This means that these data processing computers can communicate with each other. After this installation, the network links are tested in terms of performance, reliability, and operation parameters.

The next step is to modify the computer applications systems to employ the network. This may mean the accommodation for electronic mail; software modification for sharing data and transmitting files to other computers; adaptation to serve different types of terminals, microcomputers, and printers.

The user terminals that currently connect directly to one computer may be rerouted to the nearest computer in the network. The changes that can occur in connections are shown in Fig. 6. The elapsed time to carry out these changes can be significant. New communications hardware must be ordered and installed, then the communications themselves have to be installed.

After installation each user site must be tested. If backup lines are installed, these must be tested as well. The testing involves physical ability to communicate, performance response time, and communications compatibility with the computers in the network.

After testing the installation, the user groups must be trained. The areas of interest in training typically involve network access to any computer in the network, use of electronic mail and any other utilities, information on the software and data bases available at other computer sites.

After training and documentation distribution,

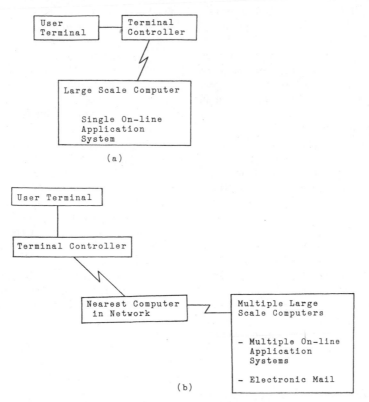

FIG. 6.   Migration from on-line system to computer network. (a) On-line environment. (b) Computer network connection.

network use can begin. Additional steps relating to the method of charging for services must be defined. It has been found in some networks that the marketplace motive will actually result in significant shifts in computer workload between sites. Thus, if a user finds that the same software is available on another machine in the network and costs 40% less to use than the current rate, the user can transfer files across the network to the cheaper computer. This can result in substantial changes and dislocations. This issue will be considered in Section VII.

VII. Computer Network Operation and Performance

After reaching a state of operation, a computer network must be measured and controlled like any other resource. This includes regular measurements of performance and cost for internal use by technical staff as well as visibility of the data to management.

A variety of data measurements can be collected for a network. Some of these include traffic on each link of the network (measured in terms of messages), end-to-end traffic across the network (measured in messages), communications response time for each link and each path in the network (time required to traverse the network), computer and communications response time between when the user inputs a message at a terminal and receives a response back from the computer, problem reports and failure incident analysis, mean time to failure and to repair components, overall availability and reliability of network, and cost reports through accounting and billing. This data is collected automatically by system logs and monitors in the network, by manual recording of problems and events, and by reports generated by the data processing computers.

From the standpoint of the technical staff, interest is centered on peak periods where response time is slowest and the number of active users is highest. This supports the technical staff in tuning the system for peak periods and in attempting to locate bottlenecks in the system. Bottlenecking can occur, for example, when too much traffic is being directed through one particular link in the network. Other links may be underutilized.

If one computer is receiving more than the anticipated workload, there are several alternatives. First, some of the workload could be transferred to a lesser-used machine. Second, the hardware for data processing or communications could be upgraded. A third alternative analyzes potential bottlenecks at the data processing site.

These examples point to the complexity that one encounters in a network versus one computer center serving many users in a network. There are more factors to consider, and with different equipment, efforts at analysis and improvement in performance are more time consuming and complex.

The technical actions based on the data include diagnosis and correction of problems, tuning for improvement in performance, and connection of additional user sites. In addition, the applications systems are not static. New releases of system software and applications software as well as new applications systems must be installed. Hardware at the data processing sites is also changing. These remarks indicate that the process of measurement and tuning is continuous.

For management there are higher level reports and data that are necessary. The goal of management is to ensure that the network is being managed and controlled properly, that overall performance and reliability are satisfactory, and that the network is delivering benefits that outweigh the costs.

Some of the commonly used management measures of a network include the following.

1. Cost per user. This is the monthly cost of the entire computer network (including the data processing hardware) divided by the number of users. This ratio gives management a measure of the productivity gain necessary to cover the cost of the network. As more users are added to the network, additional incremental hardware and communications are necessary. However, with economies of scale in support and improved hardware performance, the cost per user should decline.

2. Cost per terminal. This is the ratio of the total monthly network cost and the number of terminals. Since there are usually more users than terminals, the ratio for terminals should be much larger than that of users. This ratio should also decline over time.

3. Response time for the network. This is portrayed in several ways. First, the response time over the day or week can be shown. This could be an average across the month. A second approach is to indicate average response time,

the worse case response time, and a standard deviation band of response time.

4. Number of transactions in total and per terminal by site and overall. After defining a transaction, the ratios include the total number of transactions divided by the number of terminals per hour and the number of transactions per user per hour. This could be shown for each data processing computer site and for the entire network.

These measures would be calculated each month and then inserted in graphs to show year-to-date and comparison with the previous year.

Another measure is to show the down time and failures. A pie chart can show the distribution of problems by source. Another chart would portray availability by day across the month. The number of terminal hours can be calculated that were unavailable as compared to the total hours. Thus, if part of a network were not working and 10 terminals were not functional for an hour, the number of nonworking terminal hours would be 10. If there were 100 terminals and the network were operational for on-line use for 10 hours per day, five days per week, then the total hours per week would be 50,000.

Most of these remarks pertain to the on-line operation and performance of the network. However, when the network is not available to end users in the evenings and weekends, it is still in use for testing and installation of new features and systems.

In some networks there is also batch processing and file transfer operations being performed. Suppose, for example, that users in one part of the country enter transactions that are stored in a transaction file in the data processing computer in the network nearest to them. During a normal business day the transaction file grows. These transactions could be orders for goods or services of various types. At the end of the day the transaction file is transmitted across the computer network to the data processing center where the master files reside and where the updating applications system is based. The transaction file is used to update the master files. A copy of a master file may then be transmitted through the computer network to other data processing computer sites.

The transfer of batch files is quite common in many fields. For financial systems a snapshot of account balances and other information can be taken and the information transmitted through the network for inquiry purposes the next day. It can occur in a network that some data processing computers support on-line work and processing and other data processing computers support batch processing with older applications systems. The various data processing computers are linked in a computer network wherein the network link between the batch oriented and on-line oriented machines is most active in the off hours when users are not active.

Computer network operations are affected by time zones and the nature of the business. In some businesses there are active users in multiple time zones. Added to this are extended hours and possible weekend on-line use of systems. This creates additional pressure on the network and the staff. The system may have to be operational and available for 18 hours or more per day for more than six days per week.

A typical schedule in an active network during a work day might be the following 6 A.M.–8 P.M., on-line operation of the network and software by users; 8 P.M.–midnight, backup of files and transmission of transaction files to other data processing computers in the network; midnight–4 A.M., updating of files received from other computers in the network; 4 A.M.–5 A.M., maintenance and analysis; 5 A.M.–6 A.M., setup of the network operation for the operation of that day. This operations cycle leaves little room for human error. Yet operations errors occur with some frequency. A network has to be reestablished. A batch job stream has to be rerun. The wrong data is processed. These and other errors add up to time to restore operations. Translated into impacts on users, this means a delay in the startup time of the network for that day. A delay of even one hour for 100 users means a substantial loss in productivity. Assuming that the users are highly dependent on the system and that the average hourly wage with fringe benefits and overhead is $15.00 per hour, the loss is $1,500 per hour. It may cost even more if overtime is necessary to make up for the lost time. This example reinforces the need for high reliability and availability.

During network operation it is necessary at times to test new technology products and services to assess the benefits and effects. A computer network may have the capability of isolating the part of the network involving the new technology from the remainder of the network. In this way if there is a problem or if response time suffers, the impacts on production are reduced. This approach is fairly typical when new releases of network software and other systems software are available. It can also be used for new applications systems. However, there may be some enhancements or new software that impacts the network overall. An example might be an electronic mail system or a system that verifies user identification and allows the user to access multiple applications systems from the same terminal without any changes or major steps. In such instances it is difficult to restrict the test and use to one segment of the network.

Attention has been centered on the network part of operations. There is also the normal day-to-day operation of the individual data processing computer centers. These obviously are undergoing maintenance and support according to their own schedules. This can have a cross impact on the network. The necessity for coordination points to the need for network management and planning so that an event in one part of the network does not impact the entire network. In addition to regular meetings, there should be communications using electronic mail and written communications. The network support group needs to be informed of any changes in a site that might impact the network.

VIII. Example of Network within an Organization

In one large organization with six locations, there were different and incompatible computers. The mainframe computers were of two different makes and models. There were also minicomputers of three different vendors. With the lack of compatibility and no network, the only way the computers could be used was to work with a terminal tied directly to one of the machines. Then if another machine was desired, the user had to move to a terminal to the other machine. Data and file transfer was either manual by human data entry or by computer tape when output and input programs would have to be written for supporting the transfer.

In addition to these problems there were limitations on the available capacity and functions supported in the melange of computer types. After analysis, the approach that was taken was to install a computer network with minicomputers at each site and with the interface of the minicomputers to the other minis and mainframe computers. The minicomputers in the network are tied together by T-1 carrier links. At each site the minicomputer is connected to other machines via a fiber optics link. User terminals at each site are connected to the minicomputer via

a coaxial cable network. The network supports a user terminal connected to any computer in the network. The network computers also provide for security and for electronic mail and word processing. The network then is not only a communications backbone, but is also a software utility for all users.

There are several benefits to this approach. First, the current hardware did not need to be altered. The software and connectivity was able to be established without major reworking of systems. Second, a common network was established for all users with benefits shortly after the network became operational. In some networks, it may take an extended time for new applications systems to be functional. In this instance, electronic mail, word processing, and connection to the data processing computers could be used immediately after an initial training period.

The computer network here resulted in the benefits discussed in Section IV. There were some additional benefits that evolved over a period of several years. One is that the existing data processing computers could continue to be employed without as much growth as earlier estimated. This was made possible because of the offloading of communications traffic from the data processing computers to the network computers. Another benefit was the encouragement of increased communications between users and managers. Electronic mail provides for a record of calls and activity. Telephone calls proved to be less than satisfactory since such a low proportion of calls were completed. Moreover, the high ratio of staff professionals to secretaries made this situation worse. Periodic management reports, administrative forms and reports, budgets, and other work integrated into the network. The most unanticipated result was that acceptance of the network was so great that additional hardware and communications facilities were required to handle the additional load.

IX. Example of an Industry Network

The automobile industry is becoming more automated, not only within the factory but also between the manufacturer, supplier of components, distributor, and dealer. Dealerships offer a wide range of services. There is the sales and leasing department, parts and accessories department, finance and insurance department, service department, the general office, and management functions.

These departments link to other external entities through networks. The sales and leasing department links to a dealer network to determine what vehicles are on the lots of other dealers and distributors. Sales and finance link to credit bureau networks to approve applications for loans and leases. The finance department links to the manufacturer for warranty information. Service links to manufacturers for warranty information. Parts must access a parts network to locate parts that are not on site at the dealership. The dealer management orders new cars and trucks through the network.

The network allows the dealer to have fewer vehicles on location. This reduces costs substantially since space, insurance, and financing costs are diminished. By reviewing credit information faster, the service to the customer is improved and the productivity of the sales force is increased. Having access to parts data can locate the needed parts faster and lead to faster repair. For service, the network can track and schedule service visits so that staff productivity is raised.

For the manufacturer and distributor, a computer network allows for greater access to order, sales, credit, and other information. This information can be used in directing sales and advertising campaigns, managing the relationship with the dealer, and keeping in touch with the consumers who purchase the products.

X. Trends in Technology

The continued improvement of hardware is one trend that will continue to contribute to the expansion of networks. The spread of fault tolerant and nonstop systems will increase reliability and availability of networks even more. Microcomputers and local area networks have shown to be factors leading to increased communications. Organizations have moved with technology to establish major data bases and files on-line for users. The computer network allows them to spread this knowledge and software capability to more users at a very low incremental cost.

Competition in the data communications industry has increased with the deregulation of communications services. There are more value-added carriers today. New hardware and technology is reaching the marketplace faster. Deregulation has also led to a faster modernization of the backbone communications networks. The use of current telephone lines to support much higher speed and multiplexed signals portends an era with greatly increased communications capacity.

BIBLIOGRAPHY

IEEE Publications on data communications and computer networks.
Lientz, B. P. (1981). ''Introduction to Distributed Systems.'' Addison-Wesley, Reading, Massachusetts.
Martin, J. (1983). ''Systems Analysis for Data Transmission.'' Prentice-Hall, Englewood Cliffs, New Jersey.
Ruskin, R. (ed.), (1974). ''Computer Networks.'' Prentice-Hall, Englewood Cliffs, New Jersey.

DATA COMMUNICATION NETWORKS

Paul S. Kreager *Washington State University*

GLOSSARY

Asynchronous: Method of synchronizing serial character-level data between sender and receiver, characterized by the addition of start and stop bits before and after the bit-string representing the character. The synchronizing bits are added before the sender transmits the information over the network, allowing the receiver to tell when the character begins and ends. Decoding of the character bits is done solely by the receiver, knowing the predetermined data transmission rate. Contrasts with synchronous data transmission.

Bandwidth: Range of frequencies that a communication line can pass. In data communications, the term is also loosely used to denote the bit transmission rate (bits per second) that a line will pass.

Baseband: Form of data communication where a communication line is turned over to the sole use of one subscriber and binary information is put directly on the line in its direct electrical current or voltage equivalent. Form of time-division multiplexing and generally in contrast to broadband communication.

Broadband: Form of data communication where several subscribers share and can simultaneously use one common communication line. Each subscriber's data are modulated over a carrier frequency, i.e., information is frequency-division multiplexed. Generally, in contrast to baseband communication.

Carrier: Company in the business of carrying data over its transmission-line facilities, e.g., telephone companies. Also, an electrical signal used to carry information by modulation of the signal. Modulating a fixed-frequency carrier amplitude with voice information in AM radio is an example.

Circuit: Generally, a communication path between two points, also referred to as a line, link, or path.

Multiplexing: Sharing of a communication path, possibly simultaneously, by several subscribers. Implemented by dividing the path bandwidth into distinct frequency bands (frequency-division multiplexing) or by rotating a time slot, in turn, among line subscribers (time-division multiplexing).

Network: Conglomeration of circuits connected together to form communication paths between subscribers with some common interest.

Protocol: Set of rules or conventions governing how data communication takes place over a network, covering a wide range of areas including physical connections, electrical signaling, data representation and packaging, and line access and utilization.

Synchronous: Method of synchronizing serial block-level data, characterized by sender and receiver being in synchronization for the duration of the data block. Accomplished by the sender incorporating timing information with the data, allowing the receiver to determine both bit and character times. Contrasts with asynchronous transmission.

The subject of data communication in its broadest sense studies the electronic transfer of information between two or more points. Distances between points can range from inches to an interplanetary scale. Here, the interest is in how the points are connected, or networked, and in how network dependencies influence information flow over the network. The technology of data communications touches a large segment of our society. Data systems are an integral part of the business world, technical people in all disciplines are dependent on computers, and even the average person is likely to have a personal computer at home. All these situations often require the communication of data from one point to another. With this widespread usage in mind, the material presented here is more global in nature, covering neither the deeper technical details of data transmission nor the more theoretical aspects.

I. Network Basics

A. NETWORK TOPOLOGIES

Connecting points together is a topological concern and a logical place to start discussing data communication networks. Figure 1 illustrates basic network topologies that may be considered primitive structures, applying to all types of networks. For the time being, a network can be viewed as a simple pathway between points over which information is conveyed. There may or may not be some control element accompanying the network that allows for proper routing of information. The multipoint topology, for instance, has several points connected by a common circuit (communication path or line), communication being between the controller and points on the network, not between points on the network. In contrast, the bus or ring structures suggest that any point can communicate with any other point on the network. These basic topologies can be linked together to form hierarchical networks of considerable complexity.

Several circuit configurations can exist within the various network topologies dictating the di-

rection of information flow. A simplex transmission is one where data flows only in one direction. Half-duplex transmission occurs when information can travel in either direction over a circuit, but in only one direction at a time. Full-duplex, or simply duplex, transmission is the simultaneous transfer of information in both directions over a circuit.

B. NETWORK COMPONENTS

The most basic component of any network is the medium over which information is carried. All networks use some type of conductive media at least somewhere in the data path. However, the conductive path may be broken by a radiative (e.g., microwave) terrestrial or satellite communication link, for example. The basic types of conductive cable are illustrated in Fig. 2. All types except fiber conduct an electrical signal; fiber is used as a "light pipe" to conduct electromagnetic energy in the light spectrum. [*See* MICROWAVE COMMUNICATIONS.]

For other common network components, Figs. 3 and 4 will serve as a reference for defining both components and some of the associated trade jargon. Starting with the local area environment in Fig. 3, note first the more global aspects. Buildings A, C, and D have a convenient interbuilding cable path within which a single bus-type topology network is installed. Building D also happens to have a ring network installed. Building B has no convenient open path to building A and therefore uses a telephone company (telco) circuit to reach the services of the large computer. Telco circuits are a very common type of a data connection, since telephone cable is the single most widely installed communication media. Note that this circuit has a modem installed at each end. Loosely, a *modem* is a device that changes the way data is transmitted when it passes from one environment to another. Here, the terminal or front end at either end of the circuit communicates in a digital fashion. However, the telephone circuit conveys

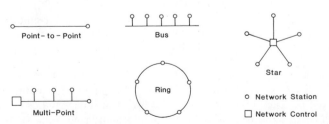

FIG. 1.   Network topologies.

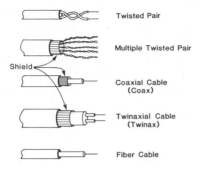

FIG. 2.   Examples of network media.

data in an analog fashion. The modem allows for the translation from digital to analog and back to digital as data goes from one end of the circuit to the other. Large computers typically have a front-end processor attached, which is really a smaller computer custom-designed to handle network traffic into and out of the large computer. Front ends are also called terminal control units. [*See* SIGNAL PROCESSING, GENERAL; TELECOMMUNICATION SWITCHING.]

Note the abundance of interface boxes. Generally speaking, all terminals, computers, or other devices accessing a common network must do so using a common electrical protocol and information format. Interfaces allow this to happen. A terminal, for instance, may not have the proper electrical signal levels to directly connect to a network. An interface will allow for the proper translation of levels. Messages traveling over networks often have other information appended to the message for proper operation. For example, the number of characters in the message or a destination address for the message may have to be added to the message itself. Interfaces can perform these functions also. These

functions may be included within a terminal or computer, thus allowing a direct device connection to the network. In any event, proper protocols are enforced somewhere, even if not visible as a separate function.

Other network topologies are evident in Fig. 3. Terminals T_5 and T_6 form a simple star with the computer in building C. A less noticeable star can be seen with terminals T_2, T_3, and T_7 connected to the large computer front end. Connections between T_2, T_3, or T_7 and the large computer, individually, are examples of simple point-to-point circuits. Note the three-level network hierarchy: T_3 in building B could conceivably reach the computer in building D via the front end to the bus network, then through the gateway to the ring network on which the building D computer is connected. One can think of a gateway as an interpreter. The bus and ring networks no doubt have different protocols of use (they speak different languages). The gateway is a device (microcomputer) that allows for a language conversion from one system to another. More formally, this process is called protocol conversion. A similar device to a gateway is the bridge. Linking together two or more distinct networks that use the same protocol is done with a bridge. One common bridge function is to change the data rate of information as it passes from one network to the other. Another function is address translation: a bridge needs to recognize internetwork addresses so that the information can be properly routed from one network to the other. Gateways also accomplish these same functions. Unfortunately, it is generally the rule that differing computers, terminals, and networks do not easily connect to one another, creating a need for a variety of devices that translate protocols, information formats, and electrical characteristics.

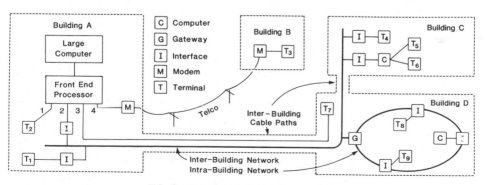

FIG. 3.   Local-area network examples.

The bus or ring networks of Fig. 3 would be suitable candidates for either coaxial or fiber cable. The reason is that for networks having to support a high aggregate data rate (many terminals and/or high-data-rate devices attached), one needs to select a medium that can support that data rate. Both coax and fiber are high-data-rate (bandwidth) cable. Telephone circuits are typically a twisted pair, having a much lower bandwidth. In Fig. 3 applications, one would typically not find radiative links, such as microwave, because of the short distance. There are exceptions, of course. For instance, two buildings may need to be connected with a high-bandwidth link, but say a 10-lane freeway exists between them. Then it may be cheaper to install a radiative-type link using microwave or infrared transmitters and receivers, for example.

Figure 4 changes the environment by increasing network distances. Long-distance data communication is often called "long-haul" in trade jargon. "Wide-area" is also commonly used. Again, take a global look at the figure. Note that Austin, Denver, and Phoenix are connected in a ring fashion. Point-to-point circuits connect Denver to Seattle and Seattle to Portland. A multipoint (or multidrop) circuit originates at the Portland computer, linking to Salem, Los Angeles, and San Diego. A satellite link connects Seattle to Honolulu. Real intercity data connections are more complex than Fig. 4 suggests. Someone desiring a circuit to another city generally leases one from the telephone company. The circuit leaves the user location often as a simple send/receive pair of wires. Since it looks the same way at the other end of the circuit, one is conditioned to think of long-haul data circuits as simple point-to-point send/receive pairs. In reality, intercity connections are often done by microwave, which is transparent to the circuit subscriber. Satellite links are also microwave, but with one major difference over their terrestrial counterparts. Satellite distances add significant delays to data transmission, which can create problems in certain situations. We will discuss more about this later.

In "short-haul" circuits, where distances are relatively small—i.e., Fig. 3 situations—media cost is often an almost insignificant portion of the overall communication system expense. Multiple connections to a computer, say, may be implemented by simply replicating circuits, one to each terminal. In long-haul situations, circuit costs can become a large part of the communications expense, requiring a different approach. Multiplexing is a technique that allows several devices (terminals, computers) to share one communication path, thus eliminating the need for costly multiple circuits. Figure 4 shows Seattle and Portland connected by one circuit,

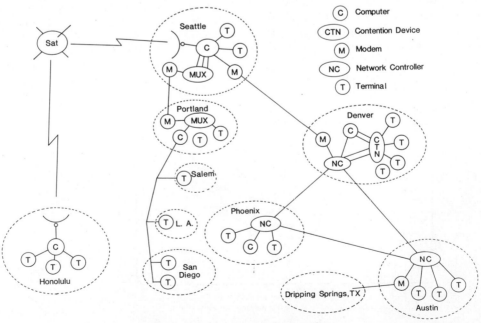

FIG. 4.   Wide-area network examples.

each end terminated with a modem, and each modem connected to a multiplexer. The modem and multiplexer may be packaged in one unit. The two Portland terminals and computer all share the same circuit via the multiplexer to access a computer system in Seattle. Several forms of multiplexing exist. Frequency-division multiplexing (FDM) takes the allowable bandwidth of a circuit and divides it into several frequency bands. For instance, each Portland terminal could be allocated a unique 300-Hz portion of the circuit bandwidth. One terminal might have 600–900 Hz, another 900–1200 Hz, etc. Time-division multiplexing (TDM) is another form where users have their own unique time slot rather than frequency slot. View the multiplexer as a switch that sequentially switches to each terminal, say every millisecond. If there were 10 terminals connected to the Portland TDM, then every 10 msec each terminal would be allocated 1 msec of time. Note that neither FDM nor TDM techniques discriminate whether or not a terminal is actually using its allotted frequency or time slot. If not in use, that portion of the information spectrum is wasted. An improvement to TDM is to dynamically allocate time slots only to active terminals. A device doing this is known as a statistical time-division multiplexer (STDM), often just called a stat mux, and is a preferred way to multiplex information. By dynamically allocating circuit bandwidth, the number of terminals connected can be increased considerably over earlier techniques, making for much better circuit utilization. There are drawbacks to using a stat mux even though it efficiently utilizes communication paths. When the number of active users increases, an information bottleneck will occur when the rate of information flow to the multiplexer exceeds the capacity (bandwidth) of the circuit. Buffering is a technique used to manage these peak instances of information flow. Input information passes through a buffer (storage area) within the stat mux on its way for transmission over the circuit. As long as the input rate does not exceed the capacity of the circuit, data flow through the mux is optimal. However, once the input information rate exceeds the line capacity, the buffer begins to fill, queuing the input for later transmission over the link. As the input rate ebbs, the buffer depletes as information is transmitted out of the mux. To visualize the buffering process, compare to a balloon (buffer) inflating and deflating as its air (information) content varies. If the buffer fills, the mux

has no choice but to throttle the input by sending a signal back to the input terminals that it cannot accept any more information. In turn, the terminal must stop its activity, potentially placing an immediate hardship on the user. Obviously, the buffer size, the aggregate input data rate, and the data bandwidth of the connecting circuit must be integrated together for proper operation of the statistical multiplexer.

Contention is another mechanism used to increase utilization of communication resources. In a manner similar to frequency or time slots not being utilized, computer or network ports (entry points to a system) are wasted when users connected to those ports are not using them. A contention device allows several ports to be shared by a larger number of users. Statistically, most of the time not all users are active; therefore, by having a larger number of users contend for a smaller number of ports, most of the time everyone will be adequately serviced, resulting in optimal use of port resources. When the condition arises that a maximum of m ports are being utilized, and the $(m + 1)$th user tries to access a system, it simply gets a "busy," i.e., all ports are busy. Figure 4 demonstrates this technique with a contention unit that allows four terminals in Denver to contend for either two ports into a computer or two other ports into the wide-area network. This particular configuration allows for switched contention, where a user can switch between two or more destinations. It is not uncommon for contention to be included within front-end or other equipment.

Finally, a word about switched networks and dial access components in a data communications environment. The occasional computer user often "dials up" the computer service desired using special equipment connected to their telephone circuit. The telephone circuit is part of the telephone company switched network for voice and therefore can reach almost anywhere in the world. The reason dial access is preferred is cost. It is very costly to have a dedicated long-haul circuit, and even more so if it is only sporadically used. A direct-connect modem allows user equipment to directly connect to a telco's switched network by plugging into the same outlet as the standard telephone. An alternate method is via an acoustic coupler into which a telephone handset is inserted. Data are then acoustically coupled via the telephone ear and mouthpieces. In either case, data are converted to audio tones for transmission over the switched network. Figure 4 gives an example.

Suppose a user in Dripping Springs, Texas, needs to occasionally access a certain computer in Seattle. To make a more interesting example, further suppose that the Seattle computer does not have a local dial access number but does have remote dial access in Austin, via a dial access modem connected to a network controller. The user in Dripping Springs can make a (relatively short distance) call to Austin and ride the network up to Seattle to the desired computer service. This is an example of how a remote computer service can offer "local" access using a communications network. Back in Fig. 3, terminal 3 could also access building A computer services via a dial connection. The telephone company switched network is commonly used for both local and wide-area data communication.

II. Network Intelligence

Up to this point a network has been considered more or less as a convenient pathway to resources such as computers. Not much has been asked of the network; indeed, once installed, it has been viewed as a passive entity. Suppose we wish information such as the amount of data traffic traversing the network or diagnostic data of the network in failure situations. How about asking the network itself to dynamically establish a circuit between a terminal and computer? Or how about asking the network to send a parcel of information to one or more entities on the network? These types of functions considerably elevate the status of a network above simple circuits connecting users to computers. To accomplish this, intelligence must be somehow added to the network itself. As an example, consider Phoenix, Denver, and Austin linked together in Fig. 4 by network controllers (NCs). Adding the NCs makes the network an intelligent entity in its own right, considerably enhancing utility of the network. NCs have other names, such as network processors, communication processors, node processors, or node controllers. They all are basically small computers tailored to perform data communication functions for the network. If a link failure between Phoenix and Denver occurs, the network dynamically adjusts, routing all Phoenix/Denver traffic via Austin. A network user would not even necessarily know of such a rerouting. Since all data must pass through the controller, statistics are available such as the amount of traffic, peak traffic times, outage information, etc. This information allows for more optimal operation of the network.

From a user perspective, the basic function of a network is to direct or switch the user's data to the desired destination. Several switching schemes have evolved as network intelligence has increased. Circuit switches were the earliest, simply establishing point-to-point paths between subscribers. Using a standard telephone amply describes this type of switching, where telephone-company equipment establishes a (point-to-point) circuit between two parties. This same concept can apply to data and is often seen in PBX (private branch exchange) where these devices can establish data circuits between subscribers (in addition to their voice functions). Circuit switching establishes dedicated paths. Once done, subscribers need not tell the network where the information is to go; the dedicated path establishes that inherently. An alternative to this scheme is to connect all network subscribers via a common (shared) circuit and switch the information itself as it flows over the network. This is the basic idea in packet switching. If all network subscribers have an assigned unique address, and if a "packet" of information has an address appended, then information can be delivered across the network between subscribers. A major difference in this switching scheme is that all subscribers time-share the network, but essentially the same end result is accomplished as for circuit switching, i.e., transferring information between points on the network. Obviously, the packet size has a large bearing on the performance of the network. At one extreme, appending 10 bits of address to 10 bits of data results in 50% overhead to get the job done. On the other hand, 10 address bits for 10,000 data bits may reduce overhead significantly, but it might take an extraordinary amount of time to assemble a packet of that size. Selecting optimal packet lengths is an art in itself. Because of the addressing capability, packets can be easily routed around a network to bypass failed elements or congested nodes. Also an art in this technology is the design of packet routing algorithms. Packets may be statically routed, which essentially means that once a path is established between network subscribers, all packets will take that path for the duration of the session. Adaptive routing algorithms allow packets to be dynamically routed over the network, not necessarily taking the same path as previous packets between the same subscribers. This scheme allows the network to adapt information flow to the current conditions of the network. In Fig. 4, if the Phoenix-to-Denver link started to become congested at some point, part of the packet traffic could be rerouted to Denver via Austin. It is also obvious that adaptive routing can become quite complicated when hundreds of users are shipping packets over a large network. Adaptive packet routing algorithms are often very complex to avoid such subtle things as loosing packets over the network or having packets ride the network forever.

As a final note on items related to network intelligence, message switching can be viewed as a network function. The key concept in message switching is that information does not necessarily have to traverse the network in real-time. Up to this point, user connections to resources on a network have been assumed to be responsive, not introducing intolerable delays. Message switching relies on the concept that a message can be stored and forwarded at a later time. Therefore, a storage element has been added, which allows the delayed transmission of the stored message. It should be noted that this function can be provided by a computer resource connected to a network and therefore may not necessarily be an attribute of the network proper.

III. Network and Information Management

The basic unit of information in the data world is the bit, having a value of either 0 or 1. Bits of information are transported in two basic ways, either serially or in parallel. Parallel transmission is used primarily in short-distance (few meters) applications such as between a computer processor and its memory or other peripheral equipment. For example, if 10 paths (wires) exist between two points, 10 bits of information can be sent simultaneously (in parallel). In trade for the high information rate allowed by parallel transmission, one pays the added cost of multiple transmission paths. Alternatively, virtually all longer-distance data communication (described here) is done serially, i.e., one bit at a time over a single communication path. The details of how 0s and 1s are sent over communication paths is the separate subject of data transmission. For the purposes at hand, all data communication between network points can be viewed as a serial stream of information bits.

This bit stream must be properly packaged and controlled as it is conveyed over the network. A message is a general name for the information package itself.

Message formats depend to a large extent on the types of data circuits over which they flow. A point-to-point circuit, for instance, does not require the sender to identify the intended receiver. A multipoint or bus network does. Errors are inevitable in data communication, especially over longer distances. The choice of a particular transmission scheme may depend on how well that scheme can detect errors. The overall flow, control, and synchronization of information over networks are important management functions.

A. Data Synchronization

To transfer information from a sender to a receiver over a network requires that the two entities be synchronized throughout the data transmission time. The most fundamental requirement for synchronization is for both sender and receiver to agree on the transmission rate, i.e., the bits per second (bps). Knowing only the basic bit time, a receiver still does not know when to sample incoming data from the network, i.e., it does not know where an arbitrary bit time starts or ends. Ideally, one would want to sample near the middle of the bit time to stay away from transition levels at the beginning or end. At a higher level, bits are often grouped together to form characters or bytes (1 byte = 8 bits), or higher yet, blocks or frames of data. A string of characters forming a data block, for instance, would need to be synchronized so that the receiver knows when the block begins. Not knowing this, it is difficult to decompose the block back into the character string. Generally, these basic synchronization problems are solved by transmitting data in one of two standard ways: asynchronously or synchronously.

Asynchronous data transmission is described logically in Fig. 5. The general problem being solved is the synchronization of a character of information between sender and receiver. Shown is the ASCII character "h" being transmitted over a communication line. In addition to the 7-bit character itself (1101000), a start bit, stop bit, and parity bit are included. Odd parity is shown in the example, which means that an odd number of 1s are in each transmitted character (ensured by proper setting of the parity bit). The parity bit is used by the receiver to detect a

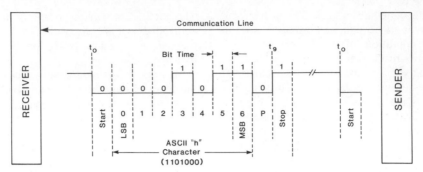

FIG. 5.   Asynchronous data transmission.

limited number of errors that can occur when the character is sent over the communication path. There are variations to the asynchronous format shown, including 8-bit character, no parity, and two rather than one stop bit. Both sender and receiver must agree on a common format in addition to a common bit rate. Once this is done, the receiver constantly looks for a start bit to occur on an idle line. When the start bits occurs, the receiver then has all the information required to determine the character being transmitted over the link. For instance, starting 1.5 bit times after the start bit transition would be an optimal time to sample the first data bit on the line. Since the sender and receiver have agreed on the basic bit time and information format, the receiver then continues to sample succeeding bit times to determine the remaining bits in the character. A stop bit, among other things, ensures that the line goes back to the idle condition, readying it for another character transmission. It is customary in asynchronous data transmission to transmit the least significant bit (LSB) of the character first and the most significant bit (MSB) last. This explains the inversion of the 7-bit character as it traverses the communication line. Note the synchronization overhead for asynchronous transmission. The start, parity, and stop bits are not part of the desired information (character) but are required to effect the transmission. For the example shown, 10 bits are required to send the 7 real information bits, representing a significant overhead, which is constant for every character transmitted. Reiterating, in asynchronous communication the sender and receiver are only in synchronization between the start and stop bits of each character transmitted. Asynchronous is meant in the sense that the sender and receiver are not in continuous synchronization.

Synchronous data transmission allows larger pieces of information to be synchronized over the network and significantly reduces transportation overhead in comparison to asynchronous transmission. Figure 6(a) shows three bit-synchronization schemes. In short-distance applications it is common to use both data and clock lines to transport information. The clock line tells the receiver exactly when to sample the data line. In the example shown, positive clock transitions inform the receiver to read out 1 0 1 1 from the data stream. For longer distance applications, data and bit timing information are combined to eliminate the costly need for a separate clocking line. Over certain telco lines, information must be transported in an analog fashion. A rudimentary example is given to demonstrate how bit synchronization can be implemented using an amplitude-modulated (AM) carrier. The carrier amplitude has one of two levels corresponding to the 0 and 1 information bit states. Bit timing is derived from the carrier frequency. In other longer-distance applications, data transmission might be done digitally, requiring a different bit-synchronization scheme. One common way is to use a Manchester code, the third example in Fig. 6(a). This code is a variation of a class of codes known as biphase codes. Notice that the digital waveform has a low to high transition for a logical 1 and high to low transition for a logical 0. Further, there is at least one transition for each bit time; in other words, the code is self-clocking, carrying both clock and data information. Several other self-clocking schemes exist. These examples illustrate how individual bits are synchronized. When bits are logically combined to form characters or bytes as information entities, a requirement is imposed to know where the first bit of a character or byte begins. Figure 6(b) illustrates one way to accom-

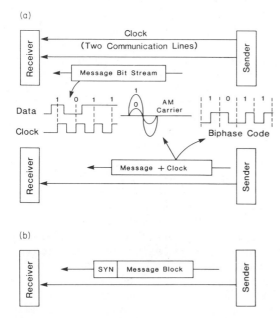

FIG. 6.   Synchronous data transmission. (a) Bit synchronization. (b) Character synchronization.

plish this. The message block is simply a concatenation of information bytes forming a serial bit string eight times the number of bytes long. To achieve byte-level synchronization, a SYN (synchronization) byte is added to the header of the message block. Remember that bit synchronization has already been accomplished. The SYN byte is a unique bit string (for example 10010110, or 96 hexadecimal) previously agreed on by sender and receiver. When the bit stream is first detected, the receiver begins a pattern-matching operation to align the incoming bit stream to the known SYN code. When the match occurs, sender and receiver are then in byte-level synchronization (sync). Once in sync, the succeeding bit stream is easily decomposed into its constituent bytes of information. Synchronization overhead is due to the added SYN byte. As the message block length increases, SYN overhead reduces. Even for short message blocks, overhead is considerably less compared to asynchronous transmission.

B. SYNCHRONOUS MESSAGE FORMAT

The preceding discussion on data synchronization was somewhat idealized to simplify explanation of basic synchronization concepts. Because synchronous data transmission is so prevalent in longer-distance data communication and closely linked to network architecture, message formatting needs to be amplified. Figure 7 illustrates a general synchronous message format, better depicting what actually occurs in practice. At first glance, considerably more overhead is apparent. However, considering that all overhead elements of the message are generally one or two bytes, overhead is still relatively small when the data block is large. Several SYN bytes usually prefix a synchronous message because the network environment can garble the beginning of a message. More than one SYN byte allows the receiver to lock into synchronization even though the first SYN byte may be distorted by transmission noise. Over a shared network, address information is necessary to direct the message to the proper destination. Address information of the sender as well might be part of the address. Uses of the control field include defining the message function, signaling the beginning or end of message blocks, acknowledge other transmissions, code switching, status signaling, and issuing commands over the network. The error field is used to detect if a transmitted message has been distorted after traversing the network.

Synchronous formats are distinguished as being either bit- or byte-oriented. Early formats, still in use, are byte-oriented in design where management of the message is done at the character level. Character-level management forces code dependency, i.e., the basic network logic depends on the particular code used to represent the characters [e.g., the American standard code for information interchange (ASCII) or the extended binary-coded-decimal interchange code (EBCDIC)]. Subtle problems are also inherent in byte-oriented formats. For example, a data block could contain a control character as a normal part of the sender's data. If not designed to handle this, network control logic could take

FIG. 7.   Basic synchronous message format.

action at the wrong time in response to the data block control character. Bit-oriented formats treat the message as a bit string, leaving character composition to higher-level network functions. This has the immediate advantage of code transparency and is the preferred method in newer format designs.

C. ERROR HANDLING

Well-designed electronic equipment is inherently reliable and provides essentially error-free performance. Within itself, the equipment does not generate noise that shows up as performance errors. The vast majority of errors occur when equipments are connected via some media. Further, errors increase as media distances increase. The common television set is a good example. The set itself generally provides good performance for years even though errors are evident almost hourly. Snow, picture distortion, and other errors are not because of the set but rather, principally, due to environmental noise impacting the communication path. Lightning or human-made electrical signal sources, for example, often inject considerable noise into transmission media.

Because the sole purpose of data communication networks is to provide communication paths between equipments, and since those paths are often very long, error handling is a common and critical function of network design. It is application-dependent. In transmitted text, for example, it is easy for a person to detect errors, and for that application, error detection may not be of paramount importance. Errors in numeric data also may be obvious, such as for a singularly misplaced point in an otherwise smooth graph. Errors in financial data can be catastrophic. Error handling is generally discussed in terms of detection and correction, with implementation scheme complexity proportional to the importance of errors being reduced. As a primitive example, a receiving station may echo back each character sent by a terminal operator. Detection and correction, in this case, may rest solely on the operator. Implementing automatic detection is the next step up in complexity, followed by automatic error correction. To add these automatic features it is necessary to add information to the data being transmitted, which is subsequently used by the receiver to sense and perhaps correct errors introduced over the link. For lower data rates, overhead is generally added to the transmitted

data to be used solely for error detection by the receiver, with correction implemented by asking the transmitting station to resend errored data. This simple scheme allows for minimal overhead to the transmitted message and concentrates that overhead at the most critical point, i.e., error detection. Adding more overhead to a message allows the receiving station to both detect and correct certain classes of errors without retransmission. This is known as forward error correction and may be desirable in higher-data-rate applications, for instance, where the added overhead can be justified. In practice, adding correction code to a message becomes increasingly more difficult as the code covers more and more error conditions. This is one reason why many strategies place emphasis on the detection code design and use the simple expedient of retransmission for correction. Another reason is the bit error rate performance of typical communication lines. Under normal conditions, 1 bit error in 100,000, or a 1 in 10^5 error rate, is to be expected even for continental distances. The rate can approach 1 in 10^8 as the distance decreases. In many situations, it is inefficient to add correction code overhead to the message when, on the average, that overhead would exceed the overhead in simply resending the message.

The simplest error-handling procedure is the addition of the parity bit, as discussed during asynchronous data transmission and elaborated on in Fig. 8(a). This scheme is called a vertical redundancy check (VRC) and has obvious limitations. Any odd number of bit errors can be detected, but any even number of bit errors cannot. Parity is generally added by the sender's equipment just before the character is transmitted and checked by the receiver's equipment upon reception. Typically, parity error detection in the receiving equipment simply flags the error and allows higher-level functions to decide what to do about it. In some cases no action is taken.

Longitudinal redundancy checks (LRC) [Fig. 8(b)] allow blocks of characters to be checked. One way to do this is to form parity over corresponding bits of each character in a block of characters. An alternate method is to use some other arithmetic operation. Summing or exclusive-or-ing the character string and using the single character result as a check character is common. The arithmetic check character is appended to the message block for transmission over the network. The receiver performs the same checking operation upon reception of the

(a)

Character Bit Positions:	P	6 5 4 3 2 1 0
Example Character (1101000):		1 1 0 1 0 0 0
Odd Parity: Σ l's = Odd	0	1 1 0 1 0 0 0
Even Parity: Σ l's = Even	1	1 1 0 1 0 0 0

(b)

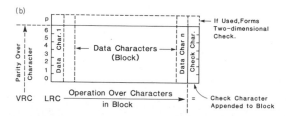

FIG. 8. Basic error detection schemes. (a) Vertical redundancy check over a character. (b) Longitudinal redundancy check over a block.

message block and compares its result to the check character sent by the transmitter. A mismatch signifies an error condition. Note that if each character in the block has an integral parity bit, a two-dimensional check is created with more powerful detection capability.

The error detection schemes considered thus far are commonly used for asynchronous transmission. For synchronous transmission, the cyclic redundancy check (CRC) is common. Errors over typical transmission links tend to occur in bursts. Further, the burst time tends to last over several bit times. CRC techniques are designed to increase the reliability of error detection in this environment. The algorithm to do this is more complex than LRC, taking the user data-bit stream (message block) and dividing it by a predetermined binary number. The quotient forms a check character(s), which is appended to the end of the message block. Figure 7 shows this error-detection information as a separate field in the synchronous message format.

D. ACCESS CONTROL

Whenever a communication line is shared among several subscribers, rules need to be established governing line access and utilization. This subject was introduced earlier when multiplexing was described. Dividing the line spectrum into frequency or time slots allows a predetermined number of subscribers to access a common communication line. The multiplexer

allows for sharing of one line but effectively provides point-to-point circuits to the individual subscribers. Also note that there is no intersubscriber communication over a multiplexed line. The basic rule to follow in this environment is to maintain a balance between the aggregate subscriber bit rate and the line bandwidth. Described now are techniques to control sharing of communication lines when the full bandwidth of the line is turned over for the sole use of one subscriber at a time in multipoint, bus, or ring network architectures. These environments are more complicated than a point-to-point topology because more rules are required to handle the protocol of turning over use of the line to the various subscribers.

An extension of the time division multiplexing concept is to time-slot the communication line and allow each subscriber sole use of the line for a predetermined amount of time. The only real difference now is that intersubscriber communication can take place. The slot holder must take care to manage its available time for transmitting one or more messages. The same problem surfaces here as for TDMs, in that time is wasted when a subscriber has nothing to transmit. Work-arounds exist to this problem, such as passing a time slot from one subscriber to another. A subscriber not needing its time allotment would then simply pass its time slot to someone else.

Polling or selection [Fig. 9(a)] was an early development in line control techniques and is still in wide use. In its infancy, data communication was mainframe-oriented where the central computer totally controlled communication line activity in a master–slave relationship with remote stations. Polling is common in wide-area multidrop network topologies. Returning to Fig. 4, as an example, the Portland computer is the master station of the multidrop circuit connecting slave stations in Salem, Los Angeles, and San Diego. The master sequentially polls the slaves to determine if they want to transmit or are ready to receive information to or from the master. The master has a poll list that can be manipulated to control polling priority and timing. Rules of this scheme govern the dialog between master and slave, a process sometimes called handshaking, which proceeds similar to the following. If the master wants to send data to station C, it sends a poll to C asking if C is ready to receive information. Station C responds with an ACK (positive acknowledgement) if ready, allowing the master to send data. Station C will

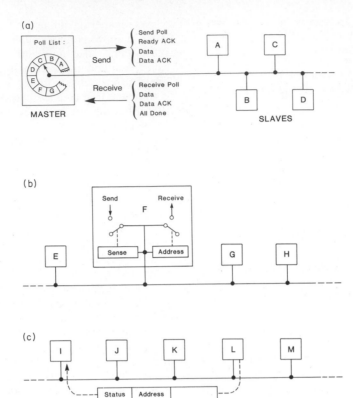

FIG. 9.   Line control techniques. (a) Polling. (b) Contention. (c) Token passing.

ACK or NACK (negative acknowledgement) each message for error control. A similar dialog takes place when the master needs to receive information from a slave station. The dialog can become fairly complex to cover the many situations that arise in practice. Polling is widely used and offers a centralized approach to line management where priorities can be established, since all is under the control of the master. Interstation (slave) communication is generally not a feature of this architecture. Another drawback is the rather high transmission overhead. Even when no actual data transmissions are taking place, polls and NAKs are continually traveling over the network. If time between polls is increased to effect a decrease in overhead, individual subscribers on the network witness increasingly long response times. Thus, both the number of polled stations allowed and the polling rate become factors in designing polled multipoint networks. Redundancy of the master is also a concern, since a failure there literally shuts down the entire polled network.

Consider next the community line in Fig. 9(b) having the rule that whoever first accesses an idle line gains control of it, i.e., stations are in contention for the line. This is also known as demand access. Station F shows the rough control mechanisms involved with contention line control. A station desiring access continually senses line status and, when idle, switches its operating mode and begins sending data over the link. In addition to sensing, stations also look at the address information of messages traveling over the network. When a station recognizes its own address, it knows to switch its operating mode to receive the message directed to it. Two problems are immediately apparent. First, a rule needs to be established to prevent line hogging. Second, a line going idle may experience several subscribers trying simultaneous access, resulting in collisions. Properly designed, contention is a viable scheme for managing the line even with the collision problem. Collisions are not as frequent as one might think, especially when designs include ways to reduce them such as having subscribers wait a random amount of time after sensing the line going idle before trying to access it. Subtle problems are inherent in this scheme due to communication line transmission delay. A particular subscriber could gain access to the line and start transmitting its data message. Another subscriber, however, could likewise think it had access simply because transmission delays gave the appearance of an idle line. These timing considerations are inherent in the design of contention networks. Because of delay, contention networks are necessarily limited to campus-level distances. Contention is a fair management approach as every subscriber has equal status and equal opportunity to the line. Fairness, however, is not always a good management approach. Some subscribers might need more frequent access or be allowed longer times for data transmission to better optimize network and subscriber performance. The inability to assign priorities is a principal disadvantage of the contention approach.

Token passing will be the last line-management technique discussed and is generalized in Fig. 9(c). This concept was introduced earlier when passing an unused time slot was used to improve efficiency of a time-multiplexed line. A token is a bit-string that is passed around the network to the various subscribers. It is an access control mechanism in that whoever receives an idle token has the right to use the network for data transmission. Initially, with no data transfers taking place, the token is idle and declared as idle in its status information field. Destination routing of idle tokens is done either implicitly or explicitly. Stations on a ring topology network inherently pass information from one station to the next in order around the ring. In this case, an idle token does not need destination address information and is said to be passed implicitly. For a bus topology, there is no inherent station ordering, resulting in random access. Here, the token is passed explicitly, or with destination address included. Note that explicit passing still orders or sequences the stations, resulting in a logical ring over the physical bus network.

A station desiring to transmit information assembles a frame consisting of the data and address information, then waits for an idle token to arrive. When it does, token status is changed to busy, the frame is appended to the token, and the message sent out over the network. The preceding, although quite general, essentially describes the access protocol involved in token passing networks. Token passing is a natural choice for a ring architecture but can be implemented over a bus as well.

Briefly, access controls can be viewed as operating in a controlled environment or a demand environment. Polling and token passing allow access rights to the network to be passed from one subscriber to another in a controlled, orderly fashion. Collision schemes are demand access situations, where access rights are demanded by each station needing access rather than having the right passed to it by some other station.

E. Flow Control

Up to this point the basic rudiments of networking have been presented. Synchronization, message formatting, error handling, and access control to the communication link are fundamental requirements in data communications. The basic tools now exist to carry on data transfers among subscribers of a common network. Much remains to improve the efficiency of data communication over the network, however. Consider the simple case of a sender that needs to transmit several messages to different subscribers. Considerable time can be wasted if the sender must wait for a positive acknowledgement to each transmitted message before issuing its next message to another station. It can take time for a receiver to check for errors and send back the ACK or NACK. A similar problem exists even if the sender needs to transmit several messages to the same receiver. Implementing controls to handle delayed acknowledgements can improve data flow over the network. Errors further complicate design of the controls. For example, say a receiver correctly reads a message from the network and sends back an ACK. However, due to link noise, suppose the ACK is lost. In this case, the sender will eventually become aware that something is wrong because of the missing acknowledgement and then has no choice but to resend the message that the receiver correctly obtained earlier. Further, if a stream of messages was sent, the ACK block in error must somehow be identified.

From the preceding, it can be seen that flow control is an important function to improved operation of the network. Networks with flow control, at the very least, incorporate sequencing information in the message. Figure 7, for instance, could have an added field identifying the message number. Other enhancements are also possible. The reader is referred to the bibliography for study of these higher-level network functions. The topics under discussion in this section generally fall under the heading of data link controls (DLC) in the literature.

F. THE ISO REFERENCE MODEL

Some of the problems and complexity inherent in data communication networking have now been surfaced. There are many ways to implement networking functions, and further, hierarchies of tasks exist as just described for flow control versus lower-level functions. Because of this diversity, an array of incompatible networking products evolved in the communication marketplace. Further, additional markets were spawned for gateway products to allow dissimilar networks to intercommunicate. While networking complexity might never allow total standardization, the incompatibility problems faced by consumers made the entire industry keenly aware that some guidance was needed.

In recognition of the basic networking incompatibilities, problems, and lack of direction, the International Standards Organization (ISO) developed their Open Systems Interconnection (OSI) reference model, shown in Fig. 10 [reference: ISO/TC97/SC16/N117 and related documents]. The intent of the model is to provide a base level of standardization for the data communications industry in terms of peer groups or common tasks involved when one subscriber talks to another over a network. The OSI model is a hierarchical layered structure with complexity and abstractness increasing from the bottom up. Each layer or level performs a reasonably well-defined network function and is dependent on the layers below to accomplish more elementary functions. The following briefly describes

each layer. (The names of some technical specifications are included for further reference.)

Level 0 is simply the physical communication media, and is often not listed as a level in itself.

Level 1 governs the physical connection of network entities in regard to basic electrical, mechanical, and signaling protocols. The basic function accomplished is to establish linkage to the physical network and pass logical bits to level 2. Examples of level 1 are EIA-232, EIA-449, and X.21. [EIA specifications are generated by the Electronic Industries Association (EIA) and X standards by the Consultative Committee in International Telegraphy and Telephony (CCITT).]

Level 2 governs much about what the previous section discussed: error handling, polling, and flow control. The idea of level 2 is to provide for reliable and efficient data transfers. Examples are BSC, HDLC, SDLC, ADDCP, and CSMA/CD (see Section IV).

Levels 3–7 cover higher network functions beyond the scope of this chapter, for which the reader is referred to the bibliography.

The idea of peer communication is now evident from the model. For instance, the data-link layers (level 2) of network subscribers communicate with each other in a common, mutually understandable language. This layer does not care about how information was physically passed over the network: that is the job of level 1. Indeed, once a common level language is established, a logical level of communication is formed shown by the dotted lines. Each level can be considered a module with prescribed ways to pass information between adjacent modules. The model is "open" in the sense that the modules are not so tightly specified that they would exclude certain vendor approaches to implementation. In short, the OSI model provides a general architecture for networking.

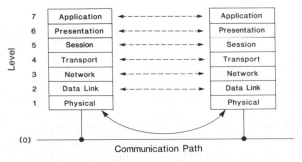

FIG. 10.   ISO reference model.

IV. Local Area Networks

As mentioned before, early networks were mainframe-oriented. This fostered the master–slave environment where slave stations were tyically dumb terminals. With all the intelligence at the mainframe, none was required at the terminal or in the network itself. Therefore, networking could be accomplished with simple, passive, point-to-point or multipoint circuits. Development of minicomputers started the decentralization of processing power from the mainframe, bringing it closer to individual users. With more intelligence at the user end, the need developed to allow equipments to communicate among themselves, in addition to the mainframe. Network architecture changed accordingly. Networks also had to support increasingly higher data rates to connect sophisticated stations found in the workplace. As the evolution continued, the interconnection of workstations tended to become clustered within a building or small group of buildings to support normal business functions. Local area networks, or LANs, became the product of this environment.

The concept of LANs was introduced in Fig. 3, which shows the connection of equipment in a local setting. LANs are not rigidly defined, but tend to share the following characteristics:

Data rates are higher, in the range of 1–50 Mbit/sec (M, mega $= 10^6$), to support high-speed communication requirements such as computer-to-computer linkage.

Network geography generally ranges from a building up to a campus-level setting. Distance limitations are primarily a result of the higher data rates.

Network architecture generally allows station-to-station communication in contrast to master–slave relationships.

LANs tend to be privately owned.

Functions other than data communication might utilize the network, such as voice and video communication.

The above characteristics of LANs have a large influence on the media. In the upper megabit range, data transmission must be done over higher-quality media such as coaxial cable or fiber. Use of a twisted pair generally requires a reduction in distance or speed, or both. A simplistic view of an LAN is a high-performance cable threaded throughout a building or group of buildings; anyone wishing connection to the LAN runs a cable from their station, tapping into the nearest run of the LAN cable.

Before discussing LANs in more technical detail, it is necessary to first define two general forms of communication which LANs typically employ: baseband or broadband. Baseband refers to dumping the raw digital signal itself onto the communication line. For instance, a station could have an electrical output of $+5$ volts for a logical 1 and -5 volts for a logical 0. In baseband, once the station had access to the line, it could literally switch the line voltage between the $+5$ and -5 levels to form the electrical analog of the logical bit stream being transmitted. A characteristic of baseband communications is that whoever has access to the line excludes all others. This creates a form of time-division multiplexing (TDM) when the line is shared among several subscribers. Broadband communication is similar to cable TV (CATV); indeed, digital broadband is implemented with many CATV industry components. On the TV cable, many channels are broadcast simultaneously by allotting each channel a certain portion of the frequency spectrum. To use broadband technology for LANs essentially only requires that a portion of the spectrum be devoted for that use and that a way be provided to convert digital data to a form compatible with analog broadband transmission. To accomplish the latter, a modem is used between LAN subscribers and the broadband cable system. Via the modem, digital data modulates the carrier within that allotted part of the frequency spectrum. Broadband, then, is done by frequency-division multiplexing (FDM). The principal advantage of broadband is its multiuse capability. It is not uncommon for a broadband network to serve digital LAN, video teleconferencing, TV broadcasts, and other functions. Broadband LANs tend to have a higher number of addressable stations allowed. In practice, however, this feature may not be realized due to the particular protocol in use. For baseband, the principal advantage is cost. If only the LAN function is required, baseband is considerably cheaper at installation and throughout the life of the LAN.

Two of the line protocols already discussed are commonly used for LANs: contention and token passing. Imbedded in these LAN protocols are also many of the other line-management techniques discussed earlier, such as error handling and message acknowledgement. Because contention and token passing are so prevalent, they are used here as case examples to further explain higher-performance local area networking. Contention is better known as CSMA/CD in the LAN world, meaning carrier sense multiple access/collision detect. This translates to a network allowing multiple subscribers that sense carrier for access and are able to detect collisions that normally occur with contention. Token-passing LANs are either token ring or token bus, corresponding to the two topologies. Both CSMA/CD and token passing are supported by the IEEE 802 standard on local networking, which is another reason for stressing these protocols here. (IEEE 802 is the Institute of Electrical and Electronic Engineer's committee 802 on local networking.)

A. CSMA/CD LANs

Figure 11 shows the approximate implementation of a commercial CSMA/CD LAN using baseband coaxial cable. As an example, a bus topology is shown that allows information to flow equally well in either direction over the bus. Although only two stations are shown for explanation, over 1000 are possible in practice. Each end of the bus is terminated in its characteristic electrical impedance. ISO levels are included to further demonstrate how the model applies to real networks. Levels 3–7 are often called the client layer in CSMA/CD.

All stations are equipped with transceivers that both transmit and receive information over the physical media. The transceivers are responsible for ISO level 1 functions, principally managing the electrical details of getting the bits over the physical media. In this case, the logical bit stream to or from the data-link control level would typically be converted to a biphase code, allowing self-clocking of the information over the bus. CSMA/CD bit rates of 10 Mbit/sec are common. Note that the bandwidth of the communication channel (baseband coax) must be capable of sustaining twice this rate, since biphase coding can have up to two transitions per bit time. With the bus topology, all stations examine information passing over the communication path simultaneously (ignoring transmission delay). This feature allows for broadcasting information to one, some, or all stations on the bus.

The information packet begins with a string of bits called the preamble. This is a string of alternating 0s and 1s used to establish bit synchronization of transceivers. Typically, the last two bits of the preamble are 1 1, signifying the end of the preamble and also establishing block synchronization with following address fields. Both destination and source addresses are included in the packet. The type field, explained shortly, is followed by the user data block. The data block is bit-oriented and of variable length. The last field, CRC, is the cyclic redundancy checksum used for error detection, as previously explained.

For certain applications, client layer functions may need to intercommunicate over the network. An example might be to establish the protocol being used at the client layer. Passage of this information is accomplished via the type field. To lower levels, the type field looks like part of the data block. Information flow proceeds approximately as the following for transmit and receive functions:

1.   Packet assembly begins in the client layer with data, type, and address information, then proceeds to level 2. The DLC calculates the CRC over the packet information at that point, appends it to the packet, and passes on to level 1. The physical layer adds the preamble, performs the biphase code conversion, and sends the packet out over the link.

2.   An incoming packet is stripped of the pre-

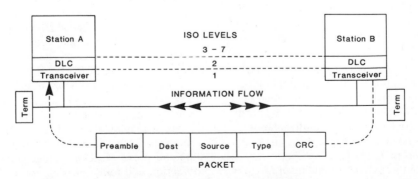

FIG. 11.   CSMA/CD bus baseband LAN.

amble by level 1, with the remaining bit stream passed to the DLC. Using the CRC field, the DLC determines if any transmission errors occurred and checks to see if the packet destination is intended for its particular station. If so, the remaining address, type, and data information is sent on to the client layer.

A shortcoming of CSMA/CD, and one often used to compare against the next discussion of token passing, is the collision problem. Collisions are generally not a problem in lightly loaded networks. As load increases, however, collisions are more frequent, which then begins to impact the throughput performance of the network. CSMA/CD generally includes a feature to help with performance degradation due to collisions, but this does not eliminate it. When the collision rate reaches a certain point, an exponential backoff algorithm is used to increase the time a station waits before trying another transmission. This is a useful approach to smoothing out times when the network becomes quite busy.

B. TOKEN-PASSING LANs

For comparison to CSMA/CD, a representative commercial token-passing LAN is approximated in Fig. 12. A ring topology is shown, which has two distinct differences from a bus. First, information flow is unidirectional over the ring, and second, any station on the ring can cause a break in the communication path. Again, only two stations are shown for simplicity, but many can exist on a single ring in practice. The ring requires the transceiver function of the bus topology to be replaced with individual receivers and drivers to receive and transmit data over the link. This arrangement effectively makes the network a circular linkage of point-to-point circuits. The ISO level functions are basically the same for token and CSMA/CD, needing no further elaboration.

Data transfer over a token ring begins when a station desiring to transmit data receives an idle token. Then the packet status is changed to busy, the information frame is added to the token, and the message is broadcast down-link to succeeding stations. The intended receiving station recognizes its address in the message, resulting in that station capturing the message as it goes by. The message goes around the network until it returns to the original sender. The sender then removes the information frame, changes token status back to idle, and sends the available token down-link to other stations. The target station that received the message performs the usual tasks of checking for transmission errors and passing the information frame on to higher-level functions.

The frame format used in token passing is similar to the packet format of CSMA/CD. The frame is divided into three general fields: the header, data block, and trailer. The header and trailer both have delimiters (DEL) that signify the beginning and end of the frame. The remainder of the header includes token status, destination and source addresses, and other control information for link and message management. In conjunction with the trailer delimiter, a modifier (MOD) field terminates the frame and is used for certain control functions, such as error detection and address recognition.

Only the two more popular higher-performance LAN approaches have been discussed. Both are baseband. In practice, token-passing

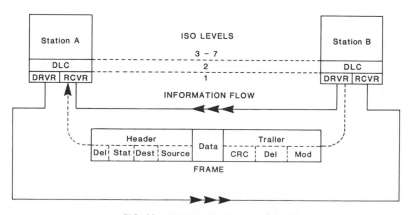

FIG. 12.   Token ring baseband LAN.

LANs tend to be baseband only, not being implemented over broadband. Both token ring and token bus are in use, both having their respective IEEE 802 standards. CSMA/CD is found in both baseband and broadband implementations.

C. Lower-Performance LANs

Several LAN strategies exist in the range of under 1 Mbit/sec. With lower data rates, twisted-pair transmission medium comes back into play because of its low cost and widespread use. As a case example for lower-performance LAN technology, digital PBX-type technology, is chosen here for several reasons. First, these devices are immediately usable for digital data transport. If an older analog-type switch exists, it can be easily replaced with a digital version using the same cable plant. Second, the existing cable plant of the PBX generally reaches everyone in the local community to service their telephone needs, and, as a result, is a significant resource when evaluating LAN needs. For example, if 1000 stations needed connection over an LAN, considerable time and expense would be devoted just to install cable. If the capabilities of the PBX and cable plant match the LAN requirements, no additional wiring may be necessary. As LAN data rates decrease, LAN functions begin to merge with data applications running over telephone circuits. It then becomes difficult to define the boundaries of an LAN. It

becomes more correct to speak of local networking in general rather than about an "LAN." Lastly, this environment also serves to introduce wide-area networking, which immediately follows this topic.

Referring to Fig. 13, a local setting is shown about a telephone-company central office. All telephone-subscriber voice and data lines for a community are connected into a central office. These lines are referred to as local loops, being simple twisted-pair circuits. At the central office, switching equipment exists that connects a subscriber to other local or distant subscribers as directed by the caller's dialed digit sequence (telephone number). Lines to the central office can functionally be classified into switched (dial) or dedicated circuits. Both are physically the same type of twisted-pair cable. A switched or dial circuit is one that is tied into the switching equipment at the central office—i.e., it relies on the caller to dial the desired connection. A dedicated circuit is one that is permanently connected: source and destination circuits at the central office are hardwired together, dedicating the path to the subscribers involved. The PBX is a smaller version of the central-office switching equipment and is commonly used in many business applications. It allows all telephones within a building, or community of buildings, to be interconnected. Geographically, this situation is precisely the same as that for an LAN. PBXs are linked to central office functions via trunk lines.

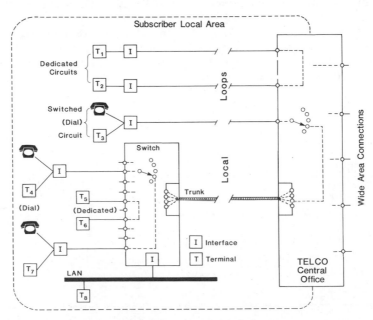

FIG. 13. Private exchanges, local loops.

Figure 13 shows a generic switch that could be one of several things, one being the PBX. At a more sophisticated level, it could be a PNX, or private network exchange, which may incorporate higher-level LAN functions. Computer base exchanges, or CBX, is also a term used to denote a more sophisticated switch. An interface is shown from the switch to a higher-performance LAN to illustrate a higher-level function than a simple PBX.

Several examples are given in Fig. 13 to illustrate lower-performance local networking. Terminals 1 and 2 are connected as a point-to-point dedicated circuit via a hardwired connection at the central office. Either terminal could just as easily have been connected to some remote city via the telephone-company wide-area circuitry. Terminal 3 is connected to an interface along with a standard telephone. The user can dial another subscriber to establish a connection, then switch over to the data terminal. There are several variations of this capability as discussed earlier. Terminal 4 is like 3, but connected to a local switch rather than directly to the central office. Terminals 5 and 6 are likewise similar to 1 and 2, being connected via the local switch. With this brief overview, it is easy to visualize many ways to make local data connections using switch devices.

V. Wide-Area Networking

Figure 4 introduced wide-area networking in general, ignoring some of the detail involved in establishing a data path. Figure 14 takes up where Fig. 13 left off in terms of wide-area connections and adds enough detail to visualize how long-haul data circuits are implemented. The more common long-haul circuits use common carrier services of the telephone company. This is an old term originating from the idea of providing a carrier service common to the population at large. Originally, telephone service was classed as a basic carrier, as were ship and rail services. Later, data services were offered over the same analog transmission services, followed later yet with digital data services. Shown in the figure are several general ways a terminal in city A might be connected to a computer in city B. If a dial connection is desired, the caller dials the destination telephone number, which results in a path being established through all the switching toll centers between the two cities. A dedicated circuit can be ordered from the carrier, established by hardwiring connections along the path between subscribers. As the distance increases between points, it is more likely that connections include high-bandwidth transmission media, rather than simple switched or dedicated copper wire connections. Terrestrial microwave is an example, providing high-bandwidth communication for voice, data, and video between cities. Microwave dishes are a common sight on top of buildings and high-terrain points, with about 30 miles between dishes being the maximum. Another example is co-ax or fiber cable, typically buried between cities. Information traffic between cities, then, is typically multiplexed over the high-bandwidth capabilities of microwave or cable. Lastly, long-haul circuits can be implemented by multiplexing a subscriber over a high-bandwidth communication satellite link.

Each of the several ways to implement wide

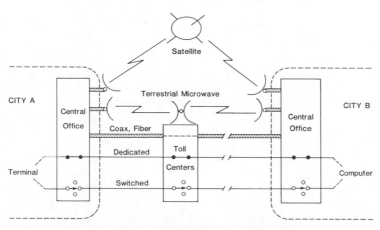

FIG. 14.   Wide-area networking.

area circuits have their pros and cons. Switched circuits tend to be noisy and are used in sporadic-need, lower-data-rate applications. Dedicated terrestrial links are of higher quality with less noise and are used in applications where high time-connectivity and low circuit delays are important. Satellite links are generally implemented via geosynchronous satellites, which are in stationary orbit relative to points on earth, about 22,300 miles in altitude. This distance requires considerable time for signal propagation, averaging about 540 msec and peaking to 900 msec in some situations. To visualize the impact of this, consider a polled circuit implemented via a satellite circuit. Short blocks of information can require only a few milliseconds of time, small compared to satellite-link transit time. In a polled environment, considerable time would be wasted waiting on satellite-link delays.

With a variety of communication services available, subscribers can order data circuits with a range of data rates and performances. Using commonly available equipment and existing subscriber cable services, it is easy to implement circuits with data rates of 50 kbit/sec (k, kilo = 10^3). With a little more effort, over a megabit per second can be provided for long-haul applications.

A. ANALOG AND DIGITAL FACILITIES

Originally, data sent over wide area links had to be converted to analog form to be compatible with common carrier equipment. Referring back to Fig. 13, the interfaces shown would be modems that modulate analog carrier with data to ship information over the network. As technology improved, common carriers implemented digital data services that allowed digital data to be directly interfaced to the carriers' equipment. A digital interface is required to multiplex the subscriber data into the carrier's bit stream, but no digital/analog conversion takes place. This reduces cost, as digital multiplexing is cheaper than modulating and demodulating data, and also benefits from the higher reliability inherent in digital equipment.

To understand some of the rationale of bit rates used in digital data transmission, it is necessary to discuss digitizing voice. Telephone voice circuits are generally analog at least up to the central office. Past there, analog signals may be digitized so that they can be transmitted over the telephone company's digital facilities. To transfer voice over telephone circuits, a band-

width of about 4 kHz is needed to carry a reasonable representation of the talker's voice characteristics. To digitize the voice occurring over the line requires that it be sampled at twice the highest frequency, or 8000 times per second. At each sample time, the signal amplitude must be converted to a binary number. Seven bits are used for the conversion, which allows discrimination of 128 different amplitude levels. Another bit is added for control, bringing the total to 8 bits for each sampled point. Twenty-four voice channels are customarily time-multiplexed (TDM) for transmission over digital facilities. The multiplexing takes place by sampling each of the 24 channels, in order, producing 192 bits of information. Another bit is added for control, bringing the total to 193 bits. Since this needs to be done 8000 times per second, a transmission rate of 1.544 Mbit/sec results. This is called a T1 carrier. Rough multiples of the T1 service are used to form a hierarchy of digital transmission, as shown in Fig. 15. Data rates increase as more subscribers are multiplexed together. A hierarchy of media also results, starting with the basic twisted pair at the subscriber end up to coax or fiber, which is required to carry the higher data rates. Basic synchronous digital data service is offered by carriers at data rates of 2.4, 4.8, 9.6, and 56 kbit/sec over standard local loops to the central office. It is also possible to access the T1 carrier for high-data-rate applications.

B. INTEGRATED-SERVICES DIGITAL NETWORK

The cost effectiveness, functionality, and inherent reliability of digital transmission result in a continued migration of digital connectivity to end-user applications. Direct digital interfaces to end-user equipment are becoming available from the carriers for a widening array of applications. A common example is the digital telephone. This type phone, within itself, converts the voice analog signal to a digital form, allowing the device to directly connect to a digital transmission channel. As an example of higher complexity, video teleconferencing is possible by bringing a digital "pipe" to a subscriber's premises of sufficient bandwidth to allow the transmission of digitized video signals. The concept of directly connecting a variety of end-user devices to digital carrier services is known as integrated-services digital network, or ISDN. ISDN is an infant industry that will mature throughout the remainder of the century.

As already discussed, digital transmission ser-

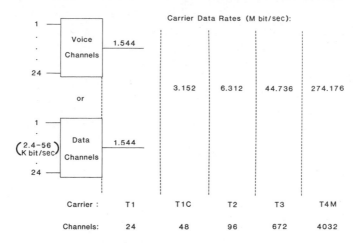

FIG. 15.   Digital data services.

vices are commonly used between carrier toll centers. Therefore, a high-speed network backbone for ISDN in some sense already exists. The difficulty is extending the digital service to the end-user over existing local loops. To bring the full capability of ISDN to the household, for instance, requires greater bandwidth than the local loop can provide. As ISDN matures, we can expect direct digital connection to the carriers to service many applications: PBX connection, data terminals, LAN connection, utility meters, and security devices, to name a few. Once a connection is made, the ISDN will literally allow subscriber access to the world. ISDNs will exist in all countries, linked together to form a worldwide digital network.

C. Other Wide-Area Services

Up to this point, only common carrier services have been discussed. Specialized common carriers (SCCs) also exist, offering only specialized services, not the wider offerings of the common carriers. For example, an independently owned, high-bandwidth terrestrial microwave could be installed with a tie-point to central office equipment (Fig. 14) leased from a common carrier. This would allow the specialized carrier (independent microwave) access to local loop service to provide its specialized services to the general population, which keeps it in the class of common carrier, but specialized to some particular application.

Another example of carrier services is the value-added carrier (VAC). As an example, someone owning a computer could implement a special computer data base containing certain information that might be of interest to a wide variety of people. One way to provide access to the information would be to install dial modems to the computer, allowing anyone with a telephone to tie into the data base. Passing an account number, say, after dialing in would allow the provider to bill for services rendered. One problem with this approach is that many potential customers might be turned away by the long-distance telephone charges. If the service is popular over a large geographical area, even globally, another approach would be to distribute the access mechanism. The service could lease long-haul circuits from the common carriers and install dial or leased line access equipment over a wide area, providing local access to residents in the cities serviced. In other words, the common carriers have been used to implement a service that added a value to the basic common carrier service.

Bibliography

Black, U. D. (1983). "Data Communications, Networks and Distributed Processing," Reston Publishing Co. Inc., Reston, Virginia.

Dixon, R. C., Strole, N. C., and Markov, J. D. (1983). A token-ring network for local data communications, *IBM Syst. J.* **22**(1/2), 47–62.

Kreager, P. S. (1983). "Practical Aspects of Data Communication." McGraw-Hill, New York.

Stallings, W. (1985). "Data and Computer Communications." MacMillan, New York.

Tanenbaum, A. S. (1981). "Computer Networks," Prentice-Hall, Englewood Cliffs, New Jersey.

A Building Planning Guide for Communication Wiring, G320-8059-1, IBM Publication. March, 1984.

DATA TRANSMISSION MEDIA

John S. Sobolewski *University of Washington*

GLOSSARY

Bit: Short for binary digit, the smallest unit of information in a binary system. A bit represents the choice between a one (mark) or a zero (space) condition.

Carrier: Continuous frequency or periodic pulses capable of being modulated or changed by an information signal.

Channel: Path for transmission of information between two or more points using some form of transmission medium.

Channel capacity: Number of information bits per second that can be transmitted over a channel.

Demodulation: Process of retrieving information from a modulated carrier wave. The opposite of modulation.

Medium, bounded: Transmission medium that constrains and guides or conducts information-carrying signals. Also called a guided transmission medium. Examples include wires and optical fibers.

Medium, transmission: Transmission medium that provides the physical channel or path used for transmission of information between two or more points.

Medium, unbounded: Transmission medium that permits signals to be transmitted but does not guide or constrain them. Also called an unguided transmission medium. Examples include radiowave transmission using space or air as the transmission medium.

Modulation: Process of varying some characteristic of a carrier wave in accordance with the instantaneous value or sample of the information to be transmitted.

Multiplexing: Division of higher bandwidth communication channel into two or more lower bandwidth channels.

Repeater: Device whose function is to retime and restore received signals back to their original form and strength and to retransmit them further down the channel.

Signal: Aggregate of waves propagated along a transmission channel toward a receiver at the destination.

Electronic communication systems depend upon transmission media for sending voice, data, or video signals from one point to another. Among the physical media are twisted pairs, coaxial cable, microwave radio, and optical fibers. The distance of transmission can be very small, as between computer memory and the central processing unit, or very large, as between a corporate headquarters and an overseas subsidiary. Without the media, communication systems could not exist.

I. Basic Signal Types

Information is transmitted over communication channels either as analog signals or digital signals. Analog signals are in the form of a continuous time-varying physical quantity, such as voltage magnitude or frequency, that reflects the variations of the information or signal source with time. Speech, radio, and television signals are examples of analog signals. Digital signals consist of a stream of discrete pulses with well-defined states. Data generated by digital computers and their peripherals or data terminals are digital in nature and, in general, consist of two discrete levels corresponding to digital ones and zeros, as illustrated in Fig. 1. [*See* SIGNAL PROCESSING, GENERAL.]

It is important to note that a communication system can be designed to carry either analog or digital signals by using appropriate signal conversion devices. This applies to systems using any type of transmission media—wire, coaxial cable, microwave radio, satellite, and others. Furthermore, since analog signals can be converted to digital signals and vice versa, any type of information can be transmitted either in digital or analog form that is compatible with the transmission medium over which it will travel with no significant loss of information. [*See* SIGNAL PROCESSING, ANALOG; SIGNAL PROCESSING, DIGITAL.]

Most communication channels today are designed for transmission of analog signals within a certain range of frequencies. If we need to transmit digital computer data for long distances over such channels, we can use a special device known as a modem that converts the digital data into an analog signal consisting of a continuous range of frequencies, as illustrated in Fig. 1. We can thus use existing analog channels for transmission of digital data. Conversely, it is possible to convert analog voice or television signals into digital signals and transmit them over channels designed for digital signals. The capability of converting analog to digital signals and digital to analog signals is important, since the need for transmission of digital data is growing rapidly

while most of the communication channels in existence are analog in nature and will remain so for many years to come because of the hundreds of billions of dollars that are tied up in this equipment. However, digital transmission has advantages over analog transmission, and consequently digital technology is evolving rapidly. Over the past two decades, millions of miles of digital transmission channels have been installed by telephone companies, mostly to carry digitized speech. This trend will continue, and communication channels installed in the future will be almost exclusively digital in nature, not only for transmission of data but also because of increased use of digital telephones and digital television. [*See* DATA COMMUNICATION NETWORKS.]

II. Characteristics of Some Common Signals

Before describing the concept of a communication channel and the various transmission media in greater detail, it is necessary to describe the characteristics of some of the most common types of signals that are routinely transmitted over existing communication channels. Even though significant economic payoff can result from designing communication channels that can carry all types of signals, some of the signal characteristics are so different that different communication architectures are needed. The sections that follow describe the characteristics of program and speech, which are analog signals, data, which is a digital signal, and video, which can be considered a hybrid between analog and digital.

A. PROGRAM SIGNALS

Program signals include radio broadcasts, the audio portion of television programs, and high-fidelity music programs that are distributed over cable systems to subscribing customers. Such signals are characterized by a greater average volume, dynamic range in volume, duration, and bandwidth (which results in greater fidelity) than typical telephone speech signals. A high dynamic range of volume is needed to reproduce faithfully the variety of program material transmitted, which may include speech, music, and special sound effects. The bandwidth required may be as high as 30–15,000 Hz for high-quality music for FM stereo or the audio portion of television programs, even though only a fraction of

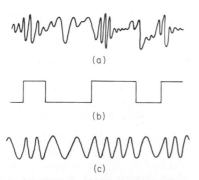

FIG. 1. Any information can be transmitted in (a) analog or (b) digital form. Conversion from one to the other is available; (c) shows digital signal converted to an analog signal using frequency modulation.

this bandwidth is used for speech during, say, a newscast. [*See* RADIO SPECTRUM UTILIZATION.]

B. SPEECH SIGNALS

Speech signals are usually transmitted over telephone channels. The telephone set transforms the voice signals into electrical analog signals which are transmitted to the local telephone exchange and then through the wide area telephone network to the receiving party. Although speech typically covers frequencies from 30 to 10,000 Hz, most of the energy is in the range from 200 to 3500 Hz. Since the human ear is not very sensitive to small changes in frequencies and since humans can correct for things such as missing syllables or words, we do not need to reproduce speech signals precisely to achieve acceptable quality of transmission. Consequently, most telephone communications are bandlimited to between 200 and 3500 Hz to save transmission costs. This savings comes about since costs are directly related to the bandwidth (i.e., the range of frequencies transmitted) and (as will be described later) since bandlimiting allows a greater number of voice channels to be multiplexed over a high-bandwidth channel. It must be pointed out, however, that the nominal bandwidth of a voice channel is defined as 4000 Hz, with the additional bandwidth allowing a guard band on either side of the speech signal to reduce interference between channels. [*See* VOICEBAND DATA COMMUNICATIONS.]

The time required to set up a telephone connection can vary from a few seconds to a minute, communication is almost always two-way, and the two parties continuously transmit (talk), listen (receive), or pause until the call is terminated. A wide dynamic range in volume is needed, although not as large as for program signals. Conversations require immediate delivery of the signal (i.e., delays are not tolerated), but communication is relatively tolerant of noise on the channel.

C. DATA SIGNALS

Computers, workstations, or data terminals usually produce digital signals consisting of streams of binary digits (bits) representing 1s and 0s. Unlike the speech signal, which contains much redundant information and is relatively tolerant of noise or errors on the channel, the digital data signal may have little or no redun-

dancy; thus, in general, errors cannot be tolerated. In fact, bits (parity or cyclic redundancy bits) are usually added to the basic digital signal to detect errors and correct them by retransmission. By addition of even more bits and use of error correcting codes, errors can be detected and corrected without retransmission of the data.

Other important characteristics are shown in Table I, which summarizes the differences between digital and telephone voice signals. It is important to recognize that those differences are significant and that telephone and computer users have fundamentally different requirements. However, because telephones have had such a dominant role in communication technology since the 1920s while data transmission is relatively new, economies of scale have dictated that much data traffic be converted to analog form so it can be transmitted over channels originally designed for voice signals. This severely limits the potential of the computer and can waste channel capacity. With the large increase in data transmission since the beginning of the computer era, and the even greater increases projected for the future, there will be dramatic progress in the development of new network architectures designed for data transmission.

D. VIDEO SIGNALS

Most video signals carried over communication channels today are color television signals. A video transmission system must deal with four important factors when transmitting images of moving objects, specifically:

1. Perception of the distribution of luminance (light and grey shade)
2. Perception of a three-dimensional image (width, height, and depth)
3. Perception of motion related to the two factors above
4. Perception of color (hues and tints)

Monochrome video transmission deals with the first three factors. In color video transmission, color is combined with the monochrome (black and white) picture. At every moment, the video signal must integrate luminance and color from a scene in three dimensions, as a function of time, and combine them into a complex electrical signal for transmission. What makes this more complex is that time itself is variable since the scene is changing all the time. The integra-

TABLE I.   Differences in Characteristics between Data and Telephone Voice Signals

Data signal	Telephone voice signal
Desirable to set up a connection in one second or less	One second to one minute to set up a connection
One- or two-way transmission	Two-way transmission in most cases
Error-free received data	Tolerant of noise and some errors on the channel
Little or no redundancy	Much inherently redundant information
Transmission usually in bursts	Transmit or receive continuously until call is disconnected
Data can usually be stored and transmitted later when convenient	Not tolerant of transmission delays
Transmission has high peaks. Peak-to-average ratio as high as 1000 to 1.	Transmission rate relatively constant
Connection may be required for 24 hr/day, 7 days/week (e.g., bank cash machine)	Duration of connection usually several minutes
May require a wide range of bandwidths, from hundreds to millions of Hz	Requires fixed bandwidth of about 4000 Hz

tion of visual information is carried out by a process called scanning.

The scanning process consists of taking a horizontal strip across the image, and scanning it from left to right beginning at the top. When the right-hand end of a strip is reached, the next lower horizontal strip is scanned. Luminance values are translated on each scanning interval into voltage or current variations and are transmitted. In the United States a total of 525 horizontal scan lines makes a whole picture frame. Scanning a whole frame is repeated at a rate of 30 frames per second. This rate takes advantage of the persistence of vision of the human eye, thus eliminating flicker and giving the perception of motion, as in motion pictures. Because of the amount of information that must be transmitted to produce flicker-free images with acceptable horizontal and vertical resolution, video signals require a bandwidth of 6 MHz in the United States, as compared with the 4-kHz bandwidth required for acceptable transmission of voice signals.

To permit decoding of the video signal at the receiver, it is necessary to transmit vertical and horizontal synchronizing pulses interleaved with the picture information. These pulses synchronize the transmitter and receiver by identifying the start of every horizontal scan line that corresponds to the top of the picture and the subsequent horizontal scan lines that make up the rest of the picture. Since these synchronizing pulses

are essentially digital signals while the picture information in each horizontal scan line is in analog form, video signals may be thought of as hybrid signals.

III. Communication Channels

A block diagram of an electronic communication system is shown in Fig. 2. It consists of an input signal, a transmitter, a communication channel, a receiver, and an output signal from the receiver. The transmitter modifies the input message signal into a form suitable for transmission over the channel, which is the transmission path for providing communication between transmitter and receiver. The purpose of the receiver is to recreate the original message signal at the output. For example, when the digital data signals must be sent over a communication channel designed primarily for analog signals, the transmitter has to convert the digital signals to analog signals by a process known as modulation. The receiver then demodulates the analog signal back to digital form and passes it on to its ultimate destination.

The communication channel can consist of various media such as wire, coaxial cable, optical fibers, or free space in which the signal is radiated as an electromagnetic wave as in television or radio broadcasting. Communication channels have practical limitations, such as bandwidth, and suffer from various impair-

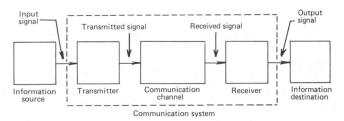

FIG. 2.   Model of a communication system.

ments, such as nonlinearities as well as noise and interference, which may be introduced from within or from outside the channel. Transmitters and receivers must be carefully designed to match the signals to be transmitted to the physical properties of the communication channel and to minimize the effects of channel impairments on the quality of reception.

We can classify communication channels as analog or digital depending on the basic signals they transmit. Analog channels can be further classified as voice, program, or video channels depending on their intended use and the kind of signals they carry. They may be characterized by their bandwidth, which is the range of frequencies they transmit. Thus a voiceband channel may be called a 4-kHz channel while broadband channels may be 48-kHz or 240-kHz channels. Digital channels are usually characterized by their bit rate. Channels with a bit rate of 64 kbps (kilobits per second), for example, are used with increasing frequency and can be used to transmit simultaneously several lower bit rate channels through multiplexing, as explained in Section VII.

Analog and digital channels can also be classified as simplex, half-duplex, or full-duplex. Simplex channels transmit in one direction only; they are used for radio broadcast but seldom for data communication, which usually requires two-way transmission. Half-duplex channels can send in both directions but only in one direction at any given time. In other words, they provide nonsimultaneous two-way communication. Full-duplex channels allow simultaneous two-way communication.

IV.   Channel Capacity

In practice, a communication channel represents a financial investment, and hence the goal of communication engineers is to derive as much as possible from that investment. In the case of digital transmission this is done by maximizing the channel capacity, which may be defined as the maximum rate at which information can be sent over the channel with an arbitrarily small probability of error.

In 1928, Nyquist showed that for binary transmission (i.e., the transmitted signal has one of two possible values at any one time) $2W$ bits/sec can be transmitted over a channel of bandwidth W in the absence of noise. In the general case, if we use M-ary ($M > 2$) rather than pure binary transmission by sending one of M different and distinguishable signal values at any one time, then in the absence of noise we have

$$C = 2W \log_2 M \qquad (1)$$

where C is the channel capacity in bits/sec and W the channel bandwidth in Hz. Thus, if we have a channel bandwidth W of 3300 Hz, then in the absence of noise we can transmit up to 6600 bits/sec using pure binary transmission ($M = 2$). However, if we use a transmission scheme that makes it possible to send one of four distinguishable values at any given instant ($M = 4$), then we increase the channel capacity to 13,200 bits/sec. Similarly, if we can send one of eight distinguishable values at any instant, then each level in effect represents $\log_2 8 = 3$ bits, and the channel capacity becomes 19,800 bits/sec using this transmission scheme.

For a channel of bandwidth W, it is highly desirable to increase M in the above equation to maximize the channel capacity. The question arises, however, as to the number of different signaling values that can be transmitted and be separately distinguished at the receiver in the presence of noise, distortion, limits on signal power, and other channel impairments that occur in practice. In 1948, Shannon proved that if signals are sent with a signal power S over a channel with Gaussian noise (amplitude of noise signal follows a Gaussian distribution) of power N, then the capacity C, of the channel in bits per second is

$$C = W \log_2(1 + S/N) \qquad (2)$$

where W is the bandwidth of the channel. This is one of the fundamental laws of communication and gives the maximum signaling rate over a communication channel in terms of three parameters that are known or measurable. We can design elaborate coding schemes or modulation techniques, but we will never be able to increase the capacity unless we increase either the available bandwidth or the signal-to-noise ratio S/N.

If we take the previous example of the channel with a bandwidth of 3300 Hz and if we assume a signal-to-noise ratio of 63, then the maximum possible rate at which data could be transmitted over this channel is $3300 \log_2(1 + 63)$ = 19,800 bits/sec, no matter how many different signaling levels our transmission scheme may have at any one time. In fact, for this particular channel, if we tried to use a transmission scheme with one of 16 different signal values at any one time to increase the channel capacity, then in the absence of noise, relation (1) gives a capacity of 26,400 bits/sec. However, with the given signal-to-noise ratio of 63, the capacity is limited to the 19,800 bits/sec, implying that 16 different values would not be distinguishable on this channel unless the signal-to-noise ratio were increased, which could be done by increasing the signal power of the transmitted signal.

Unfortunately, as explained in Section VIII, practical communication channels include impairments other than Gaussian noise, and the limit given by relation (2) is very difficult or costly to achieve in practice. However, it clearly identifies the two resources that designers can use to increase the channel capacity, namely, channel bandwidth and transmitted signal power. Sometimes these resources are limited, in which case the channels may be classified as bandlimited, as in the case of telephone voice channels, or power-limited, as in the case of, say, a satellite channel.

V. Baseband Transmission

Signals can be transmitted over a communication channel using either baseband or broadband transmission. In baseband transmission, the information signal is sent over the channel directly without modification. In broadband transmission, the information signal is modified by superimposing it on a higher-frequency signal, called the carrier, which "carries" the information over the channel. Broadband transmission—or carrier transmission, as it is sometimes called—uses modulation techniques that are described in the next section.

Because there is no modification of the information signal, baseband transmission systems can be used for digital or analog signals. In practice, they require an electrical or light (e.g., optical fiber) conductor to carry the baseband signal and are used primarily for transmission over short distances. Separate transmitters and receivers are seldom required (e.g., the source and destination act as transmitter and receiver respectively), and in those cases in which they are required, they are relatively simple devices. The simplicity of baseband transmission makes it very common; in fact, it is used for short-distance signal transmission amongst most components of electronic devices or systems. Transmission of signals between components within a computer, a computer and its local peripheral devices, and an audio amplifier and the speakers are all examples of baseband transmission. Others include transmission over local area networks in a building or factory and over telephone subscriber loops between a telephone set and the local exchange. Longer telephone links such as those between exchanges do not usually carry baseband signals but use higher carrier frequencies to carry many signals simultaneously.

Because of the relatively short distances involved, the bandwidth of baseband channels is relatively high. The major limitations of such channels include stray capacitance, inductance, and resistance which cause distortion, especially of the leading and trailing edges of digital signals. These limitations will be described in greater detail in Section IX and XII.

VI. Broadband Transmission

In broadband transmission, information signals are processed and superimposed on a carrier that is more suitable for transmission over a particular channel. This is known as modulation, which is defined as the process by which some characteristic (e.g., amplitude, frequency, or phase) of a carrier wave is varied or modulated in accordance with the instantaneous value or samples of the information signal to be transmitted. The information signal, which can be analog or digital, is called the modulating wave while the result of the modulation process is called the modulated wave. Demodulation restores the modulated signal to its original form at the receiving end of the communication channel.

At first, modulation techniques were used primarily for radio transmission; here a relatively low-bandwidth speech or music program signal was used to modulate a much higher frequency carrier, resulting in a modulated signal whose frequency spectrum was suited for radio transmission. Subsequently these techniques were used for telephone transmission, because it was realized that telephone lines had a bandwidth greater than that needed for speech. By using several voice signals to modulate carriers of different frequencies, several voice signals could be transmitted (or multiplexed) over a single telephone line, thus reducing the cost per voice channel.

Modulation is used for two other important reasons. First, it is used to convert digital signals for transmission over analog channels that would otherwise destroy the digital signal. Second, it is used to convert voice or other analog signals to digital form to achieve lower noise or less distortion in addition to economies in system implementation.

Modulation, detection, and demodulation are complex topics, and only the most basic principles can be described here. A more complete treatment is given in Schwartz (1980) and Proakis (1983) given in the Bibliography.

A. ANALOG MODULATION

In analog modulation, a sine-wave carrier is modulated by an analog signal such as a voice or a program signal. The sine-wave carrier $c(t)$ may be represented by

$$c(t) = A_c \sin(2\pi f_c t + \theta_c) \qquad (3)$$

where A_c is the carrier amplitude, f_c the carrier frequency, and θ_c the phase. The values of A_c, f_c, and θ_c can be varied by the message or modulating signal $m(t)$ to form the modulated wave $s(t)$. In all cases, the carrier frequency f_c must be much greater than the highest frequency component of the modulating signal $m(t)$.

In amplitude modulation (AM), the amplitude of the carrier is varied in accordance with the message signal $m(t)$. This is done by adding the product of $c(t)$ and a fraction k of $m(t)$ to the carrier $c(t)$, yielding

$$\begin{aligned} s(t) &= c(t) + km(t)c(t) \\ &= A_c(1 + km(t)) \sin(2\pi f_c t + \theta_c) \end{aligned} \qquad (4)$$

The absolute value of $s(t)$ has the same shape as $m(t)$ provided that the absolute value of $km(t)$ is always less than 1.

By taking the Fourier transform of $s(t)$, it may be shown that it consists of two symmetrical sidebands each of bandwidth m on either side of the carrier frequency f_c, where m is the bandwidth of the message signal $m(t)$. Consequently, one of the sidebands is sometimes removed through filtering to conserve bandwidth, and the carrier is suppressed to reduce transmission power. This variation of AM is known as single sideband suppressed carrier (SSB/SC) modulation and is commonly used for voice and data transmission over telephone circuits.

In frequency modulation (FM), the amplitude of $s(t)$ is constant, but the instantaneous frequency $f_i(t)$ of the modulated wave $s(t)$ changes linearly with the amplitude of the message signal $m(t)$ and is given by

$$f_i(t) = f_c + k_f m(t) \qquad (5)$$

which yields

$$s(t) = A_c \sin(2\pi(f_c + k_f m(t))t + \theta_c) \qquad (6)$$

where k_f is a constant and determines the maximum frequency deviation that can occur. In phase modulation (PM), the amplitude and frequency are maintained constant, but the instantaneous phase of the modulated wave $s(t)$ changes linearly with $m(t)$, yielding

$$s(t) = A_c \sin(2\pi f_c t + k_\theta m(t)) \qquad (7)$$

where k_θ is a constant and determines the maximum phase shift that can occur.

Amplitude and frequency modulation are used widely for radio program signals. An important feature of frequency and phase modulation is that they are more immune to noise and interference than amplitude modulation. This advantage, however, is achieved at the expense of a more complex receiver and greater bandwidth requirement. In fact, frequency and phase modulation provide a convenient means for obtaining better noise performance at the expense of bandwidth. Figure 3 illustrates the three modulation techniques in the case where the modulating signal is a digital rather than an analog signal.

B. DIGITAL–ANALOG MODULATION

The three analog modulation techniques just described can also be used for digital signals. In the case of binary digital signals, the modulation process involves keying (switching) the amplitude, frequency, or phase of the carrier between two different values according to the message signal. This is a form of digital-analog modulation and results in three distinct techniques known as amplitude shift keying (ASK), frequency shift keying (FSK), and phase shift keying (PSK). They are illustrated in Fig. 3 and are extensively used in modems (modulator–demodulators) which allow digital signals to be transmitted over analog channels. Figure 4 illustrates a simple demodulation (or detection) scheme for amplitude shift keying. The received signal is rectified and smoothed using a low-pass filter, and then the message signal is reproduced by passing the smoothed signal through a trigger circuit with the thresholds indicated.

C. MULTIPLE LEVEL MODULATION

The preceding discussion assumed binary signals. With multiple level (*M*-level) modulation, it is possible to transmit more than two amplitudes, frequencies, or phases. If four levels of amplitude are used, as shown in Fig. 5, each distinguishable level can represent a dibit (two bits) of information. The number of bits carried by the signal can be twice as great in theory but only at the expense of greater susceptibility to noise. If eight levels were used, a tribit (three bits) could be sent at any one time, but the noise susceptibility would be even greater.

The noise susceptibility of *M*-level amplitude modulation can be overcome by using *M*-level

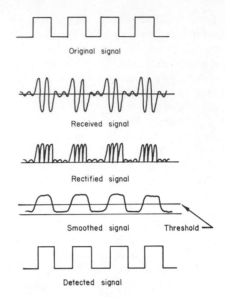

FIG. 4.   Demodulation used with ASK.

phase modulation. Thus dibits 00, 01, 10, and 11 could be represented by phase shifts of 90°, 0°, 180°, and 270°, respectively. *M*-level amplitude and phase modulation can be combined, and the resultant is sometimes known as quadrature amplitude modulation (QAM). Figure 6(a) shows a signal-space diagram (also known as the signal constellation) for a QAM system in which a combination of two levels of amplitude and four levels of phase shift (eight possible combinations) are used to transmit a tribit of information in one of eight possible states. Figure 6(b) shows a more complicated modulation technique in which a quadbit (four bits) of information can be transmitted in one of 16 possible combinations of phase and amplitude modulation. Note that

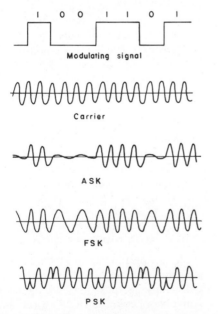

FIG. 3.   Three basic techniques of modulating a sinusoidal carrier using a digital modulating signal: ASK, FSK, and PSK.

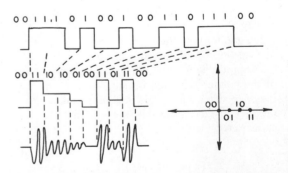

FIG. 5.   Amplitude modulation with four states and the corresponding signal-space diagram showing how the four amplitudes are coded in terms of magnitude and phase.

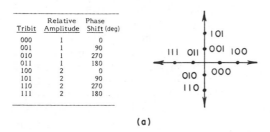

Tribit	Relative Amplitude	Phase Shift (deg)
000	1	0
001	1	90
010	1	270
011	1	180
100	2	0
101	2	90
110	2	270
111	2	180

(a)

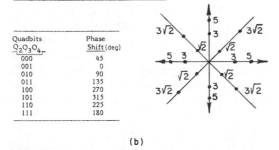

Quadbit Q_1	Relative Amplitude	Absolute Phase (deg)
0	3	0 , 90 , 180 , 270
1	5	0 , 90 , 180 , 270
0	$\sqrt{2}$	45 , 135 , 225 , 315
1	$3\sqrt{2}$	45 , 135 , 225 , 315

Quadbits $Q_2Q_3Q_4$	Phase Shift (deg)
000	45
001	0
010	90
011	135
100	270
101	315
110	225
111	180

(b)

FIG. 6. *M*-ary modulation using a combination of amplitude and phase modulation. (a) Two levels of amplitude and four levels of phase shift; (b) a more complicated technique.

this scheme is much more complex than others described previously. A synchronizing signal first establishes the absolute phase. The first bit Q_1 of each quadbit determines the signal element amplitude to be transmitted. The next three bits Q_2, Q_3, and Q_4 determine the phase shift or change relative to the absolute phase of the preceding element.

M-level modulation techniques such as QAM are finding increasing use in modern data transmission systems. A 9600-bit/sec QAM modem, for example, maps a quadbit of information into one of 16 (2^4) possible combinations of amplitude and phase, as shown in the space-signal diagram of Fig. 6b and transmits 2400 quadbits/sec.

A related modulation technique is trellis code modulation (TCM), in which redundant code bits are added and transmitted with the data. A 14,400-bits/sec TCM modem, for example, assembles six data bits and generates a redundant seventh bit using two of the data bits and a binary convolutional encoding scheme. The resulting seven bits are mapped into one of 128 (2^7)

possible combinations of amplitude and phase, much like QAM, and 2400 of these seven-bit symbols are transmitted each second to obtain the 14,400-bits/sec effective data transmission rate. The 14,400 comes about because only six of the seven bits represent actual data.

At the receiving end, the three encoded bits (the redundant bit and the two data bits used to generate it) select one of eight (2^3) subsets, each consisting of 16 amplitude–phase combinations, while the remaining four bits select one of the 16 combinations from the selected subset. The redundant seventh bit ensures that only certain sequences of amplitude–phase combinations are valid and that any two valid sequences of combinations are far apart on the signal-space diagram to help in the decoding.

The major differences between QAM and TCM are the encoder that generates the redundant bit and the resulting signal-space diagram which, for TCM, has twice as many amplitude–phase combinations because of the introduction of the redundant bit. By careful choice of the signal-space diagram and the convolution code to generate the redundant bit, TCM can have a noise tolerance of twice that of QAM [see Payton (1985)] which translates to higher transmission rates over bandwidth-limited channels. Typical modems using QAM are limited to 9600 bits/sec while newer TCM modems can operate reliably at speeds up to 14,400 bits/sec over telephone channels. This improvement implies that TCM will most likely replace QAM as the principal modulation technique for high-speed modems.

D. PULSE MODULATION

All the modulation techniques described previously use a continuous sinusoidal carrier wave that is modulated by analog or digital signals. Because of the continuous carrier wave, these are sometimes referred to as continuous-wave (CW) modulation techniques. In pulse modulation, the carrier is not a continuous wave but a periodic pulse train whose amplitude, duration, or position is varied in accordance with the message. Pulse amplitude (PAM), pulse duration (PDM), and pulse position (PPM) modulation are illustrated in Fig. 7. Note that PPM consists of equal width pulses derived from the trailing edge of PDM pulses. PPM has an advantage over PDM since the latter can require significant transmitter power for transmitting pulses of long duration.

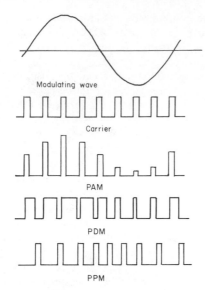

FIG. 7. PAM, PDM, and PPM using a sinusoidal modulating wave and a pulse carrier.

E. PULSE CODE MODULATION

In pulse amplitude modulation, the amplitude of the pulse can assume any value between zero and the maximum value. Pulse code modulation (PCM) is derived from PAM but is distinguished from the latter by two additional signal processing steps, called quantizing and encoding, that take place before the signal is transmitted. Quantizing replaces the exact amplitude of the samples with the nearest value from a limited set of specific amplitudes. The sample amplitude is then encoded, and the codes are transmitted typically as binary codes. This means that, unlike other modulation techniques described so far, in PCM both the sampling time and the amplitude are in discrete form.

Representing an exact sample amplitude by one of 2^n predetermined, discrete amplitudes introduces an error called quantization noise that can be made negligible by using a sufficiently large number of quantizing levels. Studies have shown that using 8 bits per sample to represent one of 256 quantizing levels provides a satisfactory signal-to-noise ratio for speech signals [see Rey (1983) in Bibliography]. The sampling rate is usually determined from the sampling theorem, which states that a baseband (information) signal of finite energy with no frequency components higher than W Hz is completely specified by the amplitudes of its samples taken at a rate of $2W$/sec. The corollary of the sampling theo-

rem states that a baseband analog channel can be used to transmit a train of independent pulses at a maximum rate that is twice the channel bandwidth W. These results are important in determining appropriate sample rates and bandwidths for conversion between analog and digital signals.

Applying the sampling theorem to speech signals that are limited to 4000 Hz, we find that they need to be sampled 8000 times/sec to be completely specified. Using PCM with 8 bits to represent one of 256 discrete amplitude samples, $8 \times 8,000$ or 64,000 bits/sec are required to transmit the 4000-Hz voice signal. If we now use the corollary to the sampling theorem, we find that a channel with a bandwidth of 32,000 Hz is required to transmit the 64,000 bits/sec needed to specify the 4000-Hz voice signal. Although it is true that PCM requires more bandwidth than the baseband analog signal (32,000 Hz bandwidth for the 4000-Hz voice signal in the above example), this is more than offset by the following facts:

1. PCM has very high immunity to noise.
2. PCM repeater design is relatively simple.
3. The PCM signal can be completely reconstructed at each repeater location by a process called regeneration.
4. PCM provides a uniform modulation technique suitable for other signals on many different types of media including wire, coaxial cable, microwave and optical fibers.
5. PCM is compatible with time division multiplexing.

F. COMPARISON OF MODULATION TECHNIQUES

The various modulation techniques that have been described have distinct advantages and disadvantages in terms of cost, immunity to noise, and other impairments commonly encountered in communication channels. Amplitude modulation is widely used for radio programs. It is easy to maintain and relatively low in cost, but it is susceptible to noise. Frequency modulation is more effective in terms of noise tolerance and more suited for data transmission than AM. Phase modulation is more complex and costly but is relatively immune to noise and theoretically makes the best use of bandwidth for a given transmission rate. Various forms of phase and hybrids of phase and amplitude modulation (QAM and TCM) are increasingly used for data communication over analog channels at rates higher than 2400 bits/sec. Although it requires higher bandwidth, PCM has the advantage that the signal is regenerative and has greatest immunity to noise. Optical fibers, with their very high bandwidths, are well suited for PCM, consequently, fibers and PCM are rapidly becoming the two leading technologies for transmission of digital data and digitized analog signals.

VII. Multiplexing

Multiplexing is a technique that allows a number of lower bandwidth communication channels to be combined and transmitted simultaneously over one higher bandwidth channel. At the receiving end, demultiplexing recovers the original lower bandwidth channels. The main purpose of multiplexing is to make efficient use of the full bandwidth of a communication channel and achieve a lower per channel cost. The three basic multiplexing methods in use are space-division multiplexing, frequency division multiplexing and time-division multiplexing.

A. SPACE–DIVISION MULTIPLIEXING

Space-division multiplexing refers to the physical grouping together of many individual channels or transmission paths to save physical space. A large number of wire pairs, coaxial cables, and/or optical fibers are usually grouped together to form a larger cable, such as the one illustrated in Fig. 11(a). Each wire pair, fiber or coaxial cable in the main cable is a communication channel giving a high total aggregate bandwidth. With each such individual channel in the cable capable of being frequency- or time-division multiplexed, such cables have enough bandwidth to carry more than 10,000 two-way voice channels in a cable diameter of under 3 in.

B. FREQUENCY-DIVISION MULTIPLEXING

As illustrated in Fig. 8, frequency-division multiplexing (FDM) divides the frequency spectrum of a higher-bandwidth channel into many individual smaller-bandwidth communication channels. Signals on these channels are transmitted at the same time but at different carrier frequencies. Guard bands are needed between the frequency channels to reduce interchannel interference.

Perhaps the most familiar example of FDM is radio broadcasting. Various stations occupy different frequencies in the radio spectrum. Program signals use amplitude or frequency modulation to modulate a carrier whose frequency is

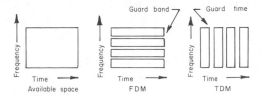

FIG. 8. Relationship between frequency and time in FDM and TDM.

allocated to that station. Tuning circuits in the radio receiver allow the signal from one station to be separated from the others. The signals are in the form of modulated electromagnetic waves using the atmosphere as the communication channel.

FDM for voice signals over telephone channels is very similar. At one end, a number of modulators with carrier frequencies differing by 4000 Hz modulate the various channels onto the higher bandwidth channel. At the receiving end, an equal number of demodulators tuned to the same frequency bands as the modulators, receive and demodulate the multiplexed signal into the corresponding channels. This is illustrated in Fig. 9, for one-way transmission for the sake of simplicity.

C. TIME-DIVISION MULTIPLEXING

Time-division multiplexing (TDM) operates by giving the entire channel bandwidth to a stream of characters or bits consisting of one bit from each of the low-speed channels for a small segment of time, as illustrated in Fig. 8. This allows many low-bandwidth channels to be accommodated on a high-bandwidth channel by interleaving them in the time domain. Guard times

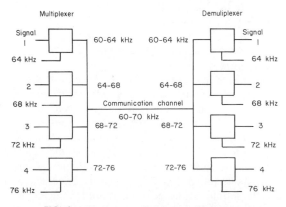

FIG. 9. Frequency-division multiplexing.

are used to separate time slices, and the transmitting and receiving ends must be synchronized. A familiar example of TDM might be the input–output bus of a computer servicing many peripherals for short periods of time, one at a time.

D. MULTIPLEX HIERARCHIES

As networks grow and get more complex, hierarchies of multiplexing are required whereby low-bandwidth channels are multiplexed onto high-bandwidth channels which in turn are multiplexed onto even higher-bandwidth channels, and so on. In the FDM hierarchy, multiplex levels correspond to increasingly higher-frequency bands. In the TDM hierarchy, multiplex levels correspond to increasingly higher pulse rates. For example, using PCM, each voice channel requires 64,000 bps. Twenty-four voice channels can be time-division multiplexed onto a Tl carrier operating at 1.544 Mbps, while four Tl carriers can be multiplexed onto a T2 carrier operating at 6.312 Mbps. A T4M carrier operates at 274.176 Mbps and can carry traffic from 42 T2 carriers or up to 4032 voice channels.

When hierarchies of multiplexing are used, it is important to have accurate carrier frequencies (pilot tones) for FDM and stable timing signals for TDM to synchronize transmitting and receiving stations.

VIII. Types of Data Transmission Media

Data transmission media provide the physical communication channel to interconnect nodes in a network. Media that constrain and guide or conduct communication signals are called bounded or guided media. Media that permit signals to be transmitted but do not guide them are called unbounded media. Bounded media include open wires or other conductors, twisted wire pairs, coaxial cables, waveguides, and optical fibers which are conductors of light. The atmosphere and outer space are examples of unbounded or unguided media for transmitting broadcast radio, television, terrestrial microwave radio, and satellite communications. Table II summarizes the various media, the main applications, the fundamental frequency bands used, and other relevant characteristics. Table II is summarized in Table III, for convenience.

It should be noted that multiplexing hierarchies described in Section VII,D imply hierarchies in the transmission media. A single intercontinental data message, for example, may use the following hierarchy of transmission media:

1. Twisted pair between source and the local telephone exchange
2. Microwave transmission between the local exchange and the central office which switches lines onto higher-capacity trunks and vice versa
3. Coaxial cable between the central office and the toll office which switches the continental trunk to an overseas trunk
4. Satellite transmission between toll offices on the two continents
5. Optical fiber from the intercontinental toll office to the overseas local exchange
6. Twisted pair between the overseas local exchange and the final destination

IX. Transmission Impairments

In practice, communication signals are degraded by physical limitations of transmission media such as bandwidth, impairments arising from within the channel such as echo, and impairments introduced from outside the channel such as impulse noise. The impact of such impairments depends on the type of signal carried and whether the channel is designed for analog or digital transmission. They can attenuate, amplify, or severely distort the signal as it passes through the communication channel and introduce errors. Impulse noise, which can be caused by electrical storms, for example, has little effect on the reception of analog speech signals because of amplitude limiters in the channel and the relative tolerance of the human ear to occasional errors. However, such noise can obliterate digital data signals.

The planning, design, installation, operation, and maintenance of communication channels for transmission of analog or digital signals depend upon an understanding of the impairments to which these channels are susceptible and the effect of these impairments on the signal. Some commonly encountered impairments are briefly described below. Other impairments related to specific transmission media will be described in later sections.

A. WHITE OR GAUSSIAN NOISE

This has a normal (Gaussian) amplitude distribution and causes the background hiss occasionally encountered over telephone channels.

TABLE II.   Transmission Media

Designation	Year in service	Operating frequency (MHz)	Modulation	1- or 2-Carrier channel			Per system			Repeater spacing (miles)	System length (miles)
				Operating band (MHz)	Voice capacity (ccts)	Binary capacity (mbps)	Carrier channels	Voice capacity (ccts)	Binary capacity (mbps)		
Open wire (2 wire, a pair of open wire)											
A(Bell)	1918		SSB/SC	0.020	4	—	16	64	—	—	—
C(Bell)	1925	0.005–0.030	SSB/SC	0.025	3	—	16	48	—	130/180	2000
J(Bell)	1938	0.036–0.143	SSB/SC	0.107	12	—	16	192	—	30/100	1500
CCITT	—	0.003–0.300	SSB/SC	0.300	28		—	—	—	—	1600
Paired cable (4 wire, two cable pairs)											
K(Bell)	1938	0.012–0.060	SSB/SC	0.048	12	0.048	48	288	1.152	17	1500
CCITT	—	0.012–0.552	SSB/SC	0.048	120	0.480	—	—	—	—	1600
Land coaxial cable (4 wire, two coaxial tubes)											
L1(Bell)	1941	0.068–2.78	SSB/SC	3.000	600	3.000	2 + 2	600	3.000	6/8	4000
CCITT	—	0.060–4.09	SSB/SC	4.000	960	4.000	—	—	—	6	1600
L3(Bell)	1953	0.312–8.28	SSB/SC	8.000	1860	8.000	6 + 2	5580	24.00	3/4	4000
L3(Bell)	1962	0.312–8.28	SSB/SC	8.000	1860	8.000	10 + 2	9300	40.000	3/4	4000
L3(Bell)	1965	0.312–8.28	SSB/SC	8.000	1860	8.000	1 + 2	16740	72.000	3/4	4000
CCITT	—	0.300–12.43	SSB/SC	12.000	2700	12.000	—	—	—	3	1600
L4(Bell)	1967	0.564–17.55	SSB/SC	18.000	3600	18.000	18 + 2	32400	182.000	2	4000
L5(Bell)	1974	3.12–60.56	SSB/SC	60.000	10800	60.000	18 + 2	97200	540.000	1	4000
Submarine coaxial cable (2 wire, one coaxial tube, except SB which is 4 wire with 2 tubes)											
SB(TAT 1)	1956	0.024–0.168	SSB/SC	0.160	48	0.160	2	48	0.160	48	2500
SD(TAT 3)	1963	0.108–1.05	SSB/SC	1.100	128	0.550	1	128	0.550	24	4000
SF(TAT 5)	1970	0.564–5.88	SSB/SC	5.900	720	3.000	1	720	3.000	12	4000
SG(TAT 6)	1976	—	SSB/SC	30.000	4000	15.000	1	4000	15.000	6	4000
Microwave radio (4 wire, two radio-frequency carrier channels)											
TDX(Bell)	1947	3700–4200	FDM/FM	10.000	60	0.240	8 + 0	240	0.960	30	4000
TD2(Bell)	1950	3700–4200	FDM/FM	20.000	600	2.400	10 + 2	3000	12.000	30	4000
TH(Bell)	1961	5925–6425	FDM/FM	30.000	1860	7.500	12 + 2	11160	45.000	30	4000
TD3(Bell)	1967	3700–4200	FDM/FM	20.000	1200	4.800	20 + 4	12000	48.000	30	4000
Troposcatter radio (4 wire, two radio-frequency carrier channels)											
BMEW	1961	400–2000	FDM/FM	10.000	240	0.960	2	240	0.960	100	900
EMT	1965	400–2000	FDM/FM	6.000	120	0.480	2	120	0.480	200	3200
DEW EAST	1962	400–2000	FDM/FM	3.000	72	0.288	2	72	0.288	300	2400
UK SPAIN	1962	400–2000	FDM/FM	1.000	24	0.096	2	24	0.096	400	1200
Communication satellites (4 wire, two radio-frequency carrier channels)											
INTEL SAT I	1965	3700–4200 5925–6425	FDM/FM	25.000	240	0.960	2 + 2	240	0.960	N/A	8000
INTEL SAT II	1966	3700–4200 5925–6425	FDM/FM	130.000	240	0.960	2 + 2	240	0.960	N/A	8000
INTEL SAT III	1968	3700–4200 5925–6425	FDM/FM	225.000	1200	4.800	2 + 2	1200	4.800	N/A	8000
INTEL ST IV	1970	3700–4200 5925–6425	FDM/FM	36.000	300	1.200	24	3600	14.400	N/A	8000
INTEL SAT IV A	1975	3700–4200 5925–6425	FDM/FM	36.000	300	1.200	40	6000	24.000	N/A	8000

B. Impulse Noise

This usually consists of a short duration (less than 100–200 milliseconds), high amplitude burst of noise energy that is much greater than normal peaks of message circuit noise. The short duration does not unduly impair analog speech signals but can obliterate digital data signals. For this reason, data signals are usually transmitted in blocks containing error detecting codes so that when errors are detected, the data can be retransmitted. Because of its relative short duration, impulse noise results in greater error rates at higher data rates since at higher rates it becomes increasingly difficult to distin-

guish between a data and noise pulse. Impulse noise is usually caused by lightning storms and voltage transients by electromechanical switching systems. The objective for telephone lines is to have no more than 15 impulse counts in 15 minutes for 50% of all telephone communications.

C. Cross-talk Noise

Cross-talk is usually caused by capacitive or inductive coupling between adjacent channels. It can be a problem in high-speed digital circuits. If it is intelligible on analog channels (i.e., a con-

TABLE III. Commonly Used Transmission Media

Medium used	Main application	Frequency range	Relative use
Open conductor	Short distances	Varies with distance	Very high
Twisted wire pair	Up to 50 miles	Varies with distance	Very high
Coaxial cable	Short or long haul	3–300 MHz	High
Waveguide	Short distances	3–12 GHz	Medium
	Long haul	16 Gbits/sec	Experimental
Optical fiber	Short or long haul	Up to 16 Gbits/sec	Increasing
Microwave radio	Short haul	10.7–11.7 GHz	Medium
	Long haul	3.7–6.425 GHz	High
Satellite	Long haul	5.925–6.425 GHz up	Medium
		3.7–4.2 GHz down	
		14.0–14.5 GHz up	Increasing
		11.7–12.2 GHz down	
Broadcast radio	Short to long	Varies with distance	Very high

versation on an adjacent circuit can be over-heard), it becomes particularly objectionable, not only because it impairs the conversation but also because it creates a loss of privacy. Methods to control cross-talk include shielding conductors, separating tightly coupled circuits, impedance matching of lines, and suppressing nonlinearities. Telephone companies have a goal of limiting the cross-talk index, which is the actual percentage probability of receiving an audible, intelligible speech signal on a call, to under 0.5% on most of their channels.

D. Quantizing Noise

This is the noise caused by errors introduced when quantizing an analog signal into one of 2^n discrete amplitudes. It can be controlled by using a code with a suitably large value of n (e.g., eight for voice signals). A codec is a coder–decoder that is used in PCM systems for analog-to-digital and digital-to-analog conversion.

E. Impedance Mismatch

If long wires, wire pairs, and coaxial cables are not terminated in their characteristic impedance, reflections can take place, which can cause high standing-wave ratios that appear as noise in the system. This can be overcome by terminating long lines by a load impedance that is equal to or very close to the characteristic impedance of the line.

F. Attenuation Distortion

Attention distortion (sometimes called amplitude/frequency distortion) occurs when the relative magnitude of different frequency components of a signal are altered during transmission over a channel. A transmission medium that is ideal has a flat frequency response for the range of frequencies over which the spectrum of the information signal is nonzero. The distortion is caused by capacitive and inductive reactances in series and in parallel with the electrical conductors comprising the channel. Attenuation distortion is much less a problem in voice transmission than in data transmission because of the redundant nature of speech signals as opposed to the small amount of redundancy in data signals, where the loss or alteration of one bit usually alters the meaning of the code word in which the bit is contained.

G. Envelope Delay Distortion

A signal that carries information has a phase component in addition to amplitude and frequency. To obtain an undistorted signal, it is required that the transmission medium have not only a flat amplitude-versus-frequency characteristic but also a linear phase-versus-frequency characteristic. As in the case of attenuation distortion, real transmission media do not exhibit ideal phase response because of distributed stray inductance, capacitance, resistance and, conductance which affect the time or phase relationship between various frequency components of a transmitted signal. This is known as envelope delay, or group distortion, and causes various frequency components of a complex signal to propagate through the transmission medium at different velocities, causing the composite signal at the receiver to be distorted. If the distortion is sufficiently large, late-arriving energy from one pulse can interfere with the start of the next pulse, resulting in a phenomenon known as intersymbol interference. Equalizers are devices that attempt to equalize the envelope delay and compensate for the different components that make up amplitude distortion.

H. Frequency Translation

This is a phenomenon that results in a frequency shift whereby all frequency components in the modulated signal are shifted by a constant amount. It is generally due to oscillator drift or frequency offset in the carrier wave equipment. Frequency translation can be especially troublesome in systems that transmit FDM signals. This type of system relies heavily on channel filters in the modulation–demodulation process. Since frequency translation causes a shift in the energy spectrum for which the system was designed, the spectrum shift causes some of the desired information-signal energy to encounter the undesirable amplitude and phase distortion characteristics that are usually found at the band edges of filters. If the frequency error is sufficiently large, the system is seriously degraded, especially for high-rate data transmission. This is because many modems use carriers that are synchronized between the receiver and transmitter, and frequency errors in the media appear as errors in the modems. On telephone channels, the objective is to hold the frequency shift to ±2 Hz for acceptable performance.

I. Phase Jitter

Phase jitter refers to small changes in the phase of the received signal and appears as a slight frequency modulation on the carrier or baseband signal. It often takes the form of multiples or submultiples of ac power frequencies. It is usually caused by reactive coupling between equipment associated with power lines or jitters in signals used for timing (clocks). Phase jitter is of little consequence in voice transmission since the human ear is not sensitive to small changes in phase or frequency. However, phase jitter can seriously deteriorate data transfer since it

causes the data pulses to jitter, and if large variations occur, then one data pulse may try to occupy the time slot of another and cause an error, especially in TDM systems. In addition, phase jitter may make synchronization of the transmitter and receiver more difficult.

J. Harmonic Distortion

Harmonic distortion is usually caused by clipping or limiting the transmitted signal, which causes higher harmonics of the transmitted signal at the receiver. In most cases this is not a serious problem.

K. Echo

Talker echo is a telephone line impairment that is caused by reflections when long lines are not terminated in their characteristic impedance. It causes part of the speaker signal to be reflected back and can seriously interfere with the talker's speaking process. If the elapsed time (i.e., the delay) is long, and the echo path loss is inadequate (i.e., there is not enough loss to attenuate the echo sufficiently), the interference may be so great that speaking is nearly impossible. Impedance mismatches are usually caused by hybrids in the signal path. Hybrids are circuits used for converting a two-wire circuit to a four-wire circuit and vice versa. Echoes can be controlled using echo suppressors, which detect echoes and automatically insert a loss over the listener's speaking path to prevent the echo from reaching the talker. When both speak at the same time, the suppressor inserts a loss in both directions, resulting in undesirable clipping of speech signals. An echo canceler, rather than inserting a loss in the return path, uses the transmitted speech signal to generate a replica of the echo and subtracts this signal from the echo, thus cancelling it. Using echo suppressors and cancellers reduces the echo to tolerable levels on 99% of all telephone connections in which this problem is encountered [Rey (1983) in Bibliography].

L. Inductive Interference

Because power and telephone companies tend to serve the same customers and share the same right of way, telephone wires are exposed to electromagnetic fields created by power distribution systems. If the telephone lines are not perfectly balanced, a voltage difference between the two conductors of a cable pair may develop in the presence of a strong electromagnetic field that can result in an audible noise level. This type of inductive interference is sometimes known as electromagnetic interference or EMI for short. Telephone sets and transmission media are designed to have high tolerance to 60-Hz hum, but higher harmonics can cause problems. This is minimized by balancing lines, grounding cable shields, and separating power and telephone media where necessary. As a last resort, magnetic core devices are used to minimize this type of noise.

M. MESSAGE CIRCUIT NOISE

This is a signal composed of noise from a number of sources such as power hum, inductive interference, Gaussian noise, central office noise, and impulse noise, which have been previously discussed. It causes impairments similar to those caused by its various component parts.

N. GROUND NOISE

This is caused by voltage differences between the receiving and transmitting ends due to current flow through the ground return path. Transmission media consisting of single wire over ground, as shown in Fig. 10(a), are especially susceptible to this type of noise.

X. Bounded Transmission Media

Bounded transmission media constrain and guide communication signals. They include single wire (or single conductor), paired cable, coaxial cable, optical fibers, and waveguides. The first three media are electrical conductors, optical fibers are conductors of light, and waveguides constrain and guide electromagnetic waves.

A. SINGLE WIRE LINES

This medium uses a single wire or some other electrical conductor to provide a path for an electric current with electrical ground providing the return path, as shown in Fig. 10(a). Since this is the medium used almost exclusively to interconnect components in all types of electronic devices over short distances (most connections between components on a circuit board or even on a silicon chip are examples of a single wire over ground transmission medium), a substantial portion of Section XII will be devoted to this medium, with emphasis on its proper use in circuits where high-bandwidths are involved.

B. PAIRED CABLE

Because of ground noise problems associated with a single conductor when transmitting signals over longer distances, two-conductor paired cable is used, with the second cable providing the return path for the signal current. The two cables can be parallel to each other (open wire pair) or twisted (twisted pair), as shown in Fig. 10(b) and 10(c). Open wires are susceptible to cross-talk and electromagnetic interference from power lines. As explained in Section XII, twisted pairs reduce the distance between the conductors and reduce the inductive cross-talk by reducing the area of the current loop. Consequently, twisted pairs have replaced open pairs for telephone transmission except in some rural areas. In many cases, space-division multiplexing is used by combining many twisted pairs into a larger cable. Twisted pairs are stranded into a ropelike form called a binder group, and several binder groups are twisted together around a common axis to form the cable core. A protective sheath is wrapped around the core, resulting in a cable like that in Fig. 11(a). Paired cable is

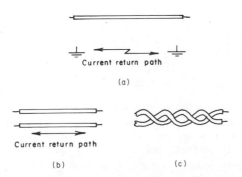

FIG. 10.   Single and paired cables. (a) Single cable over ground; (b) open pair; (c) twisted pair.

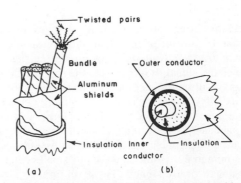

FIG. 11.   (a) Multipair cable and (b) coaxial cable.

made in a number of standard sizes and may contain from 2 to 3600 wire pairs.

Paired cable is the main type of medium for local telephone and data transmission. It can be installed easily using commonly available tools. Today, it is typically used for relatively short distances (normally less than 10 miles). It has bandwidth limitations and is susceptible to noise. As frequency or bit rates of transmission increase, the repeaters needed to amplify or regenerate the signals must be placed closer together to overcome the increased effect of the transmission impairments. With proper design, twisted pair cables can carry about 12 voice channels per pair with repeaters every 2–4 miles.

C. COAXIAL CABLE

Coaxial and twinaxial (two coaxial cables in a single jacket) cables are used for high-frequency analog or digital transmission requiring wide bandwidth. As shown in Fig. 11b, a coaxial cable consists of an inner conductor completely surrounded by an outer conductor, with high-quality insulation between the two. The outer conductor is surrounded by a protective sheath. As with twisted pairs, coaxial cable can be bundled together with twisted pairs to form a larger cable with very high total bandwidth.

Coaxial cables have various characteristic impedances, with 50, 75, and 93 ohms being the most common. They have the advantage of operating at very high frequencies, which allows them to carry a very large number of analog or digital channels using FDM. Since the outer conductor is grounded and provides an effective shield that improves with increasing frequencies, the cross-talk between adjacent coaxial cables decreases with frequency, rather than increasing as is the case with cable pairs.

In the telephone network, coaxial cable is used primarily on intercity routes for the long haul network where heavy traffic exists. The Bell L5 coaxial cable, for example, has a capacity of 10,800 voice circuits per carrier channel with repeaters every mile. In these applications, the installed cost per voice circuit is lower than for paired cable. Intercontinental submarine cables also use coaxial cables, but in this environment the design, operational, and reliability requirements are different from cables used on land.

Coaxial cables are also used extensively in community antenna TV (CATV) or cable TV systems for distribution of TV and music programs.

More recently, coaxial cables are being used for local area networks within or between buildings. Such networks interconnect a large number of terminals, workstations, and computers using baseband transmission with TDM or broadband transmission with FDM. They have a low incidence of errors, support high data rates, can be tapped into easily, and devices such as taps, controllers, splitters, couplers, multiplexers, and repeaters are readily available.

D. WAVEGUIDES

A waveguide is a rectangular or circular pipe, usually made of copper, that confines and guides very high frequency radio waves between two locations. Compared with coaxial cable, it has a very low attenuation at microwave frequencies, which is its main advantage. Its main disadvantages are that it must be manufactured with extreme uniformity to achieve the low attenuation, and great care must be taken during installation to minimize sharp bends which also increase attenuation. These disadvantages have limited the application of waveguides to carry 4-, 6-, and 11-GHz range signals from the base of microwave radio towers to the dish antennas at the top. More recently, waveguides are being used in an experiment to help determine feasibility of their use as a viable medium for very high capacity transmission of 16 Gbits/sec over long distances. A circular waveguide is used whose diameter is chosen to have a considerably lower attenuation per unit length than a rectangular waveguide at the above transmission rate.

E. OPTICAL FIBERS

Optical fibers hold great promise as a transmission medium capable of carrying up to several gigabits of information per second over short or long distances. The basis of fiber-optic, or lightguide, transmission systems is an optical fiber, which is a thin, flexible glass or plastic fiber through which light is transmitted. An optical fiber is actually a waveguide that guides the propagation of optical frequency waves through total internal reflection at the fiber boundaries. A fiber-optic transmission system is shown in Fig. 12 and is similar to a conventional transmission system, except that the transmitter uses a light emitting diode (LED) or a laser diode to change electrical signals into light signals while

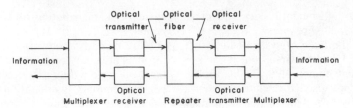

FIG. 12.   Typical optical fiber communication system.

the receiver uses a photodiode to convert the light signals back into electrical signals.

Compared with wire, twisted pairs, or coaxial cable, optical fibers offer advantages so great in terms of information-carrying capacity and signal protection that they will start replacing many of these older media over the next decade. These advantages include the following:

1.   Large bandwidth. Optical fibers have bandwidths of several GHz compared with an upper limit of 500 MHz for coaxial cable. These bandwidths are achievable because the information-carrying capacity of a transmission medium increases with the carrier frequency, and for optical fibers the carrier is light which has frequencies several orders of magnitude higher than the highest radio frequencies. With currently available systems, a 144-fiber cable can accommodate 250,000 voice frequency circuits, and this will increase rapidly with the introduction of new optical fiber technology.

2.   Low losses. As shown in Fig. 13, the attenuation of an optical fiber is essentially independent of the modulation frequency whereas copper wires and coaxial cables exhibit increasing attenuation with increased frequency. The advantage of the flat attenuation response of low-loss optical fibers is that repeaters need not be placed so close together. While a 44.7 Mbit/sec. T3 carrier in a digital telephone system has a standard repeater spacing of 1.1 km for coaxial cable and 6.4 km for existing optical fibers, optical systems with repeater spacing of more than 20 km are being planned using new, low-loss fibers. This means that interoffice trunks will require no repeaters since the distances involved are usually shorter than the above repeater spacing for new fibers.

3.   Electromagnetic interference (EMI) immunity. Because optical fibers are made of materials like glass and plastic which are insulators, they are not affected by ordinary electromagnetic fields. With conductors such as copper wire, shielding is required in many cases to reduce the effects of stray magnetic fields. The EMI immunity of optical fibers also offers the potential for signal protection through low error rates of 1 per 10^9 bits or even better. This has the potential for reducing the cost and overhead associated with the complex error checking and error correction mechanisms that have to be used with other transmission media.

4.   Small size and light weight. A fiber with a

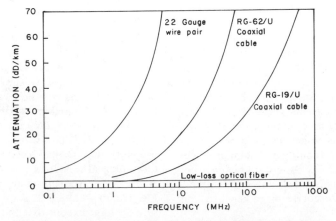

FIG. 13.   Attenuation as a function of frequency for some commonly used media.

0.125-mm diameter and a 3.5-mm protective jacket has the same information-carrying capacity and can replace a cable consisting of 900 twisted copper wire pairs with an 8-cm outer diameter. The smaller size for a given capacity and the fact that glass and plastic weigh less than copper mean that fibers are lighter by a factor of 10 to 100 than copper cables of equivalent capacity. Both size and weight are important factors when considering overcrowed conduits running under city streets.

5.   Security. Unlike electrical conductors, optical fibers do not radiate energy; consequently, eavesdropping techniques that can be used on the former are useless with fibers. Furthermore, it is virtually impossible to tap an optical fiber since, in general, this affects the light transmission enough to be detected. Since data are increasingly perceived as a valuable asset, the security offered by fibers is important because it may reduce data encryption costs.

6.   Safety and electrical isolation. Because fiber is an insulator, it provides electrical isolation between data source and destination. This avoids ground noise often encountered when systems are interconnected by electrical conductors. Furthermore, fibers present no electrical spark hazards and can be used in applications where electrical codes and common sense prohibit the use of electrical conductors.

Despite their many important advantages, optical fibers have several disadvantages. They are a relatively new technology and not yet understood well enough by designers, who are comfortable designing with electrical conductors and who base their solutions to problems on what they know best. This situation should improve as industry-wide standards related to this new technology continue to emerge; the costs of light sources, detectors, fibers, and other related hardware fall with increased use, and as technicians are trained to handle, install, and maintain optical fiber systems. Since it has been established that this technology can outperform cable as a transmission medium in many situations, the above disadvantages will be overcome, and the next decade should see a great increase in its use.

1. Types of Fibers

Optical fibers are made of plastic, glass, or silica. Plastic fibers are the least efficient but tend to be larger, more economical, and more rugged. Glass or silica fibers are much smaller,

and their lower attenuation loss makes them more suited to high-capacity channels. A third type of fiber, with size and performance intermediate to the other two types, has a glass or silica core with a plastic cladding.

The basic optical fiber consists of two concentric layers, the inner core and the outer cladding which has a refractive index lower than the core. The construction is such that light injected into the core always strikes the core-to-cladding interface at an angle greater than the critical angle and is therefore reflected back into the core. Since the angles of incidence and reflection are equal, the light wave continues to zigzag down the length of the core by total internal reflection, as shown in Fig. 14a. In effect, the light is trapped in the core, and the cladding not only provides protection to the core but also may be thought of as the "insulation" that prevents the light from escaping. The exact characteristics of light propagation depend on fiber size, construction, and composition, as well as on the nature of the light source. Although fiber performance can be reasonably approximated by considering light as rays, more exact analysis must deal in field theory and solutions to Maxwell's equations, which show that light does not travel randomly through a fiber but is channeled into modes that represent allowed solutions to the field equations. Consequently, it is more informative to classify fibers by refractive index profiles and the number of modes supported.

The two main types of refractive index profiles are step and graded. In a step index fiber, the core has a uniform refractive index n_1 with a

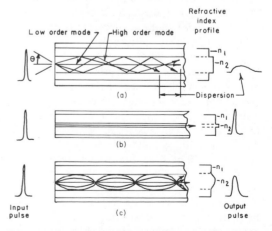

FIG. 14.   Characteristics of common optical fibers. (a) Multimode step index; (b) single mode step index; (c) multimode graded index.

distinct change or step to a lower index n_2 for the cladding. In a graded index fiber, the refractive index of the core is not uniform but is highest at the center and decreases until it matches that of the cladding. The index profiles and the light wave propagation supported by these different fibers are illustrated in Fig. 14.

The simplest type of fiber is the multimode step index fiber, which usually has a core diameter from 0.05 to 1.0 mm. The relatively large core allows many modes of light propagation; and since some rays follow longer paths than others, their original relationship is not preserved. The result is that a narrow pulse of light has a tendency to spread as it travels down the fiber, as shown in Fig. 14a. This spreading of light pulses is known as modal dispersion and is 15–40 nsec/km for typical multimode step fibers. Consequently, such fibers tend to be used for transmission over short to medium distances.

Modal dispersion can be reduced or eliminated by using a fiber with a core diameter small enough that the fiber propagates efficiently only the lowest-order mode along its axis, as shown in Fig. 14b. Single mode step index fibers typically have core diameters between 0.002 and 0.01 mm and are very efficient for very high-speed, long-distance transmission. Their small size, however, makes them relatively difficult to work with, especially when fibers must be linked.

Modal dispersion can also be reduced by using graded index fibers, as shown in Fig. 14c. Their core is a series of concentric rings, each with a lower refractive index as we move from the center of the core to the cladding boundary. Since light travels faster in a medium of lower refractive index, light further from the fiber axis travels faster. This means that in a graded index fiber, high-order modes have a faster average velocity than lower-order modes; therefore all modes tend to arrive at any point at nearly the same time. Light is refracted successively by the different layers of the core, with the result that the path of travel appears to be nearly sinusoidal. In such fibers, modal dispersion is typically well under 10 nsec/km and sometimes even less than 1 nsec.

As in the case of electrical conductors, optical fibers are usually cabled by enclosing many fibers in a protective sheath, which is usually made of some plastic material such as polyvinyl chloride or polyurethane. Because the fibers are so thin, the cable is usually strengthened by add-

ing steel wire or, more commonly, Kevlar aramid yarn with a high Young's modulus to give the cable assembly flexibility and tensile strength. Cables are also available containing both optical fibers and electrical conductors, with the latter usually being used to provide power for remotely located repeaters.

2. Signal Degradation

Signal degradation in optical fiber systems is caused by one or more of the following:

(a) Attenuation. Attenuation, or transmission loss (dimming of light intensity) is caused by absorption and scattering. Absorption is the optical equivalent to electrical conductor resistance and is usually caused by fiber impurities that absorb light energy and turn it into heat. The amount of absorption by these impurities depends on their concentration and the light wavelength used. High-absorption regions, such as that occurring in the 950-nm wavelength range due to hydroxyl ion (OH^-) impurities, should be avoided, for example. Scattering results from imperfections in fibers. Rayleigh scattering comes from the atomic and molecular structure of the core and from density and composition variations that are introduced during manufacturing. Unintentional variations in core diameter, microscopic bends in the fiber, and small discontinuities in the core-to-cladding interface also cause scattering loss. For commercially available fibers, attenuation ranges from about 1 db/km (decibel per kilometer) for premium small-core fibers to over 2000 db/km for large-core plastic fibers. Since attenuation losses vary with the light wavelength used, it is important for optical performance to match carefully the characteristics of the light source transmitter and the fiber.

(b) Dispersion. Dispersion, a measure of the widening of a light pulse as it travels along the fiber, is usually expressed in nsec/km. Dispersion limits the bandwidth or information-carrying capacity of a fiber, since input pulses must be separated enough in time that dispersion does not cause adjacent pulses to overlap at the destination to ensure that the receiver can distinguish them. Dispersion can be considered as the optical analog of envelope delay distortion in electrical conductors. Dispersion can be of two types. Modal dispersion arises from the different length of paths traveled by the different modes. Material dispersion is due to different velocities of different wavelengths. In single mode° fibers, which exhibit no modal dispersion, material dispersion is the sole frequency-limiting mechanism. High-quality optical fibers have a dispersion of about 1 nsec/km or even less.

(c) Other losses. Interconnection is a critical part of optical fiber communication links. Fibers must be connected or spliced to provide low-loss coupling through the junction. Precise alignment results in low loss, but the small size of fiber cores together with dimensional variations in core diameter and alignment mechanisms make this a formidable task. Factors affecting these losses include:

(i)   Differences in core diameters of fibers to be joined

(ii)   Differences in the numerical aperture of fibers to be joined. The numerical aperture (NA) is a measure of the light-gathering capability of a fiber and is equal to $\sin \theta$ where θ is the half-angle of the cone within which all incident light is totally internally reflected by the fiber core, as shown in Fig. 14(a).

(iii)   Core-to-cladding eccentricity, which can cause misalignment

(iv)   Core ellipticity, which can reduce the contact area

(v)   Lateral misalignment of the axes of the two cores to be joined

(vi)   Angular misalignment of the two cores

(vii)   Fiber-end separation

(viii)   Tilting angle between the two cores if the fiber is not cut on an axis that is perfectly perpendicular to the core axis

A good splice, which refers to a permanent interconnection, has a typical loss of about 0.1 db. A connection refers to detachable interconnections, which in practice have less precise alignments, resulting in a loss of up to 4 db using inexpensive connectors for large fibers.

3. Light Sources and Detectors

In an optical fiber communication system, the light source must efficiently convert electrical energy (current and voltage) into optical energy in the form of light. A good source must be

(a)   small and bright to permit the maximum transfer of light into the core of the fiber,

(b)   fast to respond to rapidly changing modulating signals encountered in high-bandwidth data systems,

(c)   Monochromatic (i.e., produce light within narrow band of wavelengths) to limit dispersion within the fiber, and

(d)   Reliable, with a lifetime in the tens of thousands or preferably in the hundreds of thousands of hours of operation.

The most commonly used light sources are gallium arsenide light emitting diodes (LEDs) and injection laser diodes (ILDs). Both devices are small, with sizes compatible with the cores of fibers, and emit light wavelengths in the range of 800 to 900 nm, where fibers tend to have relatively low loss and dispersion. LEDs are not monochromatic in that they emit light with a relatively broad spectrum of 40 to 50 nm, which translates to an upper bit-rate capacity of 200 Mbits/sec, with 50 Mbits/sec being a more conservative figure. With present technology, LEDs can transmit about 100 μW of optical power into the core of a fiber and have typical lifetimes of 100,000 hr. They have a light output power characteristic that is almost linear over a large range of driving current and hence can transmit analog signals using amplitude modulation. ILDs produce semimonochromatic light with a spectrum 1–2 nm wide and can transfer up to 10 mW into the core of a fiber, or about two orders of magnitude more than LEDs. Their light power output characteristic is linear over a limited region and hence they are more suited for digital applications. This combination of characteristics allows ILDs to be driven at Gbit/sec rates. Unfortunately, they are not quite as reliable as LEDs, and their characteristics vary with temperature, resulting in more complex driver circuitry.

Optical detectors convert optical energy into electrical energy. Devices most commonly used for this purpose include PIN and avalanche photodiodes (APDs). A PIN diode is a specially made diode with intrinsic material between the P and N materials to give a faster response time to incident light energy. The photodiodes are usually made of silicon because of their sensitivity in the 750–950 nm wavelength region. Silicon PIN diodes convert light power input to electrical current output with a quantum efficiency of over 90%, a response time on the order of 1 μsec, and low noise. APDs have much faster response times than PIN diodes and can be used at higher bit rates, but they display temperature instabilities that require more complex detector circuitry to compensate. They also require high voltages (100 and sometimes even 200 V), which also tends to complicate repeater and receiver design.

4. Modulation in Optical Fiber Systems

Optical fiber communication systems usually use a form of amplitude modulation called intensity modulation in which the intensity of the light carrier is changed by the modulating signal. Both LED and ILD light sources can be conveniently modulated in the intensity mode. The PIN diode as well as the APD responds to intensity modulation by producing a photocurrent proportional to the incident light intensity. Although the light output power of LEDs and ILDs is relatively linear over a wide range of drive currents, indicating they are suitable for continuous-wave modulation, today's optical fiber communication systems are more suitable to digital rather than analog operation. In the digital mode, the light source is switched on and off, which greatly simplifies the detection process at the destination.

The selection of line signaling format is an important consideration in optical communication systems. It is desirable to use a format that is self-clocking at the receiver and conserves output power at the source, especially when using ILD sources since they should be driven to high power output for only short intervals to improve their lifetimes. Some of the popular digital signaling formats are illustrated in Fig. 15. They include the following:

(a) Nonreturn to zero (NRZ) in which a change of state occurs only if there is a 1-to-0 or a 0-to-1 transition. A string of 1s is a continuous "on" condition while a string of 0s is a continuous "off" condition.

(b) Return to zero (RZ) format in which there is a complete pulse transmitted for each logic 1 condition. Note that the pulse width must be less than the bit interval to permit the return to zero condition.

(c) Bipolar RZ format in which a pulse is transmitted for each logical 1 and 0 condition. Again the pulse width must be less than the bit interval to permit the return to zero condition.

(d) Manchester code format in which, by convention, a logic 0 is defined as a positive-going transition while a logic 1 is defined as a negative-going transition occurring at times other than bit time boundaries.

For optical systems, the RZ and Manchester code formats are good candidates because they are self-clocking and because the relatively short pulses, even for a continuous string of logical 1s, conserve the life of ILDs. Note, however, that these formats require at least twice the bandwidth of the NRZ format for a given bit rate. Pulse modulation techniques such as PDM and PPM are also sometimes used.

F. OUTLOOK FOR BOUNDED MEDIA

Wire and its variants, such as conductors over ground, open wire, and twisted pairs, provide the universal method of connecting components within equipment and of interconnecting equipment to make up systems. Of all transmission media, they are the most pervasive and will remain so since they are by far the least costly, especially over short distances. Compared with other transmission media, they are also by far the most limited in terms of bandwidth, distances spanned, and susceptibility to interference. Because they radiate electromagnetic energy, they can interfere with other equipment and should not be used in an environment where data security is important. Although they will be increasingly replaced by other media for high transmission rates and distances greater than a kew kilometers, they will remain the most pervasive medium for use over short distances because of their simplicity and low cost.

Coaxial cable has made inroads in telephone transmission, television, distribution (CATV), and short run system interconnects at relatively high frequencies. It will continue to be widely used for the latter two applications, but it will be increasingly replaced by optical fibers for transmission over longer distances. However, over the next decade, local area networks based on coaxial cable will be increasingly common to interconnect terminals, workstations, and computers at bit rates up to 50 Mbits/sec. Such local

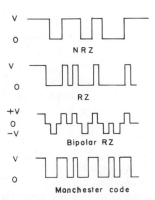

FIG. 15.   Examples of binary signal formats for optical fibers.

area networks could range from tens to hundreds of meters for intrapremises distribution and up to 3 km for interbuilding connections within a campus or industrial complex.

Even though major improvements have been made over the past few years, optical fibers are still an immature technology when compared with wires and coaxial cable. Nevertheless, major improvements in light sources, detectors, and splicing technology, and the emergence of standards and reduction in optical component costs, are expected greatly to increase the use of this transmission medium. The dominant application areas are expected to include telephone trunk lines, local area and local distribution networks, system interconnects requiring very high bandwidths, and in high-interference environments where electrical conductors would be too costly to shield. In the more distant future, optical and electro-optical technologies may be used to solve interconnection problems in high-speed digital systems, even for very short distances [Goodman et al. (1984)]. Today, electrical conductors are used for this purpose almost exclusively.

XI. Unbounded Transmission

Bounded media require a physical connection between two points to conduct or guide current, electromagnetic waves, or light. In unbounded media no such physical connection is required. Space or air is the transmission medium for electromagnetic waves, which are therefore unbounded by the medium. Today, this medium is used almost exclusively for some forms of radio transmission, which include broadcast, microwave, and satellite transmission. In fact, it is estimated that 60% of today's long-distance communication transmission is provided by some form of radio transmission, as measured in circuit miles. In industrialized nations, the probability is very high that radio is involved in some portion of a communication circuit that goes beyond the local level. The major advantages of radio transmission include the following:

1. No physical path is needed between the source and destination. This is extremely important since providing a voice line or cable between two points may be economically infeasible or physically impossible, as in the case of communication with a mobile terminal.
2. The source (transmitter) and destination (receiver) can both be mobile.

3. Transmission can be point-to-point or point to many points simultaneously (multiaccess) as in broadcast radio.
4. Broad spectrum from low to high bandwidth is available.
5. Can be quickly implemented and no right of way is required for the transmission path.

Radio transmission also has some disadvantages. It is subject to interference and a number of propagation anomalies that may require considerable engineering effort to overcome. It also requires space in the crowded radio transmission spectrum and therefore requires licensing and design approval from regulatory agencies.

A. RADIO FREQUENCY SPECTRUM

In radio transmission, electromagnetic radio wave power from a transmitter is coupled by the transmitter antenna into air or free space. In radio reception, electromagnetic radio waves are intercepted by a receiving antenna and coupled into a receiver for detection. The antennas can be omnidirectional as in broadcast radio or highly directional as in microwave radio transmission. With an omnidirectional antenna, a receiver can receive signals (as well as noise and interference) from other transmitters located anywhere within receiving distance. With directional transmitting and receiving antennas, signals are transmitted in a very narrow angular sector (e.g., 1° for microwave transmission), and similarly, a receiver can only detect signals emanating from a narrow sector. If the transmitting and receiving antennas are aimed at each other, good communication can be established with relatively low transmitted power and little interference to other radio links, which is a key point in conservation of the limited radio frequency spectrum shown in Table IV. Such point-to-point radio communication means that frequencies can be reused in the same general vicinity without interference. This should be contrasted with omnidirectional broadcast systems where frequencies can be reused only if transmitters using the same frequencies are located sufficiently far apart so as not to cause mutual interference at local receivers. Table IV shows the radio frequency bands in current use and their typical uses, characteristics, and principal modulation methods.

B. SIGNAL DEGRADATION

Radio waves, like light waves, are subject to reflection and refraction. They are also subject

TABLE IV.   Radio Frequency Spectrum

Frequency	Name	Characteristics	Principal modulation methods		Typical users
			Analog	Digital	
3–30 KHz	Very low frequency (VLF)	Very noisy, very limited bandwidth, very large	Seldom used	ASK, FSK, PSK	Sonar, long-range navigation
30–300 KHz	Low frequency (LF)	As for VLF	Seldom used	ASK, FSK, PSK	Navigationál aids, radio beacons
300–3000 KHz	Medium frequency (MF)	Noisy, limited bandwidth, large transmitting antenna	AM	ASK, FSK, PSK	Commercial AM radio, maritime radiotelephone, distress calls
3–30 MHz	High frequency (HF)	Noisy, long distance but subject to fading, subject to interference, moderate antenna size	AM	ASK, FSK, PSK	Shortwave radio, CB radio, ship to coast, point to point
30–300 MHz	Very high frequency (VHF)	Line-of-sight range, moderate noise and interference, small antenna size	AM, FM	FSK, PSK, others	Télevision, FM radio, land and air mobile traffic, police
300–3000 MHz	Ultrahigh frequency (UHF)	Line of sight, high bandwidth, low noise, high congestion in some areas	FM	FSK, PSK, others	UHF television, radar, space telemetry, microwave links
3–30 GHz	Superhigh frequency (SHF)	Line-of-sight range, bandwidth to 500 MHz, low noise, narrow antenna beams	FM	FSK, PSK, others	Microwave links, radar, satellite communication
30–300 GHz	Extremely higher frequency (EHF)	Line of sight, bandwidth to 1 GHz, low noise, high attenuation, very small antenna	FM	FSK, PSK, others	Radar landing systems, experimental

to attenuation losses due to atmospheric and natural phenomena such as rain, snow, and fog. These result in three major types of signal degradation: multipath interference, fading, and attenuation losses.

Reflections from the earth surface, the ionosphere, natural or manmade objects, and atmospheric refraction can create multiple paths between the transmitting and receiving antennas. Depending on the relative path distance, the reflected wave is shifted in phase with respect to the original wave, which can cause interference at the receiver, called multipath interference. Since the amount of phase shift is frequency dependent, the combined received signal is also frequency dependent, which can lead to serious problems in wideband transmission.

Fading is caused by abnormal changes in the refractive index of the atmosphere. Normally, the atmosphere refracts or bends radio waves back toward the surface of the earth. However, abnormal distribution of temperature, humidity, and heavy ground fog can cause radio waves to be bent toward the surface much more than normal, so that they never reach the receiving antenna, causing changes in received signal strength or even complete loss. Variation or, specifically, reduction of received signal strength at different periods in time is called fading.

As transmission frequencies increase, path attenuation losses due to the atmosphere also increase. More serious losses are caused by fog, snow, and especially rain, which becomes very significant at frequencies above 4 GHz. The effect of these losses is to cause fading and increase error rates. They are usually allowed for during the design process by using published meteorological data of the region in which a radio link is located.

C. TERRESTRIAL MICROWAVE RADIO

Terrestrial microwave transmission uses highly directional antennas for line of sight propagation paths using frequencies in the 4–12 GHz range. The antennas are usually parabolic with diameters ranging from 12 in. to several feet, depending upon their spacing. The spacing between repeater stations depends upon the geographical terrain over a given route, the technology used in the terminal equipment, and the transmitter power permitted by the local regulatory agency. For long-distance transmission, typical repeater spacings are 20–30 miles, but longer spacings are possible in areas where atmospheric conditions result in little fading.

Microwave links can carry several thousand voice channels using frequency-division multiplexing. They are an important source of competition with coaxial systems and have the advantages that no physical facility is required to guide the microwave energy between separate stations and that these stations can normally be 20–30 miles apart as opposed to about 1 mile for coaxial systems.

Using the earth's atmosphere as the transmission medium, however, results in multipath interference and fading. This can be minimized using frequency diversity, which means the transmission of the same radio signal over different microwave frequencies at the same time. Since for a given set of atmospheric conditions radio signals at different frequencies experience different degrees of multipath interference and fading, the strongest received signal is selected for retransmission or detection. Frequency diversity has the disadvantage of using more of the available radio frequency spectrum, consequently space diversity is sometimes used as an alternative approach, especially for paths that experience severe fading. Space diversity uses two receiving antennas, usually mounted on the same tower and separated vertically by several wavelengths. By switching from the regular to the diversity antenna whenever the signal level drops, the received signal can be maintained nearly constant. Neither frequency nor space diversification is effective in countering attenuation caused by rain; the only remedy for this is the use of greater transmission power or shorter repeater spacing.

D. SATELLITE TRANSMISSION

Satellite transmission consists of a line-of-sight propagation path from a ground station to a communications satellite (up link) and back to an earth station (down link). The satellite is usually placed in a geosynchronous orbit about 22,300 miles above the earth so that it appears stationary from any point from which it is visible and acts like a repeater in the sky. The ground station includes the antennas, buildings, and electronics necessary to transmit, receive, multiplex, and demultiplex signals. The frequency spectrum used is similar to that used for terrestrial microwave radio. The ground station antenna is usually highly directional, while the satellite antenna has a larger beam width to cover a larger portion of the earth's surface and be able to communicate with many widely separated earth stations simultaneously.

The capability of a satellite to act as a repeater for many different earth stations is called multiple access (MA). Three main methods are currently in use to accomplish this: frequency-division multiple access (FDMA), time-division multiple access (TDMA), and code-division multiple access (CDMA). In FDMA, circuits between different earth stations are assigned different frequency bands within the allowable bandwidth. In TDMA, the entire bandwidth is allocated to each earth station for a short time, just like in time-division multiplexing. CDMA (also called spread spectrum multiple access) uses a pseudorandom code and operates in both time and frequency domains. It is effective against jamming techniques and is used primarily in military satellite communications.

Satellites have had a dramatic impact on communication topologies and pricing. They can be used for transmission of program, video, voice, or data signals almost anywhere on earth, no matter how remote the location or whether it is fixed or mobile like a ship. They provide multichannel capabilities, wide bandwidths, and high data rates. The transmission cost is independent of the distance between the source and destination. Furthermore, only one satellite repeater is required for most transmissions. This characteristic of satellite transmission may make it superior to terrestrial systems such as microwave or coaxial cable, since the latter systems require many repeaters in tandem to cover long distances, and amplification of the signal by each repeater tends to increase the effects of distortion and noise.

Because of the greater distance between the earth and the satellite repeater, attenuation and transmission delays can cause problems. Attenuation can be overcome by using high gain, narrow beams, and path elevation angles greater than 20° for the ground antennas. The high path elevation reduces the distance the signals travel through the atmosphere to reduce attenuation and fading. However, rain attenuation can still be a problem, especially at the higher carrier frequencies, but it can be minimized by a form of space diversity since rain is unlikely to occur at two widely separated ground stations. The total transmission delay is approximately 0.5 sec, which is much higher than for terrestrial transmission media and is due to the much longer distance the signal must travel (a minimum of $2 \times 22,300$ or 44,600 miles). This delay can impair the quality of voice communication, but it has the greatest detrimental effect on data transmission unless communication protocols are designed to match the characteristics of this transmission medium.

E. Other Unbounded Media Transmission Systems

Communications satellites and terrestrial microwave radio are by far the most frequent systems using unbounded media for transmission of data. However, many other systems using unbounded media are available and are commonly used, especially for transmission of voice and program signals. A partial listing and a brief description of these systems is given below. A more detailed description may be bound in Freeman (1981) and Pooch et al. (1983).

1.   Radio and television broadcast. These systems use omnidirectional antennas to broadcast program signals to receivers in the local area: AM radio uses the band from 500 to 1300 KHz, while FM radio uses from 90 to 110 MHz. Television uses the very high and ultrahigh frequency bands to provide the bandwidth required for acceptable transmission of video program signals. A receiver can select one of many local stations by tuning to the frequency allocated to the desired station, which is a form of FDM.

2.   High-frequency radio. As little as thirty years ago, high-frequency (HF) radio carried almost all transatlantic telephone traffic. Today it is used primarily for short-wave radio and for telephone communications to ships at sea and to countries not connected by cable and with no satellite antenna. Long-distance communication is possible with HF radio since the frequencies employed allow the radio waves to be reflected by the ionosphere back to earth until they reach their final destination in one or more hops. For this reason it is sometimes called ionospheric radio transmission. Because of the movement and changes of the ionosphere, HF radio is subject to fading, interference, distortion, and periodic blackouts. Thus it is rarely used for data transmission unless extreme precautions are taken for detection and correction of errors.

3.   Tropospheric scatter radio transmission. In tropospheric scatter radio transmission, the troposphere is used to scatter and reflect a fraction of the incident radio waves back to earth. With two highly directional antennas pointed at the troposphere (which is about 6 miles up, as opposed to 30 miles for the ionosphere), enough radio wave energy is reflected back to result in very reliable over-the-horizon communication systems with effective ranges of 100 to 600 miles. Such systems are used where it is not possible or economical to use land lines or even microwave radio. Island chains or very rugged mountain ranges are good examples of terrain

where such systems are used. The carrier frequencies are in the 900–5000 MHz range and can carry several hundred voice circuits over short distances of 100 miles, although 72 is more typical over longer distances. It is subject to fading, which is usually overcome using space diversity with two receivers and transmitters at each end of the link with the antennas separated by at least 30 wavelengths to achieve reliable transmission.

4.   Packet radio. Packet radio can be used to allow terminals to communicate with a computer that is connected to an omnidirectional home base transmitter. Data to be transmitted to a terminal are formatted into fixed length packets which include the address of the destination terminal at the beginning and cyclic redundancy check characters at the end of the packet. Each terminal receives all the transmitted packets and discards those that are not addressed to it. Packets with detected errors are not acknowledged by the terminal, which causes the home base to retransmit the message after a certain time. When terminals transmit to the computer home base, it is possible that two or more terminals may transmit simultaneously, in which case the messages are not acknowledged. If a terminal does not receive an acknowledgment within a suitable time; it reschedules the packet for retransmission at a random time. Eventually, all packets will be received correctly. Different frequencies can be used for the computer-to-terminal and the terminal-to-computer links to allow concurrent reception and transmission. A variation of packet radio can allow any workstation or computer to communicate with any other without the master–slave relationship described above.

5.   Light transmission systems. Light transmission systems use a light-emitting or laser diode as a transmitter and a photodiode or phototransistor as the receiver. As in microwave radio, the transmission path must be line of sight, and air is the transmission medium. The problem with such systems is that extreme care must be taken to align the light source and detector. Even then, problems occur in heavy rain, fog, or snow, which attenuate the transmitted light signal and impair error-free reception. Bounded communication using optical fibers overcomes these problems.

F. Outlook for Unbounded Transmission Media

The ease of establishing communication links will remain the primary attraction of links using unbounded transmission media. Multiple access capability of some systems using these media is also an important advantage. New developments in systems such as packet radio are of particular importance since they promise inexpensive communication between fixed or mobile workstations and computers. With satellite communications significantly reducing the cost for long-distance transmission, the main competition to systems using unbounded transmission media is coaxial cable and optical fibers for the shorter distances. Perhaps the greatest obstacle to the growth of use of unbounded media is the limited frequency spectrum, but with improvements in antenna design, integrated circuits operating at high frequencies, and advances in modulation techniques to use bandwidth even more efficiently, systems using unbounded media will always play a major role in communications.

XII. Interconnection Considerations for Electrical Conductors

It has been mentioned in Section X that electrical conductors are and will continue to be the predominant transmission medium for interconnecting components of a system. As bandwidths and bit rates increase, it becomes increasingly important to take into account the stray reactances that influence the performance of these media. If we have a circular conductor of diameter d whose center line is a distance D over a ground plane, the expressions for the stray reactances are

$$L = 0.46 \log_{10}(4D/d) \ \mu\text{H/m} \qquad (8)$$

$$C = \frac{24.1}{\log_{10}(4D/d)} \ \text{pF/m} \qquad (9)$$

where L is the stray inductance and C the stray radial capacitance, assuming air is the dielectric. If the same conductor is shielded by a concentric shield of diameter D that is grounded, the stray reactances become

$$L = 0.46 \log_{10}(D/d) \ \mu\text{H/m} \qquad (10)$$

$$C = \frac{24.1}{\log_{10}(4D/d)} \ \text{pF/m} \qquad (11)$$

Equations 8–11 indicate that the stray reactances are strongly dependent on the geometry, conductor length, and proximity of the ground plane. By carefully positioning the conductor we can reduce one stray component, but only at the expense of increasing the other. Because of the relative slow rate of change of the logarithmic terms, the conductor length is the major factor determining the stray reactances.

The current or voltage propagation velocity v_p in a conductor is given by

$$v_p = \frac{1}{\sqrt{LC}} \ \text{m/sec} \qquad (12)$$

while its propagation delay T_d due to the finite velocity of propagation is given by

$$T_d = l/v_p = l\sqrt{LC} \ \text{sec} \qquad (13)$$

where l is the length of the conductor in meters, L its distributed inductance, and C its distributed radial capacitance per unit length. The conductor is said to be "short" if its propagation delay T_d is much less than the rise and fall times of the pulses it is expected to transmit. But if its length is such that its T_d is comparable to or longer than the rise and fall times or greater than $0.35/f$, where f is the greatest frequency per second for which it will be used, the conductor is termed as being "long"; then transmission line theory should be used to predict its characteristics.

The transmission line parameters relevant to our discussion are the characteristic impedance and the reflection coefficient. The characteristic impedance Z_0 is the impedance seen at one end of a infinitely long line and is given by

$$Z_0 = \sqrt{\frac{R + sL}{G + sC}} \qquad (14)$$

where s is the Laplace operator, R the resistance, L the inductance, G the conductance, and C the radial capacitance of the conductor per unit length. Sometimes R and G can be neglected, in which case,

$$Z_0 = \sqrt{\frac{L}{C}} \qquad (15)$$

The reflection coefficient r is the ratio of the reflected voltage v_r to the incident voltage v_i if a conductor is not terminated in its characteristic impedance and is given by

$$r = \frac{v_r}{v_i} = \frac{Z_L - Z_0}{Z_L + Z_0}. \qquad (16)$$

where Z_L is the terminating or load impedance. The propagation velocity and propagation delay are given by Eqs. (12) and (13). If a long conductor, as explained above, is not terminated or matched in its characteristic impedance, reflections will be present, and ringing will occur if pulses with sharp leading and trailing edges are transmitted over the line. Such reflections and ringing increase the inherent electrical noise of the system in which they occur. The impedance ranges of transmission lines that are usually encountered in practice are given below. Note that a strip line is a rectangular conductor over ground with the width of the conductor much greater than its thickness. This type of conductor is encountered in printed circuits, for example.

Type of transmission line	Characteristic impedance (Ω)
Wire over ground	80–400
Twisted pair	80–200
Coaxial pair	40–120
Strip line	20–140

The above relations are useful in determining the length and placement of lines. In high-impedance circuits, it is important to minimize capacitance by making the line as short as possible and placing it as far as possible from earth. In low-impedance circuits, it is important to minimize the inductance by making the line as short as possible and placing it as close to earth as possible. For long lines, the lines should be terminated in their characteristic impedance whenever possible to minimize reflections and hence circuit noise.

The presence of stray electromagnetic fields in a system causes cross-talk voltage or current to be induced in neighboring conductors. This cross-talk may be predominantly capacitive or inductive. Capacitive cross-talk is due to stray capacitance between conductors and is predominant in circuits with large voltage swings and small currents (i.e., high-impedance circuits). The capacitive cross-talk voltage v_c is given by

$$v_c = k_c \varepsilon (\text{voltage swing} \times \text{length of line}$$

$$\times \text{ function of spacing})/\text{rise time of voltage} \quad (17)$$

where ε is the dielectric constant of the medium between the conductors and k_c a constant. In general, the closer the lines the greater the v_c. Spacing the lines further apart and away from ground reduces v_c but has the effect of increasing inductive cross-talk. When the voltage swings are small and the currents are large (i.e., low-impedance circuits), inductive cross-talk usually predominates. It arises chiefly because of mutual inductance coupling between conductors and is generally given by

$$i_i = k_i (\text{current swing} \times \text{length of line}$$

$$\times \text{ function of spacing})/\text{rise time of current} \quad (18)$$

where i_i is the cross-talk current, which can be reduced if the lines are further from each other and as close as possible to earth.

If more than one line is driven at the same time, the inductive currents in the passive line are additive. This is not the case with capacitive cross-talk. An upper bound is reached for v_c if the passive line is surrounded by active lines. Both types of cross-talk may be reduced by using special lines. Twisted pairs, coaxial lines, and multiple shielded lines all have zero mutual impedance in theory and therefore zero inductive cross-talk. In practice, the following numbers are useful:

Type of wiring	Mutual inductance (μH/m)
"Neat" in bundles	1–7
Point-to-point over ground	0.6
Twisted pairs	0.06
Coaxial cable	0.006
Multiple shielded lines	As small as necessary

In summary, it is impossible to eliminate all the stray inductance and capacitance of conductors. It is possible only to trade one against the other by varying the characteristic impedance. In interconnecting components or systems, the goal should be to use point-to-point connections over a ground plane. All lines should be coaxial and matched at least at the receiving end. Practical considerations may not permit this, but the more we approach this type of interconnections, the less trouble will be experienced with system noise and cross-talk and therefore with getting a high-speed system to operate successfully.

XIII. Standards

Modern communication systems represent a complex and rapidly changing technology. Effective communications on a local, country, and worldwide basis require that communications equipment and transmission media work together in harmony. Thus there is need for widely accepted standards in the communications field. Such standards are established by a number of organizations including the International Telegraph and Telephone Consultative Committee (CCITT), the International Organization for

Standardization (ISO), the American National Standards Institute (ANSI), the Electronic Industries Association (EIA), the European Computer Manufacturers Association (ECMA), the National Communication System (Federal Standards—Telecommunications), and the National Bureau of Standards (NBS). Standards most relevant to data communication are compiled and summarized in Folts (1982).

BIBLIOGRAPHY

AMP (1982). "Designer's Guide to Fiber Optics." AMP Incorporated, Harrisburgh, PA.

Folts, H. C., ed. (1982). "McGraw-Hill's Compilation of Data Communications Standards." McGraw-Hill, New York.

Freeman, R. L. (1981). "Telecommunication Transmission Handbook." John Wiley & Sons, New York.

Goodman, J. W., Leonberger, F. I., Kung, S.-Y., and Athale, R. A. (1984). *Proc. IEEE, Special Issue on Optical Computing* **72**(7), July, pp. 850–866.

Payton, J., and Quréshi, S. (1985). Trellis encoding: What it is and how it affects data transmission. *Data Commun.*, May 1985.

Pooch, U. W., Greene, W. G., and Moss, G. G. (1983). "Telecommunications and Networking." Little, Brown and Co., Boston.

Proakis, J. G. (1983). "Digital Communications." McGraw-Hill, New York.

Rey, R. F., ed. (1983). "Engineering and Operations in the Bell System." 2nd ed. AT&T Bell Laboratories, Murray Hill, NJ.

Schwartz, M. (1980). "Information Transmission, Modulation and Noise." 3rd ed. McGraw-Hill, New York.

DIGITAL SPEECH PROCESSING

L. R. Rabiner

J. L. Flanagan   *AT&T Bell Laboratories*

GLOSSARY

Adaptive coder: Coder that changes (adapts) its encoding process in response to characteristics of the encoded signal.

Complexity: Measure of the computation required to implement a digital algorithm.

Excitation model: Computational characterization of the properties of the human voice source.

Formant: Resonance of the vocal tract manifested acoustically as a concentration of energy in the frequency spectrum.

Lexical access: Process of looking up an item in a dictionary or table.

Predictive coder: Coder that uses a method of predicting the value of a signal from previous (known) values of the signal.

Speech coding: Process of converting a speech signal into a digital format for transmission at a specified information rate.

Speech synthesizer: Device that generates approximations to speech signals from information about natural speech sounds.

Subjective quality: Perceived quality of a speech-processing system, typically as judged by a group of listeners.

Digital speech processing is the science and technology of transducing, representing, transmitting, transforming, and reconstituting speech information by digital techniques. The technology serves needs for communication between humans as well as between humans and machines. The former embraces digital encoding and decoding of speech for efficient transmission, storage, and privacy. The latter embodies digital synthesis of speech, to give machines a "mouth" with which to speak information to humans, and automatic recognition of speech, to give machines an "ear" with which to listen to human-spoken instructions.

I. Introduction

The field of speech processing includes the topics of speech coding, speech synthesis, and speech recognition. Speech coding is concerned with communication between people and therefore deals with techniques of speech transmission. Relevant issues are conservation of bandwidth, cost and design of terminal equipment, techniques for voice privacy and secure communications, and tandeming of voice-coding systems. Speech synthesis, or computer voice response, is concerned with machines talking to people. Thus, synthesis deals with systems as simple as announcement machines to those as complicated as printed-text-to-speech converters. Speech recognition is concerned with people talking to machines. Speech recognizers range in sophistication from the simplest isolated word or phrase recognition systems to fully conversational recognizers that attempt to deal with vocabularies and syntax comparable to natural language. Also included in the broad area of speech recognition is the topic of speaker recognition (i.e., verification or identification of the individual talker).

Digital speech processing has advanced in the past decade for several reasons. One key reason is the explosive growth in computational capabilities, supported by economical very large scale integrated (VLSI) hardware. General-purpose signal-processing computers exist today that can run ordinary FORTRAN code and, with minimal user assistance, execute algorithms at rates of the order of 50 to 100 megaflops. Such machines are classified as minisupercomputers and cost less than the mainframe machines of the 1970s. Similarly, VLSI digital signal processor (DSP) chips now exist that do calculations in floating point arithmetic at an 8-megaflop rate. Thus, even a 100-megaflop algorithm can potentially be realized with about a dozen DSP chips on a single circuit board.

Another reason for the progress in digital speech processing is the improvements that have been made in speech-processing algorithms. Speech coding has benefited significantly from the introduction of multipulse linear predictive coding (MPLPC) and code-excited linear prediction (CELP); the field of text-to-speech synthesis has seen major improvements due to the introduction of large pronouncing dictionaries; and the field of speech recognition has seen the maturity of algorithms for recognizing connected words (e.g., level building) and the widespread acceptance of statistical modeling techniques, namely, hidden Markov models (HMMs).

Finally, perhaps the greatest recent impetus to advancing digital speech processing has been the growing need for products that serve real-world applications. Since the mid-1970s there has been major growth in the utility of voice products for at least four market sectors: telecommunications, business applications, consumer products, and government. In the telecommunications sector, voice coders are used for reduced bit-rate transmission and privacy; repertory name dialers are used for hands-free dialing; announcement systems provide information to customers; and a wide variety of operator and attendant services depend on recognition and synthesis for increased utility. In business applications, voice mail and store-and-forward services are already in widespread use, and voice interactive terminals and workstations are beginning to appear on the market. In the consumer products and services sector, toys using either synthesis and/or recognition have been available for several years, and residence communication systems and alarm announcement systems have started to appear. In the area of government communications, anticipated uses include coding for secure communications

and voice control of military systems. These examples, by no means exhaustive, illustrate the burgeoning applications of speech processing and point to a growing market in the coming years.

The purpose of this article is to review the main issues in speech coding, synthesis, and recognition, to indicate where progress has been made, and to point out areas where new research is necessary to achieve desired goals. [See SIGNAL PROCESSING, GENERAL.]

II. Speech Coding

Research in speech coding aims to achieve high-quality speech transmission at digital data speeds. An important relation is the subjective quality obtained for a given transmission rate. One widely used measure of subjective speech quality is the so-called mean opinion score (MOS), based on a five-point judgmental scale ranging from 1 (unacceptable) to 5 (excellent). An MOS score of 4 or above is typically regarded as high quality. An MOS score of the order of 3 is sometimes considered useful "communications" quality.

To illustrate the current status of speech coding, Fig. 1 shows plots of subjective quality (MOS) versus transmission rate for high-, medium-, and low-complexity coders. A low-complexity coder (e.g., μlaw pulse code modulation (PCM) encoding) requires hardware capable of $\sim 10^5$ multiply–addition operations per second.

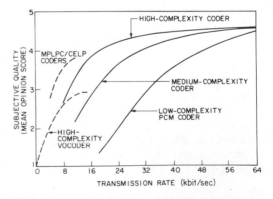

FIG. 1.   Curves of subjective quality (in terms of mean opinion score) versus transmission rate for three types of waveform coder (solid curves), for vocoders (dashed curve at lower left), and MPLPC and CELPC coders (dashed curve at upper left). (Courtesy of N. S. Jayant.)

The curve of MOS versus bit rate for the low-complexity coder shows that transmission rates of 48 kbit/sec and above typically preserve high-quality coding. By moderately increasing the complexity of the coder (by a factor of about 4 to 10), a more sophisticated algorithm for coding, such as adaptive differential PCM (ADPCM), can be used, resulting in improved subjective quality. Using a moderate-complexity coder, high quality can be maintained down to ~32 kbit/sec and communications quality can be maintained down to ~16 kbit/sec.

A further increase in complexity of the coder (typically by a factor of 4 to 10) again leads to improved coder performance. Typical coders of this complexity include adaptive predictive coders (APCs), adaptive transform coders (ATCs), and subband coders (SBCs). Such high-complexity coders achieve high-quality coding down to 16 kbit/sec and communications quality down to 9.6 kbit/sec.

The problem faced by all three classes of coders is that subjective quality falls rather precipitously below ~9.6 kbit/sec. As a result, at the traditional data rates of 4.8 and 2.4 kbit/sec the standard speech-coding algorithms presently used cannot achieve high quality.

The outlook for improved-quality speech transmission at data speeds seem good as a result of two developments. First, MPLPC attempts to bridge the gap between conventional LPC vocoders (voice coders) and conventional waveform coders. By representing spectral information in a standard vocoder manner and by representing the excitation for the vocoder by a fixed number of excitation pulses per analysis interval, the subjective quality of MPLPC speech is judged to approach high quality at 9.6 kbit/sec. More recently, the fixed-excitation model of MPLPC has been replaced by a stochastic code book excitation model in the CELP coder. For this coder, the subjective quality of the coded speech at 4.8 kbit/sec is judged to be at or above communications quality. Further research on CELP coders aims to achieve subjective quality approximately as shown by the dashed curve at the upper left of Fig. 1. Achieving this performance will be a well-defined challenge in digital speech coding in the next several years. If such performance is achieved, applications of 2.4 and 4.8 kbit/sec may become widespread in telecommunication networks. Electronic voice mail and store-and-forward systems also will become enhanced for a wide range of services and systems.

III. Speech Synthesis

A principal objective in speech synthesis is to produce natural-quality synthetic speech from unrestricted text input. The goal is to provide great versatility for having a machine speak information to a human user in as natural a manner as possible. Useful applications of speech synthesis include announcement machines (e.g., weather, time), computer answer back (voice messages, prompts), information retrieval from databases (stock price quotations, bank balances), and reading aids for the blind and vocally handicapped.

At least three major factors influence the performance of speech synthesizers. The first is the quality (or naturalness) of the synthesis. It is often possible to trade between quality and message flexibility. For example, announcement machines often use the best speech-coding methods to give high-quality speech, since the messages to be spoken are fixed in context and limited in number. However, text-to-speech systems must provide complete flexibility, and thus the speech signal must be synthesized from fundamental units.

The second factor is the size of the vocabulary. Again, if a relatively small vocabulary is required, (e.g., 100–500 words), it is possible to custom-adjust the synthesis for improved naturalness. However, for vocabularies of more than 1000 words, customized tuning is inappropriate.

The third factor affecting speech synthesis is the cost (or complexity) of the system. The cost includes hardware required for the storage of words, phrases, and production rules, as well as hardware required for speech signal generation (e.g., coder, synthesizer). The cost of synthesis systems has been falling rapidly with advances in VLSI, so this factor is no longer so important.

Synthesis techniques are distinguished by their message versatility or flexibility. Restricted techniques use prerecorded and coded speech, based on words, phrases, and sentences. The system then concatenates these units into messages. More versatile text-to-speech systems accomplish synthesis from units smaller than words (e.g., dyads, fractional syllables, or phonemes) and compute the pronunciation factors for these units.

Figure 2 shows a block diagram of the concatenative type of synthesis system. The storage consists of a fixed set of words, phrases, and sentences, which have been encoded by any of the coders discussed in Section II. An input

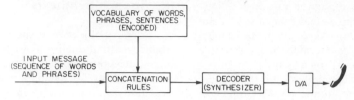

FIG. 2.   Block diagram of a concatenation type of speech synthesizer.

message—a sequence of words, phrases, and sentences—is converted to the appropriate sequence of units, which are retrieved and concatenated (usually with some type of smoothing at the junctions between units). The concatenated units are sent to a decoder (synthesizer) and to a digital-to-analog converter for transmission, playback, or both. The concatenative type of synthesis is used primarily in announcement machines and for applications such as automatic intercept of incorrectly dialed telephone numbers, where only a small vocabulary is required and a limited set of output sentences is needed [13]. [*See* DATA COMMUNICATION NETWORKS.]

A text-to-speech (TTS) synthesizer is shown in Fig. 3. In its most general form, the input to the system is a message in the form of unrestricted ASCII text, and the output of the system is the continuously spoken message. The system has three major modules: letter-to-sound conversion, sound-to-parameter assembly, and synthesis from a parametric description of the text. The letter-to-sound conversion can utilize either a set of programmed pronouncing rules or a stored pronouncing dictionary (which provides the phonetic spelling of every word in the text message) or a mixture of these two techniques. Even with dictionaries of several hundred thousand words, there will be cases where the words of the ASCII text are not always found (e.g., proper names, cities, specialized terminology), and for such cases programmed pronunciation rules are mandatory. In addition to deriving the phonetic symbols that correspond to the text of the input message, the first module must also provide prosody markers (pitch, duration, intensity) for the message to be spoken.

The second stage of the TTS system performs the conversion from phonetic symbols to continuous synthesis parameters, based on the set of subword units used to represent the speech. Thus, if dyad units are used, a conversion from phonetic symbols to dyads is required, followed by retrieval and smoothing of the synthesis parameters corresponding to the dyads in the message. Continuous contours for pitch and timing are also computed in this stage. The final stage is the synthesis of speech from the parametric representation of the subword units. Typically an LPC, MPLPC, or formant synthesizer is used.

TTS synthesizers can be used for database access, such as stock price quotations and bank balance checking, for access to voluminous amounts of text material over telephone lines (e.g., medical or legal encyclopedias), and as reading aids for the visually handicapped. Current TTS synthesizers produce speech that ap-

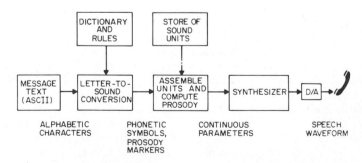

FIG. 3.   Functional diagram of a text-to-speech synthesis system based on stored subword units.

proaches the word intelligibility of natural speech but that is distinctly synthetic sounding. They perform with very large vocabularies and great flexibility and at relatively low cost. The challenge in speech synthesis in the next several years will be to improve voice quality and increase flexibility by providing a wide range of voice styles (male, female, child) and voice characteristics (Southern drawl, New England accent, etc.). In this manner TTS systems can be tailored both for the application and for the intended group of users.

IV. Speech Recognition

The objective of research in speech recognition is the automatic recognition of continuous speech for unrestricted vocabulary, context, and speakers. The goal of the research is to provide enhanced access to machines via voice commands. Some obvious applications include a voice typewriter, voice control of communication services and terminals, financial services, order entry, and aids for the hearing and motion handicapped.

Unlike each of the two preceding areas of speech processing, speech recognition research is far from achieving its ideal. To understand why this is the case, it is worthwhile discussing several of the factors that make automatic speech recognition by machine an exceptionally difficult problem. One important factor is the speech unit used for recognition. Units as large as words and phrases are acceptable for limited task environments (e.g., dialing telephone numbers, voice editors), but are totally intractable for large vocabularies and continuous speech recognition. In this case a subword recognition unit (e.g., dyad, diphone, syllable, fractional syllable, or even phoneme) is required, and techniques for composing words from such subword units are needed (such as lexical access from stored pronouncing dictionaries).

Another important factor affecting performance is the size of the user population. So-called speaker-trained systems adapt to the voice patterns of an individual user via a training or enrollment procedure. For a limited number of frequent users of a particular speech recognizer, this type of procedure is reasonable and generally leads to good recognizer performance. However, for applications where the user population is large and the usage casual, it is infeasible to train the system (e.g., users of automatic number dialers or order entry services). In such cases the recognizer must be speaker independent and able to adapt to a broad range of accents and voice characteristics.

Other factors that affect recognizer performance include vocabulary complexity, transmission medium over which recognition is performed (e.g., telephone line, airplane cockpit), task limitations in the form of syntactic and semantic constraints on what can be spoken, and cost and method of implementation.

These factors (which are only a partial list) illustrate why automatic speech recognition, in the broadest context, remains a challenging problem and is likely to remain so for some time. Currently, speech recognition has achieved modest success by limiting the scope of its applications. Thus, systems required to recognize small- to moderate-size vocabularies (10–500 words) in a speaker-trained manner, in a controlled environment, with a well-defined task (e.g., order entry, voice editing, telephone number dialing), and spoken in either an isolated or connected word format have been reasonably successful. Such systems have been, for the most part, based entirely on the techniques of statistical pattern recognition (Fig. 4). In the simplest context, each word in the vocabulary is represented as a distinct pattern (or set of patterns) in the recognizer memory. (The pattern can be either a sample of the vocabulary word, stored as a temporal sequence of spectral frames, or a statistical model of the spectrum of the word as a function of time.) Each time a word or sequence of words (either isolated or connected) is spoken, a match between the unknown pattern and each of the stored word vocabulary patterns is made, and the best match (or sequence of matches) is used as the recognized string. In order to do the matching be-

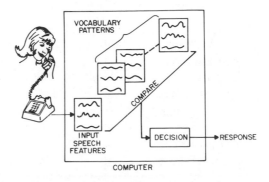

FIG. 4.   Automatic speech recognition using a pattern recognition framework.

tween the unknown speech pattern and the stored vocabulary patterns, a time alignment procedure is required to register the unknown and reference patterns properly. Several algorithms based on the techniques of dynamic programming have been devised for optimally performing the match. [*See* PATTERN RECOGNITION.]

A summary of typical recognition performance for systems of the type shown in Fig. 4 is given in Tables I and II. Table I shows average word error rate for context-free recognition (i.e., no task syntax or semantics to help detect and correct errors) of isolated words for both speaker-dependent (SD) and speaker-independent (SI) systems. It can be seen that, for the same vocabulary, SD and SI recognizers *can* achieve comparable performance (if the SI system is allowed to do an order of magnitude more computation by using multiple patterns for each vocabulary word). It is further seen that performance can be more sensitive to vocabulary complexity than to vocabulary size. Thus, the 39-word alpha digits vocabulary (letters A–Z, digits 0–9, three command words) with highly confusable word subsets such as B, C, D, E, G, P, T, V, Z, 3 has an average error rate of 20%, whereas a 200-word vocabulary of Japanese city names (all polysyllabic and highly distinctive) has an average error rate of ~3%.

Table II shows the typical performance of connected-word recognizers applied to tasks with various types of constraints. The performance on connected digits (speaker trained) reflects significant advances in the training procedures for recognition.

TABLE I.   Performance of Isolated Word Systems

Vocabulary	Mode[a]	Error rate (%)
10 Digits	SI	1.8
37 Dialer words	SD	0
39 Alpha digits	SD	20.5
	SI	21.0
54 Computer terms	SI	3.5
129 Airline words	SD	12.0
	SI	9.0
200 Japanese cities	SD	2.7
1109 Basic english	SD	20.8

[a] SI, speaker independent; SD, speaker dependent.

A system of continuous speech recognition using subword units and vocabulary sizes greater than 1000 words uses phonemelike units in a statistical model to represent words, where each phonemelike unit is a statistical model based on vector-quantized spectral outputs of a speech spectrum analysis. A third statistical model represents syntax; thus, the recognition task is essentially a Bayesian optimization over a triply embedded sequence of statistical models. The computational requirements are very large, but a system has been implemented at IBM research laboratories using isolated word inputs for the task of automatic transcription of office dictation. For a vocabulary of 5000 words, in a speaker-trained mode, with 20 min of training for each talker, the average *word* error rates for five talkers are 2% for prerecorded speech, 3.1% for read speech, and 5.7% for spontaneously spoken speech.

The challenges in speech recognition are many. As illustrated above, the performance of current systems is barely acceptable for large vocabulary systems, even with isolated word inputs, speaker training, and favorable talking environment. Almost every aspect of continuous speech recognition, from training to systems implementation, represents a challenge in performance, reliability, and robustness.

V. Speaker Verification

The objective of speaker verification is the authentication of a claimed identity from measurements on the voice signal. The applications of speaker verification include entry control to restricted premises, access to privileged information, funds transfer, voice banking, and similar transactions.

Figure 5 illustrates the components of a speaker verification system. The storage for this

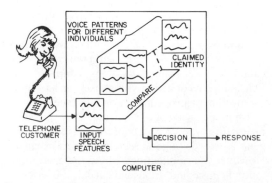

FIG. 5.   Automatic speaker verification based on pattern comparisons for individual voice features.

TABLE II.   Performance of Connected-Word Recognition Systems on Specific Tasks

Vocabulary	Task	Syntax	Mode[a]	String error rate (%)
10 Digits	1- to 2-digit strings	None	SD	2 (Unknown string length)
				1 (Known string length)
	2- to 5-digit strings	None	SI	9 (Unknown string length)
				5 (Known string length)
26 Letters of the alphabet	Directory listing retrieval	17,000-Name-list syntax	SD	4
			SI	10
129 Airline terms	Airlines reservation and information	Explicit state diagram syntax	SD	13
			SI	26

[a] SD, speaker dependent; SI, speaker independent.

system consists of voice patterns for all valid users. The voice patterns can be in the form of models of average spectral behavior, or they can be temporal sequences of spectral frames similar to those used in the speech recognizer of Fig. 4. For verification, the talker must make an identity claim and speak a test phrase (which can be a key phrase or arbitrary speech), and the system compares the input speech features with the stored model or pattern for the claimed identity. On the basis of a similarity score and a carefully selected decision threshold, the system can accept or reject the speaker. The features useful for verification are those that distinguish talkers, independent of the spoken material. In contrast, the features useful for speech recognition are those that distinguish different words, independent of the talker.

The decision threshold is often made dependent on the type of transaction that will occur as a result of the verification process. Clearly, a more stringent acceptance threshold is required for the transfer of money from one account to another than is required to report a current balance in a checking account.

Key factors affecting the performance of speaker verification systems are the type of input string, the features that characterize the voice pattern, and the type of transmission system over which the verification system is used. Best performance is achieved when sentence-long utterances are used in a relatively noise free speaking environment. Conversely, poorer performance is achieved for short, unconstrained spoken utterances in a noisy environment. Table III summarizes the current performance of several types of speaker verification systems.

The challenge in speaker verification is to build adaptive talker models, based on a small amount of training, that perform well even for short input strings (e.g., one to four words). To achieve this goal, more research is needed in the

TABLE III.   Performance of Speaker Verification Systems

Input	Mode	Acoustic signal processing	Feature pattern	Verification performance (equal-error rate, %)
Sentence-long utterances	Text dependent	Tenth-order cepstral analysis	Time contours of cepstral coefficients	1 (Recorded telephone) 4 (Live telephone)
Isolated word strings	Text independent	Eighth-order LPC analysis	Speaker-dependent word templates Speaker-independent template distance	4 (Recorded telephone) 8 (Recorded telephone)
Isolated word strings	Text independent	Eighth-order cepstral analysis	Vector quantization code book; talker models	1 (Recorded telephone)

area of talker modeling as well as in the area of robust analysis of noisy signals.

VI. Concluding Comment

This cursory overview of digital speech processing has aimed to highlight recent advances, current areas of research, and key issues for which new fundamental understanding is needed. Future progress in speech processing will surely be linked closely with advances in computation, microelectronics, and algorithm design.

BIBLIOGRAPHY

Flanagan, J. L. (1972). "Speech Analysis, Synthesis, and Perception," Springer-Verlag, New York.

Jayant, N. S., and Noll, P. (1984). "Digital Coding of Waveforms," Prentice-Hall, Englewood Cliffs, N.J.

Rabiner, L. R., and Gold, B. (1975). "Theory and Application of Digital Signal Processing," Prentice-Hall, Englewood Cliffs, N.J.

Rabiner, L. R., and Schafer, R. W. (1978). "Digital Processing of Speech Signals," Prentice-Hall, Englewood Cliffs, N.J.

INTELLIGENT NETWORKS

Syed V. Ahamed   *City University of New York*

GLOSSARY

B (bearer) channel: Information bearing channel that provides a transparent digital path between one customer and another. In the context of integrated services digital network, the B channel has been standardized at 64,000 bits/sec.

Carrier serving area: Designation of a particular geographic area whose communication needs can be served by a carrier system between a central distribution point within the geographic area and an exchange facility.

Carrier system: System for modulating a periodic carrier signal with the information of one or more channels to be able to transmit the combined signal over any specific transmission medium.

Channel: Logical connection between any two points in the network to exchange information. For the channel user, the exact physical path or the type of information carrying media is inconsequential. For the network, the physical and the logical address of the channel play an important role. The network may allocate, switch, reallocate, monitor, transfer, multiplex, or accomplish any function to convey the user information optimally.

Circuit switching: Mode of interconnecting logical channels between nodes of the network, maintaining the digital transparency for the duration of the use, and finally releasing the channels to return to their idle state.

D(delta) channel: Supporting channel that carries out-of-band signaling information for the information-bearing B channels.

Digital hierarchy: Ordering of the various lower capacity local channels that permits easy multiplexing and demultiplexing to and from higher capacity long-distance channels.

Electronic switching systems (ESS): Integrated facilities for switching channels within the network by electronic devices operated by stored program. Intelligent network switches also perform administrative, interfacing, and service functions for the local and remote switching modules, various networks, and carrier systems.

Network intelligence: Network capacity to adaptively switch, seek, monitor, and interconnect the information bearing channels, information sources, and information users.

Network services: Class of customer services provided by the intelligent networks above and beyond those readily available from conventional networks, such as the standard telephone network.

Node: Physical location or logical address in the network at which specific network functions, such as switching, relaying, tapping, and monitoring, can be performed.

Packet: Fixed and maximum length unit of information (data or control) that is sent from a source to a destination. A packet may be a totally independent unit of communication with all the routing information necessary for transmission. A packet may also be one of a series of packets enroute to the destination. The entire series will then complete the communication function across a preestablished path. The packet contents may be data or the control information to establish and inform the features of the path.

Packet switching: Mode of relaying information in a packet form with complete identification for the packet to reach its final destination and for the network to complete the transmittal, recovery, billing, and other associated network functions.

Stored program control: Capacity to control the network functions by programs or microcode generated and stored as software or utilities.

Subscriber loop carrier: Carrier system used in the loop plant of existing telephone network that is used to carry numerous telephone conversations between an exchange and a remote terminal. Such carrier systems are used in relatively few countries and are not universally prevalent.

Time division multiplexing: Concept of allocating finite time slots to individual channels and thus share a high-speed digital medium among two or more low-speed users.

Vendor services: Services provided by source of information to the customers at no additional cost or for a fee. The network then communicates the information necessary to complete the transaction.

An information society has started to emerge. The availability of the right information at the right time becomes the necessity for survival and excellence. In view of providing key information to a very large community of users with quick access to and dependability of the computer systems, the telecommunications scientists have conceptualized global information networks. These networks are preprogrammed to be adaptive, algorithmic, resourceful, responsive, and intelligent.

Intelligent networks (IN) are defined as the carriers of information with distinct algorithmic adaptation. In the context of hardware, intelligence resides in the customized integrated circuit chips, their sophisticated layout, and their interconnections. In the context of software, intelligence is coded as programs, utilities, or modules. It may reside in the active memories of computers during the execution phase. In the context of firmware, intelligence is microcoded into the control memories of the monitoring computers. Thus, the basic computers (the hardware, the software, and the firmware), which control, monitor, and process the information, become an integral part of intelligent networks. The flow of the information takes place over appropriate channels within the network and also within a diversity of participating networks, as shown in Fig. 1. These chan-

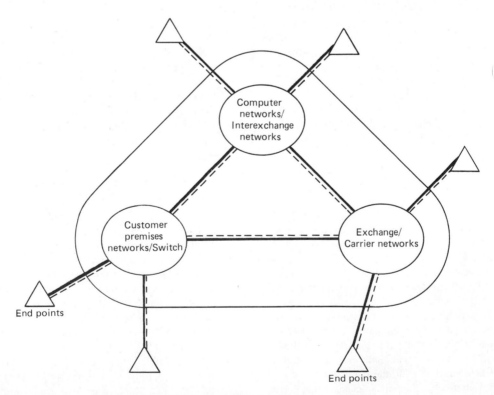

FIG. 1.   Conceptual representation of an IN: data, ———; control, - - - -.

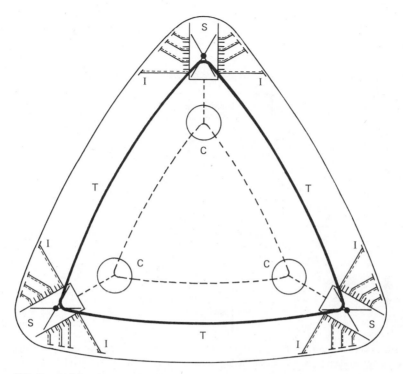

FIG. 2. Building blocks of an IN showing the flow of control and data: data, ——; control, – – – –; c, controlling computers, S, switches; I, network interfaces; T, transmission path and facilities.

nels are dynamically assigned, switched, and reallocated to carry the information from node to node. The facilities that actually switch channels may be central offices, switching centers, private branch exchanges, or even satellites that relay information. The actual transport of the information is carried out over transmission facilities of the network. Such facilities may span a small laboratory or a nation or the whole world. The size of the network or its geographical expanse is inconsequential to its nature. The four essential building blocks (see Fig. 2) of any intelligent network become the interface for the flow of information in and out of the network, the monitoring computers, the switching systems, and the associated transmission facilities.

Large amounts of intelligence are encompassed in the conventional analogue networks, such as the plain old telephone service (POTS) network. However, in more recent context, intelligent networks refer only to digital networks, even though there are analogue components that exist within the network.

I. Basic Concepts

The capacity to adapt to the extensive and dynamic network environment may change because of a large number of internal and external conditions. The network may become overloaded or faulty; it may experience switching delays or inadequate standby channel capacity or any other network condition. Further, the user or the source and the destination of the information may lead to extraneous searching before the right information is conveyed to the customer. It is here that the built-in algorithmic intelligence should monitor the network performance without the user or operator or any other human intervention. In essence, the network adaptively responds to the commands that control and execute the entire range of communication functions.

Intelligent networks perform in two distinct directions. First, they have to actively process information to respond to the queries of the user. Second, they have to adapt and fulfill the switching and transmis-

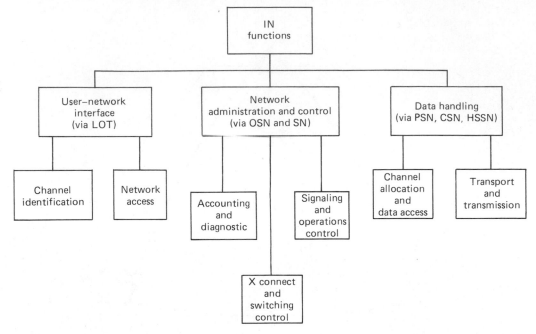

FIG. 3. Organization of IN functions: LOT, local office termination; OSN, operations systems network; SN, signaling network; PSN, packet switched network; HSSN, high-speed switched network; CSN, circuit switched network.

sion requirements to convey information from its source to its destination, wherever either one may be geographically or computationally located. The network intelligence can thus be grouped into its information processing aspect and its switching and transmission aspect. In Fig. 3, the functional organization of an intelligent network is depicted.

A. INFORMATION PROCESSING ASPECTS

Networks are designed to serve a large number of users with a large number of queries, seeking a wide variety of answers. The queries may be in-depth or peripheral, they may have constraints and modifiers, they may seek solutions and modify the subsequent queries depending upon the previous answers, etc. For this reason, the network functions require a certain amount of sophistication to comprehend the query, seek the answer, and convey the right answer to the right user within a reasonable amount of time.

The front-end processor of the computers which handle the user queries needs natural language processing capabilities. The software for processing natural language queries would exist at the user network interface shown in Fig. 3. Such processors have embedded elaborate rules of grammar to comprehend a complex query. Next, the information sought needs

to be identified and accessed. The information may be available locally within the local data base environment of the computer handling the user query or may be available elsewhere in the network. In the earlier case, a search within the local data base is initiated. In the latter case, a query is dispatched to another node where the answer is available. Information that is supplied to the customer through a public domain intelligent network (PDIN) may be from an information vendor. Such information vendors are abundant. For example, a hospital may be searching for an organ donor. In this case, a national data base with a list of donors may be maintained by a private facility, which dispenses its information for a fee. In this case, the user would incur a cost in receiving the information. The network has to determine the appropriate information needed by both parties (the customer and the vendor) so that the transaction may take place.

Such transactions have to be monitored by the unattended network via its own follow-up sequence of programs. The functions of locating the information sought by the user, the choice of the programs to execute in the network functions, etc., become the software environment that drives these networks.

Thus, the nature of intelligence required in the networks to handle information access and retrieval, the comprehension of the user query, and the integration of the steps necessary in the procurement of the information sought becomes a necessary subfunction of an IN. Concepts evolved from formal language theory provide the software tools to comprehend the user queries. Concepts evolved from knowledge engineering provide software tools to tackle the retrieval, the design, and the fabrication of the answers sought by the users.

B. SWITCHING AND TRANSMISSION ASPECTS

The switching of information-bearing channels is essential in order to use a specific channel to perform a variety of service functions for the customer for a finite length of time. Switches perform in three distinct ways (see Fig. 4). First, the channel may be circuit switched, that is, switched by the network and allocated to a certain user for as long as the user may need it. Second, the channel may be packet switched, that is, the information may be packetized and dispatched to the appropriate destination, either as an individual packet or as a series of packets in an appropriate sequence to complete an informational transaction. Third, the switching may be performed by the channel itself; in this case, the information is known to be channel switched. The third type of switching generally occurs in private networks, and the customer (private branch exchanges, PBXs) exchange carries out the localized switching. In the first and second cases, the external network has the option to choose and switch between various channels depending upon the availability, type of service, and source and destination of information. This calls for intricately designed switching systems, which are discussed in Section VI.

The transmission of the information also needs intelligence. The network can choose between a large number of paths and circuits. The network conditions strongly influence the path selected. These paths may be quite physical, such as a pair of wires, or they may be one or more channels multiplexed over one physical media, such as a wire pair, a coaxial cable, an optical fiber, a digital radio circuit, or any viable medium for data transmission.

Fortunately, a large number of adaptive algorithms have been developed during the evolution of the conventional telephone network. The newer and truly sophisticated networks offer a new breed of intelligence. The capacity to incrementally modify the network at the customer command would be impossible by the older POTS network hardware and its rudimentary software. The choice of the path may also satisfy certain other user-defined constraints,

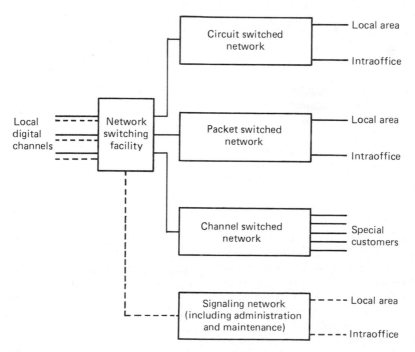

FIG. 4.   Integration of various networks for the transport of data (via the circuit switched network, packet switched network, and channel switched network) and for the flow of control (via the signaling network.) Data, ———; control, - - - -.

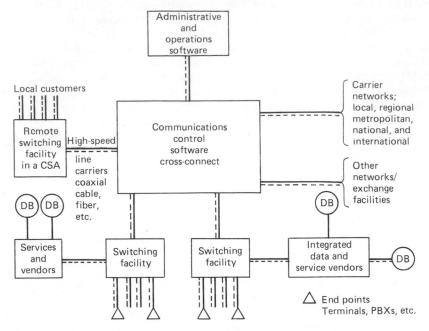

FIG. 5. Architectural layout of an exchange facility for an IN (CSA, carrier serving area; DB, data base) Data, ——; control, - - - -.

such as minimal cost, minimal delay, and low error. The path routing intelligence is handled by a different layer of the widely accepted network model (see Section IV). Processing of address information for communication also needs certain specialized capabilities. For example, the intelligence to wake up an idle channel and to test it before the actual communication needs very different procedures than those for cost minimization. Thus, it becomes necessary to classify the types of intelligence, necessary for the smooth, error-free transmission of information. Functioning of these networks without chaos becomes a fundamental requirement in light of the possible faulty conditions that can exist in the network, the transmission, and the associated control computers.

Intelligent networks have widely dispersed intelligence modules (software routines, microcoded read only memories, programmable read only memories) and information modules (data bases). It becomes necessary to provide access to the right module at the right time. Hence, some of the questions addressed by the network architects pertain to the types of networks and their environments, local and global intelligence, local and global data bases, and accessibility in the communications, interfaces, sources, and destinations of the information communicated. In Fig. 5, a typical layout of an exchange facility within an

IN is depicted. This leads to the architectural issues which govern the network design.

II. Intelligent Network Architecture

A. BASIC COMPONENTS

Intelligent networks consist of sources and destinations of information, network nodes, transmission facilities, and networks to control the switching and transmission through the network. The need for a common channel signaling network to support and control the information-bearing channels has been recently recognized, and most intelligent networks have this additional network component in their architecture. These controlling networks adopt out-of-band signaling. It should be recognized that this mode of signaling (as opposed to in-band signaling) is only one of the possible ways to control intelligent networks.

B. DATA SOURCES AND DESTINATIONS

Customers, vendor data bases, billing systems, public domain information banks, and any information storage or handling facility are data sources, provided they can be interfaced with the network. The sources of intelligence, such as programs, routines, and modules, also become part of the control information. This information may reside as data bases, programs, or operating systems for the interconnected computers within the network. However, the difference between information that passes through the network and the information to control and monitor the flow of customer information remains fundamental. Such a difference also exists in computer systems. The flow of data signals is distinct from the flow of the control signals within the computer architecture.

C. NETWORK NODES

Nodes of an IN perform a wider variety of sophisticated functions in comparison with the functions of the nodes within the telephone or computer network. Nodes may be designed as one of four types: (1) nodes in which no through data traffic is permitted (e.g., the end office or a computer port); (2) nodes through which only through traffic is advised (e.g., a satellite relay station or a digital repeater bank); (3) nodes through which both through and terminal traffic are allowed (e.g., a metropolitan switching office that forwards information to other nodes and has spare capacity and links terminating at local area customers); and (4) nodes where numerous links meet to permit local and global exchanges of channels and where local and through traffic is permitted. See Fig. 6 for the distribution of these nodes in a typically mountainous region of a national network.

In addition to the switching of information-bearing channels, a typical node in a PDIN manages and controls the flow of information in context of the services it is providing. Vendor services need the correct querying, retrieval, and follow up. Administrative, maintenance, and billing information is appropriately exchanged. Diagnostics are administered and follow-up operator actions are communicated. In the context of the integrated services digital network (ISDN), the nodes serve to manage the B channels. As presented earlier, the control and signaling information for switched channels is received on the D channel or in the appropriate X.25 protocol for the packet switched channels. The network control module responds to this information and acts to establish the digital connectivity of the B channels via the switching or the connection control module. Further, the node may be called upon to send signaling information to the next ISDN central office regarding the activated and active B channels.

The node also provides access to the information vendors to supply the necessary information to the network users. It serves to alert the office administration and management network regarding the call pro-

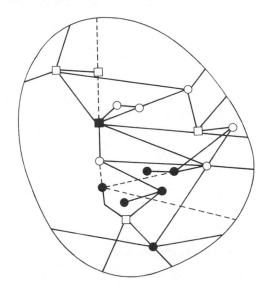

FIG. 6. Common types of nodes and links within a regional network. High-capacity coaxial carrier systems, ———; radio links (depending upon the terrain), – – – –; solid circle, no through traffic (e.g., end office in a small town); open circle, only through traffic (e.g., a repeater station); open square, through and terminal traffic; and solid square, through traffic switching and switching facility for local traffic.

gress, charges, rates, billing information, etc. The node can serve as a gateway office by providing access to other networks, such as packet switched or private networks. Very specialized interfacing functions and, thus, resident modules are necessary to permit a dependable flow of voice, data, or video signals through the node. Diagnostic and maintenance functions are also undertaken by nodes to check out the network modules and their functionality. In case the gateway office serves numerous countries, the node functions to send control signals to access foreign networks through the international signaling interface.

In the United States, where the subscriber loop carrier systems (see glossary), private networks, and carrier serving areas (see glossary) are plentiful, the nodes of the intelligent networks serve a greater diversity of functions. Private network interfacing calls for certain specialized functions from the node. Such functions are individually tailored to the specific node or the central office. The more recent electronic switching systems (ESS, see glossary) have made such customization possible by the modular hardware and software approach used by the ESS architecture designers. In pursuing a modular approach, these nodes are divided to pursue four basic functions. First, the interfacing with a 4-wire or 2-wire digital subscriber loop at any one of the basic or wide-band ISDN rates is accomplished by the peripheral switching modules. Second, the switching function for the numerous data channels is carried out by the main switching modules within the node for the local traffic or the call is forwarded via the interexchange carrier network and the interoffice signaling network. Third, the localized communications between the various switching modules is carried out by the communications module that provides a cross-connect facility between the switching modules. Finally, the office administration and network diagnostic and routine housekeeping functions are carried out by the administrative module (see Fig. 5).

D. TRANSMISSION FACILITIES

Digital information may be transmitted over any number of physical media. As shown in Fig. 7, the line coder, the transmit filter, the channel media (with or without repeaters), the regenerator, and the line decoder constitute an entire digital facility. In the regional, national, and global networks, a large number of digital carrier systems are in place.

Metallic media has been used extensively to carry digital data in most of the local, metropolitan, and regional networks. Transmission rates depend upon the distances and the coding techniques used. Local loops to the subscribers can carry data effectively at

the basic access rate of 144 kb/sec. The line rate can be as high as 192 kb/sec. With the proposed 2B1Q (2 binary bits converted to 1 quaternary level) code with refined digital echo cancellation techniques, the line rate can approach 1 Mb/sec over the shorter loops that span the carrier serving area (CSA) location and subscribers at a maximum distance of 12,000 ft. The T1 carrier facility carries digital information at 1.544 MB/sec extremely dependably. Digital carriers at the T1C rate (3.152 MB/sec in the United States) and at the DS2 rates (6.312 Mb/sec in United States and 8.448 Mb/sec in Europe) are also readily available.

Other rates for digital transmission facilities (DS3) are 44.736 Mb/sec in the United States and 34.368 Mb/sec in Europe. The physical media at this DS3 rate are usually light-wave transmission facilities and free space with the 11-GHz digital radio carrier facilities. Higher bit rates at 274.176 MB/sec (United States) and 139.264 MB/sec (Europe) are also in service. The intelligent networks can be interfaced with some or all of these digital carrier systems depending upon the type of the node and its switching facility. These data rates are derived from the digital hierarchies of the different countries around the world.

These carrier systems are used in context of the intelligent networks to carry digital information over any given distance from one node to the next. The switching and transmission facility does not consti-

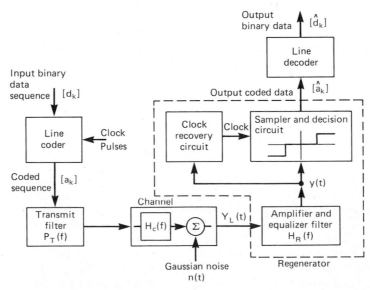

FIG. 7. A typical digital transmission facility for an IN. [From Miller, M. J., and Ahamed, S. V. (1987, 1988)].

tute an intelligent network unless it is architecturally controlled by its own control packets or its control network.

E. NETWORK ARCHITECTURE ISSUES: NETWORK ARCHITECTURE AND COMPUTER ARCHITECTURE

Network architecture has evolved from computer architecture. The functionality of the digital network influences its architecture. Networks have immensely distributed processing, memory, administrative, and data-base capabilities. Conventional computer architecture addresses issues pertaining to the hardware modules (central processors, memories, input/output processors, and their bus structures). Privileged instructions available to the operating system and the basic utility programs perform the start-up and core functions that monitor the functionality of the computer system. The PDINs are continuously active and have to be functional under far more stringent conditions. The monitoring and fault tolerance of networks is more crucial in the public domain. Hence, their architecture reflects these additional features.

Network architecture addresses a far wider spectrum of issues. For the participating computers with any IN, the hardware, their localized privileged instructions, the local software modules, and priority instructions become localized issues. On a global basis, network architecture addresses the issues dealing with the communication between the various nodes (which may be of various types, see Section II, C), the channel capacities, the data bases available and their access, the types of interface, and the compatibility and control of the network layers functions (see Section IV) and the channel banks (which separate out the channels and lead them to the appropriate destinations). Fortunately, considerable standardization has taken place in the terminology and design of network structure and the associated protocols by the International Telegraph and Telephone Consultative Committee (CCITT).

III. Network Types

Public domain networks and totally private networks constitute the extremes of network ownership and thus its organization and management. Whereas there is considerable freedom in the architecture and operation of the private network, the public domain networks' architecture and operation tend to be standardized and streamlined (see Section IV). In the discussion presented here, acknowledged international standards are used in context to the PDIN, especially for the ISDN.

There are two major architectural variations in the networks used in the public domain. First, in the switched architectures (connection orientation), there are five stages for data transfer through a switched channel. The sequence may be summarized as follows: an idle channel is identified and tagged, a connection is established, the data are transferred, the channel is released, and the channel resumes to be idle again. This sequence of stages closely parallels the well-established network steps in a typical voice call, which follows the network functions (call setup, alert, connect, disconnect, and release). In context to the ISDN, the information necessary to accomplish these individual steps is incorporated in the CCITT Q.930/931 protocol at the network layer used over the D channel to set up the B channel.

Second, in the packet switched architectures (connectionless orientation), a packet of information is assembled with its own address so that it can be forwarded to its destination. The packet may transit through any number of nodes until it reaches its destination. The exact physical routing is not known; but the innate functioning of the participating nodes through the network assures the correct transmission of the packet. (In many ways the packet can be compared to a letter dropped in a mailing system, which assures the delivery of the letter to the right destination.) In context to packet switching, CCITT has specified the widely accepted X.25, X.75, X.28, and X.29 protocol at the network layer (see Section IV).

The two architectures can work cooperatively, and one of the objectives of the intelligent network switches and their nodes is to assure the compatibility of the two major network ideologies. Hence, there are differences in functioning but not in integration toward being one intelligent network in the public domain. Concepts culminating in the present standardization and implementation of the IN and the ISDN have initiated the evolution of intelligent networks.

A third supporting network that has emerged because of the need to control and monitor the flow of information is called the common channel signaling (CCS) network. Signaling is essential to the functionality of any IN. In context to ISDN, the signaling is carried out by the standard CCS7 network adopted in United States and Europe. This network uses the out-of-band information (i.e., the information on the D channel) to control and signal the various switches to complete and monitor the B channels. Figure 8 depicts the role of the signaling and control of the information-bearing B channels. This network becomes essential in the circuit switched context because the B channels provide transparent end-to-end

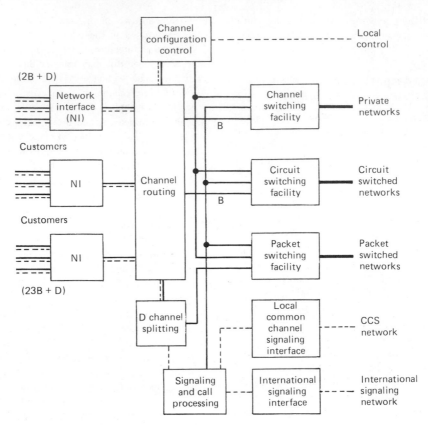

FIG. 8.   Architecture of a gateway node showing the need for a common channel signaling network to control and monitor the B channels via the D channel in context to the ISDN environment.

digital connectivity for the network users. If and when a transition to the packet mode is to occur, the CCS information is used to provide the transition and vice versa.

Private and semiprivate networks (sometimes called channel switched networks) may also use the standard components and the associated interfaces. Generally, private and dedicated lines, and digital distribution facilities with any amount of computerized information handling capacity may be encountered in private (intelligent) networks. These networks are emerging with great amounts of sophistication and intelligence. Typically, such networks are found in the scientific community (e.g., National Science Foundation's NSFNET, Defense Advanced Research Projects Agency's ARPANET, and Carnegie Mellon and Bell of Pennsylvania's Metropolitan Campus Network (MCN). Other examples from industry (e.g., VLSI vendors' networks) and national laboratories (e.g., Lawrence Livermoore National Laboratory's private network for their 12,000 employees, spanning 500 buildings) also prevail.

IV. ISO Model and Intelligent Networks

The open system interconnect (OSI) standard (International Standard 7498) proposed by the International Organization for Standardization (ISO) and the identical CCITT recommendation (Recommendation X.200) are internationally accepted. Most networks follow these standards closely for the ease of interconnection to other networks and the flexibility of using different product vendors and standard software modules to work with these products. A conceptual representation of the OSI environment is shown in Fig. 9.

A. LAYERS OF THE OSI MODEL

Germane to the OSI concept are the seven layers for its implementation. The individual layers, shown in Table I, perform very specific functions to maintain the network coherent functions. The intelligence (hardware, software, firmware) to perform the vari-

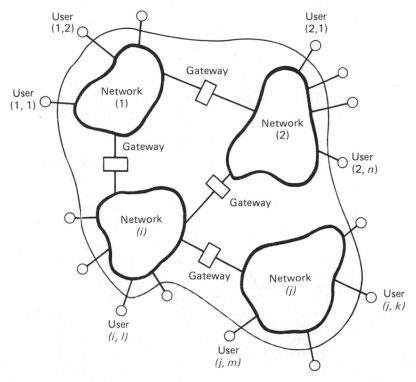

FIG. 9.   A conceptual representation of the open system interconnect (OSI) environment.

ous functions at any one of the layers has to be provided in specific or common devices that perform these necessary functions.

The top three layers are specific to the users of the data and information services of the network. The fourth layer, dealing with the transportation of data that is very critical in matching the user specific requirements to the network function, becomes a liaison between the network service users and the network service providers. The lower three layers are specific to the actual network providing the network services.

In the application layer (AP), the intelligence modules assure the user that the service provided is consistent with the type of application. In the presentation layer (PRL), the intelligence modules interpret the information exchanged between the AP and the network. The communication syntax is appropriately negotiated, chosen, and translated. In the sessions layer (SL), the intelligence modules bind and unbind various users and their applications into logical units for dispatch and recovery to and from the network during a session. In the transport layer (TL), the modules control the end-to-end information exchange with the required degree of reliability. In the network layer (NL), the intelligence modules provide, maintain, and terminate the connections in a

switched environment of the end systems. Routing and addressing information is also provided by these modules. The input and the interface to these modules is from the TL and does not depend upon the lower two levels (DL and PHL) of the network. Because of the need to carry out very critical functions, most intelligent networks follow the CCITT recommendations for the interface and modularity of the functions. In the data link layer (DL), the modules perform the error-free transmission of bits. Two basic functions, synchronization of the data streams and error control, are essential at this layer. In the physical layer (PHL), the intelligence modules provide the functional and procedural steps to activate, maintain, and deactivate the physical connection. Both the electrical and mechanical characteristics of the interface with the external transmission media are considered by this layer.

It is important to note the intelligence modules may be in the form of software, hardware, or firmware. The actual form of these modules is unimportant as long as the functions are dependably carried through at the various layers of any IN. The standards established by CCITT in unison with ISO and those accredited by the Accredited Standards Committee (ASC) of the American National Standards Association (ANSI) have been crucial in establishing the

TABLE I.   The Open System Interconnect Model

OSI layer	Type of functional intelligence[a]
Application	AL: Facility to serve the end user. Provision of the distributed information service. Communication management between the AL and PRL. Service to the user, the application (SASE), and the group of applications (CASE) served. Authentication of user IDs, destination IDs, authority to exchange information. Determination of service quality from the lower layers. Data integrity, error recovery, and file transfers are also assured. (See ISO DIS 8649, protocol ISO 8640, CCITT X.400 message handling system.)
Presentation	PRL: Assures the delivery of information to the end users in a form that is usable and understood. The information content (semantic) is retained, even though the presentation format and language (syntax) can be altered to suit the source(s) and destination(s) of the information. (See ISO 8824 for abstract syntax notation ASN.1, and ISO 8825 for encoding, CCITT X.409.)
Session	SL: Provision to transfer data and to transfer control in an organized and synchronized manner. User may define the degree of control and synchronization that the session layer will provide. (See CCITT X.215, ISO 8326; X.225, ISO X.8327, T.62.)
Transport	TL: Selection of network service. Evaluation of the need for multiplexing. Selection of the functions from the lower layers. Optimal data size decisions. Mapping of transport addresses to network addresses or the end-point users and negotiated. Data flow regulation between the end users. Segmentation and concatenation. Error detection and its recovery. (See ISO 8072, CCITT X.214; ISO 8073, ISO DIS 8602-connectionless, CCITT X.224.)
Network	NL: Establish, maintain, and terminate switched connections. Addressing and routing functions. Service TL, independent of DL and PHL. (a) Connection: network connection, data transfer, optional expedited data and receipt transfers, reset, and connection release. (b) Connectionless: UNITDATA. (See ISO 8348, CCITT X.213; protocols CCITT X.25, 1984, packet, ISO DP 8878 with X.25 for connection oriented network service; ISO 8473 for connectionless internetworking; CCITT Q.930/ Q.931 ISDN.)
Data link	DL: Synchronization/framing, error detection and recovery, and flow control for information transmitted over the physical link. (See ISO DIS 8886, 1745, 2111, 2628, 2629; CCITT X.212, X.21 for basic mode; ISO 3309, 4335, 6159, 6256; CCITT X.25, X.75, X.71 for HDLC; CCITT Q.920/ Q.921 for ISDN.)
Physical	PHL: Activation, maintenance, and deactivation of the physical connection. The electrical and mechanical characteristics for physical interface and the transmission media. (See ISO 2110, CCITT V.24, V.28 or EIA RS-232-C, also EIA RS-449, CCITT X.21, V.35; EIA RS-422A, CCITT V.11, X.27 for balanced voltage; EIA RS-423A, CCITT V.10, X.26 for unbalanced voltage; CCITT I.430 for (2B + D) ISDN and CCITT I.431 for 24 B or 30 B channels.)

[a] The nature and amount of intelligence required at each level and the established standards for specifications and protocol.

functions of different layers for the open systems interconnect facilities for most of the intelligent networks. The standards specify the information and the functions at each of the seven layers of the OSI reference model. If the designers of any particular network architecture decide to follow these standards, the distribution of the intelligence in the network also get firmly distributed for the control and monitoring of the network functions. The distribution of these functions is better established for the lower three layers (network, data link, and physical). The guide lines are summarized as follows. For the packet mode, the X.25 protocol is the CCITT standard at layer 3. This protocol is implemented together with link access protocol for the B channel (LAPB) at layer 2, the data link. The X.25 protocol at layer 3 is implemented with link access protocol for the D

channel (LAPD) at the second layer. For the switched mode, the Q.931 standard has been established at layer 3 between the user and the exchange network interface. At layer 2, the data link layer, LAPD format is used again to carry the control information to the network call control module. This information is carried by the D channel to facilitate the network in serving the associated B channels. Typically, the Q.931 call setup message has very specific information (e.g., caller identification), the B channel service desired, the B channel identification, and the destination address. The call control network module uses this information to monitor the call setup for the B channel. The D channel message changes if the user wishes to disconnect the B channel, and the network control module responds accordingly.

The OSI layered structure also facilitates the segregation of the functions performed by individual layers. The intelligence required to perform the functions within any one layer can be localized to the hardware and software particular to that layer. The protocols that facilitate the transfer of information between the lower layers of the OSI model are undergoing standardization. The protocols at the lowest (physical) layer have been established and documented.

B. X.25 PACKET SWITCHED NETWORK STANDARD

The best known standard for interfacing with the network is based upon CCITT recommendation X.25 in 1980 and revised in 1984. It spans the lower three layers of the seven-layer OSI network model and permits full network layer service by the network services vendors throughout the globe. The recommendations of 1980 (X.25 version) have been widely accepted and deployed by most network services vendors. The latter recommendations of 1984 (X.25 versions) are being rapidly implemented.

The OSI standard (or the CCITT X.200 recommendation) permits any layer to communicate with a corresponding layer at any other node in the network.

V. Example of an Intelligent Network

The variety of the functions offered by the mature ISDN are both intelligent and adaptive. For this reason, the proposed ISDN architecture and administration can be adapted to any IN. In more highly specialized private networks (such as industrial, corporate, scientific, military, and governmental), the methodology for implementation may tailored to suit and serve the priorities of the individual network. The basic concepts for the design and management prevail in most INs. Another example of the PDINs is the 800 network, which incorporates call management data-base facilities. Flexible networks and private virtual networks offer customer control on the configuration of the network for call forwarding, vendor selection, and other local modifications to the network, which influence customer activity. Such intelligent networks are regularly offered by the Bell Operating Companies in the United States. American Telephone and Telegraph Co. is in the process of installing its worldwide intelligent network, which is capable of carrying out most of the intelligent functions discussed here.

The French have also introduced the Biarritz network, which is capable of producing a wide range of intelligent functions, such as a home video library, home shopping, video games, an on-line videotex,

an encyclopedia, and graphic services. Capabilities of other national intelligent networks are also increasing. For the rest of the discussion, we refer to the evolution of the ISDN.

A. BASIC RATE ISDN

ISDN basic access consists of providing the customer with two circuit or packet switched B channels at 64 kb/sec and one D channel at 16 kb/sec. The two B channels provide bidirectional digital connectivity, that is, the facility to see and use a clear transparent digital channel at 64 kb/sec bearing the information that the user wishes to communicate between the digital device and the distant customer. The two B channels may be individually and independently switched and provide digital communication with any other customer or source in the network. Voice, data, or video may be transmitted as digital information over the B channel. Initially, the circuit switched B channel will provide 56 kb/sec throughput capacity, and eventually it will provide the full 64 kb/sec capacity. The packet switched B channel will provide a maximum throughput of 48 kb/sec at a line rate of 64 kb/sec, (Layer 2—LAPB, subscriber digital line carrier or SDLC)

The D channel serves a variety of functions (see Fig. 8), including signaling and control information across the interface between the network and the customer. This signaling and control information manages the information-carrying B channels. This signaling is done out of band from the B channels thus facilitating the B channels to be clear and transparent. Further, the control of the B channels via the D channel facilitates robust distributed processing across the ISDN interface. Displaying the call origination address, call transfer, call forwarding information, connected address, etc., would be facilitated by the D channel. Call progress information, such as idle, dial, alert, and connect, will also be carried by this channel. Finally, the signal information identifies the type of tone being applied to the B channel. These tones are necessary to alert the digital facility at the end of the connection. These tones may also differ or be interchanged by the network or the user terminal.

The D channel may also serve two other functions: telemetry and low-capacity packet switched transport. The services included in the telemetry type of capabilities consist of burglar and fire alarms, energy management, and emergency services. For the low-capacity packet switched transport, the maximum throughput rate is limited to 9.6 kb/sec at a line rate of 16 kb/sec.

B. Broadband ISDN

A variety of higher rates are also available in the context of the ISDN. In North America, the H_0, H_{11}, and H_{12} channels support clear access at 384, 1536, and 1920 kb/sec, respectively. A separate D channel is used for signaling (control and maintenance). These channels may be circuit switched or packet switched.

In North America and Japan, the ISDN primary access takes place at (23B + D) rate. In Europe, the rate is (30B + D) at the primary rate interface (PRI). At this interface, the D channel is at 64 kb/sec. The single D channel can carry signaling information for as many as 40 B channels. The D channel protocol is defined by CCITT Q.921 specification or LAPD (link access protocol for D channel). This protocol is used for signaling and information transfer. It is similar to the protocol defined for packet data in the X.25 (LAPB). However, it allows for more than one logical link between end points because of the difference in the nature of the D channel information as opposed to the nature of the B channel information, which is targeted to one end point. In instances where the D channel carries X.25 data together with signaling, the LAPD protocol is used (at the data link layer, i.e., layer 2 of the 7-layer ISO model) for both logical channels—one reserved for signaling and one for X.25 data. At the network layer (i.e., layer 3), the protocol reverts to X.25 for the packet switched data and to CCITT Q.931 protocol for the signaling.

C. Rates and User Characterization

The use of the intelligent network capacities differs widely between the categories of users. Businesses (ranging from very large to small) are expected to use a large proportion of the services and features offered by intelligent networks. Hence, the higher rates are generally intended for the larger businesses for their interactive and batch modes of data transport needs, large-scale data-base access, and machine and intelligent terminal access of the network. Interfacing with the business's private networks and the digital PBXs will also become necessary at the higher network rates.

The lower standard for accessing the network at the basic rate is intended for small businesses and the residential market. The type of network and vendor services are of a different category in this sector. Home information, computers, vending, electronic directories, demand news, facsimile, low-rate video, etc., are expected to become standard features offered by the newer intelligent networks as the information society becomes better established. For most of these services, the basic (lower, 2B + D) rate is expected to suffice.

Wide-band access at a higher bit rate for specialized customers for large, high-speed data is also envisioned. Metropolitan area networks (MANs) with specialized and dispersed computational facilities may call for one intelligent network interfacing with another intelligent network. These types of high-capacity intelligent networks have been installed around scattered campuses and the nation by academic institutions and the National Science Foundation. The emergence of the optical DS3 using high-speed synchronous optical fiber networks (SONET) with data rates in excess of 2 Gb/sec will bring about the truly integrated optical switching and transmission in the architecture of intelligent networks.

VI. Intelligent Network Switches

The switching function is essential to the functioning of an intelligent and dispersed network. The same physical medium has to perform a large variety of functions, such as carrying voice, data, and video information from any other location bidirectionally. It must also be able to monitor the network response by program or by user interaction and seek and respond with the appropriate responses to system and user queries. Packets of information have to be switched within the network to appropriate modules that can respond to the appropriate user or network commands. Hence, the architecture of a dispersed intelligent network is not complete without an intricate fabric of remotely controlled switching elements.

A. Switching Facilities

In this context, switching of digital information takes place in three types of switches: channel (private), digital (circuit), and packet. The B channel information, and any X.25 packet in the D channel, generally flows through these switches. The control and signaling information in the D channel, which supports the various B channels, flows into the common channel signaling facilities. It is here that most of the intelligence and control reside in the intelligent networks. The signaling information to and from the customers, vendors, networks, etc., converges in these facilities and gets intelligently interpreted to interconnect the bearer channels and dispatch appropriate commands and control information throughout the network.

At the information gateway (typically an ISDN office) interfacing the intelligent network with the customer terminal, the D channel information is separated from the B channel information so that the individual B channels that are served can be individually controlled and signaled. The signaling and call processing information is generated and dispatched to the common channeling signaling interface. This control information flows through a common channel signaling network to the appropriate gateway (another ISDN office) where other functions, such as call setup, call completion, and call monitoring, take place. The B channels are set up for the flow of information between the customers, vendors, service providers, etc.

As is usual in the architecture of computer systems, the flow of control information is separated out from the flow of data. In intelligent networks, the concept is carried through by separating out the common channel signaling facilities from the switching facilities. The concept of distributed processing becomes applicable in the intelligent information networks, as it is in distributed computing networks. To implement the entire system, enormous amounts of software modules and algorithmic steps become essential. For this reason, the use of stored program control (SPC) becomes necessary. Fortunately, over the last three decades, the switching and the computing technologies have been synergetic. Most of the SPC concepts in computing have been implemented in the electronic switching systems (ESS).

B. Electronic Switching in Intelligent Networks

Over the last decade, more and more of the telephone switching environments have been under the control of stored programs residing within the realm of ESS. These switching systems are controlled by intricate software modules exactly as any other sophisticated computer. The reprogramming of these switching systems is relatively straightforward in light of the requirements of the intelligent functions expected from the network. For the circuit switching part of the facility, the existing #1ESS (introduced in 1965 to local switching and to toll switching in 1970) and #1AESS (introduced in 1976 to local switching) have the basic switching capacity to switch and be programmed accordingly. However, the common channel signaling aspects need to be upgraded to accommodate the intelligent network functions, especially those of the ISDN. The flexible and highly programmable digital switches are key components in the distributed and intelligent networks.

In North America, the advent of the more sophisticated ESS facility (#5ESS) was a major step in the realization of intelligent networks. Its modular hardware and software building blocks permit the capacity to customize these switching systems to perform the flexible and programmable functions with inherent mechanized intelligence. The more recently introduced #5ESS switch will meet the CCITT requirements for the (2B + D), (23B + D), and the (30B + D) rate interfaces.

In Europe, the steady evolution of the switching systems (Ericsson, AXE10) for intelligent network functions has been well along its way since 1984. The modularity for customization and for CCITT signaling system No. 7, which facilitates the customer services integrated in the network functions, has been incorporated into the newer switching systems being built in Europe. These switches also provide for the (2B + D), (23B + D) and (30B + D) rates.

VII. Trends in Intelligent Networks

Public domain intelligent networks can only evolve because of the massive capital expense necessary to standardize, design, and construct national and international networks. Small, private networks with any reasonable amount of sophistication can be built with the existing technology. Between these two extremes lie a group of specialized, high-capacity, sophisticated networks, which can be classified as highly intelligent in their own right. Typical examples of these are military, industrial, computer, campus, and private networks. However, if one specifies the public domain networks, their vendors, services, and specialized features, a few general remarks may be made.

The PDINs invariably have to employ the most economical of the possible modes for data switching and transmission. The well-established concept of stored program control of electronic switching has been implemented by most of the major vendors for digital switches. There is a rapid trend to replace the electromechanical switches by digital switches. The analogue switches are being replaced more slowly because of the continued demand for the POTS by a majority of the rural and suburban customers.

The impact of optical switching will have its own effect on the switching functions of the intelligent networks. Optical switching holds enormous potential for speed of and access to the time division multiplexing of the various channels. Integrated optoelectronics offers features that the SPC switching systems accomplish now with the programmable software modules. Typical functions that can be performed are customized services, selected data-base access, the interface with the packet switched networks, common channel control interface, fiber-optic media, and the large number of digital carrier systems that are in use.

Optical media for the transmission of high-speed digital information are other major forces that shape the access possible in the PDIN environment. Optical fiber and optical communication systems are being introduced in most digital networks. The flexibility, economy, and rate of transmission (up to 1.7 Gb/sec with 20-mile repeaters and 295 Mb/sec overseas intercontinental ranges) make these systems "move" gigabits with extreme flexibility and accuracy. Coupled with optical switching facilities and optical computing techniques, the intelligent networks will enter their next generation.

Information vendors play a critical role in supplying customer information via the network. There is a definite need for these vendors to seek and supply information from their own data bases. The queries can be complex and also in a natural language. For this reason, the front-end processors of the vendor data-base systems need the recently introduced natural language processors.

In summary, the three emerging technologies (transmission, switching, and data-base) unified into one composite science of network architecture are modifying the course of IN evolution. At the present time, intelligent networks are in the stages of conceptual evolution that computers were during the sixties. The impact of very large scale integration (VLSI) was felt by the computing industry. Some very positive events, such as massive mainframes, chip computers, computer graphics, and artificial intelligence, were yet to occur. Intelligent networks will find their direction from the growth of ISDN, massive database technology for information vendors, optical switching, integrated optical transmission, integrated computing and switching in the communications arena, and worldwide standardization of signals and protocols. Software and knowledge engineering will also facilitate the final implementation of the global, public domain intelligent networks.

BIBLIOGRAPHY

Ahamed, S. V. (1982). Simulation and design studies of the digital subscriber line. *Bell Syst. Techn. J.* **61,** 1003–1077.

Bell Telephone Laboratories (1982). "Transmission Systems for Communications." Western Electric Company, Winston-Salem, North Carolina.

Miller, M. J., and Ahamed, S. V. (1987). "Digital Transmission Systems and Networks, Vol. I: Principles." Computer Science Press, Rockville, Maryland.

Miller, M. J., and Ahamed, S. V. (1988). "Digital Transmission Systems and Networks, Vol. II: Applications." Computer Science Press, Rockville, Maryland.

MICROWAVE COMMUNICATIONS

U. S. Berger[†]

GLOSSARY

Bandwidth: Frequency occupancy needed for the satisfactory transmission of useful information. Very often the frequency spectrum is greater than that theoretically needed.

Baseband: Band of frequencies, usually the lowest frequencies in the microwave system, where basic information is assembled; this spectrum is generally that provided to the microwave system to be delivered to a distant point in the same format and information content.

Carrier: Used with two widely different, yet unambiguous meanings: (1) in business, organizational entity to render communications to customers or users; (2) electrical wave which by itself does not carry information but is generally modulated by another wave which does contain desirable information.

Compandor: In telephony, use of a compandor (a device that reduces the amplitude level of strong talk spurts) at the head end of a system permits the average level of the voice channel to be raised without exceeding the system design level; this automatically raises the level of the weak talk spurts in relation to the stronger speech signals. At the receiving end, the inverse is done (expandor) to restore the original relationship of the strong spurts to the weak ones. Overall effect is a very substantial improvement in the signal-to-noise ratio on voice circuits.

Dielectric resonators (YIG): Magnetic devices that have certain desirable characteristics, primarily magnetic resonances, which can be used for frequency selective elements in microwave networks (filters, oscillators, equalizers, etc.).

Digital transmission: Signals sent as a series of pulses of a prescribed amplitude and received and detected in predetermined time slots. Digital signals for microwave transmission are usually encoded and processed in such a way that the spectral bandwidth is minimized.

Downconverter (or mixer): In a superheterodyne receiver, the circuit that accepts the SHF signal and the channel carrier to produce a difference frequency known as the intermediate frequency.

Drop and add: In a long, multichannel microwave communication facility not all of the information is conveyed from one end of the system to the other: signals may be dropped to customers along the way (e.g., network television). Similarly, the channel may accept new (added) signals at intermediate points. Access to the system baseband is necessary for this function.

Hot standby: Second repeater tuned to the same frequency as another and available for substitution into the system for the working equipment that has failed. Substitution is usually made automatically after a failure has been detected.

Intersymbol interference: In digital systems, distortion can cause a part of one pulse to

---

[†] Retired from Bell Telephone Laboratories.

overlap other pulses so that it may be difficult or impossible to identify the presence or absence of a specific pulse of interest.

Isotropic antenna: Theoretical antenna of infinitesimal size in which it is assumed that all of the energy is radiated (point source). It has no visual shape or form and has no practical use except as a basis for reference for other antennas of finite dimensions.

Intermediate frequency (IF): Frequency, in the vicinity of 70 MHz (which is a common frequency for all channels) at which considerable amplification takes place, interconnections are made, automatic gain adjustment is provided, and channels are disabled upon command (squelch). Use of the relatively low frequency simplifies the design of these functions compared to performing them at SHF. IF is derived by demodulating the SHF signal with a channel carrier.

Intermodulation: Heterodyning or undesired mixing of two signals in a nonlinear device to produce an undesired product.

Klystron: Electron tube that can operate in the SHF bands. Some klystrons are designed as oscillators with a single cavity (reflex klystrons). Multicavity klystrons can amplify or oscillate, depending upon the configuration of the external components used with the device.

Modulation and demodulation: Process whereby useful information is translated from one portion of the frequency spectrum to another. Demodulation is the inverse of modulation. Modulation is sometimes used interchangeably with detection. Modulation is frequently used to connote the mixing of two frequencies to derive a third frequency.

Negative resistance: Only positive resistances exist as discrete passive components that can be purchased to specific values from a vendor. Positive resistances obey Ohm's law, where $R = E/I$. This means that the component is linear; higher voltage across it will result in higher current through the device. In a negative resistance, $R = -E/kI$, which states that over some range of values an increase in voltage will result in a reduction in current. This relationship will maintain only over a limited range of the $E-I$ characteristic.

Noise figure: Measure of the excess noise in a device over the theoretical or ideal conditions. Very useful measure of the quality of an active component in a radio system. The lower the noise figure, the better the performance and the less contribution the component will make to system noise.

Polarization: Relation of the electromagnetic-wave vectors to some reference. Polarization may be vertical (wherein the electric vectors are in a vertical plane), horizontal, circular (right or left handed), etc. Coupling between H and V polarizations can be substantially minimized with careful design, and this can be helpful in reducing adjacent channel couplings in multichannel systems. 30–35-dB cross-polarization discrimination can be obtained in some cases.

Repeater: Radio station capable of receiving a weak signal on a specific frequency and substantially increasing the power for transmission on another frequency within the system spectrum.

Superheterodyne: Type of radio equipment in which an incoming frequency is converted to another frequency for purposes of amplification and for providing selectivity against unwanted frequencies. A carrier frequency differing from the incoming signal by a specific amount, the IF frequency, and coupled into a nonlinear device produces the modulation function.

Superhigh frequency (SHF): Wavelength shorter than 10 cm; frequencies above 3 GHz.

Waveguide: Usually a hollow metal structure intended to guide an electromagnetic wave along its length. Internal dimensions are related to the efficacy of transmission at specific frequencies. Cutoff frequency is the frequency below which a particular waveguide cannot satisfactorily transmit the wave.

Waveguide modes: Depending upon the shape and size of the waveguide with respect to the length of the wave traversing the guide, one or more patterns of wave propagation may exist within the guide. Each of these modes will have a specific topology, different velocity, and different energy distribution along and across the guide cross section. Both desirable and undesirable modes can exist under specific conditions.

The transmission of television, telephone, and data services via superhigh radio frequencies is known as microwave communications and has been available since the late 1940s. Terrestrial microwave systems consist of a series of radio stations, generally located within line of sight of the adjacent stations, where the signals may be retransmitted repeatly to achieve a system of substantial length. This has frequently been referred to as a microwave radio relay system or a microwave repeater system. Usually, these facilities are capable of carrying a number of TV programs or thousands of telephone or data circuits simultaneously. Microwave communication systems have become highly reliable, efficient, and economical in handling communications of all types, and a vast network of circuits has been provided over the past 40 years. Satellite microwave systems have supplemented the terrestrial facilities in providing worldwide communications.

I. Introduction

Communication via microwave radio is not a recent development for theoretical studies of radio transmission in the superhigh frequency (SHF) band (3 GHz and above) and was carried out in the late 1930s. Throughout the world, scientists and engineers were pushing the frontiers of all radio transmission higher and higher in frequency. Well before World War II radio communication was being vigorously pursued in the frequency region from 50 to 500 MHz for frequency modulation (FM) broadcasting, point-to-point transmission, television broadcasting, and vehicular communication (including aircraft). Military aircraft radio communication in this portion of the spectrum was highly successful in the 1940s and aided in the defeat of the Japanese in the Pacific because they were unaware of our substantial communication capabilities at these higher frequencies. Television broadcasting, which requires substantial bandwidths in the radio spectrum, was demonstrated to the public at the New York World's Fair in 1939. All of these were part of the evolution of the field of communication that started in the 1930s and has continued until the present with communications people today working at lightwave frequencies with glass fibers as the transmission medium. [See RADIO SPECTRUM UTILIZATION.]

II. Early Work

Researchers in many countries in the 1930s were exploring radio at even higher frequencies, and many basic concepts pertaining to SHF were well understood before World War II. The lack of adequate components and test equipment slowed the progress of these studies, and it was often necessary to pause and invent or make the parts that were needed.

At frequencies above ~3 GHz, some of the components and subunits within a microwave station are interconnected with waveguide, but in the early days the waveguide itself was equally unknown and had to be thoroughly investigated. One of the research pioneers in this field was George Southworth of Bell Laboratories. He and his research team evaluated the loss and other characteristics of a radio wave as it was propagated through a smooth metal pipe. To obtain meaningful results without a prohibitively long section of pipe, a very short pulse of radio energy was injected into the pipe and allowed to traverse back and forth many times until the loss in the walls of the waveguide attenuated the signal below a detectable level. Counting the detected pulses, measuring the time required for successive pulses to pass the detection point, and observing the amplitude of the pulse each time it appeared at the detector, the characteristics of the medium could be determined. Other equally significant studies of microwave signals were conducted by many people during this period.

As a result of the work done before 1940, it was possible to develop a four-channel microwave system for the Armed Services operating in the 4.3–4.8-GHz band using a reflex klystron tube with an output power of 0.1 W. This system was capable of carrying initially 8 voice circuits using pulse-position modulation with time multiplexing of the 8 circuits in the pulse stream (before the end of the war the capacity had been increased to 16 circuits). This early system was used in the field in Europe from 1943 through 1945 and provided communication over impassable terrain. With the pulse signal format, this system was the forerunner of what today is one of the newest and most important forms of communication, namely, digital transmission. During and after the war, tests were made in the field to study propagation characteristics as they might be experienced in actual communication systems. A 40-mile link in the New York–New Jersey area was established, and propagation tests were made at a number of frequencies over a frequency range of 1 to 24 GHz. These results were extremely important in determining how a relay system might perform in a field environment and laid the basis for planning long-repeatered systems. [See DATA COMMUNICA-

TION NETWORKS; DATA TRANSMISSION MEDIA.]

III. Early Needs and Applications

At the close of World War II, there was a backlog of needs for all forms of consumer goods, one of which was television. The previously mentioned demonstrations at the World of Tomorrow Exhibition in New York and the promise by the television people that their offering was about ready for widespread release to the public raised the question as to how the program material could be distributed to local broadcasters. While television could be transmitted via coaxial cable, the limited amount of cable available in the ground, the high cost of laying additional cable, and the length of time required to put new facilities in place throughout the country made microwave radio a very attractive alternative. The high cost of a cable system was due to (1) the need for rights-of-way for the entire length of the cable trenches; (2) the high construction costs of routes through rocky terrain, particularly in the East, the Northeastern United States, and the Rocky Mountain region; and (3) the cost of the cable itself. Radio sites, with average spacings of about 25 miles on high hills and often in remote areas, were much more readily obtainable and at lower initial cost.

In 1944, plans were made for the development of an experimental microwave radio relay system between New York and Boston that could provide television service between the two cities, but more importantly, could also answer a myriad of very important questions on economics, expected quality of transmission, reliability, maintenance requirements, public acceptance, entrance into cities, and relations with the entertainment industry. Earlier studies had already addressed the question of what frequency band should be used for this type of service, and it was concluded that somewhere between 3 and 10 GHz would be acceptable from a propagation standpoint. From the component point of view, the lower end of this range would be preferable for the first system. Below 3 GHz, components tend to become large and cumbersome, whereas frequencies above 9 or 10 GHz are subject to severe attenuation problems in areas of high rainfall such as in the Gulf Coast region. Negotiations with the FCC led initially to a common carrier frequency allocation of 3.9 to 4.2 GHz, but this was later expanded to 3.7 to 4.2 GHz.

Other bands at 6 and 11 GHz were explored but developed later.

The four 2-way channels with seven intermediate repeater stations and the terminal stations in New York and Boston represented a substantial amount of microwave hardware that had to be manufactured and installed. Including all of the ancillary items such as antennas, buildings, and power and site preparation, the aggregate added up to a rather expensive experiment. The equipment units in the New York–Boston system included klystron oscillators and klystron amplifiers to boost the power to approximately 0.5 W. Narrowband FM was selected for the modulation plan to avoid the nonlinear amplitude distortion effects of the klystron amplifiers, which would introduce intolerable distortions in the television picture and in the telephony circuits that were to be studied during the early testing. The New York–Boston SHF radio link was completed in November of 1947 and was turned up for use by the television industry for their experiments. In May of 1948, commercial service between the two cities was inaugurated.

While the New York–Boston system was being constructed, plans were being made to build a much longer 4-GHz facility to serve a larger portion of the country. New York to Chicago was the first phase to be built through New York State, along the Great Lakes in Ohio and Indiana, and into Chicago. This flat farm country was easy terrain on which to build a radio system. The route was expedited for the 1950 World Series, and this was a treat for those television viewers along the way who previously had only local programs or film presentations to watch. It would not be correct to say that there were no troubles in the New York–Chicago route: equipment failures, particularly failures of electron tubes, were numerous and a major source of trouble, but this may not be surprising when the number of components operating in tandem is taken into account. There were about 20 electron tubes in each repeater and about 35 repeaters in the route for a total of 700 tubes, many in unattended locations. Component life then was not good, but even with a mean life of 5000 hr for the tubes, a failure could be expected every 7 hr. Not knowing which tube would fail next, the total downtime of the system was appreciable and the maintenance effort was high. These were the growing pains of a new venture.

Despite these kinds of reliability problems and some performance difficulties, the second phase of the transcontinental route was being planned before the troubles in the first section were all solved. The Chicago–San Francisco route was more difficult to build through the mountainous areas, but sites could be readily found in the farm belt of the Midwest and the mountaintops of the Rockies. Service dates were targeted for the San Francisco peace talks with Japan in September of 1951, and with about 65 repeater sites to be built, installed, and tested, the task was enormous, but again the deadline was met. Reliability continued to be a problem, and like any first experience, many of these difficulties were overcome by careful study, analysis of the failure information, redesign of parts that were causing trouble, and improving manufacturing techniques of many components. In Section VIII the subject of reliability is addressed in greater detail. While the backbone of a network was being formed, side branches were also being planned and constructed in many parts of the country, and this has continued over the years until the present; virtually every locality is being served by microwave radio.

IV. Microwave System Design

Some of the considerations that go into the design of a microwave system include: (1) the service needs, (2) the initial size of the route in terms of length and channel cross section, (3) the potential for growth, (4) the number of drop-and-add points where service is to be supplied, (5) the availability of frequencies in the vicinity of the proposed route, (6) the availability of suitable sites for the repeater stations, (7) the noise and reliability objectives, (8) the availability of power and access roads, and (9) not the least, cost.

There are two general types of systems in use: (1) the long-haul system, which may have only a modest number of drop-and-add points but may have many repeater stations and may go a long distance end to end, and (2) the short-haul system, which may consist of a relatively small number of repeaters with frequent drop-and-add points. Both of these systems use a superheterodyne type of receiver with an intermediate frequency of 65 or 70 MHz.

Three common carrier bands, 4, 6, and 11 GHz, have been used for communication purposes, and generally, the best overall use of these bands has been made by the communications industry. The early use of 4 GHz for long-haul service, such as New York to San Francisco, more or less established a pattern that 4

GHz would be used for long haul. Similarly, the 11-GHz portion of the spectrum, which experiences problems in heavy rainfall areas, would not be used in long-haul facilities. As might be expected, 6 GHz has been employed for both applications, and because short, light route sections of 6 GHz were built in a great many places in the United States, full development of long-haul 6-GHz routes was not always possible.

A. Short Haul

Short-haul systems are generally used where traffic loads are relatively light or the length of the route is not great, for television pickup from remote locations, for intrastate facilities for the common carriers, for control communications for gas and electric utilities, for telemetry control signals for dams, and for some government communications. Because the distances may not exceed a few hundred miles and the traffic load may be relatively light, the performance requirements have been set slightly below long-haul standards. The need for frequent access to and from the system has mandated detection and modulation capabilities in the repeater, and when these functions are provided, some distortion occurs. This type of repeater is known as a remodulating repeater, and the connection between the receiver and the transmitter within the repeater is made at baseband frequencies (0–5 MHz) through a level-adjusting network to assure that the proper deviation of the transmitter is obtained.

B. Long Haul

Long-haul systems, on the other hand, may have a number of repeaters that require no dropping or adding, primarily because they may be located in open country away from cities or towns where no baseband access to the system is required. It would be unwise to use remodulating equipment in these situations because the distortion would be too high and the deviation of the transmitters would be difficult to hold to the proper value. This leads to a second type of repeater, namely, the heterodyne repeater, wherein the interconnection from the microwave receiver to the microwave transmitter is made at the 70-MHz IF rather than at baseband.

C. Growth

The service needs of a system, both initially and in the future, determine the number of radio channels that must be included in the design plans. Experience of almost 35 years has shown that growth of microwave use in virtually all areas of the United States has occurred but not always at the same rate. In many cases, the original construction took care of only the initial requirements, and later, building and equipment additions were made to accommodate the growth needs. The 4-GHz band that was first developed, in many areas, has long been filled, and the facilities cannot be expanded further on those routes. The same can be said of the 6-GHz band; so, if these frequencies are to be reused for additional growth, new routes have to be constructed. This may be feasible in the open areas of the country, but eventually the routes must converge in the cities, where the service needs exist. The entrance problem has become an ever-increasingly serious one. In the period from the early 1950s until the present, microwave systems have grown into vast networks carrying all forms of communication services to all parts of the nation. Some telephone and data circuits traverse 2500 miles or more with little or no intermediate demodulation.

D. PERFORMANCE

Noise and reliability objectives for the carriers are invariably set high, because many customers are dependent upon these facilities for important communications. A short "hit" or bat of noise may be of no consequence to a home viewer of a television program, for the eye is highly forgiving of a minor signal imperfection, or to a telephone user, because a slight click in the telephone receiver is ignored, but the same "glitch" to an unforgiving computer may raise an order for 5 tons of steel to 5000 tons. (It was for this reason that data systems introduced error detection capabilities.) Noise and reliability enter into such crucial system design factors as transmitter output power, receiver noise figure, and repeater spacing.

Repeater spacing is determined from many considerations. Microwave transmission is almost always "line of sight," because this results in by far the best performance, and therefore high locations are almost always selected for the repeater station. Clearance over hills, trees, buildings, and other obstructions is sometimes difficult to obtain at a given point, and the designer may have to reexamine the terrain to get a radio path that is acceptable. If the distance between repeaters is too great, the received signal

may be too low and atmospheric fading may be excessive. However, route cost savings can result if repeaters are located far apart because less equipment will be needed. Conversely, short hops can improve both noise performance and reliability, but the cost will increase because more equipment will be required. Economics play an important role in system design. AT&T has a vast microwave network with an average hop length of about 26 miles. There are some hops as short as 1 mile because of terrain factors, for example, providing service into a valley from a repeater on a nearby mountain ridge. Repeaters over 50 miles apart to span some territory, such as an Indian reservation in the Southwest, are also found. A few long spans can be accommodated by the statistics of the performance of the system or by taking special steps, such as providing space-diversity antennas, to overcome serious fading phenomena.

E. SYSTEM CAPACITY

In terrestrial FM microwave systems, only one television program can be carried on a single radio channel. There are a number of reasons for this: (1) the 20-MHz channel width at 4 GHz will only accommodate one television program; (2) the stacking and unstacking of a second television program above the first one in the baseband makes for very complex FM terminals; (3) interference between the two television programs on the same radio channel may put bars in the television pictures, and if the two pictures are not synchronized, the patterns will move; and (4) if the two signals are not coterminous, a television network management problem arises. Accordingly, television transmissions via microwave relay facilities are handled as a complete channel load and can be switched at IF if necessary.

Telephony, on the other hand, started with 480 circuits on a fully loaded 4-GHz radio channel in 1950, and as the technology developed, more and more circuits were added until, a quarter of a century later, the 4-GHz channels could each carry 1500–1800 4-kHz voice-frequency circuits (for either voice or data). The 6-GHz FM channels are about 30 MHz wide and can carry 1800–2400 voice-frequency circuits on each SHF channel.

When single-sideband (SSB) amplitude modulation (AM) at 6 GHz was turned up for service in early 1981, the voice circuit capacity was increased to 6000 circuits. It is quite apparent that if the basic components of a system, such as the antenna, waveguide, tower, basic and emergency power, station alarms, and land, can be used with a new system of greater capacity per radio channel, the individual voice circuits will be more economical and the development of a new and different kind of system can be justified.

F. FREQUENCY COORDINATION AND CHANNELIZATION

Frequency coordination with other services and users in a specific area must be done to avoid serious interference problems and to provide for future growth for all users. The long-range plans of a carrier are taken into consideration when another service seeks to operate in the same band in the same area. Occasionally, the problem is adjudicated with the assistance of the FCC.

Very early in the history of microwave communication systems, extensive studies were made on frequency plans and channelization schemes and recommendations were made to the FCC for industry-wide acceptance. Without standardization of a plan and the subsequent licensing of stations to use specific frequencies, utter chaos would exist. After the New York–Boston development was started, a frequency plan for the 3.7–4.2-GHz band was established using 24 20-MHz-wide channels with 12 transmitter frequencies interleaved with an equal number of receiving frequencies across the 500-MHz band. This scheme spread the high-level transmitting signals over the entire authorized 4-GHz band, and at that time, it was felt that this spreading would ease the receiver design problem. This channelization arrangement provided 12 usable 20-MHz frequencies in each of two directions of transmission. (Only a very small percentage of the licensed channels are used for one-way service. Studio transmitter links and remote television pickups are two examples.) An IF of precisely 70 MHz was chosen somewhat arbitrarily and, once so selected, was retained for all 4-GHz equipment. This selection was satisfactory because, at that time, all repeater equipment used self-contained microwave frequency generating equipment.

When the 6-GHz common carrier band was developed starting about 1950, the plan placed eight 30-MHz transmitting channels in one-half of the 5.925–6.425-GHz spectrum and eight 30-MHz receiving channels in the other half, with a small guard band between. This was felt to be an advantage in systems using separate antennas for transmitting and receiving because the interferences would be less. It has turned out that these two plans are "fielder's choices" because, in later years, the same antenna has been used for both 4 and 6 GHz simultaneously. Hindsight after 30 years of experience indicates that only one scheme should exist, but unfortunately, once a decision is made and then is accepted as the standard, any deviation from it is impossible to introduce. The present arrangement is workable as attested to by the large numbers of 4–6-GHz joint routes, but without knowledge of the original arguments for one frequency plan versus the other, the justification for two plans to exist is virtually impossible to comprehend.

The first 6-GHz long-haul system used a common source of microwave frequency generation for all of the repeaters in a relay station to minimize intermodulation tone problems. In essence, all channels were "synchronized" to a common source. To take full advantage of this asset, the IF was established at 74.1 MHz and the channel spacings were set at 29.6 MHz. This has become the industry standard on frequency assignments in this band and has also been adopted for the newest SSB AM radio systems.

V. Waveguide

Reference has already been made to the theoretical studies done on waveguide and its use in microwave systems. Waveguide is a component that, as the name implies, conducts or guides the microwave energy. Electromagnetic theory indicates that if the frequency is high enough the center conductor of a coaxial structure is no longer needed and the energy can be propagated on the inner walls of the medium. The cross section of the guide can be square, rectangular, circular, or elliptical. Each one of these configurations supports propagation in one or more modes, a mode being the way the electromagnetic energy is distributed within the pipe. Discrete modes or wave topologies are often used to obtain some specific result. The waveguide dimensions are somewhat critical for they have to be large enough to support the desired modes at the lowest frequency of interest, yet not be so large that undesired modes are also propagated. The waveguide size determines the cutoff frequency, a frequency below which satisfactory propagation cannot be achieved. Therefore, systems operating in different bands use different sizes of guides.

A. RECTANGULAR WAVEGUIDE

Rectangular waveguide usually has a cross section with an aspect ratio of $1:2$, the width being about twice the height. For operation in the 3.7–4.2-GHz band, the large inside dimension is 2.29 in. For the 10.7–11.7-GHz band, the wide dimension is 0.9 in. Waveguide is coded WR- (waveguide, rectangular) with numbers following to indicate the large dimension (i.e., WR-229). The numbers commonly used are the width dimension with the decimal point omitted.

The dominant mode in a rectangular guide is the TE_{01}, the transverse electric mode, where the electric vector is perpendicular to the direction of propagation in the guide. In this case, the electric wave vector is from one wall of the large dimension to the other.

The loss of waveguide is somewhat dependent on the smoothness of the guide walls and also on the material from which it is fabricated. Generally, the waveguide is carefully drawn from billets of copper, brass, or aluminum. The loss for copper is less than 2 dB per 100 ft at 4.2 GHz. Pieces of waveguide or components fashioned from waveguide are joined together by carefully machined flanges on mating parts.

B. ROUND WAVEGUIDE

Round waveguide is also widely used for connections from the electronic equipment to the antenna system because of its lower loss. Round waveguide can support several modes, depending upon the frequency and how the waves are launched into the guide. Being a symmetrical structure, two polarizations can be carried, and by careful design, it can be used for several frequency bands simultaneously. For example, WC-281 (2.81 in. in diameter) is used for systems in the 3.7–4.2-GHz, 5.925–6.425-GHz, and 10.7–11.7-GHz bands. Waveguide components that are complicated in mechanical design or are difficult to manufacture by casting or machining techniques are often electroplated using precision mandrels to control the size and shape of the part.

C. ELLIPTICAL WAVEGUIDE

Waveguide with an elliptical cross section is often made from an intricately formed metal ribbon that is wound on a shaped mandrel in such a fashion that the finished product is semiflexible and can be bent on a large radius without deforming the cross section of the guide. This structure will support only a single polarization in the TE_{01} mode. Its use has been limited and is chosen primarily for its lower cost and the ability to manufacture it in long sections and to curl it up in large coils for shipping. Field flanging techniques have been developed and installation is relatively simple.

D. SQUARE WAVEGUIDE

Square waveguide is rarely used but has some applications in waveguide components that require the propagation of two polarizations at the same time and for which circular guide is unsuited. An example is a case in which two polarizations of a signal from the round antenna waveguide must be converted into two rectangular waveguides at the equipment location.

VI. Antennas

One of the most important components of a microwave system is the antenna. It is important because (1) system performance is highly dependent upon its characteristics, (2) the antenna and its supporting structure (tower) are physically the largest part of the station, and (3) the antenna tower structure and the ancillary components, such as the antenna feed line and supporting hardware, represent a sizable portion of the cost of the repeater installation. The microwave antenna is physically large because it must provide high gain to make up for the huge propagation loss that exists between the two repeater sites. The loss between two isotropic antennas operating on a frequency F (in gegahertz) with a spacing of S (miles) between the sites is

$$\text{loss} = 20 \log 71,000 \times F \times S$$

At 4 GHz with $S = 25$ miles, the loss is 137 dB.

Antenna design, predicated upon the principles of reflection and refraction of electromagnetic waves, has led to three basic types of structure. The first and most widely used is the parabolic reflector or some portions of a parabola, followed by the lens, and then by the passive reflector or periscope. Two types of lens antennas have been used, namely, the accelerator lens and the delay lens, each acting on the emerging wavefront. These designs are inherently frequency sensitive and therefore cannot be used in multiband systems. The New York–Boston route used the accelerator-type lens, while the early transcontinental 4-GHz stations employed the delay lens structure. Neither of these antennas is used to any extent today because the parabolic reflector design operates effectively over wider bandwidths and with two degrees of polarizations. [*See* ANTENNAS.]

A. PARABOLIC REFLECTOR ANTENNA

A parabolic reflector, with an antenna feed at the focal point, will radiate energy from the reflector with parallel rays that can be focused. Similarly, parallel rays entering a receiving antenna reflector will concentrate the energy at the focal point. In the case of a transmitting antenna, a beam is formed which can be concentrated and directed toward a distant target, and theoretically, no energy is transmitted backward from the reflector. Similarly, in the case of a receiving antenna, little energy will be received from the backward direction. A microwave antenna is stated to have gain compared to an isotropic radiator; yet it is a passive device and no external power is added. This so-called gain is achieved by the high directivity obtained from the large diameter of the reflector compared to the wavelength of the signal and the use of a signal launcher that illuminates the reflector from the focal point. The larger the area of the parabolic surface, the greater the directivity and, hence, the gain. For example, an antenna 3 m in diameter has a gain at 4.2 GHz of approximately 40 dB. One watt of power from the antenna feed at the focal point will provide an effective radiated power of 10,000 W and will have a beam width of approximately 3° ($\pm 1.5°$) at the 3-dB down point of the major lobe. This requires high accuracy in aiming the antenna toward a target 25 or so miles away. The gain of the antenna in decibels is equal to 20 log 7.8FD, where F is the frequency in gigahertz and D is the diameter in meters referred to an isotropic radiator. [*See* ANTENNAS, REFLECTOR.]

In terrestrial microwave systems, the antenna is generally pointed at or near the horizon, and for a parabolic antenna on the top of a tall building where the SHF equipment may be nearby, rectangular waveguides can be used to connect to the antenna feed elements. Two polarizations will require two waveguide runs. If the antenna is on a tall tower and the equipment is at ground level, the two rectangular guides needed may have excessive losses. It is difficult to connect a direct radiator parabola with circular guide because of the vertical-to-horizontal transition needed at the top of the tower. The horn reflector antenna, invented by H. T. Friis in the mid-1940s and in wide use today, uses only a section

of a very large parabolic surface as the radiator and is illuminated from the focal point by a tapered horn of either round or square cross section in such a way that the vertical-to-horizontal transition takes place within the antenna itself. The horn has a carefully shaped transition taper from round to square to connect the WC-281 circular waveguide to the square throat of the antenna with minimum mode conversion effects. This use of circular guide allows the microwave equipment to be located at the base of the tower without incurring prohibitively high losses. This antenna has metal sides for structural reasons and, therefore, has excellent side-to-side shielding, permitting several antennas to be mounted on the same tower platform. It has excellent radiation patterns, high gain, and a large front-to-back loss ratio. This antenna can be identified in the countryside by its appearance, which resembles a cornucopia or a large ice-cream cone atop the tower, and frequently there will be a cluster of four or more horn reflector antennas at large repeater stations. In recent years, this antenna, made by a number of vendors, has been altered to include a conical cross section from the circular guide to the parabolic surface.

B. RECEIVE TELEVISION ONLY

The parabolic antenna is gaining wide acceptance for receive-only earth stations to provide home reception from television satellites. Many households today display a dish in the yard, which is not easy to hide and still have it function efficiently. It may become a status symbol despite its unaesthetic appearance in some neighborhoods. Most home antennas are ~3 m in diameter and, as previously stated, have about 40 dB of gain, which usually provides an acceptable picture for home viewing.

Satellites for commercial applications such as telephone, data, and network television transmission use much larger parabolic antennas; some are as large as 10 m in diameter and have gains approaching 60 dB at 6 GHz. This is necessary to provide the performance needed for these services. Harvard University has a radio astronomy parabolic antenna 25.6 m (84 ft) in diameter with a gain of 58 dB at the frequency of interest, which is 1.25 GHz. At Arecibo, Puerto Rico, the world's largest antenna is 1000 ft in diameter and has 60 dB of gain at 430 MHz. In some areas of the country, large dish antennas must be heated to prevent ice and snow from collecting on the parabolic surface and distorting the wave pattern, which would otherwise introduce excessive losses.

C. PASSIVE REFLECTOR

The last type of antenna that has been mentioned, and which was in widespread use in the 1960s for short haul, is the passive reflector. In this case, the usual configuration is a parabolic antenna mounted at ground level and radiating vertically. A nearby tower supports a large metallic plate, usually at about 45° with respect to the ground, which serves as the radiating element. This arrangement saves the expense of the waveguide feed line up the tower, but the reflector is difficult to adjust because two angles are involved (the angle of incidence and the angle of reflection), and it also has the tendency to allow energy to spill past the reflector and interfere with other services, including orbiting satellites. This antenna has been "outlawed" for new construction, although many still exist under a grandfather clause.

VII. Components

Many components go into a transmitter–receiver unit and the ancillary subsystem parts needed to make up a total transmission facility. The evolution of microwave components from the days of New York to Boston to the present day is a saga of the work of thousands of individuals and groups that honed the technology from a crude beginning to a high degree of sophistication for a wide variety of communication uses and needs. As new technology, materials, processes, and inventions come into being, microwave system designers keep a careful eye on how these can be used to improve or expand the capability of the systems at hand or how they can be used for the next system to be developed. Such scrutiny in the past has paid huge dividends, and today's advanced designs are far, far ahead of those of a generation ago. The array of components and subsystems that make up the transmitting and receiving portions of a microwave system and the integration of this equipment into the repeater installation requires meticulous attention in the design and the manufacture of each of the individual parts.

A. KLYSTRONS

The source of microwave energy for the New York–Boston system, and many other early systems, was the klystron electron tube, a device that, because of its internal construction, can be made to amplify or oscillate at SHF. These tubes were invented prior to World War II and became the first available for either laboratory experimentation or working systems. A reflex klystron has a resonant cavity that can be tuned over a small frequency range by exerting pressure on the cavity wall to change the internal dimensions of the tube and, therefore, the frequency of oscillation. Multicavity klystrons can be used as amplifiers by injecting an input signal into one cavity near the gun structure and extracting the amplified signal from another cavity further along the electron stream. A two-cavity klystron was used in the transmitter of the New York–Boston system.

B. PLANAR TRIODE

It was learned even before the New York–Boston experiment that the klystron tube would not be satisfactory for long systems, and in the mid-1940s work had started on a three-element electron tube that would amplify at 4.2 GHz. J. A. Morton and others conceived a high-performance vacuum tube that had internal electrode sizes and spacings only a fraction of what had been used before in other electron devices. The tube that evolved had a grid structure whose wire was only 0.0003 in. in diameter and which was tension wound 1000 turns to the inch. The spacing between the grid and the cathode is 0.0005 in., which is much less than the diameter of a human hair. The transconductance of this triode is 50,000 μmho. This tube is known as a planar triode tube because the grid is a flat, meshed disk sandwiched between a flat cathode and a flat plate. Operation of this early microwave triode in a grounded-grid, tuned cavity structure had a power output of only 0.5 W across the 3.7–4.2 band. The power limitation was one of poor heat dissipation of the glass used as insulation in the tube envelope. The input power to the tube had to be kept low enough to avoid damage and early failure of the device.

This tube has turned out to be an extremely important component for one very large system, and development efforts to improve its performance, life, and output power and reduce its cost continued for 25 yr. Replacement of the glass insulation with beryllia (a ceramic with excellent insulating and thermal conductivity properties) in the late 1960s, in part, contributed to an increase in the safe operating power level of the tube. Improved versions have provided 5 W of output power in some applications. Today, this tube has been replaced by solid-state gallium arsenide field-effect transistor (GaAs FET) amplifiers in virtually all applications.

C. TRAVELING-WAVE TUBE

About the time that Morton started work on the planar triode, Rudolph Kompfner of England conceived a device that became known as the traveling-wave tube (TWT) which consisted of a gun structure to focus an electron beam down the center of a small-bore helical winding and a collector to receive the high-speed electrons at the far end. The microwave signal to be amplified is coupled into the helix at the gun end, and the electron beam is focused along the tube length by a precision magnetic field around the outside of the tube envelope. The signal propagates down the helix, gaining energy from the electron beam. Output is coupled out of the tube near the collector. Traveling-wave tubes for transmitter applications have been made with power outputs ranging from a few watts to several hundred watts in the SHF band. One virtue of the TWT is its very large bandwidth.

The first application of a TWT amplifier in a long-haul system was in a 6-GIIz application in 1960, and since that time a number of other applications have used this basic design. Unlike the planar triode, where the life is 15,000 hr or less, the TWT life is generally on the order of 100,000 hr or more, approaching what can be achieved with solid-state devices. The cost of a TWT is at least 10 times that of the triode, but the TWT can operate at much higher frequencies and at higher power levels. One disadvantage of the TWT is that for 10 W of output, the power supply voltages are on the order of 3000 V, necessitating elaborate safety precautions to protect maintenance personnel.

Other components for microwave systems include high-frequency diodes, rf converters (mixers), wave filters, equalizers, isolators, circulators, and generators of microwave power. Over the years, many other systems have evolved in other frequency bands, and in each case the components have been tailored for the specific applications. All, however, require components and subsystems to perform essentially the same functions.

D. MICROWAVE NETWORKS

Wave filters are needed to select specific frequencies and to reject others. In low-frequency communication applications, frequency selectivity is achieved by the use of discrete capacitors and inductors arranged in resonant circuits. At SHF, these discrete components cannot be used effectively; rather, distributed elements, usually in waveguide or strip-line structures, must be employed. In a waveguide filter, irises in the guide and posts across the guide perform the resonant functions that coils and capacitors achieve at lower frequencies. Over the years, an entire technology has been developed for ready computation of various characteristics of waveguide networks. Originally, these involved computations were performed on slide rules and mechanical calculators requiring a substantial staff of people spending many hours in the design process, but today the electronic computer and the many programs available have reduced this design effort to routines rapidly carried out by relatively nontechnical personnel.

Recently, a new material known as yttrium–indium–garnet (YIG) has become available which has ferroresonant properties with high stability that are almost ideal for microwave filters. Its great virtue is that filters made from this material are very much smaller and lower in cost than conventional filter structures. Both of these factors are highly significant.

E. FERRITES

About 1960, a new material called *ferrite* became available which has proven to be highly useful in a number of microwave components. Ferrites can vary in composition, depending upon the specific results desired, but most contain a mixture of the oxides of nickel, iron, copper, and aluminum. Physically, ferrites are a form of ceramic for they are hard and brittle and must be either molded in the green state and kiln fired or ground to size after firing. These materials, when used in a waveguide structure with associated magnets, can form a nonreciprocal (one-way) transmission path, that is, microwave energy can propagate in one direction through the device, but energy flowing in the reverse direction is absorbed in the material.

F. ISOLATORS

Depending upon the composition of the ferrite material and the shape of the piece, it can exhibit either field displacement or Faraday rotation characteristics. The field displacement material is often used for isolators, which are made by mounting the ferrite slab lengthwise in the guide and perpendicular to the wide walls. An external magnet coupled into the edge of the slab provides the polarization field necessary to achieve directivity.

G. CIRCULATORS

Circulators use the Faraday rotation type of material with the ferrite mounted in the center of three converging microwave paths, each 120° to the adjacent path. The ferrite in a circulator is usually in the form of a small disk with the external magnet directly over the area of the ferrite. Depending upon the polarity of the magnet, the energy can be circulated from one of the three ports to an adjacent port either to the right or left. If circulated to the right, energy cannot go to the left port. If the polarity of the magnet were reversed, the circulation would be in the opposite direction. The circulator provides great flexibility in the design of microwave integrated circuits.

H. DOWNCONVERTERS

In the 1950s, the conversion from SHF to the 70-MHz IF was accomplished with a point-contact silicon diode and a microwave source as a local oscillator (at that time the oscillator was a reflex klystron). This mixer or converter, as it is sometimes called, has loss, and the point-contact diode introduces noise which can result in a receiver noise figure over 12 dB. In the late 1960s, the Schottky or hot-carrier diode became available, and because of the way the junction is formed, it is an ideal mixer at SHF. A 4-GHz Schottky diode and associated structure can have a noise figure of less than 8 dB. This was a major step forward, and in one case, it resulted in a very sizable increase in the number of voice and data channels that could be carried over a single radio channel. The Schottky diode is rugged, it can sustain substantial overload from the local oscillator and the signal, and it can be manufactured with relative ease.

I. TUNNEL DIODE AMPLIFIER

Solid-state diodes have been used to provide amplification by using the negative resistance property of the device. The details of diode characteristics, how doping in the course of manufacture can alter their properties, and how negative resistance can occur can be found else-

where. By embedding a specially doped diode in a strip-line circuit and applying an appropriate bias to the diode, an amplifier can be designed. One type is the tunnel diode amplifier (TDA), so named because of the special fabrication of the semiconductor junction, which can provide low-level gains up to 20 dB with noise figures as low as 4 dB. The bias voltage on the diode to optimize the negative-resistance characteristic is very low, on the order of 0.4 V, and to avoid swinging outside this region, the input signal must also be kept low. Due to this tendency to readily overload, the TDA has found little use in terrestrial systems even though its gain and noise figure are quite good and the design is not very complicated. Tunnel diodes have been used effectively in some phase-array radar systems primarily because the radar return signal is usually not high enough to cause overload difficulties. Germanium diodes are frequently used in TDAs at frequencies up to almost 30 GHz.

J. PARAMETRIC AMPLIFIERS

Another type of negative-resistance amplifier for receiver applications which is more complicated than the TDA but which does not suffer from the low-level limitation of the TDA is the parametric amplifier. This type of circuit requires that the diode be "pumped" or driven by a local oscillator source at several times the operating frequency. For example, a 4-GHz amplifier is pumped by a 12-GHz source that must be very carefully stabilized in amplitude.

Parametric amplifiers were thoroughly evaluated for widespread use in the long-haul environment in 1968, but it was found that the circuit complexity needed to keep the pump signal at the optimum value was impractical and unduly expensive. This amplifier is no longer in use in long-haul terrestrial common-carrier systems, although they have been used in some satellite applications.

K. IMPATT AMPLIFIERS

Impact avalanche transit time (IMPATT) solid-state power amplifiers also use the negative resistance of the diode in specially designed circuits that include circulators and strip line to connect the devices together. Both silicon and gallium arsenide materials have been used, depending upon the power level at which the amplifiers are to operate. In relatively low-power amplifiers (1–2 W), a single diode can provide the required output, but for higher powered applications, several devices must be cascaded.

IMPATT amplifiers are basically reflection amplifiers. The input signal from one port of a circulator is terminated with an IMPATT diode that is suitably biased into the negative-resistance region. Amplification takes place because negative resistance has been introduced into the circuit which cancels out the losses so that the output signal is higher than the input. The output is returned to the same port, circulated to an adjacent port, and then delivered to the system output in the case of a single-diode unit or to the next stage of a multistage amplifier. The output cannot be conducted toward the input because of the high reverse loss of the circulator.

In the case of single-diode amplifiers, approximately 20 dB of gain can be realized at frequencies in the 6–11-GHz range. Tuning procedures, while not complicated, frequently require caution by the operator to avoid improper matching of the diode into the circuit. If this should happen, there is a possibility that the diode will be permanently damaged. Electron tubes are much more tolerant of misadjustment than IMPATT diodes. Noise figures of IMPATT amplifiers are generally high; some typical power amplifier noise figures range from 30 to 50 dB, which is higher than those for electron devices.

In multistage amplifiers, each diode must be individually biased and it is necessary to carefully regulate the voltage to each semiconductor. The IMPATT amplifier can be highly nonlinear, and therefore it is not well suited for AM systems. These nonlinearities do not degrade the performance of angle-modulated systems using either FM or PM. In some applications requiring high gain, the IMPATT diode may function as a locked oscillator, where the input signal causes the oscillator to instantaneously follow the input frequency exactly. This is the equivalent of an amplifier, for the output signal is a replica of the input but the amplitude is the equivalent of that of the oscillator. Precautions must be taken to assure that in case of failure of the input source the amplifier is adequately squelched to prevent interference.

The IMPATT amplifier has been successfully used in 6- and 11-GHz systems operating in the 1–3.5-W power range with quite satisfactory results. These have been single-diode units of both silicon and gallium arsenide. As the output power level requirements rise, the maintenance difficulties, circuit complications, and costs accelerate rapidly. A five-stage amplifier requires five voltage regulators, output monitors on each of the five stages, an intricate strip-line circulator structure, and meticulous circuit adjustment. The power efficiency, that is, the rf output power versus the overall dc input power, is considerably lower for the IMPATT amplifier as compared to the TWT. In large installations where many amplifiers are used, these power differences may be significant.

The component costs of the 10-W IMPATT amplifier, which can provide the same gain as a TWT (about 32 dB), are not substantially lower than the TWT and the high-voltage power supply that it was to replace, and while it has been demonstrated that solid-state 10-W amplifiers are feasible, they are not used in long-haul terrestrial common-carrier systems today.

L. Gunn Diode Amplifiers

The Gunn diode has been used in low-power microwave systems, often with the diode as a locked oscillator, as described earlier for the low-power IMPATT. The Gunn diode has lower noise than the IMPATT and is ideally suited to low-power applications (up to about 2 W). The Gunn diode can also serve as the receiver local oscillator when suitably stabilized in frequency from an external source.

M. Transistor Amplifiers

In the early 1970s, sufficient exploratory work had been completed on microwave transistors to indicate clearly that the GaAs FET offered intriguing possibilities in power amplifier, receiving amplifier, and frequency converter applications. The performance of this class of transistor was a result of the high electron mobility of the class III–V elements in the periodic table and the ability to design more nearly planar transistor geometries. These transistors had less stringent spacing requirements than the equivalent bipolar units and, as a result, could offer better high-frequency performance for a given degree of manufacturing complexity.

Circuit development ensued, and for the first time, active microwave circuits were being developed using an equivalent-circuit design approach that was sufficiently detailed to accurately predict performance and, at the same time, simple enough to lend itself to direct calculations of the device geometry. Amplifier circuits can be designed that cover 10% bandwidths, and therefore frequency adjustment is not required in the field. This is a distinct advantage over the application of the planar triode tube and the IMPATT diode amplifiers. A 2-W, 4-GHz amplifier was designed to replace the planar triode amplifier in many applications, and this was followed by a 5-W version that offered the possibility of increasing the capacity of older 2-W relay systems. Field experience with GaAs FET power amplifiers has indicated that after a short burn-in period failures are rare and performance is sufficiently stable that routine maintenance is not required. This results in great economic savings to those who own and operate these systems.

The low inherent noise of the GaAs FET transistor is used to increase the sensitivity of the receiver section of the radio repeater. Amplifiers with a noise figure of 2 dB have been built and provide an improvement by a factor of at least two in receiver sensitivity. To obtain the best noise figure, the gate lengths and widths in the transistor have to be kept to a minimum. The use of a common transistor low-noise amplifier in the rectangular waveguide run to a number of repeaters in tandem provides a low noise figure to all units at a considerable cost savings. A common amplifier in the receiver input, if not designed to have low intermodulation characteristics, can result in undesirable interchannel cross talk during heavy selective fading conditions. Therefore, linear amplifier designs are sometimes necessary in low-level applications in the receiver input.

In the early 1980s, further semiconductor developments had a continuing impact on the capability of components for microwave radio relay systems. The development of the field-effect transistor with two gate electrodes made possible the design of a new group of frequency converters and gain control circuits with sufficient gain to mask the noise from subsequent circuit elements. Special doping formulations, the introduction of indium, and improved semiconductor geometry gave rise to higher electron mobility, closer spacings, thinner conductors, and better grounds. The ability to add circuit-matching elements on the semiconductor chip has increased microwave performance. Monolithic circuits with one or more stages on a single chip and high-electron-mobility devices are examples of these improvements. Circuit developers have been able to introduce meaningful computer-aided design programs that enable more complete evaluation of circuit and environmental options in the design stage.

Typical performance capabilities now include power levels as high as 100 W at low microwave frequencies (1 GHz) and as high as 8 W at 18 GHz. Bandwidths with relatively constant gain have been extended to an octave or more. Amplifiers for quadrature amplitude modulation (QAM) digital modulation have power outputs of 4 W or more at saturation and gains and linearities that equal or exceed those of TWTs. Low-noise amplifiers have noise figures of 1 dB at 1 GHz, 1.3 dB at 6 GHz, and 3 dB at 20 GHz. By refrigerating to 20 K, noise figures as low as 0.1 dB at 1 GHz, 0.25 dB at 6 GHz, and 0.9 dB at 20 GHz are possible. Using high-electron-mobility transistors (HEMTs), a room-temperature noise figure of 1.4 dB at 11 GHz has been obtained. If refrigeration is used, the noise figure drops to 0.35 dB.

Circuits that combine a dielectric resonator and a GaAs FET transistor in an oscillator configuration can provide a stable frequency source with a stability of a fraction of a part per million per degree Celcius, which would satisfy short-haul requirements. For more critical operations, a transistor operating in a phase-locked loop with a low-frequency reference offers even better frequency stability.

N. STRIP LINE

The invention of the transistor, which is considerably smaller than the electron tube, required new methods of mounting it and the other electronic components needed to form a complete circuit. For many years, printed circuits have been routinely employed where the electrical and thermal demands are not great and where the use of a phenolic or epoxy glass board does not affect the circuit performance. At SHF, these materials are not thermally stable and dielectric losses can be excessive. An improved technology is available for these high frequencies called *strip line*, which is a process that deposits a conductive path on a high-quality ceramic substrate. The pattern is usually of copper or gold, which provide high conductivity to keep the circuit losses low. The pattern is actually the electrical circuit with inductors, capacitors, resistors, and the interconnection paths formed by the deposited film on the substrate. At these frequencies, inductors and capacitors are physically small, and in a strip-line design an inductor may be only a narrow strip of metal on an otherwise plain section of the ceramic sheet. Intricate circuits are possible; the semiconductor active components can be appliquéd onto the circuit. By this means, the leads can be kept very short, which is a necessity at these frequencies. Manufacturing reproducibility is very good with strip line because the metal deposition onto the ceramic is extremely accurate with the photographic techniques used.

Strip line is used for many SHF circuits in transmitters and receivers. By integrating strip-line circuits with ferrite circuits, very elaborate microwave subsystems can be devised. For example, the entire SHF portion of a transmitter or receiver can be designed as a single unit.

VIII. Reliability

Generally four factors can impair the reliability of a microwave system, and each requires specific action to minimize its impact upon service quality. These four causes of degradation are (1) propagation effects, (2) equipment failure, (3) power failure, and (4) maintenance errors and/or personnel intervention. All of these factors have been minimized by additions and improvements to the systems and by adequate personnel training in the operation and maintenance of the facilities.

In the case of multipath fading or equipment failures described in the next two sections, circuit outage can be greatly overcome by the rapid substitution of a good channel for the one that has failed. Protection switching is described in Section VIII,E.

A. Propagation

Long ago it was observed that the input signal from the antenna into the receiver was not always constant. When more information was obtained from various parts of the country, it showed that some sections were more prone to fading than others; the months of July and August were worse than the other months from the standpoint of the severity and frequency of occurrence, and the time of day had an effect on the constancy of propagation. The flat plains areas, particularly in the time period from midnight until after sunrise, had more fading than the irregular topography of the Rocky Mountain region. Reflections from flat ground with peculiar temperature conditions and still air enhance the fading. The locale around the Gulf Coast area, Florida, Georgia, and Alabama is also a high-fade region that is caused by atmospheric conditions. [See RADIO PROPAGATION.]

Propagation phenomena have had, and continue to receive, a large amount of theoretical study and field verification. Crawford and Jakes in 1952 reported on selective fading at SHFs and DeLange made tests at these frequencies using very short pulses. Kaylor in 1950 studied fading at 4 GHz by very elaborately instrumenting (for that period) one 30.8-mi hop between Lowden and Princeton, Iowa. These tests showed some remarkable effects that led to corrective steps to minimize the outages due to fading. This is covered in Section VIII,E on protection switching.

Fading can be of two types: a partial or total obstruction of the path that might exist for some period of time (from minutes to hours). This has the same effect as if a "bulge" in the earth had risen up to produce a path blockage. Earth bulge fading affects all channels at the same time for it has no frequency dependence (flat fade). Fortunately, this kind of fading is relatively rare in both location and frequency of occurrence.

Multipath or selective fading occurs when the atmospheric conditions cause the various rays of the radio wave to be reflected and/or refracted over the path until some arrive at the receiving antenna out of phase with the main or direct wave. The cancellations at the receiver, either in total or in part, affect the received signal level. This partial cancellation in a single 20-MHz channel can be highly selective; it has frequently been observed that a fade can be much more severe at one part of a 20-MHz channel than at another. This leads to serious distortions of the signal and often makes the channel unusable. At other times, the fade will affect the entire channel about equally and the channel will become noisy. This is frequently observed on a television broadcast that is being carried by terrestrial radio facilities. Selective fading is very rapid, often measured in terms of a few seconds per occurrence. A great wealth of published material on atmospheric fading at SHF exists for the interested reader.

It has already been mentioned that at frequencies above about 10 GHz rain affects the transmission as though the transmitter power had been lowered or an attenuator had been inserted into the input of the receiver. This rain attenuation is caused by the dispersion of the radio wave by the individual rain drops as they fall to earth. In normal or dry circumstances, the wave is directed to the target (receiver) by the transmitting antenna and the path, so established, is then considered the norm. When falling rain is interspersed over the path, the rain drops act as small refractors and divert the beam off the norm. More rain, from either higher precipitation or longer paths (more rain drops), will produce higher attenuation. Rain effects are frequency sensitive only to the extent that all frequencies above about 10 GHz will be affected, and the higher the frequency the less rainfall is needed to produce a given amount of attenuation. Similarly, the greater the rainfall, the higher the attenuation, and the longer the path over which the rainfall exists, the greater the attenuation. For this reason, the 11-GHz band is usually used in short-haul systems only, and then only when rain effects will not be serious. However, the 11-GHz band is not a total loss even in high rain areas because acceptable performance can always be achieved by reducing the distance between the repeaters. By doing this, the cost of the system will increase due to the additional sites and equipment that are needed. Instead of the 26-mile average spacing that prevails at 4 and 6 GHz, the 11-GHz systems average about 15 miles between repeaters. At frequencies in the region of 18 to 20 GHz, the repeater spacings may average only 2–4 miles because rain attenuation will exist at more locations and at lower levels of precipitation.

B. Equipment Failure

Microwave equipment, like all electrical and electronic equipment, will ultimately fail, and when this occurs, the remainder of the system will not be usable until repairs are made or the system is restored with an automatic protection system. Over the years, every subsystem, equipment unit, and component has been studied and the industry has spent great effort and expense in redesign and in introducing new technology to make the individual parts of a system more reliable. The results have been very effective, and the life of the components, including the electron tube devices, has been greatly extended over the years. Manufacturers of components have also made improvements by studying their processes and making changes whenever needed, and tests have been added at the time of manufacture to assure that early failure in the field will not occur. Only two communication facilities are given more attention and scrutiny in design and manufacture than microwave relay: they are (1) undersea cable systems and (2) the orbiting equipment units of satellite systems.

Since solid-state components have become available, many circuits of existing systems have been redesigned to take advantage of both improved performance and higher reliability. Of the latter, it can be said that very often a redesigned circuit with solid-state devices can pay for itself in the reduction of maintenance costs in a short while. For example, in one large system that had used the planar triode tube, redesign to introduce the GaAs FET power amplifier has replaced virtually all of the 135,000 tubes (system wide), which had required maintenance at least once a year. Considering the locations of all the repeaters throughout the United States, a great amount of time was required to get to the site, make the repairs, and return to the home location. Studies show that the average maintenance call from a center to a remote repeater location is 1.5 h. New designs, therefore, are not always used for the manufacture of equipment of new systems; frequently, they are retrofitted into existing systems to save money.

C. Power

Home electronics equipment usually operates from commercial alternating current, and when a power failure occurs, the other uses of electricity are also affected. Generally, the user associates the loss of television or radio reception with the local power failure, and when the lights come back on, usually the television is again operating normally. The home consumer accepts this situation as an "unavoidable" event, even though an automobile crashing into a utility pole has severed the power line. When the same conditions occur a hundred miles away and telephone, data, and television circuits are interrupted, the facility users are less tolerant of a power failure and they demand that the carrier do better. To overcome system failure of this type, the carriers generally have established emergency power facilities at each repeater site. Battery standby is by far the simplest and cheapest, and on light route systems, this may be the only backup. Capacity is frequently set for 8 to 24 hr with the expectation that the commercial power will be restored by that time. Oftentimes means are provided for bringing in a portable engine generator set, towed by a truck, to connect into the station to recharge the batteries if it appears that the loss of alternating current will be prolonged.

In the larger stations, the battery backup and the engine alternator are a part of the permanent station installation and automatic startup of the engine is provided. With such an arrangement, there is no interruption at all in the continuity of service. With added control and command facilities at each location, maintenance people can periodically "exercise" the reserve power system remotely to be assured that the equipment is in proper working order.

D. Maintenance Errors

In large systems, such as a transcontinental route, many service control points are involved and a large number of people are charged with the maintenance of the system. It has been learned from experience that even a single operating error can effect a serious outage; a single erroneous push of a switch button can cross CBS with NBC. Operating errors can happen, but the carriers have laid great emphasis on personnel training and the importance of avoiding operating errors. In general, these people are well informed and perform at a high level of competence. By making the equipment more reliable and by making many of the functions automatic, people are less involved and the number of man-made interruptions are reduced accordingly.

E. Protection Switching

Fading of the microwave signal can produce interruptions in service from momentary outages to prolonged intervals when the radio channels are unusable. Long outages are usually the result of earth bulge atmospheric effects and almost always degrade all channels simultaneously. The other type of fading is caused by multipath transmission, and on any specific channel the disturbance is only momentary. All channels in a system might experience serious fading in the same time period but not at the same instant. Equipment failures can also cause circuit outage, but with modern-day equipment, this is no longer the problem that it once was with short-lived components. Unfortunately, specific equipment failures cannot be anticipated and such events always cause an interruption in service.

For both multipath fading and equipment failures, microwave systems have rather sophisticated schemes for using one of several authorized channels to provide protection for the remaining ones in the group (as many as 11 channels). This channel substitution (channel switching) is done very quickly, and even with unanticipated equipment failures, service is restored in less than 50 msec. Protection arrangements of this type are ineffective for earth bulge fades because one protection channel cannot protect a multiplicity of channels that have failed simultaneously and in which there is a high probability that the protection channel in the same band would also have failed under these circumstances.

Protection switching in long-haul systems is done by sectioning the route into what is known as a switching section, and each segment is operated independently of the other sections for protection purposes. Telemetry functions are provided between the ends of the section to control the switches. In practice, a switching section may be a single hop or it may be as long as about 12 repeaters (longer sections become impractical for a number of reasons.) The average section length in the United States is three hops, which is predicated on the need to provide protection switching wherever the baseband signal is to be delivered to the user. Without this requirement, many longer switching sections would exist.

Channel switching is done at the IF with solid-state switches. A switching matrix is arranged to substitute a channel known to be good (the protection channel) for the failed (working) channel, and in a multichannel system this requires a switch at both the head end and the tail end of the section. The quality of each channel, including the protection channel, is monitored for noise and the presence of a pilot tone. When the detector at the tail end of the section senses an abnormality (high noise or lack of pilot or both), a switch is initiated by sending a control signal to the head end (transmitting end) to make a bridge between the working channel and the protection channel. The monitor on the protection channel recognizes that the head-end switch has been made and proceeds to connect the output of the protection channel to the working channel. These precautions, namely, verifying that the service is on the protection channel before the final switch is made, give assurance that a switch to a dead channel will not be completed.

The failed working channel is monitored continuously for noise and presence of pilot even though the channel is disconnected from the system. This is done to determine when the noise performance is such that the working channel can be returned to the system so that the protection channel can be released and made ready to serve some other channel that requires protection.

Current systems use microprocessors for logic control of the switches that can be used for more than the protection of failed channels. Switches can be made manually or automatically for maintenance checking and for circuit rearrangements either on demand or on a predetermined schedule, such as television feeds of a network program to a local outlet or a television pickup of some event of interest that is to be connected to a television network.

In short-haul systems where baseband signals are available at each repeater site, switching is usually done at the baseband frequencies. Depending upon the number of working channels in the system, head–end switching at baseband may or may not be used. In light systems, two repeaters operating on the same frequency may be used, one being a hot standby for the other, with a simple transfer switch at the tail end to make the transfer when ordered by the switch-initiating circuits.

F. Space Diversity

The large amount of attention given to microwave propagation many years ago established the fact that if two receiving antennas are located on the same support but separated vertically by 35 ft or more, the output of the two antennas, under fading conditions, is almost totally uncorrelated. This says that at any given instant, if one antenna output is faded and unusable, the likelihood that the output of the other antenna is normal or quite acceptable is very high. Successful reception virtually 100% of the time then depends only on providing the means to select the right antenna at the right time. Over the years, many clever schemes have been devised for detecting and selecting the good signal. An arrangement for combining the outputs of both antennas, with appropriate phase shifters to account for the differences in the two signal path lengths, has been devised so that a good signal is available essentially all of the time. In many systems, such as single-sideband and digital systems, space-diversity arrangements are necessary.

IX. Radio Terminals

The information-bearing signal that is to be transmitted via microwave facilities is always generated at a lower frequency and is then translated (modulated) up in one or more steps to a discrete frequency in the SHF band. This lower frequency is called the *baseband* and may extend from virtually direct current to ~8–10 MHz. Television baseband signals generally will be originated by the broadcaster and will contain the sound as well as the picture information. It is delivered to the carrier in a very standard format; usually the sync pulse information will be more negative than the black or blanking level, and the picture information will be positive relative to the blanking level. Voice-frequency telephony, on the other hand, will be delivered to the carrier as a broadband signal resembling random noise after the individual voice-frequency circuits have gone through several steps of translation to place each 4-kHz circuit in a discrete 4-kHz slot in the baseband spectrum. Digital signals may either be handled as baseband signals or by special signal processing to get a suitable IF format to be applied to the radio. More information on these terminals can be found in Section XIII on digital radio.

A. FM Radio Terminals

For FM microwave systems, the FM terminal equipment consists of an FM transmitter and an FM receiver. The FM transmitter contains an oscillator circuit that is capable of having the rest frequency deviated above and below this value by the baseband signal. The greater the amplitude of the baseband signal the greater the FM deviation from the center frequency. In recent years, voltage-controlled diode capacitors have been used in the frequency-determining circuits of the oscillator to produce the instantaneous changes in frequency, thereby generating frequency modulation. Over the years, considerable attention has been given to the means for controlling and stabilizing the average or rest frequency at the desired value. This is important because the IF signal from the FM transmitter is added directly to the channel local oscillator source to produce the SHF signal that is radiated from the antenna. If these signals are not accurately controlled, interferences in multichannel systems can be severe.

At the receiving end, the SHF signal of each radio channel is demodulated with the use of a discrete local oscillator frequency to produce the IF signal. This FM IF signal is then detected to baseband in an FM discriminator quite similar to that used in commercial FM radio receivers except with much wider bandwidth. The baseband signal is then delivered to the end user, the television broadcaster, or the telephone company.

FM transmission consists of carrier and FM sidebands above and below the carrier and, as such, is a reasonably rugged signal format. Frequency shifts through repeaters and through the many stages of frequency translations that occur in long systems do not alter the frequency of the recovered baseband. This is because the discriminator in the FM receiver uses the carrier frequency to demodulate the FM sidebands. All frequencies are shifted the same amount in going through the system and any frequency shift is canceled out. This is not true in radio terminals used in SSB systems.

Baseband circuits in the terminals are used to adjust the levels of the signals and can provide means for equalizing the gain–frequency response characteristics of the various parts of the system. Because of the wide variety of signals that must be accommodated, the baseband circuits must be broadband.

B. AM Radio Terminals

Single-sideband AM radio systems are used to transmit information by using the spectrum in a more efficient manner than that of FM systems. Where FM equipment transmits information using two sidebands around the carrier that on the average require appreciable power for the carrier itself, SSB not only uses just one sideband but also the carrier is suppressed, thereby saving the power of the carrier. This higher efficiency in the usage of the spectrum is obtained at the expense of a very elaborate and precise control of all the frequencies in the system, including those used in the terminal equipment.

In telephone service applications, it is absolutely mandatory that virtually no frequency shift take place from end to end; otherwise, the voice quality is impaired and the customer will complain. Data systems produce errors and telephoto circuits produce distorted pictures if the frequency shift is more than 20 Hz. Frequency translation throughout the multiplexing hierarchy must be strictly controlled by locking to a high-precision standard. More information on frequency control can be found in the section on SSB AM radio.

X. System Testing

When microwave systems were first used for communication purposes, it was believed that a considerable amount of testing would be required to assure that reliable service could be provided at all times. It was felt that in a long chain of repeaters performance would not be stable, that frequent failures, especially of the electron tubes, would occur, and that the equipment would degrade with time. It was assumed that by frequently visiting the repeater sites and testing the electronic circuits, failures could be anticipated to some degree. Routine measurements for a number of key characteristics were made on many parts of the repeater with the expectation that they would essentially eliminate the equipment as a cause of trouble. These routine tests were made for many years and at considerable cost, and in retrospect, one might question whether this practice was cost effective. As equipment performance and reliability improved with a maturing technology with more reliable components, it became apparent that less and less maintenance was really necessary. Eventually, maintenance was done only on a demand basis (i.e., when something either failed completely and station alarms pinpointed the location or end-to-end performance had degraded beyond acceptable limits and terminal alarms called attention to the problem).

It should not be concluded that the carriers no longer test systems, repeaters, and parts. Very elaborate test facilities are available to diagnose trouble and locate the defective elements, but tests are no longer performed on subunits of a repeater in a system that is in no apparent trouble.

In television systems, the carrier's "customer" is technically competent in that the people who take the television signal from the carrier are critical observers who have the means for determining the quality of the circuit. Therefore, imperfections in the television signal can result in a complaint from the broadcast people to the common-carrier operators. Television system testing is exacting and detailed. The broadcasters and the carriers use essentially the same kind of equipment and one can verify the observations of the other. The broadcaster uses a multiburst signal wherein a number of discrete sine-wave signals, all at equal amplitude, are transmitted sequentially and at the receiving end are displayed on an oscilloscope. By observing the height of the display of each of these tones, which appear as a series of vertical bars, the response characteristic can be evaluated. If the high-frequency bars are lower than the low-frequency bars, the high-frequency response rolls off and the picture will lose fine-grain detail. As an example, the picture of the shirt of a football referee standing directly in front of the camera will show large and distinct black and white stripes; as he moves out onto the field and away from the camera, the stripes lose their distinct pattern and smear into a grey smudge. Differential gain and phase, a measure of the amplitude and phase of a reference test signal as the input amplitude is varied, gives an indication of the stability of the color of the picture. Poor differential gain will cause the intensity of the color to change when the scene changes from light to dark. Similarly, poor differential phase characteristics will actually cause the hue of the color to change with the intensity of the general picture content. Noise measurements are simple to perform but are important to assure the broadcaster that the picture being sent out over the air is not snowy or flecked with noise bursts.

A number of tests can be performed on a microwave system that carries telephony signals. When many voice circuits are multiplexed together, the total baseband spectrum has almost the same characteristics as an equal band of random or "white" noise. Tests made by using a random noise source can determine the amount of cross talk between voice circuits that is being experienced. The noise source, connected in place of the telephone circuits, is equipped with narrow band-rejection filters tuned to selected portions of the spectrum. At the receiving end, a sensitive detector is connected to the output of the corresponding bank of bandpass filters. Since no noise transmission exists at the selected frequencies at the sending end because of the reject filters, any detected output at these frequencies at the receiving end is caused by intermodulation in the various parts of the microwave system. This is a very powerful tool in determining the amount of intercircuit cross talk present.

It is also important that the level of all of the voice circuits of the radio channel be essentially the same in a microwave system used for telephony. To check this, a variable-frequency oscillator at the head end and an oscilloscope at the receiving end will show if the frequency characteristic is "flat" (i.e., signals at all frequencies are of the same amplitude). This measurement is made by automatically "sweeping" the frequency from the low end of the band to the high end. The operator can readily see the actual performance on the oscilloscope.

In data systems, the all-important characteristics are the error rate and the degree of intersymbol interference that exists. To check the error rate, a signal of known characteristics is transmitted over the facility and an error detector is connected to the receiver output. By observing the count on the detector, the quality of the system can be easily determined. A typical number frequently quoted is a bit error rate of one in a thousand bits (error rate of 1 in 10^3). The intersymbol interference can be evaluated by sending a random data pattern and allowing the detector (oscilloscope) to examine all values of the wave. The pattern so displayed will appear as a series of traces that ideally will all cross the zero axis at the same time and all rise to maximum at the same time. With distortion in the system, these conditions will not be met and what should be a wide-open "eye" will begin to show more and more closure as the distortion increases. Since the detector in the system must accurately determine whether a pulse is or is not present at the time of decision, an open "eye" is highly desirable if low error rates are to be achieved.

The specialized test equipment used in current test programs, both at terminal locations and at repeater sites, is not inconsequential in quantity, degree of sophistication, or cost. Furthermore, it is mandatory that the personnel maintaining large microwave systems be adequately trained to use this array of precision test equipment, to interpret the results of the measurements made, and to effect repairs when tests indicate that the system is no longer meeting specifications. In order to maintain long-haul systems to the degree needed to meet present-day service requirements, a uniform set of procedures and practices must be provided to the maintenance forces assigned to these tasks. Such practices must be easily performed and understandable and provide unambiguous results.

XI. Single-Sideband AM Radio

It was noted in a previous section that a decision was made in the late 1940s to use FM rather than AM for the first application of microwave radio. It was the consensus of the decision makers then that the distortions arising from the anomalies of the phase characteristics of the components and subsystems could be more readily handled than those that would be associated with the nonlinear amplitude characteristics of these same elements. It turned out to be a wise decision since SHF FM repeater designs are much less complex and less costly than AM repeater designs. In the search for ways of adding voice circuit capacity to the larger systems, attention was given in the late 1960s to reexamining the SSB question in light of the advancements in the field of component technology and in the technical understanding of propagation characteristics, system design, and circuit capabilities. An intensive study indicated that an AM design was feasible and also economical when fully loaded.

The major deterrent that kept the industry from going the AM route in the 1940s and the one characteristic that had to be overcome in the 1970s was that of amplitude linearity. Amplitude linearity in an electronic circuit is the degree to which the amplitude of the output signal is a replica of the amplitude of the input signal. If the output is not identical (except for scale) to the input, this is equivalent to an output that is identical plus a component that is equal to the difference or the distortion. The latter is not a usable component of the signal and is actually an undesired perturbation of the signal. The various circuits of a SSB AM repeater were studied to see where nonlinearities existed and what might be done to reduce them. These undesirable characteristics are most predominant in those circuits that must deliver substantial amounts of power, and in the case of a microwave repeater, this is the transmitter output device. In 6-GHz systems where the output power must be on the order of 10 W, the TWT is the logical choice for the output stage. In studying the TWT from a linearity standpoint, some internal design changes in the tube were made to improve the linearity, but this was still not sufficient to meet long-haul system requirements. A number of circuit innovations were studied and tested before a decision was made to employ a technique of predistortion to solve the nonlinearity problem.

Predistortion, as the word implies, is a means of altering the signal into the TWT in a predictable fashion so that this input, when passed through the TWT amplifier, will emerge at a higher level but with the undesired distortion canceled out. Such a scheme is not new per se, but it has not been used previously at SHF. It is necessary that components, including the TWT, be very uniform in their characteristics so that replacement tubes and other parts can fit the predistortion circuitry with only slight adjustment and so that a single predistorter will fit all equipment units.

All SSB circuits must be designed with this one characteristic at the top of the requirements list. The IF amplifiers (preamplifiers and main IF amplifiers) have to provide sufficient gain to boost the low received level to that required for the upconverter and to accommodate an atmospheric fade of approximately 40 dB. Automatic gain control (AGC) to compensate for this depth of fade must be provided and at the same time must not introduce amplitude distortion. The high degree of linearity required in SSB AM is one of the costs of getting substantially greater system capacity.

Most of the circuits of the SSB repeater are quite similar to their counterparts in FM systems, but because the SSB signal does not contain a rugged carrier within its format, the SSB signal can suffer distortion from selective fading more readily than FM. Equalization of the amplitude–frequency response characteristic is achieved by pilot-controlled regulators that increase or decrease the amplification at certain frequencies relative to the other frequencies in the channel spectrum. More precise equalization of the signal is another cost in getting the benefits of SSB's higher capacity.

The SHF signal that is radiated from the antenna at each repeater is derived by modulating the IF with a precise carrier frequency. In FM systems, this carrier is obtained from a stable crystal oscillator and harmonic generating equipment that is independent of equipment outside the repeater site. Frequency instabilities or offsets from the norm, while small, do not affect the demodulated baseband signal, because of the nature of the FM receiver as mentioned in the section on radio terminals. These same frequency instabilities and departures from the exact design values would be intolerable in SSB AM systems. To get these oscillators at precisely the right frequencies, synchronization to a nationwide frequency standard is necessary and this standard must be available at every repeater location. This may mean that the frequency standard must be provided separately by wire or cable facilities or be locally derived from the difference between two precision pilots transmitted over the radio. To get an appreciation for the frequency accuracy required, a 20-Hz frequency error in a voice channel would be noticeable by almost every listener. Other services may be even more demanding because some customers may depend upon the telephone system to transmit accurate signals for medical or scientific applications. Therefore, it is probably necessary to have an accuracy of ~1–2 Hz from end to end. The SHF signal at 6 GHz is 6 billion Hz, and 1 Hz is an accuracy of 6×10^{10}. This frequency control requirement is another cost attributable to the use of SSB AM.

The disbursive medium, caused by selective fading, requires that in the areas where this phenomenon is severe, space-diversity antennas must be used to ease the amplitude equalization problem. This is still another cost that must be borne for the use of SSB AM. Despite the costs and added complexity of (1) overcoming component nonlinearities, (2) providing more complex amplitude equalization, (3) obtaining increased precision of beating oscillator signals at each repeater point, and (4) in some locations providing space-diversity antennas, the overall economics of SSB systems are still favorable and allow the maximum usage of the assigned frequency spectra.

XII. Satellite Communication

Satellite radio transmission plays a very important role in providing international communication facilities (overseas circuits) that, at the present time, can otherwise be provided only by ocean cable systems. The latter have limited capacity, are used only for telephony and data applications, and are not used for television signal transmission. Satellite channels are also widely used for domestic telephony, data and network television long-haul transmission, as well as television distribution to local outlets. Satellite systems compete favorably with established terrestrial systems for some services. Initially, satellite systems used the 4- and 6-GHz common-carrier bands, sharing frequencies with the terrestrial common-carrier users. The practice has been to use the 6-GHz band to transmit from earth to the satellite and to use the 4-GHz band from the satellite to the earth. In order to have the satellite in a stationary position with respect to points on the earth and be available on a continuous basis, the vehicle must be accurately positioned at a distance of 23,600 miles above the equator. In an earlier section it was mentioned that the path loss between antennas increases as the path length increases. At 4 GHz, two isotropic antennas separated 25 miles have a path loss of 137 dB. At 23,600 miles the path loss is 196 dB! [See SATELLITE COMMUNICATIONS.]

Satellite systems in some respects are quite similar to terrestrial systems. There are radio terminals, transmitters, receivers, and facilities to switch inputs and outputs to and from the various working channels. Specialized power supply equipment that is highly reliable is needed as in terrestrial systems. Because only two ground stations are required to make up a long-haul facility, the financial investment in ground station equipment can be quite high. In enumerating the basic subdivision of satellite system equipment and earth system equipment, the similarities with terrestrial systems end. Specific subunits for satellite systems are quite unique.

In the satellite unit, extreme care in design must be exercised to obtain the ultimate in performance and reliability. Here, two items that are of little or no consequence on earth but are of great importance aloft are weight and power efficiency. Weight is a very important concern in placing the vehicle in orbit, and power efficiency determines how much size, weight, and cost must be dedicated to the solar cell power source. Any trade-off that can be made with the ground station equipment is made, and accordingly, high power in the transmitters, large ground station antennas, and carefully designed high-sensitivity receivers (low-noise figure) are always used. In the past, large klystrons, operating at output levels well over 1000 W, have been used to obtain sufficient power to put an acceptable signal into the satellite receivers. Similarly, large antennas are used to obtain high gain and high directivity of the beam. The latter is important because it is mandatory that a sharp beam be used if other satellites in space are to reuse the limited number of SHF frequencies available. Receiver sensitivity is optimized in the ground station of high-capacity systems usually by refrigerating the front end and achieving the lowest noise figure possible. The importance of the cooling is mentioned in Section VII,M, where it was shown that appreciable improvement in receiver sensitivity (reduction in noise figure) can be achieved using this technique.

Satellites are launched in such a way that, after stabilization in the orbit over the equator, each is assigned a specific longitudinal position that must be maintained. This is called "station keeping" and small jets with on-board gas reservoirs can make minor adjustments of the satellite's position to keep it "on station."

Modern satellites are becoming more sophisticated and include very advanced electronics other than those needed to transmit and receive a SHF signal. Switching of the signals within the satellite permit the routing of circuits on command and provide other operational functions that previously would have been done at ground terminals with much more electronic equipment.

Power in the satellite vehicle is derived from solar energy impinging upon the solar cells mounted on the surface of the orbiting station. Some reserve is stored in the batteries to provide power during earth shadows, but the bulk of the power from the solar cells is used directly in the satellite electronics.

Some of the problems that plague terrestrial systems are much less severe or nonexistent in satellite systems. Fading caused by atmospheric conditions along a 25–40-mile path in a terrestrial system is virtually nonexistent at 4 and 6 GHz in a satellite path that traverses only about 10 miles of the earth's atmosphere. Therefore, the range of the AGC in the receiver can be much less than in earth systems. Also, the level of the various input signals into the satellite receivers from the earth station transmitters are essentially at equal levels, making the intermodulation problems in the satellite less severe.

The satellite positions (stations) in space are optimized to obtain the maximum number of locations with acceptable levels of interferences from cochannel ground stations. Present practice is to locate satellites at 5° intervals; however, it is proposed that this spacing be reduced to as low as 3° longitude. This results in a finite number of systems that can be deployed using the 4- and 6-GHz bands. Quite a few years ago it was recognized that using only these frequency bands would be inadequate in the years ahead and that systems would have to be developed at higher frequencies, where propagation problems and electronic component design complexities would be more severe. Today, satellite systems operate at 4, 6, 11, and 12 GHz. Some consideration has been given to operating at 25 to 30 GHz.

To achieve the maximum performance of a satellite facility from the standpoint of system capacity and signal-to-noise ratio, the baseband signal is often processed in very specific ways. In the case of television transmission, the electronic elimination of redundancy in the picture allows satisfactory transmission with less baseband occupancy. Compandors (compressors/expandors) are widely used in telephony (but not data) terminals to improve the signal-to-noise ratio in a voice channel by effectively increasing the modulation level on the satellite link. At the present time, frequency modulation of the rf carrier is used in these systems. SSB technology has not yet been employed in satellite systems, but it is now under consideration. The sharing of frequencies with terrestrial systems makes it necessary to coordinate their use in each instance. A terrestrial system normally points the beam very near to the horizon, but if a satellite earth station is in line with this beam, mutual interference can (will) exist. It will be necessary to relocate one of the sites so that the three stations will not be in this unfavorable alignment. The high power of the earth station transmitter and the high sensitivity of the earth station receiver aggravate the problem in both directions.

XIII. Digital Microwave Radio

In previous sections, the history of microwave has shown virtually a total use of analog technologies for the transmission of television, telephony, and voiceband data including telegraph. First, FM was used, and more recently, SSB AM has been introduced primarily to obtain better utilization of the facilities and of the frequency spectrum. To be able to put 6000 voiceband circuits onto a 6-GHz channel when only about 2000 could be accommodated with FM results in a considerable reduction in cost for each circuit carried by the SSB channel. Over the years, it was the primary aim of system designers to increase the load-carrying capacity of the radio channels while still maintaining the quality of the services being rendered.

By contrast, the trend in the local telephone system for the past quarter century has been to provide interoffice connections of lengths greater than about 5 miles by digital means. This has been an ever-accelerating trend, and the reason for this is the rather large difference in cost of analog multiplex terminals versus digital terminals. What gives the digital terminal the edge over analog techniques is the common use of a large number of components for the digital terminal, whereas an analog multiplexer uses components on a per-circuit basis. Digital facilities are taking over rapidly, and in many instances analog circuits are being removed only to be replaced by digital equivalents. This is done partly to update the equipment in the plant, partly to provide better circuits with lower noise and less maintenance, and partly to be able to handle voice and data services without concern of the mix over the same facility. Once a signal is encoded in digital form, the various kinds of signals are generally indistinguishable from each other and the routing of signals via a digital switch is greatly simplified.

In the mid-1970s, considerable thought was given to whether digital signals could be transmitted economically over microwave channels. It was a well-known fact that a digital voice channel using pulse-code modulation (PCM) requires substantially greater bandwidth than a SSB channel. In the basic PCM terminal, the normal 4-kHz voiceband requires 64 kHz of bandwidth, an increase of 16 times what is required for SSB. It has been stated several times herein that SSB is the most efficient use of the rado spectrum, so the question can be raised, "Why should digital transmission via radio be considered at all?" It has become apparent that the costs of long-haul systems are not only the costs of the radio line but of all of the equipment, including multiplex terminals. More and more switching in the telephone system is being done on a digital basis, and for distances less than about 400 miles, experience shows that digital radio will be less expensive overall than an analog system. (Beyond about 400 miles, the line costs will predominate and it will be more economical to convert digital signals to analog for transmission purposes only.)

Originally, digital modulation schemes were considered wherein phase-shift keying (PSK) of the carrier would convey the information on the channel. Unfortunately, these simple schemes do not provide the bit speeds needed for high system capacity without using excessive bandwidth. A number of other enhancements of the modulation process must be used to obtain higher bit rates. More phases of the carrier can be used, and amplitude shifts plus phase shifts and the modulation of two coherent carriers in QAM can be employed in combination to obtain higher speed of transmission. All of these cause the detection process to become more critical and fragile and require more precise control of all system parameters. Since digital systems are all predicated upon the principle that a pulse will be present or absent at a particular instant, a reference must also be present at the detector so that the received pulse can be recognized when it should be present. Unlike SSB, which requires a precise frequency but not a precise time or phase relationship for proper detection, digital systems must extract precision timing information from the received signal itself. All of this has altered the once rugged PCM signal to one which is less robust and sensitive to perturbations of the signal than is FM.

With the vast microwave facilities already in place, using well-established channel assignments and channel bandwidths in presently authorized common-carrier bands, digital microwave must fit precisely into the analog microwave world. Furthermore, the FCC has established a rule that a system operating on a 20-MHz-wide channel, as at 4 GHz, shall have the capability of carrying at least 1150 4-kHz voiceband circuits. This requires transmission at a rate of about 78 Mbit/sec or higher. The digital signal must not interfere with adjacent channels that might be operating on other modes of modulation such as FM and SSB AM. Current systems use 90 Mbit/sec for systems with 20-MHz channel bandwidth and ~140 Mbit/sec for systems with 30-MHz channel bandwidths.

It is highly likely that no phase of microwave radio technology has received more attention from the standpoint of high-level mathematical analysis of (1) the frequency spectrum, (2) the error rate performance, and (3) the sensitivity to distortion effects as has digital modulation of SHF waves. Modern computer programs make possible system simulations and performance evaluations without constructing models and field testing an actual system. This has speeded up the development process and has resulted in more optimum designs.

One of the serious problems encountered in digital radio systems is the dispersive nature of a channel undergoing selective fading. The inter-symbol distortion that results when selective fading is present requires the use of space-diversity antennas with combining arrangements or adaptive equalization control of the two paths into the receiver to meet performance objectives. This adds to the complexity of the system and to system costs.

When modulated onto the SHF carrier, a digital signal can produce a frequency spectrum quite different from and at times far broader than either a television signal or a multiplexed telephony signal. The higher the bit rate, the wider the spectrum spread. To keep this under control when operating in a multichannel environment or in a multichannel digital band and to meet FCC spectrum occupancy rules, sophisticated filtering in the IF portions of both the transmitting and receiving equipment of the repeater is used. Again, computer analysis and synthesis help to determine the design of these filters.

XIV. Conclusion

After more than 35 yr of continuous development of FM analog microwave systems, the law of diminishing returns has set in and further improvements are not likely to be earthshaking. SSB, while much more recent, has already maximized the frequency utilization of 4-kHz occupancy for a 4-kHz voice channel, and this, too, is not likely to show astounding change in capacity. Digital transmission, however, is relatively new, and powerful analytical forces are being brought to bear on all facets of the subject by literally hundreds of researchers and engineers, and it would be highly dangerous to speculate where this fast-moving technology will be in another decade. The future, however, may be made still more uncertain by the rapid growth of fiber guide systems which operate at frequencies very much higher than SHF but which do not experience the anomalies of atmospherically induced dispersion of the transmission medium. This would be a definite advantage for that facility. The past 35 yr have made permanent contributions to the quality and the quantity of communications in the United States and throughout the world, and microwave radio has played a large part in bringing about this communications revolution.

BIBLIOGRAPHY

Borgne, M. (1985). "Comparison of high-level modulation schemes for high-capacity digital radio systems." *IEEE Trans. Commun.* **COM–33,** No. 5.

Fenderson, G. L., Parker, J. W., Quigley, P. D., Shephard, S. R., and Siller, C. A. (1984). "Adaptive Transversal Equalization of Multipath Propagation for 16–QAM 90 Mb/s Digital Radio," *Bell Syst. Tech. J.* **63,** No. 10. John F. Kennedy Parkway, Short Hills, N.J. 07078.

Jakes, W. C. (ed.) (1983). "The AR6A single-sideband microwave system," *Bell Syst. Tech. J.* **62,** No. 10.

Leclert, A., and Vandamme, P. (1985). "Decision feedback equalization of dispersive radio channels," *IEEE Trans. Commun.* **COM–33,** No. 7.

Moridi, S., and Sari, H. (1985). "Analysis of four decision-feedback carrier recovery loops in the presence of intersymbol interference." *IEEE Trans. Commun.* **COM–33.**

Tattleman, P., and Grantham, D. (1985). "Minute rainfall rates for microwave attenuation calculations," *IEEE Trans. Commun.* **COM–33,** No. 4.

Vigants, A. (1981). "Microwave radio obstruction fading," *Bell Syst. Tech. J.* **60,** No. 6.

Vigants, A., and Pursley, M. V. (1979). "Transmission unavailability of frequency–diversity protected microwave FM radio caused by multipath fading." *Bell Syst. Tech. J.* **58,** No. 8.

Yeh, Y. S., and Greenspan, L. J. (1985). "A new approach to space–diversity combining in microwave digital radio." *Bell Syst. Tech. J.* **64,** No. 4.

MODULATION

Rodger Ziemer   *University of Colorado at Colorado Springs*

GLOSSARY

Analog Modulation: Process whereby a message signal, which is the analog of some physical quantity, is impressed on a carrier signal for transmission through a channel.

Carrier Signal: Signal, usually a sinusoid or a pulse train, which is used to carry the message signal by means of several different types of modulation techniques.

Coherent demodulation: Method of recovering a message signal from a modulated signal that makes use of a reference signal at the receiver that is phase coherent with the received carrier signal.

Correlation detector: Receiver for digitally modulated signals that correlates a replica of the transmitted signal with the received signal in order to recover the data.

Demodulator: Device used to recover a message signal from the received modulated carrier.

Digital modulation: Technique for placing a digital data sequence on a carrier signal for subsequent transmission through a channel.

Envelope detector: Device that has as its output the envelope of a modulated carrier signal at its input.

Linear modulation: Modulation technique in which the modulated carrier is a linear function of the message signal.

Lower sideband: Portion of a modulated signal spectrum with frequency components having frequencies less than the carrier frequency.

Message signal: Signal that contains the information to be transmitted through the channel through modulation of a carrier signal.

Modulation: Process whereby a message signal is impressed on a carrier signal for subsequent transmission through a channel.

Modulator: Device for impressing a message signal on a carrier signal through the process of modulation.

Multiplexing: Modulation process that makes it possible to transmit several message signals through the same channel on a single carrier.

Noncoherent modulation: Digital modulation scheme for which it is possible to recover the digital data from the modulated carrier signal without the use of a local reference signal at the receiver, which is in phase coherence with the received carrier signal.

Nonlinear analog modulation: Modulation scheme for which the modulated carrier is a nonlinear function of the message signal.

Probability of error: Measure of performance for a digital modulation scheme; essentially the ratio of symbols received in error to the total number received over a time period that is sufficiently long so that the ratio is essentially constant.

Sampling theorem: Theorem that relates the minimum number of samples of an analog signal that must be taken per second to the signal's bandwidth such that the signal can be recovered from its sample values. The low-pass sampling theorem states that this minimum number of samples per second must be at least twice the bandwidth of the low-pass signal.

Spectrum: Function that gives the relative magnitudes and phases of a signal (phasor components) versus frequency such that the signal can be reconstructed through vector addition of these phasor components.

Upper sideband: Portion of the spectrum of a modulated signal whose frequencies are greater than the carrier frequency.

Modulation is the systematic variation of some attribute of a carrier signal in accordance with a function of the message signal. Examples of carrier signals are sinusoids or pulse trains. Various attributes of sinusoids that can be varied in accordance with the message are amplitude, phase, or frequency, while those of a pulse train are pulse amplitude, pulse width, or pulse position with respect to a fixed reference. Message signals can be the analog of some quantity, such as a speech wave, in which case the modulation is referred to as analog. If the message is chosen from a discrete set, such as the English alphabet, and a carrier attribute, such as phase, is varied in a one-to-one manner in accordance with this discrete message, the modulation is referred to as digital. Modulation techniques can be further categorized as linear or nonlinear, depending on whether the modulated signal varies linearly with the message (i.e., superposition holds) or whether it varies nonlinearly with the message. Examples of the former are amplitude modulation, a term usually applied to analog modulation, or amplitude-shift keying, which is the term used when the modulation is digital. Examples of nonlinear modulation schemes include phase and frequency modulation, which are usually the terms applied to nonlinear analog modulation, or phase-shift keying and frequency-shift keying in the case of digital modulation. Reasons for modulation include (1) easing radiation of the signal by means of an antenna; (2) reducing noise and interference; (3) assigning channels; (4) multiplexing, in order to transmit many messages through the same channel at the same time; and (5) overcoming equipment limitations. The process of modulation is carried out by means of a modulator, and the inverse operation of recovering the signal from the modulated signal is carried out by a demodulator.

I. Modulation Basics

A. SIGNALS AND SPECTRA

The study of modulation methods is facilitated by taking both a time and frequency domain viewpoint. A signal's frequency domain description, or spectrum, is provided by its Fourier transform, defined as

$$X(f) = \int_{-\infty}^{\infty} x(t)\, e^{-j2\pi ft}\, dt \triangleq \mathcal{F}[x(t)] \qquad (1)$$

The time domain representation of the signal, $x(t)$, is obtained from its spectrum by the integral

$$x(t) = \int_{-\infty}^{\infty} X(f)\, e^{j2\pi ft}\, dt \triangleq \mathcal{F}^{-1}[X(f)] \qquad (2)$$

which is called the inverse Fourier transform of $X(f)$. These integrals provide what is referred to as a Fourier transform pair of any signal. Not only can tables of Fourier transform pairs be derived, but theorems relating important properties of Fourier transforms can be obtained and give much insight into how various operations on signals affect their spectra. Important theorems for the consideration of modulation techniques are the frequency translation and modulation theorems, which state the following. Given a signal $x(t)$ with Fourier transform $X(f)$, then

$$\mathcal{F}[x(t)\, e^{\pm j2\pi f_0 t}] = X(f \pm f_0)$$

(frequency translation theorem)

or

$$\mathcal{F}[x(t)\cos(2\pi f_0 t)] = \tfrac{1}{2}[X(f - f_0) + X(f + f_0)]$$

(modulation theorem)

These theorems simply state that multiplication of a signal by a sinusoid of constant frequency f_0 shifts the spectrum of the signal from being centered around zero frequency to being centered around frequency f_0.

Convenient Fourier transform pairs for the sequel are

$$\mathcal{F}[\cos(2\pi f_0 t)] = \tfrac{1}{2}[\delta(f - f_0) + \delta(f + f_0)] \qquad (3)$$

and

$$\mathcal{F}[A] = \delta(f) \qquad (4)$$

where $\delta(f)$ is the unit impulse or delta function and A is a constant. These transform pairs as well as the theorems given previously will prove useful in the description of modulation systems.

B. ANALOG MODULATION TERMINOLOGY AND METHODS

In analog modulation systems, the transmitted carrier is modulated by a signal that is the analog of some physical quantity, such as a speech wave, liquid level, or velocity. The message can be a function of a continuous-time variable, or it can be a function of a discrete-time variable. Continuous-time signals are sometimes sampled to produce discrete-time signals, also referred to as sample-data signals. Analog message signals are often used to modulate sinusoidal carriers of the form

$$x_c(t) = A_c \cos(2\pi f_c t + \theta_c) \qquad (5)$$

Chief examples of analog modulation of sinusoidal carriers include double-sideband modulation (DSB), amplitude modulation (AM), frequency modulation (FM), and phase modulation (PM). In the former two cases, it is the amplitude, A_c, of the carrier that is varied in accordance with the message signal. In the case of FM, it is the instantaneous frequency of the carrier that is varied linearly with respect to the message signal, while in the case of PM, it is the instantaneous phase of the carrier that is varied linearly with respect to the message signal.

Sample-data message signals are often used to modulate pulse train carriers of the form

$$p_c(t) = A_c \sum_{n=-\infty}^{\infty} \Pi[t - nT_s - t_0)/\tau] \qquad (6)$$

where $\Pi(t) = 1$ for $|t| \leq \frac{1}{2}$ and is zero otherwise. Depending on whether the amplitude, A_c, pulse position, T_s, or pulse width, τ, is varied in accordance with the message sample value, the result is pulse amplitude modulation (PAM), pulse position modulation (PPM), or pulse width modulation (PWM), respectively.

C. DIGITAL MODULATION TERMINOLOGY AND METHODS

Digital modulation results if a new message is chosen each T_s seconds (referred to as the symbol interval) from a discrete set, this discrete set is mapped into a discrete set of voltage levels in a one-to-one fashion, and the voltage levels are used to modulate some attribute of a carrier. If only two voltage levels are used, corresponding to two possible messages, the resulting modulation is referred to as binary. If $M > 2$ voltage levels are used, the modulation is referred to as M-ary. If the modulation is binary, it is convenient to denote the message set as $\{0, 1\}$, whereas, if the modulation is M-ary, it is convenient to denote it as the integer set $\{0, 1, 2, \ldots, M-1\}$.

If the carrier is sinusoidal, the amplitude, phase, or frequency can be varied in accordance with the message set in which case the modulation is referred to as amplitude-shift keying (ASK), phase-shift keying (PSK), or frequency-shift keying (FSK), respectively. Other higher order types of digital modulation can also be employed. For example, if both amplitude and phase are varied, with the latter varied in a quadrature manner, the result is referred to as quadrature-amplitude-shift keying (QASK). If PSK is used, but the shift in phase between one symbol interval and the next occurs in such a fashion that the phase change is continuous, the result is referred to as continuous-phase modulation (CPM). A special case of binary CPM results in minimum-shift keying (MSK). More complex CPM modulation schemes can be used to provide the simultaneous desirable attributes of constant amplitude, bandwidth efficiency, and power efficiency and are at this time in the developmental stage.

With the advent of laser sources and low-loss optical fibers, digital communications in the optical portion of the spectrum are becoming more widespread. The initial approach to optical communications made use of photodetectors, wherein a burst of light energy is directly detected and output from the detector is in the form of a baseband voltage pulse. Useful modulation methods for such optical modulation schemes can be viewed as the modulation of a baseband pulse train carrier, that is, the pulsed output of a laser. In such applications, the modulation technique can be viewed as the variation of some attribute of a pulse train, for example, amplitude, pulse width, or pulse position, by the digital message.

II. Analog Modulation

A. LINEAR ANALOG MODULATION SCHEMES

1. Double-Sideband and Amplitude Modulation

Linear analog modulation results from causing the amplitude of the carrier signal to depend linearly on the analog message signal. Three examples of this type of modulation are DSB, AM, and PAM. Pulse amplitude modulation will be discussed later with the other possible analog pulse modulation methods. In the case of AM and DSB, the carrier is sinusoidal as expressed by Eq. (1), and the modulated signal can be written as

$$x_{c, \text{DSB}}(t) = A_c\, m(t)\, \cos(2\pi f_c t) \qquad (7)$$

for DSB, and

$$x_{c, \text{AM}}(t) = A_c[1 + \alpha\, m(t)]\, \cos(2\pi f_c t) \qquad (8)$$

for AM, where $m(t)$ is the message signal, f_c is the carrier frequency, and α is a parameter called the modulation index. The implementation of a modulator for either DSB or AM is accomplished simply by a multiplier.

If the Fourier transforms of Eqs. (7) and (8) are taken, the resulting spectra are

$$x_{c, \text{DSB}}(f) = \tfrac{1}{2}A_c[M(f - f_c) + M(f + f_c)] \qquad (9)$$

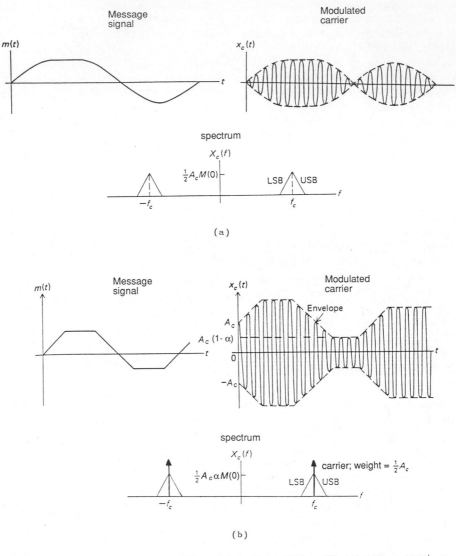

FIG. 1. Waveforms and spectra for (a) DSB modulation [note: $M(f) = \mathscr{F}[m(t)]$; $M(0) = M(f)|_{f=0}$] and (b) AM modulation.

for DSB, and

$$X_{c,\ AM}(f) = \tfrac{1}{2} A_c\{\delta(f - f_c) + \delta(f + f_c)]$$

$$+ \tfrac{1}{2} A_c\alpha[M(f - f_c) + M(f + f_c)] \tag{10}$$

for AM. The main difference between DSB and AM is the presence of a carrier component in AM, which mainfests itself as the term $A_c \cos(2\pi f_c t)$ in the time domain and as the pair of delta functions located at frequencies $\pm f_c$ in the frequency domain. Waveforms and spectra for DSB and AM modulation are compared in Fig. 1. Note that in both cases, the spec-

trum corresponding to the message signal consists of a mirror image about the carrier (the carrier is absent in the case of DSB). These message spectral components are known as the upper and lower sidebands and carry essentially duplicate information.

The demodulation of either DSB or AM can be accomplished by a coherent demodulator shown schematically in Fig. 2(a). The implementation of such a demodulator presupposes the availability of a coherent carrier reference at the receiver. Two circuits for achieving this are the squarer-frequency-divider circuit of Fig. 2(b) and the Costas loop of Fig. 2(c).

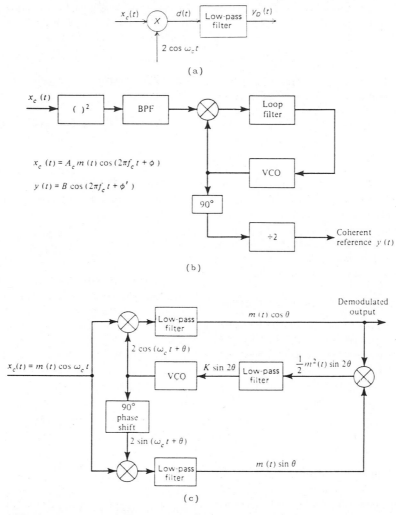

FIG. 2.   Demodulator circuits for double sideband modulation. (a) Coherent demodulator; (b) squarer-frequency-divider circuit; and (c) Costas loop.

The use of coherent demodulation is necessary for the successful demodulation of DSB because of the 180° phase reversals of the carrier that occur each time $m(t)$ changes sign. Because of the insertion of a carrier component in AM, these phase reversals do not occur, and it is possible to use an envelope detector. Such a demodulator is illustrated in Fig. 3. Its operation depends on the rectification of the positive halves of the modulated carrier pulses and the subsequent smoothing of these pulses into a waveform that closely resembles the modulating signal by the low-pass RC filter and DC blocking capacitor, which follow the rectifier. The simplicity of this scheme for demodulation of AM is one reason for its use in commercial broadcast radio.

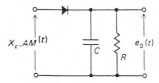

FIG. 3.   Envelope detector circuit.

2. Single-Sideband and Vestigial-Sideband Modulation

Two modulation schemes that are derivatives of DSB modulation are single-sideband (SSB) and vestigial-sideband (VSB) modulation. The SSB modulation is produced by filtering out all of one sideband of a DSB signal, while VSB is produced by leaving a vestige of one sideband and all of the other sideband of a DSB signal. Such sideband rejection can be effected by a sideband filter at the output of the DSB modulator, which consists of a multiplier, as pointed out previously. The reason for using SSB is that the information in the message signal is essentially duplicated by the upper and lower sidebands of the DSB signal; rejecting one sideband conserves bandwidth while preserving the information content of the signal. The reason for leaving a part of one sideband in VSB is to allow for easier sideband filter realization. Both SSB and VSB can be demodulated using the coherent demodulator shown in Fig. 2. It is also possible to demodulate both modulation schemes by reinserting a carrier component and passing the sum through an envelope detector. VSB is the modulation technique used for transmitting the video signal in commercial television broadcasting.

B. NONLINEAR ANALOG MODULATION SCHEMES

Examples of nonlinear analog modulation schemes are FM, PM, PWM, and PPM. The PWM and PPM examples will be discussed later under the heading of analog pulse modulation. For the sinusoidal carrier given by Eq. (4), the instantaneous phase is

$$2\pi f_i(t) = \frac{d\theta_i}{dt} = \frac{d}{dt}[2\pi f_c t + \theta_c(t)]$$

$$= 2\pi f_c + \frac{d\theta_c(t)}{dt} \qquad (11)$$

where $\theta_c(t)$ and $d\theta_c/dt$ are referred to as the phase and frequency deviation, respectively. If the phase deviation is made proportional to the message signal, the result is phase modulation, and if the frequency deviation is made proportional to the message signal, the result is frequency modulation. Thus, for phase modulation, the resultant modulated signal is of the form

$$x_{c, PM}(t) = A_c \cos[2\pi f_c t + k_p m(t)] \qquad (12)$$

while, for frequency modulation, the resultant modulated signal is of the form

$$x_{c, FM}(t) = A_c \cos[2\pi f_c t + k_f \int^t m(\alpha)\, d\alpha] \qquad (13)$$

where k_p and k_f are proportionality constants. Both FM and PM are referred to as angle-modulated signals because of the dependence of the modulated signal on the message signal through the angle, or argument, of the carrier.

Because of the nonlinear dependence of the modulated signal on the message signal for FM and PM, it is possible to obtain results for the spectrum only in certain special cases. One such case is that of a si-

nusoidal message signal of amplitude A_m and frequency f_m. For this situation, both Eqs. (12) and (13) can be written in the form

$$x_c(t) = A_c \cos[2\pi f_c t + \beta \sin(2\pi f_m t)] \qquad (14)$$

where β is referred to as the modulation index. It is given by

$$\beta = \frac{A_m f_m}{f_d} \quad \text{for FM where} \quad f_d = k_f/2\pi$$

and

$$\beta = k_p A_m \quad \text{for PM}$$

Equation (14) can be expanded in the trigonometric series

$$x_c(t) = \sum_{n=-\infty}^{\infty} A_c J_n(\beta) \cos[2\pi(f_c + nf_m)t] \qquad (15)$$

where $J_n(\beta)$ is a Bessel function of order n and argument β. Thus, for sinusoidal modulation, the spectrum of an angle-modulated signal is infinite in extent and consists of lines at the frequencies f_c, $f_c + f_m, f_c - f_m, f_c + 2f_m, f_c - 2f_m$, etc.

Figure 4 shows amplitude spectra for a sinus-

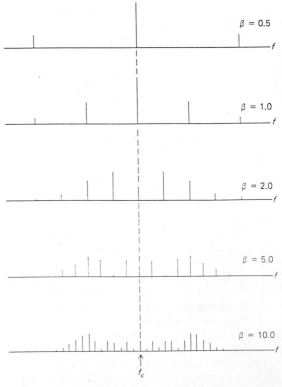

FIG. 4.   Amplitude spectra for a sinusoidally modulated FM signal (βf_m held constant).

oidally modulated carrier for various values of the modulation index, β. It is noted that the spectrum is essentially negligible beyond a certain frequency separation from the carrier. A rule of thumb often used to estimate the transmission bandwidth, B_T, required for an angle-modulated signal is Carson's rule, which is

$$B_T = 2(\beta + 1)f_m \qquad (16)$$

This rule is heuristically applied to the case of non-sinusoidal modulating signals by replacing β with a deviation ratio, D, defined as

$$D = \frac{\text{peak frequency deviation}}{\text{bandwidth of } m(t)}$$

$$= \frac{f_d}{W}[\max|m(t)|] \qquad (17)$$

where W is the bandwidth of $m(t)$.

The demodulation of FM can be accomplished by a *slope detector,* which, in its simplest form, can be modeled as a differentiator followed by an envelope detector. That such a circuit can be used for demodulation of FM can be seen by differentiating Eq. (13), which gives

$$y(t) = -A_c[2\pi f_c + k_f m(t)] \sin[2\pi f_c t + k_f \int^t m(\alpha)\, d\alpha] \qquad (18)$$

The envelope of Eq. (18) is

$$e(t) = A_c[2\pi f_c + k_f m(t)] \qquad (19)$$

which is proportional to the message, $m(t)$. In actuality, any circuit with a frequency response that is approximately linear with frequency over a range about f_c will do in place of the differentiator. To demodulate PM, one can follow the envelope detector with an integrator circuit. A second method that can be used to demodulate FM is by means of a phase-lock loop (PLL). It is, in fact, the preferred method nowadays because of the availability of inexpensive PLL integrated circuits.

Frequency modulation is used for commercial broadcast radio with a carrier frequency in the range of 88.1 to 107.9 MHz and a channel spacing of 200 kHz. In this application, a technique known as pre- and de-emphasis is used to combat the effects of interference, which is most detrimental in FM at the higher frequency end of the message spectrum. At the transmitter, a pre-emphasis filter is used to boost the high-frequency end of the message spectrum. At the receiver, after demodulation, a de-emphasis filter is used to bring the high-frequency end of the message down to what it was at the transmitter while simultaneously attenuating noise at frequencies at the high-frequency end of the message spectrum where

they are most severe because of the demodulation process.

C. ANALOG PULSE MODULATION

Three types of modulation using a pulse-type carrier of the form in Eq. (6) can be employed. These are pulse amplitude modulation, pulse width modulation, and pulse position modulation.

In PAM, the amplitude of the pulse-type carrier is varied in accordance with samples of the message taken at T_s-second intervals. If $m(t)$ is the message signal, the modulated waveform for PAM can be expressed as

$$x_{c,\text{ PAM}} = A_c \sum_{n=-\infty}^{\infty} m(nT_s)\, \Pi[(t - nT_s)/\tau] \qquad (20)$$

In PAM, the pulse width τ and pulse position t_0 (taken as zero here) are held constant. The PAM signal defined by Eq. (20) is also known as flat-top sampling, and the message can be recovered by low-pass filtering with a filter bandwidth greater than W, the message bandwidth, but less than $T_s^{-1} - W$. Ideally, the low-pass filter should be followed by an equalizer, which removes the distortion introduced by representing each sample by the rectangular pulse shape in Eq. (20) [in this idealized case, the equalizer has a frequency response that is the inverse of the Fourier transform of the rectangular pulse, which is τ sinc $(f\tau)$]. The error in ignoring the equalizer is small, however, if the pulse width is small compared with the sampling period.

The other two types of analog pulse modulation, PWM and PPM, are nonlinear modulation techniques as contrasted to PAM, which is linear. As such, their baseband spectra are wider than that of the modulating message signal, $m(t)$. This apparent disadvantage is counterbalanced by the advantage that these two modulation techniques are more immune to the detrimental effects of noise than is PAM. The message signal can be recovered from PWM by low-pass filtering, just as for PAM. A similar procedure can be used for recovering the message signal from PPM once it is converted to PWM. A disadvantage of PPM is that it requires a reference clock signal at the receiver for demodulation.

D. PULSE CODE AND DELTA MODULATION

Representation of analog messages in digital format is effected by first sampling the message signal, $m(t)$, in accordance with the sampling theorem and then converting the samples to binary number representation through the process of analog-to-digital

TABLE I. Example of PCM Waveform Representation

Sample time	Sample value	Rounded value	Binary representation
T_s	1.1	1.0	001
$2T_s$	2.4	2.0	010
$3T_s$	3.1	3.0	011
$4T_s$	1.9	2.0	010
$5T_s$	5.1	5.0	101
$6T_s$	6.6	7.0	111

(A/D) conversion. For example, for the message samples given in Table I, the corresponding A/D converted representation is given in the last column (rounding to the nearest integer value is used). The binary representation of each sample value is then transmitted through the channel by a sequence of on–off rectangular pulses in the T_s-second interval corresponding to that sample value. This procedure is referred to as pulse-code modulation (PCM).

Another way of representing the message sample values in binary form is by means of a technique

TABLE II. Coherent Digital Modulation Schemes

Modulation type	Signal set	Bandwidth efficiency (bits/sec/Hz)
Binary ASK	$s_1(t) = 0$ $s_2(t) = \sqrt{\dfrac{2E_b}{T_b}} \cos \omega_c t \quad 0 \le t \le T_b$	0.5
Binary PSK	$s_1(t) = \sqrt{\dfrac{2E_b}{T_b}} \cos \omega_c t \quad 0 \le t \le T_b$ $s_2(t) = -\sqrt{\dfrac{2E_b}{T_b}} \cos \omega_c t \quad 0 \le t \le T_b$	0.5
Binary FSK	$s_1(t) = \sqrt{\dfrac{2E_b}{T_b}} \cos \left(\omega_c + \dfrac{\Delta\omega}{2} \right) t \quad 0 \le t \le T_b$ $s_2(t) = \sqrt{\dfrac{2E_b}{T_b}} \cos \left(\omega_c - \dfrac{\Delta\omega}{2} \right) t \quad 0 \le t \le T_b$	0.33
QPSK & OQPSK	$A_c[d_1(t) \cos 2\pi f_0 t + d_2(t) \sin 2\pi f_0 t]$ d_1 and d_2 offset by $\dfrac{T_s}{2}$ for OQPSK	1.0
MSK	$A\left[d_1(t) \cos \dfrac{\pi t}{2T_b} \cos 2\pi f_0 t + d_2(t) \sin \dfrac{\pi t}{2T_b} \sin 2\pi f_0 t \right]$ $d_1, d_2 = \pm 1, \quad 0 \le t \le T_s$	0.67
M-ary PSK	$s_i(t) = \sqrt{\dfrac{2E_s}{T_s}} \cos \left[\omega_c t + \dfrac{2\pi(i-1)}{M} \right]$ $0 \le t \le T_s; \quad i = 1, 2, \ldots, M$	$0.5 \log M$
M-ary FSK	$s_i(t) = \sqrt{\dfrac{2E_s}{T_s}} \cos 2\pi [f_0 + (i-1)\, \Delta f]$ $0 \le t \le T_s; \quad i = 1, 2, \ldots, M$	$(\log_2 M)/(M+1)$
16-QASK (QAM)	$s_i(t) = \sqrt{\dfrac{2}{T_s}} (A_i \cos \omega_c t + B_i \sin \omega_c t) \quad 0 \le t \le T_s$ $A_i = \pm a, \pm 3a$ $B_i = \pm a, \pm 3a$	2.0

known as delta modulation. This technique uses a feedback modulator structure that compares a binary representation of the current sample value with the actual value. If greater than the actual value, a logic-1 is sent through the channel; if less than the actual value, a logic-0 is sent through the channel. This example is the simplest verson of delta modulation, known as zero order. Other higher order versions can be implemented, which take into account the slope of the message signal at the sampling instant, its curvature at the sampling instant, etc.

III. Digital Modulation

A. Performance Measures

Two primary considerations when choosing a digital modulation scheme are bandwidth occupancy, measured in terms of information rate per unit bandwidth (bits/sec/Hz), and error-rate performance, measured in terms of the signal energy-to-noise–spectral-density ratio (E_b/N_0), required to give a specified probabilitiy of bit error, $P_b(\varepsilon)$. The first

Probability of error	E_b/N_0 for $P_{e,bit} = 10^{-6}$
$P_b(\varepsilon) = Q\left(\sqrt{\dfrac{E_b}{N_0}}\right)$	13.6 dB
$P_b(\varepsilon) = Q\left(\sqrt{\dfrac{2E_b}{N_0}}\right)$	10.6 dB
$P_b(\varepsilon) = Q\left(\sqrt{\dfrac{E_b}{N_0}}\right)$	13.6 dB
$P_s(\varepsilon) \simeq 2Q\left(\sqrt{\dfrac{E_s}{N_0}}\right),\ P_b(\varepsilon) = Q\left(\sqrt{\dfrac{2E_b}{N_0}}\right)$	10.6 dB
$P_b(\varepsilon) = Q\left(\sqrt{\dfrac{2E_b}{N_0}}\right)$	10.6 dB
$Q\left(\sqrt{\dfrac{2E_s}{N_0}}\sin\dfrac{\pi}{M}\right) \le P_s(\varepsilon) < 2Q\left(\sqrt{\dfrac{2E_s}{N_0}}\sin\dfrac{\pi}{M}\right)$	$M = 4;\ 10.3$ dB $M = 8;\ 13.7$ dB $M = 16;\ 18.2$ dB
$P_s(\varepsilon) \le (M-1)Q\left(\sqrt{\dfrac{E_s}{N_0}}\right)$ $E_b/N_0 \triangleq E_s/N_0 \log_2 M$	$M = 4;\ 10.8$ dB $M = 8;\ 9.3$ dB $M = 16;\ 8.2$ dB
$P_s(\varepsilon) = 1 - [\tfrac{4}{16}P(C\mid I) + \tfrac{8}{16}P(C\mid II) + \tfrac{4}{16}P(C\mid III)]$ $P(C\mid I) = \left[1 - 2Q\left(\sqrt{\dfrac{2a^2}{N_0}}\right)\right]^2$ $P(C\mid II) = \left[1 - 2Q\left(\sqrt{\dfrac{2a^2}{N_0}}\right)\right]\left[1 - Q\left(\sqrt{\dfrac{2a^2}{N_0}}\right)\right]$ $P(C\mid III) = \left[1 - Q\left(\sqrt{\dfrac{2a^2}{N_0}}\right)\right]^2,\quad \dfrac{\overline{E}_b}{N_0} = \dfrac{5a^2}{2N_0}$	14.3 dB

presupposes that some means of specifying bandwidth has been chosen. This can be in terms of an energy containment bandwidth, which is the bandwidth within which a specified percent of the total energy is contained, or the bandwidth required for the main lobe of the signal spectrum. The former is more precise, and the latter easier to apply. Main lobe bandwidth will be used in the following discussion when bandwidth efficiencies are quoted for the various signaling schemes even though the numbers given are somewhat pessimistic.

B. COHERENT DIGITAL MODULATION SCHEMES

A coherent digital modulation scheme is one which requires a reference signal at the receiver that is in phase coherence with the carrier of the transmitted signal (which may be suppressed), accounting for the phase shifts introduced by the channel. Noncoherent schemes require no such reference signal, but at the expense of poorer error-rate performance. The price paid for the improved performance of coherent digital modulation schemes is the complication of having to acquire the reference at the receiver.

The characteristics of several coherent digital modulation schemes are summarized in Table II. In this table, the following symbol definitions apply: E_b is the symbol energy for a binary scheme; E_s is the symbol energy for an M-ary scheme (see Section I, C for an explanation for binary versus M-ary); T_b is the symbol duration for a binary scheme (bit period); T_s is the symbol duration for an M-ary scheme; $P_b(\varepsilon)$ denotes bit error probability; $P_s(\varepsilon)$ denotes symbol error probability; $Q(x)$ denotes the Q function, which is defined as

$$Q(x) = \int_x^\infty \frac{\exp(-x^2/2)}{\sqrt{2\pi}} \, dx \qquad (21)$$

$d_1(t)$ and $d_2(t)$ denote data sequences having values of ± 1 in T_s-second intervals; ω_c denotes carrier frequency or nominal center frequency in rad/sec; $\Delta\omega$ denotes frequency shift in rad/sec; f_c denotes carrier frequency or nominal center frequency in Hz; and M denotes the number of possible transmitted symbols in a symbol interval ($M = 2$ is a binary system). In Table II, the signal set in each case is defined for the interval $0 \leq t \leq T_s$; it is understood that the symbols follow in sequence so that the ith signaling interval is $(i - 1)T_s \leq t \leq iT_s$.

For the M-ary schemes listed in Table II, the symbol energy can be related to the bit energy by

$$E_s = (\log_2 M) \, E_b \qquad (22)$$

The bit error probability can be related to the symbol error probability in one of two ways. If a procedure

known as Gray encoding is used to ensure that an error in mistaking a symbol for its nearest neighbor results in only one bit error, then

$$P_b(\varepsilon) = P_s(\varepsilon)/\log_2 M \qquad (23)$$

If orthogonal signal sets are used (e.g., M-ary FSK), then

$$P_b(\varepsilon) = \frac{M}{2(M - 1)} P_s(\varepsilon) \qquad (24)$$

Some observations about the results in Table II are as follows. Quadrature phase-shift keying (QPSK) and offset quadrature phase-shift keying (OQPSK) offer a good compromise between bandwidth efficiency and error-rate performance, with MSK following close behind. M-ary PSK allows an increase in bandwidth efficiency with M at the expense of error-rate performance, and this increase in bandwidth efficiency goes only as the logarithm of M. M-ary FSK allows an increase in error-rate performance with M at the expense of a decrease in bandwidth efficiency. Another modulation scheme not shown, known as vertices-of-a-hypercube, allows essentially an even trade-off between these two performance measures.

Not addressed in Table II are other considerations, such as complexity of hardware and constant versus nonconstant envelope, to name only two such considerations. It is assumed that the demodulators/detectors are optimum in the sense that they achieve the minimum error probability in additive, white Gaussian noise (AWGN) of all possible detectors. They are generally known as matched filter or correlation detectors. Figure 5(a) shows the correlation demodulator/detector for binary PSK, and Fig. 5(b) shows the correlation demodulator/detector for M-ary PSK or 16-QASK (the only difference is in the threshold/decision logic box).

C. NONCOHERENT DIGITAL MODULATION SCHEMES

A noncoherent digital modulation scheme is one not requiring a reference signal at the demodulator in phase coherence with the received carrier. Two examples of such schemes are differential phase-shift keying (DPSK) and noncoherent frequency-shift keying. These are discussed briefly in the following.

In DPSK, the previous symbol period is used to provide a phase reference for the current symbol period. This procedure presupposes that phase perturbations introduced in the channel vary slowly compared with a symbol period and that there is a known phase relationship between adjacent bits. The latter is provided by using a procedure known as differen-

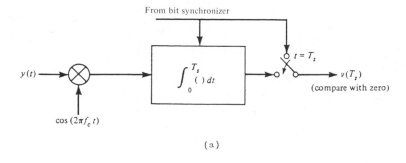

(a)

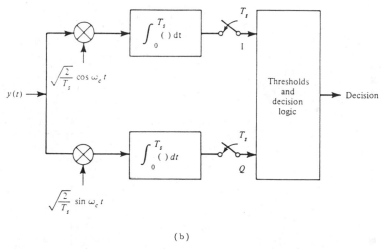

(b)

FIG. 5. Coherent detectors for digitally modulated signals: (a) binary PSK; (b) QASK.

tial encoding on the data before PSK modulation of the carrier. Although differential encoding can be carried out for M-ary alphabets, it is easiest to explain it in terms of binary data. Let the data sequence from a binary source be denoted as $\{D_n\}$ and the differentially encoded data sequence be denoted as $\{C_n\}$. The operation used to produce the nth bit of the differentially encoded sequence is

$$C_n = D_n \oplus C_{n-1} \qquad (25)$$

where $\oplus$ denotes a binary add without carry (also called an Exclusive-OR). For example, the sequence 10111001010 is differentially encoded as 100101110011, where the first bit is an arbitrarily selected reference bit. The differentially encoded sequence is then used to PSK modulate a carrier, which results in the DPSK modulated signal. Figure 6(a) shows a suboptimum detector for demodulating and detecting DPSK modulated data. That the original

source data sequence emerges from this detector can be checked by considering the result of passing successive symbols of the DPSK modulated signal through the detector. The error-rate performance of an optimum detector for binary DPSK can be shown to be

$$P_b(\varepsilon) = \exp(-E_b/N_0)/2 \qquad (26)$$

When compared with binary PSK, the difference in E_b/N_0 between DPSK and PSK is less than 1 dB. This is a small price to pay for not having to establish a coherent carrier reference at the receiver. Disadvantages of DPSK are that the phase characteristics of the channel must be stable, particularly at high data rates, and that the receiver is locked to a particular data rate because of the necessity of having to compare the current symbol with the previous symbol in the receiver.

Noncoherent FSK uses the same signal set as for

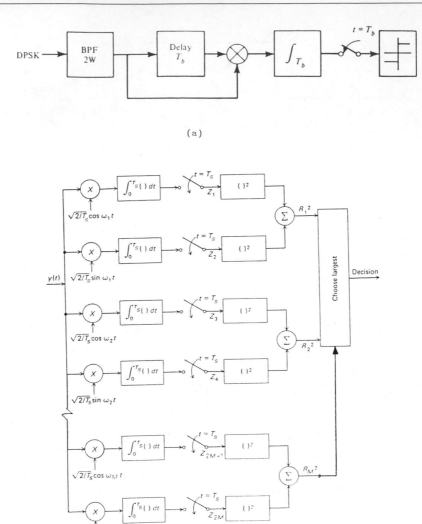

(a)

(b)

FIG. 6. Detectors for digitally modulated signals not requiring coherent references: (a) DPSK; (b) M-ary FSK.

TABLE III. Performances of Noncoherent and Coherent FSK Compared

Number of symbols per signaling interval, M	E_b/N_0 for $P_{e,bit} = 10^{-6}$	
	Noncoherent	Coherent
2	14.2	13.6
4	11.3	10.8
8	9.7	9.3
16	8.7	8.2

coherent FSK. The difference is that the receiver does not make use of the phase in demodulation/detection of the received signal. A demodulator/detector for noncoherent M-ary FSK is shown in Fig. 6(b). The symbol error probability can be shown to be

$$P_s(\varepsilon) = \sum_{k=1}^{M-1} \binom{M-1}{k} \frac{(-1)^{k+1}}{k+1}$$
$$\times \exp\left(-\frac{k}{k+1}\frac{E_s}{N_0}\right) \qquad (27)$$

The bit-error-rate performance of noncoherent FSK is compared with that for coherent FSK in Table III.

As with DPSK, the penalty in performance for using noncoherent FSK instead of coherent FSK is small.

BIBLIOGRAPHY

Haykin, S. (1988). "Digital Communications." Wiley, New York.

Sklar, B. (1988). "Digital Communications: Fundamentals and Applications." Prentice-Hall, Englewood Cliffs, New Jersey.

Wozencraft, J. M., and Jacobs, I. M. (1965). "Principles of Communications Engineering." Wiley, New York.

Ziemer, R. E., and Peterson, R. L. (1985). "Digital Communications and Spread Spectrum Systems." Macmillan, New York.

Ziemer, R., E., and Tranter, W. H. (1985). "Principles of Communications: Systsems, Modulation, and Noise," 2nd ed. Houghton, Boston, Massachusetts.

MULTIPLEXING

Rodger Ziemer *University of Colorado at Colorado Springs*

GLOSSARY

Asynchronous multiplexing: Sharing of a common channel resource by several message sources using time division multiplexing whereby the use of a common clock for all sources is not required.

Binary digit (BIT): Symbol that conveys one of two possible messages through a communications system.

Character (WORD): Sequence of digital symbols (bits) that conveys a particular piece of information (e.g., letter of the English alphabet) in addition to including certain overhead information such as start bits, stop bits, and parity check bits.

Code division multiple access (CDMA): Sharing of remote channel resources through the assignment of a unique code to each user, which is used to encode the information transmitted by that user. This encoding is usually done through a phase modulation process of each bit using the code or through using the code to frequency hop the carrier in a unique fashion for each user.

Demand assigned multiple access (DAMA): Multiple access system whereby communication resources are assigned as demanded by the users.

Fixed assigned multiple access (FAMA): Multiple access system whereby communication resources are assigned on a fixed basis to the users.

Frame: Sequence of data from several different sources. The information can be interlaced on a bit-by-bit basis or on a character-by-character basis and often includes overhead information for synchronization purposes, error correction purposes, etc.

Frequency division multiple access (FDMA): Remote sharing of communication channel resources through allocation in frequency in a similar fashion to FDM. The assignment of these resources may be fixed or varied with demand.

Frequency division multiplexing (FDM): Scheme that uses common communication channel resources by fixed allocation in frequency.

Multiple access: Method that allows the use of a remote communication resource, such as a relay satellite, by several different users.

Slot: Portion of the total time in a frame assigned to a given user in a time division multiple access system.

Synchronous: In reference to a time division multiplexed system, this term implies that all sources are kept in step from the same clock.

Time division multiple access (TDMA): Sharing of a remote communication resource, such as a relay satellite, on a time division basis.

Time division multiplexing (TDM): Scheme that uses common communication channel resources by allocation on a time interval basis.

Multiplexing is the use of some means of interleaving narrow-band or slow-speed data from multiple sources to make use of a wide-band or high-speed transmission resource (channel). Depending on whether a master system clock is used or not, multiplexing is either synchronous or asynchronous. Closely related to multiplexing is the idea of multiple access. Multiplexing presupposes fixed assignment of channel resources at a local level to the separate users, whereas multiple access schemes make use of a remote channel resource, such as a communications satellite, with a possibly dynamic resource assignment. The two most common means of multiplexing are time division multiplexing (TDM) and frequency division multiplexing (FDM). The former interleaves data from separate sources in time while the latter does so in frequency. When applied to multiple access, these respective methods are referred to as time division multiple access (TDMA) and frequency division multiple access (FDMA), respectively. There is a third method for multiple access—code division multiple access (CDMA), which makes use of a unique code assigned to each user in order to separate each user's transmission from the others.

I. Multiplexing of Analog Sources

A. FREQUENCY DIVISION MULTIPLEXING

In the article entitled "Modulation," it was shown that by multiplying a low-pass signal by a sinusoid of f_0 Hz, the spectrum of the low-pass signal was translated to f_0 Hz. This principle can be used to advantage when it is desired to transmit several low-pass message signals through the same channel. A frequency division multiplexed transmitter for n messages, $m_1(t), m_2(t), m_3(t), \ldots, m_n(t)$, with bandwidths $W_1, W_2, W_3, \ldots, W_n$, respectively, is shown in block diagram form in Fig. 1(a). Each message is used to modulate a subcarrier, assumed in this case to be by means of double-sideband modulation. The outputs of the subcarrier modulators are summed and possibly used to modulate a carrier in order to move the entire spectrum to some preassigned fre-

quency location. This step may not be required in some applications.

It is assumed that the subcarrier frequencies are separated such that $f_2 - f_1 > W_1 + W_2, f_3 > W_2 + W_3, \ldots, f_n - f_{n-1} > W_{n-1} + W_n$ so that the separate message signal spectra are not overlapping. It is clear that if the minimum separation between subcarriers is used, the transmission bandwidth required is given by

$$B_T = 2 \sum_{i=1}^{n} W_i \qquad (1)$$

The demodulation of FDM at the receiving end, as shown in Fig. 1(b), is accomplished by first going through a carrier demodulator if carrier modulation was used at the transmitting end and subsequently through a band of bandpass filters centered on the appropriate subcarrier frequencies in order to separate the various subcarrier modulated signals from each other. The final step is to pass each subcarrier modulated signal through a subcarrier demodulator. In this case, coherent demodulation is used because of the double-sideband modulation used at the transmitting end. A coherent carrier reference is required, and a convenient way to achieve this is by sending a carrier reference through one of the subchannels, usually the highest frequency one, and then frequency dividing it down to the appropriate subcarrier frequency.

Although the technique was illustrated using double-sideband modulation for the subcarriers, any type of subcarrier modulation could be used. If single-sideband modulation is used, then the required transmission bandwidth is the minimum possible and is given by

$$B_{T, \text{ min}} = \sum_{i=1}^{n} W_i \qquad (2)$$

The demodulator for each of the subcarrier channels can again be implemented as coherent demodulators.

Frequency division multiplexing is a simple multiplexing scheme to implement. Its major disadvantage is that any channel nonlinearities will generate harmonic and intermodulation components that may interfere with the desired messages. This is evident in some long-distance telephone conversations when one can hear cross talk from another voice channel.

B. TIME DIVISION MULTIPLEXING

Time division multiplexing can be applied in instances where the messages are represented in pulse modulation format. Figure 2 illustrates such a scheme, where pulse amplitude modulation is used

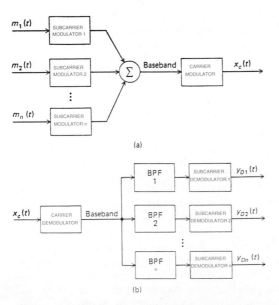

FIG. 1.   (a) An FDM modulator and (b) a demodulator.

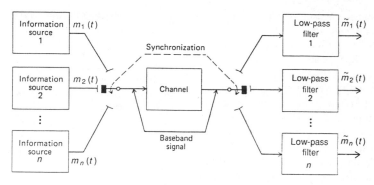

FIG. 2.   A TDM system.

to represent the messages $m_1(t)$, $m_2(t)$, $m_3(t)$, . . . , $m_n(t)$, which are now assumed to all be of bandwidth W Hz. A commutator samples each message in turn, which according to the sampling theorem must be at a minimum rate of $2W$ samples per second per message. At the receiving end, a second commutator, which is synchronized with the one at the transmitting end, deinterleaves the samples corresponding to the respective messages. A low-pass filter then allows the recovery of each message from its sample values. Using the sampling theorem in reverse, it can be shown that the theoretical minimum transmission bandwidth requirement for TDM is exactly the same as for FDM.

Any analog pulse modulation format can be used in TDM, and in fact, there is no restriction to the pulses representing analog messages. However, additional considerations, such as framing, are necessary when the messages are digital (to be discussed later). Time division multiplexing avoids the cross talk problems of FDM. However, it is necessary to maintain synchronization of all message samples and to synchronize the decommutation at the receiving end with the commutation at the transmitting end. Time division multiplexing can also accommodate messages of unequal bandwidths through the use of creative sampling. For example, for three messages m_1, m_2, and m_3 of bandwidths W, W, and $2W$, the commutator would consist of one contact for m_1, one for m_2, and two for m_3 in the sequence $m_1 m_3 m_2 m_3 m_1 m_3 m_2$ . . . .

II. Time Division Multiplexing of Digital Data

A. INTRODUCTION

Digital data can originate in two ways: (1) it may come from a source that is inherently digital, such as a computer or messages consisting of written text, or (2) it can result from sampling an analog message, such as voice. In the latter case, an intermediate step, known as analog-to-digital (A/D) conversion is necessary wherein the analog message is sampled in accordance with the sampling theorem and the samples converted or encoded into digital words. Encoding is usually binary and can be accomplished with any of several different types of encoders, such as ramp, feedback, or flash. It is not the purpose of this discussion to enlarge upon encoder types. When analog signals are sampled and encoded into digital format using an A/D converter, the process is referred to as pulse code modulation (PCM). It is often necessary to time division multiplex PCM signals to fully utilize the channel. Such time division multiplexing of PCM data can be accomplished in various ways as discussed next.

B. SYNCHRONOUS TIME DIVISION MULTIPLEXING OF PCM DATA

In synchronous multiplexing of PCM data, it is necessary that the rates of all sources be derived from a common clock. This guarantees high efficiency, but at the expense of greater complexity. For example, long-range plans for the U.S. telephone system include plans for a completely synchronous TDM system with a master clock located at Hillsboro, Missouri.

An advantage of a synchronous TDM PCM system is that it is completely feasible to intermix analog and digitial sources. For example, consider the time division multiplexing of two 2-kHz analog sources, a 4-kHz analog source, and eight 8-kb/sec digital sources, with 4-bit representation used for each analog sample. By virtue of the sampling theorem, each analog source must be sampled at a rate at least twice its bandwidth. If the analog sources are sampled at their minimum rates of 4, 4, and 8 kHz, respectively, the samples combined, and the combined samples converted to 4-bit digital format, then the rate of samples taken from the analog sources is 64 kb/sec. These bits, representing the analog sources, can then be interlaced with the bits from the digital sources to produce a combined bit stream of 128 kb/sec. Letting a_j represent the jth bit from the combined outputs of the analog sources and d_{ij} represent the jth bit from the ith digital source, the bit stream would be $a_1d_{11}a_2d_{21}a_3d_{31}a_4d_{41}a_5d_{51}a_6d_{61}a_7d_{71}a_8d_{81}a_9d_{12}$ . . . . This example illustrates bit-by-bit TDM. If the various bit streams are not synchronous, but nearly so, a technique known as pulse stuffing can be used to make them synchronous. This will be discussed later.

An alternative technique to this example is to interlace words or characters from different sources. A complete sequence of characters along with overhead bits for purposes such as synchronization and error correction constitute a frame of data. For example, the T1 carrier system for TDM telephone transmission in the U.S. consists of 24 channels, each channel consisting of 8-bit words representing samples taken at 8-kHz rates (one sample per 125 μsec) from each telephone conversation. Thus, in each frame there are 192 bits representing telephone conversations plus 1 bit added for framing. If b_{ij} represents the jth bit in the ith channel and f represents the framing bit, the frame sequence is as follows:

$$fb_{11}b_{12}b_{13}b_{14}b_{15}b_{16}b_{17}b_{18}b_{21}b_{22}b_{23}$$
$$b_{24}b_{25}b_{26}b_{27}b_{28}b_{31}b_{32}b_{33}b_{34}b_{35} \ldots b_{24,8}$$

Each bit is (125 μsec)/(193 bits/frame) = 0.6477 μsec in duration and is represented in bipolar format (A volts for half the bit duration, and $-A$ volts for the other half) so that no DC component exists on the line. The framing bit for the even numbered frames follows the sequence 001110, and for the odd numbered frames, it follows the sequence 101010. This is so that the signaling information, which consists of the eighth bit of each of the 24 channels on every sixth frame, can be identified. The data rate corresponding to a 0.6477-μsec bit time is 1.544 MB/sec, requiring a minimum channel bandwidth of 1.544 MHz. When compared with the bandwidth requirements of a 24-channel FDM system, it is seen that TDM requries much more—roughly a factor of 11 more. Yet, TDM systems are more efficient in transmitting information from digital sources than FDM systems. With the advent of fiber-optic transmission media, TDM will become even more attractive for voice and data transmission.

C. ASYNCHRONOUS TIME DIVISION MULTIPLEXING OF PCM DATA

For asynchronous multiplexing, the information bits are sent in blocks, called characters or words. Each character begins with a start bit followed by several information bits and finally by one or two stop bits. The receiver clock starts with the arrival of the start bit. As a result, no synchronization with a master clock is necessary. Keyboard terminals are typical asynchronous sources and operate with a 7-bit ASCII code with each character having appended to it a start bit, a parity bit, and one stop bit for a total character length of 10 bits. In asynchronous multiplexing, the data from multiple sources is interleaved on a character-by-character basis instead of on a bit-by-bit basis. Because of the required start and stop bits, asynchronous multiplexing is less efficient than synchronous multiplexing, although distribution of a clock is required for the latter.

A system that lies between synchronous and asynchronous multiplexing is an almost synchronous multiplexing system wherein the individual clocks of the various data sources are nearly the same frequency but not exactly. Because of this, it is necessary to occasionally "stuff" bits into the multiplexed bit stream if a data bit is unavailable when required. When demultiplexing takes place, these stuffed bits are identified and removed.

III. Multiple Access Techniques

Multiple access (MA), like multiplexing, involves sharing of a common communications resource between several users. However, whereas multiplexing involves a fixed assignment of this resource at a local level, MA involves the remote sharing of a resource, and this sharing may under certain circumstances change dynamically under the control of a system controller.

There are three main techniques in use for utilizing the bandwidth resources of a remote resource, such as a relay satellite. These are

1. frequency division multiple access (FDMA),
2. time division multiple access (TDMA), and
3. code division multiple access (CDMA).

The first two are analogous to FDM and TDM, respectively. Other means by which the bandwidth resources of a relay satellite can be extended are spatial frequency reuse and polarization frequency reuse.

When using FDMA, TDMA, or CDMA, the satellite resources may be utilized in a fixed assigned

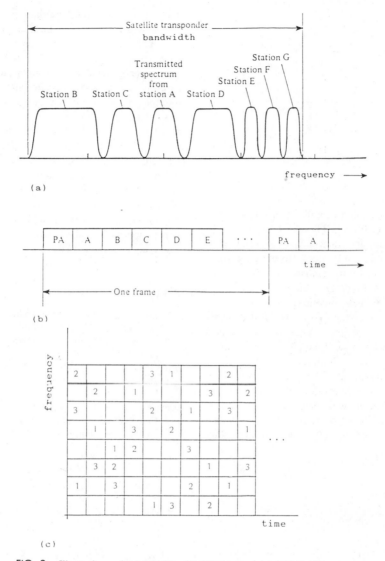

FIG. 3. Illustrations of (a) FDMA, (b) TDMA, and (c) CDMA-FH systems. Numbers in (c) denote hopping sequences for channels 1, 2, and 3.

multiple access (FAMA) mode or in a demand assigned multiple access (DAMA) mode. In the FAMA mode, the assignment format does not change even though the traffic load of the various earth station users may change. In the DAMA mode, the assignment format is changed as needed depending on the traffic load of the various users.

Figure 3 illustrates these three accessing schemes. In FDMA, signals from various users are stacked up in frequency, just as for FDM, as shown in Fig. 3(a). Guard bands are maintained between adjacent signal spectra to minimize cross talk between channels.

If frequency slots are assigned permanently to the users, the system is referred to as FAMA. If some type of dynamic allocation scheme is used to assign frequency slots, it is referred to as a DAMA system.

In TDMA, the messages from various users are interlaced in time, just as for TDM, as shown in Fig. 3(b). As illustrated in Fig. 4, the data from each user is conveyed in time intervals called slots. A number of slots make up a frame. Each slot is made up of a preamble plus information bits. The functions of the preamble are to provide identification and allow synchronization of the slot at the intended receiver.

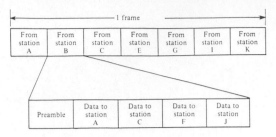

FIG. 4.   Detail of a TDMA frame format.

Guard times are utilized between each user's transmission to minimize cross talk between channels. It is necessary to maintain overall network synchronization in TDMA. If, in a TDMA system, the time slots that make up each frame are preassigned to specific message sources, it is referred to as fixed assignment TDMA. If the traffic is bursty, fixed assignment TDMA systems are not very efficient. Thus, it may be preferable in such situations to implement a method whereby the slots can be assigned dynamically. Such DAMA schemes required a central network controller and a separate channel (low information rate) between each user and the controller to carry out the assignments.

In CDMA, each user is assigned a code that ideally does not correlate with the codes of other users. The codes of various users are then used to separate one user's transmissions from those of the other users. This can be accomplished in various fashions, one of which is to divide the time–frequency space occupied by all users into several nonoverlapping subblocks as shown in Fig. 3(c). Each user then makes use of a unique set of subblocks to transmit its data. Such a technique is called frequency hop (FH) multiple access. Another technique is for the ith user to send a 0 bit as $-c_i(t)$, where $c_i(t)$ is its code, and a 1 bit as $+c_i(t)$. Such a technique is an example of direct sequence (DS) multiple access. Although CDMA schemes can be operated synchronously, they can be designed so as not to require synchronization, and this is the preferred mode. When operated without synchronization, one must account for multiple access noise, which is a manifestation of the partial correlation of a desired user's code with all other users' codes present on the system.

BIBLIOGRAPHY

Couch, L. W., (1987). "Digital and Analog Communications Systems," 2nd ed. Macmillan, New York.

Haykin, S. (1988). "Digital Communications." Wiley, New York.

Sklar, B. (1988). "Digital Communications: Fundamentals and Applications." Prentice-Hall, Englewood Cliffs, New Jersey.

Ziemer, R. E., and Tranter, W. H. (1985). "Principles of Communications: Systems, Modulation, and Noise," 2nd ed. Houghton, Boston, Massachusetts.

OPTICAL CIRCUITRY

H. M. Gibbs, U. J. Gibson, N. Peyghambarian, D. Sarid, C. T. Seaton,
G. I. Stegeman, and M. Warren   *University of Arizona*

GLOSSARY

Exciton: Excitation of a crystal in which an electron is excited with insufficient energy to place it in the conduction band but with that energy minus the binding energy to the hole left in the valence band. By staying together in an exciton as they move through the crystal, the electron and hole need less energy than if they are free of each other. This situation is very analogous to a hydrogen atom or a positronium atom.

Fabry–Perot etalon: Optical interferometer consisting of parallel reflecting surfaces, which cause the light to reflect back and forth. When the optical length of the etalon equals an integral number of wavelengths of the light, the multiple beams interfere constructively, resulting in high transmission and a higher light intensity inside the etalon than incident. Otherwise, the beams interfere destructively, most of the light is reflected from the etalon, and the intensity inside is lower than that incident.

Heterostructure: Structure consisting of two or more kinds of crystals (i.e., GaAs and AlGaAs).

Multiple-quantum-well: Many quantum wells can be grown on top of each other by growing alternating layers of GaAs and AlGaAs.

Nonlinear Fabry–Perot etalon: Fabry–Perot etalon whose transmission properties change with light intensity. The most promising all-optical logic devices so far are based on an intensity-dependent index of refraction.

Nonlinear refractive index: Intensity-dependent index of refraction. Close to an optical resonance of the material, the intensity dependence is often very complicated. Far from resonance, the total index n can be expanded in powers of the intensity (i.e., $n = n_0 + n_2 I + \cdots$ where n_0 is the weak-intensity limit of n).

Nonlinear waveguide: Device that confines light relative to free-space propagation and whose properties (guiding ability, phase shift, etc.) depend on the light intensity in the device.

Optical bistability: Characteristic of a system if its output intensity can have two different values for the same input intensity. This implies a hysteresis in output versus input, no matter how slowly the input is varied.

Optical nonlinearity: Characteristic of a material if its index of refraction and/or absorption depend on the intensity of the light beam propagating through the material.

Optical pathlength: Product of the material's index of refraction and its physical length.

Quantum well: When a thin layer of GaAs is grown with AlGaAs below and above it, the carriers and excitons excited by absorption of light in the GaAs are confined to the GaAs layer. This is because AlGaAs has a larger energy gap, so a carrier in the GaAs sees a potential barrier at the interfaces with AlGaAs (i.e., it is trapped in a well). When the GaAs thickness is of the same size or less than the separation between the electron and hole in an exciton (280 Å in bulk GaAs), the exciton wave function and binding energy are changed by quantum effects.

Superlattice: Properties of excitons in a quantum well can depend on neighboring quantum wells if the AlGaAs barrier is thin enough ($\sim$50 Å or less) so that the wave function can penetrate into one or more adjacent wells. Thus a superlattice is a multiple-quantum-well structure with barriers so thin that the properties cannot be described by a single quantum well.

Optical circuitry is a generic term used to include a wide range of optical devices in which a light signal is used to control another signal (electrical or optical) and to communicate between devices (electrical or optical). This definition encompasses practically all optical devices: all-optical devices for logic operations, preprocessing of optical images, and optical computing; optical interconnections between electronic processors and interfaces between electrical and optical arrays; and light-controlled electrical switches for ultrafast oscilloscopes and slow high-voltage or high-current control applications.

I. Introduction

We are in the midst of a communications and information expansion that is placing ever higher demands for faster and higher-capacity signal processing. The development of rapid transportation, instant telecommunication, and high-speed computers has greatly increased the desire for high-volume data transmission. More data can be utilized by supercomputers, and they, in turn, produce information and control decisions at an ever-increasing rate. Most of this expansion has been made possible by the electronic transistor and its offspring up through the present very-large-scale integrated circuits. This progress is beginning to run into fundamental limitations. Optical circuits have the potential to surpass electronics in speed, with interestingly low switching energies.

The potential of optical circuitry has been discussed for many years, but the excitement is growing rapidly on the strength of the success of optical fibers for optical transmission, the generation of subpicosecond optical pulses, and the development of promising optical logic elements, namely optical bistable and gate devices. Much research remains to be done to discover the best nonlinear optical materials and fabrication techniques. [*See* OPTICAL FIBER COMMUNICATIONS.]

Optical transition energies are on the order of 1 eV, permitting room-temperature and very-high-frequency operation, but requiring more energy per bit than Josephson devices. Clearly, the speed and packing density of integrated electronics are impressive. However, in the generation and transmission of picosecond and femtosecond pulses, optics is far ahead of electronics. Optical circuitry is also much more resistant to electromagnetic interference and to cross-talk than is electronics. If suitable all-optical control and logic devices can be developed, then optics should supplant electronics for very-high-speed decision making. Recently developed nonlinear optical devices have the desired characteristics for constructing all-optical systems. However, the nanosecond switching times must be greatly reduced before an all-optical serial computer should be contemplated. More immediate development will likely occur in two areas where optics offers unique advantages, namely, fast switching and parallel processing. Switching in the 0.1- to 10-psec regime is already (and only) optically possible; it could permit military signal encryption or high-speed multiplexing and demultiplexing to utilize the large bandwidth of optical fibers. Parallel processing takes advantage of the two-dimensional nature of light propagation; entire arrays can be interconnected by light beams that can even pass through each other.

Perhaps even more exciting is the possibility of constructing optical computers that "think" much more like the human mind than like an electronic digital computer. The potential for massive optical interconnections can mimic the brain's thousands of connections to a single neuron. Approximate but very rapid pattern recognition via associative reasoning processes would complement capabilities of electronic computers.

Emphasis in this article is upon all-optical devices: how they work and what they can do for signal processing (Section II); origins of the light dependence of the index of refraction on which they are based (Section III); the growth of the nonlinear thin films and their fabrication into devices (Section IV); and their potential application to parallel optical computing (Section V). Optoelectronic devices that are hybrids between optics and electronics are not treated here. Devices that control light with electrical signals (i.e., modulators) are especially important for optical communications; this is often achieved by pulsing the laser itself. Devices that use optical sampling techniques to analyze ultrafast electrical signals underscore the fact that ultrashort pulses can be generated and handled much more easily optically.

The optical circuitry devices treated here are divided into waveguide and etalon devices. Waveguide devices confine the light over distances much longer than free-space propagation would permit. This lowers the input power required to obtain the same nonlinear effect needed for a logic operation. Losses in the material must be minimized for this advantage to be realized. This extended interaction length also results in a longer transit time, which may not be acceptable in some cases but is no problem for pipeline processing. An optical waveguide is the optical analog of an electrically conducting wire; the latter can be made smaller than an optical waveguide whose guiding dimension must be at least half an optical wavelength in the material. Consequently, guided-wave optical devices are likely to find applications in high-speed serial applications such as optical communications, data encryption, convolvers, etc., rather than in computing. Guided-wave interconnections are permanent and shock-resistant, but they do not easily utilize the unique capabilities of light for massive parallelism and interconnectivity. A whole array can be imaged onto another array with only a lens in between. Light beams can pass through each other with no interaction; even the direction of a beam can be changed. No doubt waveguiding will play an important role in many optical control devices, but unguided devices that employ propagation of many beams perpendicular to the plane will be likely to dominate in parallel processing applications. Etalon devices can perform picosecond logic operations on light beams focused onto the etalon. They have the potential for massive parallelism and multiple interconnections. In addition, their short transit time makes them ideal for ultrafast computations in which the outputs of one decision plane are needed as inputs to the next decision plane.

Note that both guided-wave and etalon devices employ thin films of nonlinear optical material. The search for better materials and for better growth and fabrication techniques constitutes a large part of current research in optical circuitry.

II. Device Principles and Applications

The propagation and spatial manipulation of light energy is a well-established technology, both in the guided-wave regime with optical fibers and planar waveguides and in free-space propagation through the use of lenses, mirrors, etc. The crucial elements needed to allow optical circuitry to perform decision making and computational tasks are components whose transmissions depend on some variable parameter of the light incident on them. The property of light used to carry information and interact with the components must be easily modified and measured. In addition, the output of one component should be suitable as an input to another component. For example, a device whose output phase depends on the input polarization might have some useful functions, but would not be a desirable component for building processing systems. Consequently, most of the work in optical circuitry so far has concentrated on intensity-dependent nonlinear materials that can be utilized in such a way as to produce an output intensity that is a nonlinear function of the input intensity.

Such nonlinear optical devices are analogous to the transistors in electronics. These devices have been designed and fabricated for use either in guided-wave applications (see Section II,B), in which the light is confined in a planar or cylindrical structure, and/or for applications in which the light is incident perpendicular to the material surface. Most, but not all, of these latter devices are based on the nonlinear Fabry–Perot etalon.

A. Nonlinear Etalons

A nonlinear Fabry–Perot etalon consists of a material with an intensity-dependent refractive index positioned between two partially reflecting mirrors of a Fabry–Perot resonator (Fig. 1a). As in an ordinary Fabry–Perot etalon, if the optical length of the cavity is equal to an integral number of half-wavelengths of the incident light, the etalon will be in resonance and highly transmitting (peaks in Fig. 1b). The presence of the nonlinear material allows the optical length of the etalon to be intensity dependent (Fig. 1c). Consequently, the transmission of the etalon can be a highly nonlinear function of the intensity incident on it. If the feedback into the nonlinear material from the partially reflecting mirrors is high enough, the device can exhibit

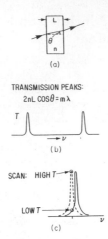

FIG. 1. (a) Nonlinear Fabry–Perot etalon, consisting of a light-intensity-dependent material between two parallel reflecting surfaces. (b) Transmission function of a Fabry–Perot etalon. (c) Shift of Fabry–Perot peak toward laser frequency ν_L with increasing light intensity.

optical bistability, in which there are two stable output intensity levels possible for a given input intensity. If the output intensity of an optical bistable device is plotted against its input intensity, a hysteresis loop appears (Fig. 2).

Various materials exist that exhibit a nonlinear change in refractive index (see Section III). Semiconductor materials have received the most attention for etalon applications because they interact efficiently with light over short distances. The semiconductors studied include GaAs, GaAs–AlGaAs multiple quantum wells, and InSb, InAs, CuCl, CdS, and CdHgTe. These materials have very large nonlinear indices of refraction due to electronic mechanisms near their absorption band edge. The GaAs-based devices are the most promising for applications. They operate at room temperature and are very fast with switch-on times of 1 psec and switch-off times of less than 200 psec.

Another group of materials exhibits thermally induced changes in both refractive index and physical thickness. Inside an etalon structure, the effects can combine to produce optical bistability. These materials also include semiconductors such as GaAs, ZnS, and ZnSe, as well as color filters and liquid dyes. Of particular interest are the ZnS and ZnSe materials. Devices are made from these materials in the form of conventional interference filters in which the semiconductor forms the spacer layer of the filters. Although much slower than the GaAs etalons,

the interference filters are more easily fabricated and operate with visible light.

Finally, there are devices that are not based on a nonlinear Fabry–Perot etalon. These, such as the SEED (self electrooptic effect device), have the advantage of not requiring the use of mirrors to provide feedback. The SEED is based on the increase of optical absorption of a GaAs–GaAlAs multiple-quantum-well (MQW) structure resulting from shifts caused by the application of an electric field perpendicular to its surface. The MQW layer is fabricated in the device as part of a PIN diode (see Fig. 3), which acts not only as the means of applying a field to the MQW layer, but also as a photodetector that allows the device to operate with electrical rather than optical feedback. The device operates at room temperature in the near infrared. By choosing positive or negative feedback modes, the SEED can exhibit either optical bistable operation or linear modulation. Although its construction is more complicated than that of the nonlinear Fabry–Perot etalon because of the additional electronic circuitry, the SEED has great potential as an optical circuitry component and as an interface between electronics and optics (spatial light modulator).

Nonlinear Fabry–Perot etalons exhibit a variety of operating modes that suggest applications as both analog and digital components in optical circuitry. Analog operations of nonlinear etalons include differential gain (Fig. 4), in which the device allows one beam to control a more intense beam. In this and in most other applications, the nonlinear etalon is operated as a three-port device, again analogous to a transistor. Another analog-type operating mode is as a limiter: a nonlinear etalon operated in the upper branch of bistability will maintain an almost constant transmitted intensity as the incident inten-

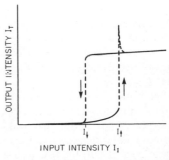

FIG. 2. Characteristic curve for an optical bistable system.

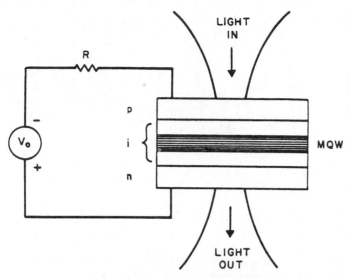

FIG. 3.   Schematic of the quantum-well SEED.

sity is varied. This is due to the shift of the etalon transmission peak with cavity intensity (Fig. 5).

Etalon peak shifts can also be used to perform gate operations for digital logic. In this case, zero, one, or two input pulses are simultaneously incident on a device at a wavelength that may be far from the transmission peak (Fig. 6). The pulses serve as the logic inputs for the gate. Another pulse with a wavelength equal to or close to the transmission peak serves as the probe pulse, and its transmission (or reflection) is the output state of the gate. The type of gate (NOR, AND, etc.) is determined by the detuning of the probe pulse from the initial transmission peak of the etalon and the energies of the short input pulses. The complete set of two-input logic gates has been demonstrated in both dye and GaAs (Fig. 7) etalons. A disadvantage of this type of logic-gate operation is that the input and output signals have different wavelengths, complicating the design of a system with sequential logic operations. This disadvantage can be eliminated by bistable operation of a nonlinear etalon. A bistable device can operate as an AND gate in transmission (I_1, $I_2 < I_\uparrow$ but $I_1 + I_2 > I_\uparrow$ in Fig. 2) and can also function in steady state as an optical-memory device. In both cases the input and output wavelengths can be identical.

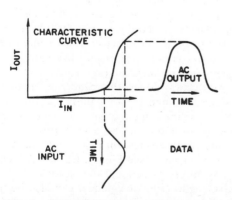

FIG. 4.   Differential gain in sodium vapor.

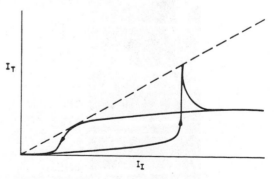

FIG. 5. Optical bistability and limiter action in a 2-mm-long etalon consisting of a polished Corning 3-142 filter with $R = 0.8$. In the on state, the output changes very little for a factor of 4 change in the input. The spike in the transmission occurs as the device turns on and the etalon's frequency of peak transmission is swept through the laser frequency.

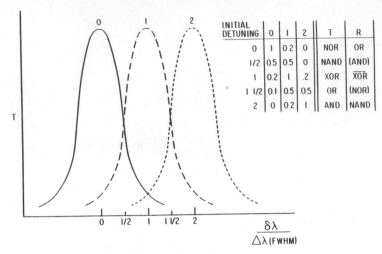

INITIAL DETUNING	0	1	2	T	R
0	1	0.2	0	NOR	OR
1/2	0.5	0.5	0	NAND	(AND)
1	0.2	1	.2	XOR	$\overline{XOR}$
1 1/2	0.1	0.5	0.5	OR	(NOR)
2	0	0.2	1	AND	NAND

$$\frac{\delta\lambda}{\Delta\lambda\,(FWHM)}$$

FIG. 6. Use of a nonlinear etalon to perform all of the logic operations. The transmission curves are labeled 0, 1, 2 to indicate the peak position seen by a probe pulse after no (0), one (1), or two (2) input pulses. With the probe wavelength at one of the five labeled values [expressed by the initial detuning in full widths at half maximum (FWHMs) of the transmission peak], the gates in the table are obtained. The fractional values in the columns below 0, 1, and 2 (of inputs) are the approximate transmissions when each input shifts the peak by 1 FWHM. In reflection, the AND and NOR have poor contrast.

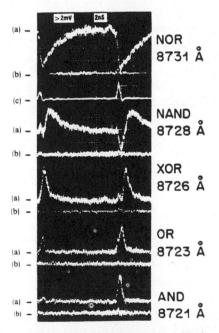

FIG. 7. Logic gate operation using a GaAs etalon. (a) Transmission of a continuous-wave (cw) probe beam. Response on the left side is due to 8 pJ (one input), while that on the right is due to 16 pJ (two inputs). (b) Probe zero line. (c) Input pulses (same for all gates).

B. NONLINEAR WAVEGUIDES

Nonlinear interactions occur whenever the optical fields associated with one or more laser beams are large enough to produce polarization fields proportional to the product of two or more fields. These nonlinear polarization fields radiate in phase with the generated field under optimum conditions (phase matching), growing linearly with propagation distance; hence the key to obtaining efficient (low-power) nonlinear interactions is to maintain high optical intensities over as long a distance as possible.

Optical beams can be confined to an optical wavelength in one dimension by total internal reflection at the boundary of a film whose refractive index is higher than its surroundings. Diffractionless propagation (in the confined dimension) occurs down the film for centimeter distances limited by absorption, scattering, or both. Nonlinear interactions can take place either in the film or in the neighboring media by means of the evanescent fields that accompany the guided wave (see Fig. 8). Rectangular channels with cross-sectional dimensions of optical wavelengths (rib waveguides) can also be used to further confine the optical beams.

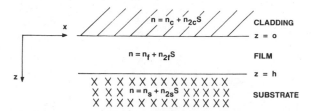

FIG. 8. Schematic of a nonlinear waveguide in which the guide, substrate, and cladding are all nonlinear (S is the intensity).

An impressive number of nonlinear optical devices such as bistable switches, and logic gates, have already been successfully demonstrated in a plane-wave context in a large variety of materials exhibiting some form of intensity-dependent index of refraction (see Section III). The goal of these, and in fact of any all-optical device, is to perform some signal-processing function with a minimum amount of power and device volume. Since it is desirable to operate with input and output beams of the same frequency, this limits the nonlinear interactions to those based on the third-order nonlinearity, for which the pertinent coefficient is usually the intensity-dependent refractive index n_{2_i} for the medium i. Most devices require an intensity-dependent phase shift accumulated over some propagation distance, making guided-wave geometries optimum for such interactions. Other waveguide devices are those based on power-dependent field distributions. Experiments reported to date on third-order guided-wave phenomena are degenerate four-wave mixing, coherent anti-Stokes

Raman scattering, nonlinear waveguide coupling, optical limiting, nonlinear coherent coupling, etc. These observations represent the initial stages of this field, and other potential applications have been proposed. Because the fabrication techniques are planar-technology oriented, such a waveguide approach to optical signal processing is primarily of interest for serial, rather than parallel, operations.

Degenerate four-wave mixing in waveguides involves the mixing of three input beams of the same frequency to produce a fourth wave of the same frequency. This interaction has potential application to real-time signal processing operations that involve the mixing of two or more waveforms, such as convolution, correlation, or time inversion, as shown in Fig. 9.

In most guided-wave cases the effect of a nonlinear refractive index is to produce a power dependence of the guided-wave wave vector that can be analyzed with coupled-mode theory. However, if, for example, the cladding medium in a nonlinear waveguide is nonlinear, then the field distributions can also be affected if the optically induced change in the cladding index is comparable to the zero-field index difference between the film and cladding. This leads to intensity-dependent field distributions, as shown in Fig. 10. Furthermore, self-focusing of the guided wave can occur in a cladding material with a positive n_2, resulting in the light creating its own waveguiding medium in the cladding. This leads

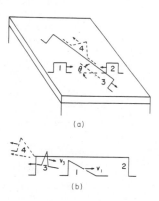

FIG. 9. Schematic diagram of degenerate four-wave mixing in a thin-film optical waveguide. (a) The convolution (waveform 4) of the two input waveforms (1 and 2). (b) Time inversion (waveform 4) of the input waveform 1 by virtue of the δ-function pulse (3) overtaking 1 in the interaction region.

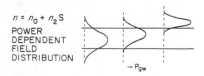

FIG. 10. Schematic of intensity-dependent guiding. At low intensities the light is guided in the central layer, but the self-focusing nonlinearity in the cladding results in guiding in the cladding at high intensities.

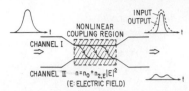

FIG. 11.   Schematic of a nonlinear coherent coupler. Schematic of two coupled waveguides with a nonlinear material in the coupling region. At low (high) intensities, the output comes out the lower (upper) guide.

to the anomalous variation in guided-wave refractive index with guided-wave power that gives rise to a number of new power-dependent guided-wave phenomena. For example, a positive-n_2 cladding medium in an asymmetric waveguide can cause a device to guide only above a minimum power threshold, which can be set by controlling the waveguide parameters. Alternatively, a limiting device arising from power-dependent control of waveguide cut-off can be obtained by using a negative-n_2 cladding medium. When a film is bounded by two nonlinear media, new power-dependent wave solutions occur with the possibility of switching between them.

The other major class of devices based on intensity-dependent refractive indices utilizes a power-dependent wave vector. Such devices are mostly a variation of the basic nonlinear coherent coupler (shown in Fig. 11), whose transfer efficiency from one channel to the other depends on the guided-wave power. Optical control of this transfer can be used to produce optical logic operations in devices such as Mach–Zender interferometers (Fig. 12) and ring-channel resonators (Fig. 13).

Another class of promising nonlinear devices is based on optical tuning of the Bragg condition of a grating embedded in a nonlinear waveguide (Fig. 14). This can lead to bistable switching for guided waves and numerous all-optical logic operations.

All of the applications and devices mentioned here are dependent on being able to utilize and

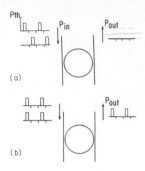

FIG. 13.   Schematic of a nonlinear ring-channel resonator operated as an AND gate. (a) The inputs are never coincident, so P_{out} is zero. (b) When the inputs are coincident, their sum exceeds P_{th} and there is output.

fabricate materials with a large nonlinearity in a thin-film format suitable for optical waveguiding. Standard thin-film fabrication techniques (see Section IV) can be used where appropriate along with in-diffusion, ion-exchange, Langmuir–Blodgett deposition, etc., depending on the materials selected for a given application.

III. Nonlinear Refractive Indices

In Section II, a number of nonlinear optical devices were described. In almost every case, the basic physical mechanism is an intensity-dependent change $\delta(nL)$ in optical pathlength nL or corresponding change $\delta\phi$ in phase shift ϕ,

$$\delta\phi = (2\pi/\lambda)\delta(nL)$$

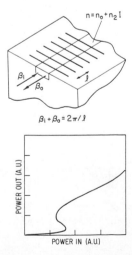

FIG. 14.   Schematic of a nonlinear grating operated as a bistable device with calculated transmission function.

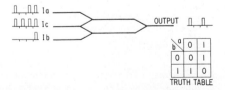

FIG. 12.   Schematic of a nonlinear optical Mach–Zender interferometer operated as an exclusive OR gate.

where λ is the wavelength, L the sample length, and n the refractive index. Far from optical resonances in the material, n can be written as

$$n = n_0 + n_2 I + \cdots$$

where n_0 is the background index and n_2 is often called the nonlinear index of refraction. The term $\delta(nL)$ may have contributions from a change δL in physical length and from a change δn in refractive index. One of the simplest origins for $\delta\phi$ is thermal (i.e., an intensity-dependent heating of the medium) resulting in both δn and δL. Bistability and logic operations have been achieved using thermal effects, but electronic effects are faster.

The Kramers–Kronig relations express the relationship between the absorption spectrum $\alpha(\omega) \propto \chi''(\omega)$ of a material and its refractive index $n(\omega) \propto \chi'(\omega)$. Consequently, if the absorption spectrum changes, so does the refractive index, as depicted in Fig. 15. An absorption resonance gives rise to the well-known anomalous dispersion with higher refractive index below the resonance and lower above. If the absorption profiles can be made to change by increasing the light intensity, then the refractive index is also intensity-dependent (i.e., nonlinear) and the material can be used for dispersive optical bistability. Contributions to the refractive index by a simple two-level transition in a collection of gas-phase atoms is an illustration of this type of nonlinearity.

Band filling can be envisaged as in Fig. 16: if

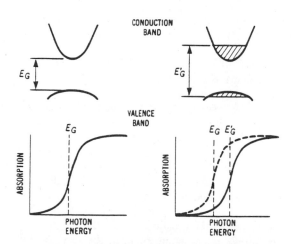

FIG. 16.   Schematic illustration of the shift in optical absorption in a direct-gap semiconductor due to band filling (excitonic effects are omitted for simplicity).

light of energy exceeding the bandgap is sufficiently intense to fill all of the lower states in the conduction band faster than they decay, the bandgap effectively shifts up to the photon energy. This shift in band edge changes the absorption spectrum by removing an absorption peak at the band edge, resulting in a reduction of the refractive index below the band edge as in Fig. 16. This band-filling saturation is also known as dynamic Burstein–Moss shifting and is the dominant band-edge mechanism in InSb ($n_2 \approx 1$ cm²/kW at 5.4 μm and 77 K) and InAs ($n_2 \approx 0.03$ cm²/kW at 3.1 μm and 77 K).

Exciton nonlinear refraction can be viewed similarly. Less energy is required to create an electron and hole that are bound to each other to form an exciton than to create a free electron and free hole. The exciton resonance appears as a sharp absorption peak at an energy slightly below the band edge. Saturation of the exciton absorption—for example, by creating so many excitons that they overlap and shorten each other's lifetimes by screening the Coulomb potential responsible for their binding—changes the refractive index as in Fig. 15 (GaAs/AlGaAs multiple quantum wells, $n_2 \approx 0.2$ cm²/kW at 0.84 μm and 300 K).

The intensity dependence of a two-photon absorption peak arising from biexcitons (i.e., excitonic molecules) gives rise similarly to a nonlinear refractive index but of the opposite sign, since the absorption increases with increased light intensity (i.e., CuCl). Nonlinear refraction can also arise from an electron–hole plasma. Free electrons and holes in semiconductors be-

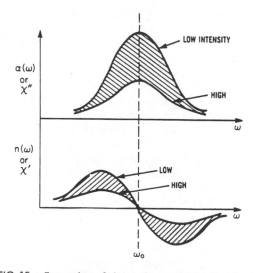

FIG. 15.   Saturation of absorption and refraction for a simple absorption line as a function of frequency.

have like mobile charge carriers that can be moved by applied electric (optical) fields. Their effective masses are often less than one-tenth of that of a free electron, so they move easily and produce large effects in the free-carrier Drude model. If one compares the values of the ratio of n_2 to the response time as a measure of the size of the nonlinearity and the speed, one finds that the only mechanisms with values as large as 0.01–10 cm^2 kW μsec are resonant nonlinearities in semiconductors. For this reason, most research directed toward fast practical devices is concentrated on semiconductors.

There are several techniques for measuring n_2. Perhaps the most popular method at present is optical phase conjugation in the form of degenerate four-wave mixing, which has been applied to the measurement of both small and very large n_2 values. Wavefront encoding leading to changes in the far-field profile has also been used, and various techniques utilize a change in interference fringes.

IV. Growth of Materials and Fabrication of Devices

The exploitation of the physical mechanisms, described in the previous section, for the development of optical circuits requires precise control over the form and quality of the materials involved. This section describes the methods employed to produce high-purity materials and combine them in forms that allow observation of their nonlinear behavior.

The two classes of devices discussed, the etalon and guided-wave configurations, both rely on the production of thin, uniform layers of materials with known properties. The thicknesses and indices of the layers must be accurately controlled, and the chemical purity must be maintained at a high level, particularly for the semiconductors. Both vacuum-deposition and bulk-modification methods are employed in the production of these devices.

A. VAPOR-DEPOSITION GROWTH

The etalon configuration and some of the waveguide structures require thin layers with optical thicknesses that vary by less than 1% of the film thicknesses. The best way to achieve this type of uniformity is by deposition from a vapor, either in vacuum or using a chemical reaction at a heated substrate. There are a number of variations possible for each of these methods,

and those applicable to the production of materials for optical circuitry will be discussed.

1. Vacuum Evaporation

This technique is used for the deposition of thin layers of ZnS, ZnSe, and CuCl, as well as for the deposition of the dielectric multilayer stacks that act as nonabsorbing reflectors in the etalon configuration. A vessel is evacuated to a pressure of about 10^{-6} Torr, and either resistive heating or electron-beam bombardment is used to heat the source material. When the vapor pressure of the material exceeds the base pressure of the system, material traverses the distance from the source to the substrate, where the vapor condenses to form a thin layer. With appropriate choice of source configuration and source-to-substrate distance, the uniformity of this layer can be excellent. The thickness may be monitored either with a quartz-crystal microbalance or by observing the quarterwave fringes optically. Impurity concentration is dependent primarily on the purity of the source and the presence of residual gases.

Films deposited in this manner are, in general, polycrystalline, but use of a single-crystal substrate, low evaporation rates, and heated substrates may result in epitaxial growth. Reduction of the base pressure and use of either elemental or compound effusion ovens as sources results in a transition from vapor evaporation to the technique known as molecular beam epitaxy.

2. Molecular-Beam Epitaxy (MBE)

This technique is employed extensively in both research and production environments for the production of high-quality GaAs. In addition to the high purity achieved by the use of extremely low base pressures, the slow deposition rates allow fine control over the thicknesses of the layers and allow the production of abrupt interfaces. These capabilities are crucial for the formation of quantum-well structures.

The principle of MBE is the same as for conventional vapor deposition, but the hardware is different. The deposition is performed in a stainless-steel chamber, with all-metal seals, which can be baked under vacuum to desorb gases from the walls. This equipment, coupled with a high-capacity pump (often of the entrainment type), allows base pressures in the range of 10^{-12}–10^{-10} Torr. The sources used are temperature-regulated ovens known as Knudsen cells, which maintain a constant flux at fixed temperatures. The cells are enclosed in a liquid-nitrogen-cooled shroud, which keeps the background pressure at the substrate low and prevents cross-talk between the sources. For GaAs, separate sources are used for the Ga and As, as the compound evaporates incongruently. Careful control is required over the arrival rates of the constituents to form stoichiometric films, as the two vapors have different sticking coefficients. The use of a hot single-crystal substrate permits epitaxial crystalline growth. Various analysis techniques are used *in vacuo* to assess the quality of the films. X-Ray photoelectron or Auger spectroscopy may be used to detect impurities, and electron diffraction probes the quality of the crystal surface. The quantum wells with GaAs–$Ga_{1-x}Al_xAs$ layers are made by adding an Al source and opening and closing the shutter as required to create the desired structure.

3. Atomic Layer Evaporation (ALE)

This is a relatively new technique, which has demonstrated promise for the production of high-quality II–VI semiconductors; it has also been applied to the production of GaAs. The technique has been performed in both ultra-high-vacuum and medium-high-vacuum conditions. It involves exposure of the substrate to alternating vapor bunches from two or more sources. A heated substrate is used to drive a chemical reaction between the two source materials. The deposit is a compound, and the reactant species, as well as any waste products, must have a much higher vapor pressure at the substrate temperature than the desired deposit. Either elements or compounds may be used as the source materials. In the case of elemental sources, no waste products would be anticipated. After each vapor pulse, which should contain in excess of the number of atoms (molecules) required for a monolayer, there is a time lag before the reactant pulse is sent in. This allows the first material to diffuse on the surface and react at each available site, and allows re-evaporation of the excess material. With this method, the thickness of the deposit is determined by the number of cycles of the two ovens. This allows the production of very-well-controlled thicknesses, and possibly superlattices. Use of a single-crystal substrate has been shown to result in epitaxial films of CdTe.

4. Chemical Vapor Deposition (CVD)

This technique is similar to ALE, except that there is no temporal separation of the reactive materials. A heated substrate is held beneath the flow of gases containing the desired deposit constituents. The reaction that is then thermally driven results in the deposition of a film with good thickness uniformity and, usually, low intrinsic stress. The use of organometallics and single-crystal substrates has been shown to result in high-quality epitaxial layers. The equipment for MOCVD (metal–organic chemical vapor deposition) is modestly priced, although the gases involved require special handling. There has been some controversy about the abruptness of the interfaces formed by this technique, but they are of sufficient quality to permit lasing in heterostructures made in this way.

B. Other Growth Techniques

1. Liquid-Phase Epitaxy (LPE)

In this method, condensation of the new layer of material occurs from a liquid solution above the single-crystal substrate. A graphite tray holds the solute, and the substrate then slides underneath it. Precise cooling of the solute allows growth of an epitaxial layer. If more than one layer, with different compositions, is required, the substrate slides underneath the solutions with different compositions. The technique is economical and satisfactory for the growth of single layers, but interdiffusion at, and characterization of, the boundaries can be a significant problem. The technique is applicable to a wide range of materials, including GaAs and compounds thereof.

2. in-Diffusion

This method is used extensively for the production of waveguides and other devices for integrated optics. Rather than growth of a new layer for waveguiding, the index of the outer region of the substrate is altered by the controlled introduction of chemical impurities. The two most used techniques are Ti doping of $LiNbO_3$ and silver-ion exchange in glasses containing an alkali. In each case, a controlled heating process, with the desired impurity close to the surface, results in a reproducible increase in the refractive index profile at that surface. Complex structures can be defined by the use of photoresist patterning techniques.

C. Patterning

A long-established need in waveguide devices, and an emerging one in etalons, is for control of the form of the layers within the plane defined by one of the above growth techniques. Most applications require pattern generation on the scale of a few micrometers. This is readily achieved using photoresist techniques adopted from the semiconductor industry.

A solution of polymer containing a photosensitive element is spread in a thin (0.5–2.0 μm) layer on the substrate by spinning at high speeds. This layer is exposed to ultraviolet (UV) light through the desired pattern, and a selective developer removes the exposed or unexposed portion, depending on the type of photoresist. The pattern is transferred to the active layer of the device either by etching or in-diffusion. In some devices, such as grating couplers, the pattern is holographically formed, and the resist may be used as a grating rather than a transferred copy.

There are several methods used for removing material through a mask, including chemical, ion, and reactive-ion etching. Chemical etching is the simplest method to implement, and is invaluable in many processes. However, undercutting of the mask, and selective directional and defect etching effects may make the process hard to control. These have led to the increased use of ion-beam techniques.

The most straightforward method of ion-beam etching is to accelerate inert gas ions to energies of the order of 500 eV and direct them toward the substrate. The ions will sputter material away wherever they strike. The photoresist protects the regions under it, and the pattern is transferred to the substrate. This method can be set up in any existing vacuum system, but the etch rates for some materials of interest are rather slow. This has led to the development of reactive-ion etching techniques. Reactive-ion etching is usually performed in a dedicated vacuum system and uses an RF (radiofrequency) plasma discharge rather than a directed beam of ions. This discharge creates fragments that react chemically with the substrate as they bombard it, thus increasing the etch rate dramatically. Very sharp etch profiles result from the appropriate choice of reactive gas.

V. Parallel Optical Computing

In contrast with the strong emphasis on serial communication in waveguide technology, two-dimensional optical circuitry has been focused on the development of parallel computing or signal-processing systems that can utilize the inherent parallelism of light. A single image formed by a simple optical system can contain millions of bits of encoded information. This information can be simultaneously communicated through space using a variety of optical systems with ease and without mutual interference. If elementary logic operations could be performed on this information at even moderate speeds, the total throughput of a parallel optical information-processing system would be orders of magnitude higher than that of any present or planned electronics system. Such optical systems are far from realization, however. Progress needs to be made in the key components of an optical parallel-processing system and in new architectures and algorithms that are needed to utilize the massive parallelism and interconnections possible with light.

Regardless of the type of parallel optical-processing system envisaged, it is at some stage necessary for lightwave signals to interact with each other so that decision making or mathematical operations may occur. In most optical processing systems attempted in the past, this required an optical–electronic conversion process, and, if further operations were needed, a reconversion from electronic signals to optical signals. This process was extremely inefficient in terms of both energy and time and has been a major obstacle to the success of optical processing systems. Only with the recent development of intrinsic nonlinear optical devices has there been a means of performing optical logic without this expensive conversion process. A two-dimensional array of optically bistable or nonlinear devices (Fig. 17) could operate as the fundamental building blocks of optical parallel-processing systems in a manner analogous to the use of arrays of electronics nonlinear devices (i.e., transistors) in electronic processing systems. The optical systems would, however, have the advantage of utilizing the previously mentioned massive parallelism of optics. The arrays of nonlinear optical devices could be nonlinear etalons operating as optical transistors, logic gates, or bistable devices, as described previously (Section II,A). The mode of operation of the arrays would of course depend on the type of processing or computation desired.

Much of the optical parallel processing performed to date has been analog processing of information encoded as a two-dimensional image and based on the Fourier-transforming capability of a lens. This kind of optical parallel processing has had a long and successful history of development and application to a variety of real-

FIG. 17.   An array of 9×9-μm GaAs pixels defined by reactive etching. Some have been operated as NOR gates.

world problems. Although these systems are not dealt with here, there are ways in which these systems can benefit from the development of two-dimensional optical-circuitry components. Many of the analog optical parallel-processing designs require the use of an optically addressed spatial light modulator. This is a two-dimensional planar device that allows a light beam or an electrical signal to control the transmission or reflection of light at each point on its surface. An array of optically bistable devices is obviously an optically addressed spatial-light modulator. Present optically addressed spatial light modulators rely on optical–electronic–optical reconversion such as the liquid-crystal light valve or utilize photographic film, which makes it impossible to reconfigure the system in real time.

Even though the development of two-dimensional nonlinear etalon arrays could be advantageous for the development of analog optical parallel-processing systems, these systems still suffer from the inherent disadvantages of analog computation: limited accuracy and computational complexity due to noise propagation and dynamic-range limitations. Recently, progress has been made in modifying this traditional approach. The new approach is the best optical scheme so far for high-speed numerical calculations, such as matrix–vector or matrix–matrix addition or multiplication. The addition process is the analog addition of light intensities by a detector, and the multiplication process is the analog multiplication of a light intensity by a transmission function. However, it is possible to

bypass the analog limitations of dynamic range by using a digital (but usually not binary) number representation scheme such as residue arithmetic to break up the numerical dynamic range requirements over a number of digits. This process also enhances the noise immunity of the processor. These systems do not, however, make full use of the parallelism of optical systems. The systems are closely coupled to an electronic processor that sends the data to the optical processor in time-multiplexed form so that the optical processor is operating on one-dimensional data arrays. This allows the use of linearly dimensioned components such as acousto–optic modulators. By sacrificing some spatial parallelism, it is possible for the controlling electronic system to keep up with the optical processor. The optical systems can, in principle, be expanded with parallel channels to handle higher-dimension arrays and achieve greater two-dimensional throughputs. However, the limitations on system performance are imposed by the large electronic input/output overhead inherent in the system design.

The highest-performance and most general-purpose optical-circuitry-based information-processing systems envisioned are digital systems based on binary number representation (usually) and two-level logic. These systems would allow full utilization of the parallel-processing capabilities of optics for image processing, numerical computation, and other applications. They are, however, the furthest from realization of any optical information-processing systems discussed.

The basic concepts of the digital optical (parallel-processing) systems are two-dimensional arrays of all-optical logic gates, with the outputs of each gate array communicating to the inputs of the next array through free-space reimaging interconnects. The free-space interconnection of the logic-gate arrays could be provided by classical optical components (prisms, lenses, etc.) for systems with simple regular interconnection patterns, as might be used for image processing applications. More complex interconnection patterns may require the use of holographic optical elements.

The memory-storage function for the system would be provided by the propagation time of the light through the interconnection systems. Since the entire two-dimensional data array that is input to the processor can be operated on in a single system (clock) cycle, it is not necessary to store data for more than one cycle.

Input/output functions will require the development or improvement of a number of optical-circuitry components. Input data could be coded onto a single wavefront with an electrically addressed spatial light modulator or by a two-dimensional array of directly modulated laser diodes. Two-dimensional laser diode arrays are not yet available but are under development. Current electrically addressed spatial light modulators lack the bandwidths needed to keep pace with proposed digital optical systems, but their development is continuing. Output signals from the optical system would be monitored by two-dimensional detector arrays, which are now commonplace. Any conceivable digital optical-processing system would be interfaced with an electronic system to manage its input–output to the real world. The bandwidth restrictions on the electronics system could cause a bottleneck affecting the overall performance of the systems. Some analog-processing systems have exhibited this problem already.

A problem as important as the components and design of a digital optical-computing system is the development of algorithms and interconnection architectures that allow the utilization of the system's parallel-processing capabilities. Optical parallel-processing systems are going to be developed as special-purpose systems. Certainly some of the first applications will be in image processing and pattern recognition. In these areas the data are already encoded in two-dimensional optical form and, for many cases, the processing required consists of a few simple, regular operations performed in parallel. These are applications that, despite their apparent simplicity, are not handled well by electronics processing systems.

Another forseeable application of optical parallel-processing systems is an attached high-speed numerical processor for electronic supercomputer-type systems. Large numerical calculations often involve operations on large matrices or data arrays. An optical parallel processor could handle simple repetitive operations on large arrays for general-purpose computers in much the same way that electronic array processors do. The free-space interconnection capabilities of optical processors would also be advantageous for Fourier transform-type calculations. Numerical calculations with optical processors may require a different number representation and arithmetic than electronic systems. Conventional binary arithmetic can be implemented in optical systems, but there are certain advantages to other number representations. The use of residue arithmetic can eliminate the need for carry operations when performing additions or multiplications, for example. Symbolic substitution has been proposed as a method in which simple logic operations on patterns of binary optical data can perform Boolean logic operations or some complex mathematical operations.

Recent research in artificial intelligence and human thought processes has contributed systems that seem particularly suited to optical implementation. In particular, associative memory systems require massively interconnected systems that can best be implemented with light.

In summary, linear optical signal-processing systems in which optical elements perform various transformations but electronics performs the decision making have been explored for a few decades. *Nonlinear* optical signal-processing systems are now being designed and tested in which nonlinear optical devices make decisions, (i.e., perform logic operations). Hopefully this will greatly increase the throughput of optical systems by reducing the rate at which the electronics must accept data. In other words, nonlinear optical signal-processing systems should supply answers, not just make transformations, leaving electronics to make the decisions required to arrive at answers.

BIBLIOGRAPHY

Caulfield, H. J., Horvitz, S., Tricoles, G. P., and VonWinkle, W. A. (eds.) (1984). Special issue on optical computing. *Proc. IEEE* **72,** 755.

Gibbs, H. M. (1985). "Optical Bistability: Controlling Light with Light." Academic, New York.

Gibbs, H. M., McCall, S. L., and Venkatesan, T. N. C. (1980). Optical bistable devices: The basic components of all-optical systems? *Opt. Eng.* **19,** 463.

Gibbs, H. M., Mandel, P., Peyghambarian, N., and Smith, S. D. (eds.) (1986). "Optical Bistability III," Proceedings of OSA Topical Meeting on Optical Bistability, Tucson, December 2–4, 1985. Springer-Verlag, Berlin.

Macleod, H. A. (1969). "Thin-Film Optical Filters." Adam Hilger, London.

Miller, D. A. B. (1983). Dynamic nonlinear optics in semiconductors: Physics and applications. *Laser Focus* **19**(7), 61.

Stegeman, G. I., and Seaton, C. T. (1985). Nonlinear integrated optics. *J. Appl. Phys.* **58,** R57.

OPTICAL FIBER COMMUNICATIONS

B. K. Tariyal   *AT&T Technologies, Inc.*
A. H. Cherin   *AT&T Bell Laboratories*

GLOSSARY

Attenuation: Decrease of average optical power as light travels along the length of an optical fiber.

Barrier layer: Layer of deposited glass adjacent to the inner tube surface to create a barrier against OH diffusion.

Chemical vapor deposition: Process in which products of a heterogeneous gas–liquid or gas–solid reaction are deposited on the surface of a substrate.

Cladding: Low-refractive-index material that surrounds the fiber core.

Core: Central portion of a fiber through which light is transmitted.

Cut-off wavelength: Wavelength greater than which a particular mode ceases to be a bound mode.

Dispersion: Cause of distortion of the signal due to different propagation characteristics of different modes, leading to bandwidth limitations.

Bandwidth: Measure of the information carrying capacity of the fiber. The greater the bandwidth, the greater the information carrying capacity.

Graded index profile: Any refractive index profile that varies with radius in the core.

Microbending: Sharp curvatures involving local fiber axis displacements of few micrometers and spatial wavelengths of a few milli-meters. Microbending causes significant losses.

Mode: Permitted electromagnetic field pattern within an optical fiber.

Numerical aperture: Acceptance angle of the fiber.

Optical repeater: Optoelectric device that receives a signal and amplifies it and retransmits it. In digital systems the signal is regenerated.

Communications using light as a signal carrier and optical fibers as transmission media are termed optical fiber communications. The applications of optical fiber communications have increased at a rapid rate, since the first commercial installation of a fiber-optic system in 1977. Today every major long-distance telecommunication company is spending millions of dollars on optical fiber communication systems. In an optical fiber communication system, voice, video, or data are converted to a coded pulse stream of light using a suitable light source. This pulse stream is carried by optical fibers to a re-generating or receiving station. At the final receiving station the light pulses are converted to electric signals, decoded, and converted into the form of the original information. Optical fiber communications are currently used for telecommunications, data communications, military applications, industrial controls, and medical applications.

I. Introduction

Since ancient times, humans have used light as a vehicle to carry information. Lanterns on ships and smoke signals or flashing mirrors on land are early examples of uses of how humans used light to communicate. It was just over a hundred years ago that Alexander Graham Bell (1880) transmitted a telephone signal a distance greater than 200 m using light as the signal carrier. Bell called his invention a "Photophone" and obtained a patent for it. Bell, however, wisely gave up the photophone in favor of the electric telephone. Photophone at the time of its invention could not be exploited commercially because of two basic drawbacks: (1) the lack of a reliable light source, and (2) the lack of a dependable transmission medium.

The invention of the laser in 1960 gave a new impetus to the idea of lightwave communications (as scientists realized the potential of the dazzling information carrying capacity of these lasers). Much research was undertaken by different laboratories around the world during the early 1960s on optical devices and transmission media. The transmission media, however, remained the main problem, until K. C. Kao and G. A. Hockham in 1966 proposed that glass fibers with sufficiently high-purity core surrounded by a lower-refractive-index cladding could be used for transmitting light over long distances. At the time, available glasses had losses of several thousand decibels per kilometer. In 1970, Robert Maurer of Corning Glass Works was able to produce a fiber with a loss of 20 dB/km. Tremendous progress in the production of low-loss optical fibers has been made since then in the various laboratories in the United States, Japan, and Europe, and today optical fiber communication is one of the fastest growing industries. Optical fiber communication is being used to transmit voice, video, and data over long distance as well as within a local network.

Fiber optics appears to be the future method of choice for many communications applications. The biggest advantage of a lightwave system is its tremendous information carrying capacity. There are already systems that can carry several thousand simultaneous conversations over a pair of optical fibers thinner than human hair. In addition to this extremely high capacity, the lightguide cables are light weight, they are immune to electromagnetic interference, and they are potentially very inexpensive.

A lightwave communication system (Fig. 1) consists of a transmitter, a transmission medium, and a receiver. The transmitter takes the coded electronic signal (voice, video or data) and converts it to the light signal, which is then

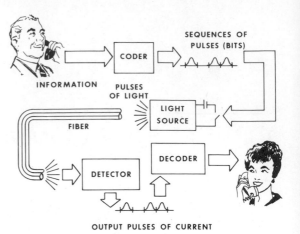

FIG. 1. Schematic diagram of a lightwave communications systems.

carried by the transmission medium (an optical fiber cable) to either a repeater or the receiver. At the receiving end the signal is detected, converted to electrical pulses, and decoded to the proper output. This article provides a brief overview of the different components used in an optical fiber system, along with examples of various applications of optical fiber systems.

II. Classification of Optical Fibers and Attractive Features

Fibers that are used for optical communication are waveguides made of transparent dielectrics whose function is to guide light over long distances. An optical fiber consists of an inner cylinder of glass called the core, surrounded by a cylindrical shell of glass of lower refractive index, called the cladding. Optical fibers (lightguides) may be classified in terms of the refractive index profile of the core and whether one mode (single-mode fiber) or many modes (multimode fiber) are propagating in the guide (Fig. 2). If the core, which is typically made of a high-silica-content glass or a multicomponent glass, has a uniform refractive index n_1, it is called a *step-index fiber*. If the core has a nonuniform refractive index that gradually decreases from the center toward the core–cladding interface, the fiber is called a *graded-index fiber*. The cladding surrounding the core has a uniform refractive index n_2 that is slightly lower than the refractive index of the core region. The cladding of the fiber is made of a high-silica-content glass or a multicomponent glass. Figure

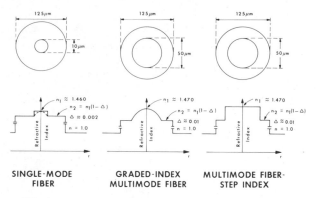

FIG. 2.   Geometry of single-mode and multimode fibers.

2 shows the dimensions and refractive indices for commonly used telecommunication fibers. Figure 3 enumerates some of the advantages, constraints, and applications of the different types of fibers. In general, when the transmission medium must have a very high bandwidth—for example, in an undersea or long-distance terrestrial system—a single mode fiber is used. For intermediate system bandwidth requirements between 200 MHz km and 2 GHz km, such as found in intracity trunks between telephone central offices or in local area networks, either a single-mode or graded-index multimode fiber would be the choice. For applications such as short data links where lower bandwidth requirements are placed on the transmission medium, either a graded-index or a step-index multimode fiber may be used.

Because of their low loss and wide bandwidth capabilities, optical fibers have the potential for being used wherever twisted wire pairs or coaxial cables are used as the transmission medium in a communication system. If an engineer were interested in choosing a transmission medium for a given transmission objective, he or she would tabulate the required and desired features of alternate technologies that may be available for use in the applications. With that process in mind, a summary of the attractive features and the advantages of optical fiber transmission will be given. Some of these advantages include (a) low loss and high bandwidth; (b) small size and bending radius; (c) nonconductive, nonradiative, and noninductive; (d) light weight; and (e) providing natural growth capability.

To appreciate the low loss and wide bandwidth capabilities of optical fibers, consider the curves of signal attenuation versus frequency for three different transmission media shown in Fig. 4. Optical fibers have a "flat" transfer function well beyond 100 MHz. When compared with wire pairs of coaxial cables, optical fibers have far less loss for signal frequencies above a few megahertz. This is an important characteristic that strongly influences system economics, since it allows the system designer to increase

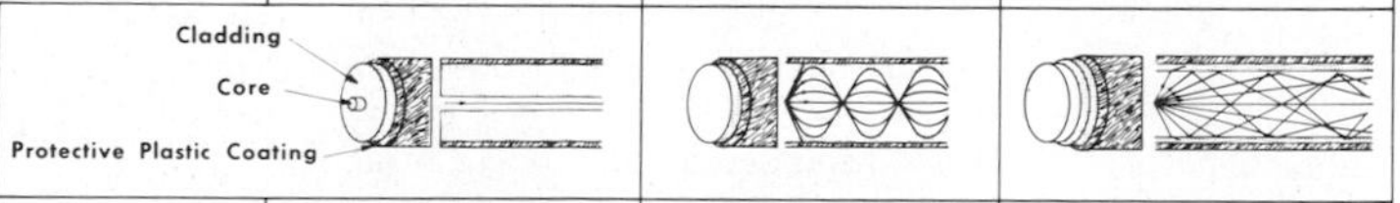

	SINGLE-MODE FIBER	GRADED-INDEX MULTIMODE FIBER	STEP-INDEX MULTIMODE FIBER
Cladding, Core, Protective Plastic Coating			
SOURCE	LASER PREFERRED	LASER or LED	LASER or LED
BANDWIDTH	VERY VERY LARGE > 3 GHz km	VERY LARGE 200 MHz 3 GHz km	LARGE < 200 MHz km
SPLICING	DIFFICULT DUE TO SMALL CORE	DIFFICULT BUT DOABLE	DIFFICULT BUT DOABLE
EXAMPLE OF APPLICATION	SUBMARINE CABLE SYSTEM	TELEPHONE LOOP DISTRIBUTION SYSTEM	DATA LINKS
COST	LEAST EXPENSIVE	MOST EXPENSIVE	EXPENSIVE

FIG. 3.   Applications and characteristics of fiber types.

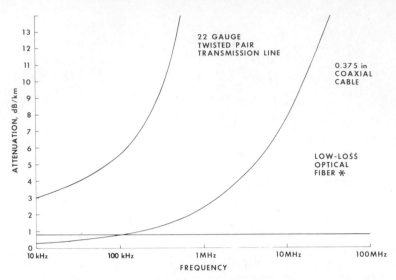

FIG. 4.   Attenuation versus frequency for three different transmission media. Asterisk indicates fiber loss at a carrier wavelength of 1.3 μm.

the distance between regenerators (amplifiers) in a communication system.

The small size, small bending radius (a few centimeters), and light weight of optical fibers and cables are very important where space is at a premium, such as in aircraft, on ships, and in crowded ducts under city streets.

Because optical fibers are dielectric wave-guides, they avoid many problems such as radia-tive interference, ground loops, and, when installed in a cable without metal, light-ning-induced damage that exists in other trans-mission media.

Finally, the engineer using optical fibers has a great deal of flexibility. He or she can install an optical fiber cable and use it initially in a low-capacity (low-bit-rate) system. As the system needs grow, the engineer can take advantage of the broadband capabilities of optical fibers and convert to a high-capacity (high-bit-rate) system by simply changing the terminal electronics. A comparison of the growth capability of different transmission media is shown in Table I. For the three digital transmission rates considered (1.544, 6.312, and 44.7 Mbit/sec), the loss of the optical fiber is constant. The loss of the metallic transmission lines, however, increases with in-creasing transmission rates, thus limiting their use at the higher bit rates. The optical fiber sys-tem, on the other hand, could be used at all bit

TABLE I.   Growth Capability, Transmission Media Comparisons

Transmission media	Loss in dB/km at half bit rate frequency (digital transmission rates)		
	T1 (1.544 Mbit/sec)	T2 (6.312 Mbit/sec)	T3 (44.736 Mbit/sec)
26-Gauge twisted wire pair	24	48	128
19-Gauge twisted wire pair	10.8	21	56
0.375-in.-Diameter coaxial cable	2.1	4.5	11
Low-loss optical fiber[a]	0.75	0.75	0.75

[a] Fiber loss at a carrier wavelength of 1.3 μm.

rates and can grow naturally to satisfy system needs.

III. Fiber Transmission Characteristics

The proper design and operation of an optical communication system using optical fibers as the transmission medium requires a knowledge of the transmission characteristics of the optical sources, fibers, and interconnection devices (connectors, couplers, and splices) used to join lengths of fibers together. The transmission criteria that affect the choice of the fiber type used in a system are signal attenuation, information transmission capacity (bandwidth), and source coupling and interconnection efficiency. Signal attenuation is due to a number of loss mechanisms within the fiber, as shown in Table II, and due to the losses occurring in splices and connectors. The information transmission capacity of a fiber depends on dispersion, a phenomenon that causes light that is originally concentrated into a short pulse to spread out into a broader pulse as it travels along an optical fiber. Source and interconnection efficiency depends on the fiber's core diameter and its numerical aperture, a measure of the angle over which light is accepted in the fiber.

Absorption and scattering of light traveling through a fiber leads to signal attenuation, the rate of which is measured in decibels per kilometer (dB/km). As can be seen in Fig. 5, for both multimode and single-mode fibers, attenuation depends strongly on wavelength. The decrease in scattering losses with increasing wavelength is offset by an increase in material absorption such that attenuation is lowest near 1.55 μm (1550 nm).

The measured values given in Table III are probably close to the lower bounds for the attenuation of optical fibers. In addition to intrinsic fiber losses, extrinsic loss mechanisms, such as absorption due to impurity ions, and microbend-

TABLE II.   Loss Mechanisms

Intrinsic material absorption loss
   Ultraviolet absorption tail
   Infrared absorption tail
Absorption loss due to impurity ions
Rayleigh scattering loss
Waveguide scattering loss
Microbending loss

TABLE III. Best Attenuation Results (dB/km) in Ge–P–SiO$_2$-Core Fibers

Wavelength (nm)	$\Delta \approx 0.2\%$ (Single-mode fibers)	$\Delta \approx 1.0\%$ (Graded-index multimode fibers)
850	2.1	2.20
1300	0.27	0.44
1500	0.16	0.23

ing loss due to jacketing and cabling can add loss to a fiber.

The bandwidth or information-carrying capacity of a fiber is inversely related to its total dispersion. The total dispersion in a fiber is a combination of three components: intermodal dispersion (modal delay distortion), material dispersion, and waveguide dispersion.

Intermodal dispersion occurs in multimode fibers because rays associated with different modes travel different effective distances through the optical fiber. This causes light in the different modes to spread out temporally as it travels along the fiber. Modal delay distortion can severely limit the bandwidth of a step index multimode fiber to the order of 20 MHz km. To reduce modal delay distortion in multimode fibers, the core is carefully doped to create a graded (approximately parabolic shaped) refractive index profile. By carefully designing this in-

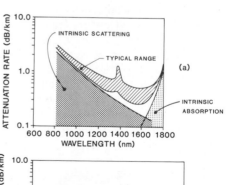

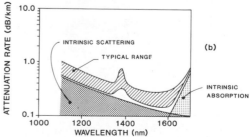

FIG. 5.   Spectral attenuation rate. (a) Graded-index multimode fibers. (b) Single-mode fibers.

dex profile, the group velocities of the propagating modes are nearly equalized. Bandwidths of 1.0 GHz km are readily attainable in commercially available graded index multimode fibers. The most effective way of eliminating intermodal dispersion is to use a single-mode fiber. Since only one mode propagates in a single mode fiber, modal delay distortion between modes does not exist and very high bandwidths are possible. The bandwidth of a single-mode fiber, as mentioned previously, is limited by the combination of material and waveguide dispersion. As shown in Fig. 6, both material and waveguide dispersion are dependent on wavelength.

Material dispersion is caused by the variation of the refractive index of the glass with wavelength and the spectral width of the system source. Waveguide dispersion occurs because light travels in both the core and cladding of a single-mode fiber at an effective velocity between that of the core and cladding materials. The waveguide dispersion arises because the effective velocity, the waveguide dispersion, changes with wavelength. The amount of waveguide dispersion depends on the design of the waveguide structure as well as on the fiber materials. Both material and waveguide dispersion are measured in picoseconds (of pulse spreading) per nanometer (of source spectral width) per kilometer (of fiber length), reflecting both the increases in magnitude in source linewidth and the increase in dispersion with fiber length.

Material and waveguide dispersion can have different signs and effectively cancel each other's dispersive effect on the total dispersion in a single-mode fiber. In conventional germanium-doped silica fibers, the "zero-dispersion" wavelength at which the waveguide and material dispersion effects cancel each other out occurs near 1.30 μm. The zero-dispersion wavelength can be shifted to 1.55 μm, or the low-dispersion

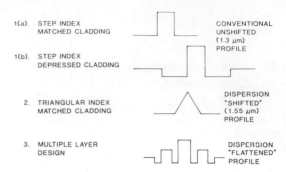

FIG. 7.   Single-mode refractive-index profiles.

characteristics of a fiber can be broadened by modifying the refractive-index profile shape of a single-mode fiber. This profile shape modification alters the waveguide dispersion characteristics of the fiber and changes the wavelength region in which waveguide and material dispersion effects cancel each other. Figure 7 illustrates the profile shapes of "conventional," "dispersion-shifted," and "dispersion-flattened" single-mode fibers. Single-mode fibers operating in their zero-dispersion region with system sources of finite spectral width do not have infinite bandwidth but have bandwidths that are high enough to satisfy all current high-capacity system requirements.

IV. Optical Fiber Cable Manufacturing

Optical fiber cables should have low loss and high bandwidth and should maintain these characteristics while in service in extreme environments. In addition, they should be strong enough to survive the stresses encountered during manufacture, installation, and service in a hostile environment. The manufacturing process used to fabricate the optical fiber cables can be divided into four steps: (1) preform fabrication, (2) fiber drawing and coating, (3) fiber measurement, and (4) fiber packaging.

A. PREFORM FABRICATION

The first step in the fabrication of optical fiber is the creation of a glass preform. A preform is a large blank of glass several millimeters in diameter and several centimeters in length. The preform has all the desired properties (e.g., geometrical ratios and chemical composition) necessary to yield a high-quality fiber. The preform is subsequently drawn into a multi-kilometer-long

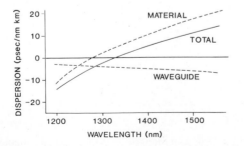

FIG. 6.   Single-mode step-index dispersion curve.

hair-thin fiber. Four different preform manufacturing processes are currently in commercial use.

The most widely used process is the modified chemical vapor deposition (MCVD) process invented at the AT&T Bell Laboratories. Outside vapor deposition process (OVD) is used by Corning Glass Works and some of its joint ventures in Europe. Vapor axial deposition (VAD) process is the process used most widely in Japan. Philips, in Eindhoven, Netherlands, uses a low-temperature plasma chemical-vapor deposition (PCVD) process.

In addition to the above four major processes, other processes are under development in different laboratories. Plasma MCVD is under development at Bell Laboratories, hybrid OVD–VAD processes are being developed in Japan, and Sol-Gel processes are being developed in several laboratories. The first four processes are the established commercial processes and are producing fiber economically. The new processes are aimed at greatly increasing the manufacturing productivity of preforms, and thereby reducing their cost.

All the above processes produce high-silica fibers using different dopants, such as germanium, phosphorous, and fluorine. These dopants modify the refractive index of silica, enabling the production of the proper core refractive-index profile. Purity of the reactants and the control of the refractive-index profile are crucial to the low loss and high bandwidth of the fiber.

1. MCVD Process

In the MCVD process (Fig. 8), a fused-silica tube of extremely high purity and dimensional uniformity is cleaned in an acid solution and degreased. The clean tube is mounted on a glass working lathe. A mixture of reactants is passed from one end of the tube and exhaust gases are taken out at the other end while the tube is being rotated. A torch travels along the length of the tube in the direction of the reactant flow. The reactants include ultra-high-purity oxygen and a combination of one or more of the halides and oxyhalides ($SiCl_4$, $GeCl_4$, $POCl_3$, BCl_3, BBr_3, SiF_4, CCl_4, CCl_2F_2, Cl_2, SF_6, and $SOCl_2$).

The halides react with the oxygen in the temperature range of 1300–1600°C to form oxide particles, which are driven to the wall of the tube and subsequently consolidated into a glassy layer as the hottest part of the flame passes over. After the completion of one pass, the torch travels back and the next pass is begun. Depending on the type of fiber (i.e., multimode or single-mode), a barrier layer or a cladding consisting of many thin layers is first deposited on the inside surface of the tube. The compositions may include B_2O_3–P_2O_5–SiO_2 or F–P_2O_5–SiO_2 for barrier layers, and SiO_2, F–SiO_2, F–P_2O_5–SiO_2, or F–GeO_2–SiO_2–P_2O_5 for cladding layers. After the required number of barrier or cladding layers has been deposited, the core is deposited. The core compositions depend on whether the fiber is single-mode, multimode, step-index, or multimode graded index. In the case of graded-index multimode fibers, the dopant level changes with every layer, to provide a refractive index profile that yields the maximum bandwidth.

After the deposition is complete, the reactant flow is stopped except for a small flow of oxygen, and the temperature is raised by reducing the torch speed and increasing the flows of oxygen and hydrogen through the torch. Usually the exhaust end of the tube is closed first and a small positive pressure is maintained inside the deposited tube while the torch travels backward. The higher temperatures cause the glass viscosity to decrease, and the surface tension causes the tube to contract inward. The complete collapse of the tube into a solid preform is achieved in several passes. The speed of the collapse, the rotation of the tube, the temperature of collapse, and the positive pressure of oxygen inside the tube are all accurately controlled to predetermined values in order to produce a straight and bubble-free preform with minimum ovality. The complete preform is then taken off the lathe. After an inspection to assure that the preform is free of defects, the preform is ready to be drawn into a thin fiber.

The control of the refractive-index profile along the cross section of the deposited portion

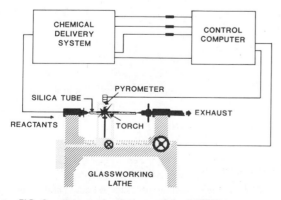

FIG. 8.   Schematic diagram of the MCVD process.

of the preform is achieved through a vapor delivery system. In this system, liquids are vaporized by passing a carrier gas (pure O_2) through the bubblers, made of fused silica. Accurate flows are achieved with precision flow controllers that maintain accurate carrier gas flows and extremely accurate temperatures within the bubblers. Microprocessors are used to automate the complete deposition process, including the torch travel and composition changes throughout the process. Impurities are reduced to very low levels by starting with pure chemicals, and there is further reducing of the impurities with in-house purification of these chemicals. Ultra-pure oxygen and a completely sealed vapor-delivery system are used to avoid chemical contamination. Transition-metal ion impurities of well below 1 ppb and OH^- ion impurities of less than 1 ppm are typically maintained to produce high quality fiber.

2. The PCVD Process

The PCVD process (Fig. 9) also uses a starting tube, and the deposition takes place inside the tube. Here, however, the tube is either stationary or oscillating and the pressure is kept at 10–15 torr. Reactants are fed inside the tube, and the reaction is accomplished by a traveling microwave plasma inside the tube. The entire tube is maintained at approximately 1200°C. The plasma causes the heterogeneous depositions of glass on the tube wall, and the deposition efficiency is very high. After the required depositions of the cladding and core are complete, the tube is taken out and collapsed on a separate equipment. Extreme care is required to prevent impurities from getting into the tube during the transport and collapse procedure. The PCVD process has the advantages of high efficiency, no tube distortion because of the lower temperature, and very accurate profile control because of the large number of layers deposited in a short time. However, going to higher rates of flow

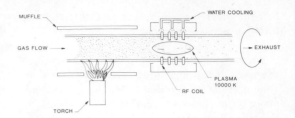

FIG. 10.   Schematic diagram of the PMCVD process.

presents some difficulties, because of a need to maintain the low pressure.

3. The PMCVD Process

The PMCVD is an enhancement of the MCVD process. Very high rates of deposition (up to 10 g/min, compared to 2 g/min for MCVD) are achieved by using larger diameter tubes and an RF plasma for reaction (Fig. 10). Because of the very high temperature of the plasma, water cooling is essential. An oxyhydrogen torch follows the plasma and sinters the deposition. The high rates of deposition are achieved because of very high thermal gradients from the center of the tube to the wall and the resulting high thermophoretic driving force. The PMCVD process is still in the development stage and has not been commercialized.

4. The OVD Process

The OVD process does not use a starting tube; instead, a stream of soot particles of desired composition is deposited on a bait rod (Fig. 11). The soot particles are produced by the reac-

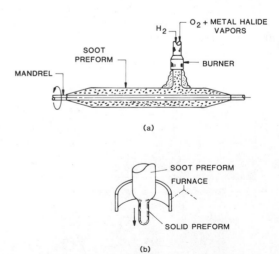

FIG. 11.   Schematic diagram of the outside vapor deposition. (a) Soot deposition. (b) Consolidation.

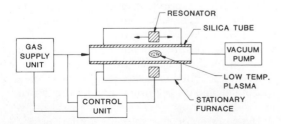

FIG. 9.   Schematic diagram of the PCVD process.

tion of reactants in a fuel gas–oxygen flame. A cylindrical porous soot preform is built layer by layer. After the deposition of the core and cladding is complete, the bait rod is removed. The porous preform is then sintered and dried in a furnace at 1400–1600°C to form a clear bubble-free preform under a controlled environment. The central hold left by the blank may or may not be closed, depending on the type of preform. The preform is now ready for inspection and drawing.

5. The VAD Process

The process is very similar to the OVD process. However, the soot deposition is done axially instead of radially. The soot is deposited at the end of a starting silica-glass rod (Fig. 12). A special torch using several annular holes is used to direct a stream of soot at the deposition surface. The reactant vapors, hydrogen gas, argon gas, and oxygen gas flow through different annular openings. Normally the core is deposited and the rotating speed is gradually withdrawn as the deposition proceeds at the end. The index profile is controlled by the composition of the gases flowing through the torch and the temperature distribution at the deposition surface. The porous preform is consolidated and dehydrated as it passes through a carbon-ring furnace in a controlled environment. $SOCl_2$ and Cl_2 are used to dehydrate the preform. Because of the axial deposition, this process is semicontinuous and is capable of producing very large preforms.

B. Fiber Drawing

After a preform has been inspected for various defects such as bubbles, ovality, and straightness, it is taken to a fiber drawing station. A large-scale fiber drawing process must repeatedly maintain the optical quality of the

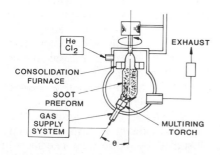

FIG. 12. Schematic diagram of the vapor axial deposition.

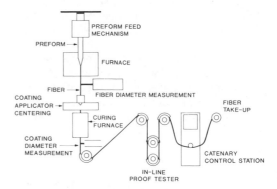

FIG. 13. The fiber drawing process.

preform and produce a dimensionally uniform fiber with high strength.

1. Draw Process

During fiber drawing, the inspected preform is lowered into a hot zone at a certain feed rate V_p, and the fiber is pulled from the softened neck-down region (Fig. 13) at a rate V_f. At steady state,

$$\pi D_p^2 V_p/4 = \pi D_f^2 V_f/4 \qquad (1)$$

where D_p and D_f are the preform and fiber diameters, respectively. Therefore,

$$V_f = (D_p^2/D_f^2)V_p \qquad (2)$$

A draw machine, therefore, consists of a preform feed mechanism, a heat source, a pulling device, a coating device, and a control system to accurately maintain the fiber diameter and the furnace temperature.

2. Heat Source

The heat source should provide sufficient energy to soften the glass for pulling the fiber without causing excessive tension and without creating turbulence in the neck-down region. A proper heat source will yield a fiber with uniform diameter and high strength. Oxyhydrogen torches, CO_2 lasers, resistance furnaces, and induction furnaces have been used to draw fibers. An oxyhydrogen torch, although a clean source of heat, suffers from turbulence due to flame. A CO_2 laser is too expensive a heat source to be considered for the large-scale manufacture of fibers. Graphite resistance furnaces and zirconia induction furnaces are the most widely used heat sources for fiber drawing. In the graphite resistance furnace, a graphite resistive element produces the required heat. Because graphite reacts with oxygen at high temperatures, an inert environment (e.g., argon) is maintained inside the furnace. The zirconia induction furnace does not require inert environment. It is extremely important that the furnace environment be clean in order to produce high-strength fibers. A zirconia induction furnace, when properly designed and used, has produced very-high-strength long-length fibers (over 2.0 GPa) in lengths of several kilometers.

3. Mechanical Systems

An accurate preform feed mechanism and drive capstan form the basis of fiber speed control. The mechanism allows the preform to be fed at a constant speed into the hot zone, while maintaining the preform at the center of the furnace opening at the top. A centering device is used to position preforms that are not perfectly straight. The preform is usually held with a collet-type chuck mounted in a vertically movable carriage, which is driven by a lead screw. A precision stainless-steel drive capstan is mounted on the shaft of a high-performance dc servomotor. The fiber is taken up on a proper-diameter spool. The fiber is wound on the spool at close to zero tension with the help of a catenary control. In some cases fiber is proof-tested in-line before it is wound on a spool. The proof stress can be set at different levels depending on the application for which the fiber is being manufactured.

4. Fiber Coating System

The glass fiber coming out of the furnace has a highly polished pristine surface, and the theoretical strength of such a fiber is in the range of 15–20 GPa. Strengths in the range of 4.5–5.5 GPa are routinely measured on short fiber lengths. To preserve this high strength, polymeric coatings are applied immediately after the drawing. The coating must be applied without damaging the fiber, it must solidify before reaching the capstan, and it should not cause microbending loss. To satisfy all these requirements, usually two layers of coatings are applied: a soft inner coating adjacent to the fiber to avoid microbending loss, and a hard outer coating to resist abrasion. The coatings are a combination of ultraviolet (UV) curable acrylates, UV-curable silicones, hot melts, heat-curable silicones, and nylons. When dual coatings are applied, the coated fiber diameter is typically 235–250 μm. The nylon-jacketed fiber typically used in Japan has an outside diameter of 900 μm. All coating materials are usually filtered to remove particles

that may damage the fiber. Coatings are usually applied by passing the fiber through a coating cup and then curing the coating before the fiber is taken up by the capstan. The method of application, the coating material, the temperature, and the draw speed affect the proper application of a well-centered, bubble-free coating.

Fiber drawing facilities are usually located in a clean room where the air is maintained at class 10,000. The region of the preform and fiber from the coating cup to the top of the preform is maintained at class 100 or better. A class 100 environment means that there are no more than 100 particles of size greater than 0.5 μm in 1 ft^3 of air. A clean environment, proper centering of the preform in the furnace and fiber in the coating cup, and proper alignment of the whole draw tower ensure a scratch-free fiber of a very high tensile strength. A control unit regulates the draw speed, preform feed speed, preform centering, fiber diameter, furnace temperature, and draw tension.

The coated fiber wound on a spool is next taken to the fiber measurement area to assure proper quality control.

5. Proof Testing of Fibers

Mechanical failure is one of the major concerns in the reliability of optical fibers. Fiber drawn in kilometer lengths must be strong enough to survive all of the short- and long-term stresses that it will encounter during the manufacture, installation, and long service life. Glass is an ideal elastic isotropic solid and does not contain dislocations. Hence, the strength is determined mainly by inclusions and surface flaws. Although extreme care is taken to avoid inhomogeneities and surface flaws during fiber manufacture, they cannot be completely eliminated. Since surface flaws can result from various causes, they are statistical in nature and it is impossible to predict the long-length strength of glass fibers. To guarantee a minimum fiber strength, proof testing has been adopted as a manufacturing step. Proof testing can be done in-line immediately after the drawing and coating, or off-line before the fiber is stored.

In proof testing, the entire length of the fiber is subjected to a properly controlled proof stress. The proof stress is based on the stresses likely to be encountered by the fiber during manufacture, storage, installation, and service. The fibers that survive the proof test are stored for further packaging into cables.

Proof testing not only guarantees that the fiber will survive short-term stresses but also guarantees that the fiber will survive a lower residual stress that it may be subjected to during its long service life. It is well known that glass, when used in a humid environment, can fail under a long-term stress well below its instantaneous strength. This phenomenon is called static fatigue. Several models have been proposed to quantitatively describe the relationship between residual stress and the life of optical fibers. Use is made of the most conservative of these models, and the proof stress is determined by a consideration of the maximum possible residual stress in service and the required service life.

C. FIBER PACKAGING

In order to efficiently use one or more fibers, they need to be packaged so that they can be handled, transported, and installed without damage. Optical fibers can be used in a variety of applications, and hence the way they are packaged or cabled will also vary. There are numerous cable designs that are used by different cable manufacturers. All these designs, however, must meet certain criteria. A primary consideration in a cable design is to assure that the fibers in the cables maintain their optical properties (attenuation and dispersion) during their service life under different environmental conditions. The design, therefore, must minimize microbending effects. This usually means letting the fiber take a minimum energy position at all times in the cable structure. Proper selection of cabling materials so as to minimize differential thermal expansion or contraction during temperature extremes is important in minimizing microbending loss. The cable structure must be such that the fibers carry a load well below the proof-test level at all times, and especially while using conventional installation equipment. The cables must provide adequate protection to the fibers under all adverse environmental conditions during their entire service life, which may be as long as 40 years. Finally, the cable designs should be cost-effective and easily connectorized or spliced.

Five different types (Fig. 14) of basic cable designs are currently in use: (a) loose tube, (b) fluted, (c) ribbon, (d) stranded, and (e) Lightpack Cable. The loose tube design was pioneered by Siemens in Germany. Up to 10 fibers are enclosed in a loose tube, which is filled with a soft filling compound. Since the fibers are rela-

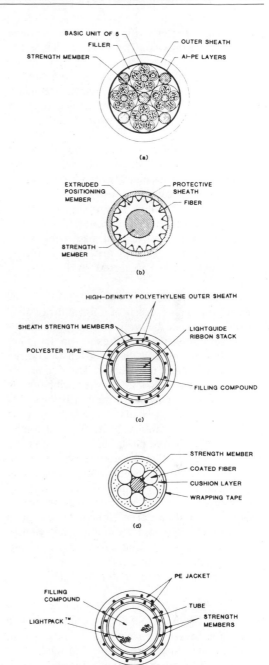

FIG. 14.   Fiber cable designs. (a) Loose tube design. (b) Slotted design. (c) Ribbon design. (d) Stranded unit. (e) Lightpack™ Cable design.

tively free to take the minimum energy configuration, the microbending losses are avoided. Several of these buffered loose tube units are stranded around a central glass–resin support member. Aramid yarns are stranded on the cable core to provide strength members (for pulling through ducts), with a final polyethylene sheath on the outside. The stranding lay length and pitch radius are calculated to permit tensile strain on the cable up to the rated force and to permit cooling down to the rated low temperature without affecting the fiber attenuation.

In the fluted designs, fibers are laid in the grooves of plastic central members and are relatively free to move. The shape and size of the grooves vary with the design. The grooved core may also contain a central strength member. A sheath is formed over the grooved core, and this essentially forms a unit. Several units may then be stranded around a central strength member to form a cable core of desired size, over which different types of sheaths are formed. Fluted designs have been pioneered in France and Canada.

The ribbon design was invented at AT&T Bell Laboratories and consists of a linear array of 12 fibers sandwiched between two polyester tapes with pressure-sensitive adhesive on the fiber side. The spacing and the back tension on the fibers is accurately maintained. The ribbons are typically 2.5 mm in width. Up to 12 ribbons can be stacked to give a cable core consisting of 144 fibers. The core is twisted to some lay length and enclosed in a polyethylene tube. Several combinations of protective plastic and metallic layers along with metallic or nonmetallic strength members are then applied around the core to give the final cable its required mechanical and environmental characteristics needed for use in specified conditions. The ribbon design offers the most efficient and economic packaging of fibers for high-fiber-count cables. It also lends the cable to preconnectorization and makes it extremely convenient for installation and splicing.

The tight-bound stranded designs were pioneered by Japanese and are used in the United States for several applications. In this design, several coated fibers are stranded around a central support member. The central support member may also serve as a strength member, and it may be metallic or nonmetallic. The stranded unit can have up to 18 fibers. The unit is contained within a plastic tube filled with a water-blocking compound. The final cable consists of several of these units stranded around a central member and protected on the outside with various sheath combinations.

The Lightpack Cable design, pioneered by AT&T, is one of the simplest designs. Several fibers are held together with a binder to form a

unit. One or more units are laid inside a large tube, which is filled with a water-blocking compound. This design has the advantage of the loose tube design in that the fibers are free of strain, but is more compact. The tube-containing units can then be protected with various sheath options and strength members to provide the final cable.

The final step in cabling is the sheathing operation. After the fibers have been made into identifiable units, one or more of the units (as discussed earlier) form a core, which is then covered with a combination of sheathing layers. The number and combination of the sheathing layers depend on the intended use. Typically, a polyethylene sheath is extruded over the filled cable core. In a typical cross-ply design, metallic or nonmetallic strength members are applied over the first sheath layer, followed by another polyethylene sheath, over which another layer of strength members is applied. The direction of lay of the two layers of the strength members is opposite to each other. A final sheath is applied and the cable is ready for the final inspection, preconnectorization, and shipment. Metallic vapor barriers and lightning- and rodent-protection sheath options are also available. Further armoring is applied to cables made for submarine application.

In addition to the above cable designs, there are numerous other cable designs used for specific applications, such as fire-resistant cables, military tactical cables, cables for missile guidance systems, cables for field communications established by air-drop operations, air deployment cables, and cables for industrial controls. All these applications have unique requirements, such as ruggedness, low loss, and repeaterless spans, and the cable designs are accordingly selected. However, all these cable designs still rely on the basic unit designs discussed above.

V. Sources and Detectors

In this section we will review the characteristic of optical sources and detectors that are used in fiber-optic communication systems.

A. Optical Sources

Semiconductor light emitting diodes (LEDs) and injection-laser diodes (ILDs) are attractive as optical carrier sources because they are dimensionally compatible with optical fibers.

They emit at wavelengths (0.8–0.9 μm and 1.3–1.6 μm) corresponding with regions of low optical-fiber loss, their outputs can be rapidly controlled by varying their bias current and therefore they are easy to modulate, and finally they offer solid-state reliability with lifetimes now exceeding 10^6 h. Although LEDs and ILDs exhibit a number of similar characteristics, there are important differences between them that must be understood before one can select a source for a specific fiber-optic communication system.

One major difference between LEDs and ILDs is their spatial and temporal coherence. An ILD radiates a relatively narrow beam of light that has a narrow spectral width. In contrast, LED sources have much wider radiation patterns (beam width) and have moderately large spectral widths. These factors govern the amount of optical power that can be coupled into a fiber and the influence of chromatic dispersion on the bandwidth of the fiber medium. The second difference between ILDs and LEDs is their speed. The stimulated emission from lasers results in intrinsically faster optical rise and fall times in response to changes in drive current than can be realized with LEDs. The third difference between the devices is related to their linearity. LEDs generate light that is almost linearly proportional to the current passing through the device. Lasers, however, are threshold devices, and the lasing output is proportional to the drive current only above threshold. The threshold current of a laser, unfortunately, is not a constant but is a function of the device's temperature and age. Feedback control drive circuitry is therefore required to stabilize a laser's output power. Table IV illustrates typical ILDs and LED characteristics found in fiber-optic communication systems.

GaAlAs devices (both ILDs and LEDs) emitting in the wavelength region of 0.8–0.9 μm are commercially available and widely used in optical fiber systems. InGaAsP devices, with their emission wavelengths in the region 1.0–1.6 μm, are available for application near 1.3 and 1.6 μm where fiber chromatic dispersion and transmission losses are minimal. The high-radiance Burrus type (surface emitting) LED is well suited for application in systems of low to medium bandwidth (<50 MHz). The power that can be coupled from an LED into a fiber is proportional to the number of modes the fiber can propagate, that is, to its core area times its numerical aperture squared. For simple butt coupling, where the emitting area of the LED is equal to or less than the core area of the fiber, presently available surface emitting GaAlAs and InGaAsP LEDs can launch about 50 μW into a graded-index fiber of numerical aperture (NA) 0.2 and core diameter 50 μm. The spectral width of an LED is a function of the operating wavelength, the active-layer doping concentration, and the junction current density. The rms spectral width of a typical 0.85-μm GaAlAs Burrus-type LED is about 16 nm, while that of a 1.3-μm InGaAsP LED is about 40 nm (spectral width is approximately proportional to λ^2). The modulation bandwidth of an LED depends on the device geometry, its current density, and the doping concentration of its active layer. Higher doping concentration yields higher bandwidth, but only at the expense of lower output power and wider spectral width. Figure 15 shows the tradeoff between output power and modulation bandwidth for a group of Burrus type GaAlSAs LEDs. Typically, a 50-μm-diameter LED that can butt couple 50 μW into a 0.2-NA, 50-μm-core graded-index fiber can be current modulated at rates up to about 50 MHz.

TABLE IV.   Optical Source Characteristics

	ILDs	LEDs
Output power, mW	1–10	1–10
Power launched into fiber, mW	0.5–5	0.03–0.3
Spectral width (rms value), nm	2–4	15–60
Brightness, W/cm² sr	~10^5	10–10^3
Rise time, 10–90%, nsec	<1	2–20
Frequency response (−3 dB), MHz	>500	<200
Voltage drop, V	1.5–2	1.5–2.5
Forward current, mA	10–300	50–300
Threshold current, mA	5–250	Not applicable
Feedback stabilization required	Yes	No

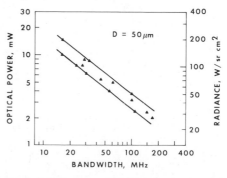

FIG. 15.   Output power versus 3-dB bandwidth for Burrus-type LED.

ILDs are well suited for application in medium- to high-bandwidth fiber-optic communication systems. Compared to LEDs, injection lasers offer the advantage of narrower spectral width (<3 nm), larger modulation bandwidth (>500 MHz), and greater launched power (1 mW). ILDs are the sources most compatible for use with single-mode fibers. However, ILDs are not as reliable as LEDs, are more expensive, and require feedback circuity to stabilize their output power against variations due to temperature and aging effects.

B. OPTICAL RECEIVERS—PHOTODETECTORS

The basic purpose of an optical receiver is to detect the received light incident on it and to convert it to an electrical signal containing the information impressed on the light at the transmitting end. The receiver is therefore an optical-to-electrical converter or O/E transducer. An optical receiver consists of a photodetector and an associated amplifier along with necessary filtering and processing, as shown in Fig. 16. The function of the photodetector is to detect the incident light signal and convert it to an electrical current. The amplifier converts this current into a usable signal while introducing the minimum amount of additional noise to corrupt the signal. In designing an optical receiver, one tries to minimize the amount of optical power that must reach the receiver in order to achieve a given bit error rate (BER) in digital systems, or a given signal-to-noise ratio (S/N) in an analog system. In this section we will describe the characteristics of the photodetectors used in fiber optic systems. Since the performance of an optical receiver depends not only on the photodetector but also on the components and design chosen for the subsequent amplifier, we will also briefly describe configurations for this amplifier and their associated resulting receiver sensitivities.

In all the installed commercial fiber optic communication systems in existence today, the photodetector used is either a semiconductor $p-i-n$ or avalanche photodiode (APD). These devices differ in that the $p-i-n$ basically converts one photon to one electron and has a conversion efficiency of less than unity. In an APD carrier, multiplication takes place that results in multiple electrons at the output per incident photon.

The reasons for choosing $p-i-n$ or APD photodetector are usually based on cost and on required receiver sensitivity. The avalanching process in the APD has a sharp threshold, which is sensitive to ambient temperature, and may require dynamic control of a relatively high bias voltage. The APD control and driver circuits are more expensive than those for the $p-i-n$ detector, and the APD itself is more expensive than the $p-i-n$ device. An APD with optimum gain, however, provides about 15 dB more receiver sensitivity than that achieved with a $p-i-n$ diode.

An excellent spectral match exists between

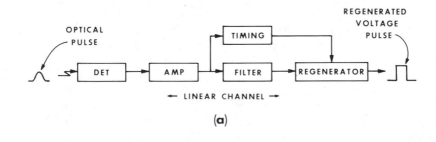

(a)

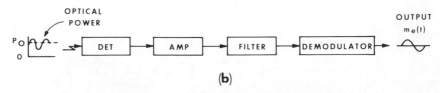

(b)

FIG. 16.   Schematic of digital receiver. (a) Digital. (b) Analog.

TABLE V.   Typical Photodetector Characteristics

Characteristic	Silicon	Germanium	p–i–n Diodes		
			InP	Silicon	Germanium
Wavelength range, μm	0.4–1.1	0.5–1.8	1.0–1.6	0.4–1.1	0.5–1.65
Wavelength of peak sensitivity, μm	0.85	1.5	1.26	0.85	1.5
Quantum efficiency, 1%	80	50	70	80	70
Rise time, nsec	0.01	0.3	0.1	0.5	0.25
Bias voltage, V	15	6	10	170	40
Responsivity, A/W	0.5	0.7	0.4	0.7	0.6
Avalanche gain	1.0	1.0	1.0	80–150	80–150

GaAlAs sources operating in the wavelength range of 0.8–0.9 μm and photodiodes made of silicon (spectral range 0.5–1.1 μm). The silicon p–i–n diode having no gain but with low dark current ($<10^{-9}$ A) and large bandwidth (1 GHz) is best suited for applications where receiver sensitivity is not critical. Silicon APDs are preferable in applications that demand high sensitivity, and those employed in presently installed telecommunication systems have current gains of about 100 and primary dark currents in the range of 10^{-10} to 10^{-11} A. Germanium photodiodes are used in the longer-wavelength region (1.3–1.6 μm), since the response of silicon decreases rapidly as λ increases beyond 1.0 μm. Germanium APDs with gain bandwidth products of approximately 60 GHz have been made, but their dark currents are high (10^{-8} to 10^{-7} A) and their excess noise factors are large. InGaAs and InGaAsP diodes are also used in the long-wavelength region. InGaAs p–i–n diodes have been made with very low capacitance (<0.3 pf) and acceptably low dark current ($<5 \times 10^{-9}$ A). However, further work is required in the area of long-wavelength APDs to reduce their high excess noise factor. Table V summarizes the characteristics of photodetectors used in fiber-optic communication systems.

To calculate the system margin for a communication system, a knowledge of receiver sensitivity is needed. Receiver sensitivity is determined primarily by the characteristics of the photodetector and the low-noise front-end amplifier, which is optimized for use with the detector. Figure 17 shows two commonly used configurations. To achieve the best receiver sensitivity, the amplifier should have a high input impedance or provide feedback as in a transimpe-

dance amplifier. The first stage can be either a GaAs field-effect transistor (FET) or a silicon bipolar transistor with a suitably adjusted emitter bias.

In the short-wavelength region (0.8–0.9 μm), silicon APDs can provide sufficiently high gain and low excess noise to overcome the input amplifier noise. In this wavelength region, design of the first amplifier stage is not very critical. A conventional silicon bipolar transistor having an emitter capacitance of 10 pF and a current gain of 150 has been used to build a digital receiver that requires only 10 nW average optical input power (-50 dBm) for a bit error rate (BER) of 10^{-9} at 100 Mbit/sec.

Since leakage currents (high noise factor) severely limit the use of avalanche gains as low-noise amplification process in the long-wavelength region (1.3–1.6 μm), p–i–n detectors are usually used. With a p–i–n diode, the microwave GaAs FET is well suited for use in the first amplifier stage because it has a low gate capacitance and high transconductance. It is typically used in a transimpedance configuration and offers wide bandwidth and good dynamic range. The best receivers have been built using InGaAs p–i–n detectors and GaAs FETs and require an

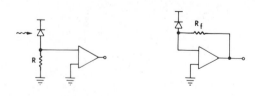

FIG. 17.   Schematic of amplifier designs.

average optical input power of 25 nW (−46 dBm) for 10^{-9} BER at 100 Mbit/sec.

Curves for receiver sensitivity as a function of bit rate (bandwidth), along with curves of power available from LED and ILD sources, will allow us to calculate the net transmission loss tolerable between regenerators (system margin) as a function of bit rate (frequency).

Figure 18 shows, for digital systems, the average optical power required at the receiver (for BER of 10^{-9}) as well as the power available from optical sources as a function of bit rate. The lower boundary for receiver performance applies for receivers using silicon APDs at wavelengths of less than 1 μm. The upper receiver curve in Fig. 18 reflects the performance of p–i–n FET receivers sensitive in the wavelength range between 1.1 and 1.6 μm. Sources made of GaAlAs and InGaAsP have similar characteristics in terms of modulation speed and power delivered into a fiber. Light emitting diodes are usually restricted to those applications where the modulation bandwidths are less than 50 MHz. The separation between the sources and receiver bands in Fig. 18 is an indication of the gross transmission margin of a system. Practical repeater spans are designed with about 10 dB subtracted from the maximum values given in Fig. 18. This will account for variation in the transmitter and receiver components due to temperature variations and aging. As a general rule of thumb, the required fiber bandwidth in a digital system is equal to or larger than the specified system bit rate. The noise penalty for using this rule is less than 1 dB. Along with this potential noise source, the 10-dB safety margin also

allows for signal degradation from various noise sources in the transmitter and receiver. Once the net transmission loss that is tolerable between regenerators is obtained, the distance between regenerators (link length) can be determined from the loss characteristics for the fibers and interconnection devices used in the system.

VI. Source Coupling, Splices, and Connectors

In this section we will investigate the coupling of energy from an optical source into a fiber, and the effects of intrinsic and extrinsic splice-loss parameters on the transmission characteristics of an optical fiber link. In addition, we will give examples of different types of optical fiber connectors and splices.

A. SOURCE COUPLING

The factors that effect the coupling efficiency of a source into a fiber can be broadly divided into two categories, as shown in Table VI. The first category, loss due to unintercepted illumination, can be caused by the source's emitting area being larger than the fiber's core area. Even if the source is smaller than the core, one can still have problems with unintercepted illumination if separation and misalignment of the source and fiber axes allow emitted light to miss the core and become lost. Coupling loss due to unintercepted illumination can be eliminated, however, if the source emitting area and the fiber-core area are properly matched and aligned. Figure 19 shows a fiber "pigtail" permanently

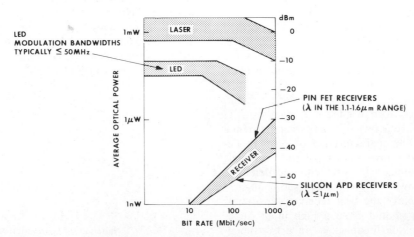

FIG. 18.   Transmission margin versus bit rate for digital transmission systems.

TABLE VI. Factors Influencing Source Coupling Efficiency into a Fiber

1. Unintercepted illumination loss
   (a) Area mismatch between source spot size and fiber core area
   (b) Misalignment of source and fiber axis
2. Numerical aperture loss, caused by that part of the source emission profile that radiates outside of the fiber's acceptance cone

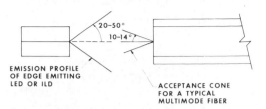

FIG. 20. Schematic showing numerical-aperature mismatch of source and fiber.

mounted and properly aligned on a Burrus-type light emitting diode. This configuration essentially eliminates loss due to unintercepted illumination. The second category of coupling loss that affects efficiency of source coupling into a fiber is due to mismatches between the source beam and fiber numerical apertures. This type of mismatch is shown schematically in Fig. 20. For fiber-optic communication systems, two types of light sources, light emitting diodes (LEDs) and injection laser diodes (ILDs), are typically used. A lambertian source whose radiant intensity varies with the cosine of the angle between a line perpendicular to it and another line drawn to an observation point is often used to model the radiation pattern of a surface emitting LED. Some sources such as edge emitting LEDs and ILDs exhibit radiation patterns that are narrower than a Lambertian source.

The radiation pattern obtained from an edge emitting LED is elliptical in cross section with half-power beam divergence angles of approximately ±60° and ±30°. The radiation pattern obtained from a double heterojunction laser diode is also elliptical in cross section but narrower in

beam width than a LED. For example, the typical half-power beam divergence angles of an ILD are ±25° and ±5° perpendicular and parallel to the junction plane, respectively (see Fig. 21).

If we consider the problem of coupling energy from an LED into a multimode graded-index fiber and assume that the LED is a lambertian source in direct contact with the fiber core and covering its entire cross section, we can calculate the source coupling efficiency to be approximately equal to $\frac{1}{2}(NA)^2$. Sources such as edge emitting LEDs and ILDs, which have more directional emitters, will have a higher coupling efficiency than a lambertian source. ILDs with a properly aligned fiber pigtail will have a source coupling efficiency of 10–20%. For the case where the emitting area of the source is smaller than that of the fiber core, imaging optics can

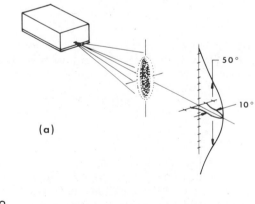

(a)

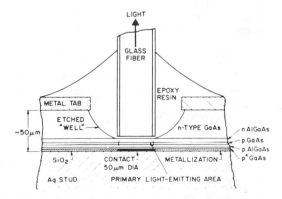

FIG. 19. Burrus-type LED with a fiber pigtail. Small-area, high radiance GaAs–Al$_x$Ga$_{1-x}$As double-hetero-structure surface emitter.

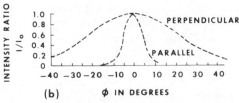

(b)   φ IN DEGREES

FIG. 21. Radiation pattern of an injection laser diode. (a) Schematic representation of far-field radiation pattern. (b) Far-field intensity pattern measured in planes parallel and perpendicular to the junction.

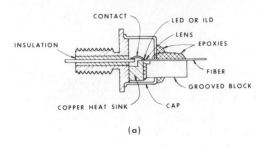

(a)

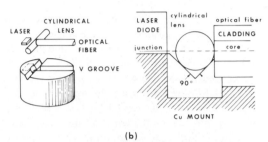

(b)

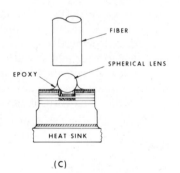

(c)

FIG. 22. Lens system to increase source coupling efficiency. (a) LED or ILD with lens incorporated in the source package. (b) ILD with cylindrical lens coupling source energy into graded index fiber ($\eta_c \approx 40–50\%$). (c) LED with spherical lens coupling source energy into fiber ($\eta_c \approx 8–10\%$).

improve the source coupling efficiency. Figure 22 illustrates a number of different lens arrangements that have been used and their associated source coupling efficiencies. Efficiently coupling a light source into a single-mode fiber is a difficult task because of its small core diameter and small difference in the core cladding refractive index. Mechanical tolerances are more stringent for small-core fibers, and careful alignment is needed between the emitting area of the source and the fiber core. Table VII is an estimate for the power launched by surface (SLED) and edge (ELED) emitting LEDs and ILDs into both single-mode and multimode fibers.

TABLE VII.   Launched Power in dB min

Diode	Wavelength (μm)				
	0.85	1.3		1.55	
	MMF[a]	MMF	SMF	MMF	SMF
SLED	−15	−17	−34	−18	−35
ELED	−9	−11	−22	−12	−23
ILD	+8	l 5	+3	+4	+2

[a] MMF, multimode fiber; SMF, single-mode fiber.

B. SPLICES AND CONNECTORS

The practical implementation of optical fiber communication systems requires the use of interconnection devices such as splices or connectors. A connector, by definition, is a demountable device used where it is necessary or convenient to easily disconnect and reconnect fibers. A splice, on the other hand, is employed to permanently join lengths of fiber together. The losses introduced by splices and connectors are an important factor to be considered in the design of a fiber-optic system, since they can be a significant part of the loss budget of a multi-kilometer communication link. In this section we will divide losses of splices and connectors into two categories, as shown in Table VIII. The first category of losses is related to the technique used to join fibers and is caused by extrinsic (to the fiber) parameters such as transverse offset between the fiber cores, end separation, axial tilt, and fiber end quality. The second category of losses is related to the properties of the fibers joined and is referred to as intrinsic (to the fibers) splice loss. Intrinsic parameters include variations in fiber diameter (both core and cladding), index profile (α and Δ mismatch), and ellipticity and concentricity of the fiber cores.

TABLE VIII.   Splice Loss Factors

Extrinsic
  Transverse offset
  Longitudinal offset
  Axial tilt
  Fiber end quality
Intrinsic
  Fiber diameter variation
  α Mismatch
  Δ Mismatch
  Ellipticity and concentricity of fiber core

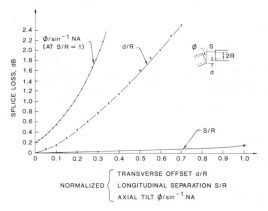

FIG. 23.   Splice loss due to extrinsic parameters.

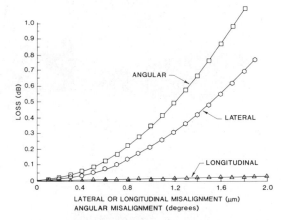

FIG. 24.   Extrinsic splice loss of single mode fibers.

Figure 23 compares the relative influence on splice loss of the major extrinsic parameters of transverse offset, end separation, and axial tilt, for multimode graded-index fibers. Splice loss is significantly more sensitive to transverse offset and axial tilt than it is to longitudinal offset. For example, a transverse offset of 0.14 core radii or an axial tilt of 1° (for a fiber with NA = 0.20) will produce a splice loss of 0.25 dB. A longitudinal offset of one core radius will produce a loss of only 0.14 dB. Fiber end quality has a minimal effect on splice loss if proper fracturing or grinding and polishing end-preparation techniques are used in conjunction with an index-matching material. A matching material with a refractive index approximately the same as that of the core is used to reduce the Fresnel reflection loss caused by the glass–air interfaces between the coupled fibers of a joint.

Figure 24 compares, for single-mode fiber, the effects of transverse offset, end separation, and axial tilt on splice loss. Notice that very small (fractions of a micrometer) transverse offsets or axial tilts (fraction of a degree) can cause a significant amount of splice loss (>0.1 dB). The mismatch of intrinsic fiber parameters can also significantly affect the loss of a splice. For multimode graded-index fibers, splice loss is most sensitive to a mismatch of core diameter or delta. A normalized core radius or delta mismatch of 0.1 will produce a splice loss of approximately 0.2 dB. For single-mode fibers, the splice is most sensitive to a mismatch in mode field diameter. A mismatch in the mode diameter of 0.5 μm can cause a splice loss of 0.05 dB. To minimize the effect of intrinsic parameters on splice loss, tight manufacturing tolerances on the fibers used in a low-loss communication system must be maintained.

There is a variety of techniques that have been developed to interconnect fibers. Table IX describes some of the salient features of three of the techniques that will be illustrated in this text.

Figure 25 is a good example of a passively aligned plug and alignment sleeve type of connector. This molded biconical plug connector is widely used, with both single-mode and multimode fibers, as part of a jumper cable for a vari-

TABLE IX.   Characteristics of Interconnection Methods

Interconnection method	Alignment technique		Single-mode fiber	Multimode fiber	Connection	
	Active	Passive			Single	Multiple
Plug and Alignment sleeve	X	X	X	X	X	
V-Groove alignment		X	X	X	X	X
Fusion	X	X	X	X	X	

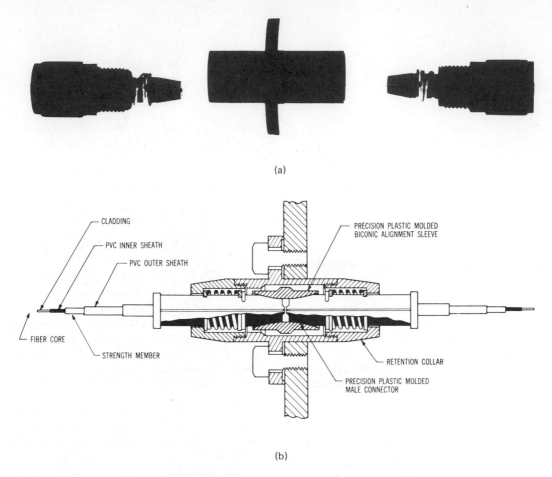

(a)

(b)

FIG. 25. (a) Single-fiber biconical connector. (b) Cross section.

ety of central-office applications. The heart of this connector is a biconical sleeve that accepts two plugs and aligns the axes of the fiber ends that are centrally located in these plugs. An inherent advantage of the conical alignment configuration is that virtually no abrasive wear occurs between the mating parts until the plug is fully seated within the biconical alignment sleeve. Losses of less than 1.0 dB are usually obtained with this type of connector. Another example of a plug and sleeve splice that illustrates the concept of active alignment of fibers is shown in Fig. 26. For single-mode splices, micrometer-type extrinsic and intrinsic tolerances must be maintained to achieve low-loss connections. To avoid putting very stringent constraints on both the fibers and the parts used to interconnect them, the technique of active alignment of the fiber cores while monitoring a signal related to transmission loss is used. Single-mode

fibers mounted in precision glass cylindrical plugs are actively aligned in an eccentric sleeve to produce the very-low-loss (<0.1 dB) rotary splices shown in Fig. 26.

The silicon chip array splice shown in Fig. 27 is a very good example of the use of V grooves as the alignment mechanism for joining fibers.

FIG. 26. Rotary splice for single-mode fibers.

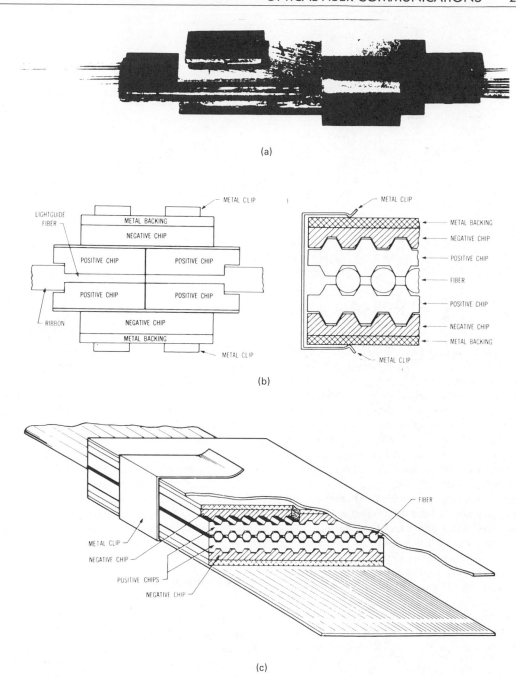

FIG. 27.   Multiple-fiber silicon chip array connector. (a) 12-Fiber silicon chip array connector. (b) Side view and partial cross section of array connector. (c) Cross section of array connector.

This splice is used in conjunction with ribbon-type structures to simultaneously splice groups of optical fibers (multiple-fiber splice). This reenterable splice consists of two array halves, two negative chips with metal backing, and two spring clips. An array half is formed by permanently affixing two positive preferentially etched silicon chips to the end of a ribbon. The fiber ends are then simultaneously prepared by grinding and polishing the end of the array. The splice halves are usually assembled on a cable prior to shipment from a factory, although they can be assembled in the field. A craftsperson assembles a splice by simply aligning two array halves with

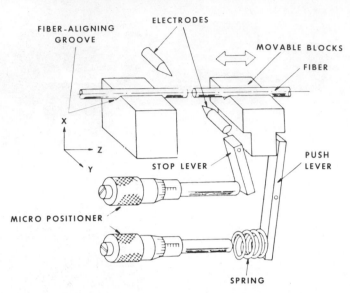

FIG. 28. Schematic showing fusion splicing using an electric arc.

two negative chips. The splice is held together with spring clips, and a matching material is inserted to complete the connection. The reenterable array splice can be disconnected and reconnected in the field, although it was not designed for use where many connect–disconnect operations are required. The average loss of the silicon chip array splice is less than 0.2 dB for multimode fibers and less than 0.6 dB for single mode fibers.

Fusion splicing, illustrated schematically in Fig. 28, is a widely used method for permanently joining individual fibers together. Fusion splicing, or welding as it is sometimes called, is accomplished by applying localized heating at the interface between two butted, realigned fiber ends, causing them to soften and fuse together. Commercially available fusion splicing machines produce splices with average losses of less than 0.3 dB (active alignment) with single-mode fibers and less than 0.2 dB with multimode fibers.

VII. Applications of Optical Fiber Communications

The unique advantages of lightwave communications—the tremendous information-carrying capacity, freedom from electromagnetic interference, the light weight of fibers, and the relative chemical and high-temperature stability of fibers—suggest a communication revolution will occur in the near future. Already the telecom-

munications market for optical fibers has exploded in the United States, and unprecedented demand for fibers has been experienced. Within a short space of 5 years, from 1980 to 1985, the total U.S. circuit mile capacity installed with optical fibers far exceeded the total existing U.S. circuit mile capacity using copper cables in 1980. So far the field of fiber optics has been technology-driven, but the future will be application-driven. Lightwave application may be broadly classified into telecommunications applications, industrial applications, military application, medical applications, and sensor applications.

A. TELECOMMUNICATIONS APPLICATIONS

By far in terms of product volume and revenues the telecommunications application has been the major market. The need for greater circuit capacity coupled with the problem of congested duct space in the cities led to the initial applications of optical fiber communication for the interoffice trunk in the big cities (Fig. 29). However due to the tremendous progress made in reducing the transmission loss and increasing the bandwidth (information carrying capacity) of the optical fibers, they are now the economic choice for most telecommunication applications. Today the biggest market for fiber is the long-distance application. Japan has already completed its long distance routes using single-mode fibers. In the United States, thousands of

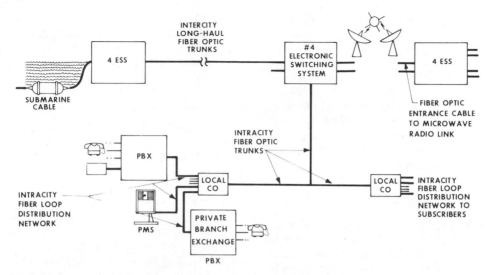

FIG. 29. Optical fiber telecommunications application.

miles of lightguide cables are being installed by AT&T, MCI, GTE, and several other long-distance companies, as well as regional Bell and independent operating companies. The majority of these companies use single-mode fibers that operate at 1300 nm wavelength, with future up gradability to 1550 nm, when reliable sources and electronics are developed. Already commercial systems have been installed with bit rates as high as 565 Mbit/sec, which is equivalent to 7680 two-way conversations over a pair of fibers. Another application for the fibers is in the feeder loop, that is, a cable that is used between a telephone central office and the local distribution interface. Since fibers have very high capacity and can transmit voice, data, and video, efforts are underway to install fibers into individual houses. AT&T has announced its intention to develop and offer a universal information service (UIS) to satisfy all the video, voice, and data communications needs of a subscriber. One of the most attractive features of the universal information system is the use of interactive systems. Systems are already in use that provide continuing interactive service for smoke and heat detectors to automatically alert fire departments of potential or active fire hazards, police alarms, and medical alert alarms to summon aid. In addition to the above, all the other information services can be obtained on demand. The potential for completely automating all the control needs of a household is there. Several trial projects have been implemented in Germany, France, Japan, the United Kingdom, Canada,

and the United States. However, the information needs of individual households have not yet reached a level where fibers can be economically justified. However, definite potential exists for an information outlet in each household, which could be used for a multitude of services such as video phone, television, news, banking, and time-share computing.

B. LOCAL AREA NETWORKS

Networks using optical fibers to transmit voice, video, and data within a building or within industrial and university campuses have been offered by several vendors. These systems are called local area networks (LANs). For economic reasons, many of today's LANs use a fiber with high (higher than that used for trunking applications) numerical aperture and operate with light emitting diodes working at 1.30 μm. Fiber-optics LANs can improve the communications inside a high-use area, reduce the bulk of copper cables, and in many cases eliminate congestion in computer rooms by remoting peripheral equipment. LANs are likely to be installed in most office buildings, industrial facilities, university campuses, and military bases. This is potentially one of the biggest future markets for fiber-optic applications.

C. INDUSTRIAL APPLICATIONS

Industrial applications can be of different types, such as process control, manufacturing automation, and energy management and application. Applications for process control are in nuclear, petrochemical, chemical, and food industries. Applications for manufacturing automation are for numerical control of machines and large data systems. Applications for control exist in airways, highways, shipping, railways, gas and oil transportation, and distribution control. Although the information rates needed for such application are low compared to fiber capacity, fiber-optic control systems have several major advantages over the electronic wire systems. These include freedom from electromagnetic interference, improved signal quality, large potential for upgrading, and safety in a hazardous environment such as oil or gas. Also, remote control can be easily achieved. In the case of power failures, fiber-optic systems can be easily maintained with auxiliary power sources.

D. COMPUTER APPLICATIONS

Computer applications are the other part of industrial use of fiber optics. Fibers are ideally suited for internal links that require very high data rates, of the order of gigabits per second (Gbit/sec). Auxiliary equipments require lower data rates and hence can be handled both by fibers or copper wires; however, fibers offer the added advantage of longer-distance network and error-free operation, because fiber transmission is unaffected by the electromagnetic noise. The fibers will gradually be used in greater volume as inter- and intracomputer links.

E. MILITARY APPLICATIONS

Three attributes of fiber optics—security, freedom from electromagnetic interference, and light weight—make fibers very attractive for use in the defense applications. Fibers can be used for data busing in aircrafts with tremendous reduction in weight; thereby increasing the range of these aircraft, along with increased information capacity and more secure and error-free operation. Such aircraft include surveillance aircraft, attack aircraft, space vehicles, and strategic air command bombers. Similar fiber applications exist for inter- and intraship communications, submarine mobile command centers, ship-to-satellite communication links, and all types of missile guidance systems. For military terrestrial applications, all communication within any base could be a local area network and could include missile-center links and

ground-to-air and ground-to-ship communications. All the above types of applications exist today, making the miliary applications a large market for fiber optics communications.

F. MEDICAL APPLICATIONS

In the medical field there are two distinctly different types of fiber-optics systems being offered. The first type takes the form of a LAN that will satisfy all the communications needs for a hospital, from keeping patient records to the energy-management and financial-management systems. The second category of medical fiber-optic systems uses fibers in diagnostic equipment and as sensors. Because of their small size and relatively benign impact upon body elements, much effort is underway to develop fiber-optic sensors for medical applications. The mechanical flexibility and small size allows the measurements of parameters inside the tissue, organs, and blood vessels. Also because of the freedom from electromagnetic interference, these instruments can be used in conjunction with other electronic monitoring systems. Three types of applications for biomedical diagnostics exists: (1) sensors for temperature, pressure, and flow of blood; (2) sensors for monitoring oxygen saturation, pH, and oxygen and carbon dioxide concentrations; and (3) sensors for immunoassay reactions. The sensors consist of fibers through which light is carried to the sample and returned back to analyzing equipment. At the end of the fiber is a suitable miniaturized transducer. The backscattered light is then analyzed to determine the velocity of moving blood cells using the Doppler shift. Spectral analysis can also be performed to measure concentration of blood constituents such as oxyhemoglobin. Concentration of gases in the bloodstream can be measured by coating the fiber end with an appropriate reagent for spectrophotometric or fluorimetric analysis. For monitoring immunoassay reactions, an antibody is attached to the surface of the fiber core from which cladding has been stripped. The reaction to the antigen or reaction product is measured at the fiber output end.

G. SENSOR APPLICATIONS

Use of fiber optics for sensors has been continuously increasing. As with other applications, the sensor applications also fall into three categories: military, medical, and industrial. The biomedical sensors were discussed in the previous section. Ultrasonic sensors use the phase shift caused in the light propagating through the core through radial stresses. The other sensors used are the pressure and temperature sensors, radiation sensors, depth sensors, acoustic sensors, sensors for color and turbidity; and sensors for underwater bubble detection. In general, the sensor system consists of an electronic control module, sensor head, and fiber-optic cable. The sensor head senses the pressure, temperature, velocity, and reaction and converts it into a change in the optical signal, which is then analyzed to measure the desired change by the electronic control unit. The field of fiber-optics sensors has expanded to the extent that separate conferences are held on fiber-optic sensors.

BIBLIOGRAPHY

Barnoski, M. K. (ed.) (1976). "Fundamentals of Optical Fiber Communications." Academic Press, New York.

Bendow, B., and Shashanka, S. M. (ed.) (1979). "Fiber Optics: Advances in Research and Development." Plenum Press. New York.

Cherin, A. H. (1983). "Introduction to Optical Fibers." McGraw Hill. New York.

Li, T. (ed.) (1985). "Optical Fiber Communications." Academic Press. New York.

Midwinter, J. E. (1979). "Optical Fibers for Transmission." John Wiley & Sons. New York.

Miller, S. E., and Chynoweth, A. G. (ed.) (1979). Optical Fiber Telecommunications." Academic Press. New York.

Suematsu, Y., and Ken-ichi, I. (1982). "Introduction to Optical Fiber Communication." John Wiley and Sons. New York.

PACKET SWITCHING

Robert W. Stubblefield *AT&T Bell Labs*

GLOSSARY

Data-circuit-terminating equipment (DCE): Point within a communications network to which a terminal outside the network is connected.

Data terminal equipment (DTE): Computer terminal (which may be a computer itself) that is connected to a communications network.

Logical channel: Virtual, rather than physical, communications channel that is created by labeling each unit of information in a physical channel with the virtual channel's name (logical channel address) to distinguish it from other units of information in that physical channel.

Packet assembler/disassemblers (PAD): Used to take information from various physical communications channels, break the information into units called packets, label the packets, and funnel them out on a different physical channel. (They are used to do the inverse for information flowing in the opposite direction.)

Permanent virtual circuit (PVC): A logical connection provided by a network between two devices that needs to be established (by administrative procedures) only once.

Protocol: Detailed conventions used by one telecommunications device communicating with another such device.

Virtual calls (VC): Temporary logical connections between two devices that a network establishes and clears when instructed to do so by the devices.

Window size: Number of units of information that a terminal may send without waiting for confirmation that they have been received or permission to send more.

Packet switching is a technique used in telecommunications in which the information to be transmitted between two points is broken into packets, which are labeled to allow them to be switched to the appropriate destination. This technique allows communications resources to be shared in a statistical fashion among several conversations. It is to be contrasted with other sharing techniques (e.g., time division multiplexing) that reserve some fraction of a physical channel to be used exclusively for one conversation. It is also to be contrasted with circuit switching, in which information on one physical circuit is switched to another physical circuit. The packet switching function might be distributed across all the communicating devices, as in packet radio or many local area networks (LANs) or the devices might send packets to a centralized device—the packet switch—to do the switching of each packet to the right physical and logical destination.

I. Motivation

A. PACKETIZATION

Data streams were packetized for other reasons before it was seen that packets could be used as a unit of switching. The value of breaking a stream of information into units is particularly evident when the stream is data being exchanged between computer devices. In voice telecommunications, the people at either end of the conversation can correct for minor losses in the signal introduced by the telecommunications medium. For example, if some noise on the line causes "The weather is great" to come across as "The weath[static]s great," the listener most likely has enough contextual information to know what was sent; in the worst case, the listener asks the speaker to repeat the message. If the information being exchanged between computers is "The new bank balance is $1,000" and it comes across as "The new bank balance is $[static]000," some equivalent error correction scheme is needed. Packetization facilitates the efficient detection and correction of data errors. The techniques discussed here allow each packet in a data stream to be checked for errors and retransmitted when necessary.

B. PROTOCOLS

When people communicate, they do so from a broad base of shared information. Essential to communication is a common understanding of some basic communication rules. For example, one convention is that both parties should not talk at the same time; another is that questions are asked (usually) with the expectation that the listener will answer. The protocols that define the rules of communicating between computers also need to be commonly understood by the programmers of each machine. The mutual agreement to use published standard protocols for packet switching is the means used to ensure such an understanding.

C. PACKET SWITCHING

Packet switching grew from a need to conserve data communications resources. It is based on the observation that reserving a communications channel for a conversation that has lapses of silence is wasteful. Packet switching allows such gaps to be filled with data of another conversation. Packetization allows the units of the various conversations to be switched to the appropriate destination. Terminals may benefit from being on the same packet switch with locations with which they do not communicate by getting economy-of-scale cost advantages on their communications expenses. Larger packet switches are less expensive than smaller ones when the cost is measured per packet switched. Another reason for accessing a common packet switch is increased flexibility: terminals that do not communicate with each other today might do so tomorrow. If they are on the same packet switch, no new physical connections need be made to allow such communication.

D. NETWORKS

Although many benefits of packet switching are obtained with a single packet switch, packet switches are frequently interconnected with communications links (called trunks) to form packet switching networks. Such a network provides an additional benefit: each device can have a shorter (and therefore less expensive) access line to the network. The purpose of a packet switching network is to make an intercon-

nected set of packet switches have the same functionality as a single packet switch at lower cost.

II. Techniques

A. UNITS

1. Bit

The atomic unit of data communications is the smallest unit of information, a binary digit or bit, which has one of only two possible values—zero or one. Although there are sophisticated multi-value transmission techniques, all data communications are measured in terms of bits. A common measure of the speed of data communications is bits per second. The sender and receiver must use the same protocol to agree on what various sequences of bits mean. Some protocols define sequences of eight bits to be a unit known as a byte. For example, the even-parity, eight-bit sequence for the letter "A" as defined by the American Standard Code for Information Interchange (ASCII) is 01000001. The first bit (the leading zero) is a "parity" bit. If the communicators agree to use even parity, this bit will be zero when the number of bits that are one is even, and it will be one if the number of one bits is odd. This algorithm means that every eight-bit byte should always have an even number of ones. Thus, if there is an error in a single bit, it can be detected.

2. Frame

Another important unit for data communications is the frame. A frame is a sequence of bits representing an integral unit of information delimited by a defined pattern. Data to be transmitted across a communications link is framed with other information according to an agreed protocol. Some of the framing data may define the length of the frame, some may be added to control the flow of information, and more may be added to detect transmission errors. The protocol defines (1) the pattern of bits to mark the start of the frame, (2) some header and trailer information that surrounds the string of data bits whose integrity is to be preserved across a single communications link between two terminals, and (3) the procedures to be followed to transmit the information reliably.

3. Packet

The basic unit for packet switching is the packet. Packets are similar to frames in that they are sequences of bits representing an integral unit of information delimited by defined patterns. The difference

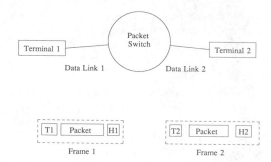

FIG. 1.   A packet is carried in different frames between two terminals. H1, frame 1 header information; T1, frame 1 trailer information; H2, frame 2 header information; T2, frame 2 trailer information.

is that frames are used to preserve the identity and integrity of information across a single communication link; packets are used across the whole communications path, end to end, from the sender to the receiver. A packet of data is contained within a single frame across any communications link; but when it goes to the next link, a different frame may carry it. The relationship between frames and packets is shown in Fig. 1.

B. Protocol Layers

The idea of protocol layers was introduced to provide a hierarchical structure for the many parameters and functions that must be defined in data communications. The International Organization for Standardization has defined a seven-layer architecture called open systems interconnection (OSI). Because the interfaces among the components (layers) are published, the architecture is open to any vendor who builds to the standards, and thus, the products of different vendors may be combined into one system. The lower four layers of the OSI architecture define how to transfer information (as a stream of bits) between remote systems. Packet switching is a particular technique that is applicable to the third layer and depends on the lower two layers accomplishing their respective functions.

1. Physical Layer

The lowest layer is the physical layer. This layer gives the functional and procedural characteristics to activate, maintain, and deactivate the physical circuit. It is the level for defining what physical manifestation is used to represent the bits of information. For example, the standard known as RS232 defines a "zero" as a voltage in the range of $+3$ to $+25$ V and a "one" as a voltage in the range of -25 to

-3 V. This same standard defines a 25-pin connector, its physical dimensions, and which leads are for signals, data, ground, etc. In the United States, the RS232 standard is commonly used for data transmission rates of up to 20,000 b/sec. Another standard, V.35, is commonly used for rates of 56,000 b/sec.

2. Data Link Layer

The second layer is the data link layer. The purpose of the data link layer is to achieve the error-free transfer of information over the physical circuit. This layer defines the patterns of bits that make up a frame of data and specifies the allowed states and transitions of the communications processes at each end of a telecommunications link. States and transitions are illustrated by the following example. When two people speak, protocol (politeness) calls for one to be in a talking state and the other to be in a listening state. The transition to a different state might be signaled by a pause by the speaker. In radio communications, the protocol has been made more explicit with the sender saying "10-4" or "over" to allow the other to speak.

One data link layer protocol defined for packet switching is link access protocol balanced or LAPB. In this protocol, each communications process has many more states and transitions than the simple speaking and listening states just described. For example, one system might be able to transmit more information than the other can receive and decode in some period of time. LAPB allows the receiver to send a special signal known as receiver not ready (RNR) to tell the sender to stop sending until the receiver sends a receiver ready (RR) signal.

Reliability of the data link layer protocol LAPB is achieved by use of a frame check sequence (FCS). This is additional information (16 bits) added to each frame that is a function of the other information in the frame. The sender computes it and inserts it in the frame; the receiver uses the other information in the frame to compute what it should be and checks to see if the computed FCS matches the sent FCS. If it does not match, it is assumed that the information was corrupted during transmission, and the receiver may request the sender to retransmit the frame. (The sender may decide to retransmit the frame because the receiver did not acknowledge receiving it within a previously specified period of time.) Packet switching services use this technique to achieve a very low error rate of one packet in error in 100 million.

The data link layer may be used to control the flow

of information between a terminal (DTE) and the network (DCE). LAPB defines a parameter the DTE and DCE must agree on to specify the maximum number of unacknowledged frames that one can send to the other. After sending that many, the sender must wait for an acknowledgment before sending another. This parameter is sometimes called the level 2 window size and is typically set to 7.

The value of a layered protocol architecture is illustrated by the fact that the same LAPB data link layer can be used for 9600-b/sec data with an RS232 physical layer interface and for 56,000-b/sec data with a V.35 physical layer interface. The designers of the data link layer functions only need to know which physical layer interface is used, not how it works. For example, they need not know what voltage ranges represent ones and zeros.

3. Network Layer

The third layer of the OSI reference model is the network layer. This layer provides the means to establish, maintain, and terminate connections that may span a network of data links. Packet switching is a specific technique for implementing the network layer with virtual, rather than physical connections. The OSI packet switching protocol commonly used for the network layer is X.25 (see Section II,c). (In fact X.25 defines a couple of data link protocols and specifies that it will work with an enumerated set of physical layer protocols in addition to defining the network layer functions.)

These lower three layers deal with the transmission of information from one device to another without any dependence on the meaning of that information. The upper three layers (session, presentation, and application) define what that information means (such as a request for an airline reservation or a computer-to-computer file transfer) and are oblivious to how the information is transmitted (whether by a direct communications line or by storing the information now and forwarding it later).

4. Transport Layer

Between the upper three and lower three layers is the transport layer. The purpose of this fourth layer is to allow the upper levels to specify the appropriate quality of service for reliable transfer of information between the devices. Thus, an airline reservation application might have the transport layer choose a network layer connection that has low delay (such as packet switching), while a file transfer application might have the transport layer choose a network layer connection with large bandwidth (such as a circuit switched network).

Because packet switching is concerned with the lower three layers of the OSI reference model, the remaining layers will be described briefly.

5. Session Layer

The fifth layer is the session layer, which manages the dialogue between communicating systems; it maintains the state of the dialogue even over data loss by the transport layer. For example, the session layer could request the transport layer to reestablish a connection after a transport layer failure.

6. Presentation Layer

The sixth layer is the presentation layer, which controls the encoding of the information being transmitted. For example, it is at this level that the ASCII coding of 01000001 for the letter A would be defined.

7. Application Layer

The seventh and uppermost layer is the application layer, which provides the means for the software application programs to access the OSI environment. An example of a function provided by this layer is a "login/password" capability used to authenticate the identity of a process.

C. X.25—A PACKET SWITCHING ACCESS PROTOCOL

The X.25 protocol is the standard protocol defined for dedicated access to a public data network by terminals operating in the packet mode. This standard is defined in documents published by the Consultative Committee for International Telegraph and Telephone (CCITT), which are updated by international agreement every four years. The X.25 protocol points to other standards to define acceptable physical layers (such as RS232 and V.35) and specifies the details of the data link protocols with which it will work (LAP and LAPB). The network layer protocol functions are achieved by X.25 at the packet level; for example, a specific type of packet is defined to signal the network that a terminal would like to establish a connection. Two types of connections are defined: virtual call service and permanent virtual circuit. Both of these services are built on the important concept of the logical channel.

1. Logical Channel

A logical channel for communicating over a given physical path is implemented by augmenting each packet that traverses the path with some additional data known as the "logical channel number." A circuit switch always switches a signal (within a single conversation) from one physical connection to a single other physical connection. Multiplexing techniques allow multiple conversations over a single physical link by using a fraction of a physical link (sometimes a range of frequencies, as in frequency division multiplexing, and sometimes a range of time, as in time division multiplexing) of the physical link for each individual conversation. Most multiplexing techniques reserve a fixed fraction of the physical link for each conversation that might take place. For example, a physical link might carry a single conversation of 48 kb/sec or 5 different conversations of 9.6 kb/sec. A logical channel is identified "logically" by the information itself carried within the channel, rather than "physically" by some predetermined fraction of the channel.

The logical channel allows conversations to be statistically multiplexed. Instead of a fixed number of conversations at fixed rates, a variable number of conversations at various rates can share the physical channel. As long as the total average data rate needed by all conversations is less than the data rate the physical channel can provide (neglecting complications of overhead and congestion that are discussed later) and each conversation can be uniquely identified by its logical channel identifier, any number of conversations can be supported. A terminal subscribes to the network for a specific number of logical channels. Some of these channels may be specified for a virtual call service and others may be specified for a permanent virtual circuit service. There may be as many as 4096 logical channels (including channel 0) defined if the network and terminal agree. (More logical channels give the benefit of more simultaneous conversations at the expense of using more computer memory in both the terminal and the network to keep track of those conversations.) The range of logical channels that can be used for virtual calls and permanent virtual circuits is shown in Fig. 2.

The data link layer technique of defining a window size to restrict the transmission of unacknowledged frames controls the total flow across a data communications link. At the network layer, this technique can be used on individual logical channels to make sure that no conversation consumes the whole capacity of the transmission facility. A typical value for the packet level (network layer) window size is two packets.

2. Virtual Call Service

The X.25 virtual call (VC) service defines procedures to establish a virtual (rather than physical) con-

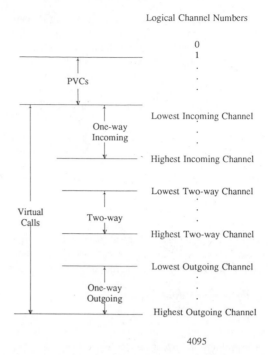

FIG. 2.   Ranges for logical channels to be used for different virtual circuit services.

nection between two terminals via a packet switched network. Four packet types are defined to establish a connection and four others to terminate the connection. For call establishment, the caller sends the network a call request packet identifying the caller terminal (using the available logical channel with the highest number in the range agreed to be used for such purposes by the network and that terminal). The network sends an incoming call packet to the called terminal (using the available logical channel with the lowest number). (Note that these logical channel numbers have only a local significance, i.e., they have meaning between a terminal and its nearest edge of the network.) The called terminal signifies acceptance of the call by returning a call accepted packet. The call connection is actually established when the network signals the caller with a call connected packet.

A terminal may terminate the connection by sending the network a clear request packet. The network acknowledges the request and notes that the associated logical channel is ready to be reused for another call by returning a DCE clear confirmation packet. The network tells the terminal on the other end of the conversation of the termination by sending it a clear indication packet. That terminal acknowledges and notes freeing of the logical channel resource by returning a DTE clear confirmation packet. The called terminal could have refused to accept the original call by answering the incoming call packet with a clear request packet. While the terminals are connected, they can exchange data by sending it to the network in DTE data packets; the network transmits the data to the other terminal in DCE data packets.

Other types of packets allow control of the flow of information on an individual virtual connection (DCE RR, DCE RNR, DTE RR, and DTE RNR), resetting of an individual virtual connection (reset request, reset indication, DTE reset confirmation, and DCE reset confirmation), restarting all virtual connections simultaneously (restart request, restart indication, DTE restart confirmation, and DCE restart confirmation), reporting of error conditions (diagnostic), and a determination of what options a terminal has registered with the network (registration request and registration confirmation).

3. Permanent Virtual Circuit Service

In addition to the virtual call service, X.25 defines a permanent virtual circuit (PVC) service. This service provides a virtual connection between two terminals that is identical to the VC except that the call establishment is not done by dynamically choosing a ready logical channel at the request of a terminal. Instead, the logical channels are assigned when the terminals subscribe to the service. The PVCs allow a packet switching service to emulate many of the features of a customer's private data line service. The PVCs have the virtue of being accessible only by the terminals defined to use them; no other terminal can inadvertently connect by "dailing a wrong number." The price a terminal pays for this privacy is the inability to establish connections with other terminals on short notice. Of course, this privacy is ineffective if that terminal has some logical channels available for virtual calls. Among the many optional user facilities defined by X.25 is a set designed to achieve privacy for virtual calls.

4. X.25 Optional User Facilities

A set of closed user group (CUG) optional facilities enables users to form groups of terminals to and from which access is restricted. Terminals may belong to multiple closed user groups and may be connected to other terminals in compatible groups.

Other optional facilities can restrict the usage of particular terminals. Incoming calls barred prohibits a terminal from receiving any calls, and outgoing calls barred prevents a terminal from originating any calls.

Some optional facilities are associated with billing for services. For example, reverse charging allows a

TABLE I.   X.25 Optional Facilities

On-line facility registration
Extended packet sequence numbering
D-bit modification
Packet retransmission
Incoming calls barred
Outgoing calls barred
One-way logical channel outgoing
One-way logical channel incoming
Nonstandard default packet sizes
Nonstandard default window sizes
Default throughput classes assignment
Flow control parameter negotiation
Throughput class negotiation
Closed user group
Closed user group with outgoing access
Closed user group with incoming access
Incoming calls barred within a closed user group
Outgoing calls barred within a closed user group
Closed user group selection
Closed user group with outgoing access selection
Bilateral closed user group
Bilateral closed user group with outgoing access
Bilateral closed user group selection
Fast select
Fast select acceptance
Reverse charging
Reverse charging acceptance
Local charging prevention
Network user identification
Charging information
Recognized private operating agency selection
Hunt group
Call redirection
Called line address modified notification
Call redirection notification
Transit delay selection and indication

terminal to request that the called terminal get the bill; reverse charging acceptance allows a terminal to acknowledge that it will accept reverse charged calls. The network user identification facility gives a network information it can use to identify the billed party. The charging information facility allows a terminal to discover what the network has charged for a call. The complete list of X.25 optional facilities is shown in Table I.

D. X.75—Packet Switching Internetworking Protocol

Just as the protocol used for terminals to communicate with a network has to be rigorously defined, so does the protocol that two networks use to communicate with one another. There are multiple X.25 networks for political and economic reasons. Canada, France, Germany, Spain, and other nations each have their own. AT&T, GTE, Tymnet, and several local telephone companies each have public X.25 networks in the United States. The X.75 protocol is used to interconnect two networks. If the interconnections exist, a terminal on one network may talk with a terminal on another network.

The call establishment procedures and the facilities of X.75 are much like those of X.25 except that the communication is between two signal terminating equipments (STEs) on behalf of DTEs connected to different networks. The STE is the CCITT name for the software and equipment in one network that is connected to the physical transmission facility that goes to the other network. The connections between two terminals that subscribe to two different networks are shown in Fig. 3. Call establishment would involve the following sequence of packets: terminal A sends an X.25 call request packet to network 1; network 1 sends an X.75 call request packet to network 2; network 2 sends an X.25 incoming call packet to terminal B; terminal B sends an X.25 call accepted packet to network 2; network 2 sends an X.75 call accepted packet to network 1; and, finally, network 1 tells terminal A the connection is established with an X.25 call connected packet.

A public data network might even be an intermediary network between two other networks; that is, neither the calling nor the called terminal is on the network. Such a network is called a transit network. Traffic carried within the network for such connections goes from one STE to another.

E. X.121—ADDRESSING CONVENTIONS FOR PUBLIC DATA NETWORKS

One of the problems an X.25 network has to solve is to locate the terminal the caller wants to call. Part of this problem is solved by the X.121 standard, which specifies identifying DTEs with a number of up to 14 digits. The first four digits identify which network the terminal is connected to, and that network uses the remaining digits to locate the terminal. Just as telephones have telephone numbers, terminals have data communications numbers; and just as you have to know a telephone number to place a call, a computer has to be programmed with the number of a terminal it will call.

F. NETWORK IMPLEMENTATION CHOICES

1. Numbering Plans

There is no standard that tells a public data network how to use the remaining network address digits to locate a terminal. This is one of the decisions a network provider must make in implementing an X.25 packet switching service. One such implementation is to use the next six digits (after the data network identification code—the four digits mentioned previously) to identify a packet switch within the network and the last four digits to identify an access line connected to a DTE. The digits identifying the packet switch might have geographic significance as telephone area codes do. The scheme for associ-

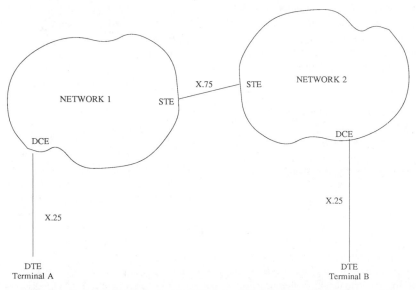

FIG. 3.   Connections between two terminals on different packet switching networks.

ating the "logical" network number with a "physical" access line is known as the numbering plan.

2. Internal Protocols

Another implementation choice involves the protocol to be used to get information across a network. The X.25 and X.75 protocols imply what has to be done but do not define the means to do it. A network must define an internal protocol to convey information between DCEs, between STEs within the network, or between a DCE and an STE. Because the logical channels used at the two ends of a conversation are independently chosen, the network must provide some means of associating them. The internal protocol can make this association by appending additional information to the packet. An internal protocol might be a switch-to-switch protocol (where the protocol work done in transferring a packet from each switch to the next is like that done in getting the packet to the network) or it might be edge-to-edge. In the latter implementation, the switches at the edges of the network have the responsibility of maintaining the integrity of the virtual connection, and the intermediate switches do the simpler work of merely relaying the packets. The various choices made in implementing a packet switch and a network of packet switches affect the performance seen by the connected terminals.

III. Packet Switching Performance

A. DELAY

One of the most important measures of packet switching performance is that of delay. Delay is defined for several different contexts. Cross-network delay is the amount of time a packet takes from the time it enters the network until the time it leaves the network. Such delays are typically in the hundreds of milliseconds; for example, AT&T's ACCUNET® Packet Service has an average cross-network delay of less than 200 msec.

The delay a packet switching customer sees is known as the response time. This delay consists of two cross-network delays plus the delay contributed by the terminal at the other end of the conversation. For most interactive applications, the response time is on the order of a few seconds. An example of such an application would be a reservation system where the response time would be the delay between the request for information (such as what rooms are available for a given date at a given location) and the display of that information on the requester's terminal.

The most stringent requirement for response time, which is used by a few applications, is known as echoplex. In such applications, a symbol typed on the terminal console does not appear on the terminal display until it is echoed by the remote terminal (a host computer) at the other end of the network. Such delays should be less than 250 msec (one-fourth of a second) to be acceptable to experienced typists. Currently, the best packet switching networks are only marginal performers based on this criterion.

The portion of response time that is within the control of packet switching networks is the cross-network delay. The components of this delay are the sum of the cross-switch delays, the propagation delay, and the sum of the insertion delays, each of which is discussed in the following.

Cross-switch delay is the amount of time a packet takes from the time the packet enters the switch until the time it leaves the switch. Cross-switch delays are typically measured in the tens of milliseconds; for example, the Western Electric 1PSS has an average cross-switch delay of 30 msec. The cross-switch delay component of cross-network delay can be reduced by having fewer (and therefore larger—if the network is to carry the same traffic) switches in the network and enough trunks between the switches so that a virtual connection between any two terminals goes through only two or three switches. The size of the packet switch has an effect on delay in another way, too. Just as your delay in getting served by a bank teller increases with the number of people who simultaneously want such service, the delay a packet sees goes up as the number of packets that must be switched increases, and just as increasing the number of tellers decreases your average wait time, increasing the packet switching capacity decreases the average packet delay.

Another component of cross-network delay is the propagation delay—the time it takes the electronic signal conveying information to get from one place to another. This signal travels at the speed of light and thus takes only about 16 msec (3000 miles/ 186,000 miles per sec) to cross the country. Network designers merely need to avoid making virtual connections by circuitous routes to keep the propagation contribution small. Note that if satellite transmission facilities are used, propagation delay becomes significant. Transmission satellites are maintained in a geosynchronous orbit 22,400 miles in altitude. Thus, the propagation delay up to and down from the satellite is about 240 msec (44,800 miles/186,000 miles per sec).

Insertion delay is the time it takes to put a packet on a transmission line. It is a function of the data rate of the physical transmission channel. A 1200-b/sec transmission line will take 1 sec to get a 2100-bit packet inserted onto it. A 9600-b/sec line will require one-eighth as long, 125 msec. The corresponding delay for a 56,000-b/sec line is about 21 msec. A network that can use trunks with a data rate of 1.536 Mb/sec could drop that delay to less than 1 msec. Network providers typically give customers several options for access line speeds. Note that insertion delays are also incurred when packets are placed on trunks between switches; therefore, networks with routing plans that involve fewer trunks and networks with higher speed trunks will have better delay performance.

B. THROUGHPUT

Another performance measure for packet switching is throughput—the amount of information that can be sent in a unit of time. For circuit switching, the throughput is the same as the data rate of the circuit being used. In packet switching, many different virtual connections may use the same physical channel; throughput is measured for each logical channel. The total throughput (the sum of that for each logical channel) will be less than the line speed because some bits are used for logical channel identification, defining packet types, error control, etc., and are not available for customer data. This header information is a smaller percentage of the data in a packet as the length of the packet increases; thus, a terminal that wants higher throughput should send larger packets.

Another effect on throughput performance is associated with the flow control parameters chosen for a virtual connection. For example, if a level three window size of 1 is chosen, 1000-bit packets are sent, and the round-trip network delay is 200 msec, then the maximum throughput on the logical channel will be 5000 b/sec.

C. RELIABILITY

Packet switching reliability refers to the ability to get uncorrupted data from one terminal to another without a loss of packets and without interruptions in the connections. The error checking techniques mentioned previously and the protocol provisions for retransmitting packets can reduce lost and misdelivered packets to less than 5 in 100 million and errored packets to less than 1 in 100 million. To prevent service interruptions from prematurely halting a conversation, a packet switching network must either restore failed transmission facilities promptly (before timers expire to bring down the virtual connection) or else be able to quickly reroute virtual connections around failed facilities. Such techniques can achieve a reset rate of less than twice per 100 million packets sent.

D. AVAILABILITY

Availability of a packet switching network means the percentage of time that one terminal connected to the network can reach any other on the network (assuming they meet any closed user group constraints and are not using all their logical channels for other conversations). To achieve high availability (for example, 24 hours a day, 7 days a week availability of 99.85%), a network must have very reliable packet switches, multiple trunks interconnecting them, and alternate routing algorithms to route around failures. If the network is always to be available, it is extremely important that new access lines and trunks (and even new switches) be able to be added to the network without taking current equipment out of service.

IV. Conclusion

Packet switching has proven to be a reliable and economic means for transmitting data among terminals connected to networks spanning a large geographic area. The same sort of techniques that allow audio signals to be stored on compact discs as digital data can be used to digitize voice telecommunications. Telecommunications service providers are interested in using their resources to provide a single communications service for both data and voice—the integrated service digital network (ISDN) concept. One means toward that goal is to use packet switching techniques for both voice and data in the future.

BIBLIOGRAPHY

CCITT Red Book (1985). "Data Communication Networks Interfaces," Vol. VIII—Fascicle VIII.3, Recommendations X.20–X.32, Geneva.

Deasington, R. J. (1985). "X.25 Explained," E. Horwood Series, Halstead Press, New York.

Schwartz, M. (1987). "Telecommunications Networks," Addison-Wesley, Reading, Massachusetts.

Seidler, J. (1983). "Principles of Computer Communications Network Design," Halstead Press, New York.

St. Amand, J. V. (1986). "A Guide to Packet-switched Value-added Networks," Macmillan, New York.

RADIO PROPAGATION

Herbert V. Hitney, Juergen H. Richter, Richard A. Pappert, Kenneth D. Anderson, and Gerald B. Baumgartner, Jr.   *Naval Ocean Systems Center*

GLOSSARY

Advection: Horizontal transport of an airmass.

Advective duct: Ducting phenomenon caused by the horizontal transport of an airmass.

Bulk meteorological measurements: Refers to the bulk aerodynamic method, which obtains low-level flux of atmospheric parameters using meteorological measurements at the surface and at a reference height (usually at 10 m).

Ducting: Channeling or guiding of electromagnetic energy.

Interference region: Region within the radio horizon where path loss is the result of constructive and destructive interference of direct and surface-reflected signals.

Inversion: Increase in temperature with altitude.

Lidar: Laser radar (acronym for light detection and ranging).

Microwave refractometer: Instrument that measures radio refractivity directly.

Multipath: Signals received along different propagation paths (and as direct and surface-reflected).

Oceanic evaporation duct: Ducting phenomenon caused by rapid decrease of humidity above the ocean surface.

Path loss: Ratio of transmitted to received power using loss-free isotropic antennas (usually expressed in decibels).

Radio horizon: Horizon for radio waves (slightly different from the optical horizon since atmospheric refractivity is different for optical and radio waves).

Radiometer: Sensitive, calibrated receiver for measuring electromagnetic emissions.

Radiosonde: Weather balloon carrying temperature, humidity, and pressure sensors, which are telemetered to a ground station.

Skywave: Propagation via ionospheric reflection (usually limited to the high-frequency band).

Subsidence: Downward transport of an airmass.

Subsidence duct: Ducting phenomenon caused by the downward transport of an airmass.

Troposphere: Lowest part of the atmosphere where temperature normally decreases with altitude.

Waveguide normal modes: Resonant solutions to the wave equation that are often useful for beyond-the-horizon field evaluations.

The propagation of radio waves for frequencies above ~30 MHz is often significantly affected by the structure of the lower atmosphere, the troposphere. Examples are "holes" in radar coverage or unusually long propagation ranges far beyond the normal radio horizon. These phenomena are caused by the vertical moisture distribution, which is often sharply layered due to temperature inversions, advection of different airmasses, subsidence, and evaporation above the water. Since nearly horizontally propagating waves intersect those layers at very small angles, small vertical moisture (or refractivity) gradients can cause significant wave front bending.

I. Introduction and Background

As soon as radars became available, unusual atmospheric propagation effects were observed. A famous example is the World War II sighting of Arabia with a 200-MHz radar from India 1700 miles away. Subsequently, extensive analytical and experimental investigations were conducted. Some of the experiments were so comprehensive and of such high quality that their results are still used today for validation of propagation models. The most dramatic propagation anomalies are encountered over water, where atmospheric ducting is more significant and consistent than over land. In addition, evaporation ducting is a persistent phenomenon found only over water. The propagation measurement and analysis effort begun in the 1940s has provided an understanding of the physical processes responsible for anomalous radio propagation. Initially, quantitative propagation assessment was limited and mathematically cumbersome. In the 1960s, increased sophistication of radars and weapons systems necessitated better ducting assessment techniques. During that period, radars were used and specially built for sensing refractive layers, their fine structure, and their temporal as well as spatial behavior. In particular, the ultra-high-resolution FM–CW radar tropospheric sounding technique settled important questions like the relative contribution of clear-air refractivity fluctuations and point targets, such as birds and insects, to the radar returns. Extensive propagation measurements, advances in the knowledge of atmospheric boundary layer processes, and the use of sophisticated mathematical modeling techniques resulted in a much improved understanding of radio propagation anomalies. The availability of small computers in the 1970s provided a near-real-time radar and radio propagation assessment capability, which in the 1980s has been expanded to include tactical decision aids for mitigation or exploitation of atmospheric propagation effects.

II. Fundamentals of Tropospheric Propagation

The mechanisms that control the propagation of radio waves in the near vicinity of the earth within the troposphere can be roughly separated into the two classes of standard and nonstandard mechanisms. Standard propagation mechanisms are those associated with an atmospheric structure that closely resembles the so-called standard atmosphere, in which the radio refractive index decreases exponentially with increasing height in a manner such that at low altitudes it decreases nearly linearly with height. The presumption of a standard atmosphere is purely for computational convenience and is based on long-term averages, usually over continental areas. A standard atmosphere should never automatically be assumed to be the most common condition, however, since there are many geographic areas where important nonstandard distributions of refractive index occur.

Spherical spreading, characterized by an inverse square falloff with range, is the only loss mechanism associated with free-space propagation. Consistent with usual practice, free-space propagation is used throughout this article as a standard against which both propagation measurements and calculations are referenced.

Within the horizon, the coherent interference of signals from the direct path between the transmitter and receiver or target with any reflected paths is very important. In a marine environment, such multipath effects are usually the result of interference of the sea-reflected path with the direct path. In this case, the effects of divergence or spreading due to reflection from the earth's curved surface and the effects of the rough sea surface on reducing the reflection coefficient become important factors in determining the total amount of interference between the two paths. The phase difference between the two paths is determined by path geometry and the phase lag angle associated with reflection from the sea surface. If there are landmasses or other large obstacles nearby, multipath from them will also be a determining factor, but these effects are beyond the scope of this article and will not be discussed further. Two-wave interference will be the dominant propagation effect when both the transmitter and receiver antennas are illuminating both direct and reflected paths and when the transmitter and receiver are reasonably well within line-of-sight. As the receiver or target approaches the radio horizon, the interference effect begins to blend with the diffraction propagation mechanism.

At ranges sufficiently well beyond the radio horizon, the dominant propagation mode is diffraction of the radio waves by the earth's curved surface. In a standard atmosphere, the diffraction mechanism is usually the limiting factor for radar applications, since the attenuation rate with range is much greater than in the interfer-

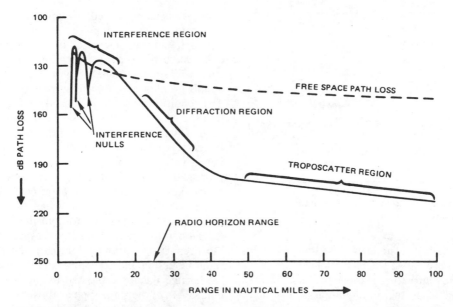

FIG. 1.   Path loss versus range illustrating standard propagation regions. The example is for terminals at 18 and 30 m and a frequency of 5.0 GHz.

ence region and the maximum propagation loss, or path loss, consistent with radar detection of reasonable size targets falls in this region. For high-power communication systems where high-gain antennas can be employed on both terminals, the dominant propagation mechanism at ranges far beyond the horizon is tropospheric scatter, or troposcatter, whereby energy is scattered from turbulent fluctuations of the refractive index in the common volume of the two antennas. The final standard propagation effect is absorption by atmospheric gases, primarily water vapor and oxygen, although the attenuation from absorption is generally negligible below 20 GHz. Figure 1 shows path loss in decibels (ratio of transmitted to received power for loss-free isotropic antennas) versus range for a 5-GHz transmitter at 18 m and a receiver at 30 m over seawater to illustrate the various standard propagation regions discussed above. [See ANTENNAS.]

Nonstandard propagation mechanisms are all associated with abnormal vertical distributions of the refractive index. If n is the refractive index, then the refractivity N is defined as $N = (n - 1) \times 10^6$. For a standard refractivity gradient, a radio ray will refract downward toward the earth's surface, but with a curvature less than the earth's radius. Standard gradients are usually considered to be from -79 to $0N$ units per kilometer of height, which are characteristic of long-term mean refractive effects for a particular area. For instance, the long-term mean gradient over the Continental United States is approximately $-39N$ units per kilometer. If the gradient exceeds $0N/km$, a radio ray will bend upwards and the layer is said to be subrefractive and have the effect of shortening the horizon. If the gradient is between -157 and $-79N/km$, the ray will still bend downwards at a rate less than the earth's curvature but at a rate greater than standard. These gradients are called superrefractive and have the effect of extending the horizon. The most dramatic nonstandard effects

TABLE I.   Refractive Conditions in Relation to N and M Gradients

Condition	N Gradient (N/km)	M Gradient (M/km)
Trapping	$dN/dh \leq -157$	$dM/dh \leq 0$
Superrefractive	$-157 < dN/dh \leq -79$	$0 < dM/dh \leq 78$
Standard	$-79 < dN/dh \leq 0$	$78 < dM/dh \leq 157$
Subrefractive	$dN/dh > 0$	$dM/dh > 157$

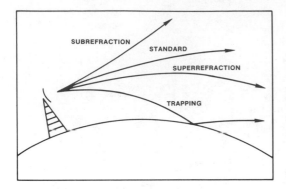

FIG. 2. Relative bending for each of the four refractive conditions.

are those caused by gradients less than $-157N/$ km, which are called trapping gradients. In this case, the ray curvature exceeds the earth's curvature and leads to the formation of ducting, which can result in propagation ranges far exceeding the normal horizon. The modified refractivity M is defined by $M = N + (h/a) \times 10^6 = N + 0.157h$, where h is height in meters and a is the earth's radius in meters. M is useful in identifying trapping gradients in that trapping occurs for all negative M gradients. Table I lists the four refractive conditions discussed above and their relation to N and M gradients, and Fig. 2 illustrates the relative curvature for each.

In a marine environment there are three distinct types of ducts caused by trapping gradients, namely, surface-based ducts, elevated ducts, and evaporation ducts. Surface-based

ducts are usually created by trapping layers that occur up to several hundred meters in height, although they can be created by a trapping layer adjacent to the surface (sometimes referred to simply as surface ducts). Figure 3 shows the N and M profiles for a typical surface-based duct. The condition for a surface-based duct to exist is that the M value at the top of the trapping layer be less than the M value at the surface. These ducts are not particularly sensitive to frequency, being able to support long over-the-horizon propagation ranges at frequencies above 100 MHz. Surface-based ducts occur with annual frequencies of up to 50% in areas such as the Eastern Mediterranean and Northern Indian Ocean and are the type of duct responsible for most reports of extremely long over-the-horizon radar detection and communication ranges.

Elevated ducts are created by elevated trapping layers of the same type as those that create most surface-based ducts. However, in this case, either the layer is too high or the M deficit across the trapping layer is too small to meet the condition previously stated to form a surface-based duct. Figure 4 illustrates the N and M profiles required for an elevated duct. The vertical extent of the duct is from the top of the trapping layer down to a height where the M value is equal to the M value at the top of the trapping layer. Elevated ducts can also affect propagation for frequencies above approximately 100 MHz but the effects are usually limited to airborne emitters or sensors located close to or above the elevated duct. The primary effects are

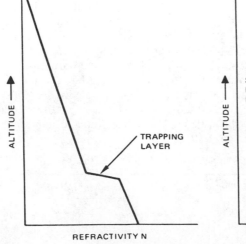
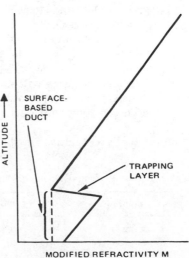

FIG. 3. N and M profiles for a surface-based duct.

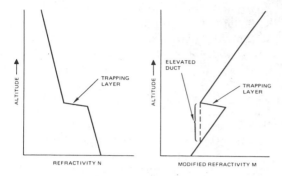

FIG. 4. N and M profiles for an elevated duct.

extended ranges for receivers or targets within the duct and radio or radar "holes" in coverage for receivers or targets at altitudes above the duct. Elevated ducts occur at altitudes up to about 6 km, although they are most common below 3 km.

The evaporation duct is a nearly permanent propagation mechanism created by the very rapid decrease of moisture immediately above the ocean surface. For continuity reasons, the air adjacent to the ocean is saturated with water vapor and the relative humidity is thus 100%. This high relative humidity decreases rapidly with increasing height in the first few meters until an ambient value is reached which depends on the general meteorological conditions. The rapid decrease in humidity creates a trapping layer adjacent to the surface as illustrated by the modified refractivity curve in Fig. 5. The height at which a minimum value of M is reached is called the evaporation duct height, which is the measure of strength of the duct. The evaporation duct itself extends from the duct height down to the surface, but because evaporation ducts are very "leaky," they affect radio and

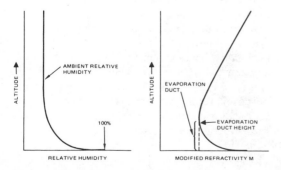

FIG. 5. Relative humidity and M profiles for an evaporation duct.

radar terminals significantly above as well as within the duct. The frequencies that the evaporation duct can affect are strongly dependent on the existing duct height, with a lower practical limit of about 3 GHz. The evaporation duct heights vary generally between 0 and 40 m, with a long-term mean value of about 8 m at northern latitudes, and up to about 30 m at tropical latitudes. The primary evaporation duct effects are to give extended ranges for surface-to-surface radio or radar systems operating above 3 GHz. The optimum frequency to achieve extended ranges via the evaporation duct appears to be around 18 GHz. Although the ducting effect extends beyond this frequency, absorption by atmospheric gases and extra attenuation due to a rough sea surface begin to counteract the benefits of the duct, as illustrated in Section VI.

III. Aspects of Refractivity

A. REFRACTIVITY-SENSING TECHNIQUES

For propagation assessment purposes the vertical profile of radio refractivity should be known to within one N unit with a vertical resolution of less than 10 m. The best instrument for measuring atmospheric refractivity is a microwave refractometer. It measures radio refractivity directly with excellent accuracy and response time. It must be carried by aircraft or balloon for vertical profiling and is relatively costly and electronically complex. Many versions are in use today including one on a U.S. Navy aircraft-carrier-based airplane. The most commonly used instrument for measuring atmospheric refractivity is still the radiosonde. It measures temperature and humidity as a function of pressure. Radio refractivity and altitude are calculated from those values. The calculated refractivity profile is generally satisfactory if sufficient care in calibration and data reduction is taken. One particular problem with radiosonde-deduced surface ducts concerns erroneous surface-based ducts resulting from limitation in the response time of the sensors and in data-reduction procedures. This problem is particularly troublesome for refractivity climatologies based on archived standard radiosonde data. The most desirable instrument for measuring atmospheric refractivity would be based on a remote-sensing technique, preferably passive. Unfortunately, water vapor profiles obtained with microwave radiometers do not have the vertical resolution needed. Active remote sensors using microwave radar sense refractivity fluctuations associated with mixing across gradients in the vertical refractivity profile. The relationship between those fluctuations and the gradients is presently the subject of intense investigation. Two lidar techniques (differential absorption, or DIAL, and Raman scattering) have been used for measuring water vapor and temperature profiles. They will, however, not work under cloudy conditions, which severely limits their general usefulness for refractivity applications. Reception of signals from known emitters either ground- or satellite-based may be used to sense the propagation medium. In this way, conditions along the entire propagation path of interest for the desired frequency can be sensed directly. However, only in rare cases will it be feasible to have a source at the right frequency and location. Inferring the vertical refractivity structure itself along the path requires multifrequency signals and complex profile inversion schemes. Use of beacons from satellites passing through the horizon has been recently analyzed. This technique relates shifts in interference null location with vertical refractivity structure. The practical utilization of this technique, however, is limited by signal fluctuations caused by random inhomogeneities of the medium. Averaging times required for precise measurements of interference patterns are too long for the relatively short time a satellite passes through the horizon.

Some qualitative radio propagation inferences may be drawn from satellite-sensed cloud patterns usually in conjunction with weather charts. Well-mixed atmospheric conditions will generally indicate standard propagation conditions while stratifications may indicate the possibility of ducting.

B. HORIZONTAL HOMOGENEITY, PERSISTENCE, AND REFRACTIVITY FORECASTING

Propagation assessment is usually based on one refractivity profile measured along a slant path in the vicinity of the propagation path as close as possible to the time of interest. The underlying assumption is that the atmosphere is horizontally stratified to justify use of a single profile for propagation calculations. This assumption is based on a physical reason since the atmosphere, in particular over ocean areas, is horizontally much less variable than vertically. Horizontal stratification also implies temporal persistence. Propagation forecasts are often based on persistence; that is, it is assumed that present conditions will not change significantly in the near future. There are, however, conditions for which horizontal inhomogeneity may be important, for example, at airmass boundaries or in coastal regions. The question is how often horizontal inhomogeneity must be considered for valid propagation assessment. For this purpose, simultaneous refractivity and propagation data were examined. It was found that calculations of propagation enhancements based on a single profile were correct in 86% of the cases. A similar conclusion has been reached from five years of shipboard experience with IREPS.

Propagation assessment would be considerably more complicated and costly if effects of horizontal inhomogeneity must be included. In this case, refractivity has to be sensed at multiple locations and more frequently since persistence is no longer a valid assumption. This would require an increase of the number of refractivity profile soundings by roughly an order of magnitude (for reliable shipboard propagation assessment, presently one to two radiosonde soundings are taken in each 24-hr period). Propagation calculations are more complex for inhomogeneous paths and are azimuth-dependent. In addition, they must be performed more frequently. One may assume that horizontal inhomogeneity is responsible for part of the 14% of incorrect assessments mentioned before and that their proper assessment would reduce this number by one-half. Then the expected improvement in routine propagation assessment would be only about 4% for those areas with the greatest occurrence of surface-based ducts of around 50% and much less for areas where ducting is uncommon. Consideration of horizontal inhomogeneity effects depends, therefore, on a tradeoff between the improvement in propagation assessment accuracy and cost and must be decided on a case-by-case basis.

Refractivity forecasting requires very accurate prediction of the atmospheric humidity profile and its dynamic behavior. Refractive boundary layer structures have been modeled by S. Burk and E. Gossard.[1] However, for operational purposes, refractivity forecasts (other than persistence) are qualitative. For example, a high-pressure ridge and associated subsidence may

---

[1] References to authors named in the text can be identified from the reference list of the Hitney *et al.* (1985) paper of the bibliography.

produce ducting layers, and a well-mixed atmosphere will probably indicate standard propagation conditions. Satellite imagery of clouds can be used for similar qualitative statements of ducting conditions.

C. REFRACTIVITY CLIMATOLOGIES

When a radar or communication system is in place, routine upper air sensing and surface meteorological observations are adequate to determine refractive effects on system performance. However, prior to installation of the radar or communication system, proper site selection must consider a statistical description of the vertical refractivity profile. Coastal and shipboard systems operating at frequencies greater than 3 GHz must also consider the evaporation duct, which may not influence site selection but may impact system design. Global climatologies of refractivity profiles and evaporation ducts exist and are described in the following paragraphs.

Upper air radiosonde observations from 921 land- and sea-based meteorological stations reporting during five selected years (1969 to 1971 and 1973 to 1974) were analyzed by L. Ortenburger. Results of the study are compiled as tables of monthly percent occurrence, surface refractivity, layer gradients, inflection heights, and trapping frequency for both surface-based and elevated ducts as well as superrefractive layers. This analysis also contains data on reliability of station reports, instrument types, winds aloft, and much more. It is the best climatological study of upper air observations known. Ortenburger is continuing to improve the utility of the database by developing a microcomputer-hosted program to access and display the statistics for a user-specified time and location. The target machine is the IBM XT personal computer, which is well known and generally available.

For the specific case of advective ducting in a maritime environment, profile statistics for 399 coastal, island, and weathership stations have been extracted. This subset was created for use with the IREPS computer program, which is discussed in Section VII. Locations of the selected stations are shown in Fig. 6. Through the IREPS program, the user is able to simulate a nonducting environment, a surface-based duct environment, an elevated duct environment, or a combination of all three for any maritime position.

Unlike advection ducts, the evaporation duct is nearly always present, although much thinner, with typical mean heights of 8 to 10 m above the

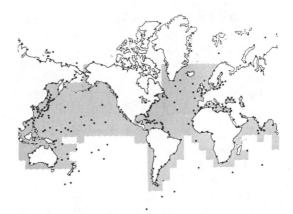

FIG. 6.   World map showing selected radiosonde locations (dots) and shipboard surface observation areas (shaded regions) contained in the IREPS climatology.

sea surface. Duct-height distributions, however, vary significantly with geographic area, season, and time of day. Therefore, the evaporation duct climatology consists of percent occurrence histograms for duct heights from 0 to 40 m in 2-m intervals. This database was developed by the National Climatic Data Center (NCDC), Asheville, N. C., which processed all shipboard surface meteorological observations reported during the years 1970 through 1979. In addition to the evaporation duct statistics, NCDC simultaneously compiled percent occurrence histograms for wind speed, wind rose, sea-surface temperature, and other parameters. All histograms were developed for ocean regions in grids of 10° latitude and 10° longitude known as Marsden Squares. A total of 213 such squares are included in the climatology and are shown as the shaded ocean regions in Fig. 6. These climatologies can be extremely useful in making long-term predictions of system performance, as the example in Fig. 18 in Section V shows, for use in system design studies.

IV. Propagation Measurements

Since World War II, there have been many radiometeorological measurement programs to investigate the effects of anomalous tropospheric propagation in a marine environment. Several of these programs will be presented in this section and are used to compare with the models presented in Sections V and VI. There may be other measurement programs of comparable quality, hence those presented here should not be considered to be comprehensive.

A 1953–1954 measurement program explored the effects of nonstandard refractivity profiles on high-altitude signal levels as a function of range, primarily within the horizon. One example of the results is shown in Fig. 7. A total of 22 such runs from 25 to 250 nautical miles were made at 218, 418, and 1089 MHz for both horizontal and vertical polarization with the transmitters in high-altitude aircraft from 8700 to 11,900 m and the receivers located at 33 m above mean sea level (msl). Refractivity profiles were measured near the receiver in all cases, which indicated both standard and nonstandard conditions. All paths were entirely over water off the San Diego coast. Overall results showed that high-altitude signals were not reduced due to the presence of low-level refractive anomalies as had been postulated, even though the null patterns could be significantly shifted.

In 1944, the U.S. Navy Radio and Sound Laboratory, a predecessor of the Naval Ocean Systems Center, carried out an 80-nautical mile one-way, over-water, surface-to-surface transmission experiment at 52, 100, and 547 MHz for 40 days. Transmitters and receivers were all located about 30 m above msl. The primary results are summarized in Fig. 12, which shows the base of the temperature inversion plotted in a time series along with the received signal levels. This experiment vividly showed how over-the-horizon signals are affected by changes in the structure of the troposphere, with signals as high as free-space levels or more associated with low-altitude temperature inversions. As can be seen from the results, the ducting effects were much more pronounced on the higher frequencies.

A massive radio meteorological data collection effort to examine low-level ducts created by advection of warm, dry air over the ocean was undertaken in 1946 and 1947 off the coast of Canterbury Province, South Island, New Zealand. In general, meteorological profiles, from which modified refractivity was computed, were recorded every 20 km in range from the beach out to a maximum of 160 km and from the surface level up to about 300 m. Simultaneous radio measurements at 100, 500, 3240, and 9875 MHz were made from aircraft-mounted transmitters to ground-based receivers. Ninety sets of observations were made over seventy-four days. Field strengths were recorded versus range out to a maximum of approximately 180 km for aircraft altitudes of about 600 m, and for ascents and descents covering altitudes up to 600 m at various ranges. All of the data are of very high quality and quite useful even now after nearly 40 years.

From 1945 to 1948, the U.S. Navy Electronics Laboratory (NEL), another predecessor to the Naval Ocean Systems Center, collected a large amount of radio-meteorological data on a 280-nautical mile over-water path between San Diego and Guadalupe Isle. An aircraft was fitted with transmitters at 63, 170, 520, and 3300 MHz and receivers and recorders were located in San

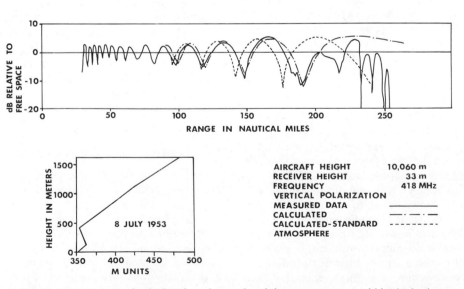

FIG. 7.   Comparison of calculated to observed path loss versus range within the horizon.

Diego. Receiver antenna heights were either 30 or 150 m and the aircraft flew a "sawtooth" flight profile varying altitude from near zero to 1200 m, out to a maximum range of 280 nautical miles. The example in Fig. 19 is for a surface-based duct that becomes an elevated duct with increasing range. The meteorological profiles are given in the figure in B units, which are related to the modified refractivity M by

$$M = B + 0.118h \qquad (1)$$

where h is height in meters.

The final measurement program to be discussed here, directed at investigating subsidence inversions and resulting surface-based or low-elevated ducts, was carried out by the Naval Electronics Laboratory Center from May to August 1974. A 3088-MHz transmitter was located in San Diego on the coast at 21 m above msl and an aircraft was instrumented with a receiver and recorded signals as the aircraft flew an over-water radial course away from the transmitter. Horizontal polarization was used. Forty-five flights from fixed level altitudes of 30 to 4500 m and to ranges up to 400 nm were made concurrently with radiosondes launched from the transmitter site. Figure 15 is an example of the measured path loss versus range for one case on May 28, 1974 with the aircraft at 914-m altitude. Radiosonde measurements indicated an elevated duct existed between about 400 and 700 m. These measurements have been the basis for several recent comparisons with modeling theory.

In 1972, the Naval Electronics Laboratory Center made a series of over-the-horizon propagation measurements to investigate the effects of the evaporation duct. Over-water paths of approximately 35 km between islands off San Diego, Greece, and Key West, Florida, were instrumented at 1, 3, 9.6, 18, and 37 GHz with transmitter antennas at about 5 m and receiver antennas at about 5, 10, and 20 m above msl. The Greek measurements were the most extensive. Measurement periods were from two to three weeks for each of the four seasons. The specific objective of the tests was to determine the optimum antenna heights to best exploit the evaporation duct effect. Figure 16a shows the 9.6-GHz measured path loss for the low receiver antenna (solid curve) and Fig. 16b shows decibel difference between the high and low antennas (higher signals on the high antenna gives positive difference) versus time for a two-week period in November 1972. The two reference levels indicate the path loss expected in free space and due to diffraction at the same range. Note the extreme variability of path loss from well below free space to greater than diffraction with the average value being just slightly greater than free space and well below diffraction. Note also that when the path loss is low, the lower antenna generally has up to only 10 dB higher signal levels than the higher antenna. In other words, when the evaporation duct is strong, the highest signals will be received at the low antenna, but the high antenna will still receive signals very much enhanced over nonducting conditions. In addition to providing a wealth of information for comparison to theoretical models as discussed briefly in Section V, the joint measurement and analytical program concluded, in part:

1. An antenna need not be in the evaporation duct to benefit from signal enhancements of the duct.
2. Signal enhancements are strongly dependent on frequency and evaporation duct height.
3. Frequencies below 3 GHz are rarely affected by the duct. Sea-surface roughness and atmospheric absorption begin to counteract the effect of the duct above 10–20 GHz.
4. For long-term, wide-area applications, the optimum antenna height for shipboard use would be as high as possible because strong evaporation ducts are not persistent in all areas.

An experiment to investigate evaporation duct effects on longer paths was performed during 1981–1982. The frequencies chosen were 3 and 18 GHz and the path chosen was between two southern California off-shore islands with a path length of 81 km. The transmitter antennas were about 20 m and the receiver antennas about 11 m above msl. Both path loss and meteorological measurements, from which the evaporation duct height could be calculated, were made. Figure 17 shows a scatter diagram of observed path loss at 18 GHz and calculated duct height for a three-week period in June 1982. The dashed reference lines in this case are free space and troposcatter levels expected at the same range of 81 km. The receiver threshold corresponds to 210-dB path loss. This experimental program clearly established the capability of the evaporation duct to support high signal levels at moderate over-the-horizon ranges and further confirmed the theoretical understanding of this important propagation mechanism.

V. Propagation Modeling in Horizontally Stratified Media

A. WITHIN AND NEAR-THE-HORIZON PROPAGATION

For ranges and altitudes well within the horizon, physical optics techniques are usually used to calculate field strength. Ray trajectories for both direct and reflected paths are calculated using piecewise-linear-with-height refractivity profiles and ray-trace procedures. When one terminal of the path is close to the surface, say within 100 m, and ranges are fairly large, the interference effects of the direct and reflected ray can be accounted for using a technique described by L. Blake, which assumes the direct and reflected paths are parallel throughout the atmosphere, except for the region close to the lower terminal. This lowest region is assumed to be characterized by a single linear refractivity layer and the path-length difference calculations become quite tractable. For shorter ranges and lower altitudes, the assumption of parallel rays breaks down. In this case, an alternate solution can be used which requires a single linear refractivity layer to exist over the entire height interval of interest. Refractive distortion can still be taken into account, however, by determining the single linear layer that best matches an actual raytrace through the entire region of interest at low elevation angles. In determining the contribution of the reflected ray, the reflection coefficient for the proper polarization as well as the spherical earth divergence factor needs to be computed. In addition, an ocean roughness factor should be included in the calculations, as discussed later in Section VI. The optical field calculations are applicable to ranges and heights such that one or both of the following conditions apply. First, the path-length difference between the direct and reflected ray must be at least one-quarter wavelength, and second, to ensure valid divergence factor calculations, the reflection grazing angle must exceed a limiting value.

Figure 7 is a comparison of measured field strength versus range data from a 1953 NEL measurement program described in Section IV to calculations based on the techniques described above. The transmitter in this case was in an aircraft at 10,060 m, the receiver was located 33 m above msl, and vertical polarization was used. The modified refractivity versus height was measured near the receiver and is shown in the figure. In this case, the limit of the optical region is based on the quarter-wavelength limit. Calculations are shown for both a standard atmosphere (dashed curve) and for the actual refractivity profile (chain–dot curve). Note how the trapping layer has stretched out the last few nulls in range.

The above-described methods work reasonably well in most cases. However, there are some cases where the actual path-length difference is too strongly dependent on the trajectories and electromagnetic path lengths of the direct and reflected rays for the assumption of an equivalent single linear layer to be valid for calculating path-length difference. Current investigations by the authors using complete raytrace methods including the calculation of integrated electromagnetic path length along each ray have shown some promise in overcoming this problem, although at the expense of much increased computing requirements. Figure 8 illustrates one example where the new full raytrace model does a far better job compared to the equivalent single-layer model. The example is for a trilinear surface-based duct with a duct height of 465 m.

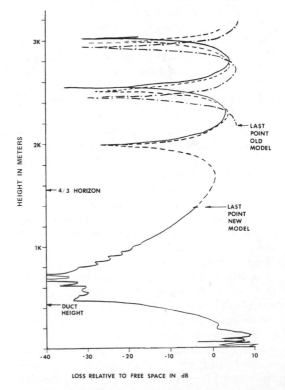

FIG. 8. Comparison of two ray-optics models to waveguide calculations within the horizon. ——, Ground truth; – · –, old model; - - -, new model.

To avoid uncertainties inherent in measured data, the figure shows as "ground truth" (solid curve) the results of a well-confirmed waveguide model program, described in this section. The frequency was 3000 MHz, the transmitter height was 30 m, the range was 100 nautical miles, and horizontal polarization was assumed. Included in the calculation were 139 waveguide modes to ensure that the results would be accurate well inside the horizon. The results from the "new model" (dashed curve) are based on the full ray-trace method and coincide exactly with the waveguide calculations below the first optical null. Note that the raytrace model not only matches the waveguide model more accurately than the "old model" (chain–dot curve) based on the single-layer assumption, but is able to calculate fields much closer to the horizon since it computes a much improved path-length difference.

As was mentioned above, ray-optic methods are limited to regions within the radio horizon. Although waveguide methods can be applied to near-horizon cases by including a sufficient number of modes, practical applications generally require that ranges and altitudes be sufficiently beyond the horizon. The time-honored method for treating the near-horizon region is based on W. Fishback's method of "bold interpolation," in which linear interpolation is used to connect the last point calculated in the optical region with the first point calculated in the beyond-the-horizon region in terms of decibel path loss versus range. As crude as this method would seem to be, experience shows it works remarkably well in matching measured data.

B. BEYOND-THE-HORIZON PROPAGATION

At large distances beyond the horizon, in normal environments, the field strengths are dominated by scattering from tropospheric refractive index fluctuations (i.e., troposcatter). The troposcatter field is generally well below radar receiver thresholds and will be touched upon only very briefly. The principal emphasis in the present section will be on anomalous beyond-the-horizon propagation produced by laterally homogeneous tropospheric layering. In particular, several case studies which have been treated by waveguide concepts are reviewed. Points of discrepancy between calculation and observation are singled out along with some problem areas worthy of further study.

Since waveguide formalism is well docu-

mented, the approaches which have been found most useful will be reviewed only very briefly. First, the waveguide developments are in terms of an earth flattened geometry where earth curvature is included in a modified index of refraction as discussed in Section II. The modified refractive index is approximated by linear segments so that the altitude dependence of the fields are, to a good approximation, expressible in terms of modified Hankel functions of order one-third or, equivalently, Airy functions. Waveguide normal modes are found subject to the boundary conditions of outgoing wave at the top and outgoing wave in the ground. One of the fundamental modal function forms can be expressed as follows in terms of plane-wave reflection coefficients,

$$F(\theta) = 1 - R_b(\theta)\bar{R}_b(\theta) \qquad (2)$$

Where R_b is the plane-wave reflection coefficient from everything above any height b within the guide with vacuum below b and $\bar{R}_b$ is the plane-wave reflection coefficient from everything below b with vacuum above; θ is the complex grazing angle of the plane wave (i.e., complement of the angle of incidence), where the real part of the cosine of the grazing angle is related to phase velocity and its imaginary part is related to attenuation rate.

Crucial to any successful waveguide program is the determination of all significant complex zeros of the fundamental mode equation, $F = 0$. One solution of the equation is $\theta = 0$. Using this knowledge and plotting the curve $G = |\bar{R}_b(\theta)R_b(\theta)| = 1$ gives one method of root extraction. An example is shown in Fig. 9. The vertical axis is the imaginary part of θ and the horizontal axis is the real part. Modes with attenuation rates less than 1.3 dB/km are shown. The inset is the modified refractivity used for the calculation. It will be observed that the inset corresponds to $M = 0$ at $z = 183$ m. In this connection, it is remarked that waveguide results are determined for practical purposes by the gradients of the refractivity profile and are quite insensitive to translational effects. The Os in Fig. 9 denote modes located on the G curve traced from the origin. The ×s represent modes located by a method discussed below. The mode not found using the G trace is a diffraction-type mode. Generally, the attenuation rates of diffraction-type modes increase approximately as frequency to the one-third power and thus tend to become less important with increasing frequency for beyond-the-horizon propagation in

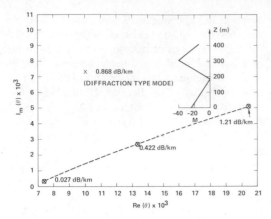

FIG. 9. Zero locations in the complex eigenangle space (frequency 65.0 MHz).

ducting environments associated with elevated layers. Most of the calculations to be presented in this section were generated using modes found by the G trace method.

A better method for finding the mode solutions is based on an ingenious algorithm developed by C. Shellman and D. Morfitt. Implementation of the method requires searching the periphery of a rectangular region of the eigenvalue space for 0° or 180° phase contours of the modal function. The modal function is required to be analytic within and on the boundary of the search rectangle. The latter requirement guarantees that the phase contours which enter the search rectangle must either terminate on zeros of the modal function or exit the search rectangle (Fig. 10). When applying the method to the modal function, given by Eq. (2), difficulties are encountered because R_b and or $\bar{R}_b$ may not be analytic in the region of interest. S. Marcus and more recently G. Baumgartner have avoided this difficulty by formulating the mode equation

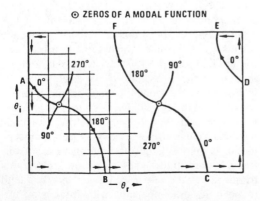

⊙ ZEROS OF A MODAL FUNCTION

FIG. 10. Schematic of root-finding method.

directly in terms of continuity of the tangential (i.e., horizontal) field components. They have put the Shellman–Morfitt root-finding algorithm to excellent use for tropospheric waveguide calculations and their methods represent a marked improvement over the G trace procedure in the sense that the method infallibly locates all nondegenerate modes within the search rectangle. That capability should be particularly useful in studies involving multiple ducts.

Once the eigenvalues are determined, the signal level relative to free space can be computed from

$$S(\mathrm{dB}) = 10 \log_{10} \left\{ [0.0207 x^2 f / \sin(x/a)] \right.$$
$$\times \left| \sum_{n=1}^{N} \lambda_n g_n(z_T) g_n(z_R) \right.$$
$$\left. \exp(-ikx \cos \theta_n) \right|^2 \right\} \quad (3)$$

where f is the frequency (MHz), x the transmitter–receiver distance (km), a the earth's radius (km), g_n the height gain function normalized to unity at level b, z_T the transmitter altitude, z_R the receiver altitude, θ_n the nth eigenangle, k the free-space wavenumber (km^{-1}), N the total number of modes used, and

$$\lambda_n = \frac{[\cos(\theta_n)]^{1/2}(1 + \bar{R}_b)^2}{\bar{R}_b (\partial F / \partial \theta)_{\theta = \theta_n}} = \text{excitation factor} \quad (4)$$

Equations (3) and (4) are consistent with the plane-wave reflection coefficient formalism. The value for N depends very much upon the height of the layer, the frequency, and the terminal locations. The height gain function for the TE wave, and to good approximation for the TM wave, obeys the equation

$$d^2 g_n / dz^2 + k^2 (m^2(z) - \cos^2 \theta_n) g_n = 0 \quad (5)$$

where the modified refractive index is given by

$$m = n + z/a \quad (6)$$

In the linearly segmented approximation, second-order terms involving the gradients of m are ignored when solving Eq. (5). The height gain functions are found subject to the conditions that they represent outgoing wave at the top and in the ground, and that the tangential field components of the RF wave are continuous at points where, in the linear segmented approximation, dm/dz is discontinuous. It is often the case that results are expressed in terms of path loss. This conversion is made by subtracting Eq. (3) from the free-space path loss L_{FS}, which is

$$L_{FS}(dB) = 32.4 + 20 \log_{10} x + 20 \log_{10} f \quad (7)$$

In the following, results will also be discussed in terms of the incoherent mode sum, which is obtained by the replacement

$$\left| \sum_{n=1}^{N} \lambda_n g_n(z_T) g_n(z_R) \exp(-ikx \cos \theta_n) \right|^2$$

$$\Rightarrow \sum_{n=1}^{N} |\lambda_n g_n(z_T) g_n(z_R) \exp(-ikx \cos \theta_n)|^2 \quad (8)$$

in Eq. (3).

To give an idea of the dynamic frequency range to which the waveguide formalism has been applied, a numerical study made of ducting in the HF band is considered first. P. Hansen reported measurements over a 235-km southern California all-ocean path in the frequency range from 4 to 32 MHz. No skywave contamination existed for the path. Hansen found that above about 20 MHz the average signal levels considerably exceeded predictions based on standard groundwave theory. During a 24-hr period of Hansen's measurements, nine refractivity profiles recorded at four different sites which were in reasonable proximity to Hansen's path were available. Each refractivity profile provided an environment for which waveguide calculations were performed. Figure 11 shows the measured and calculated path losses. The average path losses are at the midpoint of the error bars, which represent one standard deviation on each side of the average. It will be seen that the calculated averages and the experimental averages are in good agreement. The disparity between the calculated and observed standard deviation is attributed to the likelihood of lateral inhomogeneity of the guide, which has not been taken into account in the calculations.

Figure 12 shows results of the 1944 San Pedro to San Diego measurements mentioned in Section IV. Shown is the base of the temperature inversion plotted in a time series along with the envelopes of the observed signals. There were about 150 vertical refractivity profiles measured during the course of the experiment of which about 80 have survived to this day. Of that number, 64 were amenable to trilinear fits. Analytical results based on the trilinear fits are shown by the dots on the signal level curves. Particularly at the higher frequencies the calculated results are in good agreement with the observations.

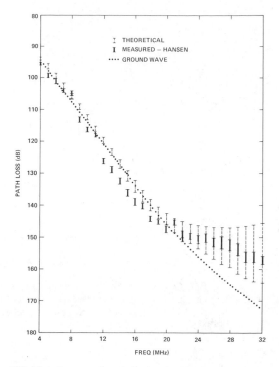

FIG. 11.   Case study of ducting at frequencies in the HF band.

Figure 13 shows height gain curves in a ducting environment and in a normal atmosphere at 3.3 GHz for a 222-km range and receiver altitude of 152 m. It applies to the Guadalupe Island study (see Section IV) where an inversion layer characterized by a 40 M-unit deficit existed between 183 and 305 m as shown on the inset. The comparison between calculated and experimental results is quite good. This case is an example of signal level calculation by waveguide concepts when the number of modes is on the order of 100. The large number of modes points out the need for approximate methods such as ray or hybrid methods. Nevertheless, the results of Fig. 13 show that with enough fortitude waveguide calculations can be carried up to at least the several gigahertz range for typical surface-based ducts. Of course, a reliable waveguide program can also be used as a tape measure to assess the accuracy of approximate methods.

Figure 14 shows results of an approximate method. It is for the same frequency and environment as those in Fig. 13. The range is 111 km with the receiver altitude at 30.5 m. Theoretical results for both the coherent- and incoherent-mode sums are shown along with field strengths for the normal atmosphere (terminated at the ho-

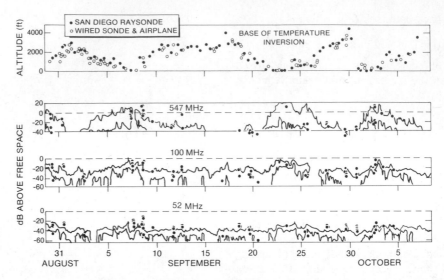

FIG. 12. San Pedro to San Diego experiment; 90-mile link; transmitter and receiver altitude 100 ft. Solid curves are envelopes of the measurement and the dots waveguide results. For comparison, troposcatter levels are −38.5 dB at 52 MHz, −41.4 dB at 100 MHz, and −48.7 dB at 547 MHz.

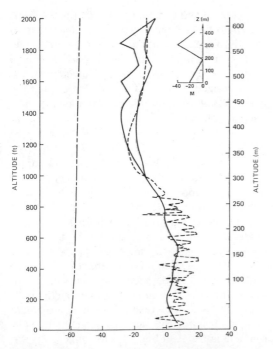

FIG. 13. Measured and calculated height gains for a surface-based duct. Frequency, 3.3 GHz; range, 120 nautical miles (222.4 km); receiver altitude, 500 ft (152.4 m); ——, experimental; ---, calculated; ---, normal.

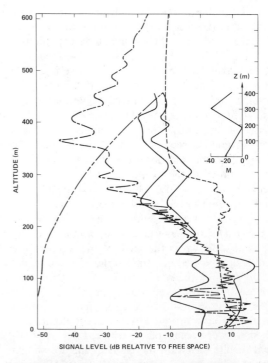

FIG. 14. Measured and calculated height gains for a surface-based duct. Frequency, 3.3 GHz; range, 111.1 km (60 nautical miles); transmitter altitude, 30.5 m (100 ft); ---, MA-coherent; ---, MA-incoherent; ——, experimental; ---, normal.

rizon). The waveguide results were obtained using asymptotic formulas, where reasonable, for the plane-wave reflection coefficients and are labeled MA in Fig. 14. Although only implemented for trilinear models, the method shows promise for speeding up waveguide calculations with relatively little degradation in accuracy from the full wave calculations. The mode phasing is such that above about 275 m the coherent mode sum considerably underestimates the measurements. If the mode phasing is destroyed by lateral inhomogeneity of the layer, index of refraction fluctuations, or surface roughness, then the expectation would be that the incoherent mode sum would be more representative of the propagation. For the case shown in Fig. 14, the coherent-mode sum gives the better agreement with measurement between about 150 and 275 m (i.e., roughly over the inversion-layer height range) and the incoherent-mode sum gives the better agreement above 275 m. Which mode sum is to be preferred, and why, under a given set of terminator conditions remains an unsolved question. It would certainly seem appropriate to direct attention to the roles played by refractive index fluctuations, layer inhomogeneity, and surface roughness in future studies.

Figure 15 shows additional results close to 3 GHz. Calculated and measured path losses as a function of range for a transmitter at about 21 m and a receiver at 914 m are shown. The measurements were taken in an off-shore San Diego en-

vironment during 1974 (see Section IV). The range covers the line-of-sight region, the diffraction region, and the region well beyond the horizon. Two waveguide results are shown. One is for the elevated layer environment shown in Fig. 15b, which was obtained from radiosonde measurements made during the period of the radio measurements and the other is for the evaporation duct shown in Fig. 15c, which meterological data indicated might have existed at the time of the measurements. The arrows are crude measures of the horizon for a direct ray (≈ 139 km) for a ray once reflected from the elevated layer (≈ 315 km), and for a ray twice reflected from the elevated layer (≈ 537 km). Observe that the waveguide-calculated path loss for the elevated layer shows a large increase in the neighborhood of the arrows consistent with what might be expected on the basis of a ray hop picture and greatly overestimates the observed path loss beyond the first horizon. The waveguide calculation for the very strong evaporation duct gives reasonable agreement with the beyond-the-horizon signal. However,

1. because of the difference between the experimental and shallow surface duct results close to the horizon,
2. because many of the 1974 measurements (for a variety of refractivity environments) showed beyond-the-horizon signal leveling off close to the 180-dB value, and

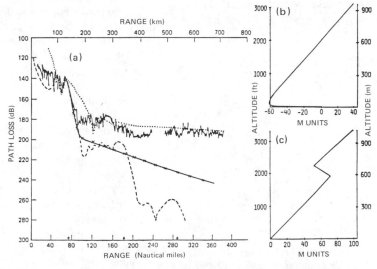

FIG. 15.   (a) Measured and calculated path loss versus range. Frequency, 3087.7 MHz; transmitter altitude, 68 ft (20.7 m); receiver altitude, 3000 ft; (914 m); ———, experimental; ⋯, evaporation duct; ---, elevated layer; ×, normal. (b) and (c) are explained in the text.

3. because an exceptional, very strong, evaporation duct is required,

it is questionable whether the shallow surface duct can explain the many observations. Why the path-loss falloff beyond the horizon is repeatedly much less in the offshore San Diego area than expected for troposcatter and the precise role, if any, played by ducting and superrefractive environments remain a mystery.

Waveguide programs have also been used extensively to model propagation effects created by the evaporation duct, whereby multilinear refractivity profiles are used to approximate the actual refractivity profiles derived from the meteorological processes. The result from one such modeling effort is shown in Fig. 16 and compared with the 9.6-GHz Greek measurements described in Section IV. Figure 16a shows calculated (dotted curve) and observed (solid curve) path loss versus time. The calculations are based on profiles derived from the bulk meteorological measurements of air temperature, relative humidity, wind speed, and sea temperature measured at one end of the path. Although the calculations are in substantial agreement with the observations, some discrepancies do exist which are probably related to an inhomogeneous path, errors in the meteorological measurements, or the presence of surface-based ducts, which were not accounted for in the calculations.

Another example of evaporation duct modeling is presented in Fig. 17 for Anderson's 18-GHz measurements described in Section IV. The figure shows path loss versus duct height,

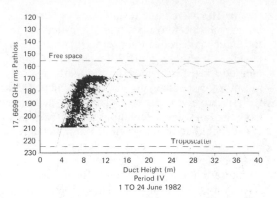

FIG. 17. Comparisons of experimental measurements to waveguide calculations at 18 GHz.

where the solid oscillating curve is the theoretical dependence based on the waveguide model and includes the effects of atmospheric absorption. The calculated fields to the peak of the first mode provide an upper bound to the magnitude of the observed fields. Since the duct heights were measured only at one end of the terminals, it is not surprising to see a spread of observed fields due to variations in the evaporation duct over the entire path. Also, the calculations were made assuming a flat ocean surface so the calculated signal levels are expected to be overestimates. The oscillatory nature of the calculated signal results from modal interference. As the duct height increases, the least attenuated mode drops out of the picture because of a decrease in its height gain function at the transmit and receive terminals. Thus although more highly attenuated than the first mode, the second mode eventually becomes dominant with increasing duct height because of the height gain effect. The process continues to repeat itself with higher order modes becoming dominant as the duct height increases further.

A final application of the waveguide model to the evaporation duct is based on the work of K. Anderson, who has recently reported results of a study on the influence of the evaporation duct on shipboard surface-search radar design. This study used the NCDC climatology discussed in Section III and the propagation models discussed previously to evaluate the probability of detecting a surface target at ranges well beyond the normal radar horizon. Four radar frequencies were considered: 3, 6, 10, and 18 GHz. Predicted radar performance is summarized in Fig. 18, which amply illustrates

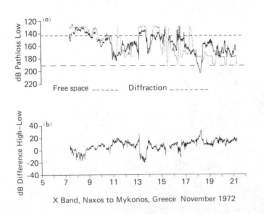

FIG. 16. (a) Calculated (dotted) and observed (solid) path loss versus time for evaporation duct conditions at 9.6 GHz. (b) Observed difference between antennas at 5 and 20 m versus time.

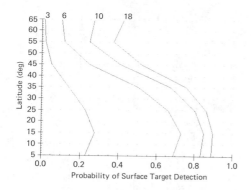

FIG. 18. Geographic influence of the evaporation duct on shipboard radar surface-target detection capabilities. The target is well beyond the normal horizon and the increased detection range is solely due to the evaporation duct. Predictions are for four radar frequencies: 3, 6, 10, and 18 GHz.

the necessity of including environmental effects in the design of radar and communication systems. For the particular case of a shipboard surface-search radar designed for long-range target detection, the added complexities of building a radar at 18 GHz may well be worth the increased probability of detection.

Finally, it is pointed out that waveguide calculations can be carried to within the line-of-sight region and can therefore be used to assist in developing empirical formulas for connecting ray theory line-of-sight fields to diffraction fields (see Fig. 8).

VI. Inhomogeneity and Roughness

A. LAYER INHOMOGENEITY

As discussed in Section III,B, for propagation prediction purposes, the assumption of lateral homogeneity appears to be adequate most of the time. Nevertheless, it is legitimate to inquire about the tools available to handle situations where the layers do have refractivity profiles which vary along the path. When this variation is sufficiently slow, the modes do not interact significantly, and thus each mode can be tracked separately. This so-called adiabatic (or WKB) approximation has been used in underwater sound studies and VLF propagation in the earth-ionosphere waveguide. If the adiabatic approximation fails, the lateral inhomogeneity is generally treated by mode-conversion techniques or by the parabolic-equation method. Both mode-

conversion (also referred to as "range-dependent mode") techniques and the parabolic-equation method have been developed to a high degree of sophistication by the underwater sound community. Despite this, the methods have been rarely used in tropospheric ducting work. The principal reason for this is probably because the detailed information relating to lateral inhomogeneity of the duct which is necessary for meaningful modeling is rarely available. Moreover, tropospheric ducts, and of course their irregularities, are deviations from the norm, whereas nonuniform ocean topology, for example, is the norm.

Figure 19, part of the Guadalupe Island data, shows measurements of height gains at 63, 170, 520, and 3300 MHz along with concurrent meterological data along the path. Also shown in the figure are experimental measurements of the refractivity in B units [see Eq. (1) in Section IV]. The receiver height is 100 ft (30.5 m) and the transmitter height is variable over the altitude range indicated by the height gain behavior of the field. The latter are in decibels relative to free space. The geometrical horizon distance is also indicated and it should be appreciated that many of the data pertain to signals received far beyond the horizon. It is clear that the layer structure varied temporally or spatially or both. These data served as the basis for a numerical modeling study of the effects of lateral nonuniformity. Only the two lowest frequencies were examined. The vertical refractivity structure was modeled by trilinear refractivity profiles as a function of range. The model properties of the layer (i.e., the middle segment of the trilinear model) are given in Table II. The levels z_2 and z_3 are the lower and upper levels of the layer, respectively, and x is the range. A single profile was assumed to apply throughout the range 0–40 nautical miles. The profile described by $z_2 =$

TABLE II. Model Parameters of the Layer

x (nautical miles)	z_2 (m)	z_3 (m)	$\Delta M/\Delta z$ (per km)
0–40	152.4	335.3	−144.4
80	213.4	457.2	−59.0
120	350.5	609.6	−49.2
160	350.5	655.4	−19.7
200	457.2	731.5	−3.3

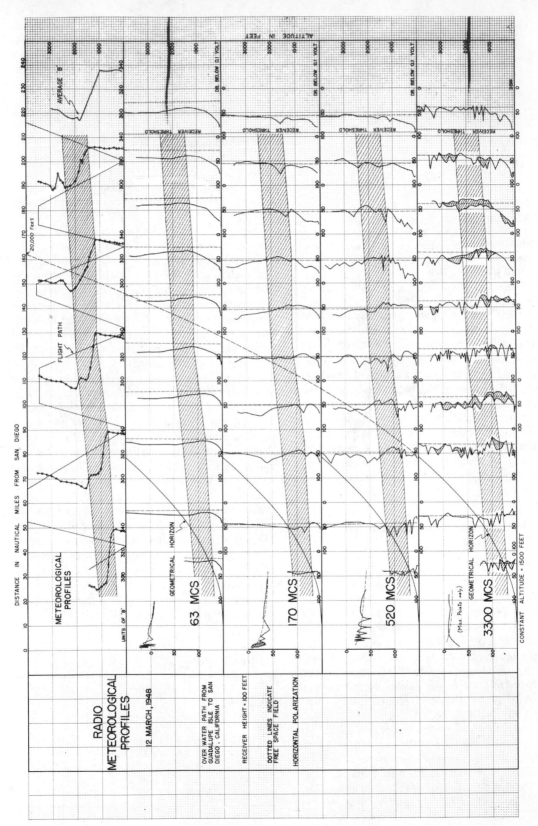

FIG. 19.   Guadalupe Island data.

213.4 m, $z_3 = 457.2$ m, and $\Delta M/\Delta z = -59.0$ is assumed to be the profile at 80 nautical miles, etc. A standard gradient of $118M$/km was assumed below z_2 in all cases. Linear interpolation of z_2, z_3, and $\Delta M/\Delta z$ was used to determine their values at ranges intermediate to 40–80, 80–120, 120–160, and 160–200 nautical miles. Figure 20 shows a comparison of measured and calculated height gains at 63 MHz. The calculated results are for the WKB (or adiabatic) method and the agreement with the measurements will be seen to be very good in this case of single-mode propagation. One manifestation of the nonuniform character of the guide is the increase with range of the height where maximum signal level occurs.

Figure 21 shows a comparison of measured and calculated height gains at 170 MHz. In this instance two theoretical curves are given. One is the WKB calculation and the other represents a slab model implementation of a mode-conversion program. Over 65 slabs were used in the implementation and a convergence study indi-

cated that was an ample number of slabs. The comparisons in this case are not nearly as good as they are for the 63-MHz case. Particularly at lower altitudes for the intermediate ranges the discrepancy between measurement and calculation is as large as about 40 dB. At the more remote ranges and at altitudes in excess of about 400 m, the WKB calculation gives the best agreement with observation. Though not shown on Fig. 20, mode-conversion results were also generated for the 63-MHz case and for an unexplained reason mode conversion again gave results at the higher altitudes which were definitely inferior to the WKB results. It can only be speculated that the mode-conversion results may be more sensitive to fine structure of the vertical and lateral refractivity profiles than are WKB results. Mode conversion is very difficult to implement as the frequency increases and the waveguide becomes more and more multi-moded. Parabolic equation methods apparently offer some promise of alleviating this shortcoming.

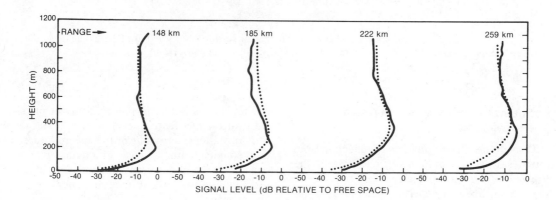

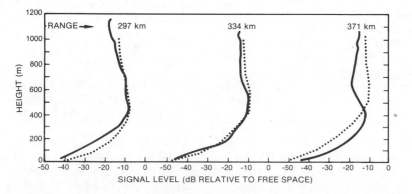

FIG. 20. Comparison of measurements and WKB results at 63 MHz. Frequency, 63 MHz; receiver height, 30.5 m; ——, measured; ⋯, WKB.

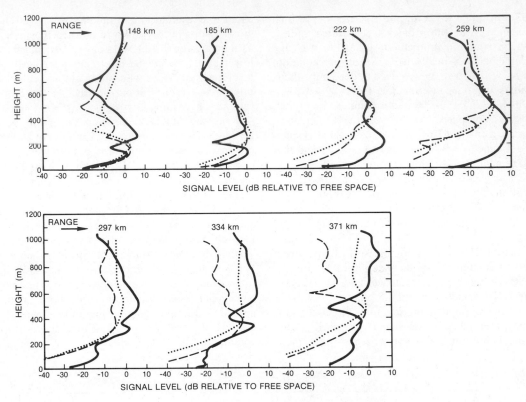

FIG. 21. Comparison of measurements, WKB, and mode-conversion results at 170 MHz. Frequency, 170 MHz; receiver height, 30.5 m; ——, measured; ⋯, WKB; ---, mode conversion.

B. SURFACE ROUGHNESS AND MILLIMETER-WAVE PROPAGATION

Though the general problem of reflection from rough surfaces is extremely complicated, the Fresnel reflection coefficient formulas for a smooth surface can be easily generalized to allow, in partial measure, for roughness of a surface obeying Gaussian statistics. This modification is developed from Kirchhoff–Huygens theory in terms of the surface rms bump height. In particular, if R is the Fresnel reflection coefficient for the smooth surface then the effective reflection coefficient R_e for the rough surface in the specular direction for the coherent part of the field can be written as

$$R_e = R \exp(-\phi^2/2), \qquad \phi = 2k\sigma \sin \theta_g \quad (9)$$

where k is the free-space wavenumber, σ is the rms bump height, and θ_g the grazing angle at the surface. C. Beard has shown this to be reasonable provided $|\phi|$ is not too large. It is also quite possible that the formula fails at near grazing angles since multiple reflections and shadowing

are not allowed for in the theory. Still, the physically attractive feature that a rough surface should scatter energy out of the specular direction coupled with its ease of implementation have made it a popular way to treat, at least in a semiquantitative way, surface roughness effects on propagation. The two following surface roughness models are commonly used to relate the rms bump height (in meters) to the wind speed u (in meters per second):

$$\sigma = 0.0051u^2 \quad (10)$$

$$\sigma = 0.00176u^{5/2} \quad (11)$$

Equation (10) is obtained from the Phillips' saturation curve spectrum and Eq. (11) is derived from the Neumann–Pierson spectrum as modified by Kinsman.

Figure 22 shows comparisons between measured and calculated line-of-sight fields. The measurements are part of the 1974 off-shore San Diego measurements discussed in Section IV. Two sets of calculated curves are shown. They represent the interference pattern between the

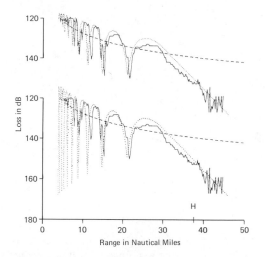

FIG. 22. Line-of-sight comparison with (upper curves) and without (lower curves) surface roughness.

direct wave and the wave once reflected from the ground. The calculation for the smooth earth gives, as expected, very deep interference nulls, while the calculation with the surface rms bump height of 0.5 m gives excellent agreement with the line-of-sight field and in particular with the depth of the nulls.

Equation (9) has also been used in waveguide studies. S. Rotheram has analyzed, with regards to roughness, evaporation duct data taken in the North Sea by researchers at Hamburg University. Figure 23 shows his results at 6.814 GHz for a 77.2-km path from Wedderwarden to Heligoland with a transmitter height of 29 m, a receiver height of 33 m, and horizontal polarization. Shown on the plot is field strength versus duct thickness, along with the number of observations divided by 5 falling in each 2-dB × 1-m element. Theoretical predictions of field strength versus duct thickness for rms surface bump heights of 0 to 1.5 m at 0.25-m intervals are shown for refractivity profiles obtained using the Monin–Obukhov similarity theory for near-neutral ($-0.01 < 1/L < 0.01$) conditions, where L is the Monin–Obukhov stability length in meters. It will be seen that the qualitative features of the data are well covered by the theoretical curves. Mediterranean Sea data provide additional support for the qualitative features predicted by the simple surface roughness model.

As an illustration of waveguide applications into the millimeter-wave region, Figs. 24 and 25 show results of a preliminary study of the modal attenuation rate associated with a moderate evaporation duct in the frequency range from 3

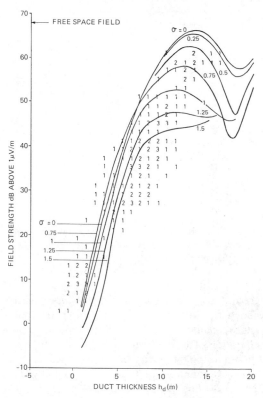

FIG. 23. Theoretical and experimental field strength–duct thickness relations for near-neutral stability. Frequency, 6.814 GHz; horizontal polarization; $D = 77.2$ km; h_T, 29 m; $h_R = 33$ m; fading types A + B; $-0.01 < 1/L < 0.01$. [Reproduced from Rotheram, S. (1974). "Beyond the Horizon Propagation in the Evaporation Duct—Inclusion of the Rough Sea," Marconi Tech. Rep. MTR 74/33.]

to 100 GHz. The results have been obtained with a multisegment waveguide program. The refractivity profile is listed in Table III.

Both Figs. 24 and 25 show the waveguide attenuation rate (i.e., attenuation rate due to

TABLE III.   Moderate Evaporation Duct

Z (m)	$M(Z)$ (M units)	DM/DZ (M units/m)
0.0	0.0	-172.6537
0.0732	-12.6383	-6.0083
0.5791	-15.6778	-1.1354
2.0940	-17.3979	-0.3099
5.1267	-18.3377	-0.0791
9.6774	-18.6977	-0.0107
11.5519	-18.7177	0.0293
19.0591	-18.4978	0.0675
30.3215	-17.7375	0.0900

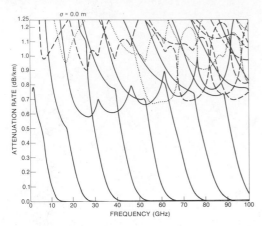

FIG. 24. Modal attenuation rate as a function of frequency for a moderate evaporation duct and zero rms bump height.

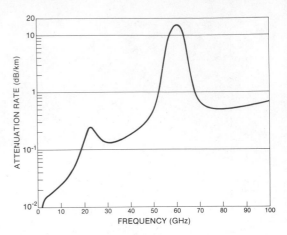

FIG. 26. Attenuation rate at the ground due to oxygen and water vapor as a function of frequency for a temperature of 15°C and 75% humidity (L_w = 0.0 g/m³).

ground loss and leakage). To this loss must be added the loss due to atmospheric absorption caused mainly by oxygen, water vapor, and precipitation. The molecular loss at the ground is shown in Fig. 26 for an air temperature of 15°C, a relative humidity of 75%, and no liquid water. The curve has been generated using H. Liebe's model.

Figure 24 is for a smooth earth and Fig. 25 has been generated for a surface characterized by an rms bump height σ = 0.32 m. Figure 24 shows the efficacy of the duct to trap modes as the frequency increases. Leakage loss is negligible for these well-trapped modes and ground loss is responsible for the residual attenuation rate which occurs after trapping. The latter slowly

increases with frequency because as the mode becomes better trapped it strikes the air–sea interface more frequently, thereby enhancing the ground loss. Modes with attenuation rates greater than about 0.6 dB/km are of the leaky variety and their proliferation with frequency is clear. For terminals within the duct and for transhorizon propagation it is likely that the signal levels will be controlled by the well-trapped modes. However, the leaky modes may play a significant role in field strength determinations outside the duct.

Figure 25 shows that the primary effect of the roughness is to increase the attenuation rate of the well-trapped modes. The distinct competition between trapping and surface scattering is clear. As a mode becomes better trapped it strikes the surface more frequently with the consequent increase in the attenuation rate. An interesting feature of the curves is that despite the frequency square dependence in the exponent of Eq. (9), the modes appear to level off at about 0.4 dB/km at the high end of the frequency band considered (indeed, it appears as though the attenuation rate is beginning to decrease for the least attenuated mode as the frequency increases).

It is known that the leaky modes are sensitive to the number of segments used for the M profile. Nevertheless, it is believed that the mode structure shown in Figs. 24 and 25 is representative of a continuous profile. Also, scattering from atmospheric turbulence has been ignored, as has been the frequency dependence of the

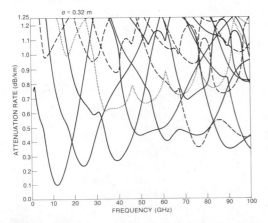

FIG. 25. Modal attenuation rate as a function of frequency for a moderate evaporation duct and an rms bump height of 0.32 m.

real part of the refractive index profile, which would be important in pulse stretching studies, particularly close to the 60-GHz oxygen absorption band. All of these effects deserve close examination in future studies.

VII. Propagation Assessment Systems

The recent advances in propagation modeling coupled with the increasing availability of inexpensive small computers has led to the development of systems that can calculate and display propagation effects in a timely fashion for use in assessing radio or radar system performance. The most advanced of these systems designed for use in a marine environment is the Integrated Refractive Effects Prediction System (IREPS). IREPS is designed primarily for use aboard naval ships to assess and exploit changes in the coverage patterns of radar, electronic warfare, and communications systems caused by abnormal refractivity structures in the lower atmosphere. The system is based around a Hewlett-Packard 9845 desk-top computer and uses the propagation models presented in Sections V and VI, except that no provision for horizontal changes in refractivity structure is included. Other limitations include the use of only ray optics for air-to-air geometries as opposed to full-wave solutions, and an approximate single-mode waveguide for surface-based ducts. The environmental input is primarily from radiosondes, consisting of pressure, temperature, and relative humidity; although the capability exists to interface IREPS with recordings from airborne microwave refractometers when they are available. The propagation models will handle a wide variety of surface-based and airborne air-search radars, some surface-search radars, and several electronic warfare and communications systems. The frequency limits of the system are 100 MHz to 20 GHz.

An example of an IREPS display is shown in Fig. 27. This example is for a hypothetical 400-MHz air-search radar with the radar antenna height being 30 m above sea level. The display shows this radar's coverage on a spherical earth range-versus-altitude diagram with three shaded regions that correspond to 90, 50, and 10% probabilities of detection of a small airborne target, based on free-space detection ranges of 46, 73, and 100 nautical miles. This example is for a refractivity structure characterized by a surface-based duct extending up to about 300 m, which

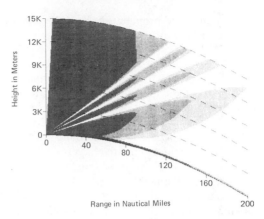

FIG. 27.   Example of a coverage display from IREPS.

results in the extended ranges indicated in the figure near the surface. At higher altitudes, the example shows three of the normal interference lobes, where their location has been modified from the standard atmosphere case based on ray traces through the actual refractivity profile. Only the first three of the lobes are shown, since higher ones get closer together and are both harder to resolve on the display and less meaningful to a user.

In addition to the coverage display, IREPS will generate path loss versus range displays that can be used in assessing maximum range performance for radar, communications, or electronic warfare applications when both terminals are located at fixed heights above the water. There are also tables that can be generated to assess maximum expected detection ranges of surface-search radars against predefined sets of ships classes or to assess maximum expected intercept ranges of predefined sets of radar emitters. Finally, there are displays that show the refractivity structure, and give general assessments of propagation conditions, based on either *in situ* measurements or the climatology described in Section II.

VIII. Conclusions

This article has reviewed tropospheric radio-wave propagation measurements and modeling applicable to radar systems operating in a marine environment. The work reported on spans five decades and has culminated in a capability to perform near-real-time assessments of radio and radar systems' actual performance based on *in situ* meterological measurements. Although there remains, and probably always will remain, justification for improvements to the models, the current IREPS capabilities appear to handle about 85% of the real-world naval applications. The largest payoff for improving performance assessment is likely to be in the area of refractivity sensor development. The ideal new sensor would be passive and remote in addition to providing sufficient accuracy to determine the presence and strength of refractive layers.

In the area of propagation model improvements, the development of approximate or ray-optics and full-wave hybrid methods aimed at reducing computation time in solutions for ducting environments would be extremely valuable. Also of importance would be an improved understanding of the interrelationship of refractivity fluctuations, surface roughness, and ducting. Recent work has shown considerable promise in developing propagation models for laterally inhomogeneous refractivity environment; however, the usefulness of such new models for assessment purposes may be severely limited by the lack of detailed refractivity measurements that will be required. Finally, there is a need to extend existing models and develop new models as necessary for millimeter-wave propagation in a marine environment, particularly with respect to evaporation ducting and surface roughness.

BIBLIOGRAPHY

Baumgartner, G. B., Jr., Hitney, H. V., and Pappert, R. A. (1983). *Proc. Inst. Elec. Eng.* **130**(pt. F), 630–642.

Gossard, E. E. (1978). *Radio Sci.* **13**(2), 255–259.

Hitney, H. V., and Richter, J. H. (1976). *Nav. Eng. J.,* **88**(2), 257–262.

Hitney, H. V., Pappert, R. A., Hattan, C. P., and Goodhart, C. L. (1978). *Radio Sci.* **13**(4), 669–675.

Hitney, H. V., Richter, J. H., Pappert, R. A., Anderson, K. D., and Baumgartner, G. B. (1985). *Proc. IEEE,* **73**(2), 265–283.

Jeske, H. (1971). *AGARD Conf. Proc.* **70**(2), 50.1–50.10.

Liebe, H. J. (1981). *Radio Sci.,* **16**(6), 1183–1199.

Pappert, R. A., and Goodhart, C. L. (1977). *Radio Sci.* **12**(1), 76–87.

Pappert, R. A., and Goodhart, C. L. (1979). *Radio Sci.* **14**(5), 803–813.

Rotheram, S. (1974). *Marconi Rev.* **37,** 18–40.

RADIO SPECTRUM UTILIZATION

William J. Weisz   *Motorola Inc.*

GLOSSARY

Allocation of a frequency band: Radio spectrum is divided into blocks of frequencies designated or allocated for use by a particular service or group of users such as public safety, police, land mobile, television. Individual users are assigned the use of a narrow band of frequencies within the allocated block of frequencies.

Band: Band is used to refer to a part of the radio spectrum; for example, the VHF band in the 150 MHz region, the television band, etc.

Bandwidth: Bandwidth consists of the range of frequencies located between two designated frequencies. One of the frequencies is the lower edge of the band and the other frequency is the upper edge of the band.

Channels: Radio channels are designated individual frequencies, or frequency bands that are assigned by a regulatory body. While television is allocated a block of frequencies between 174 and 216 MHz, there are seven television channels in this band; that is, channel 10 is assigned to occupy the frequencies between 192 to 198 MHz. Much of the radio spectrum has been assigned channel numbers; therefore, bands and channels are often synonymous.

Coverage: Coverage is the geographical area in which the user can transmit and receive the desired signal at a level that provides adequate quality of the received information. For example, the coverage of a broadcast station is the area in which a user can receive a strong enough signal to provide a clear picture and sound. This can range from a few kilometers radius to several hundred kilometers radius. In some radio systems, coverage is extended by using several transmitters placed throughout the coverage area.

Radio frequency signal: Signal on a specified designated frequency that transmits impulses or fluctuating electrical quantities, such as voltage, current, or electric field strength, the variations of which represent coded information.

Spectrum allocation: Process of assigning a certain part of the spectrum to be used for a particular radio service. It is common to refer to the entire set of assignments as the spectrum allocation.

Wavelength: Speed of light divided by the radio frequency. A working approximation is 300 m divided by the radio frequency in MHz.

The radio spectrum is that portion of the electromagnetic spectrum that can effectively be electronically radiated from one point in space and received at another. Presently this includes frequencies from 9 kHz (kilohertz = 10^3 Hz) to 400 GHz (gigahertz = 10^9 Hz). While there are international allocations for the radio spectrum up to 275 GHz, most of the commercial use takes place between 100 kHz and 20 GHz and very few experimental systems have been operated above 100 GHz. Signals using these same frequencies can also be generated and conducted via wires, coaxial cables, and glass fibers extending from transmitter to receiver, but since such signals are contained and not intended to be radiated, they are not part of the radio spectrum. There are many sources that unintentionally radiate, for example, the sun, power lines, computers, and automobile engines, and these must be considered because they impact radio spectrum usage. Since the same radio spectrum is shared by all within a given coverage area, it constitutes a common "natural resource." Allocations of portions of the radio spectrum are set worldwide by periodic international conferences and administered by the International Telecommunications Union (ITU), between the various national telecommunications authorities, who determine national policy for each country. Spectrum usage, or allocation, is determined by type of use, or "service," and is managed so as to optimize usage.

I. Introduction

Radio spectrum usage includes all of the transmissions made and signals that are received. The utilization of the radio spectrum is a measure of how effectively it is used. Efficient utilization of the radio spectrum depends on optimizing its use, but because of the various types of services using the spectrum, there is no universally accepted single method of measuring utilization and comparing the efficiency of one type of use versus another.

In the broadcast service, high utilization comes from providing as many broadcast channels as possible in a given area without causing destructive interference to the listening public. This is done by establishing frequencies or channels and controlling characteristics of the broadcast equipment used. Each channel has only one transmitter operating in any given coverage area, and many receivers. [See COMMUNICATIONS SYSTEMS, CIVILIAN.]

In services such as Land Mobile (police, fire, business use, etc.), high utilization comes from providing service to a large number of roving and fixed users in a given area, while providing a coverage over the working area that will give the quality of service required without causing destructive interference between users.

In some services, 100% receipt of signals is mandatory. Normal statistical methods of measuring spectrum utilization do not apply. In these instances, utilization may be very high despite the fact that there are very few transmissions. The distress calling channels used by ships and airplanes are examples of such cases. Here, the emphasis is on keeping all unnecessary transmissions off these channels so that the distress calls can be received, even under adverse conditions.

II. Radio Spectrum as a Natural Resource

Radio spectrum is universally considered a natural resource. In the industrialized areas of the world this resource is extensively used and there is constant pressure to regulate radio spectrum to make room for additional users. In contrast, in many developing countries, and in areas of low population density, portions of the radio spectrum are virtually unused.

Nevertheless, radio spectrum usage is increasing everywhere. Not only are more users seeking to use existing services, but also there are new services that require portions of the radio spectrum. These new and different services will place future demands for radio spectrum greater than those previously envisioned. [See DATA COMMUNICATION NETWORKS.]

The radio spectrum is a special type of natural resource that can instantly be renewed. It is not permanently depleted like oil. If everyone using the radio spectrum were to stop transmitting, the entire spectrum would be instantly available for other uses. This has only happened, however, during wartime, or when a regulatory body changes a group of users from one part of the radio spectrum to another.

On the other hand, the radio spectrum resource, or parts of it, can be effectively depleted if the number of transmissions exceed the capacity of the band. When this happens, mutual interference becomes so great that the amount of usable transmission is reduced below the point of usefulness. Where this has occurred, the cause can be traced to the national regulatory body having permitted too many users access to a band, or the lack of either clear operating rules or operator discipline. For this reason, national governments grant users the privilege of using radio spectrum in exchange for agreements to abide by usage rules. In addition, the users often agree among themselves on informal additional procedures to ensure that the spectrum resource is not overloaded.

In time of disaster, man-made or natural, the emergency demand for additional transmissions rises to several times normal. At these times even the best disciplined services may overload if radio frequency allocations have not been carefully made to accommodate increased usage needs. The regulatory body must always keep this potentially large demand in mind when considering allocations. The radio spectrum resource reaches its maximum value during and just after such an event.

As with many natural resources, the radio spectrum resource has no real value until it has been developed through generally interference-free use, that is, until equipment for the specified use has been designed, built, and placed in operation. Radio spectrum then acquires the value of the equipment and the service that it provides. As a result, radio spectrum allocations, once made, tend to be extremely resistant to change.

The potential value of a portion of the radio spectrum is a function of the ability to use it economically and efficiently. This depends on many things such as frequency band service, and the locale etc. The market value of a single television channel in a major city is in the hundreds of millions of dollars. The worldwide capital investment of the public in over 480 million television sets and broadcast equipment of thousands of TV stations provides great incentive to retain the current allocation for television. The same can be said, to some degree, for every established service in the world. Where the radio spectrum is used by commercial businesses as a tool to increase productivity, cost saving can generally be accurately determined. [*See* TELECOMMUNICATIONS.]

In contrast, spectrum used to transmit timely and accurate information to and from policemen, firemen, and soldiers may have a value not measurable in direct monetary terms. However, a large investment must nonetheless be made in equipment and training to provide for this capability. This investment can be lost if the service is required to change to different frequencies or transmission types.

National governments undertake the task of evaluating the needs of government services (either local or national), together with the needs of commercial services, in order to appropriately allocate and enforce specific radio spectrum uses. Their task is complicated by requests for spectrum for new services resulting from technology advances as well as maximization of return on the existing users' investments. Most countries have formed an agency (in the United States, this is the Federal Communications Commission) to administer the radio spectrum. However, some countries rely on a state communications company to allocate the radio spectrum. Other countries give the responsibility to the military.

In the lower frequency bands, for example, below 20 MHz, and in certain higher frequency bands, such as those used for satellites, the signals can at times reach around the world across national borders. The maritime and aeronautical use of radio spectrum requires that the transmitters aboard airplanes and ships, moving from one country to another, must operate under the jurisdiction of the country in which they are operating. It is, therefore, necessary that there be worldwide agreement on the frequency allocations in many bands. Because radio frequency signals often cross national borders where they could interfere with radio frequencies allocated in another country, nations have additionally had to find ways to coordinate their individual allocation procedures. The International Telecommunication Union (ITU), an organization of the United Nations, has been formed for this purpose. The ITU seeks and obtains agreement on the broad categories of services to operate in specified radio frequency bands. The nations belonging to the ITU accept these agreements as treaties. The failure of governments to abide by international allocations, or to enforce these agreements, could destroy the value of the service using that portion of the radio spectrum.

III. Regulatory Organizations and Procedures

A. INTERNATIONAL REGULATORY ORGANIZATIONS

The International Telecommunications Union (ITU), part of the United Nations, is the international body that determines the worldwide radio spectrum allocations. It does this through World Administrative Radio Conferences, where ITU member nations participate in the decision making. The consensus allocations developed at these conferences are ratified by member nations by treaty. The resultant tables, published by the ITU, are known as the Radio Regulations.

The Radio Regulations indicate which service is to use a given frequency band. For some bands, the rules indicate that two or more services may share a given band. Each country then makes its own detailed frequency allocation plan consistent with The Radio Regulations tables.

Many fixed service systems are intended to transmit information internationally. This requires that the equipment be mutually compatible. The International Radio Consultative Committee (CCIR), a part of the ITU, is organized to study technical and operating questions in radio communications and to issue recommendations for equipment and signal specifications. Among such recommendations are those on transmission specifications that will provide the required intercountry compatibility. This committee also provides recommendations for radio system technical requirements that are used by many countries to help them in planning their own internal allocation of the radio spectrum.

Many services use a channel designation to identify the authorized transmitter frequency. For example, television stations are known to operate on channels rather than frequencies. The CCIR makes recommendations for channel frequency spacing and numbering plans, which, when adopted, aid aircraft and ships that may have to operate worldwide.

History shows that once a service has been established and uses a particular portion of the spectrum it is seldom changed. When a new service or use for the spectrum is developed, the regulatory agency normally assigns it to high frequencies. This is the most prevalent reason why an examination of the allocation table shows that many services have allocations widely scattered over the spectrum.

B. National Regulating Agencies

In the United States, the regulation of radio spectrum is done by two agencies of the federal government. They are responsible not only for spectrum allocation, but also for any technical equipment specification necessary. The Federal Communications Commission (FCC) regulates radio spectrum for all non-federal government use, including the aeronautical and maritime mobile services. The National Telecommunications and Information Administration (NTIA) regulates spectrum used by the agencies of the federal government, including the military.

The NTIA does not have to make public the reasons for its allocation of spectrum. Its task is complicated by the necessity of coordinating with all countries in which the United States has received permission to operate governmental agencies using radio spectrum (e.g., embassies and military bases).

On the other hand, the decisions of the FCC are based on public participation where any interested party can make proposals and comment on any and all proceedings. Because the radio spectrum is a valuable resource, there are many vested interests, both personal and institutional, that impinge on any allocation changes. The U.S. system, which has by and large worked well, is by its very nature a political process. After receiving all of the proposals and comments, both written and oral, the FCC commissioners must often make difficult allocation decisions. These decisions are subject to judicial review. FCC actions fall under the *Telecommunications Act of 1934*, and all subsequent Federal laws. One advantage of the U.S. system is that the public's needs are constantly reviewed and are a part of every decision. An advocate for change can advance his cause at every level of the process. One disadvantage is that the system may not be as quick to respond to technology changes or to new services.

As a matter of policy, the FCC has not regulated receiving equipment performance specifications to the same degree as regulatory organizations in other countries. It does, however, regulate transmissions that directly affect spectrum usage. This is because a receiver owner can choose his own degree of communication selectivity and quality without causing interference to other users or affecting spectrum utilization.

In many European countries, the responsibility for radio spectrum allocation is given to state communications monopolies that may also operate national telephone systems. Some of these organizations operate on the principle that they have the responsibility and the right to provide any and all of the communication needs of their citizens. Therefore, they allocate the radio spectrum in ways that best allow them to provide services as they see fit.

In general, these organizations have exercised strong system engineering influences, and regulate more aspects of the user's radio system, including both receiver and transmitter performance specifications.

Every year there are new requests for services that require radio spectrum as well as an increased demand for the existing services. While these demands can easily be met in less developed parts of the world, it is a real problem in metropolitan areas of developed countries. A great deal of pressure is placed on regulatory agencies in developed countries to find new and better ways of allocating or reallocating the sought after radio spectrum resource.

C. ALLOCATION PROCEDURES

Frequently, more people seek to be licensed to use a frequency band than there are frequency allocations available. In many countries, only government agencies can provide the communications services. In these cases there is no problem in deciding who should be licensed to use radio frequencies. However, more countries are putting the broadcast and communications business in the hands of private companies. Therefore a government-issued license to use a frequency or band of frequencies is necessary for an organization that wants to operate a radio system, be it for public safety, business, or other use.

In the United States, the Federal Government encourages the development of many radio services by individuals or companies. As a result, broadcast licenses, for example, are transferred (with the FCC's approval) when a broadcast station is sold. This license is a tangible asset of great value. When new allocations are available, the FCC first determines which applicants are capable of providing the service. Where there are more qualified applicants than there are available allocations, the FCC can license those it feels are best capable of providing the service. Financial capability is a major factor when the service requires a big investment in equipment and personnel.

In recent years, there have been more highly qualified applicants than allocatable frequencies. In an attempt to be fair, the FCC has turned to the use of lotteries to choose among qualified applicants.

While the U.S. Government has never attempted to obtain significant revenue from the licensing of radio spectrum—as it does from the leasing or selling of other natural resources—there have been a number of proposals to do this:

1. To sell the license to the highest bidder. This is the simplest proposal and in the broadcast service is the de facto way that existing licenses are transferred.

2. To lease frequency bands to the highest bidder for fixed periods. When the lease expires that portion of the spectrum would again be leased to the highest bidder. In this proposal, the spectrum would not be allocated to any specific service, and the use of the spectrum could change each time the lease expired.

While the proposed allocation systems might determine the value of the natural resource and shift the usage to that part of the spectrum that could produce the highest dollar return (some people believe this is maximum spectrum utilization), it does not provide equal access to the spectrum by those who may need it the most and are not necessarily able to acquire the necessary financial base to compete for the use of the spectrum.

These allocation processes are only proposals in the United States. Many other countries face the same problems and could adopt one of these processes or develop other ways to extract revenue from the use of radio spectrum.

IV. Spectrum Allocation Principles

In most countries, the priority followed by the regulators for allocating radio spectrum is as follows.

1. Military.
2. Aeronautical and maritime emergency communications, radio navigation, and radio location.
3. Public safety functions, for example, police, fire, and other emergency services.
4. National telecommunication companies, for example, microwave and satellite telephone relays.
5. Broadcast radio and television.
6. Private users, for example, private mobile systems and private fixed microwave relays.
7. Others.

As an example, the Congress of the United States recently recognized the importance of the Private Land Mobile Radio Services to the public welfare and economy when it adopted the Communications Amendments Act of 1982, which cites *inter alia,* that

> "In taking actions to manage the spectrum to be made available for use by the private land mobile services, the Commission (FCC) shall consider, consistent with section 1 of this Act, whether such actions will—
> 1. promote the safety of life and property;
> 2. improve the efficiency of spectrum use and reduce the regulatory burden upon spectrum users, based upon sound engineering prin-

ciples, user operational requirements, and marketplace demands;

3. encourage competition and provide services to the largest feasible number of users; or

4. increase interservice sharing opportunities between private land mobile services and other services.''

For locations where there is high spectrum usage, regulatory agencies must determine if services can be provided using means other than radio, such as cable or telephone networks. They must also consider the relative cost of using alternative methods of providing the service. For example, many studies have been conducted to determine the feasibility of providing radio and television broadcast service by cable. Such a change would allow the regulators to reallocate much valuable spectrum to other services. Such changes, however, must also consider the cost of scrapping existing equipment. While public necessity is the first priority (see above), the public interest and convenience is a factor in many services such as broadcasting, the private use of mobile radio systems (including radio paging), the amateur and citizen radio services, and microwave ovens. While these services are not necessary to save lives, they do provide for an improved quality of life and are desired by the public.

V. Use of the Radio Spectrum

For regulatory purposes, uses of the spectrum are classified by the ITU by type of service provided. The major services and some subclassifications are as follows.

A. Broadcast Service

The radio transmissions in this service are intended for direct reception by the general public and generally are radiated from one antenna location to cover a wide area around the transmitter where thousands of randomly located fixed or moving receivers operate. The familiar AM and FM radio, as well as television transmissions, are in this service.

B. Fixed Service

This service constitutes radio communications between fixed points. The most common equipment used in this service is telephone microwave systems used to interconnect switching centers. Low-frequency radio telegraph systems capable of transmitting data long distances between countries and across oceans are also in the fixed service.

C. Mobile Service

The radio communications in this service are between and among automobiles, trains, aircraft, person-carried portable units, etc. and fixed stations.

1. Aeronautical Mobile Service: Radio transmissions to and from airplanes and including emergency radio beacons.

2. Land Mobile Service: Radio transmissions to and from land vehicles and portable units.

3. Maritime Mobile Service: Radio transmissions to and from ships.

D. Satellite Service

The broadcast, fixed, and mobile services have corresponding satellite services. In addition, there are satellite services for amateur, meteorological, standard frequency, time, earth exploration, and radio navigation. The transmissions in this service generally start from earth stations and are relayed by a satellite station back to earth. In some uses the satellite may originate communication. [See SATELLITE COMMUNICATIONS.]

E. Standard Frequency and Time Signal Service

A radio communication service for scientific, technical, and other purposes, providing transmission of specified frequencies, time signals, or both, of stated high precision, intended for general reception.

F. Radio Navigation Service

Transmissions in this service are for navigation and obstruction warning.

G. Radio Location Service

In this service the radio transmissions are used to determine position, velocity, and other characteristics of objects. Radars and radio altimeters are examples.

H. Radio Astronomy Service

The transmitters in this service are of cosmic origin such as stars. Because the signals received have very low power, many regulators have provisions in their allocations to protect certain radio frequency bands in locations that are near radio telescope sites. They do this by prohibiting or restricting the power of transmitters that could radiate power in the protected radio frequency bands.

I. Amateur Service

The amateur service is for self-training, intercommunication, and technical investigations carried out by duly authorized persons interested in radio technique solely with a personal aim and without pecuniary interest. Almost all countries have recognized that by encouraging radio amateurs they create a group of people trained in the use, maintenance, and design of radio equipment. Amateurs have consistently shown the practical use of the higher radio frequencies before there were commercial applications at these frequencies.

J. Industrial, Scientific, and Medical (ISM)

Many machines use radio frequency power for purposes other than transmission of information. Such equipment is used to generate heat, control chemical reactions, and make measurements. Microwave ovens are one example. The very nature of ISM equipment mandates that it radiate large amounts of radio frequency power. While there are no regional allocations for this service, the International Radio Regulations have specified that ISM equipment radio frequencies be restricted to certain bands. Because of widespread ISM usage, these bands are generally not suitable for other radio uses.

VI. Users and Usage

A. Broadcast Service

Broadcasting is the most widely used radio service. Its allocations include a variety of frequency bands with various types of transmission capabilities. While licenses to operate broadcast transmitters are granted to individuals or institutions (see Table I), the real beneficiary is the general public. The oldest broadcast service uses amplitude modulated signals in a frequency band that extends from 535 to 1605 kHz (the AM broadcast band). In most countries, this is the predominant method of broadcasting, although many countries still broadcast using single sideband in the lower portions of the spectrum. At

TABLE I.   Stations Authorized by the FCC for the Year 1982

Service	Number of stations
Broadcast	
AM	4,800
FM	5,000
TV	1,280
Fixed	28,000
Mobile	
Aeronautical	238,000
Maritime	435,000
Land mobile[a]	864,000
Satellite	
Earth stations	6,000

[a] While exact records are not available from the FCC, it is estimated that there are about 7,000,000 Land Mobile units in operation.

certain times, such AM broadcast signals can propagate unusually large distances, especially after sunset. Such signals can cause destructive interference in broadcast receivers tuned to receive other desired weak signals. International cooperation in frequency allocation, regulating power and determining the daily hours of broadcast has been extensive, and service to the public has been excellent.

FM radio and television are also widely used for broadcast. These systems use frequencies above 54 MHz and are much less susceptible to the long distance "skip" interference characteristic of AM broadcast above. In order to provide the users with high quality signals, stations with overlapping coverage areas are not assigned the same frequencies.

FM broadcast is resistant to interference to strong signals by weaker signals, so the same channels can be reassigned at distances of 150 km or more separation between stations.

Television, however, is much more susceptible to weak signal interference. To provide protection to these users, television transmitter frequencies cannot be duplicated with separation distances of less than 300 km. Interference may also result from transmitters operating on adjacent frequencies so that in a given location, only one transmitter is allocated for every six UHF television channels.

B. Fixed Service

The fixed service is least known to the general public, but is an essential part of the radio industry. Systems in the fixed service are, for the most part, operated by communications companies or by companies that have a need for large communications networks—for telephone or data communications. Since these systems transmit information between fixed points, there is a constant reevaluation to determine whether coaxial cables or glass fibers can more effectivcly be used. In the urban areas of the industrialized countries, most fixed services use microwave frequency bands above 5 GHz, 5 to 20 MHz in width. Relay stations are used for long distance microwave communications to receive and retransmit information. These relay stations are spread 40 to 50 km apart along a line between the end points of the network. Because of the wide bandwidth, microwave systems have large capacity and are capable of simultaneously transmitting several television signals and thousands of telephone messages. Relay stations can cost well over a million dollars and the cost of the equipment at each end can cost tens of millions of dollars.

In less developed countries and areas of low population density, equipment using lower frequencies (150–500 MHz) and narrower bandwidths (25–100 kHz) are also used for the fixed service. In these areas the fixed service is used to extend telephone services beyond the wire network into rural areas.

Frequencies below 20 MHz are also used in the fixed service to provide both telephone and telegraph service to isolated areas such as remote islands.

There has been a long history of regulatory changes that move the fixed service out of the mid-frequency bands (50 MHz to 1 GHz), and to allocate these frequencies to the mobile services. Today, more of the functions that are provided in the fixed services are being transferred to the satellite services which are more reliable and often less expensive.

C. MOBILE SERVICE

The mobile service consists of three different services: aeronautical, land mobile, and maritime. While there are possible alternatives to radio for providing the fixed and broadcast requirements, radio is the only means of providing communication to units or people in motion. The need to communicate with moving vehicles and people and to coordinate the movement of materials has made the mobile service the fastest growing segment of all users of spectrum.

While these services have dissimilar needs, they have the following characteristics in common:

1. They provide communication between and among land-based, fixed locations and airplanes, vehicles, ships, or people.
2. They began as a means of providing emergency communication, and still retain that as a first priority.
3. The communications requirements of this service have expanded until today most of the information transmitted is for administrative and operational purposes. The users of this service (including business users) are constantly finding ways to use this transmitted information to make their mobile fleets more efficient and effective. Fleets that have used radio extensively have shown significant improvements in productivity for both personnel and equipment. The use of radio is now necessary to be competitive.
4. Their transmissions are mostly of short duration and any one user may only need to transmit for a few seconds at a time, adding up to only a few minutes per hour. This is in contrast to the broadcast and fixed services where transmitters are on continually or for long periods of time. This intermittent use characteristic allows many users to share the same frequency in the same location on a time-division basis, if they follow good radio usage practices.
5. They use relatively narrow bandwidths (2 to 16 kHz) to transmit the necessary information, compared to broadcast, fixed, and radio location (television, microwave relays, and radar) which use bandwidths of 5 to 50 MHz.

An increasing amount of frequency usage in the mobile service is for transmissions between people in motion carrying small portable radio equipment. The need for this type of communication in all of the mobile services is growing at a rapid rate, especially in urban areas.

The aeronautical mobile service is used to communicate with airplanes. The familiar ground control communications with aircraft is allocated frequencies between 108 and 144 MHz. Lower frequencies (between 4 and 6 MHz) are used to maintain contact with transoceanic flights as they move out over the oceans.

Mobile telephone services are the most widely used land mobile applications in many countries. This extension of the telephone service uses many frequencies between 400 MHz and 1 GHz. In the United States, one variation of this service, called the Cellular Telephone System, uses frequencies above 800 MHz. This service provides high quality mobile telephone service to hundreds of thousands of private and business people and adds greatly to radio spectrum requirements of the mobile service.

The demand for inexpensive equipment to communicate to and from the automobile has resulted in small spectrum allocations for "citizens' services." In North America, the 27 MHz citizens' band service has been established where one can operate low power equipment without formal licensing. A great many radios have been sold for this service. In Japan, a similar "personal radio service" at a frequency near 900 MHz has attracted thousands of users. The usefulness of this type of service depends a great deal on the users' willingness to conform to good radio usage practices and to tolerate large amounts of interference from others operating on the same channel.

Personal paging is a one-way land mobile service. The paging service users wear small receivers and the system provides for the transmission of a message to the user from one or more locations, and, in some systems, from any telephone. The message may be a simple "beep," indicating that the user should contact a certain person, or it can be a short voice message or a message that can be read from a visual display. These systems operate on frequencies up to 1 GHz.

The maritime mobile service has allocations throughout the radio spectrum. This service has two types of operations. When the ships are in or near ports, they communicate on frequencies above 146 MHz. At sea, they use frequencies below 26 MHz. More than in the other mobile services, most of the information transmitted to ships is data. Radio teletype and other automatic data systems are used. Although the speed of these systems is slow, it is well adapted to the propagation characteristics of the low frequencies used. An increasing number of the larger ships now use satellite relays for transoceanic communication. These satellite systems are more reliable and allow for higher rates of information transmission.

D. Satellite Service

Satellite systems are in service, or have been proposed, for each of the other services. Most satellite usage, however, is for the fixed services where satellite relay systems are operated by international or national communications companies. These systems transmit much the same information that is transmitted by fixed service microwave systems (e.g., television programs and telephone messages). Some national satellite systems now operating provide telephone service to outlying regions of less developed countries.

There are also national systems in the Satellite Broadcast Service that provide television broadcast to remote areas. Many specialized broadcast features, including movie channels, are available to those willing to incur the cost of large antennas and program fees. As the power available in the satellite relay is increased, broadcast systems that can use inexpensive receivers and antennas will become possible. Allocations for direct home broadcast have been made.

Most of the fixed, broadcast, and maritime mobile services use synchronous orbits and frequencies above 1.5 GHz.

E. Radio Navigation Service

Radio navigation services serve aeronautical and maritime users. Radio navigation systems consist of networks of transmitters in strategic locations that allow the user to locate his position in relation to the network, and hence themselves. In some navigational systems the transmitters broadcast sufficient information so that the location can automatically be calculated and displayed in the ship, airplane, or vehicle. In addition, there are specialized beacons and location markers, like those used at airports, to provide guidance for an aircraft.

Most navigational systems use narrow bandwidths, but systems allocations range from 9 kHz to well over 1.3 GHz. Satellite radio navigation systems are increasingly used in many applications.

F. Radio Location Service

Most frequency usage in this service is for radar. Since radar obtains its directional information from the use of very narrow antenna patterns, and since such narrow antenna patterns are more easily achieved at higher frequencies, most of the radars operate at microwave frequencies above 2 GHz. Radar obtains its distance information by measuring the time for a radio pulse to travel to and from the target. To improve the accuracy of this measurement, the radio pulses are made very short, thus increasing the bandwidth occupied by the radar transmitter. The bandwidth can be several MHz wide.

G. AMATEUR SERVICE

Radio amateurs (or "hams") were the first users of the radio spectrum and originally had use of the entire spectrum. The amateur service is the only service that had continuously given up spectrum to other services. As amateurs demonstrated the practical use of the higher frequencies, they were found to have commercial value and reallocated to new uses. While much amateur spectrum has been transferred, there are still allocations for amateur bands that extend from 1.8 MHz to 250 GHz. Most countries certify a user by means of a technical examination before he is allowed to use the spectrum. Over the years, the amateur radio operator has demonstrated a high degree of technical and operational discipline.

VII. Usable Radio Spectrum and Radio Equipment

To understand the factors that must be considered in order to make optimum use of the radio spectrum, it is necessary to understand a little about spectrum and radio equipment characteristics that impinge on such judgments. Because spectrum allocations are often affected by the cost of equipment to produce and receive transmissions, some brief comments on such equipment will also be made. The most basic characteristic of the radio spectrum is the location of a frequency in the spectrum. There are many factors to be considered that are a function of frequency when determining the use that will be made of a specific band of frequencies. In general, the relationship with frequency that exists is shown in Table II.

A. RADIO FREQUENCY NOISE

Noise is present at all frequencies in the radio spectrum and consists of signals having random amplitude fluctuations. Sources of noise include the sun, the stars, the atmosphere, for example, lightning, and man-made equipment, including radios. Both naturally generated and man-made noise is greatest at the low frequencies and decreases at the higher frequencies, and man-made noise is greater in urban than in rural areas. Below about 1 GHz, noise is the limiting factor in the effective sensitivity of a receiver in urban areas.

The amount of noise a receiver is exposed to is proportional to the bandwidth of the receiver. While a well-designed receiver having a bandwidth that is typically found in the land mobile service requires a signal only slightly greater than the noise, a television receiver requires a signal that is much larger than the noise.

B. RADIO TRANSMITTER POWER

The cost of radio transmitters increases with both frequency and power. At the low frequencies, most transmitters have a low power, stable, accurate frequency source that may be modulated with the intelligence to be transmitted. This signal's power is amplified 1,000,000

TABLE II.   Radio Characteristics versus Frequency

Characteristic	Lower frequency	Higher frequency
Amount of radio frequency noise	Greater	Less
Ease of generating radio frequency power	Easier	Harder
Ease of radiating radio frequency energy	Harder	Easier
Ease of directing radiated radio frequency energy	Harder	Easier
Propagation loss of radio frequency energy over land or sea	Lower	Higher
Multipath degradation of the received signal	Less	Greater
Ease of rejecting unwanted signals	Easier	Harder
Ease of keeping transmitter or receiver on assigned frequency	Easier	Harder

times before it is applied to the antenna. The transmitter radio frequency output power can range from a few hundred milliwatts to hundreds of thousands of watts. At the low frequencies, amplifiers that produce a few watts are not generally expensive, but higher power amplifiers are costly. At higher frequencies, the cost of amplifiers increases and output radio frequency power decreases. At frequencies above 2 GHz, it is very expensive to use amplifiers that produce more than a few watts.

While it is possible to configure a transmitter to generate radio frequency power in a single device to be useful for community information, the frequency of the device must be controlled accurately and the device modulated by the intended intelligence. This takes additional, often expensive, circuitry.

One well known noncommunication appliance, the microwave oven, uses the single device method and is able to generate hundreds of watts. However, the oven frequency is not modulated and its frequency is allowed to vary over a considerable band of frequencies.

C. Radiating Radio Frequency Energy

In order for the radio frequency generated by a transmitter to be useful, it must be transformed from electrical energy in the transmitter to electromagnetic fields in space. This is the function of the antenna. Effective antennas are one-fourth to one-half of the wavelength of the frequency being used. Antennas can be improved by making them larger, but the size of such antenna structure may be impractical for long wavelengths. Effective antennas for AM broadcast stations may have a height of over 200 m. Some antenna structures for very low frequencies (e.g., 200 kHz) have lengths of over 1 km. Geographical coverage range in the higher frequencies can generally be extended by mounting the antenna on a tall tower or on the roof of a building. [See ANTENNAS.]

At the high frequencies (above 2 GHz) the antenna dimensions become very small and difficult to work with. At these frequencies a properly flared opening (horn-type antennas at the end of the transmission line) can provide effective radiation of these frequencies.

Generally, it is easier to radiate radio energy at higher frequencies because of the small size of the antenna structures required.

While the above statements have been directed at transmitting antennas, they apply equally to receiving antennas as well.

D. Directing Radiated Radio Frequency Energy

In addition to radiating radio energy, antennas have directional characteristics. The directivity of an antenna is a very important system parameter. For many services, antennas should radiate equally in all directions of the compass with as much of the energy directed at the horizon as possible. Broadcast, maritime, and land mobile stations are a few examples of such services. In such cases, the coverage area is a circle centered at the antenna.

On the other hand, most stations in the fixed service transmit almost all their energy in a narrow beam only in the direction of the intended receiving station. There are several system benefits that result from this: all the transmitter's power is focused in one direction and therefore less power is required to deliver a usable signal level to a receiver. At the receiver, interference is reduced if it originates from a different direction than the intended transmitter. Because the transmitter's power is focused in one direction, this allows the same radio frequency to be used in the same geographical area several times by other unrelated fixed stations whose transmitter power is focused in other directions. The impact on the allocation of frequencies depends on the degree of directivity that can be obtained.

The degree of directivity that an antenna can have is a function of the dimensions of the antenna. The dimensions of directional antennas range from 1 to 1000 wavelengths. Television receiving antennas are about two wavelengths long. Fixed station antennas for microwave stations can be up to 1000 wavelengths across. While it is possible to build directional antennas for the very low frequencies, the antenna structures become very large, possibly up to several kilometers long.

For some services, it is desirable to point the directional antenna in different directions at different times. Examples of this are amateur stations and television receivers. Radars in the radio location service use antennas that are highly directional, but are constantly rotated. Most satellite, earth, and space stations use directional antennas and many are steerable.

Whenever an antenna has directivity, it almost always has an effective gain. This is measured as the effective field strength radiated by a directional antenna in a given direction, compared to the field strength produced by an omnidirectional antenna that radiates the same power equally well in all directions. Typical gains range from 2, for some television antennas, to over 1000 for some fixed station antennas.

It is easier to obtain high directivity at the higher frequencies because of the shorter wavelengths and hence smaller antenna dimensions required.

E. Propagation Loss of Radio Frequency Energy over Land or Sea

At the low frequencies (below 500 kHz), radio frequency energy propagates through earth, sea, and the atmosphere. At these frequencies, propagation loss over sea is about three times less than that over land and communication circuits of over 1500 km are readily established. This has made these frequencies very valuable to the maritime service for communication to ships at sea and to the fixed service for communication to remote islands. [See RADIO PROPAGATION.]

At higher frequencies (2–20 MHz) the land and sea propagation losses increase at a rate faster than the atmospheric losses. At times, longer communication circuits can be established using atmospheric radio wave reflections, but the loss varies with time and such communication circuits cannot be reliably maintained for long periods of time.

Although the loss at one frequency may increase, the loss at another frequency may decrease. The allocation of these frequencies is made in such a way that communication systems can switch from one band to another to find a usable frequency. Because these frequencies are so very useful, many international agreements exist for the allocation of these frequencies.

It is of interest to note that the information bandwidth of a single satellite relay station, which operates at a frequency around 20 GHz, is about as great as the entire spectrum from 9 kHz to 20 MHz and such a satellite can provide communication service to ships and remote islands more reliably.

In the frequency range between 30 MHz and 1 GHz the radio propagation approaches that of light. At 30 MHz there is enough refraction of the radio waves that transmissions are possible beyond the horizon by as much as 100%. As frequency increases, transmission becomes more line of sight and is limited more and more to the horizon. Increased propagation loss is also incurred in places that are shadowed from the transmitter by hills or buildings. For frequencies greater than 1 to 2 GHz the propaga-

tion loss is very great between locations that are not line of sight.

F. Multipath Degradation of the Received Signal

At low frequencies (500 kHz) the energy propagated through the earth and the energy propagated through the atmosphere travel different distances and can arrive at the receiving antenna out of phase, and, to some degree, can cancel one another. This can have the effect of increasing the propagation loss. If the antenna is moving, the cancellation will slowly change to reinforcement and back to cancellation. The distance between cancellation may be several kilometers. While the effect is very noticeable, the slow rate of change minimizes the degradation to communications.

At higher frequencies (e.g., 50 to 1000 MHz), the antenna may receive signals from many reflectors, hills, buildings, overhead wire—which can combine and cause cancellation, and the effect is the same as for the low frequencies, except that the cancellations occur every 6 to 0.3 m. The effect is most noticeable when the magnitude of the direct signal is about the same as the reflected signals. For the land mobile service, this can be the case over much of the coverage area. If the antenna is situated on a fast moving vehicle, it can pass through many cancellations in a second, disrupting the communication link. The degree of degradation of the received signal increases with the frequency of the signal and the speed of the vehicle.

To overcome this degradation, systems are designed to have a higher field strength in the coverage area. This affects the distance between systems before the frequency can be used again and must be accounted for in the spectrum allocation process.

G. Rejecting Unwanted Signals

Every receiver has some capability to select a desired signal and reject all others. This is called receiver selectivity. The better the selectivity, the greater the receiver cost. If, in a given coverage area, AM broadcast transmitters are assigned to adjacent broadcast channels, receivers would have to have a high degree of selectivity in order to clearly receive a remote station if the receiver was physically close to the adjacent channel broadcast station. In practice, this is never the case, since spectrum allocation regulations regarding geographic spacing of adjacent channels assure that in any location there are no broadcast stations assigned closer than 4 or 5 channels to each other. This reduces the required selectivity of broadcast receivers, with correspondingly lower cost, at the expense of using additional spectrum. A good rule of thumb is that better spectrum utilization in any given frequency band generally requires costlier equipment.

The frequency band between the assigned channels is called the "guard band," and almost all allocations have provision for some amount of guard band. The guard bands can be reduced if transmitters are not physically closely spaced. This, however, is not practical in the mobile services where the mobile transmitters may transmit from any location. While it is not possible to build receivers with perfect selectivity, professional receivers are available which can reject unwanted interference signals only 25 kHz away from the desired signal and which are 100,000 to 1,000,000 times greater than the desired signal.

VIII. Technology and Usable Radio Spectrum

Commercial use of spectrum is limited by the cost of generating the frequency being used with sufficient power and of receiving it. Almost without exception, the cost of devices to achieve transmitter power increases with the frequency and the power level required. For these reasons, the first commercial application of a higher frequency is usually for system applications that can more readily support the use of the costly devices. When the technology progresses and the cost of the devices is reduced, other system applications for that frequency are found. For example, frequencies between 500 and 1000 MHz were first used for the fixed services where the cost of the devices could be spread across many telephone channels. The devices used were low power—less than a watt. When the technology allowed one to build costly high power devices (greater than 20,000 W) at these frequencies, television transmitters were built. When the technology advanced further so that solid state devices of moderate power (up to 50 W) were available, they found use in the land mobile services in applications such as the cellular mobile telephone system and in private land mobile systems where the cost of these devices was a significant part of the mobile radio's cost.

IX. Technology and Spectrum Utilization

A. HIGHER FREQUENCY PRACTICALITY— AN EXAMPLE: THE LAND MOBILE

The land mobile service began as a police radio service using compatible modulation at 1600 kHz. In 1940, FM systems were permitted and began operating in the 30 MHz area. In its early years the police band was allocated on the basis of a scaled down version of the FM broadcast system. Channels were spaced every 120 kHz. The FM system proved to be superior and soon was the most popular two-way police radio system. This popularity, and subsequent advances in technology, resulted in several new allocations for police use higher in the radio frequency spectrum. In 1985, the United States, channels are allocated for police use in the frequency bands of 30 to 50, 150 to 170, 450 to 512, and 806 to 866 MHz.

B. NARROWER BANDWIDTHS

Increasing advances in the technological state-of-the-art enabled 120 kHz channel spacing in the land mobile service to first be split to 60, then 30, or 25 and 20 kHz depending on the frequency band. This allowed four or more additional usable channels to be placed in the same 120 kHz that previously accommodated only one. Hence, four times as many users could be handled, thus dramatically increasing radio spectrum utilization.

In parts of Europe, channel spacing has been further reduced to 12.5 kHz. This narrowing of channels has required that the allowed peak deviation had to be reduced, receiver bandwidth greatly reduced, receiver selectivity substantially increased, and the frequency stability of both the receiver and transmitter greatly improved. The reduction in FM modulation deviation resulted in degradation of the quality of transmission, but enabled some additional users (other than police) to use mobile radio. While voice transmissions over these systems have been found to be satisfactory, some amount of user learning is required to acclimate to the characteristics of the system.

C. DIGITAL SIGNALS

Until a few years ago, almost all the services used analog modulation to transmit information. One characteristic of this type of transmissions is that every time the signal is processed or relayed, some noise and distortion is added. While this can be minimized, it cannot be eliminated. Some signals may be processed 30 to 40 times as they move from source to destination. Digitizing, wherein the value of the signal is represented by a series of symbols, usually on or off pulses, allows systems to be designed that eliminate reprocessing errors. While the process of digitizing signals has many advantages in wire, coaxial cable, and radio systems, it often requires the use of a greater bandwidth in order to utilize the advantages. [*See* SIGNAL PROCESSING, DIGITAL.]

The trend in digital systems is to quantize the signal at its source and not to decode the signal until it is at the location where it will be used. The digitized signal can be transmitted over many media such as wireline, microwave, and satellite relays.

Because digital signals offer the user higher quality transmissions and greater freedom in processing the signals, many of the new services are now digital. In addition, there are trends to convert existing systems to digital modulation. Some countries are allocating large bands of frequencies to be used by systems that will use only digital signals and will provide both fixed and mobile services. While digital systems generally use a greater bandwidth than analog systems, the digital systems tend to be less degraded by low level unwanted signals. In crowded areas this may allow digital systems to achieve a utilization equal to or greater than the analog systems.

D. ADDITIONAL INFORMATION BY EXISTING TRANSMITTERS IN ALLOCATED BANDWIDTH

Many radio systems, when first developed, use more bandwidth than necessary due to technological limitations or equipment cost limitations. These systems are given allocations based on the original equipment's bandwidth needs. As stated before, once an allocation is made, it is difficult to change. On the other hand, as technology advances, it may be possible for additional usage to be made of the same bandwidth. Well-known examples of this include the addition of color to television, the addition of stereo to FM broadcasting, and, more recently, the addition of stereo to AM and television broadcasting.

More such cases can be expected. There are some successful experiments utilizing digital technology to provide new services using existing broadcast transmitters. Radio paging and telex service use one or more of the subcarrier channels of an FM broadcast transmitter.

Text can be transmitted by a television transmitter using the brief time interval while the transmitter is not sending necessary picture information, for example, between frames or lines. This type of shared use of equipment and spectrum is certain to increase.

E. HIGHER QUALITY TRANSMISSIONS

As the mobile services are developed for use by the general public, higher quality voice transmission over that which is acceptable for communication between more highly trained users, such as policemen or truck drivers where the radio is a working tool is necessary. For mobile telephone service (air, ship, or land), the goal of those providing the system is to achieve a voice transmission quality equal to the existing wireline telephone network, characterized by having a narrow bandwidth, 3400 Hz, and low noise.

Telephone-quality mobile radio systems that use reasonable transmitter power generally use modulation that requires a wider radio system bandwidth. The Cellular Mobile Telephone System uses FM modulation with peak deviation that is 240% of what is allowed in the public safety and business land mobile services. The voice transmission achieved very nearly equals the quality of the best telephone networks and the user does not have to become accustomed to a different transmission quality.

There are currently proposals for mobile telephone systems that use digital modulation. To achieve the same voice quality, the designers are proposing use of the same digital rate in radio systems that is used in the telephone network. Since the current channel allocations will not allow the radio systems to pass these high digital rates, wider channels would have to be allocated, or narrower bandwidth digital methods developed.

F. ALLOCATION ALTERNATIVES TO ACCOMMODATE GROWTH AND NEW USES AND THE DILEMMA OF NARROWBAND VERSUS WIDEBAND CHANNELS: OPPOSING OBJECTIVES

Many radio services have been, and will continue to be, faced with two opposing trends. The first is an apparent need to reduce channel width to permit more users to be served in a limited amount of bandwidth. The second is a desire to increase channel bandwidth to accommodate an enhanced type of service such as data transmission in the land mobile service, or better picture resolution in the television service.

1. How Growth Has Been Handled— an Example

The land mobile radio service is probably the prime example of a service where continual growth in number of users and type of service is a characteristic. There is continual growth in the number of users in the land mobile service which provides communication for essential services such as police and fire and is an important efficiency producing tool for businesses of all types in the United States. In 1958, there were about 1,000,000 transmitters licensed, and in 1975 there were 7,400,000.

As the U.S. economy continues to expand, and as new applications for the use of two-way radio continue to emerge, this growth is expected to continue into the foreseeable future. Land mobile channels have historically been made progressively narrower to provide additional channels to relieve congestion. For example, channels in the 150 MHz band, which once were 60 kHz wide, have been reduced to 30 kHz, and with geographical coordination, are now used with 15 kHz spacing. At 450 MHz, a similar evolution resulted in reduction of 200 kHz channels to 50 kHz, then 25 kHz, and with special restrictions to 12.5 kHz. This phenomenon is called channel splitting.

A similar pattern is now beginning in the 800 MHz band, which is in its infancy compared to the other lower frequency bands. The Federal Communications Commission, faced with the pressure of land mobile spectrum congestion, is contemplating a reduction in width in the current 25 kHz channels. Other alternatives include additional geographical sharing with other services such as television, relocating some existing service allocations to different frequencies to provide spectrum for land mobile use, and attempts to develop frequency spectrum above 400 MHz for land mobile service use. Historically, channel splitting and each of the other alternatives has been used extensively to support land mobile growth.

2. The Dilemma of New Uses—an Example: The Trend to Digital

a. Digital Voice Security.   The general procedure for radio communications has been to send all voice transmissions using one of the standard modulation types, AM, FM, or single sideband. While countries do have laws directed against the use of intercepted information by unintended recipients for personal gain, it is nearly impossible to keep people who have suitable receivers from listening. [*See* VOICEBAND DATA COMMUNICATIONS.]

Much of the information now transmitted is considered to be sensitive and so the trend is to encrypt the signal transmissions. The most secure encryption method is to convert all the transmitted information to digital form and then scramble the digits.

Digital voice security techniques currently in widespread use in the nongovernment markets require channel bandwidths in the order of 20 kHz or greater. Thus, the current bandwidth of 25 kHz will have to be maintained.

b. Data Transmissions.   In the fixed and mobile services, a rapid increase in the transmission of nonvoice or data messages is taking place. The new users of this service include computers transferring files among themselves. The trend is to do this at higher and higher information rates, which require wider bandwidth communication channels. This trend is changing the pattern of system usage as well as creating a need for more spectrum for these services.

c. Integrated Services Digital Network.   The new wireline Integrated Services Digital Network (ISDN) is digital, so it seems reasonable that many mobile systems of the future will also be digital to permit compatibility with the ISDN. In turn, this suggests that narrower channels may not be better and could be worse because of greater "overhead" considerations for data operation on narrower channels, as compared to wider channels.

Digital Voice Security, ISDN, and the requirement for transmitting ever more data faster illustrate some of the driving forces for wider channel bandwidths. Time Division Multiplex (TDM) techniques may be readily applied to digital systems. Therefore, the classical view of narrowband channels permitting more users per megahertz is no longer necessarily correct. Perhaps wideband channels using TDM techniques may result in greater efficiencies than narrower channels not using TDM. It should be pointed out that the use of digital techniques for land mobile is, in some respects, still a future issue, whereas narrower channels can be effectively used for voice communications today. Thus, the regulatory agency must balance the potentially significant benefits of what seems to be a relatively certain future trend against a short-term need that can be accommodated by current capability.

d. Shared Frequency Bands. Current policy calls for only one television channel allocated in any geographical area where spectrum is authorized for six; five TV channels of spectrum are unused. Under carefully controlled conditions, land mobile channels are now being assigned in this unused space. This technique is presently limited to the UHF television band. Cases of interference to television reception have not been apparent.

In many locations, microwave systems have been using the same frequencies that are used by satellite systems. This is possible because of improvements in directional antennas and careful allocation procedures. [*See* MICROWAVE COMMUNICATIONS.]

Another proposed method for sharing frequency bands is to use a modulation system known as spread spectrum. This system causes the transmitter power to be uniformly spread over a very wide band of frequencies, 10 to 50 MHz. The spread spectrum receiver, which is tuned to the desired signal only, receives other signals as noise. Thus it might be possible to use spread spectrum in the same frequency band with other types of radio systems.

e. Satellites. As the cost of placing satellites into orbit drops, more and more countries will want to use them to provide broadcast, fixed, and mobile services. The use of a satellite relay allows modern, high quality service to be provided without the need for the intercity and rural communication networks that have been built in the industrialized countries. While most of the satellite services today operate below 20 GHz, it is expected that devices will soon be available to generate sufficient radio frequency power at frequencies above 20 GHz. There are propagation problems to be resolved, but studies are under way that will hopefully provide practical answers.

f. Multiple Frequency Trunking Systems. Most radio communication allocations reserve a block of frequencies for use by the same or similar users. Except for police, fire, and certain

other land mobile users, no one is given a ''private'' channel. Multiple users share a single channel by voluntary, disciplined time sharing. Each user restricts his transmission time to periods when the channel is not being used by others.

Trunking systems, on the other hand, consist of many channels and utilize a single centrally located base station. Individual users are automatically shifted to any unused channel in the system by the base station rather than being restricted to a single one. This allows more users per channel than in conventional single channel systems. These systems are common in many land mobile services including the cellular mobile telephone system.

g. Digitized Time Sharing. Telephone network designers have found that it is more efficient to digitize the communications and combine many users' communications into a single high-speed serial digital signal stream. In this type of system, a time segment corresponds to a frequency channel in a conventional radio system. In telephone systems, it is less expensive to switch, combine, separate, and direct digital signals. Although the bandwidth of this digital technique is greater than the combined bandwidths of the individual communications, this method, now used in some existing radio systems, may be more widely used in the future.

The use of this technology is now common in the fixed service bands where systems can be allocated the required wide channels. It is also being used in the mobile satellite services where the wider bandwidths are available. Central controllers in these systems are complex, but for the most part are similar to small computers. Their cost, fortunately is decreasing rapidly. It can be expected that the telephone network technology will find application in other services, and is being proposed for land mobile services.

BIBLIOGRAPHY

Federal Communications Commission (FCC) (1983). 48th Annual Report, FY 82.
Federal Communications Commission (FCC) (1984). General Docket No. 84-1233, RM-4829.
IEEE Communications, Vol. 23, No. 1, January 1985, ISSN 0163-6804.
ITU General Secretariat, Radio Regulations, Edition of 1982, Vol. I, Geneva 1982, ISBN 92-61-01221-3.

SATELLITE COMMUNICATIONS

S. J. Campanella   *COMSAT LABORATORIES*

GLOSSARY

Code division multiple access (CDMA): Process that shares time and frequency among many users by assigning to each an orthogonal code pattern that uniquely identifies each users signal.

Equivalent isotropic radiated power (EIRP): Measure of the power radiated by an antenna in the direction of a receiver expressed as the equivalent power that would have to be radiated uniformly in all directions.

Erlang: Measure of telephone traffic load expressed in units of hundred call seconds per hour (CCS). One erlang is defined as the traffic load sufficient to keep one trunk busy on the average and is equivalent to 36 CCS.

Frequency division multiple access (FDMA): Process that shares a spectrum of frequencies among many users by assigning to each a subset of frequencies in which to transmit signals.

Time division multiple access (TDMA): Process that shares the time domain of a single carrier among many users by assigning to each time intervals in which to transmit signal bursts.

From its position in orbit at an altitude of 22,300 miles above the earth, a single geostationary satellite views almost half the earth's surface. Geostationary satellite communications has grown from a limited efficiency, simple, single-wideband-microwave-transponder operation to a high-efficiency, advanced multiple-transponder, multiple-base, microwave-switched operation.

I. Introduction

INTELSAT, an international communications satellite consortium managed by COMSAT between 1962 and 1978, and since than by INTELSAT's Director General and his staff, has been a world leader in commercial satellite system development. These satellites are illustrated in Fig. 1. Its first two commercial satellites, INTELSAT I and INTELSAT II, used omnidirectional antennas, which while inefficient did serve to initiate multidestinational international communications with considerably greater ease than with either submarine cable or high-frequency radio. Each of these satellites accommodated as many as 240 telephone circuits by using frequency division multiplexed, multiple-access (FDMA) C-band carriers with FDM–FM modulation to achieve a fully interconnected network between earth stations located in North America and Europe.

However, for two reasons, these satellites were relatively inefficient. Their energy was directed not only to earth but to open space where it was useless, and the number of telephone channel carriers possible was limited by transponder power and intermodulation distortion. Initially, satellite network control consisted simply of the assignment of telephone channels to frequency positions in FDM basebands carried on the frequency modulated (FM) FDMA carriers, which were received by all stations, as illustrated in Fig. 2.

This relatively simple FDM–FM–FDMA modulation and access technique combined with

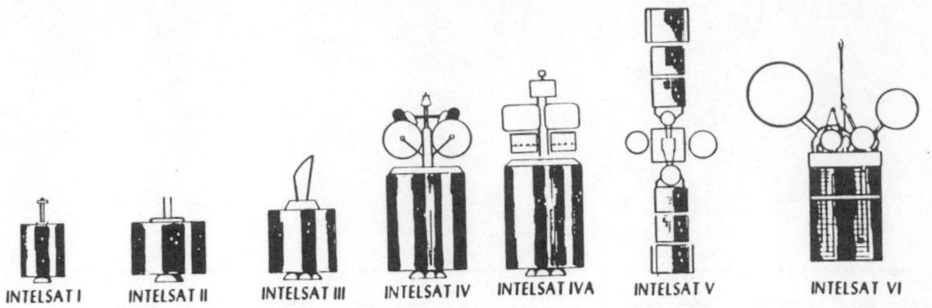

	INTELSAT I	INTELSAT II	INTELSAT III	INTELSAT IV	INTELSAT IVA	INTELSAT V	INTELSAT VI
TRANSPONDERS	2	1	2	12	20	27	48
BW/TRANSPONDER (MHz)	25	130	225	36	32, 36	40, 80, 240	40, 80, 160
COVERAGE	N. HEMISPHERE	GLOBAL	GLOBAL	GLOBAL & SPOT BEAMS	GLOBAL & SPOT BEAMS	GLOBAL, REGIONAL, & SPOT BEAMS	GLOBAL, REGIONAL, & SPOT BEAMS
e.i.r.p. (dBW)	11.5	15.5	23	22.5 (GLOBAL) 33.7 (SPOT BEAM)	22 (GLOBAL) 29 (SPOT BEAM)	22 - 29 (4 GHz) 44 (11 GHz)	23 - 31 (4 GHz) 41 - 44.4 (11 GHz)
NO. OF TEL. CIRCUITS	240 (NO MULT. ACCESS)	240	1200	4000 (AVG.)	6000 (AVG.)	12000	30000
LIFETIME (YR)	1.5	3	5	7	7	7 +	10
SATELLITE COST/CIRCUIT YR ($K)	30	10	2	1	1	0.9	0.5

FIG. 1.   The INTELSAT satellites from 1964 to 1986.

the illumination of the entire region by a toroidal antenna pattern helped realize one of the principal advantages of the geostationary satellite: it permitted full mesh interconnectivity among N stations distributed over half the earth to be accomplished with only N carriers, one carrier originating from each station, as illustrated in Fig. 3. To obtain the same flexibility using point-to-point terrestrial methods would require $(N/2)(N - 1)$ terrestrial links. Thus, for example, to fully interconnect a mesh of six nodes requires only six carriers in a satellite system, to accomplish the same mesh interconnectivity in a terrestrial system, 15 terrestrial carrier routes are required. [See TELECOMMUNICATION SWITCHING.]

As satellites evolved, the power transmitted toward the earth (EIRP) and bandwidth were increased. INTELSAT III, with increased EIRP due to deployment of the first earth coverage antenna and employing two 250-MHz transponders, achieved increased capacity but still was power limited. INTELSAT IV channelized the transponders to 40-MHz bandwidth to increase power density, and in addition to earth coverage beams, used two 4-GHz, 3-1/2 degree spot beams to enhance down-link EIRP by about 8 dB, but was capacity limited by intermodulation products and the total spectrum available. INTELSAT IV-A was the first satellite to use multiple beams to achieve spectrum reuse by antenna pattern control.

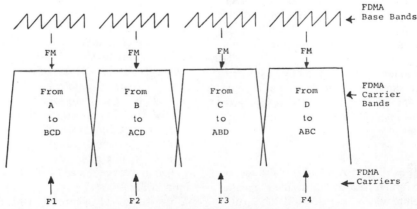

FIG. 2.   The FDM–FM–FDMA transmission format.

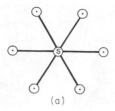

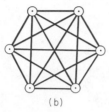

FIG. 3. Comparison of (a) satellites (N multidestination carriers) and (b) terrestrial full mesh connectivity [$N(N-1)/2$ links].

INTELSAT V and VI have expanded use of multiple beams, including reuse by dual orthogonal polarizations, as well as antenna pattern control at C band and the addition of K_u-band beams to provide reuse of the spectrum by fourfold and sixfold, respectively. However, multiple beams tend to isolate communities of earth stations thus eliminating or impeding the worldwide connectivity advantage of the stationary orbit satellite. Cross-strapping beams by interconnection through static switches is used to ameliorate this situation but becomes cumbersome for a large number of beams. Introduction of satellite-switched time division multiple access (TDMA) restores full connectivity by permitting dynamic time-domain switching of traffic between multiple beams. On-board baseband processing combined with baseband switching in a multiple-beam antenna environment achieves the same result and adds new dimensions for on-board signal processing and beam hopping.

Accompanying the evolution of the INTELSAT satellites has been the development of many other regional and domestic satellite systems operating at either C band or K_u band or a combination of both. These include COMSTAR, ANIK, PALAPA, SATCOM, GALAXY, INSAT, ARABSAT, AUSSAT, TELESAT, and others. Whereas the INTELSAT satellites have been traditionally designed in the early generations to provide low values of satellite G/T (-15 to -10 dB/K) and EIRP (20 to 28 dBW) with wide earth coverage beams requiring them to operate with large 30-m and 10-m earth stations at C band, most of the domestic and regional satellites have been designed to operate with such smaller coverages. This confined coverage has permitted them to use more directional satellite antennas as K_u-band, resulting in higher antenna gains and consequently higher satellite G/T (-5 to 2 dB/K), and accompanied by increases in satellite TWTA power output (10- and 20-W tubes) to significantly increase EIRP. (35 to 48 dBW). As a consequence, it is possible to operate communications services with earth segment antennas of considerably reduced size (1.8 to 5.5 m) and power (1 to 300 W) depending on the type of service and earth station capacity. Thin-route data services using small earth station nodes communicating to large hubs and a single channel per carrier typically require the smaller antennas and power, while high-volume trunk telephone traffic carried by TDMA requires the larger antennas and power. [*See* RADIO SPECTRUM UTILIZATION.]

II. Satellites

Communications satellites are radio-relay stations in space. They serve much the same purpose as the microwave towers one sees along the highway. The satellites receive radio signals transmitted from the ground, amplify them, translate them in frequency, and retransmit them back to the ground. Since the satellites are at high altitude, they can "see" across half of the earth. This gives them their principal communications advantage; insensitivity to distance. Another major advantage is that they can see all the microwave transmitters and receivers (earth stations) on almost half of the earth. Thus they can connect any pair of stations or provide point-to-multipoint services, such as television. Ground links, such as microwave relays and cables, are inherently limited to providing single-point to single-point services. [*See* MICROWAVE COMMUNICATIONS.]

All communication satellites operate in synchronous equatorial orbits, which makes possible uninterrupted reception and transmission with minimal tracking by earth stations. The satellites receive signals transmitted from the ground on frequencies in the 6 GHz (6 billion cycles/sec) or 12-GHz band and retransmit them back to the ground on frequencies in the 4- or 11-GHz band. Some satellites are spin-stabilized, so that they maintain their orientation in space by means of a rotation of the body of the satellite

FIG. 4.   INTELSAT VI spin stabilized communications satellite.

along an axis parallel to the earth's axis (Fig. 4), and others are platform stabilized and maintain their orientation by small jets and momentum wheels (Fig. 5). All are powered by arrays of silicon solar cells mounted on the spinning body.

The satellite body serves as a platform for the communications payload. In the case of a spin-stabilized satellite, this platform is "despun" in order to maintain its earth orientation. The communications payload includes the receivers, amplifiers, transmitters, and antennas. The structure carries with it the electric-power system, the stabilization system (including its fuel tanks), and a solid-fuel rocket engine to provide the necessary "kick" at the apogee of the satellite's transfer orbit to place it in a circular equatorial orbit with essentially zero inclination (Fig. 6). The apogee engine and its propellant weigh about as much as the rest of the satellite components put together.

The stabilization system depends on infrared sensors to determine the direction of the earth and on small thrusters to align the satellite axis properly and keep the antenna beams pointed earthward. For orbital station-keeping, it is necessary to minimize the satellite's tendency to drift along its orbit because of the gravitational perturbations due to the oblateness of the earth.

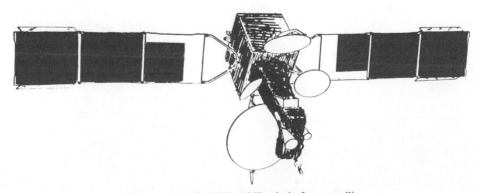

FIG. 5.   INTELSAT V stabilized platform satellite.

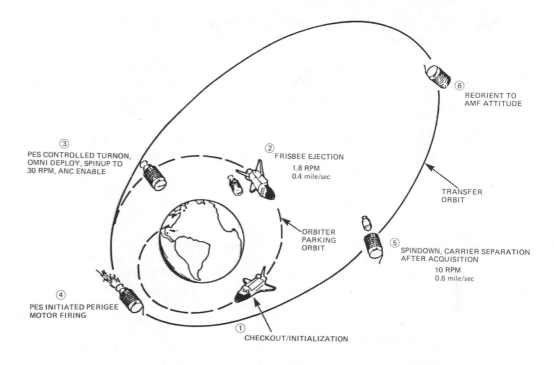

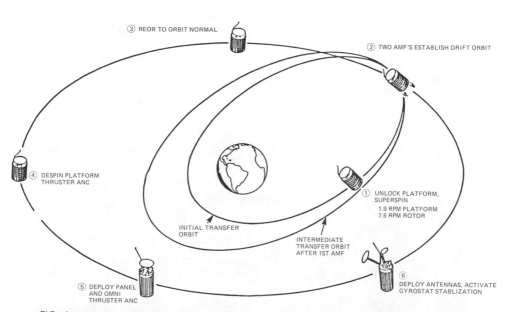

FIG. 6.   (a) The STS launch to transfer orbit and (b) transfer orbit to synchronous orbit.

The satellite is carefully tracked by ground stations, the propulsion vector needed to compensate for drift is computed, and the appropriate thrusters are fired by radio command. The thrusters, fueled by hydrazine, generate the reaction force necessary to adjust the satellite's position. With them the stabilization system can control the attitude of the platform, move the satellite in the east–west direction, and counteract the gravitation influence of the sun and moon, which tends to increase the inclination of the satellite. The correction for inclination consumes a major fraction of the hydrazine fuel and is therefore closely related to the effective life expectancy of the satellite.

It is also necessary to monitor the condition of the various systems and devices in the satellite and to take corrective action in the event of a malfunction. This is accomplished through telemetry in which a variety of sensors detect temperatures, voltages, and other conditions that are transmitted to special earth stations through a radio link. The information is subsequently sent to the control center, where it is processed for evaluation and displayed. If any out-of-tolerance condition is detected in the satellite, corrective action can be taken by radio command from the same special stations.

The effectiveness of a communications satellite is measured not only by its lifetime but also by its capacity, which in turn depends on radiated power and bandwidth. As the technology developed during the past decade, satellites were improved in all three areas; lifetime, power, and bandwidth.

More powerful launch vehicles are able to place bigger satellites in orbit, providing a greater area for solar cells and thereby more electric power for radio transmission. With increased power and more sophisticated transponder design, more of the allocated bandwidth can be utilized. Improved components, devices, and design, together with more effective quality assurance techniques, all improved satellite reliability and operating lifetime.

III. Earth Stations

An earth station transmits powerful radio signals to a satellite and receives weak signals from it. Earth stations serve to connect the satellite network to the ground communications network. They may interface into terrestrial telephone systems or private business terminations. The earth station must reorganize the various incoming signals from ground networks for voice, television, and data communications into suitable groups for satellite transmission. Each is modulated onto a radio-frequency carrier wave. The power needed for transmitting the signal to the satellite is developed by an amplifier that can produce from 1 W to several kilowatts depending on system requirements before it is fed to a parabolic antenna ranging from 1.8 to 30 m in diameter, depending on the application. The function of the antenna is to confine the radiated energy beam aimed at the satellite. The value of power radiated by the antenna is the product of antenna gain and high-power amplifier (HPA) power and is referred to as the equivalent isotropically radiated power. A typical large diameter earth station used in the INTELSAT trunk telephone system is shown in Fig. 7, and a smaller diameter station used in INTELSAT's International Business Service is shown in Fig. 8. [See COMMUNICATION SYSTEMS, CIVILIAN.]

After traveling 22,300 miles through space, the signal is received at the satellite. The satel-

FIG. 7.   Earth station for international satellite communications at Roaring Creek, Pennsylvania.

FIG. 8. The COMSAT international earth station—New York Teleport.

lite amplifies this signal, translates it to the down-link frequency band, boosts it up to an equivalent isotropically radiated power of 500 to 50,000 W, and beams it back to the ground. At the earth station, the antenna collects the energy and feeds it into a low-noise receiver. There it is amplified and interfaced into the local distribution network.

IV. Transmission

Satellite communications systems have evolved over the past two decades using principally FDMA techniques that involve multiple carriers per transponder to achieve point-to-point and point-to-multipoint telephony, television, and business services. High- and medium-capacity telephone trunk services are carried on wide bandwidth FDM/FDMA analog carriers, while low-capacity thin-route services are carried by single-channel-per-carrier (SCPC) FDMA. Digital versions of these FDMA services have also been introduced. In recent years, TDMA systems have appeared operating at transmission rates sufficient to fill a transponder and hence achieve highly efficient single-carrier-per-transponder operation. Such systems operate from as low as 25 Mbit/sec to as high as 120 Mbit/sec. Also, to achieve improved connectivity in multibeam satellites, SS-TDMA systems are about to make an appearance. A further development has been the appearance of low-bit-rate TDMA systems that operate with several TDMA carriers per transponder. These systems appear to provide a better compromise between earth station power, complexity, and cost and space segment efficiency than SCPC or high-bit-rate TDMA for medium-capacity traffic routes. Demand assignment of space segment capacity has been introduced in the SCPC and TDMA systems to permit highly efficient use of space segment capacity. The development of limited beam coverage for regional beam and domestic applications, as well as extensive use of the K_u band, has increased the satellite G/T and EIRP, making it possible to significantly reduce the power, size, and cost of the earth stations for all classes of service.

V. Multiple Access Networks

Satellites offer unique opportunities to develop new types of networks that cannot be achieved by terrestrial transmission means because of their omniview of almost half the earth from their vantage point in space. Two types of networks are prevalent, preassigned multiple access and demand assigned multiple access.

Preassigned multiple access networks are established by assigning fixed subsets of telephone channels to various destinations on FDMA carriers or TDMA burst sized to meet traffic requirements. When such subsets are large, load variation between destinations is negligible making preassigned networks control efficient. A single transmitted carrier permits each station to reach all others within the view of the satellite's downbeam. Each station can receive carriers transmitted by all others and pick out those channels it desires. Today most telephone traffic carried on both international and domestic satellite systems uses this simple, preassigned network control method. Preassignment will continue to be used for most high-volume traffic carried on the INTELSAT V and INTELSAT VI satellites; it will be extended to similar services carried by TDMA and satellite-switched TDMA.

Although preassigned operation is well suited to high-volume trunk telephone service, it is inefficient for medium and thin-route telephone service because of large fluctuations in traffic load that occur when the number of circuits on a route is relatively small. The need for thin-route telephone service was identified early in the evolution of satellite systems, and to meet this need SPADE was developed by COMSAT Laboratories to establish economical satellite communications for use by developing nations. It was found that the average traffic on such links was only in fractions of an erlang, not enough to justify full-time assignment of a channel. This fractional use of a channel can be efficiently accom-

Table I. Communication Paths in the INTELSAT System

	Atlantic	Pacific	Indian	Total
1973	156	55	82	293
1979	370	101	286	757
1984 (MAR)	660	163	392	1215

modated by allowing an earth station to have access to one channel of a pool of space segment channels when a call is placed, and to return the channel to the pool when the call is completed. This method, generally referred to as demand-assigned multiple access (DAMA), is a unique type of service that is the exclusive domain of satellites. Table I shows the channel loading advantage possible for various DAMA pool sizes. The loading efficiency approaches 100% as the pool size increases.

The SPADE system provides a pool of up to 800 single-carrier-per-channel (SCPC) FDM channels in a transponder of 40-MHz bandwidth from which assignments can be made on demand by any earth station in the system. After call termination, the channel is returned to the pool. The carriers are voice activated to achieve a 4-dB carrier-to-noise ratio improvement. Voice channels are digitized by 56-kbit/sec pulse-code modulation (PCM) and quadraphase shift keyed (QPSK) modulated at 54 kbit/sec onto the carrier. The SPADE frequency assignments are shown in Fig. 4. Whenever a circuit is set up, the forward channel is picked from a lower band, the return channel from the upper band.

To control demand assignment of the network circuits, a demand-assigned switching and signaling (DASS) system is used. The method uses a distributed DAMA network controller. Each SPADE terminal maintains a complete assignment map for the entire system. Before a station selects a channel, it first analyzes the assignment map, chooses an unused channel, and indicates its choice on the DASS network. The associated return channel completes the circuit. When the call is completed, a message is sent over the DASS system to all of the stations, which indicates that the channel is no longer in use, thereby releasing it for another call. The call assignment and releases messages must be received by all individual stations so that each can maintain its assignment map.

Other versions of SCPC demand assignment

have been introduced and are currently used in many developing countries throughout the world. These systems use a centrally controlled demand assignment, not the distributed control demand assignment of the SPADE system. A centrally controlled demand assignment scheme encounters a double-hop delay between a request for circuit assignment and the actual assignment response, distributed demand assignment, on the other hand, allows for immediate assignment. Central demand assignment costs less because all processing is performed by a single central facility shared by all users, and for this reason it is being chosen for implementing most current DAMA systems. An example of such an SCPC is that produced by Sky Switch of Boulder Colorado. DAMA is also applied to TDMA by adjusting the number of channels carried in a TDMA traffic burst to meet traffic fluctuations and also by transmitting single channel bursts to provide rapid response to demand.

VI. Satellite Access Methods

It is common practice to divide the radio spectrum assigned to satellite service into subbands, which are referred to as channels, and to equip each with a transponder that receives weak up-link signals arriving from ground stations, amplifies them and retransmits them back to earth. The block diagram of a typical transponder,

shown in Fig. 9, comprises an input low-noise amplifier to amplify the weak up-link spectrum, a frequency converter to translate the channel to its appropriate down-link frequency assignment, an output multiplexer filter to define the channel in the down-link spectrum, and an output power amplifier to supply the high down-link power needed to overcome the losses and interference encountered on the down-link. Transponder channels at C-band frequencies typically are spaced nominally 40 or 80 MHz apart and have a useful bandwidth of 36 or 72 MHz, respectively. At K band, transponder channels with frequency spacings ranging from 40 to 250 MHz are in use.

The frequency plan of the transponders used on the INTELSAT V communications satellite is shown in Fig. 10. A satellite will carry enough transponders to occupy the full radio frequency spectrum space assigned to it. The 500-MHz spectrum assigned to fixed satellite communications at C band can accommodate 12 40-MHz transponders or 6 80-MHz transponders. By frequency reuse of the same spectrum, accomplished by using different polarizations and spacial separations in the form of beams, the available spectrum can be multiplied. For example, INTELSAT V reuses the 500-MHz C band four times and its successor, INTELSAT VI, will reuse the spectrum six times. Beam coverages for the INTELSAT V are shown in Fig. 11.

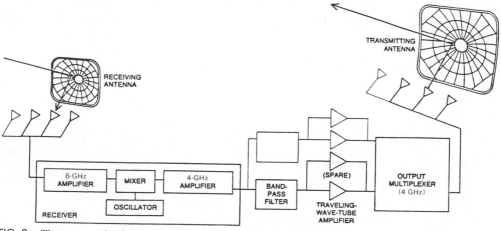

FIG. 9. The communications system of INTELSAT IV-A includes 20 transponders, or separate radio-frequency channels. Each channel has a bandwidth of 36 MHz, sufficient to accommodate 2 television channels or 600 two-way telephone circuits. Here the path through one transponder is shown. The receiving antenna on the satellite picks up the signal from the earth and feeds it into a receiver, which amplifies it and then translates it from 6 GHz to 4 GHz, the frequency used for retransmission to the earth. The band-pass filters keep the 20 transponder frequencies separated from one another. The signal is boosted by a traveling wave-tube amplifier and passed to one of the output multiplexers, where signals from several transponders are combined and fed to the appropriate antenna.

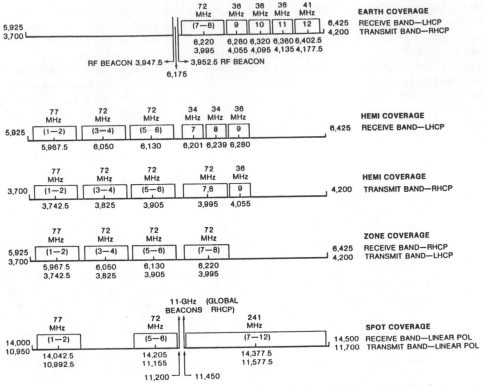

FIG. 10.   INTELSAT V transponder center frequencies bandwidths and polarizations.

Each transponder receives signals sent to it on its up-link channel frequency (in the 6-GHz C band or in the 12-GHz K_u band) from many earth stations, translates these to the appropriate down-link channel frequency (in the 4-GHz C band or in the 11-GHz K_u band), and after amplification, retransmits them back to the earth.

The signal sent from any one of the earth stations on the up-link can be received by all of the stations in the down-link. This is commonly referred to as point-to-multipoint or multidestinational operation and is an important property unique to satellite systems that cannot be easily matched by terrestrial telecommunications services and is responsible for the great utility of satellites in telephone and data telecommunications applications. Also, a multiplicity of earth stations can share each transponder channel by transmitting signals on different frequencies or at different times or with different codes. These different ways of multiple accessing are referred to as frequency-division multiple access (FDMA), time-division multiple access (TDMA), or code-division multiple access (CDMA), respectively.

A. FDMA

Each earth station accessing the satellite with FDMA is assigned its own exclusive carrier frequency and bandwidth within a particular transponder channel on both the up-link and down-link along with the carriers of several other stations. Typical FDMA carrier assignments in a typical transponder channel are shown in Fig. 2. Any station wishing to receive another station's

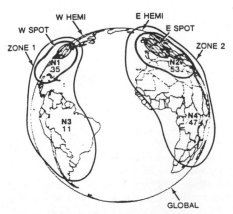

FIG. 11.   INTELSAT V beam coverage.

transmitted signal will tune its receiver to that station's down-link frequency. Stations accessing a transponder with FDMA will be assigned different bandwidths according to their traffic capacity. The carrier frequencies must be adjusted to ensure that the bands occupied by each carrier do not overlap the bands of the other carriers and cause excessive adjacent carrier interference. Because of this requirement, any readjustment of the capacity assigned to one of carriers requires corresponding adjustments to the others and can entail a considerable coordination effort. Typically, the power amplifiers used in the transponder are nonlinear and because FDMA operates with several carriers sharing a transponder channel, distortion products due to intermodulation create noiselike interference, which limits the fraction of power that can be used. Transponders operating with FDMA are typically backed off to 40% or less of their full power output to achieve near linear operation and reduce the interference. Even so, FDMA has been the dominant method for the delivery of satellite services during the first two decades of satellite communications and is expected to continue to be so in the future. FDMA is used for carrying analog trunk telephone on both domestic and international satellites, thin route single-channel-per-carrier (SCPC) telephony and data, demand-assigned multiple access (DAMA) SCPC, broadcast television distribution for major television networks, cable television distribution, and is currently being introduced for international digital trunk telephone distribution. Typically, the capacity achieved for FDMA operation in a 40-MHz transponder channel carrying several carriers in the INTELSAT system is 500 analog telephone channels. This increases to 1000 analog telephone channels if a single carrier occupies the transponder.

B. TDMA

Each earth station accessing the satellite with TDMA occupies its own exclusive fraction of time (time slot) in a time frame (TDMA frame) shared with several other stations that occupy other time slots. A typical TDMA frame is shown in Fig. 12. The TDMA frame is periodic with a period that may range from as short as 125 μsec to as long as 30 msec. During its assigned time fraction, a station transmits a short burst at an assigned carrier frequency that contains its traffic in time compressed form. On the

transmit side of a TDMA terminal, the signal presented to the input during a period equal to the TDMA frame is stored in a memory and is transmitted at an assigned time as a short burst in the next TDMA frame. Because this memory compresses the time interval containing the signal, it is called the compression memory. Usually, the compression memory consists of two memories operating in a ping–pong fashion, in which one stores the current input signal while the other transmits the previous frame's and in the next frame the roles are reversed and so on. Any station wishing to receive a station's transmitted signal will adjust its receiver to open a time window at the expected time of arrival of the transmitting station's traffic burst in the TDMA frame. On the receiver side of the terminal the received burst is stored in a memory operating at the high transmission rate and is read from this memory at the normal signal rate during the next frame. Because this memory expands the time interval of the short burst back to its normal duration, it is called the expansion memory. It is usually implemented in the same ping–pong fashion as the compression memory.

All stations transmit their traffic bursts on the same carrier frequency using a high signaling rate that completely fills the transponder channel. Typically, the signaling rate is 60 Mbit/sec for 40-MHz transponders and 120 Mbit/sec for 80-MHz transponders. The high bit rate carrier is shared by several stations accessing the TDMA frame in their assigned time fractions. An input signal arriving at a rate of 64 kbit/sec and served by a terminal operating at a transmission rate of 60 Mbit/sec will be transmitted in a burst having a duration that is $T/938$, where T is the duration of the TDMA frame period. (Note that the transmission time is compressed by the ratio of the transmission rate to the input signal rate.) A few additional bits are added to the transmitted burst as a preamble to provide information needed for carrier frequency and bit

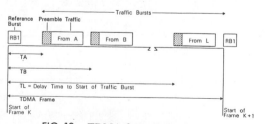

FIG. 12.   TDMA frame organization.

timing recovery, acquisition, synchronization, burst position control, and network control. TDMA traffic bursts may be either preassigned, in which case they maintain an assigned location and duration in the frame for long periods of time, or demand assigned, in which case the position and duration of a burst may be varied each time a new call arrives or an old call departs the system. Because each receiving station must have a timing reference with which to gauge the time of arrival of other stations' traffic burst in the TDMA frame, the TDMA frame contains a reference burst that is transmitted from a reference station. The reference station may be one of the traffic stations that is nominated to be the reference or it may be a special station intended to perform only the reference station function. For redundancy, some systems have two reference bursts.

The traffic carrying capacity of a TDMA station is adjusted simply by changing the duration and location of its traffic burst. A process called a coordinated burst time plan change has been devised for TDMA systems, which permits the reallocation of capacity among stations to be carried out synchronously by changing the time locations and durations of the traffic bursts of all stations in the same TDMA frame without interruption of ongoing traffic. The key to performing such synchronous burst time plan changes is to establish a fixed round trip time delay equal to an integer number of TDMA frames between the time a reference burst leaves the satellite and the time a response from any earth station synchronized to that burst returns to the satellite. This time delay, called the coordination time, is the same for all stations and is selected to be greater than the round trip delay between the satellite and the farthest station on the earth. To accomplish burst time plan synchronization, each station inserts a time delay, D_N, between the time a reference burst marking the start of a received TDMA control multiframe is received and the start of the station's transmit control multiframe. This value is selected to be equal to the difference between the actual round-trip delay for the station and the coordination time. The relationship for determining D_N is

$$D_N = MT_F - 2d_N/c$$

where T_F is the duration of the TDMA frame, M is the number of TDMA frames sufficient to cause the product MT_F to exceed the round-trip propagation time between the satellite and the farthest station, d_N is the actual distance between satellite and station N, and c is the speed of light. The derivation of the above expression is illustrated in Fig. 13. D_N is a function of the station's distance from the satellite and is continuously adjusted by feedback or open loop control to assure that the coordination round-trip delay is always maintained. Because the maximum distance between the satellite and an earth station is 26,000 miles, the coordination time must be at least 280 msec.

The ease with which TDMA system capacity can be reassigned is one of its principal advantages. Also, its burst transmissions are ideally

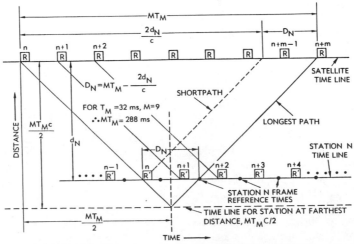

FIG. 13. Time–space graph for determination of synchronization delay D_N for traffic station N.

suited to introduction of satellite switching of traffic among multiple beams to achieve full network connectivity and to use with advanced communications satellites that employ large antennas in space to generate high radiated power hopping pencil beams that are capable of accessing widely distributed low-cost earth stations that use small antennas and low-power amplifiers.

TDMA is being introduced in both domestic and international satellite communications systems. INTELSAT initiated operation of its 120-Mbit/sec international digital trunk telephone TDMA system using the 80-MHz C-band transponders of the INTELSAT V satellites in the Atlantic and Pacific Ocean Regions of its worldwide system. In the United States, Satellite Business Systems (SBS) introduced a digital telephone and data TDMA system, operating at 48 Mbit/sec in the 43-MHz-bandwidth transponders of its own K_u-band satellites for service to business and domestic telephone users. Also in the United States, 60-Mbit/sec TDMA systems using COMSAT DST 1000 TDMA terminals have been introduced in satellite telephone systems for ARCO and GT&E and a demand-assigned DST1100 TDMA system is being intro-

duced into the Federal Republic of Germany's DFS Kopernicus satellite system. In France, a demand-assigned TDMA system operating at 25 Mbit/sec has been introduced on the TELE-COM-1 satellite for use in business communications applications and in Italy work is underway on a multibeam satellite baseband switched TDMA system operating at 147.5 Mbit/sec for experimental evaluation on the ITALSAT. Also, low-bit-rate TDMA systems operating at rates from 3 to 15 Mbit/sec able to bring the flexibility advantages of TDMA to small earth station users are becoming available from several sources. Several of these low-bit-rate TDMA carriers can be supported in a single transponder channel. Such low-bit-rate TDMA systems have economic advantages for serving medium capacity stations (up to 100 circuits) as illustrated by the breakeven revenue requirement curves plotted in Fig. 14.

When TDMA systems use only one carrier per transponder in multiple-access satellite service, they avoid the intermodulation distortion problem encountered with FDMA. They can operate very near the saturated power output of the traveling wave tube (TWT) used in the transponder provided that care is taken in selecting

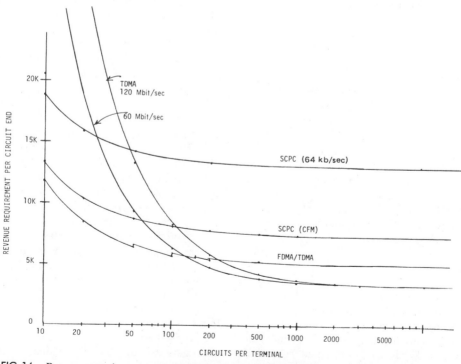

FIG. 14.   Revenue requirement per circuit end (in 1986 dollars) for TDMA at 60 and 120 Mbit/sec, SCPC carrying 64 kbit/sec and SCPC CFM carrying analog voice.

the filters used through the system. In the IN-TELSAT TDMA system the TWT can be operated 1 dB below its saturated power output. In the SBS TDMA system the TDMA carrier channel filters are significantly wider than the TDMA carrier's bandwidth. This permits the TWT to be operated several decibels in excess of saturation without serious impairment, a condition that is very desirable in combating up-link fades at its K_u-band operating frequency.

When TDMA is used in low rate applications and several TDMA carriers are carried in the same transponder, the space segment power advantage of single-carrier-per-transponder TDMA is lost because of multiple carrier inter-modulation distortion. This results in loss of space segment capacity; however, such low bit rate operation greatly reduces the size of the earth station antenna or power amplifier because the power radiated from the earth terminal's antenna is typically reduced by approximately 8 to 10 dB and the channel noise is reduced due to the reduced channel bandwidth.

TDMA systems operating in the single-carrier-per-transponder mode achieve very high traffic carrying density. The capacity of a single 80-MHz transponder in the INTELSAT system using a 120-Mbit/sec TDMA carrier is conservatively 16,000 full-time 64-kbit/sec digital PCM telephone channels, which with digital speech interpolation is increased to 3200 telephone channels. The 64-kbit/sec PCM technique for carrying telephone signals is now being replaced by a 32-kbit/sec adaptive differential pulse coding modulation (ADPCM) technique referred to as low rate encoded (LRE), which when used jointly with digital speech interpolation will double the telephone channel capacity, thus yielding 6400 telephone channels per transponder. Systems operating in the low bit rate FDMA/TDMA manner achieve approximately half of the single-carrier-per-transponder capacity, but the stations using the system are significantly smaller, lower in power, and lower in first cost and operations and maintenance costs.

C. CDMA

In a CDMA system, the individual bits of a message are encoded in terms of unique digital codes. Many stations, each having its own unique code, can simultaneously transmit their signals and each can be discriminated from all of the others. The discrimination is accomplished by means of a cross correlator, which uses as a reference the code corresponding to that of the desired transmission. Each transmitted symbol of binary information is expressed in the form of a coded sequence of shorter duration symbols called chips. The number of chips per symbol, G, is an important CDMA parameter referred to as the processing gain or bandwidth spreading factor because the bandwidth of the transmitted signal is expanded by the factor G. In a typical application for 9.6-kbit/sec transmission, a code having 7000 chips per symbol may be used. Thus, the 9.6-kbit/sec signal, which when transmitted in conventional quarternary phase shift keyed (QPSK) form requires a bandwidth of approximately 4.8 kHz, when coded for CDMA will require a QPSK bandwidth of 33.6 MHz. Because of this increased bandwidth imparted by CDMA, it is frequently referred to as a spread spectrum technique.

The codes used for CDMA are specially selected to have good cross-correlation properties so that interference from simultaneously present but unwanted transmissions of other signals can be significantly reduced. A family of psuedo-random (PR) sequence codes called GOLD codes are considered to have suitable properties. A desirable code is one which, when multiplied by itself, symbol by symbol, and summed over all of its symbols, yields an output N, where N is the number of symbols in the code, but when shifted yields very near $N/2$. Each receiver uses a unique PR code; hence, a station wishing to correspond with a particular receiver simply transmits the message using the destination receiver's code. In order to receive a desired CDMA signal, a receiver stores the PR code of the desired signal into its cross-correlator and correlates it with the total received signal. When the chips of the received desired signal align with the stored code, the cross-correlator puts out a positive or negative impulse depending whether a binary 0 or 1 was transmitted. The cross-correlation process is able to select the desired signal code even though it is buried in the composite of many other coded transmissions and thermal noise. Implementation of the cross-correlator is typically accomplished by means of a surface acoustical wave (SAW) correlator.

The number of simultaneous transmissions that can be supported using CDMA is limited by the self-generated random signal background caused by the sum of the individual signal transmissions plus the system's thermal noise background. Each unwanted signal arriving with carrier power C will contribute a noiselike interference of C/G. Thus, in a system having M simultaneous arriving signals of which only one is wanted, this noiselike interference will be MC/G. To this must be added the thermal noise contribution N. If the objective received bit energy to noise ratio is E_b/N_o, then for QPSK modulation, the number of simultaneous received signals must not exceed the limit given by the expression

$$M \leq G[(2E_b/N_o)^{-1} + (C/N)^{-1}]$$

For a bit error rate of 10^{-6} requiring an E_b/N_o of 12 dB for a practical demodulator with 1.6 dB of implementation margin and assuming a carrier to thermal noise ratio, C/N, of 20 dB, the limit is $G/131.7$. For the example considered previously using $G = 7,000$, the limit is 291 simultaneous signals. Since each signal can support 9.6 kbit/sec, the system can carry a total bit rate of 2.79 Mbit/sec. Introduction of channel coding can improve the total data capacity of the system as can be illustrated by the following example. Assume that each message is rate 1/2 encoded, which yields a 5-dB reduction in the E_b/N_o needed to achieve the same BER objective. To accommodate the doubling of the rate per channel, the value of G must be halved to 3,500. The net result is to increase the number of simultaneous signals to 384 and the total channel capacity to 3.69 Mbit/sec. An attending advantage of DCMA resides in the action of the receive side correlator processing, which causes the undesirable impact of narrow band interfering signals to be greatly reduced.

CDMA can be used to access a satellite system in two different ways. One is to identify each station in terms of its transmitted code, in which case a receiving station must use a separate correlator for each station it wishes to receive. The other is to identify each station in terms of its reception code, in which case each transmit station must transmit on different codes for each receiver. This requires only one receive side correlator. One application of CDMA used in satellite systems is to transmit low data rate messages from a centrally located station to many remote, small diameter, receive only stations. In this case CDMA is used because of its capability to selectively address the remote stations by use of different codes and to significantly reduce the interference from other satellite systems by virtue of the spectrum spreading property. Another application of CDMA currently being considered is in the GEOSTAR sys-

tem in which a centrally located station is used to establish two-way communications to remote mobile receivers and in addition by use of the precise timing that is inherently available in the CDMA method, combined with the use of two separated GEOSTAR satellites, to determine the location of the mobile station.

VII. Link Budgets

The various parameters involved in calculating a satellite link budget are illustrated in Fig. 15. Let the power transmitted in the direction of the satellite from the earth station be designated as P_E. This is determined from the product of antenna gain in the direction of the satellite and the power delivered to the antenna feed flange from the earth stations high-power amplifier (HPA). This results in a power flux density of $P_E/4\pi R_U^2$ at the satellite, where R_U is the distance from the earth station to the satellite. Assuming that the satellite has a gain G_S in the direction of the transmitting earth station and the transmission is at a wavelength L_U, the antenna has a receive aperture of $G_U(L_U)^2/4\pi$ and receives an up-link carrier power level C_U, equal to $(P_E)(G_S)(L_U/4\pi R_U)^2$. The thermal noise on the up-link is given by the product of Boltzmann's constant k, the satellite system noise temperature T_S, and the carrier bandwidth W, which is $(k)(T_S)(W)$. Hence, the up-link carrier to thermal noise ratio when expressed in decibel form is

$$(C/N)_U = P_E + 20 \log(L_U/4\pi R_U) + G/T_S$$

↑   ↑   ↑
Transmit   Path loss   Satellite
power G/T

$$+ 228.6 - 10 \log W$$

↑   ↑
Boltzmann's   Noise
constant   bandwidth

This result must be modified to account for the influence of up-link interference from other earth stations pointed at other satellites and from other beams of the same satellite and intermodulation distortion. This is done by calculating the carrier to up-link interference ratio, $(C/IF)_U$, from knowledge of the up-link interference environment and the carrier to up-link intermodulation ratio, $(C/IM)_U$, from the knowledge of the intermodulation distortion generated in the earth station's HPA. These interferences are then summed on a power basis with the ther-

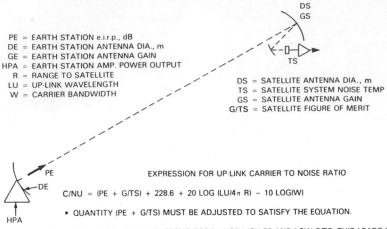

PE = EARTH STATION e.i.r.p., dB
DE = EARTH STATION ANTENNA DIA., m
GE = EARTH STATION ANTENNA GAIN
HPA = EARTH STATION AMP. POWER OUTPUT
R = RANGE TO SATELLITE
LU = UP-LINK WAVELENGTH
W = CARRIER BANDWIDTH

DS = SATELLITE ANTENNA DIA., m
TS = SATELLITE SYSTEM NOISE TEMP
GS = SATELLITE ANTENNA GAIN
G/TS = SATELLITE FIGURE OF MERIT

EXPRESSION FOR UP-LINK CARRIER TO NOISE RATIO

$$C/NU = (PE + G/TS) + 228.6 + 20 \, LOG \, (LU/4\pi R) - 10 \, LOG(W)$$

- QUANTITY (PE + G/TS) MUST BE ADJUSTED TO SATISFY THE EQUATION.

- MOST COMMERCIAL SYSTEMS TODAY USE HIGH PE AND LOW G/TS. THIS LEADS TO EXPENSIVE EARTH STATIONS.

- INCREASED SIZE OF SATELLITE APERTURE INCREASES G/TS, REDUCING PE AND HENCE EARTH STATION SIZE.

- INTERFERENCE, C/IU, MUST ALSO BE FACTORED IN

$$\therefore C/NUT = \left[(C/NU^{-1} + C/IU^{-1}) \right]^{1}$$

(a)

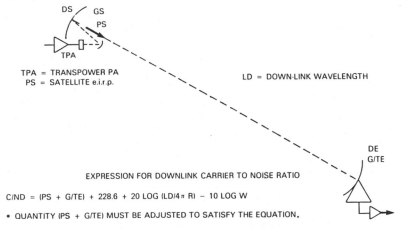

TPA = TRANSPOWER PA
PS = SATELLITE e.i.r.p.

LD = DOWN-LINK WAVELENGTH

EXPRESSION FOR DOWNLINK CARRIER TO NOISE RATIO

$$C/ND = (PS + G/TE) + 228.6 + 20 \, LOG \, (LD/4\pi R) - 10 \, LOG \, W$$

- QUANTITY (PS + G/TE) MUST BE ADJUSTED TO SATISFY THE EQUATION.

- MOST COMMERCIAL SYSTEMS TODAY USE RELATIVE LOW PS AND HIGH G/TE. THIS LEADS TO EXPENSIVE EARTH STATIONS.

- INCREASED SIZE OF SATELLITE APERTURE INCREASES GS AND PS, REDUCING G/TE AND HENCE SIZE OF EARTH STATION.

- INTERFERENCE, C/ID, MUST ALSO BE FACTORED IN

$$\therefore C/NDT = \left[(C/ND^{-1} + C/ID^{-1}) \right]^{-1}$$

(b)

FIG. 15. (a) Satellite up-link budget calculation and (b) satellite down-link budget calculation.

TABLE II.   Satellite Link Budget Calculation

Up-link

P_U, up-link radiated power	85.5 dB W
Up-link spreading loss, $4\pi/R_U^2$	163.4 dB/m²
Flux density at satellite for 2.5-dB IBO	−77.9 dB W/m
Up-link path loss for $L_U = 0.05$ m, 6 GHz	200.4 dB
Satellite Rcv G/T	−14.5 dB/K
Boltzmann's constant	228.6 dB/Hz K
Noise bandwidth, $W = 30$ MHz for QPSK	74.8 dBHz
$(C/N)_U$	24.4 dB
$(C/IF)_U$	27.0 dB
$(C/IM)_U$	∞ dB
$(C/N)_{UT}$	22.5 dB

Down-link

P_D, satellite down-link power at beam edge	21.8 dB W
Satellite down-link path loss, $L_D = 0.075$ m	197.0 dB
Earth station G/T, 30-m INTELSAT std A	39.0 dB/K
Boltzmann's constant	228.6 dB/Hz K
Noise bandwidth, $W = 30$ GHz	74.8 dBHz
$(C/N)_D$	17.6 dB
$(C/IF)_D$	27.0 dB
$(C/IM)_D$	∞ dB
$(D/N)_{DT}$	17.1 dB

End-to-end link

$(C/N)_{UD}$	16.0 dB
Objective, link BER = 10^{-6}	12.5 dB
Margin	3.5 dB

mal noise using the following relation expressed in decibel form:

$$(C/N)_{UT} = -10 \log(10^{-0.1(C/N)_U}$$
$$+ 10^{-0.1(C/IF)_U} + 10^{-0.1(C/IT)_U})$$

The same expressions can be developed for the down-link in terms of the following variables: P_D for the power radiated in the direction of the earth station from the satellite, L_D for the down-link wavelength, R_D for the down-link distance between the satellite and the earth station, G/T_E for the earth station's antenna gain to system noise temperature ratio, C/N_D for the down-link carrier power to thermal noise ratio, $(C/IF)_D$ for the down-link interference from other beams on the same satellite plus adjacent satellites, and $(C/IM)_D$ for the intermodulation distortion generated in the satellite transponder. The expressions for the down-link case are the same as those given above for the up-link with obvious variable substitutions. From the above relations the individual up- and down-link carrier to thermal noise plus interference and intermodulation

distortion ratios are determined. These are then combined using noise power addition to determine the overall end-to-end carrier to noise plus interference and intermodulation ratio. The expression for calculating this overall result in terms of the up-link and down-link contributions is

$$(C/N)_{UDT}$$
$$= -10 \log(10^{-0.1(C/N)_{UT}} + 10^{-0.1(C/N)_{DT}})$$

The value of $(C/N)_{UDT}$ thus calculated is compared to that needed to accomplish the desired performance objective and the difference is the margin. It is typical to require an overall margin of 2 dB for C-band operation and 8 dB for K_u-band operation. It is convenient to develop a tabular format for calculation of link budgets. One example is shown in Table II.

VIII. Transmission Impairments

A. Path-Induced Impairments

The most prevalent form of path-induced impairments that influence satellite communications at the C- and K_u-band frequencies are those due to rain. Rain causes three types of path impairment to appear: attenuation, increased noise temperature, and scattering. Regarding attenuation, the presence of water in the path between the earth station and the satellite causes energy absorption due to the excitation of molecular resonances. These are present to some extent at the C-band frequencies and to a much large extent at the K_u- and K_a-band frequencies. Data on such absorption is plotted in terms of the probability of exceeding a given level of path attenuation for an assumed weather model in a given region. A typical attenuation probability distribution applicable to the United States is shown in Fig. 16 for K_u-band frequencies. Also, increased noise temperature accompanies the path attenuation impairment. Rain raises the noise temperature of the system since it is at 290 K ambient, which is much higher than the background of space. The resulting noise temperature can be expressed in terms of the path attenuation a by

$$T_s = [(a - 1)/a]T_0 + T_{s0}$$

where T_s is the system noise temperature with rain, T_{s0} is the clear sky system noise temperature, and T_0 is the ambient temperature, which is nominally 290 K. The attenuation a is related to the rain attenuation L expressed in decibels by the expression

$$L = 10 \log a$$

Thus, for example, if the attenuation due to rain is 6 dB, the value of a is 4 and the resulting system noise temperature becomes $3T_0/4 + T_{s0}$. For a typical clear sky system noise temperature of 200 K, the rain system noise temperature is 417.5 K. The noise temperature increase is not apparent on the up-link because the satellite is looking at the earth, which is of course always at an ambient temperature of 290 K. Thus, in calculating link budgets it is necessary to take into account both the path loss and the noise temperature increase on the down-link but only the path loss on the up-link. [*See* RADIO PROPAGATION.]

Another form of atmospheric impairment that must be accounted for is that due to depolarization. Most satellite systems in service today use two senses of polarization, either horizontal and vertical or right- and left-hand circular, to achieve frequency reuse of the assigned frequency spectrum. Antenna systems, using modern feed technology, are able to discriminate between these polarizations to the extent that the unwanted polarization is attenuated by 30 dB or more relative to the wanted polarization under

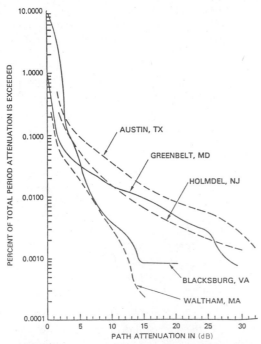

FIG. 16.   Comparison of down-annual 11.7 GHz attenuation distribution of measurements at five locations adjusted to 30° elevation angles.

clear sky conditions. However, under heavy rain conditions this discrimination can drop to as little as 15 dB and this fact must be accounted for in calculating the link availability. This is accomplished by including an appropriate carrier to interference ratio component in the calculation of the link budgets applicable to the system under study.

B. MODULATION TECHNIQUES

Numerous modulation techniques are used in accomplishing signal transmission over satellite links. Only the principal ones are discussed in the following and these are amplitude modulation (AM), frequency modulation (FM), and phase modulation as it applies to PSK digital transmission. [*See* DATA COMMUNICATION NETWORKS.]

1. Amplitude Modulation

Amplitude modulation may be expressed by the relation,

$$A_{AM}(t) = (1 + M(t)) \cos(2\pi f_c t)$$

where $M(t)$ represents the baseband modulating signal as a function of time t scaled so that its peak value does not exceed 1, and f_c the carrier

$$A_{AM}(t) = (M(t)/2) \exp(j2\pi f_c)$$
$$+ (M(t)/2) \exp(-j2\pi f_c t)$$
$$+ \cos(2\pi f_c t)$$

and in the frequency domain as

$$S_{AM}(f) = M(f - f_c)/2 + M(f + f_c)/2 + \partial(f_c)$$

where $M(f)$ is the spectrum distribution of the modulating baseband signal. This shows that the AM process generates an upper and lower sideband represented by the first two terms, respectively, and a carrier component represented by the last term. Single sideband AM transmission is frequently used in satellite systems for companded voice transmission, a process referred to as companded single sideband (CSSB), in which case only one of the sidebands is transmitted. Demodulation of such single sideband transmission requires precise reinsertion of the carrier at the receiver and an associated pilot transmission system from which the proper carrier frequency can be derived must be provided. In such CSSB transmission the individual voice channels are first processed by means of an amplitude compressor that greatly reduces the range of amplitude variation of the signal. This compression action provides a signal that uses the power of the earth station and satellite amplifiers more efficiently. The individual voice channel signals are then stacked in frequency, each assigned its own 4-kHz slot to form a multichannel baseband signal comprising several thousand voice channels. In telephone system language this is referred to as a frequency-division multiplexed (FDM) baseband. This composite baseband signal is amplitude modulated onto a carrier and transmitted over the satellite system along with a reference pilot carrier. At any receiving station the signals are received and used to recover the multichannel baseband and then the individual companded voice channels. Each such signal is next processed by an expander, which restores the signal to its original amplitude range and in the process inherently suppresses noise encountered in the transmission link. The amount of noise suppression achieved is referred to as the compander advantage and can range from 15 to 18 dB.

2. Frequency Modulation

Frequency modulation is a process in which the frequency of a carrier is deviated by a modulating signal and is represented mathematically by

$$A_{FM}(t) = A \cos 2\pi[f_c + kM'(t)]t$$

where $M'(t)$ is the time derivative of the modulating signal and k is the modulation index parameter that governs the frequency range over which the carrier is deviated. When the range of deviation Δf is large compared to the bandwidth of the modulation B, the process is referred to as wideband FM and the signal to noise ratio existing on the transmission link is increased by a multiplicative factor $(\Delta f/B)3$. This is a property of FM that is very important to satellite communications since it permits a trade between bandwidth and power. Satellite transponders are generally of low power as necessitated by the economies of satellite design and because FM permits this trade of bandwidth for power it has become the preferred modulation technique. Also for the same reasons FM is attractive to small earth station users who wish to establish satellite communications links at the minimum cost.

Frequency modulation is used in satellite systems to transmit wideband television and telephony FDM baseband signals. It has become the workhorse modulation technique of satellite communications. In the typical wideband international television transmission the nominal NTSC 6-MHz bandwidth signal is FM modulated onto a carrier with a deviation that results in a transmission bandwidth of 36 MHz at C band, which is just sufficient to occupy the entire bandwidth of a single transponder. This signal is received by many stations in the beam of the transponder using earth stations equipped with 15-m-diameter antennas. For domestic television similar FM transmission is used at K_u-band satellite transmission and since these domestic satellites transmit higher power intensities, these television signals can be received with antennas having diameters as small as 2 m.

For FDM telephone, groups of 4-kHz telephone channels, stacked in frequency on 4-kHz frequency spacings to form an FDM baseband, are FM modulated onto a carrier at an earth station and transmitted to the satellite. At the satellite, similar carriers from other earth stations but on differing frequencies are simultaneously amplified in a transponder and retransmitted to earth for reception by any properly equipped station. Thus the telephone FDM basebands transmitted by all stations are received by any station and the traffic destined to a station can be selected from among the received basebands. This transmission process is known as FDM/FM/FDMA, where FDM refers to the FDM telephone baseband format, FM to the modulation process, and FDMA to the use of frequency-division multiple access of the satellite transponders by the earth station carriers. Individual carriers may carry as few as 24 voice channels to over 1000. Satellites that serve many low and medium capacity carriers and interconnect networks of many small and medium capacity users are referred to as primary path satellites and those that carry a small number of high capacity telephone trunks and interconnect networks of large capacity users are called major path satellites.

Frequency modulation is also used to carry single-channel carriers in what is referred to as SCPC transmission. This mode of transmission is able to support a typical 3-kHz-bandwidth amplitude companded analog telephone voice channel signal in as little as 18 kHz of carrier bandwidth and to carry over 1000 such carriers in a single satellite transponder. The SCPC FM is used in domestic and international systems jointly with demand assignment of the K_u-band space channels to provide telephone communications among thousands of very low capacity users equipped with earth stations using antennas as small as 1.8 to 3.8 m and power amplifiers of only a few watts.

3. Digital Modulation

Use of digital signal transmission is rapidly growing in the satellite systems due to the current rapid migration of telephone service to use digital transmission media. Digital signals are now being used for all forms of telephone service formerly carried by analog means. Telephone voice is carried by means of pulse code modulation (PCM) at digital rates of 64 kbit/sec and most recently by adaptive differential pulse code modulation (ADPCM) at 32 kbit/sec. Data services over the public switched telephone network in both the domestic and international applications are rapidly growing due to the pressures of deregulation in the United States and the now rapid growth of the worldwide integrated services digital network (ISDN). There is also a rapid growth of packet switched computer communications networks operating solely on satellites that connect thousands of very small, low-cost earth terminals to concentrating hub stations. Teleconferencing using low bit rate digital video codecs operating at bit rates from as low as 64 kbit/sec to as high as 2048 kbit/sec is enjoying rapid growth. All of these forces are creating a rapidly increasing demand for digital satellite transmission.

Digital modulation techniques extensively used for supporting digital satellite transmission are BPSK (binary phase shift keyed) and QPSK (quadrature phase shift keyed). BPSK can be expressed as

$$B(t) = \cos(2\pi f_c t + i\pi)$$

where i is the binary state sequence of 0's and 1's representing the digital signal. QPSK can be expressed as

$$Q(t) = \cos(2\pi f_c t + j\pi/2)$$

where j is determined by the states of pairs of bits in the bit sequence according to the table

Bit	
Pair	j
00	0
01	1
11	2
10	3

The bits are transmitted in terms of symbols of duration T. In the case of BPSK, each symbol period of the modulated carrier identifies one bit in terms of carrier phases of 0 and π, while in the case of QPSK each symbol period identifies two bits in terms of carrier phase angles of 0, $\pi/2$, π, and $3\pi/2$. The signal occurring in each symbol period is a sinusoid burst of frequency f_0 and duration T having a power spectrum distribution given by the relation

$$S(f) = \{\sin[\pi(f - f_0)T]/\pi(f - f_0)T\}2$$

This power spectrum theoretically extends from minus infinity to plus infinity; however, the Nyquist sampling theorem shows that the entire information content of the digital stream is contained in a bandwidth of $W = 1/t$. In practice, the bandwidth may be made slightly greater (about 20%) to allow for imperfections in realizing the ideal filter. BPSK and QPSK are used extensively to carry satellite digital communications from rates of a few bits per second to several hundred megabits per second. In certain applications where the objective is to carry a greater density of communications in the frequency spectrum, octal phase shifted keyed (OPSK) modulation is being investigated; however, because of its sensitivity to phase distortion coding must be introduced to maintain low bit error rate. A number of other modulation techniques have also been introduced, each with the intention to improve the quality of digital communications via satellite principally by improved resistance to the influence of nonlinearities encountered in the high power amplifiers used in the earth stations and on-board the satellites. Some of these are offset QPSK, frequency shift keyed (FSK), minimum frequency shift keyed (MSK), serial MSK (SMSK), and quadrature and amplitude modulation (QAM). Also both BPSK and QPSK can be modulated in terms of the phase difference between symbols, a technique that avoids the necessity to recover the carrier to perform demodulation. In this case they are called DBPSK and DQPSK, respectively. Two versions of such modulation used in the French TELECOM 1 satellite system are 2-2PSK and 2-4PSK. Many other modulation techniques are possible and are likely to be under investigation. Thus far the phase shift keyed based techniques predominate digital satellite communications.

C. SYSTEM-INDUCED IMPAIRMENTS

System-induced impairments are those that cause the transmission to depart from theoretically perfect performance due to imperfections in the components used to construct the system. They reside in the modulators and demodulators and the filters and amplifiers used in the earth stations and on-board the satellite. These imperfections exist for all types of transmission, analog or digital. In general it is possible to calculate an ideal performance function of a modulation–demodulation process by assuming perfect arithmetic precision in the modulation and demodulation functions, and calculation of the baseband signal to noise ratio for analog signals or bit error rate for digital signals as a function of the ratio of carrier to thermal to noise ratio on a linear, memoryless transmission link. Such ideal performance curves become the theoretical target that the equipment designer attempts to approach. The imperfections listed above cause the performance to depart from this ideal performance and the departure encountered is called an impairment.

1. Modem Impairments

Modem impairments are those caused by inaccuracies and imperfections of the modulation process to generate the ideal transmission signal for the applied baseband signal and the demodulation process to perfectly extract the baseband signal at the receiver. For simple AM and FM the modulation and demodulation processes depend on linear arithmetic relations that are usually easy to realize in today's technology and near perfect modulation and demodulation implementation can be expected. Any impairments that do exist are principally due to nonlinearity in realizing the functional implementation. Only in the case of extremely wide bandwidth FM, where imperfection may be encountered in deviating the frequency of the carrier at high baseband frequencies, may some impairment be encountered in the modulation process. In digital modulation, similar constraints are at play. Digital modulation can be accomplished by methods too numerous to discuss in detail in this article. The most popular are frequency shift keyed (FSK), phase shift keyed (PSK), amplitude shift keyed (ASK), and combinations of these. Impairments in their implementation are experienced in the accuracy with which values of frequency, phase, and amplitude can be maintained through the transmission system.

2. Filter Impairments

Impairments to achieving accurate transmission are induced in the filters that the information bearing signals must pass. These filters are needed to minimize the interference from other signals that share the transmission facility and noise from natural origins. The filters themselves can produce impairment if they cause disorder in the reconstruction of the wanted signal. A typical filter characteristic is illustrated in Fig. 17. The shape of the filter in terms of its amplitude and phase response has been selected to reproduce the signal at the receiver with minimum contamination from noise and interference. The shapes of the desired amplitude and phase responses are bounded by tolerance margins within which deviations are allowed for the manufacturing process and system operation of the filter. These margins are determined by both analytic and experimental methods to produce an acceptable level of impairment in transmission of the wanted signal. Figure 18 illustrates a two-dimensional scatter diagram for the near ideal transmission of quadrature phase shift keyed (QPSK) digital transmission. Each of the states of the transmission are clear and distinct, indicating that data transmission would be accomplished with high accuracy and hence low error rate. Figure 19 is the same diagram but contaminated by noise or error due to improper filtering. The target areas are now dispersed and blurred and the ability to select a particular state accurately is impaired and the error rate significantly increased. Figure 20 shows the performance of QPSK modulation in terms of bit error rate.

3. Transmission Impairments

In addition to impairments due to implementation, impairments can also be caused by imperfections in the transmission path. These are caused by a number of factors. The transmission path includes nonlinear devices such as traveling wave tubes (TWTs) in both the earth stations and in the satellite transponders, which induce several classes of transmission impairments in satellite systems. The TWT introduces amplitude distortion, in the form of AM to AM, and phase distortion, in the form of AM to PM. Typical TWT amplitude and phase transmission characteristics are illustrated in Fig. 21. The AM-to-AM and AM-to-PM interactions cause the components of the transmission to intermodulate, resulting in the generation of in-band intermodulation products. When the transmission is composed of many independent signals, as is the case of FDMA transmission, the intermodulation products create a random signal that can

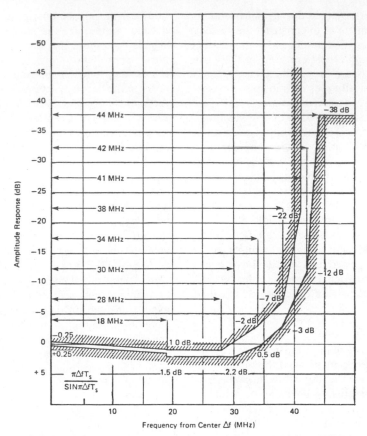

FIG. 17.   Modulator filter amplitude mask for 120 Mbit/sec QPSK.

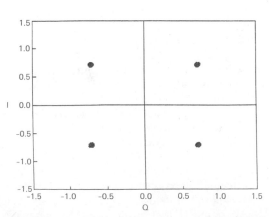

FIG. 18.   Two-dimensional scatter diagram; ideal transmission.

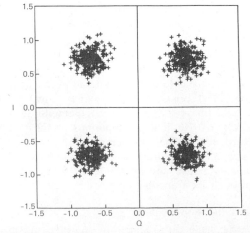

FIG. 19.   Two-dimensional scatter diagram; impaired transmission.

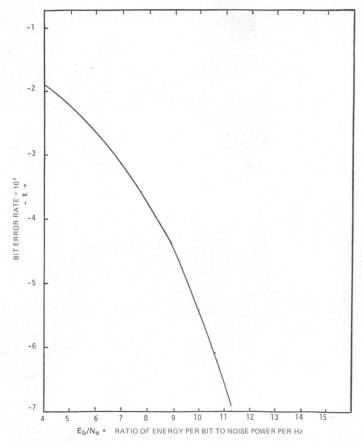

FIG. 20. Theoretical performance of QPSK modulation in terms of bit error rate versus E_b/N_o.

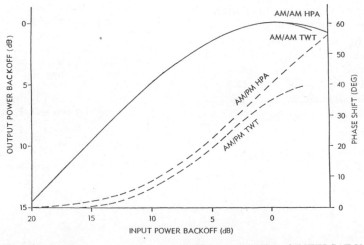

FIG. 21. Nonlinear response curves for HPA and TWTA of INTELSAT V.

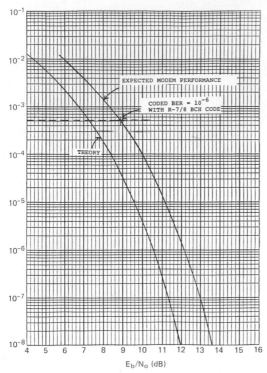

FIG. 22.   Modem performance over typical nonlinear satellite links.

the impairment can be much less depending on the bandwidth used to confine the channel. For digital transmission, if the bandwidth is equal to that of the symbol rate of the modulated signal, the TWT can be operated to within 1 dB of its saturated power output. If the bandwidth is approximately 50% greater than the symbol rate, the TWT can actually be operated beyond saturation. The TDMA network operated by SDS operates under this condition. Figure 22 shows the performance of QPSK digital modulation with error correction coding in terms of the bit error rate (BER) versus the ratio energy per bit to noise per unit bandwidth, E_b/N_o, as experienced over a typical satellite transmission link.

BIBLIOGRAPHY

Campanella, S. J. (1981). *In* "Digital Communications—Satellite/Earth Station Engineering" (K. Feher, ed.), pp. 336–404. Prentice-Hall, Englewood Cliffs, New Jersey.

Campanella, S. J. (1985). Satellite communications—challenges of the future. *Technical Marketing Society of America, Los Angeles, California, July 1985*, pp. 1–24.

Campanella, S. J. (1986). Communications satellites of the future. *Am. Astronaut. Soc., 33rd Annu. Meet., Boulder, Colorado, October 26–29, 1986*, pp. 1–21.

Campanella, S. J., and Harrington, J. V. (1984). Satellite communications networks. *Proc. IEEE*, **72** (11), 1506–1519.

Campanella, S. J., and Pontano, B. (1986). Advantages of TDMA and satellite-switched TDMA in INTELSAT V and VI. *COMSAT Tech. Rev.*, **16** (1), 207–238.

Chang, P., Fang, R. J., and Jones, M. (1981). Performance over cascaded and band-limited nonlinear satellite channels in the presence of interference. *ICC Conf,, Denver, Colorado, June 1981*, pp. 47.3.1–47.3.10.

Edelson, B. I. (1973). Satellite communications for information networking. *J. Am. Soc. Inf. Sci.* **9**, 175–187.

Edelson, B. I. (1977). "Global Satellite Communications." *Sci. Am.* **236** (2), 58–73.

be treated like another source of random noise. However, when the transmission is a single signal, as is the case for single-carrier-per-transponder TDMA, it appears as a distortion of the signal wave shape that is dependent on the history of the signal, which when averaged over all possible states of the signal can be equated to an impairment equivalent to additive random noise or to an equivalent degradation in the performance of the digital demodulator. In general, to maintain the impairment resulting from TWT nonlinearity to the equivalent of a carrier to noise ratio of 20 dB for FDMA transmission, it is necessary to operate the TWT such that its output power is 4 dB less than its saturated power output. However, for single-carrier-per-transponder operation such as that used for TDMA,

SIGNAL PROCESSING, ANALOG

W. K. Jenkins   *University of Illinois*

GLOSSARY

Active filter: Class of frequency-selective analog filters constructed from resistors, capacitors, and operational amplifiers.

Analog computer: Programmable analog circuit simulator that consists of high-quality operational amplifiers, high-precision resistors and capacitors, and analog multipliers that can be conveniently patched together to produce real-time solutions for both linear and nonlinear differential equations.

Chirp waveform: Sinusoidal pulse with a quadratic phase (linear frequency) modulation that is often used in pulsed-Doppler radars.

Compression filter: Matched filter for a prescribed waveform such as a chirp, often used as detection filters in radar receivers.

Homogeneous differential equation: Unforced differential equation that results from setting the forcing function to zero in a linear constant-coefficient differential equation.

Ideal reconstruction formula: Infinite summation of weighted $(\sin x)/x$ basis functions capable of perfectly reconstructing an analog signal from a set of samples taken at or above the Nyquist sampling rate.

Nonhomogeneous differential equation: Linear constant-coefficient differential equation with the forcing function activated, that is, not set to zero.

Operational amplifier: Differential voltage amplifier characterized by a very high linear voltage gain, a very high input impedance, and a very low output impedance.

Passive filter: Class of frequency-selective analog filters constructed from resistors, capacitors, and inductors.

Post-filter (reconstruction):   Analog filter cascaded with a digital-to-analog (D/A) converter to smooth the staircase reconstruction error and compensate for the $(\sin x)/x$ distortion inherent in the D/A reconstruction process.

Pre-filter (antialiasing):   Analog lowpass filter used before an analog-to-digital (A/D) converter to remove all frequency content above the half-sampling frequency and prevent aliasing distortion.

Synthetic aperture radar: Microwave radar carried on an airplane or orbiting satellite which collects radar returns from many different spatial locations along the flight path and produces a high-resolution terrain image by combining many returns through signal processing.

Switched capacitor filter: Class of frequency-selective analog filters constructed from capacitors, transistor switches, and operational amplifiers (op amps) and that are particularly well suited for monolithic fabrication.

Zero-input response: Response due to initial energy storage in the system of a linear system with a zero input applied.

Zero-state response: Response of a linear system to an input applied when the system is in the zero state (no initial energy storage).

In recent years the term *analog signal processing* has been used to distinguish between traditional continuous-time signal-processing techniques and the more recently popularized techniques in digital signal processing. In this chapter, the term *analog* is used synonymously with *continuous time*; a *continuous-time* (CT) signal is a function $s(t)$ that is defined for all time t contained in some interval on the real line. In this context the independent variable is usually called time, because in many signal-processing applications the functional relationships of interest are indeed dependent on time as the independent variable. However, in other applications the independent variable may represent different physical quantities such as spatial displacement or environmental temperature. The term analog signal processing refers to the representation and processing of all signals that satisfy the usual continuous-time definition, regardless of the specific physical interpretation of the independent variable.

I. Introduction

Analog signals are processed by specially designed devices, circuits, or systems in order to extract parametric information or to alter the characteristics of the input signal in some prescribed way. Spectral shaping is probably the most common form of analog signal processing and is typically done with either passive (containing resistors, capacitors, and inductors) or active [containing resistors, capacitors, and operational amplifiers (op amps)] analog filters. The active filter is probably the most common analog signal processor in use today. During the past few years, switched-capacitor filters have become popular as high-precision replacements for certain types of active filters that are to be realized in monolithic form as part of a more extensive integrated system. Switched-capacitor circuits, which contain capacitors, transistor switches, and operational amplifiers, have been successfully used for analog pre-post-filters required at the interface between analog and digital systems. This subject is treated in more detail in Section VI. Both active filters and switched-capacitor filters are closely related historically to the analog computer, from which the term analog first originated.

Analog signals can be processed by many different types of devices, some of which are surface acoustic wave (SAW) filters, charge-coupled devices (CCDs), high-frequency distributed-parameter circuits, or optical circuits, to name only a few. Since it will not be possible to discuss all the different classes of analog processing techniques and devices, and because many of these are treated elsewhere,

attention will be devoted to reviewing important mathematical tools used in analog signal analysis, discussing the connection between modern analog signal processing and its origins in the analog computer, and examining several particularly important historical topics in analog signal processing. These special topics include switched-capacitor filters, electronically tunable analog filters, nonlinear modeling by means of the analog computer, and image reconstruction for synthetic aperture radar by optical processing techniques. [*See* SIGNAL PROCESSING, GENERAL.]

II. Background Mathematics

Linear analog signal processors are characterized by the familiar mathematics of differential equations, continuous convolution operators, Laplace transforms, and Fourier transforms. Consider the single-input, single-output system shown in Fig. 1, where $y(t) = S\{x(t)\}$, $x(t)$ is the analog input signal, and $y(t)$ the analog output.

The following definitions are commonly used to define the properties of linearity, time invariance, causality, and stability:

1. *Linearity.* $S\{\cdot\}$ is linear if and only if $S\{ax_1(t) + bx_2(t)\} = aS\{x_1(t)\} + bS\{x_2(t)\}$, where a and b are scalar constants and $x_1(t)$ and $x_2(t)$ are two arbitrary input functions.

2. *Time-invariance.* $S\{\cdot\}$ is time-invariant if and only if $y(t - t_0) = S\{x(t - t_0)\}$ for all real values of t_0.

3. *Causality.* $S\{\cdot\}$ is causal if and only if $y(t)$ depends only on values of $x(s)$ at times $s \leq t$.

4. *Stability (BIBO).* $S\{\cdot\}$ is stable in the bounded input–bounded output (BIBO) sense if and only if the output $y(t)$ remains bounded for all inputs $x(t)$ that are bounded.

5. *Stability (asymptotic).* $S\{\cdot\}$ is stable in the asymptotic sense if and only if whenever $x(t) = 0$ for $t \geq t_0$, $y(t)$ approaches zero as t approaches infinity.

Whenever S is linear and time-invariant, it is possible to express the zero-state response $y(t)$, that is, the response when the system has no initial energy storage, due to an arbitrary input $x(t)$ in terms of the convolution of $h(t)$ with $x(t)$

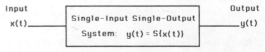

FIG. 1.   Single-input, single-output system $S\{\cdot\}$.

as expressed by Eq. (1), where $h(t) = \mathbf{S}\{\delta(t)\}$ is an analog impulse function (Dirac delta function) and $h(t)$ is the impulse response of $\mathbf{S}$.

$$y(t) = h(t) * x(t) = \int_{-\infty}^{+\infty} h(t)x(t - \tau)\, d\tau \quad (1)$$

Note that Eq. (1) characterizes only the zero-state component of the response, that is, only that part which is excited by the application of the input. In general, the zero-input response (i.e., that part due to initial energy storage in the system) must be accounted for in a different manner.

If $\mathbf{S}$ is causal, $h(t) = 0$ for $t < 0$, and the lower limit in Eq. (1) becomes zero. Also, if $x(t) = 0$ for $t < 0$, then the upper limit on the convolution becomes t. The convolutional property of linear time-invariant systems is an important fundamental property that can be summarized by the following theorem:

Fundamental Theorem of Linear Systems. The zero-state output of a linear time-invariant analog system in response to an arbitrary input signal $x(t)$ can be expressed as the convolution of $x(t)$ with the impulse response $h(t)$ of the system, that is, $y(t) = h(t) * x(t)$.

The fundamental theorem of linear systems guarantees that as long as the impulse response of the system is known and as long as the system is linear and time-invariant, then the output of the system can be computed for any input signal without having any detailed knowledge about the internal structure of the system.

A. Characterization by Differential Equations

There are important subclasses (such as passive and active analog filters) of linear time-invariant systems that can be characterized by an Nth-order differential equation expressed in terms of the input $x(t)$ and the output $y(t)$, as illustrated by

$$d^N y(t)/dt^N + \cdots + a_1 dy(t)/dt + a_0 y(t)$$
$$= b_M d^M x(t)/dt^M + \cdots + b_1 dx(t)/dt + b_0 x(t)$$
$$(2)$$

Time-invariance guarantees that the coefficients in Eq. (2) are constants. In certain situations it may be difficult to describe a system with a single differential equation, but a set of lower-order coupled differential equations may be more convenient. The traditional normal-form state equations of the form $d\mathbf{X}(t)/dt = \mathbf{A}\mathbf{X}(t) + \mathbf{B}\mathbf{U}(t)$ are simply a standard set of coupled first-order dif-

ferential equations used to describe an arbitrary linear time-invariant system in a standard language. Once a system has been characterized in a standard way, then standard mathematical techniques can be used to find solutions for time response, frequency response, transfer functions, stability, etc.

Fortunately, the vast majority of analog filters can be described with differential equation approaches, and hence they can be analyzed rigorously with the above approach. This is true because linear capacitors and inductors satisfy strict first-order differential laws of the form $y(t) = K\, dx(t)/dt$ between appropriate sets of terminal variables. The inductor law is a result of Faraday's natural law of induced voltage, while the capacitor law is due to the fact that current is the rate of flow of charge in a device. Since resistors, op amps, and ideal transformers all satisfy algebraic relationships among their terminal variables, it follows that all ideal passive and active analog filters can be characterized by appropriate sets of differential equations, because they are represented by collections of these elementary properties whose natural form is preserved by Kirchoff's laws of interconnection.

Although the entire subject of differential equations cannot be adequately reviewed here, there are several issues that should be mentioned. Linear constant coefficient differential equation can be solved either by classical techniques or by more modern transform techniques. The classical approach separates the problem into homogeneous and nonhomogeneous equations, finds solutions to these individually, and then combines them to produce the complete solution. The classical approach is intuitive and convenient in that it breaks one large problem into several smaller ones, but it also involves a certain degree of "guesswork" that the neophyte always finds disturbing. Also, the boundary values, or initial conditions, must be patched into the general solution at the very end in order to finally get a specific solution. In contrast, the transform method is a powerful approach that includes initial conditions at the very beginning of the procedure and produces a final result through a sequence of mechanical steps that do not require much insight or intuition. In other words, a solution can often be obtained by sheer persistence and hard work. The disadvantage of the transform approach is that, although the mathematical manipulations may be straightforward in principle, the actual computations often become extremely tedious, and it is not always easy to justify the result intuitively without further work. The experienced analyst finds it useful to keep both the classical and transform methods nearby and to use them interchangeably as necessary. It is simply misleading to claim that any one method of solution will result in the best approach when applied over a broad class of problems. [*See* DIFFERENTIAL EQUATIONS, ORDINARY.]

There is also a common misconception that the complementary solution (obtained from the homogeneous equation) and the particular solution (obtained from the nonhomogeneous equation) in a classical solution are the same as the zero-input response and the zero-state response from a modern state-space solution. That this is not true can be seen in Eq. (1), where it is apparent that the application of an input signal to a system that was initially in zero state may very well lead to the excitation of transient natural response terms. In the state-space approach, these transients arise as part of the zero-state response, since they do not depend in any way on initial energy storage in the system. However, in the classical solution, these terms turn up as part of the complementary solution, since the complementary solution contains all natural response terms, regardless of whether they are excited by the input or are due to initial energy storage. Another way to see this is to note that the complementary solution does indeed depend on the particular solution, because the boundary values are used in the total solution after the various components have been added together.

B. THE ROLE OF THE LAPLACE TRANSFORM IN ANALOG SYSTEM ANALYSIS

In the preceding section there was a discussion of transform methods for solving differential equations, without explicitly mentioning which transform could be used. The most familiar transform used in analog signal analysis is the Laplace transform, as defined by

$$F(s) = \mathscr{L}\{x(t)\} = \int_0^{+\infty} f(t)e^{st}\, dt \tag{3a}$$

$$x(t) = \mathscr{L}^{-1}\{F(s)\} = \frac{1}{2\pi}\int_{\sigma-j\infty}^{\sigma+j\infty} F(s)e^{-st}\, ds \tag{3b}$$

The region of convergence of $F(s)$ is the set $R = \{s \mid F(s) \neq \infty\}$ (i.e., it is the set of all s such that Eq. (3a) converges). In Eq. (3b), the contour of integration must be selected to lie entirely within R. Since the lower limit on the integral in Eq. (3a) is zero, this form is known as the one-sided

(unilateral) Laplace transform to distinguish it from the two-sided (bilateral) form. In analog system analysis, the Laplace transform is typically used to convert differential equations into algebraic equations, which can then be solved by algebraic techniques. The time-domain solution can then be found by computing the inverse Laplace transform by means of Eq. (3b) or by the use of a Laplace transform table. A list of Laplace transforms is given in Table I for some common functions used frequently in analog signal processing.

Linearity is an important property of the Laplace transform, a property that it inherits from the linear property of the integral operator used in its definition. Two other important properties are the following:

Integral Theorem of the Laplace Transform. If $f(t)$ is a Laplace transformable function with transform $F(s)$, then

$$\mathscr{L}\left\{\int_0^t f(\tau)\, d\tau\right\} = \frac{F(s)}{s}$$

Derivative Theorem of the Laplace Transform. If $f(t)$ is a Laplace transformable function with transform $F(s)$, and if $f(t)$ is continuous at $t = 0$, then

$$\mathscr{L}\{df(t)/dt\} = sF(s) - f(0)$$

The above properties, together with the linearity of the Laplace transform, allow the transform to be applied to any differential equation, or to any set of coupled equations that contain derivatives or integrals, such as those that result from circuit analysis. Note that initial conditions enter the analysis at the time the transform is applied, so the final solution automatically contains the component of the complete response due to initial conditions.

Since the one-sided Laplace transform does not consider negative time, and since initial conditions are automatically taken care of at the origin, Laplace transform techniques are extremely well suited to handle problems that contain switches that are turned on at the origin. Many analysts consider it the method of choice for circuit problems and differential equation problems in which a complete solution, including transient response terms, is required.

Finally, it should be noted that if the Laplace transform is applied to either the convolutional expression in Eq. (1) or to the differential equation in Eq. (2), the ratio of the transforms of the input and output can be found, that is, $H(s) =$

TABLE I. Laplace Transforms of Some Common Functions[a]

$f(t)$[b]	$F(s)$
$u(t)$	$\dfrac{1}{s}$
t	$\dfrac{1}{s^2}$
$\dfrac{t^{n-1}}{(n-1)!}$ $n = $ integer	$\dfrac{1}{s^n}$
e^{at}	$\dfrac{1}{s-a}$
te^{at}	$\dfrac{1}{(s-a)^2}$
$\dfrac{1}{(n-1)!}\, t^{n-1}e^{at}$	$\dfrac{1}{(s-a)^n}$
$\dfrac{1}{a-b}(e^{at}-e^{bt})$	$\dfrac{1}{(s-a)(s-b)}$
$\dfrac{e^{-at}}{(b-a)(c-a)}$ $+ \dfrac{e^{-bt}}{(a-b)(c-b)} + \dfrac{e^{-ct}}{(a-c)(b-c)}$	$\dfrac{1}{(s+a)(s+b)(s+c)}$
$1 - e^{+at}$	$\dfrac{-a}{s(s-a)}$
$\dfrac{1}{\omega}\sin \omega t$	$\dfrac{1}{s^2+\omega^2}$
$\cos \omega t$	$\dfrac{s}{s^2-\omega^2}$
$1 - \cos \omega t$	$\dfrac{\omega^2}{s(s^2-\omega^2)}$
$\sin(\omega t + \theta)$	$\dfrac{s\sin\theta - \omega\cos\theta}{s^2-\omega^2}$
$\cos(\omega t + \theta)$	$\dfrac{s\cos\theta - \omega\sin\theta}{s^2-\omega^2}$
$e^{-\alpha t}\sin \omega t$	$\dfrac{\omega}{(s+\alpha)^2+\omega^2}$
$e^{-\alpha t}\cos \omega t$	$\dfrac{s-\alpha}{(s+\alpha)^2-\omega^2}$
$\sinh \alpha t$	$\dfrac{\alpha}{s^2-\alpha^2}$
$\cosh \alpha t$	$\dfrac{s}{s^2-\alpha^2}$

[a] From Van Valkenburg, M. E. (1974). "Network Analysis," 3rd ed., © 1974, p. 192. Reprinted by permission of Prentice-Hall, Inc., Englewood Cliffs, New Jersey.
[b] All $f(t)$ should be thought of as being multiplied by $u(t)$ (i.e., $f(t) = 0$ for $t < 0$)

$Y(s)/X(s)$, where $H(s) = \mathcal{L}\{h(t)\}$ is the transfer function of the system. The transfer function $H(s)$ completely characterizes a linear system, in much the same way as for the impulse response. In particular, stability and natural response behavior of the system can be determined directly from the poles and zeros of $H(s)$. Also, the frequency response is easily found by evaluating $H(s)$ at $s = j\Omega$. If a sinusoid of the form $x(t) = A \cos(\Omega_0 t + \phi)$ is applied to a linear, time-invariant, and causal system with transfer function $H(s)$, then the sinusoidal steady-state response will be $y_{ss}(t) = A|H(j\Omega_0)| \cos[\Omega_0 t + \phi + \arg H(j\Omega_0)]$.

C. Role of the Fourier Transform in Analog System Analysis

A close relative of the Laplace transform, which is also very useful in system analysis, is the Fourier transform, defined as

$$F(j\Omega) = \mathcal{F}\{x(t)\} = \int_{-\infty}^{+\infty} f(t)e^{j\Omega t}\, dt \quad (4a)$$

$$x(t) = \mathcal{F}^{-1}\{F(j\Omega)\} = \frac{1}{2\pi}\int_{-\infty}^{+\infty} F(j\Omega)e^{-j\Omega t}\, d\Omega \quad (4b)$$

where the region of convergence is the entire $j\Omega$ axis whenever the transform exists. Note that the Fourier transform is identical to the two-sided Laplace transform with $s = j\Omega$. This implies that the Fourier transform of a causal system with impulse response $h(t)$ is identical to the frequency response of the system. Since the Fourier transform is two-sided, it contains information about behavior of the system for $t < 0$, and hence initial conditions do not arise in the same way as they do in applications of the Laplace transform. In general, analysts use the Fourier transform for sinusoidal steady-state analysis, although, theoretically, it also has the necessary properties to solve differential equations and obtain transient solutions.

Although they are closely related, it must be emphasized that the Laplace and Fourier transforms are quite different in nature, and hence they are both very useful in applications. Some of the differences are seen in Table II, which lists the Fourier transforms of some common functions, many of which we listed previously in Table I. Many of the Fourier transforms contain impulse functions that were not seen in the corresponding Laplace transforms. These impulse functions arise from the two-sided nature of the

TABLE II. Fourier Transforms of Some Common Functions[a]

Time functions, $f(t)$	Fourier transform, $F(j\omega)$
$k\,\delta(t)$	k
k	$2\pi k\,\delta(\omega)$
$\xi(t)$	$\pi\,\delta(\omega) + \dfrac{1}{j\omega}$
$\mathrm{sgn}\,(t)$	$2/j\omega$
$\cos\omega_0 t$	$\pi[\delta(\omega - \omega_0) + \delta(\omega + \omega_0)]$
$\sin\omega_0 t$	$j\pi[\delta(\omega + \omega_0) - \delta(\omega - \omega_0)]$
$e^{j\omega_0 t}$	$2\pi\delta(\omega - \omega_0)$
$t\xi(t)$	$j\pi\delta^{(1)}(\omega) - \dfrac{1}{\omega^2}$
$\displaystyle\sum_{k=-\infty}^{\infty} \delta(t - kT)$	$\omega_0 \displaystyle\sum_{n=-\infty}^{\infty} \delta(\omega - n\omega_0), \quad \omega_0 = 2\pi/T$
$\displaystyle\sum_{n=-\infty}^{\infty} F_n e^{jn\omega_0 t}$	$2\pi \displaystyle\sum_{n=-\infty}^{\infty} F_n\,\delta(\omega - n\omega_0)$
$\dfrac{d^n\,\delta(t)}{dt^n}$	$(j\omega)^n$
$|t|$	$\dfrac{-2}{\omega^2}$
t^n	$2\pi j^n \dfrac{d^n\,\delta(\omega)}{d\omega^n}$

[a] Reprinted with permission from Gabel, R. A., and Roberts, R. A. (1980). "Signals and Linear Systems," 2nd ed. p. 326. Copyright © 1980 by John Wiley & Sons, Inc., New York.

Fourier transform and often complicate solutions in simple problems. Note also that the Fourier transform of the sine and cosine functions are pure impulse functions, a result that coincides with our intuitive interpretation of frequency content of these signals.

III. Analog Filters Required in A/D and D/A Conversion

The traditional analog Fourier transform can be applied to an ideal sampled (analog) signal $s^*(t)$ as

$$\mathcal{F}\{s^*(t)\} = \sum_{n=-\infty}^{+\infty} s(nT)\mathcal{F}\{\delta_a(t - nt)\}$$

$$= \sum_{n=-\infty}^{+\infty} s_a(nT)e^{-j^n T\omega}$$

$$= \mathrm{DTFT}\{s_a(nT)\} \quad (5)$$

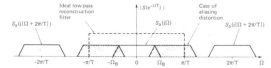

FIG. 2. Relationship between the spectrum of an analog signal and the spectrum of the ideally sampled signal.

where DTFT{·} denotes the familiar discrete-time Fourier transform of the discrete time sequence $s_a(nT)$, which consists of samples of the analog signal $s_a(t)$. This verifies that the traditional Fourier transform of $s^*(t)$ is identical to the DTFT of $s(n) \equiv s_a(t)|_{t=nT}$, that is, the sequence $s(n)$ is derived from $s_a(t)$ by ideal sampling, and it shows that $s^*(t)$ and $s(n)$ are really different models of the same phenomenon, since their spectra (as computed with appropriate transforms) are identical. Suppose that $S_a(j\Omega)$ is the spectrum of $s_a(t)$, and $S(e^{i\omega})$ is the spectrum of $s(n)$. It can be shown that

$$S(e^{j\omega}) = \frac{1}{T} \sum_{r=-\infty}^{+\infty} S_a\left(j\left[\Omega - \frac{2\pi r}{T}\right]\right) \quad (6)$$

where $\omega = \Omega T$ is often referred to as the normalized digital frequency. Equation (6) shows that the DT spectrum is formed from a superposition of an infinite number of replicas of the analog signal, as illustrated in Fig. 2. As long as the sampling frequency $\Omega_s = 2\pi/T$ is chosen so that $\Omega_s > 2\Omega_B$, where Ω_B is the highest-frequency component contained in $s_a(t)$, then each period of $S(e^{i\Omega T})$ contains a perfect copy of $S_a(j\Omega)$, and $s_a(t)$ can be recovered exactly from $s(n)$ by ideal low-pass filtering. Sampling under these conditions is said to satisfy the Nyquist sampling criterion, since the sampling frequency exceeds the Nyquist rate $2\Omega_B$. If the sampling rate does not satisfy the Nyquist criterion, the adjacent periods of the analog spectrum will overlap, causing a distorted spectrum as illustrated in Fig. 2. This effect, called aliasing distortion, is rather serious because it cannot be corrected easily once it has occurred. In general, an analog signal should always be prefiltered with an analog low-pass filter prior to sampling so that aliasing distortion does not occur.

Figure 3 shows the frequency response of a fifth-order elliptic analog low-pass filter that meets industry standards for prefiltering voice signals, which are subsequently sampled at an 8-kHz sampling rate and transmitted digitally across telephone channels. The passband ripple

is less than ±0.01 dB from dc up to a frequency of 3.4 kHz (too small to be seen in Fig. 3); the stop-band rejection reaches at least −32.0 dB at 4.6 kHz and remains below this level throughout the stop band.

When the Nyquist sampling criterion is satisfied, it is always possible to reconstruct an analog signal from its samples according to

$$s_a(t) = \sum_{k=-\infty}^{+\infty} s_a(kT) \, \text{sinc}\left[\frac{\pi}{T}(t - kT)\right] \quad (7)$$

This classical reconstruction formula can be derived by filtering $s^*(t)$ with an ideal low-pass analog filter with a bandwidth of $\Omega_B = \pi/T$, which removes the copies of the analog spectrum that were created by the sampling process (see Fig. 2). In general, exact reconstruction requires an infinite number of samples, although a good approximation can be obtained using a large but finite number of terms in the summation of Eq. (7).

Most practical systems use D/A converters for reconstruction, which results in a staircase approximation to the true analog signal, that is,

$$\hat{s}_a(t) = \sum_{k=-\infty}^{+\infty} s_a(kT)\{u(t - kT)$$
$$- u[t - (k + 1)T]\} \quad (8)$$

It can be shown that $s_a(t)$ is obtained by filtering $s_a^*(t)$ with an analog filter whose frequency response is

$$H_a(j\Omega) = 2Te^{-j\Omega T/2} \, \text{sinc}(\Omega T/2) \quad (9)$$

The approximation $s_a(t)$ is said to contain (sin x)/x distortion, which occurs because $H_a(j\Omega)$ is not an ideal low-pass filter. $H_a(j\Omega)$ distorts the signal by causing a droop near the passband edge,

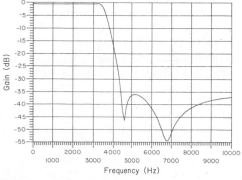

FIG. 3. A fifth-order elliptic analog antialiasing filter used in the telecommunications industry with an 8-kHz sampling rate.

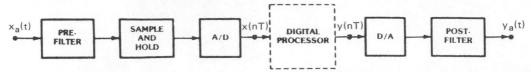

FIG. 4. Analog pre- and postfilters required at the A/D and D/A interfaces.

as well as by passing high frequency distortion terms, which leak through the sidelobes of $H_a(j\Omega)$. Therefore a practical D/A converter is normally followed by an analog postfilter.

$$H_p(j\Omega) = \begin{cases} H_a^{-1}(j\Omega) & 0 \leq |\Omega| \leq \pi/T \\ 0 & \Omega \text{ otherwise} \end{cases} \quad (10)$$

which compensates for the distortion and produces the correct $s_a(t)$, that is, the correctly constructed analog output. Unfortunately, the postfilter $H_p(j\Omega)$ cannot be perfectly implemented, and therefore the actual reconstructed signal always contains, in practice, some distortion that arises from errors in approximating the ideal postfilter. Figure 4 shows a digital processor, complete with A/D and D/A converters, and the accompanying analog pre- and postfilters necessary for proper operation.

IV. Analog Computing

Analog signal processing has many of its origins (even its name!) rooted in the historical development of the analog computer. Circuit designers found that with the aid of high-impedance, high-gain, vacuum-tube amplifiers they could design very-high-quality operational amplifiers by cascading numerous tube stages. These circuits inherited the name operational amplifiers because they could be used to realize accurately the mathematical operations of summation, scaling by constants, and integration. Given the capability to implement these operations, it was but a short time before engineers began to solve differential equations by building electrical circuits that would find solutions by electrical analogy. Simply stated, a circuit was built whose behavior is characterized by the same differential equation as the one to be solved. Then the physical behavior of the analogous circuits can be used to produce graphical solutions to many different excitations. Although the analog computer is very effective in producing real-time solutions to high-order linear differential equations, its real advantage is in the solution of nonlinear differential equations for which analytical techniques

are sometimes unknown. As long as the nonlinearity can be modeled with an electronic circuit (such as a simple analog multiplier or saturation characteristic), the analog computer is quite effective in obtaining a solution.

Figure 5 shows the modern symbol for an operational amplifier. Although state-of-the-art op amps are now rather complicated solid-state circuits, they are characterized by a small number of rather simple properties at the input and output terminals. The standard op amp is a differential amplifier that has a very high voltage gain, typically of the order of 10,000 or more. It is characterized by a very high input impedance (it does not load circuits to which it is attached), a very low output impedance (it will drive any circuit that is attached to its output), and the differential gain is linear between the saturation limits of the amplifier. The high gain implies that as long as the amplifier is operating in its linear region, the difference in voltage between the positive and negative input terminals must be extremely small. This condition, referred to as a virtual short circuit at the input of a nonsaturated op amp, is used to simplify analysis of circuits that contain these types of amplifiers.

Figure 6 shows an op amp configured as a summer, where the input–output relation is given by

$$y_0(t) = -(R_F/R_1)x_1(t) - \cdots - (R_F/R_N)x_N(t) \quad (10)$$

Similarly, Fig. 7 shows an op amp configured as an integrator, where the input–output relation is given by

$$y_0(t) = -\int_0^t [(1/R_1C)x_1(\tau) \\ + \cdots + (1/R_NC)x_N(\tau)]\, d\tau \quad (11)$$

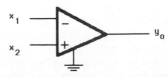

FIG. 5. Modern symbol for an operational amplifier.

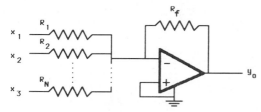

FIG. 6.   An op amp configured as a summer.

By combining these basic circuits, it is possible to solve by analog simulation a wide class of linear or nonlinear differential equations.

A nonlinear analog problem will be illustrated by using the Bonhoeffer–Van der Pol model of a nerve membrane. The electrochemical response of a nerve cell is known to be a highly nonlinear function that displays an analog threshold phenomenon. The theoretical foundation for the Bonhoeffer–Van der Pol nerve model was established many years ago by pioneers in the field, who had derived equations for the ion transfers that take place in stimulated nerve membrane. From the ion-transfer equations, a four-dimensional model was developed in terms of a set of differential equations. The model is based on the classical nonlinear Van der Pol equation, which will be shown to form an interesting model of nonlinear nerve response.

The solution to the following second-order differential equation is a damped sinusoid of the form $x(t) = e^{-kt}(A \sin bt + B \cos bt)$:

$$d_y^2(t)/dt^2 + K \, dy(t)/dt$$
$$+ \, x(t) = 0, \begin{Bmatrix} \text{initial} \\ \text{conditions} \end{Bmatrix} \quad (12)$$

If the damping constant k is replaced by a nonlinear term $c(x^2 - 1)$, the resulting equation, called the Van der Pol equation, is a nonlinear differential equation that describes the behavior of a large class of oscillators known as relaxation oscillators.

$$d^2y(t)/dt^2 + c(x^2 - 1) \, dy(t)/dt$$
$$+ \, x(t) = 0 \quad \text{ICs} \quad (13)$$

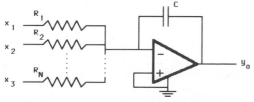

FIG. 7.   An op amp configured as a summing integrator.

By letting $y = (1/c) \, dx(t)/dt + x^3/3 - x$, Eq. (13) can be transformed into two couple first-order differential equations, which can then be programmed on an analog computer:

$$dx(t)/dt = c(y + x - x^3/3) \quad (14a)$$
$$dy(t)/dt = -x/c \quad (14b)$$

In Eq. (13), if $c = 0$, then a plot of the phase plane x versus $dx(t)/dt$ is a circle. If c is a positive constant, during a portion of each cycle $x(t)$ will be small so that $c(x^2 - 1) < 0$, while during the remaining portion of the cycle $c(x^2 - 1) > 0$. Obviously, during part of the cycle the solution is decaying and the phase plot spirals toward the origin. During the other portion it grows and spirals away from the origin. An equilibrium condition is established between these opposing forces, and a nonlinear oscillatory condition is reached known as a limit cycle. In a limit cycle the phase plane trajectory appears as a distorted circle, the distortion arising from the nonlinearity of the system.

The Bonhoeffer–Van der Pol neuron model is obtained by modifying Eqs. (14) as

$$dx(t)/dt = c(y + a - x^3/3 + z) \quad (15a)$$
$$dy(t)/dt = -(x - a + by)/c \quad (15b)$$

The quantity z is a stimulus, while the quantity x is related to membrane potential.

Figure 8 shows the analog computer circuit diagram for this neural model as it was programmed on an analog computer with $a = 0.7$, $b = 0.8$, and $c = 3.0$. The stimulus z was chosen to be a pulse, ramp, and step for experimental purposes. Pulses of width 0.1 sec were externally generated with a signal generator, with the height of the pulses carefully controlled by a potentiometer on the computer. The pulse generator was externally triggered by using a function switch and a positive reference voltage of 0.5 V. A function relay in series with a reference voltage served as a step input, while a ramp was created by integrating a reference voltage and using the comparator-hold circuits on the computer to stop the ramp and hold its final value. The pulse, step, and ramp stimuli were then used to investigate the important properties of this nonlinear model.

Limit-cycle behavior in the phase plane (x versus y) is shown in Fig. 9, with the different trajectories resulting from the application of different height pulses. The smaller trajectories about the resting point simulate subthreshold response. The neuron has not fired in this region,

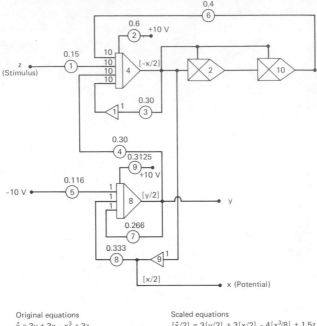

Original equations
$\dot{x} = 3y + 3x - x^3 + 3z$
$\dot{y} = -0.33x - 0.266y + 0.233$
$x(0) = 12.0$
$y(0) = 6.25$

Scaled equations
$[\dot{x}/2] = 3[y/2] + 3[x/2] - 4[x^3/8] + 1.5z$
$[\dot{y}/2] = -0.33[x/2] - 0.266[y/2] + 0.116$
$[x/2(0)] = 6.0$
$[y/2(0)] = 3.125$

FIG. 8.   Analog computer diagram for the Bonhoeffer–Van der Pol neuron model.

but there has been a disturbance of potentials due to limited ion exchange. The "enhanced" and "depressed" regions indicate areas on the phase plane where, if the model is in one of these states when a stimulus is applied, the response is correspondingly affected.

Superthreshold neural response is simulated by the larger trajectories, along which regions have been labeled to correspond to physiological nerve states. One interesting feature of this model is that it is truly an analog model that exhibits threshold-type behavior. All states within the phase plane can be reached, but the central region is nonattractive and is rarely tra-

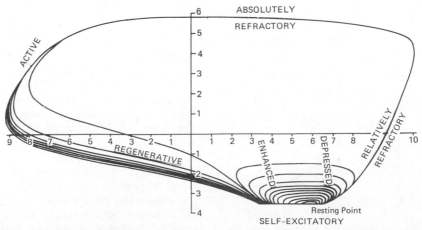

FIG. 9.   Limit cycle behavior of the Bonhoeffer–Van der Pol neuron model in the phase plane (x versus y).

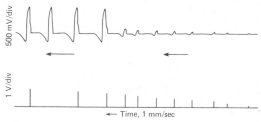

FIG. 10.   Illustration of threshold behavior.

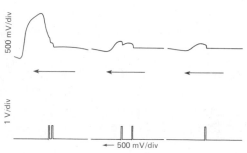

FIG. 11.   Response to subthreshold stimuli.

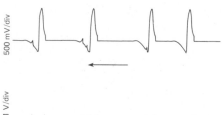

FIG. 12.   Response to stimuli during refractory period.

versed. This grouping of responses into large and small trajectories, called a quasi-threshold phenomenon in the literature, seems to describe the physical nerve behavior better than a digital description.

The time-domain operation of the Bonhoeffer–Van der Pol model illustrated in Figs. 10–13. These figures illustrate many interesting properties of the model, including threshold behavior (Fig. 10), firing by rapid subthreshold excitation (Fig. 11), response to stimuli in the refractory state (Fig. 12), and continuous firing caused by persistent stimuli (Fig. 13). It is apparent that many of these interesting properties demonstrated by analog computer simulation are a result of the nonlinear properties of the underlying Van der Pol model.

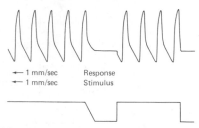

FIG. 13.   Continual firing due to persistent stimuli.

V. Characteristics of Analog VLSI Circuits

In recent years there has been an increasing emphasis on the development of very large scale integrated (VLSI) circuits for use in analog signal processing, as opposed to the more traditional discrete types of circuits constructed from collections of many small scale or medium scale integrated (SSI or MSI) circuits. The VLSI circuits are physically smaller, require less silicon area, consume less power, are more reliable, and result in reduced costs in general. However, VLSI circuits are much more difficult to design and test due to the large number of components on the chip and the limited number of terminals that can be brought out to the external environment for testing. Computer-aided design, layout, and testing are very important for VLSI

TABLE III.   Analog versus Digital Design: Important Characteristics[a]

Analog design	Digital design
Signals have a continuum of values for amplitude and time	Signals have only two states
Irregular blocks	Regular blocks
Customized	Standardized
Components have a continuum of values	Components have fixed values
Requires precise modeling capability	Optimizes VLSI technology
Difficult to use with CAD	Amenable to CAD tools
Performance optimized	Programmable by software
Designed at the circuit level	Designed at the system level

[a] From Allen, P. E. (1986). Analog VLSI design tutorial. (1986 Int. Symp. Circuits and Systems, Presymp. Workshop Analog and Digital VLSI, San Jose, California, May 1986.)

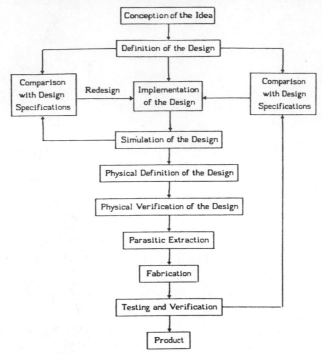

FIG. 14.   Overview of the IC design process. [From Allen, P. E. (1986). Analog VLSI design tutorial. (1986 Int. Symp. Circuits and Systems, Presymp. Workshop Analog and Digital VLSI, San Jose, California, May 1986.)]

TABLE IV.   Summary of Bipolar Integrated Circuit Technology[a]

1–3 μm, Trench isolation techniques ±3 to ±5 V power supplies
Up to four levels of metal
Bipolar structures and processing becomes simpler as the device is scaled
   down
Capacitors are not easily implemented
Low noise
High transconductance
Fast switching
Stabilized technology

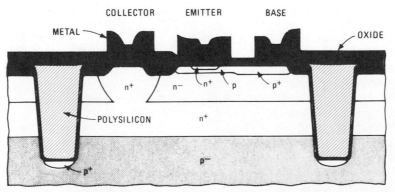

Fujitsu's trenched oxide and polysilicon isolation BJT.

[a] From Allen, P. E. (1986). Analog VLSI design tutorial. (1986 Int. Symp. Circuits and Systems, Presymp. Workshop Analog and Digital VLSI, San Jose, California, May 1986.)

TABLE V.   Summary of CMOS Integrated Circuit Technology[a]

1–3 μm, Single well
±3 to ±5 V power supplies
Double polysilicon
Several layers of metal
Substrate BJT
Good capacitor
Good switch
Low power
Problems with scaling—second-order effects, electron effects, etc.
Mature but still growing technology

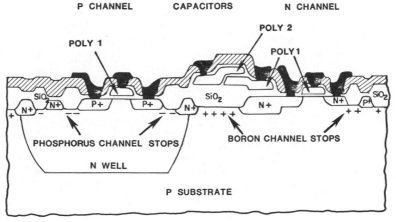

An example of state-of-the-art CMOS technology.

[a] From Allen, P. E. (1986). Analog VLSI design tutorial. (1986 Int. Symp. Circuits and Systems, Presymp. Workshop Analog and Digital VLSI, San Jose, California, May 1986.)

circuits because of the complexities that result from their size.

This section reviews analog and digital VLSI. Figure 14 shows an overview of the design process for a generic integrated circuit, as representative of the current state-of-the-art in industry. The overall design process is a complicated interaction between design, simulation, and fabrication. Since the overall process is tedious and expensive, the use of computer programs for design, analysis, and testing is extremely important. Table III summarizes the key similarities and differences between analog and digital design. In general, it is fair to conclude that analog design is more difficult and expensive than is digital. It is not as well suited for computer-aided design (CAD) tools, and it is not able to take as full advantage of VLSI capabilities as it is for digital design. Tables IV–VI summarize the essential characteristic of the three major

integrated circuit (IC) technologies that are in use today: (1) bipolar, (2) complementary metal–oxide–silicon (CMOS), and (3) gallium arsenide (GaAs). Both bipolar and CMOS are the workhorses of the integrated circuit industry today, while GaAs seems to hold a promise for the future, due to its potential for wide-band analog signal processing capabilities. This is more easily seen in Fig. 15, which compares approximate signal bandwidths for many different IC technologies. Note that GaAs (analog) technology lies at the upper extremes of signal processing bandwidths, in the same region as surface acoustic-wave devices and optical processors. Another innteresting feature that can be seen in Fig. 15 is that although the analog and digital categories overlap in the regions from 100 Hz to 10 MHz, digital is not able to compete well in the upper range of 10 MHz to 100 GHz. Therefore, in spite of the favorable position occupied by

TABLE VI.   Summary of Gallium Arsenide Integrated Circuit Technology[a]

0.5–3 μm, MESFET
± 4 to ± 6 V power supplies
$f_T \simeq 10 f_T$ (CMOS)
Poor hole mobility, only n-type devices
Wider temperature range
Not yet efficient for VLSI—conductor crossing, models, high power dissipation, etc.
Immature technology
Wide-band analog signal processing (op amp: 60 dB and 1000 MHz unity-gain bandwidth)

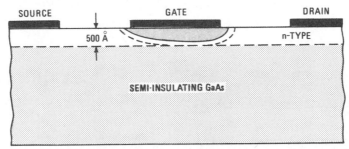

Example of GaAs MESFET technology.

[a] From Allen, P. E. (1986). Analog VLSI design tutorial. (1986 Int. Symp. Circuits and Systems, Presymp. Workshop Analog and Digital VLSI, San Jose, California, May 1986.)

digital VLSI in Table III, in the forseeable future, analog VLSI will be the choice of circuit designers for signal processing in the very high frequency ranges.

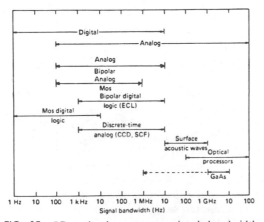

FIG. 15.   IC technology versus signal bandwidth. ("Design Methodology for VLSI Circuits," Sidhoff and Nordhoff, eds. Nato Advanced Study Institute on Design Methodologies for VLSI Circuits, Louvain La-Neuve, 1982.)

VI. Switched Capacitor Circuits

In the mid-1970s switched capacitor (SC) circuits became popular with circuit theorists and IC technologists when it was discovered that SC techniques could be used in high-precision monolithic filters. Driven by expectations of high economic payoff, SC circuits became a popular topic at workshops and conferences and in trade magazines and scientific journals, as well as throughout the entire IC industry. The explosion of activity in SC circuits tended to obscure the years of "awakening." Similarly, the frenzy of commercial development, with its focus on new products, tended to overshadow some of the important theoretical advances that were occurring at this time. This section is not meant to be a comprehensive historical survey, but rather an attempt to identify some of the important developments that marked key events in the evolution of the subject.

Prior to the discovery of SC circuits, the high-precision monolithic frequency-selective circuit was a tantalizing concept that remained elusive in practice. This was partly due to the fact that ICs cannot contain high-Q inductors, since the Q of an inductor decreases with decreasing size.

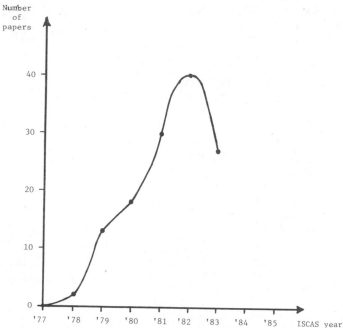

FIG. 16. Number of papers on SC circuits presented at the IEEE International Symposium on Circuits and Systems from 1977 to 1983.

Integrated resistors are nonlinear, occupy large silicon areas, and their tolerances and temperature coefficients do not closely track variances in the capacitors on the same chip. This results in RC time constants that cannot be controlled accurately and produces monolithic circuits with large variances in the frequency characteristics. In contrast, highly accurate capacitors can be realized quite easily in metal–oxide semiconductor (MOS) technology. Since different capacitors on the same chip track each other with variations in processing, capacitor ratios can be realized very accurately. Switched capacitor circuits are designed to depend only on capacitor ratios and thereby take advantage of the property to realize high-precision circuits.

After some initial publications of SC circuit design concepts circa 1977, a great amount of research began in the design and analysis of more general classes of SC circuits. Some of the initial enthusiasm decreased as the subject evolved into maturity. This observation is supported by the data presented in Fig. 16, which show the number of papers presented at the International Symposium on Circuits and Systems from 1977 to 1983 on the subject of SC circuits. This number increased rapidly from 1977 to 1981, peaked in 1982, and then fell significantly in 1983. This does not mean that the subject de-

creased in importance, or that all important problems were solved by 1983, but rather that the initial excitement had passed, the easy problems had been solved, practical limits had been identified, and circuit designers were beginning to face much more difficult problems.

A. HISTORICAL PERSPECTIVES ON SC CIRCUITS

Early researchers recognized that linear analog systems that contain switches represent an important class of linear time-varying (LTV) systems for which there exists a theory of analysis and design. If the switches are operated periodically, the system becomes linear periodically time-varying (LPTV), which is the closest relative to ideal linear time-invariant (LTI) systems. The inclusion of switches in an otherwise LTI network creates a more general structure with enhanced capabilities, although the mathematics required for analysis and design become more complicated.

Much of the early work on switched systems was done in Western Europe for telephone communication systems. In notable works on the theory of resonant transfer structures, capacitances are made to exchange charges periodically through appropriate interconnections with resonant circuits. Although the original resonant transfer circuits were largely replaced later by digital pulse-coded modulation (PCM) techniques, they provided a catalyst for the development of a more general theory of LPTV systems that become useful for the beginning studies of SC circuit analysis.

A second important development in the early years was the N-path filter structure. In its basic form this structure consists of N LTI systems arranged in parallel and switched into the system sequentially so that only one subsystem operates during a specified time interval. A number of authors used the N-path structure as a generic model for SC circuits, where each subsystem represents the SC circuit during one phase of the switching sequence. Also, the N-path model has been useful in SC circuit analysis programs because it decomposes the analysis into a sequence of LTI subsystem analyses in which each subnetwork provides initial conditions for the subnetwork that follows in the switching sequence.

The most recent flurry of SC activity began in the mid 1970s. By 1967 the idea had already been established that a filter characteristic could be specified entirely by capacitor ratios. It now seems clear that the burst of new activity in the mid-1970s was not a result of new advancements in either circuit theory or IC technology alone but rather was the result of coupling these two areas in an effective way.

The following rush of excitement was, to a great extent, led by IC technologists, that is, by researchers with the experience, facilities, and initiative to fabricate and test the first generation of experimental SC circuits. The group at the University of California at Berkeley reported success with fabricating low-pass antialiasing filters to be used as transmit prefilters and receive postfilters in telephone channel coder–decoders (CODECs). The first SC filter designs were derived from active filters in the state variable configuration by means of resistor replacement. (Note the analog computer influence in this development!) For example, the second-order filter shown in Fig. 17 is a state variable biquad in which the two resistors were replaced by switched capacitors $\alpha_1 C$ and $\alpha_2 C$. The first resistor is replaced by a "toggled" switched capacitor, whereas the second was replaced by a "parallel" switched capacitor. It was determined experimentally that the use of the toggled SC in the input path created a sharper rolloff than it would have had if a parallel SC been used in that branch also. However, in the early days of 1978, it was not clear how to choose the arrangement

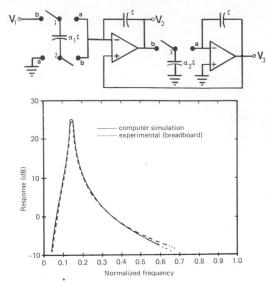

FIG. 17.   State-variable SC biquad by resistor replacement.

of switches to produce the best filter. This started a new movement to analyze SC circuits as discrete-time (but not digital) filters rather than as approximate active filters. Although they were not sophisticated filters to the circuit theorist, circuits of this type were fabricated at Berkeley, representing a first generation of circuits to demonstrate the feasibility of the SC concept.

When it became clear that resistor replacement techniques are inaccurate for switching rates near the Nyquist limit, designers began to characterize charge transfer with difference equations, derive transfer functions in the Z domain, and apply the theory of discrete-time systems to analyze performance. It was learned that if the output voltage is sampled on the "a" switch phase, then the network shown in Fig. 18a behaves as a forward Euler integrator, the network of Fig. 18b is a backward Euler integrator, and that of Fig. 18c as a trapezoidal integrator. Very-high-quality SC filters can be derived from the "leapfrog" active filter structure, which is a configuration of a doubly terminated LC ladder using these types of integrator replacements. The switch phasing between integrators in ladder networks also greatly affects performance. By simply alternating the phases between integrators, the ladder was converted from an Euler-sampled data design to a lossless discrete integrator (LDI) structure that is well known in the design of digital ladder filters. This was an important result because it showed that the use of discrete-time system theory can in certain circumstances lead to greatly improved performance without any increase in cost.

In 1978 a discovery was made that initially excited circuit theorists a great deal but that later led to considerable frustration because of practical limitations. It was discovered that a toggled switched capacitor, similar to the one used to replace the resistor R_1 in the state variable biquad of Fig. 17, transforms the analog structure into an SC structure according to a bilinear transformation that is well known as a design technique for digital filters. If each phase of the biphase clock has period T, the effective sampling rate becomes $2/T$ because the toggled switched capacitor has identical behavior on each phase. At this time, many circuit theorists thought that SC filter design theory was complete with this discovery, in that a good SC filter could be obtained from an active RC filter by simply substituting a toggled switched capacitor for each resistor. However, there are practical limitations that prevent this approach from working well in practice.

In MOS transistors there normally is a rather large nonlinear parasitic capacitance C_s between the bottom plate and the substrate, as well as a considerably smaller parasitic capacitance C_p between the top plate and the substrate. The value of C_s can be as much as 20% of the fabricated capacitor and is rather difficult to predict or compensate. The value of C_p is normally an order of magnitude smaller, although its presence can disturb a sensitive design significantly. It is important for SC circuits to be designed with the bottom plate of all capacitors grounded or switched between a voltage source and ground, so that the excess charge accumulated by the parasitic during the charging phase is harmlessly discharged to ground and does not affect the characteristics of the circuit. It is clear that a toggled switched capacitor cannot be made parasitic-insensitive in general, since the top and bottom plates must perform the same function on alternate phases. Despite efforts to compensate for parasitics in the bilinear switched capacitor, it was largely rejected by IC technologists as impractical because it led to too much uncontrolled variation in the characteristics of the final SC circuits.

Note that the forward and backward Euler integrators of Fig. 18 are insensitive with respect to both the top plate and bottom plate parasitics. For this reason these types of integrators became more successful than the bilinear integrator, and were subsequently used in many design methods that were eventually published. In 1979, a set of practical considerations for MOS implementation of SC circuits was published:

1. Switched capacitor "resistors" should not close an op amp feedback path.
2. There should be no floating nodes.
3. At least one plate of every capacitor should be connected to a voltage source or switched between voltage sources.
4. The noninverting op amp input should be kept at a constant voltage source (usually grounded) or switched between voltage sources.

Although items mentioned in this list should be treated as guidelines rather than as strict rules, they seem to have withstood the test of time in the sense that most IC technologists tend to reject designs that violate these conditions. In 1979 a design approach was published that incorporated these rules of thumb a priori and represented a considerable change of philosophy from the resistor and integrator replacement techniques that dominated earlier thinking. A network topology was chosen for a general biquad that incorporated the above guidelines. Biphase switching was selected that eliminates continuous feed-through paths and design equations were derived that express capacitor ratios in terms of desired pole locations. This represents one of the most successful attempts at de-

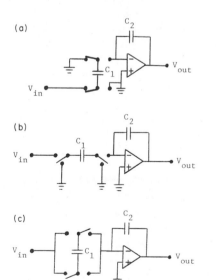

FIG. 18.   SC integrators: (a) forward Euler integrator, (b) backward Euler integrator, and (c) trapezoidal integrator.

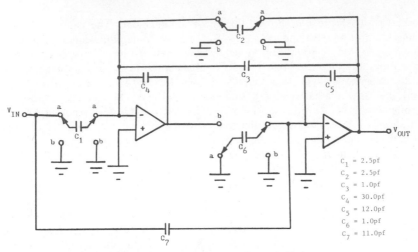

FIG. 19.   A second-order SC notch filter.

signing an SC biquad directly from $H(z)$. The approach has the added advantage that the network topology is predetermined so that the IC fabrication process can be standardized to a considerable degree.

For example, the circuit of Fig. 19 is a second-order low-pass notch filter designed by this technique. The notch was placed at 1.8 kHz, as seen in the frequency response of Fig. 20. Note that the frequency analysis shows a somewhat different magnitude response in the notch on the two phases, although one can easily determine by inspection that the magnitude response of this biquad is theoretically the same on each phase.

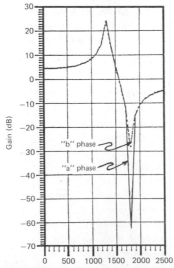

FIG. 20.   Frequency response for the notch filter of Fig. 19.

Since it was suspected that the different notch responses were due to computational inaccuracy in the frequency analysis routine, a sensitivity analysis was performed for the capacitors, the most sensitive of which are shown in Fig. 21. The sensitivity is two orders of magnitude larger in the notch in comparison with other frequencies in the pass region. It is interesting to note that the notch could not be obtained experimentally with average-quality discrete components; it is not clear whether the integrated form of this network would achieve satisfactory rejection at the notch frequency.

There have been many more recent attempts to establish more general theories of analysis and design of SC circuits. Also, there have been numerous studies of sensitivity noise and op amp finite-gain bandwidth effects. In general there are remarkable similarities in the properties and performance merits between active filters and their corresponding SC realizations. In particular, the active structures and their SC relatives track closely with respect to sensitivity and noise performance.

B. COMPUTER-AIDED ANALYSIS OF SC CIRCUITS

In the late 1970s there appeared to be two approaches to the computer-aided analysis of SC circuits; these were, in retrospect, temporary solutions. The first was the direct use of the well known SPICE program, although users quickly discovered that this approach is very expensive in terms of computer time and really generated much more information about transients between switching states than was re-

Frequency (Hz)	Magnitude Sensitivity		
	C_1	C_6	C_7
1500.0	0.3280E + 01	−0.1204E + 01	−0.2280E + 01
1550.0	0.3879E + 01	−0.1942E + 01	−0.2879E + 01
1600.0	0.4782E + 01	−0.3521E + 01	−0.3781E + 01
1650.0	0.6291E + 01	−0.7217E + 01	−0.5290E + 01
1700.0	0.9322E + 01	−0.9870E + 01	−0.8328E + 01
1750.0	0.1851E + 02	−0.3058E + 02	−0.1752E + 02
1800.0	−0.5189E + 03	−0.5163E + 03	−0.5201E + 03
1850.0	−0.1739E + 02	−0.1229E + 02	−0.1838E + 02
1900.0	−0.8669E + 01	−0.4809E + 01	−0.9668E + 01
1950.0	−0.5721E + 01	−0.2641E + 01	−0.6720E + 01
2000.0	−0.4241E + 01	−0.1692E + 01	−0.5241E + 01

FIG. 21.   Sensitivities for the most critical capacitors in the notch filter of Fig. 19.

quired. The second approach was to model an SC circuit as a digital filter with adders, multipliers, and delay registers, and then use a digital filter analysis program such as DINAP for time, frequency, and noise analyses. While this approach was effective for simple circuits with two-phase switching, the modeling proved to be too time-consuming for more complicated circuits, especially those with multiple clock phases.

Figure 22 gives a good perspective of the state-of-the-art in of SC CAD programs 1981. There clearly was an evolution from programs that dealt with specific topology and two-phase switching, to programs that were completely general with regard to topology, number of switching phases, phase intervals, and input–output (I–O) coupling. It seems that no one program has achieved the type of widespread acceptance that was extended to SPICE previously by the scientific community, although there are several that have been used considerably outside the institutions where they were developed. SWIITCAP, developed at Columbia University, is a user-oriented program that allows general network topology, general N-phase switching sequences, and continuous I–O coupling. It performs time-domain and frequency-domain analyses while permitting continuous signal paths. A particular feature included in this program is a compaction algorithm that reduces the dimension of the system equations prior to executing the sparse matrix solution. The equations are essentially nodal equations formed from appropriately defined closed surfaces that represent conservation of charge

relations. DIANA, developed at Katholieke Universiteit Leuven, is a well-known CAD package that was modified to admit SC circuits in addition to its original mixed-mode simulation capabilities. DIANA has a wide range of capabilities that lean heavily on the routines that previously existed for sparse matrix solution, sensitivity analysis, etc. DIANA is the only program that was available on a commercial basis at the time of this study. Equation formulation in DIANA is based on an adapted form of modified nodal analysis and the sensitivity analysis takes advantage of adjoint techniques. SCAPN, developed at the University of Illinois, is a third-generation SC program (following SCAPI and SCAPII of Fig. 22) that also accommodates SC networks with general topology and arbitrary switching sequences. It is a stand-alone program like SWITCAP in the sense that it was developed specifically for SC circuits and does not exist as a subprogram of a more general software package, as in DIANA. SCAPN uses modified nodal analysis directly in the Z domain, as well as sparse-matrix storage, adjoint techniques for sensitivity analysis, and a special pre-LU factorization algorithm that conserves computation during frequency updates of the LU factors. SCAPN factors a frequency-domain nonlinear distortion-analysis routine for nonlinear capacitors and op amp gains, although in its present form it does not permit continuous I–O coupling.

There are many other analysis programs and problem formulation philosophies described in the literature, and it is regrettable they cannot all be discussed here. In the earlier sections of this

Source of the program	A. Ganesan Bell Labs U.S.A.	H. De Man et al. Kath. Univ. Leuven Belgium	R. Schweer et al. Univ. Dortmund Germany	U. Kleine et al. Univ. Dortmund Germany	H. Gutsche Siemens Germany	A. Knob and R. Dessoulavy EPFL Switserland
Name of the program	BSCAP	DIANA	DOSCA	DONES		
Features						
Time–domain analysis	Yes	Yes	Yes		Yes	
Frequency–domain analysis	Yes	Yes	Yes		Yes	Yes
Aliasing effect	Yes	Yes				
Sensitivities		Yes	Yes			
Group delays	Yes	Yes	Yes			
Noise analysis		Yes	Yes			
Nonlinearities		Yes	Yes			Yes
Resistor or op amp poles	Yes	Yes	Yes		Yes	Yes
More than 2 phases	Yes	Yes	Yes		Yes	
Continuous I–O coupling	Yes	Yes				
Topological restrictions			Yes			Some
Design synthesis				Yes		
Optimization			Yes	Yes		
Mixed mode and top down design		Yes				
Symbolic transfer functions						
Multiple clock frequencies						
User's manual available			Yes			
Availability	No	Yes	Yes	No	No	Yes
						Implement in SPI, CE, ASTP, etc.
Performance						
Accuracy problems					Small and large resistance	
Size of network					Degree <16	
Speed						

FIG. 22.   CAD programs for SC circuits. [Reprinted with permission from Vandewalle, J. (1981). *Proc. 1981 Eur. Conf. Circuit Theory and Design, The Hague, August 1981*, p. 1072. Copyright © 1981 by North-Holland Publishing Company, Amsterdam.]

article emphasis was placed on approaches that emphasize the discrete-time nature of the SC circuit. There is also a different school! of thought represented in the literature, in which the problem of modeling the discrete-time behavior of the circuit is approached by developing analog equivalent circuits so that the subsequent analysis can be done with existing analysis programs. For example, this approach has been popular within Bell Laboratories, where the circuit analysis is ultimately carried out using the well-known analog analysis pro-

gram CAPECOD. It seems that both approaches to SC computer-aided circuit analysis result in valid simulation of circuit behavior, so it is likely that both will be continued in future applications.

C. SWITCHED CAPACITOR PRE- AND POSTFILTERS: A CASE STUDY

One successful application of switched capacitor circuit technology is in the realization of the analog pre- and postfilters required for digital

E. Simonyi et al. TKI Budapest Hungary	C. Lee and W. Jenkins Univ. of Illinois U.S.A.	C. Lee and W. Jenkins Univ. of Illinois U.S.A.	R. Plodeck et al. ETH-Zürich Switserland	M. Bon, CNET France A. Konczykowska T. Univ. Warsaw Poland	Y. Tsividis et al. Columbia Univ. U.S.A.	T. Rübner-Petersen Lyngby Tech. Univ. Denmark
PRSC	SCAPI	SCAPII	SCANAL	SCYMBAL	SWITCAP	SCNET
	Yes			Yes	Yes	
Yes	Yes	Yes	Yes	Yes	Yes	Yes
			Yes		Yes	
	Yes	Yes		Yes		Yes
				Yes		
						Yes
Yes			Soon	Yes	Yes	Yes
		Yes				
	Some	Some				
Yes						
Yes						
			Yes	Yes		
					Yes	
Yes	Yes	Yes	Yes	Yes	Yes	Soon
Yes	Yes	Yes	Yes	Yes	Yes	Yes
On-desk-top computer						
		Some at zero frequency				
Small				Small		
	Fast	Slower				

voice transmission in modern telephone channels. The A/D and D/A conversion, as well as some additional signal conditioning (companding, etc.), is done with a type of IC referred to as a CODEC (for coder–decoder). One IC contains the transmit filter, to handle transmission in the forward direction, the receive filters, to handle the returned voice signal, and any other circuits required for signal conditioning. The industry standard for the digital sampling rate is 8 kHz, which is designed to prevent aliasing on a 4-kHz telephone channel. Prior to SC circuit technol-ogy, the transmit and receive filtering was done with an active filter that was implemented by hybrid fabrication techniques, which added considerable cost to the system. With switched capacitor circuit techniques, the goal of integrating the transmit and receive filters together with the CODEC, while meeting the rather strict system specifications, became reachable. The philosophy is to switch the SC filters at a very high rate (ususally 256 kHz) relative to the 8-kHz sampling rate of the CODEC so that sampled data effects in the SC circuits are negligible, and for

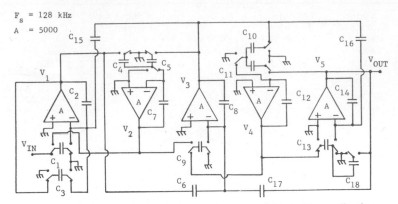

FIG. 23. A fifth-order elliptic low-pass state-variable ladder realization.

all intents and purposes the SC pre- and postfilters function as analog filters. The transmit filter must have a low-pass characteristic within a tolerance of 0 ± 0.01 dB in the pass-band up to a ripple cutoff frequency of 3.4 kHz. At 4.6 kHz the filter rejection must be at least -32.0 dB and the response must remain below this level for all frequencies above 4.6 kHz. In addition, the transmit filter must have (at least) a -20 dB notch at 50 to 60 hz to reject 60 (50) cycle noise that is usually present in the environment. In the receive channel, a postfilter is required following the D/A converter to smooth the staircase waveform that results from the reconstruction process (see Section III for a discussion of the theory). The receive filter is a low-pass filter with a cutoff at 3.4 kHz, which is essentially the same as the antialiasing transmit filter, although there is an additional requirement for a gentle 1.7- to 1.9-dB gain in the pass-band to compensate for the $(\sin x)/x$ distortion characteristic of the D/A reconstruction process. The example designs shown here are for the idealized low-pass filter alone; that is, they do not include the transmit notch and receive $(\sin x)/x$ compensation. It is well known in the telecommunications industry that systems specifications can be met with a fifth-order elliptic design for the ideal low-pass filters.

The three SC ladder structures shown in Figs.

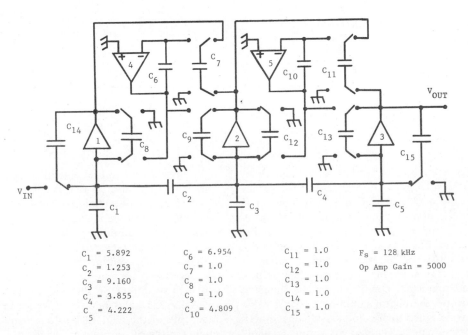

$C_1 = 5.892$	$C_6 = 6.954$	$C_{11} = 1.0$	$F_s = 128$ kHz
$C_2 = 1.253$	$C_7 = 1.0$	$C_{12} = 1.0$	Op Amp Gain = 5000
$C_3 = 9.160$	$C_8 = 1.0$	$C_{13} = 1.0$	
$C_4 = 3.855$	$C_9 = 1.0$	$C_{14} = 1.0$	
$C_5 = 4.222$	$C_{10} = 4.809$	$C_{15} = 1.0$	

FIG. 24. A fifth-order elliptic low-pass ladder realization based on inductor simulation.

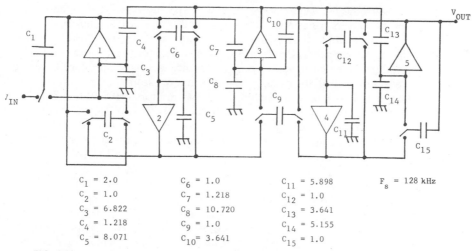

$c_1 = 2.0$ $c_6 = 1.0$ $c_{11} = 5.898$ $F_s = 128$ kHz
$c_2 = 1.0$ $c_7 = 1.218$ $c_{12} = 1.0$
$c_3 = 6.822$ $c_8 = 10.720$ $c_{13} = 3.641$
$c_4 = 1.218$ $c_9 = 1.0$ $c_{14} = 5.155$
$c_5 = 8.071$ $c_{10}= 3.641$ $c_{15} = 1.0$

FIG. 25.   A fifth-order elliptic low-pass ladder realization based on unit-gain buffers.

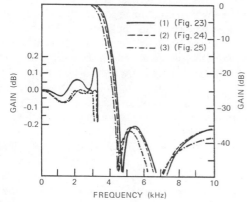

FIG. 26.   Frequency response for the three lowpass ladders of Figs. 23–25.

23–25 are all capable of realizing the required fifth-order elliptic low-pass transmit and receive filters just described. The circuit of Fig. 23 is a variation of the state-variable ladder derived by an LDI (lossless discrete integrator) transformation, although certain switches have been rearranged to guarantee a parasitic-insensitive circuit. Figure 24 shows a second design based on inductor simulation, while Fig. 25 shows a third design based on an approach using all unit gain buffers, rather than op amps. As can be seen from the frequency analyses in Fig. 26, all of these designs are capable of approximating the desired response very accurately (the Q enhancement visible in the state variable design

	Output magnitude sensitivity with respect to P_k		
Circuits	0 to ± 0.3	0 to ± 0.6	0 to ± 1.5
State Variable Ladder Filter	C2, C6, C7, C8, C9, C12, C14, C15, C16, C17, and OP AMP 1–5	C1, C3, C4, C5, C10, C11, C13, and C18	
Inductor simulation Ladder Filter	C1, C2, C3, C4, C5, C6, C7, C10, C11, and OP AMP 1–2	C8, C9, C12, C13, C14, and C15	Buffer 1–3
Unit gain Buffer Ladder Filter	C1, C2, C3, C4, C5, C6, C7, C8, C9, C10, C11, C12, C13, C14, and C15	Buffer 3 and buffer 5	Buffer 1–2 and buffer 4

FIG. 27.   Sensitivities of the most critical components in the ladder realizations of Fig. 23–25.

can be corrected by adjusting the values of the terminating switched capacitors if desired). To compare these structures further, sensitivities were calculated for every capacitor, op amp gain, and unity buffer gain. The elements were classified as not very sensitive (0.0 to ±0.3), moderately sensitive (±0.3 to ±0.6), or very sensitive (±0.6 to ±1.5) over the pass-band frequencies 0.0–2.0 kHz, as tabulated in Fig. 27. The unity gain buffers are among the most sensitive elements. Although it is difficult to make definitive claims for a superior circuit, it appears that with regard to sensitivities the original state-variable ladder is the preferred circuit because it contains no element in the most sensitive category.

VII. Switched Analog Filters

The biggest problem with the existing monolithic integrated-circuit technology for frequency-selective filters is the wide variation of the relative resistor values in a single integrated circuit. Variations in processing parameters (temperature, impurity dose, time, etc.) can cause the resistivity of diffused resistors to vary more than 1% on the same chip. Furthermore, this variation is uncorrelated to the variation of capacitance per unit area. Together these two effects cause RC products to vary by a factor of 2 when using the best processing available.

The growth of silicon dioxide on silicon is subject to similar processing variations (temperature and time), but in this case impurities are not added. For this reason the thickness of thermally grown oxide layers on a wafer does not change very much across the wafer surface. In fact, capacitors in MOS technology can be matched within 0.1%. Thus any filter topology with a transfer function dependent on capacitor ratios is ideal for unipolar (MOS) IC fabrication.

Researchers have introduced techniques that eliminate resistors by replacing them with field-effect transistors (MOSFETS) biased in the linear (ohmic) region of operation. Future development of this technique, or similar techniques that eliminate diffused resistors from filter circuits, raises the possibility that integrated filters containing resistors, capacitors, op amps, and switches may offer a simple and inexpensive means of digitally tuning state-variable analog filters. At switching rates greater than 1 MHz, MOS switched capacitor filter performance begins to degrade, which makes the use of switched capacitor filters impractical above the audio range (20 kHz). Also, the noise associated with all MOS switches and MOS amplifiers makes the dynamic range too small for demanding high-fidelity audio applications. Another dynamic range limitation is the fundamental low breakdown voltage of MOS transistors. Current technologies do not allow signals above approximately 8 V peak-to-peak. For these reasons, this more general class of switched resistor–capacitor filters may become more practical in the future.

Figure 28 shows the prototype of a switched resistor–capacitor biquad that has been studied extensively at the University of Illinois. It is essentially a state-variable biquad active filter structure, with one analog switch added to control the bandwidth (S_b) and two analog switches added to control the center frequency (S_f). Note that the filter has both a bandpass output (V_{BP})

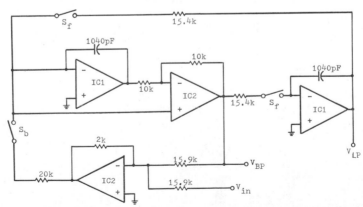

FIG. 28.   Experimental prototype for a tunable switched resistor–capacitor filter. IC1 = IC2 = LF353. All switches are part of CD4016 CMOS switch. All resistor values are in ohms.

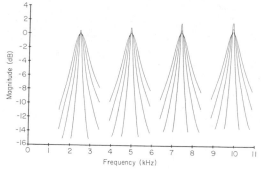

FIG. 29. Demonstration of electronic tunability of the center frequency and bandwidth for the filter in Fig. 28.

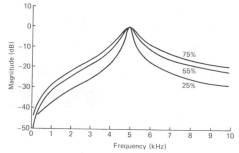

FIG. 31. Experimental bandwidth tuning of bandpass frequency response (% indicates the switch duty cycle).

and a low-pass output (V_{LP}). One of the important features of this circuit is that the bandwidth and center frequency are tuned by the duty cycles on the switches, not by the switching frequencies. This is important in practical integrated circuits because it is a much simpler matter to adjust duty cycles by proper gating, whereas it is a more costly matter to change the clock frequencies over a continuous range. This prototype was built from discrete components and has not, to the author's knowledge, ever been implemented in integrated form.

Figures 29–31 show experimental results that illustrate the electronic tuning capability of this circuit. Figure 29 shows 16 different frequency responses superimposed on the same axis: A, B, C, and D are four different center frequencies, and the four curves around each of these center points represent filters with four different bandwidths. There is a slight asymmetry visible in these curves due to nonideal effects of the discrete components used in the construction. Figure 30 shows tuning for three different center frequencies with the Q of the filter held constant (i.e., both S_b and S_c at 25, 50, and 75% duty),

and Fig. 31 shows tuning for three different bandwidths (S_b at 25, 50, and 75% duty) with the center frequency held constant. Experiments showed that a very wide range of center frequencies and bandwidths could be obtained by the proper choice of the switch duty cycles.

This example illustrates the tremendous flexibility that can be achieved with analog filters that contain resistors, capacitors, op amps, and switches. The combination of electronically tunable analog circuits with digital computer controls results in flexible and cost-effective solutions to practical signal processing problems faced by designers of electronic instrumentation.

VIII. Optical Signal Processing

Radar, or radio detecting and ranging, has been used for decades to detect distant objects, to measure their velocity, and, more recently, for terrain mapping. The basic theory first appeared in the early 1900s and was demonstrated in practice during the mid-1930s. Since its early use in World War II to detect airplane bombers, it has been refined and used in hundreds of applications, such as space exploration, aircraft avoidance systems, insect and bird tracking, weather observation and prediction, and, of course, highway speed enforcement.

When radar is used to simply detect the range to a distant object, a series of electromagnetic pulses is transmitted in the direction of the object and the two-way time delay of the reflected pulses is measured. Because it is known that the pulses propagate at the speed of light, the range to the object can easily be calculated from this delay. If it is desired to measure the radial velocity of an object, a continuous-wave (cw) electromagnetic signal (a simple sinusoid) is transmitted and the frequency shift of the returned

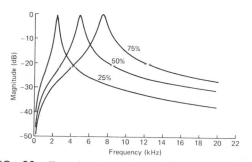

FIG. 30. Experimental constant-Q bandpass frequency response for the filter of Fig. 28 (% indicates the switch duty cycle).

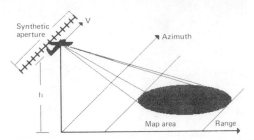

FIG. 32. Physical geometry of a strip-mapping SAR system.

signal, induced by the moving object, is measured. It is a straightforward procedure to calculate the velocity of the object from the Doppler shift and the known propagation velocity of the signal.

When the azimuthal position (perpendicular to range) is also desired, the transmitted signal must be sent as a narrow beam that is no wider than the desired azimuthal resolution. It is easily shown that at a range R and carrier ω_c, an antenna with an aperture size D will have an azimuthal resolution on the order of $R\omega_c/D$ m. For a radar operating in the conventional microwave region (10^9 Hz), this implies that very small resolution requirements dictate a physical antenna of very large size, that is, several hundred feet across. For an airborne antenna, this is impractical; however, very high azimuthal resolution can be achieved by synthesizing a large antenna through recording and signal processing.

Synthetic aperture radar (SAR) is a means of obtaining high-resolution terrain maps (complex reflectivity maps) by coherently processing the backscattered radar signal's phase (referred to sometimes as a phase history). In 1953, the University of Illinois experimentally demonstrated strip-mapped SAR. This was the earliest form of SAR in which the airborne radar antenna is held at a fixed squint angle relative to the terrain

patch of interest. Coherent processing of the returned radar signal produces an effective aperture many times the size of the physical antenna, thus creating an azimuthal beam width proportionally narrower, resulting in finer azimuthal resolution.

A higher-resolution form of SAR is called spotlight-mode SAR. This mode requires that the antenna continually be steered toward the center of the terrain patch during the aircraft flight. The same ground area is effectively pictured from many different angles and when coherently processed, produces an image of much higher than that which could be obtained from a single observation angle. Both strip-mapping and spotlight-mode SAR can also operate in space with a low-altitude satellite orbit; however, in this case the curved shape of the orbital trajectory becomes a significant factor and must be compensated as part of the signal processing required to produce the final image.

Synthetic aperture radar image processing can be done by either optical or digital signal processing. The following discussion describes the fundamentals of optical SAR processing, a subject that represents a successful application of two-dimensional analog signal processing. Figure 32 depicts a radar platform traveling with velocity V at altitude h and carrying a side-looking antenna that illuminates an area on the ground to be imaged (map area). It is assumed in this discussion that the radar is operating in the strip-mapping mode, so the antenna is held fixed on the aircraft and the ground strip is mapped out by the forward motion of the aircraft. The radar transmits a linear FM pulse

$$s(t) = \begin{cases} \cos(\omega_c t + \nu t^2/2), & |t| \leq T/2 \\ 0, & t \text{ otherwise} \end{cases} \quad (16)$$

at N distinct points along the flight path. At each location the radar receives the echo form the

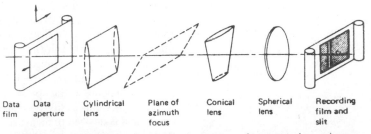

| Data film | Data aperture | Cylindrical lens | Plane of azimuth focus | Conical lens | Spherical lens | Recording film and slit |

FIG. 33. Elements of a coherent optical system for processing strip-mapping SAR imagery. [From Harger, R. O. (1970). "Synthetic Aperture Radar Systems: Theory and Design. Academic Press, New York.]

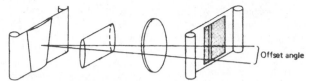

FIG. 34. Consolidated optical system for SAR processing. [From Harger, R. O. (1970). "Synthetic Aperture Radar Systems: Theory and Design." Academic Press, New York.]

Map Area and records this return for later processing. Ideally, the returned signal is demodulated into a base-band (complex) representation and then recorded for later processing. If digital processing is to be used, the I and Q channels of the base-band representation are sampled and the complex samples are stored in a digital memory. If optical processing is used, the demodulation does not translate the signal completely into the base-band, but rather the signal is left on a small offset frequency so it can be represented with real analog levels and recorded on film. This is done by using the demodulated result to modulate a cathode ray tube (CRT) sweep, which writes the signal across a recording film. After the data recording is developed it becomes the input data into an optical correlator, which match-filters (compresses) the signal in both the azimuth and range dimensions to produce a high-resolution (fully focused) image.

The essential components of the coherent optical processor used to process strip-mapping SAR images are shown in Fig. 33. The film on the left is the raw recording, which contains range sweeps written side by side from top to bottom (vertically) in the figure. The cylindrical lens removes the quadratic phase modulation in the range dimension, which was placed in the transmitted waveform by the transmitter. The conical lens performs the range-dependent quadratic-phase correction in the azimuth dimension so that all range cells are brought simultaneously into focus. The spherical lense produces a Fourier transform in the back focal plane, were the second recording film is exposed through a vertical slit as it scrolls in synchronism with the raw data film. The combination of the cylindrical and spherical lenses implements the quadratic phase-matched filter in range, while the conical and spherical lenses implement the range-dependent quadratic phase correction and integration in azimuth. The vertical slit is required to pass only the dc component of the spectrum, which means that the spherical lens performs a simple integration in azimuth when its output is viewed through the slit. Figure 34 shows the final configuration for an optical system to process SAR images.

One of the attractive features of optical signal processing for SAR is the extremely high processing rates that can be achieved with very wide band data. Equivalent digital systems require large memories and very high speed digital processors to accomplish the same task. Unfortunately, the fact that data films must be exposed and developed, together with the fact that the optical processing equipment is not very portable, has resulted in more attention being devoted to digital processing for SAR in recent years.

BIBLIOGRAPHY

Broderson, R. W., Gray, P. R., Hodges, D. A., Allstott, D., and Jacobs, G. MOS switched capacitor filters. *Proc. IEEE* **67**, 61–75.

Embree, P. M. (1983). The analysis and performance of a tunable state variable filter employing switched resistor elements. *Proc. 1983 Int. Symp. Circuits and Systems, Newport Beach, California, May 1983*, pp. 857–860.

Fang, S. F., Tsividis, Y. P., and Wing, O. (1983). SWITCAP: A switched-capacitor network analysis program—Part I: Basic features. *Circuits Systems Mag.* **5**(3), 4–10. Part II: Advanced applications, *Circuits Systems Mag.* **5**(4), 41–46.

Fettweis, A. (1981). Switched-capacitor filters: From early ideas to present possibilities. *Proc. 1981 Int. Symp. Circuits and Systems, Chicago, April 1981*, pp. 414–417.

Fried, D. L. (1972). Analog sample-data filters. *IEEE J. Solid-State Circuits* **SC-7**, 302–304.

Gabel, R. A., and Roberts, R. A. (1980). "Signals and Linear Systems," 2nd ed. Wiley, New York.

Harger, R. O. (1970). "Synthetic Aperture Radar Systems: Theory and Design." Academic Press, New York.

Jacobs, G. M., Allstot, D. J., Broderson, R. W., and Gray, P. R. (1978). Design techniques for MOS switched capacitor ladder filters. *IEEE Trans. Circuits Systems* **CAS-25,** 1014–1021.

Jenkins, W. K. (1983). Observations on the evolution of switched capacitor circuits. *Circuits Systems Mag.* **5**(4), 22–33.

Laker, K. R., and Fleischer, P. E. (1979). A family of active switched capacitor biquad building blocks. *Bell System Tech. J.,* **58,** 2235–2269.

Vandewalle, J. (1981). Survey of computer programs for CAD of switched capacitor circuits. *Proc. 1981 Eur. Conf. Circuit Theory and Design, The Hague, August 1981,* pp. 1071–1072.

Van Valkenburg, M. E. (1974). "Network Analysis," 3rd ed. Prentice-Hall, Englewood Cliffs, New Jersey.

SIGNAL PROCESSING, DIGITAL

Fred J. Taylor   *University of Florida*

GLOSSARY

Aliasing: Two or more distinctly different signals having identical or impersonating sample values.

Convolution: Mathematical process, appearing in linear or circular form, that models the input–output filtering process.

Digital filter: Digital device that is capable of altering the magnitude, frequency, or phase response of a digitally encoded input signal.

Digital signal processing chip: Electronic chip or chip set designed to perform the digital signal processing operations of transforming and filtering.

Discrete Fourier transform: Mathematical method of transforming a time series into a set of harmonics in the frequency domain and vice versa.

Fast Fourier transform: Popular algorithm used to implement discrete Fourier transforms.

Finite word length effects: Filter or transform behavior that can be attributed to the rounding or truncation of data found in finite-length register machines.

Impulse response: Response of a filter to a single input sample. If of finite duration, it is said to be produced by a finite impulse response filter, otherwise an infinite impulse response filter.

Limit cycling: Filter behavior that results in unwanted oscillations due to data rounding, truncation, or saturation.

Nyquist sample rate: Sample rate above which a bandlimited signal can be reconstructed from its sample values.

Quantization: Process by which a real variable is converted into a digital word of finite precision.

Sample value: Number or value associated with a single distinct sample; it is real for a discrete system and of finite precision in a digital signal processing system.

Time series: Continuous string of sample values.

Window: Process of isolating and modifying data to the exclusion of all others.

z-Transform: Mathematical method of analyzing and representing discrete signals.

Digital signal processing (DSP) is a relatively new technical field that is concerned with the study of systems and signals with respect to the constraints and attributes imposed on them by digital computing machinery. It differs from other signal processing technologies, namely analog and discrete, in terms of signal definition, computational procedures, and performance limitations. Because of the organization of a digital computer and the finiteness of digital calculations, the field of DSP has developed a unique set of analysis and synthesis procedures. A DSP system will accept a string of digitally encoded samples, called a digital time series (or time series), and modify, manipulate, classify, or quantify them using DSP algorithms. The tools used to implement these DSP algorithms are digital computer software, hardware, or their mix. The digitally processed data can then be used or analyzed by humans or machines. Digital signal processing has become an essential element in the study of communication, control, medicine, speech, vision, radar, sonar systems, plus a host of other applications where the high speed and high precision of digital computers can be applied to signal and system synthesis or analysis.

I. Origins of Signal Processing

The foundations of digital signal processing (DSP) are interwoven with the origins of its counterparts, analog signal processing and algorithms. An algorithm is a computing procedure that is generally well suited for execution on a digital computer. The theory of algorithms can be traced back to the seventeenth and eighteenth century work of the celebrated mathematicians Sir Isaac Newton and Karl Freidrich Gauss. An important signal processing tool now called the discrete Fourier transform (DFT) may have been derived by Gauss as an algorithm as early as 1805. This would predate Fourier's discovery in 1807 of the infinite harmonic series, which was published in 1822. In fact, many early digital signal processing successes were simply classic numerical analysis algorithms programmed for execution on a general-purpose digital computer, but modified to exhibit special signal processing attributes. [See SIGNAL PROCESSING, GENERAL.]

The other foundation on which digital signal processing is built is its predecessor, analog signal processing. An analog signal would have a graph that is continuously defined along both the independent (e.g., time) and dependent (e.g., amplitude) axes. This science goes back to the very early days of electronics, when now-primitive vacuum tube amplifiers, separately packaged capacitors, inductors, resistors, and the like were used to filter signals on the basis of their frequency content. These analog systems were often modeled in terms of ordinary linear differential equations. During the 1940s it became popular to study such systems with a differential equation analysis tool called the Laplace Transform. Using these classical analog methods, a number of important filter forms were developed. Those that have had a major impact on the design of digital filters have been the maximally flat (Butterworth) filter and the equal-ripple filters referred to as Chebyshev and elliptic filters. In the late 1940s and 1950s a new dimension in analog signal processing evolved, which was called sampled data. A sampled data

signal would have a graph whose dependent axis is continuously defined but whose independent axis is discrete. This is the result of sampling a continuous signal, say $x(t)$, at periodic intervals of time and then saving the resulting sampled values. For example, if a ± 10-volt signal is sampled at the sample rate of 1000 times per second, the resulting discrete signal could look like $x(n) = \{..., +1.23334, -0.342344, +7.24324, ...\}$, where the three displaycd *real* sample values represent the actual value of $x(t)$ at times $t = (k - 1)t_s, kt_s, (k + 1)t_s$ where $t_s = 1/1000 = 10^{-3}$ sec. These early discrete systems were used in low-frequency control applications, such as autopilots, where conventional analog filters were too bulky, unreliable, and expensive to be considered viable design options. Mathematically, the designers of discrete systems used essentially those techniques developed for linear continuous system analysis by modifying the Laplace transform to a form we now call the z-transform. [See SIGNAL PROCESSING, ANALOG.]

During the late 1940s another major milestone was achieved. Claude Shannon, of Bell Laboratories, and his associates noticed that if a signal $s(t)$ contains no frequency components at or above f_N Hz [$s(t)$ is then said to be bandlimited to f_N Hz], then $s(t)$ can be completely reconstructed (interpolated) from its sample values provided the samples are taken at a rate equal to or in excess of $f_s = 2f_N$ samples per second. This condition is known as the Nyquist sampling theorem, f_s is referred to as the Nyquist sample frequency and F_N is sometimes called the Nyquist frequency. In the time domain, the samples must be spaced no further apart than $t_s = 1/f_s$ sec, where t_s is the sample period. If a bandlimited signal is sampled at a rate slower than the Nyquist sample frequency, a type of error called aliasing can occur. As the name suggests, an aliasing signal impersonates another signal in such a manner that they cannot be discriminated from each other. This problem is suggested in Fig. 1. For example, if two signals $s_1(t) = 1$ and $s_2(t) = \cos(2\pi t)$ are sampled once per second, their resulting sample series would both be given by $\{1, 1, 1, ...\}$. The reason for this undesirable circumstance is that $s_2(t)$ is bandlimited to 1 Hz and therefore must be sampled at a rate in excess of two samples per second. Once properly applied, the sampling theorem had an immediate impact on the fields of communication, control, as it later did on the new field of digital signal processing.

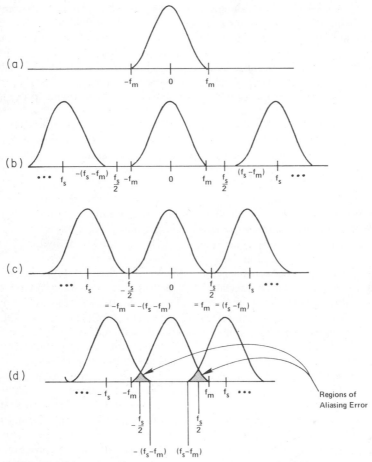

FIG. 1.   Example of aliasing: graphical interpretation of spectral isolation and overlap as a function of sample rate. (a) Signal spectra. (b) Spectra for sampling beyond the Nyquist sampling rate; $f_s > 2f_m$ (f_m is the maximum frequency). (c) Spectra for sampling at the Nyquist rate ($f_s = 2f_m$). (d) Spectra for sampling below the Nyquist rate ($f_s < 2f_m$).

II. Origins of Digital Signal Processing

The study of computing algorithms predates the advent of the digital computer. However, once digital computing machines became available, there was immediate interest in implementing a host of computing algorithms (some dating from the seventeenth and eighteenth centuries). These early attempts were, in a large part, extensions to the then maturing field of numerical analysis. Because the digital computer was arithmetically fast and could be programmed to iteratively solve a complex problem, computing algorithms were extensively used to statistically analyze and mathematically manipulate signals. During the 1960s the race into space accelerated

this type of signal processing and gave rise to a host of estimation, smoothing, and prediction algorithms for navigation, control, and space communication applications. The fields of pattern recognition and statistical hypothesis testing (e.g., radar signal detection) were also greatly enriched with the availability of general-purpose computers.

During the transition period, ranging through much of the 1960s, the general-purpose digital computer was too expensive and unreliable to be used as an on-line signal processing tool. But that began to change. In the mid-1960s the first DSP milestones were achieved in the form of fast transforms and filters. One of the attempts to use the computer as a signal processing tool was in the area of spectral analysis using Fourier transform and its variations. A discrete Fourier transform accepts as input a set of discrete samples in the time domain (i.e., a time series) and transforms them into information in the frequency domain. Unfortunately, because of its high multiplication budget, the DFT formula was not amenable to efficient digital computer data processing. It was Cooley and Tuckey who popularized the so-called fast Fourier transform (FFT), which revolutionized the field of the DFT. The FFT achieved two milestones. First, it provided a service, namely faster spectral analysis. Second, it was designed with an awareness of the strengths and limitations of digital computers, namely memory utilization and data flow. Because of this, the FFT altered the attitude of digital signal processing engineers and scientists, who charted a new direction thereafter. Refinements and design procedures that optimize computer execution rates and resource utilization were sought. Digital computer architectures have also been developed that reflect the requirement of DSP algorithms. Finally, many other fast digital algorithms and transforms have since been discovered.

The other major branch of DSP theory has concerned digital filters, which can replace many of the analog electronic filters that have been in daily use for over half a century. Passive analog filters are designed with resistors (R), capacitors (C), and inductors (L) only. Active analog filters add electronic amplifiers to the R, L, C parts list. However, these classic filters must be designed with realistic values for resistance, capacitance, and inductance. Because of the physical limitations of passive components, analog filters have a limited frequency response. They also perform poorly at very low or very high frequencies. Analog filters often require extensive and costly alignment and adjustment and require maintenance throughout their useful life span. Linear phase behavior is difficult to achieve, and the fact that analog filters cannot be programmed limits their utility and design flexibility. The alternative digital filter technology overcomes these limitations. Digital filters are a by-product of the great advances in solid-state electronics during the 1970s, which produced a wealth of high-performance, low-cost electronic devices capable of manipulating signals in an algorithmic manner not possible with analog units. Advances in digital technology such as the microprocessor have created the new field known as digital signal processing.

From a theoretical viewpoint, digital filters provide a rich field for study. Because of their robust theoretical base, one would expect more advanced digital signal processing systems to be developed in the future that are based on mathematical abstractions. The level of mathematical consciousness in digital processing is attributable to the nature of the hardware in which these systems reside. Whereas analog systems are defined over a *real coefficient field,* the data and coefficients associated with a digital filter are defined with finite precision and stored in registers of finite word length. The levels of mathematical skill required to work algebraically in this area are more advanced than those required in the field of real numbers.

The advantages of digital filters are numerous:

1. They can be fabricated with available high-density, low-cost digital hardware. This should continue to improve as digital signal processing assimilates the continuing advance in the digital electronics industry.
2. Certain classes of digital filters have guaranteed stability.
3. They are free of the impedance-matching problems associated with analog filters.
4. Digital filters can work at extremely low frequencies that cannot be supported by analog filters.
5. Digital filters are programmable if realized with a programmable processor.
6. They efficiently support computer-aided data analysis [simple input–output (I/O) logging and transfer] and can often be interfaced to the power supplies and mechanical structure of existing digital processors.
7. They can work over a wide range of critical frequencies, which is a difficult task for analog filters.
8. They can be used in conjunction with data compression (i.e., input and/or output) schemes.
9. Certain digital filters have outstanding phase linearity.
10. They can have high accuracy and precision. The precision of an analog filter generally does not exceed 60 to 70 decibels. The precision of a digital filter can be extended simply by increasing word length. For example, an n-bit digital word defines a dynamic range approximated by $6.0206n$ dB.
11. Digital filters do not require periodic alignment, which is necessary for analog filters.

Furthermore, they do not "drift" with parameter aging or environmental changes.

12. Because of their digital nature, digital filters are less sensitive to certain classes of noise that corrupt analog filters (e.g., line frequency noise).

The disadvantages of digital filters are few:

1. They are subject to quantization errors.
2. They may exhibit some form of limit cycling.
3. Hardware development time may be longer.
4. Their design and synthesis may be more difficult algebraically unless a general-purpose digital computer is used to support the design task.

Many of these disadvantages can be overcome through good digital design practices and familiarity with hardware, analysis protocols, and procedures.

General-purpose computers became available to scientists during the 1950s. At first, primitive filters were designed and implemented as computer programs. Some simply emulated the calculus operation of integration by use of iterative data processing techniques. An iterative calculation is one in which the output is a function of both the input and past outputs. Because of the recurrence relationship, these algorithms would now be called recursive filters. If the output can be written as a function of input or independent variables only, a filter would be called nonrecursive or, sometimes, a transversal filter. Digital filters can be used to smooth, interpolate, condition, or analyze data. For example, the simple recursive filter given by $y_{n+1} = -0.9y_n + x_n$ amplifies signals of alternating sign and attenuates signals of constant sign. Starting at $y_0 = 0$, the signal $x_n = (-1)^n$ produces $y_1 = 1$, $y_2 = -1.9$, $y_3 = 2.71$, and so forth, while $x_n = 1$ yields $y_1 = 1$, $y_2 = 0.1$, $y_3 = 0.9,...$ Scientific data analysis tools of this type operate "off-line" and use floating-point arithmetic. However, they do not represent true DSP statements since they were not designed with an awareness of the imprecision of digital arithmetic, memory limitations, computational complexity, data and control flow, and the like. But it was not long before engineers and scientists were developing computer routines and algorithms that did lend themselves to fast and efficient digital processing. This began the era of digital signal processing.

III. Signal and System Representation

Signal processing is concerned with the problem of manipulating and managing information. The information source (signal) may appear in one of several forms. A continuous signal process is one that is continuously resolved in both its independent and dependent axes. A discrete signal process is discretely resolved (quantized) only in the dependent variable. A digital signal process is discretely resolved in both variables. Periodic (uniform) sampling is the most common method of converting an analog signal to a discrete format.

Shannon's theorem (*ca.* 1948) states that if $x(t)$ is a signal whose highest-frequency component is bounded by f_N, and if $x(t)$ is periodically sampled so that the sample period t_s is less than or equal to $t_s < 1/2f_N$ (referred to as the Nyquist sample rate), then $x(t)$ can be interpolated from its time series by the rule

$$x(t) = \sum_{n=-\infty}^{\infty} \frac{x(nt_s)\, \sin[\omega_c(t - nt_s)]}{\omega_c(t - nt_s)}$$

However, use of this interpolating equation is not practical, and the following simpler approximations to the equation have been developed:

The zeroth-order hold (interpolator)

$$x(t) = x(n), \qquad t \in [nt_s, (n + 1)t_s]$$

The first-order hold (interpolator)

$$x(t) = x(n) + t[x(n + 1) - x(n)]/t_s,$$
$$t \in [nt_s, (n + 1)t_s]$$

The zeroth-order hold is often called a sample and hold circuit, while the first-order is called a linear interpolation.

The analysis of a discrete system can be performed using transform methods. One of the popular discrete transforms is the z-transform. It belongs to a class of algebraic operations referred to as impulse-invariant transforms, which means that the inverse transform gives back the original sample values. The z-transform is related to the Laplace transform variable s by $z = \exp(st_s)$. If $\{x(n)\}$ is a time series consisting of sample values $x(n) = x(nt_s)$, then its z-transform is given by

$$x(z) = \sum_{n=-\infty}^{\infty} x(n)z^{-n}$$

Some of the basic properties of the z-transform are:

1. The z-transform of $\{x(i)\}$ exists if and only if $\{x(i)\}$ is unique and bounded for all n.

2. Two z-transforms $X(z)$ and $Y(z)$ are equal if and only if $\{x(i)\} \equiv \{y(i)\}$.

3. The z-transform of $x(t/T)$ is independent of the sampling period T.

4. The z-transform is a linear operator in that if $x(n) \rightarrow X(z)$ and $y(n) \rightarrow Y(z)$, then if $z(n) = x(n) + y(n)$ it follows that $Z(z) = X(z) + Y(z)$.

5. The shifting property holds in that if $y(n) = x(n + m)$ and $x(n) \rightarrow X(z)$, then $Y(z) = z^m X(z)$.

6. The initial and final value theorem states that if $x(0) = \lim X(z)$ as $z \rightarrow \infty$, $x(\infty) = \lim[1 - z^{-1})X(z)]$ as $z \rightarrow 1$.

7. The linear convolution property holds in that if $x(n) \rightarrow X(z)$ and $y(n) \rightarrow Y(z)$, then $z(n) = x(n)*y(n)$ implies that $Z(z) = X(z)Y(z)$. A more formal mathematical statement is

$$z(n) = \sum_{k=-\infty}^{\infty} x(k)y(n - k)$$

$$= \sum_{k=-\infty}^{\infty} x(n - k)y(k) \quad \overset{Z}{\longleftrightarrow} X(z)Y(z)$$

8. Parseval's theorem becomes

$$\sum_{k=-\infty}^{\infty} x(n)y(n) = \frac{1}{2\pi j}\int X(z)Y(z^{-1})z^{-1}\,dz$$

9. Stability properties in the s domain transfer into the z domain. The locations of stable s-domain pole values are given by $s = \sigma + j\omega$ for $\sigma < 0$. Substituting this condition into the defin-

ing equation for a z-transform, it follows that $|z| = |\exp(\sigma t_s) \times \exp(j\omega t_s)| \leq |\exp(\sigma t_s)| < 1$ if $\sigma < 0$. That is, the stable complex values of z are bounded to be within the periphery of the unit circle given by $|z| = 1$. This condition is often referred to as the circle criterion and is summarized in Fig. 2.

Properties 7 and 8 make the z-transform well suited for representing systems as well as signals.

There are some restrictions on the use of the z-transform. From the definition of z [i.e., $z = \exp(st_s)$] it is seen that the z-transform is unique if and only if the modulus of st_s is bounded by $\pm\pi$ (i.e., $|st_s| \leq \pi$). For $s = j\omega = j2\pi f$, it follows that $|2\pi ft_s| \leq \pi$ or $|f| \leq 1/2t_s$, which is the Nyquist sampling frequency. It is sometimes useful to interpret the z-transform terms of parameters such as those shown in Table I.

Finally, the inversion of a given $X(z)$ is formally defined in terms of the contour integral

$$x(n) = Z^{-1}[X(z)] = \frac{1}{2\pi j}\oint_C \frac{X(z)z^n\,dz}{z}$$

where C is a restricted closed path found in the z plane. However, one rarely approaches the problem of inversion through the use of this difficult integral equation. Instead, more simplified methods have been developed that expedite the inversion process. The most popular of these methods are

1. Long division.
2. Partial fraction or Heaviside expansion.
3. Residue theorem of complex variables.

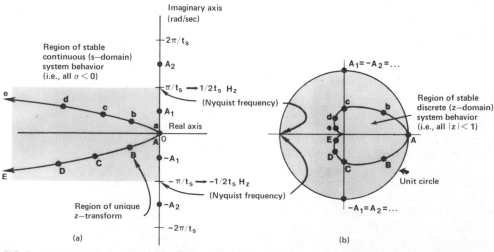

FIG. 2.   Relationship between s and z transforms in the complex plane. (a) Laplace or s domain, $s = \sigma + i\omega$; (b) z-transform domain, $z = e^{st}$. The shaded s plane in (a) is the region that results in a unique and stable z-transform. The shaded area of the z plane in (b) is the region of stable z-transforms.

TABLE I. Parameters Used in Interpreting the z-Transform

Parameter	z operator	Range
f, Hz	$z = e^{j2\pi f t_s}$	$0 \leq f \leq f_s$
w, rad/sec	$z = e^{j\omega t_s}$	$0 \leq \omega \leq 2\pi f_s$
$f' = f/f_s$ normalized to f_s	$z = e^{j2\pi f'}$	$0 \leq f' \leq 1$
$f^* = f'/0.5$ normalized to f Nyquist	$z = e^{j f^*}$	$0 \leq f^* \leq 2$
θ, rad	$z = e^{j\theta}$	$0 \leq \theta \leq 2\pi$

The standard z-transform is by no means the only mapping of the s domain to the z domain in practical use today. Many others have been considered, and have enjoyed a certain degree of popularity such as (1) the bilinear z-transform and (2) the matched z-transform.

The *bilinear z-transform* is defined by

$$s = \frac{2(z - 1)}{t_s(z + 1)}$$

A strength of this transform is its ability to map a given continuous signal having a Laplace representation $H(s)$ into the z domain without the algebraically complex operations associated with the standard z-transform. However, the bilinear z-transform is not impulse invariant and, as a result, is not necessarily the transform of choice for time domain analysis or synthesis. However, the frequency response modeling ability of a bilinear transform system is superior to that of the standard z- transformed system. Here the metric of comparison is assumed to be the similarity of the frequency response of an analog filter $H(s)$ and its discrete model $H(z)$. The spectral properties of the bilinear transform are due to its ability to distort the frequency axis by a property known as warping. If Ω represents the frequency axis of a continuous system whose Laplace transform is $H(z)$ and ω is that for the discrete system $H(z)$, where $H(s) \rightarrow H(z)$ by using the bilinear transform, then the warping equation is

$$\Omega = (2/t_s) \tan(\omega t_s/2),$$

$$\omega = (2/t_s) \tan^{-1}(\Omega t_s/2)$$

The warping of the continuous frequency axis is shown in Fig. 3. If the design of a discrete filter is predicted on knowledge of its continuous filter model $H(s)$, then the final design specifications given along the discrete frequency axis are prewarped into the continuous frequency axis.

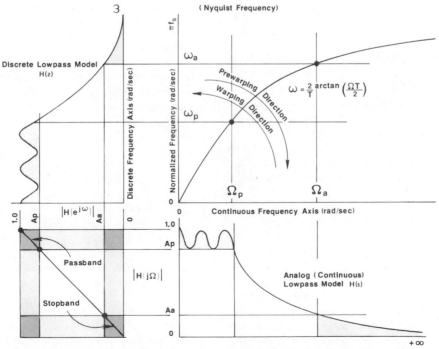

FIG. 3. Frequency and prewarping relationship between continuous (Ω) and discrete (ω) frequency domains.

From this, the resulting continuous filter $H(s)$ is designed.

The matched z-transform is normally used to model narrowband frequency-selective filters and is given by

$$H(s) = \frac{\prod_{i=1}^{N}(s - a_i)}{\prod_{i=1}^{N}(s - b_i)},$$

$$H(z) = \frac{\prod_{i=1}^{N}[1 - \exp(a_i T)z^{-1}]}{\prod_{i=1}^{N}[1 - \exp(b_i T)z^{-1}]}$$

However, in most frequency domain design applications the bilinear z-transform is used, and when time domain response is the issue, the standard z-transform is preferred.

IV. Digital Transforms

The analysis of a time series need not be performed only in the time domain. Other options such as the frequency domain are often used. The mapping of a signal from one domain (e.g., time) to and from another (e.g., frequency) is called a transform. Sometimes the transform has physical significance; at other times it is simply an artifact to support efficient analysis. In the field of digital signal processing, transforms are studied in the context of their problem-solving capability as well as their digital computational efficiency.

Historically, transforms have been a very rich area of study for engineers and scientists. Some, such as the Laplace transform, are used to solve complicated mathematical problems; others, like the Fourier transform, are used to establish a new analytic database in which the signal can be analyzed. The Fourier transform converts a signal into its frequency domain image called a spectrum. Analysis of the spectral signature of a signal is critical in many problems of communications, control, defense, pattern and voice processing, medicine, and other technical areas. The exponential form of the Fourier transform of a signal $x(t)$ is given by

$$X(\omega) = \int_{-\infty}^{\infty} x(t)e^{-j\omega t}\, dt$$

$$x(t) = \frac{1}{2\pi}\int_{-\infty}^{\infty} X(\omega)e^{j\omega t}\, d\omega$$

However, because of the finite storage and speed capability of a digital computer, infinitely long real or complex signals are replaced by their finite time series approximation. If the original signal is bandlimited to f_N Hz and is sampled at or above the Nyquist rate of $t_s < 1/2f_N$ samples per second for N contiguous samples, then

a discrete Fourier transform can be defined and used to approximate a Fourier transform. The DFT maps an N-point time series into the frequency domain by the rule

$$X(k) = \sum_{n=0}^{N-1} x(n)W_N^{nk} \qquad (DFT)$$

and the inverse DFT (IDFT) returns it to the time domain and is given by

$$x(n) = \frac{1}{N}\sum_{k=0}^{N-1} X(k)W_N^{-kn} \qquad (IDFT)$$

where W_N is a complex exponential $W_N = \exp(-j2\pi/N)$ and is sometimes referred to as the twiddle factor. A few of the more important DFT properties are

1. Linearity: $\mathrm{DFT}[ax(n) + by(n)] = a\mathrm{DFT}[x(n)] + b\mathrm{DFT}[y(n)] = aX(k) + bY(k)$.

2. Cyclic time and frequency shifting: $\mathrm{DFT}[x(n - r)] = X(k)W_N^{rk}$ and $\mathrm{IDFT}[X(k - r)] = x(n)W_N^{nr}$.

3. Time reversal: $\mathrm{DFT}[x(n)] = X(-k) = X(N - k)$. This provides a mechanism by which negative frequencies can be interpreted.

4. Symmetry for a real input: for $x(n)$ real, $X(k) = X^*(-k)$. Two special cases can be considered. If $\{x(n)\}$ is a real even time series, $X(k)$ is real (i.e., a cosine series). If $\{x(n)\}$ is odd, $X(k)$ is imaginary (i.e., a sine series).

Some of the more important DFT milestones are shown in Table II.

The DFT process of converting an N-point time series into a spectrum is indicated in Fig. 4. If $x(t)$ is a bandlimited signal whose maximum frequency component is f_M Hz, then the continuous signal must be sampled at a speed greater than the Nyquist sample rate of $2f_M$ samples/sec if aliasing errors are to be avoided. The resulting time series is then processed by the DFT formula to produce a set of N harmonics denoted by $\{X(k)\}$, where $X(k)$ is the kth harmonic and is located at frequency $f_k = k/T$ Hz. The term $T = Nt_s$ is called the fundamental period, and $t_s < 1/2f_M$ is the sample rate. The N harmonics produced by a DFT range from $f_0 = 0$ Hz to $f_{N-1} = (N - 1)/Nt_s$ Hz and are discretized into N harmonics equally spaced on f_1 Hz centers. The term $f_1 = 1/T = 1/Nt_s$ is called the fundamental frequency, and $f_0 = 0$ Hz is referred to as the dc component. If $x(t)$ is a real signal, the $N/2$ most significant harmonics can be physically interpreted as negative frequency spanning the range $[-f_s/2, 0]$.

TABLE II. Discoveries of Efficient Methods of Computing the DFT[a]

Researchers	Date	Sequence lengths N	Number of DFT values	Application
C. F. Gauss	1805	Any composite integer	All	Interpolation of orbits of celestial bodies
F. Carlini	1828	12	—	Harmonic analysis of barometric pressure
A. Smith	1846	4,8,16,32	5 or 9	Correcting deviations in compasses on ships
J. D. Everett	1860	12	5	Modeling underground temperature deviations
C. Runge	1903	$2^n k$	All	Harmonic analysis of functions
K. Stumpff	1939	$2^n k, 3^n k$	All	Harmonic analysis of functions
G. C. Danielson and C. Lanczos	1942	2^n	All	X-ray diffraction in crystals
L. H. Thomas	1948	Any integer with relatively prime factors	All	Harmonic analysis of functions
I. J. Good	1958	Any integer with relatively prime factors	All	Harmonic analysis of functions
J. W. Cooley and J. W. Tukey	1965	Any composite integer	All	Harmonic analysis of functions
S. Winograd	1976	Any integer with relatively prime factors	All	Use of complexity theory for harmonic analysis

[a] After Herdeman, Johnson, and Burris, (Oct. 1984). *IEEE ASSP Magazine*.

The Fourier transform is defined over an infinitely long data record spanning $t \in (-\infty, \infty)$ and is represented in Fig. 4 as $X_{FT}(f)$. In contrast, the DFT transforms a time series of finite duration defined over the time instances $t = 0, t_s, \ldots, (N-1)t_s$. Furthermore, the DFT incorporates the mathematical assumption that the trans- formed time series periodically repeats itself for all time with a period T. That is, if $\{x(n)\}$ is transformed into $\{X(k)\}$ the inverse transform is actually $\{x_p(n)\}$, where $x_p(n) = x(n)$ for $n = 0, 1, \ldots, N-1$; otherwise $x_p(qN + n) = x((qN + n) \bmod N)$ for q an integer. Since in general $x_p(qN + n) \neq x(qN + n)$, the spectra in question may differ.

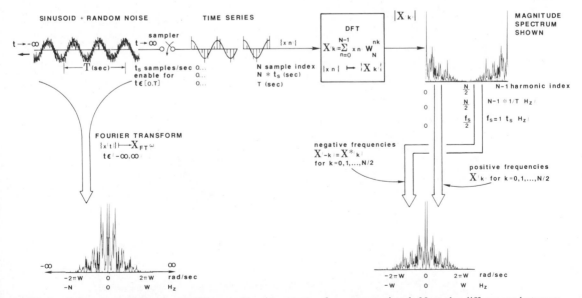

FIG. 4. Comparison of Fourier and discrete Fourier spectra of a common signal. Note the differences between DFT and Fourier transforms due to finite aperture effects in the spectra at the bottom.

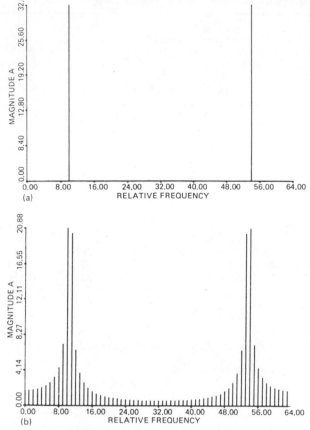

FIG. 5. Spectra representing the DFTs of a sinusoid completing (a) 10 full cycles (no leakage) and (b) 10.5 cycles (leakage).

This is called the finite aperture effect. If the original signal was, in fact, periodic with period T_p, then it is possible to achieve agreement between the DFT and the Fourier transform, which is now reduced to a Fourier series. Disagreements, if any, would now be attributed to aliasing or a phenomenon called leakage, as suggested in Fig. 5. However, leakage does not occur when the signal period and sampling interval are related by $T = mT_p$, where m is an integer.

TABLE III.   Commonly Used Window Types

Window type	Definition
Rectangular or uniform	$w(n) = \begin{cases} 1, & \|n\| \leq (N-1)/2 \\ 0, & \text{otherwise} \end{cases}$
Hamming ($a = 0.54$), Hann ($a = 0.50$)	$w(n) = \begin{cases} 1 + (1-a)\cos(2\pi n/N), & \|n\| \leq (N-1)/2 \\ 0, & \text{otherwise} \end{cases}$
Kaiser	$w(n) = \begin{cases} I_0(\beta\{1 - [2n/(N-1)]^2\}^{1/2}), & \|n\| \leq (N-1)/2 \\ 0, & \text{otherwise} \end{cases}$
	where I_0 represents a zeroth-order Bessel function given by $I_0(x) = 1 + \sum\limits_{k=1}^{\infty} \left[\left(\frac{1}{k} \frac{x}{2} \right)^k \right]^2$

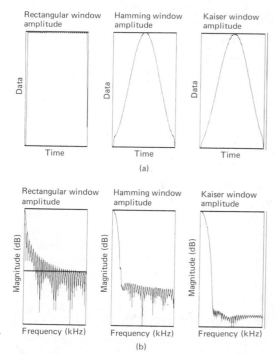

FIG. 6. Examples of standard windows (rectangular, Hamming, and Kaiser) showing (a) time series envelope and (b) log frequency spectra. (Courtesy of the Athena Group.)

Finite aperture effects can often be minimized with the use of data windows. A data window modifies a raw time series $\{x(n)\}$ by using the memoryless rule $x_w(n) = w(n)x(n)$, where $w(n)$ is the window function. The duality theorem of Fourier transforms states that multiplication in the time domain is equivalent to convolution in the frequency domain, so it follows that $X_w(\omega) =$ $W(\omega) * X(\omega)$. The object of a window is to place $X_w(\omega)$ in maximum agreement with $X(\omega)$, which implies that ideally $W(\omega)$ is given by the Dirac delta distribution $\delta_D(\omega)$. Since the convolution of anything with a δ_D is simply a copy of the original, it would follow that $X_w(\omega) \equiv X(\omega)$. Unfortunately, if $W(\omega) = \delta_D(\omega)$, then $w(t) \equiv 1$ for all $t \in (-\infty, \infty)$, which is unrealistic (i.e., $T \rightarrow \infty$). If T is to be of finite duration, a compromise is required. It is desired to find a finite-duration window whose spectral shape is an approximation to $\delta_D(\omega)$. Standard window functions commonly in use, which achieve this objective to various degrees, are summarized in Table III. Subjectively, the spectra produced from a windowed time series are cosmetically altered to be smoother and more interpretable than thin unwindowed counterparts. Some of these windows are shown graphically in Fig. 6.

Two-dimensional data sets are often encountered in picture and scene analysis, medicine, seismology, and so forth. A 2-D time series can be denoted by $x(n_1, n_2) = x(n_1 t_s, n_2 t_s)$. The sample from a 2-D z-transform of the time series $\{x(n_1, n_2)\}$ is given by $X(z_1, z_2) = \Sigma\Sigma x(n_1, n_2)z_1^{-n_1}z_2^{-n_2}$, with the 2-D DFT satisfying

$$X(k_1, k_2) = \sum_{n_1=0}^{N_1-1} \sum_{n_2=0}^{N_2-1} x(n_1, n_2) W_{N_1}^{n_1 k_1} W_{N_2}^{n_2 k_2}$$

By rearranging the DFT summation to read

$$X(k_1, k_2) = \sum_{n_1=0}^{N_1-1} W_{N_1}^{n_1 k_1} \sum_{n_2=0}^{N_2-1} x(n_1, n_2) W_{N_2}^{n_2 k_2}$$

$$= \sum_{n_1=0}^{N_1-1} W_{N_2}^{n_1 k_1} X'(n_1, k_2)$$

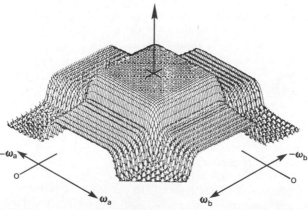

FIG. 7. Two-dimensional DFT of a lowpass filter. The ideal lowpass filter would appear as a rectangle having infinitely steep skirts in the frequency domain.

it can be seen that a 2-D DFT can be computed by N_1 applications of an N_2-point DFT. The 2-D IDFT would be analogously defined. Also, 2-D windows are simply symmetric extensions of the 1-D case and can be directly applied to $x(n_1, n_2)$. The 2-D DFT approximation of an ideal lowpass filter is shown in Fig. 7.

V. The Fast Fourier Transform

The fast Fourier transform has become a major DSP tool since being popularized by Cooley and Tuckey in 1965. In the FFT formula, the DFT equation $X(k) = \sum x(n)W_N^{nk}$ is decomposed into a number of short transforms and then recombined. The basic FFT formulas are called radix-2 or radix-4 although other radix-r forms can be found for $r = 2^k$, $r > 4$. In a radix-r implementation a problem having $N = r^n$ points is decomposed into a DFT algorithm having n/r levels, each consisting of N/r smaller r-point DFT mappings. To appreciate how the FFT algorithm works, consider the case $k, n < N = 2^n$. Using the basic DFT equation, N complex multiplications are required to compute one harmonic $X(k)$. Since there are N harmonics, N^2 complex multiplications are needed. Suppose the time series $\{x(n)\}$ is decomposed into two $N/2$-point time series of even and odd samples. The complex multiplication count for each small transform is $(N/2)^2 = N^2/4$. The total complex multiplication count would then become $(N^2/4) + (N^2/4) = N^2/2$, or half the number previously required. Continuing this partitioning policy, the complex multiplication budget becomes $(N/2) \log_2(N)$, which is significantly less than the DFT direct mechanization. This reduced multiplication budget translates into speed, hence the term *fast* Fourier transform.

If the decomposition of an N-point time series takes place as an input operation, it is called decimation in time (DIT) and the DIT–FFT output spectra are sequentially ordered. If the input is left sequentially ordered, then the output is reordered in a nonsequential manner and is called decimation in frequency. For a radix-2 FFT, the decimation rule is known as bit reversing. The FFT computer code is compact and easy to implement in software. Hardware FFT realizations are ubiquitous and are generally designed for FFTs of length $N = 2^{10} = 1024$ or less. For longer transforms, block transform methods are used that synthesize a long transform by combining the outputs of a set of short transforms.

VI. Other Transforms

The DFT belongs to a class of transforms called orthogonal transforms, and it is not the only member of this class used in DSP applications. Some of the more popular are the Walsh, slant, and COSINE transforms. Another class of transform, called the number theoretic transform, can be applied to fast DFTs or convolutions. A number of DFT formulas are based on optimized short transforms (e.g., Winograd DFT), which are used to grow longer transforms. The problem with these schemes is that they typically have a very complex data management structure. Finally, based on Fermat and Mersenne primes, multiplier-free transforms and convolution algorithms have been developed, but they are valid only for restricted transform lengths.

VII. Discrete Systems

Linear system theory is important in contemporary engineering. Linear systems may be characterized by their impulse response $h(n)$, which is the system's output to an input impulse stimulus. The output of a discrete linear system, say $y(n)$, is given by the linear convolution equation

$$y(n) = \sum_{i=-\infty}^{\infty} h(n - i)x(i)$$

$$= \sum_{i=-\infty}^{\infty} h(n)x(n - i) = h(n) * x(n)$$

where $x(n)$ is the input. The analysis of the continuous linear convolution process can be replaced with a set of Fourier transform operations. However, an N-point DFT assumes that the transformed signal is periodic with period N samples. If $\{x(n)\}$ and $\{h(n)\}$ are two N-point time series with periodic extensions $x_p(n) = x(n \bmod N)$ and $h_p(n) = h(n \bmod N)$, then their circular convolution, denoted $x(n) \circledast y(n)$, is given by

$$x(n) \circledast y(n) = \sum_{i=0}^{N-1} x(n)y((n - i) \bmod N)$$

$$= \sum_{i=0}^{N-1} x((n - i) \bmod N)y(n)$$

In general, $\text{DFT}(x(n) * y(n)) \neq \text{DFT}(x(n) \circledast y(n))$. Nevertheless, there are times when one would like to use the computational power of the FFT to do linear convolution. When this is the case, a technique called zero filling can be used to create two new zero-filled $2N$-sample time series $\{x_Z(n)\}$ and $\{h_Z(n)\}$. Each consists of exact images of the original N-sample time series $\{x(n)\}$ and $\{h(n)\}$, which are appended with blocks of N zeros. Using $2N$-point DFTs and an IDFT, it is known that $\text{IDFT}(X_{ZF}(k)H_{ZF}(k)) = x(n) * h(n)$ (linear convolution) over N contiguous samples. The other N sample values of the $2N$-point IDFT are artifacts and are discarded. To do convolution over a KN-sample time series with N-point DFTs, zero filling techniques are used in conjunction with data partitioning methods called overlap and save or overlap and add.

The DFT is used extensively to estimate the transfer function $H(k)$ of a linear system. The transfer function is given by the ratio $H(k) = Y(k)/X(k)$. The shape of the magnitude profile $|H(k)| = \text{SQRT}[H(k)H^*(k)]$ defines the magnitude frequency response of the filter, while $\phi(k) = \arctan\{\text{Im}[H(k)]/\text{Re}[H(k)]\}$ establishes its phase reponse. The derivative of the phase profile, approximated by $\Delta\phi(k) = \phi(k) - \phi(k - 1)$, is called the group delay and is often minimized as a design criterion in digital communication/telemetry applications.

The similarity-measuring autocorrelation function, with delay or lag parameter k, can be defined in terms of the DFT as follows:

$$R_x(k) = \frac{1}{N} \sum_{i=0}^{N-1} x(i)x(i + k)$$

$$= \text{IDFT}[X(k)X^*(k)] = \text{IDFT}[|X(k)|^2]$$

provided the zero filling and data partitioning methods required for linear convolution are also employed. The term $X(k)X^*(k) = |X(k)|^2$ is also referred to as the power spectrum of $\{x(n)\}$. The average value, on a per harmonic basis, of a set of power spectra is called a periodogram.

Cross-correlation is similarly defined by

$$R_{xy}(k) = \frac{1}{N} \sum_{i=0}^{N-1} x(i)y(i - k)$$

$$= \text{IDFT}[X(k)Y^*(k)]$$

and the cross-correlation coefficient at delay k, denoted by $\rho_{xy}^{(k)}$, is given by

$$\rho_{xy}^{(k)} = \text{IDFT}[X(k)Y^*(k)]/(|X(0)||Y(0)|)^{1/2}$$

The coherence function is also defined in terms of the DFT and can be used to answer the following questions:

1. Is the system linear?
2. Is the initial state of the system zero?
3. Is the system forced by a single input rather than multiple inputs?

In addition, a number of other specialized system-oriented tests have become practical due to the advent of the DFT (i.e., FFT) and DSP procedures. All of the methods have evolved because of the development of efficient and fast DSP computation tools.

VIII. Digital Filter Design

One of the major branches of study in the digital signal processing field is that of digital filter design and analysis. The design objective is to convert a discrete system, given in terms of its impulse response $[h(n)]$ or transfer function $H(z)$, into a hardware or software statement. A filter's hardware design is said to be its architecture. Filter analysis is concerned with issues of speed, precision, and complexity. It was noted earlier that if a digital filter is designed without using signal feedback, it is called a nonrecursive or transversal filter and can be mathematically represented as

$$H(z) = \sum_{n=0}^{N-1} h(n)z^{-n}$$

if the impulse response is given by the N-element time series $\{h(0), h(1), ..., h(N - 1)\}$. Some special types of recursive filters also satisfy this condition, but they are rarely seen in practice. Because the filter exhibits a finite impulse response, it is also called an FIR filter. The frequency domain response can be computed by substituting z with $\cdot z = \exp(j\omega)$. In the frequency domain the FIR transform becomes $H(\omega) = |H(\omega)| \exp(\phi(\omega))$, where $|H(\omega)|$ and $\phi(\omega)$ are called the magnitude frequency and phase response, respectively. The principal advantages of FIR filters are twofold: (1) they are simple to analyze since they are defined by a finite linear equation, and (2) they can be designed to have a phase response that satisfies the linear equation $\phi(\omega) = a\omega + b$, where $b = 0$, $\pi/2$, or $-\pi/2$. Linear phase filters are important in applications where nonlinear phase distortion can degrade a database or system performance. Some examples are digital communication line equalization, speech and image processing, and antialiasing filters to precondition the data sent to a DFT. For a FIR filter to achieve linear phase performance, the filter must have a symmetric or antisymmetric finite impulse response. It is also common to find the amplitude profile of

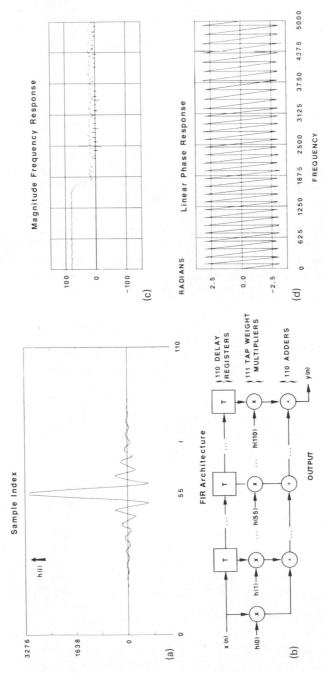

FIG. 8.   Direct implementation of a linear phase FIR lowpass digital filter using shift registers, multipliers, and adders. Phase is plotted between the principal angles $\pm\pi$. (a) Impulse response; (b) architecture; (c and d) frequency domain performance.

a FIR filter overlaid by a symmetric data window such as those introduced in the discussion of DFT. In such cases the original impulse response $\{h(n)\}$ is combined with a window function $\{w(n)\}$ to form a new FIR given by $\{h'(n) = \{h(n)w(n)\}$.

An FIR filter can be designed to be frequency selective as well. The four basic frequency selection options are low pass, high pass, bandpass, and bandstop (or bandreject). A popular design procedure approximates the shape of an ideal filter with a Chebyshev polynomial, which is used to define the individual values of $\{h(n)\}$. The ideal filter is assumed to have unit gain in its passband(s) and zero gain in the stopband(s). This design process is now highly automated and is found in many software packages used commercially and in the public domain. Finally, FIR filters also possess a simple but possibly high-order hardware architecture and software code. A direct implementation of the transfer function is suggested in Fig. 8.

If the impulse response of a filter is of infinite duration, the filter is called an infinite impulse response (IIR), and the response is given by

$$H(z) = \sum_{n=0}^{\infty} h(n)z^{-n}$$

which can often be factored into the form

$$H(z) = \left(\sum_{i=0}^{M} b_i z^{-i} \right) \bigg/ \left(1 + \sum_{i=1}^{N} a_i z^{-i} \right)$$

$$= N(z)/D(z)$$

The feedforward information paths are characterized by $N(z)$, while the feedback paths are specified by $D(z)$. The solutions to $N(z) = 0$ are called the zeros of filters; the poles are given by $D(z) = 0$. An IIR is said to be bounded if $\sum |h(n)|$, over all n, is finite. If the poles of an IIR are bounded within with the unit circle, then the filter's forced response is also known to remain

bounded or stable. Many IIR filters are specified in terms of their optimal analog counterparts, which generally take one of the following forms:

Type	Design optimality criteria
Butterworth	Maximally flat
Chebyshev	Equal-ripple passband
Elliptic	Equal-ripple passband and stopband

It is assumed that the ideal filter model has unity gain in the passband and zero elsewhere. The maximally flat criterion refers to the slope (derivative) being zero at a known critical frequency. The equal ripple (or equiripple) is sometimes referred to as a minimax criterion. In a minimax design, the maximum deviation from the ideal model is minimal. The design of a Butterworth, Chebyshev, or elliptic filter $H(s)$ is now a classic study and has been computerized. Examples of classically defined IIRs are presented in Fig. 9.

Infinite impulse response digital filters are designed by reflecting the desired filter parameters [i.e., $H(z)$] into the continuous-frequency domain [i.e., $H(s)$], where the filter equations are well known. This can be accomplished by using one of two procedures, both of which begin with definition of a desired Butterworth, Chebyshev, or elliptic analog prototype filter, denoted $H_p(s)$ in Fig. 10. For prespecified passband and stopband gains A_p and A_a, the prototype filter satisfies $|H(s)| = A_p$ at $s = j1$ (normalized frequency in radians per second) and has sufficient order to ensure that the gain at the normalized stopband frequency $\Omega_a' = \Omega_a/\Omega_p$ satisfies $|H(\Omega_a')| = A_a$. The frequency terms, denoted Ω in this formula, are in fact the prewarped version of the critical discrete frequencies ω_p and ω_a as shown in Fig. 3. From that point on, the method proceeds as shown in Table IV (see Fig. 10).

Compared to an FIR, some general conclu-

TABLE IV.   Design of IIR Digital Filters

Step	Method 1	Step	Method 2
1. Design prototype filter	$H_p(s)$	1. Design prototype filter	$H_p(s)$
2. Convert prototype into a design filter $H(s)$ using a continuous frequency-to-frequency transform	$H(s)$	2. Convert prototype into a bilinear transform prototype	$H_p(z)$
3. Convert $H(s)$ into $H(z)$ using the bilinear z-transform	$H(z)$	3. Convert prototype $H(s)$ into $H(z)$ using discrete frequency-to-frequency transform	$H(z)$

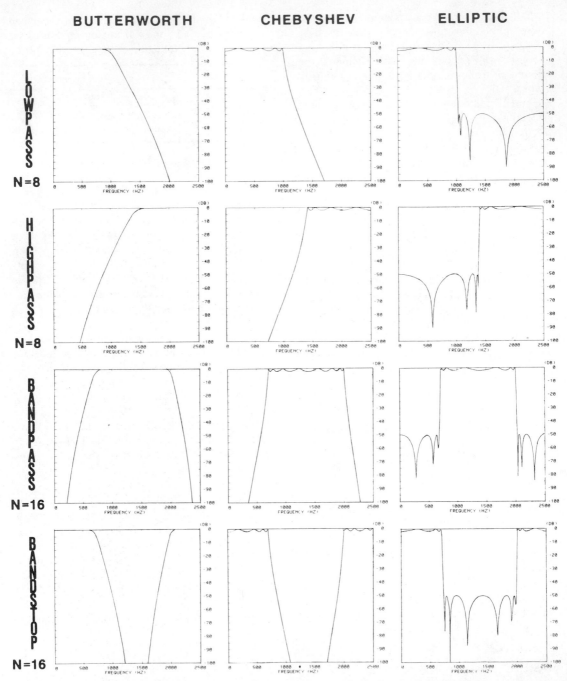

FIG. 9.   Examples of Butterworth, Chebyshev, and elliptic IIR digital filters.

sions may be drawn about on IIR. First, if magnitude frequency response is to be sharp or abrupt in terms of a transition between passbands and stopbands (called the filter skirt), the IIR is the design of choice. If phase performance is the design objective, an FIR should be chosen. The IIRs are generally of low order ($N \le$ 16), while the FIRs are usually of high order ($N \ge$ 16). For the IIR case, the derived transfer function can be factored into a number of recognizable forms. If $H(z) = \prod H_i(z)$, a serial or cascade form is realized. If $H(z) = \sum H_i(z)$, a parallel design is achieved. The individual subsystems, denoted $H_i(z)$, are normally designed

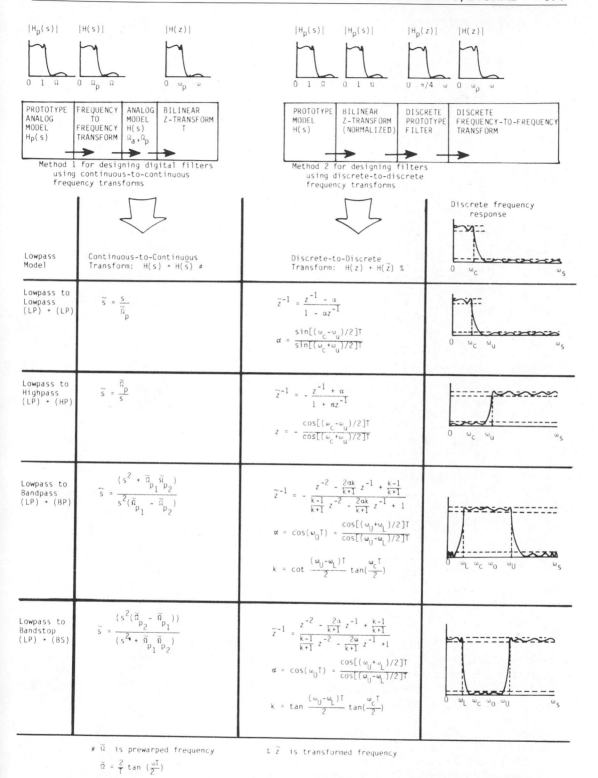

FIG. 10. Design procedures for Butterworth, Chebyshev, and elliptic IIR filters.

BUTTERWORTH

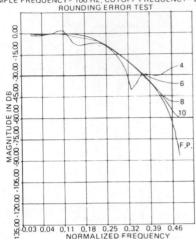

CHEBYSHEV

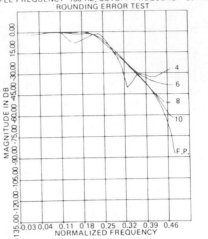

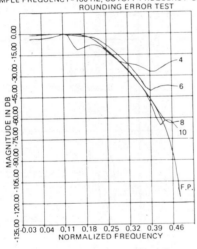

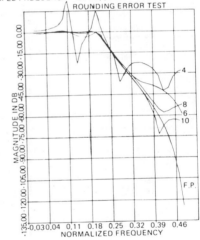

FIG. 11. Effect of finite arithmetic on the frequency response of digital filters. F.P., Floating-point response, which is equivalent to an infinite precision filter.

as first- and second-order filters having only real coefficients. These unique structures are called architectures. Other commonly found architectures are the Direct II, canonic, ladder, Gray–Markel, wave, and continued fraction, to name but a few.

Digital filters differ from their discrete counterparts $H(z)$ in that they are designed with digital hardware, arithmetic, and an n-bit data word. If floating-point arithmetic is used, the dynamic range and precision are high but the speed suffers. If speed is an important design issue, fixed-point arithmetic is usually adopted.

A high-speed fixed-point design with a signed $(n + 1)$-bit format is subject to quantization errors. Quantization is a process by which a real variable is converted into a finite-precision digital word. Errors introduced by approximating a real variable with its nearest binary-coded decimal (BCD) value manifest themselves as input, coefficient quantization, or arithmetic errors. If a real input is defined over $\pm V$ volts and is presented to an analog-to-digital (ADC) converter for quantization, the quantization step size is given by $Q \leq 1/2^n$ V/bit. The value of the quantization error ε is normally defined as the difference between a real variable X and its nearest binary-valued representation, say X_B. That is, $\varepsilon = (X - X_B)$ and is usually modeled to be of zero mean for rounding and $\pm Q/2$ for truncation. The error variance in both cases is $Q^2/12 = 2^{-2n}/12$. The error can be reduced by increasing n, but this results in a slower design.

Arithmetic errors are the result of truncation or rounding. For example, the multiplication of two n-bit numbers is a full-precision product of length $2n$ bits. This $2n$-bit product is then rounded or truncated to its nearest n-bit value. The errors associated with this operation can be modeled in the manner used for input or coefficient error analysis. However, in an IIR arithmetic errors can recirculate within the filter through its feedback paths. As a result, once inserted, these errors may never completely leave the filter. The effect of such errors on the frequency response of a Butterworth or Chebyshev lowpass filter, as a function of word length n, is interpreted in Fig. 11. Many of the IIR forms developed over the past several decades represent attempts to reduce the effects of these circulating errors and are called low-sensitivity filters. The error performance of some of the more commonly designed filters is shown in Table V, where complexity can be translated into added cost or slower throughput.

Besides the computational inaccuracies associated with finite-precision effects, other phenomena have been observed. The most important of these is limit cycling, which appears in zero input limit cycling or overflow limit cycling. Zero input limit cycling is a granularity problem in which the natural exponential decay of an impulse response wants to move the output close to zero, but this can never be achieved because of quantizing. For example, if $h(n) = 1/a^n$, then for $a = -1/2$, $y(n) = 1, -1/2, 1/4, -1/8, 1/16, -1/32, \ldots$. However, if a signed 4-bit data word with format $\pm xxx$ is used, the response is $y_4(n) = 7/8, -1/2, 1/4, -1/8, 1/8, \therefore, (-1)^n/8, \ldots$.

This low-amplitude oscillation of $\pm 1/8$, for all $n > 3$, is called zero input limit cycling, and it can result in annoying clicks in speech or telephone applications. Some filter architectures have been developed that trade off additional hardware for suppressed limit cycle behavior. Overflow limit cycling can occur whenever a signal goes into a register overflow condition, which occurs when the signal exceeds the admissible dynamic range of the numbering system. For example, adding 1 to an n-bit maximal binary value $[11 \cdots 1] = 2^n - 1$ (decimal) would send the sum back to zero. Such a rapid and unexpected change in a variable's value may cause bizarre behavior. This effect can be reduced by using saturating arithmetic, which clamps a variable to a maximum (or minimum) admissible value.

Two-dimensional digital FIR and IIR filters can also be designed. If their impulse response $h(n_1, n_2)$ can be factored as $h(n_1, n_2) = h^{[1]}(n_1)h^{[2]}(n_2)$, the filter is said to be separable. Separable filters lend themselves to one-dimensional analysis. For example, if $h(n_1, n_2)$ is separable, the 2-D convolution sum can be expressed as

$$
\begin{aligned}
y(m_1, m_2) &= \sum_{n_1} \sum_{n_2} h^{[1]}(n_1) h^{[2]} \\
&\quad (n_2) x(m_1 - n_1, m_2 - n_2) \\
&= \sum_{n_1} h^{[1]}(n_1) \sum_{n_2} h^{[2]} \\
&\quad (n_2) x(m_1 - n_1, m_2 - n_2) \\
&= \sum_{n_1} h^{[1]}(n_1) y^{[2]}(m_1 - n_1, m_2)
\end{aligned}
$$

Finally, the other 1-D concepts, such as windows, can be transferred to the 2-D case in a straightforward manner.

A filter that has the ability to adjust its coefficients to account for changes in the environment or signal space is called an adaptive filter. Such filters use an algorithm or rule-based decision law to couple measured data back into the filter

TABLE V. Survey of IIR Filter Architectures

Architecture type	Coefficient sensitivity	Round-off hardware error sensitivity	Complexity
Direct II	High	High	Low
Continued fraction	High	High	Low
Cascade	Low	Higher than parallel	Low
Parallel	Low	Lower than parallel	Low
Wave	Low	Good	Significant
Gray–Markel (ladder)	Low	Moderate	Moderate

as a set of superior coefficients. Often this information source is an error signal obtained by actually measuring the difference between a system variable and its adaptively predicted value. By its nature, an adaptive filter must be capable of altering its coefficient set to compensate for change. Therefore, general-purpose multipliers must be used to implement the filter's coefficient. Fixed (nonadaptive) filters, in contrast, often store their coefficients in read-only memory (ROM) and use special wiring techniques to replace expensive general-purpose multipliers with special-purpose ones.

IX. Implementation

Digital processing filters and algorithms are computationally intensive. Arithmetic is performed in a method that is consistent with chosen number systems. In general, three types of numbering systems are used in DSP applications:

1. Weighted (e.g., decimal).
2. Unweighted (e.g., residue).
3. Homomorphic (e.g., logarithmic).

Unsigned weighted numbers take the general form

$$X = \sum_{i=0}^{N-1} a_i r^i, \qquad 0 \le X \le r^N - 1,$$

$$0 \le a_i \le r - 1$$

where r is the radix and a_i the ith significant digit, with a_0 and a_{N-1} being the least and most significant, respectively. Typical radices found in DSP applications are $r = 2$ (binary), $r = 8$ (octal), $r = 10$ (decimal), and $r = 16$ (hexadecimal). If signed information is required, an additional sign bit is added to the code. In this model, sign-magnitude, 1's complement, or 2's complement fixed-point representations can be defined. The 2's complement fixed-point code is the most popular one because it can be used to add a column of numbers without suffering from otherwise fatal intermittent adder overflow errors, provided the sum has a valid representation.

A fractional number system can be defined by using negative (vs. positive) exponents. In either case, additional precision can be gained by adding more digits (i.e., increasing N). However, there is a practical limit to this beyond which the floating-point representation, given by $X =$ $\pm m(x) r^{\pm e(x)}$, where $m(x)$ is a mantissa and $e(x)$ the exponent, should be used. Implementation of floating-point arithmetic, though considerably more complex than that of fixed-point arithmetic, is often required in scientific calculations, where large dynamic ranges of data may be encountered. Finally, residue arithmetic (unweighted) is a method of doing carry-free highly parallel arithmetic that dates back to 500 B.C. In logarithmic systems (homomorphic) multiplication is performed by adding exponents and is therefore fast. However, addition in this system is slower and more complex than multiplication.

For hardware implementations, the primitive computational unit is the two-input half adder or modulo-2 adder. If this concept is extended to accept a third carry-in input, then a full adder results. A group of full adder cells can be integrated onto a single semiconductor chip to form longer word length adders typically having widths of 4, 8, and 16 bits. If additional hardware is traded off for higher add speeds, then carry lookahead adders or CLAs are used. Otherwise, slower ripple-type adders are chosen because of their simplicity and low cost. Multipliers can also be classified in terms of their speed–complexity metrics. The simplest, but slowest, is the shift-add multiplier, which simply emulates the manual multiplication methods learned in elementary school. Fast but more complex units can be designed by using either a multibit technique called Booth's algorithm or a specially modified full adder called a carry save unit. Examples of some commercially available multipliers are summarized in Table VI.

These relatively short word length add and multiply units are conveniently packaged, low in cost, and readily available. Longer word length fixed-point units may be designed by interconnecting the smaller units into arrays. Floating-point units can be similarly designed with separate mantissa and exponent subsystems plus logic to interconnect them.

Dedicated hardware arithmetic units are typically expensive and complex. Slower but more flexible alternatives are micro-, mini-, or maxicomputers. Besides providing arithmetic support, programmable computers are capable of performing operations such as control, storage, and logic. Where speed is a secondary issue, they have become very popular cost-effective DSP design media.

To overcome the latency problem of wide-wordlength fixed-point multipliers, various data

TABLE VI.   Some Commercially Available Multiplier Units

Vendor[a]	Size	Pins	Speed (nsec)	Power (W)	Coding[b]	Algorithm
AMD	2×4	24	25	0.6	2s	Booth
MMI	8×8	40	100	1.0	2s/uns	Modified Booth
MMI	16×16	24	800	1.0	2s	Modified Booth
TI	4×4	20	50	0.5	uns	Table lookup
TRW	8×8	40	45	1.0	2s	Carry-save
TRW	12×12	64	80	2.0	2s	Carry-save
TRW	16×16	64	100	3.0	2s	Carry-save
TRW	24×24	64	200	4.0	2s	Carry-save

[a] AMD, American Micro Devices; MMI, Monolithic Memories; TI, Texas Instruments.

[b] uns, unsigned; 2s, 2s complement code.

compression schemes have been developed. It is assumed that the speed at which the operations of addition, subtraction, multiplication, or division can be performed is inversely related to wordlength. Many of these methods emulate long-standing communication modulation techniques advanced in the 1950s. Most digital filters employ some form of pulse code modulation (PCM), proposed in 1938 by A. H. Reeves. Differential PCM began its commercial life in 1955 with Bell's T1 inverse DPCM operation, denoted $(DPCM)^{-1}$. If the DPCM is modeled as a differential process, reconstruction would require discrete integration of the form $x(n) = y(n) + y(n - 1)$. Explicitly, the FIR–DPCM convolution sum can be expressed as

$$[PCM]\ y(n) = \sum a_i x(n - i):$$
$$[DPCM]\ y(n) - y(n - 1) = \sum a_i[x(n) - x(n - 1)]$$

which can be simplified to read $\Delta y(n) = \sum a_i \Delta x(n)$. Under the proper set of conditions, the incremental quantities $\Delta x(j)$ and $\Delta y(j)$ can be amplitude compressed to n_c bits for $n_c \leq n$. For example, if $n = 16$ and $n_c = 8$, a filter of conventional architecture would require 16×16 multipliers, compared to 8×8 for the DPCM. However, to ensure that Δx does not exceed its allocated budget of n_c bits, the sampled values of $x(j)$ and $x(j - 1)$ must be similar. If they are not, an error-inducing phenomenon called slope overload will occur. This means that the sample rate t_s must often be increased to a value much greater than the original t_s to ensure that a sample $x(j)$ has insufficient time to change to a distinctly different value $x(j + 1)$. This is called

oversampling. For example, an audio filter having a Nyquist frequency of 25 kHz, with a 16-bit PCM architecture, may prove to be an inferior design based on an 8-bit microprocessor being clocked at 100 kHz.

A very popular form of DPCM is delta modulation (DM), which uses a one-bit $\{+1, -1\}$ DPCM code. Delta modulation can be traced back to a 1946 French patent. The DM systems are generally inexpensive but may suffer from severe slope overflow errors. As a result, high oversample rates must be used. These systems also suffer from a phenomenon known as *idle noise,* which is caused by asymmetry in the two current source encoders (i.e., +1 and −1) usually found in DM systems and often results in an audible disturbance.

An adaptive DPCM (ADPCM) policy is sometimes used to counter the slope overload problem found in DPCM systems. In an ADPCM system the quantization step size Q is adjusted in real time to reduce the occurrence of slope overload. As a result, the ADPCM system oversample rates are typically lower than those found in DPCM configurations. However, the price paid for those added features is greater system complexity and cost.

Beginning in the mid-1970s, high-speed semiconductor memory-intensive filters became an alternative to the more traditional multiplier-intensive designs. This development was motivated by the fact that fixed-point multiplication can be a rather time-consuming operation compared to other digital computer operations. Further, many digital filters are specified in terms of a set of fixed coefficients. As a result, scaling rather than general multiplication is required. The distinction is that multiplication involves two variables, while scaling involves one. By combining these operations, a DSP unit known as the distributed arithmetic filter was developed. This filter is also referred to by the more confusing title of a bit-slice digital filter. The distributed filter concept is used to realize a fixed-coefficient linear shift-invariant filter of the form

$$y(n) = \sum_{i=0}^{L} [a(i)x(n-i) + \sum_{j=1}^{M} b(j)y(n-j)$$

where the $\{a\}$ and $\{b\}$ are known and $L + M + 1$ general multiplications are required of traditional designs. If the variables are given an N-bit binary representation $q(t) = \sum_{k=0}^{N-1} q(k, t)2^k$, $k = 0, 1, ..., n - 1$, where $q(k, t)$ denotes the kth bit of q at time t, then

$$y(n) = \sum_{i=0}^{L} \sum_{k=0}^{N-1} a(i)x(k, n-i)2^k$$

$$+ \sum_{j=1}^{M} \sum_{k=0}^{N-1} b(j)y(k, n-j)2^k$$

Distributed arithmetic owes its name to the fact that the order of the above addition is *redistributed* as

$$y(n) = \sum_{k=0}^{N-1} 2^k \phi(n; k),$$

$$\phi(n; k) = \left(\sum_{i=0}^{L} a(i)x(k, n-i) \right.$$

$$\left. + \sum_{j=1}^{M} b(j)y(k, n-j) \right)$$

The function ϕ is defined in terms of the known $\{a\}$ and $\{b\}$ plus the $L + 1$ and M binary-valued elements $x(k, s)$ and $y(k, s)$, respectively. These $L + M + 1$ binary values can be concatenated into a single $L + M + 1$-bit word and used as an address to a ROM that contains the precomputed values of ϕ for all possible 2^{L+M+1} addresses. The sequence of table lookups of ϕ are then simply scaled by 2^k (binary shift) and sent to an adder for accumulation. Since $L + M + 1$ multiply and $L + M - 1$ add cycles have been replaced by n fast table lookups and $(n - 1)$

shift-adds, the distributed filter can often run faster. However, it is nonprogrammable, and to change the filter a new memory table lookup unit (i.e., ϕ) would have to be programmed and installed. For example, if $y(n) = 4x(n) + 3x(n-1) + y(n-1)$, $x(n) = x(n-1) = 1 \rightarrow 0001$, $y(n-1) = 2 \rightarrow 0010$, $n = 3$, then $y(n) = 9$, as shown in the insert at the bottom of this page.

Whether dedicated or programmable arithmetic logic units (ALUs) are used, a hardware DSP system would also consist of memory and peripheral devices. Memory can be broadly classified as read/write random access memory (RAM) or read-only memory (ROM); ROM is often used to store control programs and fixed coefficients, and RAM to store data and variables. Most digital filters also require some sort of peripheral data acquisition support. Analog-to-digital (ADC) and digital-to-analog (DAC) conversion can be quantified in terms of

1.   Accuracy: absolute accuracy of conversion relative to a standard voltage or current.
2.   Aperture uncertainty: uncertainty in sampling rate.
3.   Bandwidth: maximum small-signal 3-dB frequency.
4.   Codes: 2's complement, 1's complement, sign-magnitude, etc.
5.   Common mode rejection: measure of ability to suppress unwanted noise at input.
6.   Conversion time: maximum conversion rate. If a separate sample-and-hold amplifier is used, the conversion rate is the maximum conversion rate of the composite system.
7.   Full scale: dynamic range in volts.
8.   Glitch: undesirable noise spikes found in A/D output.
9.   Linearity: linearity across dynamic range.
10.   Offset drift: worst-case variation due to parametric changes.
11.   Precision: repeatability of measurement.
12.   Quantization error: $\pm(1/2)$ least significant bit (LSB).
13.   Resolution: 2^{-n}.

Analog-to-digital and digital-to-analog converters are also classified in terms of word length and speed as summarized in Table VII.

ADDRESS = $[x[i, n]$		, $x[i, n-1]$, $y[i, n-1]] \rightarrow \phi(\text{ADDRESS})$		$s \leftarrow s + 2^i\phi$
$i = 0(\text{LSB}) \rightarrow [1,$	1	, 0]	(4)1 + (3)1 + 1(0) = 7	$7 \leftarrow 0 + 2^0(7)$
$i = 1 \quad\quad \rightarrow [0,$	0	, 1]	(4)0 + (3)0 + 1(1) = 1	$9 \leftarrow 7 + 2^1(1)$
$i = 2(\text{MSB}) \rightarrow [0,$	0	, 0]	(4)0 + (3)0 + 1(0) = 0	$9 \leftarrow 9 + 2^2(0)$

TABLE VII.   Classification of Analog-to-Digital and Digital-to-Analog Converters

Analog to digital		Digital to analog	
Type	Conversion rate	Type	Settling time
Ultra high speed	>3 MHz	Ultra high speed	<100 nsec
Very high speed	300 kHz ← 3 MHz	Very high speed	100 → 1 μsec
High speed	30 kHz ← 300 kHz	High speed	1 → 10 μsec
Intermediate	3 kHz ← 30 kHz	Intermediate	10 → 100 μsec
Low speed	<3 kHz	Low speed	>100 μsec

A useful variation of the standard ADC is the data compression or robust converter. With such a device, data in a large dynamic range can be compressed into a smaller range. A commonly used data compression routine is the mu-law compander, given by

$$y(x) = V \log(1 + \mu x/V)/\log(1 + \mu)$$

Note that for $\mu = 0$ the converter is linear; the value $\mu = 255$ has been found useful in telephone applications.

Digital-to-analog converters (DACs) are of major importance as well. Some DAC units have a three-wire configuration that corrects for a difference between the ground potential of the DAC and the system it is driving. DACs can be extremely useful in supporting test, calibration, and signal synthesis tasks. The monolithic DAC units are basically low in cost and simple to interface. Some standard applications of DACs are in data display, signal generation, and servomotor control.

Software and computer programs are also important tools in the DSP design process. A number of commercially available DSP system design and analysis packages are marketed to do DFT and digital filter tasks. In developing custom code for a DSP application, software engineering techniques are used that maximize cache memory hits (if cache is present) and reduce the execution latency (delay) of a code.

X. Hardware

The technical revolution of the mid 1970s produced the tools needed to affect many high-volume real-time DSP markets. These include medium, large, and very large integrated circuits (MSI, LSI, VLSI) plus DSP metal oxide semiconductor (MOS) and bipolar devices. The ubiquitous microprocessors, with their increasing capabilities and decreasing costs, can provide control and arithmetic support to a plurality of signal processing designs. There has been a trend toward developed single-chip dedicated DSP units. Modern semiconductor technology is being used to implement high-performance real-time dedicated systems for special purposes. Many of these applications are military in nature. Nonmilitary applications, such as computer vision, speech processing, and pattern recognition, have similar performance requirements. Such systems are being developed that make use of pipelined or distributed architectures to increase the speed and capacity of a DSP design. Data flow and systolic structures are also being developed for very high speed applications; they typically trade additional complexity and data paths for speed and the simplicity of asynchronous control. For those designing high-volume telecommunications, primitive speech synthesis, and low-order digital filters, a plethora of DSP chips are available. Many of these are hybrid (both analog and digital on a common silicon foundation), charge-coupled device (CCD), or switched capacitance devices. Other semicustom strategies, such as gate arrays, standard cell, or master cell methods, can be used to fill in the gaps between custom and generic devices.

One of the technology-driven developments in DSP is the DSP chip. Perhaps the most salient characteristic of a DSP chip is its multiplier. Since multipliers normally consume a large amount of chip real estate, their design has been constantly refined and redefined. The early AMI2811 had a slow $12 \times 12 = 16$ multiplier, while a later TMS320 had a $16 \times 16 = 32$ 200-nsec multiplier that occupied about 40% more area. These chips include some amount of on-broad RAM for data storage and ROM for fixed coefficient storage. Since the cost of these chips is very low (tens of dollars), they have opened many new areas for DSP penetration. Many factors, such as speed, cost, performance, software support and programming language, debugging and emulation tools, and availability of peripheral support chips, go into the hardware design process. The 2920 chip is notable for having on-board ADC and DAC converters, but, like the 2811, it has a relatively short word length and therefore less precise calculations. The principal shortcoming of another DSP chip, the NEC7720, was the absence of saturating arithmetic, which could induce limit cycling.

Although the DSP chip concept is exciting, it cannot address all the important DSP problems. The DSP chips are still too slow and imprecise to do some real-time processing. Special processors are often required to do fast "number crunching" DSP operations over a large data base. Sometimes fast arithmetic chips (e.g., TRW-VLSI 8-, 12-, 16-, or 24-bit multiplier/accumulators) can be integrated into the design to achieve the needed throughput. In more demanding applications such as speech, geophysics, medicine, pattern recognition and so forth, array processors may be needed. Since most DSP operations are arithmetic intensive, the 10- to 1000-fold increase in speed offered by an array processor can easily justify the substantial cost of these units. Finally, newer generations of DSP hardware designs are concentrating on parallel or pipelined structures. These architectures are tuned to specific applications to achieve maximum throughput.

XI. Applications

A. RADAR AND SONAR SIGNAL PROCESSING

Radar is an important technology that affects all phases of civilian and military aviation as well as space exploration. Whether the radars are pulsed or continuous wave, DSP has become an important design and analysis tool. Because of their speed and reliability, DSP methods have found a welcome home in Doppler and moving target processors. In addition, DSP can serve in target signature analysis applications, where a target's spectral image is analyzed to provide target-type information. The heart of many radar systems is a device called a matched filter. A matched filter, under some stationary signal assumptions, maximizes the output signal-to-noise ratio. From this cleaner database, radar decisions are made. The matched filter convolution operations can be replaced by FFT operations. To achieve very high speeds, pipelined FFTs are often used. These units are generally designed to have a high real-time bandwidth (i.e., transform-per-second rate) and they operate as fast fixed-point or hybrid number system processors

rather than slower but more precise floating-point units.

There is some commonality between the principles of sonar and radar. Sonar is based on acoustic waves and appears in both active and passive forms. As with radar, DSP tools can be used to design high-performance correlators and matched filters. Time domain digital filters or FFT methods can be used to implement sonar signal processes. Often the data are collected from an array of sensors, so a plurality of processing units would be integrated into the design.

Since digital shift registers can be used effectively to design delay lines, the physical design of correlators and delay estimators is a manageable task. Because the delayed and manipulated data remain in a digital format, they can be directly passed to a digital processor for high-level analysis. The DSP hardware can also be used to build beam-forming and beam-steering FIR-like filters for sonar and radar applications.

B. TELECOMMUNICATIONS AND AUDIO

The telecommunications industry relied almost exclusively on analog devices prior to the 1960s. Systems were designed with amplifiers, capacitors, and inductors. Beginning in the early 1960s, the industry began to convert to a digital format. Digital telecommunication systems have several advantages in terms of speed and bandwidth over their analog ancestors. Early voice grade telecommunication systems used pulse code modulation. In a PCM code each bit represents a fixed amount of signal voltage. If the least significant bit or zeroth bit has a weight of V volts, the nth bit has a weight of $2^n V$ volts. To achieve recognizable voice quality speech, sampling at rates of 8000 or more samples per second over a 13-bit or greater dynamic range must often be used. To reduce the dynamic range requirement, a logarithmic $\mu - 255$ law data compander can be used to compress speech intelligence into an 8-bit word. Because of the high-volume need for voice grade telecommunications equipment, the commonly used companding ADCs are often found integrated onto low-cost highly reliable semiconductor chips. Other DSP devices have found their way into the telecommunications industry in the form of signal generators, decoders, and voice synthesizers. By the mid-1970s more than 65 million miles of digital telecommunications circuits were in operation. Because of the high volume of components needed in the telecommunications industry, a number of custom integrated-circuit chips have been developed for use as multiplexers, digital filters, and processors known as echo cancelers. The echo canceler removes the imbalance along a telephone circuit, such as a satellite link, that can cause the phenomenon known as double-talk (an overlapped incoming and outgoing signal or echo). It uses an adjustable or adaptive FIR filter to reconstruct a copy of the unwanted reflected signal, which it subtracts from the outgoing signal before transmission.

Historically, the principal concern of the pre-1970 audio engineer was sound reproduction. This field has now expanded to include digital recording and reproduction, plus digital information storage and transmission. Digital audio is somewhat distinct from digital speech processing in that speech is a telecommunications problem in which the applied criterion of success is intelligibility. In digital audio, the target is fidelity. In many cases, DSP technology has exceeded the fidelity capability of analog recording and reproduction equipment and is virtually noise-free. Also, digital audio designs have eliminated some of the electromechanical components that deteriorate with age—the phonograph stylus and records, magnetic tape and heads, and so forth. Digital equipment has also been introduced into the recording studio, since it needs much less manual attention and adjustment than analog units and is more versatile.

Audio signal processing begins with an acoustic wave, which is manipulated and possibly archived. It is returned to an acoustic form for listing. As a result, an ADC and DACs are used at the input–output level. A 16-bit ADC conversion of an analog signal, when played back through a DAC, introduces no detectable noise (distortion) to a human listener. However, the longer the required wordlength, the more expensive the DSP system and storage medium. As a result, data compression techniques are often applied to the digitized database. These include PCM encoding and delta modulation techniques. Between the 1-bit DM protocol and the n-bit PCM is the hybrid differential pulse code modulation, or DPCM. Here the difference between two successive samples is coded as a low-order PCM word and reconstructed by a DM-like incremental addition/subtraction rule. For simple speech-type signals, these methods can achieve a good degree of data compression. For complex music, their use is more difficult.

As analog tapes are played the quality of the recording degrades. Therefore, for studio as well as listener use, storing audio information on digital tape has advantages. Digital storage of audio information may also include some redundant information for use in error correction and detection.

C. Speech Processing

A milestone in speech processing was the development of the vocoder, or voice coder, by Homor Dudley in 1939. His basic speech synthesis model is still in use today. In 1946 the advent of the South spectrograph for short-term speech spectrum display marked the beginning of extensive use of spectral analysis for speech processing. This field has been growing rapidly ever since. Although they are not yet able to perform as well as humans, speech processing algorithms and machines are becoming commercially available.

Human speech is a rather short-term phenomenon that changes dynamically. As a result, its spectral signature is often studied with short-time Fourier transforms. Various windowing and data segmentation methods have been developed to serve this need. Modified transforms such as the Wigner, cepstrum, and smoothed cepstrum can also be used. If $X(k)$ is the DFT of $x(n)$, then the complex cepstrum is given by $x'(n)$, where $x'(n) = \text{IDFT}[X'(k)]$ for $X'(k) = \log X(k) = \log|X(k)| + j \arg X(k)$ ($j = \sqrt{-1}$) and arg $= \arctan[\text{Im } X(k)/\text{Real } X(k)]$.

The design of integrated voice/data networks is predicated on minimizing the channel bandwidth (number of bits per second) necessary to transmit voice, while maintaining high speech quality and intelligibility. Typically, the transmission of speech with pulse code modulation requires about 64,000 bits/sec. Specialized adaptive predictive coding (APC) and adaptive transform coding (ATC) techniques can reduce this to about 16,000 bits/sec with no loss in speech quality. Coding techniques below 4000 bits/sec are based on a parametric speech synthesis models such as linear predictive coding (LPC) vocoders and channel vocoders. A further reduction in a bit rate is possible with vector quantization rather than scalar quantization techniques. In the latter, each parameter to be transmitted is quantized independently of all other parameters. In the former, a number of parameters are quantized together as a vector into a multidimensional space. Pattern recognition methods are then used to code and decode the intelligence.

Speech synthesis refers to two distinct types of speech generation by computer and may take the form of stored-parameters algorithms (e.g., the Speak & Spell educational toy using LPC parameters) or wave form coding. The first is text-to-speech synthesis and requires that a computer phonetically "read" a scanned or stored text. The second is speech recognition, which refers to the ability of a machine to recognize or understand human speech. Most commercial and research speech recognition systems today use a pattern-matching approach. In terms of current technology, speaker-dependent continuous speech rates in the high 90th percentile can be achieved only over relatively small vocabularies (tens of words). Large vocabulary, continuous speech, and speaker-independent operation lie somewhere in the future. Finally, speaker recognition systems attempt to recognize people from their voice prints. Present-day techniques can verify a speaker with high accuracy only under controlled conditions.

D. IMAGE PROCESSING

Image processing attempts to extract or modify information found in an image-dependent signal space. The military is a long-time user and promoter of image processing theory, practice, and hardware. In medicine, many diagnostic tools rely heavily on image processors to provide fast analysis of a graphical database. Other image processing applications are in the consumer electronics industry, law enforcement, geophysics, weather prediction, and the compression, transmission, and reconstruction of television-like signals. In the growing field of robotics, computer vision will play an important role. [See ROBOT VISION.]

Image processing can be partitioned into four subordinate areas known as image restoration, enhancement, coding, and analysis. If one assumes that an image has been degraded due to transmission noise, poor optics, geometric distortion, movement, and so forth, restoration methods are used to repair as much of the damage as possible.

Detectors can convert the optical data found in the image plane to an electronic format, which is then digitized. The point spread function models the optical distortion in the image plane and can be represented as a transfer function.

The point spread function shows how much of the information at an image plant "spreads" into neighboring areas in the image plane. Distortion can also be introduced by the recording device, and noise can be added to communication channels.

A basic image restoration task is deblurring. An inverting filter is sought to reduce the effect of noise and produce a facsimile of the source. A difficult nonblurring restoration problem is that of deconvolution, which attempts to recover an image from a set of projections. Often, as in the case of medical x-ray, electromagnetic interference (EMI), and similar scanners, the signal represents the amount of energy detected in a lower-dimensional image plane. Reconstructing the three- or two-dimensional image is analogous to sketching a human face from its shadow. The energy in a two- or one-dimensional plane obtained by illuminating an absorbing three- or two-dimensional object is a projection that can be expressed in terms of its spectral signature. To reconstruct the image with some degree of definition, multiple projections are required. The projection slice theorem states that by collecting a group of one-dimensional projections at various angles a two-dimensional image can be constructed, and from the slices of two-dimensional images, a three-dimensional images can be formed, and so forth. Fourier and other types of fast transforms are generally used to do the data analysis because of the enormous amount of data that must be assimilated to reconstruct an image from its projections.

Enhancement is used to improve the ability to interpret an image database. This is often a subjective measure and relies on human perception. It differs from restoration in that restoration seeks to rebuild the original image from its distorted image, while enhancement seeks to improve on the original. Often, due to transmission bandwidth limitations, an image may lose some of its definition. When assumptions must be made about the shape of an object, enhancement methods return some of the original picture quality.

Image coding is used to compress the representation of an image into as few bits as possible while maintaining a specified degree of image intelligibility. For example, a simple 1024×1024 image, resolved to 16 bits, requires a 2^{24}-bit data field. Such figures can overtax a practical communication channel or cause a memory buffer to overflow when large or multiple images are being processed. Therefore, much attention has been given to image data compression. Fortunately, a typical image contains a large amount of redundant or correlated information. Redundant data can be eliminated by using one of a number of techniques that attempt to make trade-offs between processing speed, memory requirements, and compression capabilities. For example, if most of an image remains unchanged from frame to frame, statistical methods suggest that the data rate can be reduced toward an average of 1 bit/pixel. Furthermore, there is usually a considerable correlation between adjacent pixels (or picture elements), which can also be compressed out of the database. With a DPCM system, a signal is replaced by an estimate of its incremental change. Slowly varying signals are therefore compressed into a small dynamic range. By using this technique, an 8-bit encoded image can often be replaced with a 3-bit DPCM equivalent without a subjective loss of quality. However, if the image consists of sharp high-contrast edges, the DPCM will perform poorly.

More mathematically intense transform-domain image compression methods have been developed which work more directly on the correlation question. However, they can be computationally intense and may degrade picture quality if not correctly applied.

BIBLIOGRAPHY

Antonious, A. (1979). "Digital Filters: Analysis and Design." McGraw-Hill, New York.

Blahut, R. (1985). "Fast Algorithms for Digital Signal Processing." Addison-Wesley, Reading, Massachusetts.

Oppenheim, A. V. and Schafer, R. (1975). "Digital Signal Processing." Prentice-Hall, Englewood Cliffs, New Jersey.

Oppenheim, A. V., ed. (1978). "Application of Digital Signal Processing." Prentice-Hall, Englewood Cliffs, New Jersey.

Rabiner, L. R. and Gold, B. (1975). "Theory and Applications of Digital Signal Processing." Prentice-Hall, Englewood Cliffs, New Jersey.

Taylor, F. (1983). "Digital Filter Design Handbook." Dekker, New York.

SIGNAL PROCESSING, GENERAL

Rao Yarlagadda *Oklahoma State University*
John E. Hershey *BDM Corporation*

GLOSSARY

A/D converter: System that accepts as input an analog signal and an analog reference and provides as output a digital signal.

Convolution: Integral (summation) expression for continuous (discrete) signals, generally used to represent an output of a linear system in terms of the input and impulse response of the system.

Deconvolution: Procedure that removes the effects of convolution.

Encoder: System that takes each quantized sample and represents it by a code word, such as a set of bits.

Quantizer: System that produces a distinct set of amplitude levels for a continuous range of amplitude levels of the input signal.

Signal processing deals with the electrical representation of the processes such as voice and imagery. It comprises those analytical techniques that can be applied to these electrical representations in order to achieve a variety of objectives. These objectives include noise cleaning for quality enhancement, redundancy removal for compression, and the estimation of basic properties of the signal as a means of analyzing the mechanism producing the signal or the characteristics of the medium or channel through which the signal is conveyed.

I. Introduction

It is very difficult to define signal processing in a meaningful and succinct way. There is an inbred desire to compare it with statistics and perhaps this is a good place to start. The discipline of signal processing and the science of statistics are very different and yet it is not at all clear what forms a clean dividing line between them. Statisticians typically try to determine which of two or several hypotheses are best supported by the outcome of an experiment. Electrical engineering, within whose bounds signal processing is generally admitted to fall, has greatly benefitted from statistical methods. Perhaps the most classic of examples is the decoding of (a priori) equally probable binary signals received through Gaussian noise. The decision region is often represented as in Fig. 1. If we decide that signal S_1 was sent if our detector input, x, is less than the threshold and that signal S_2 was sent if x is equal to or greater than the threshold, then we see that an error is possible as the channel noise can cause x to lie in S_2's region even though S_1 has been sent and vice versa. If the signals are not equally (a priorly) probable, statistical methods that use knowledge of the channel's noise statistics will allow us to pick a threshold that minimizes the probability of incorrect decoding.

Signal processing is a bit more "heuristic" and it is also often a diagnostic process. Instead of trying to decide which of several sources originated a signal, signal analysis will often be used to try to diagnose properties of the processes originating the signals or the characteristics of the medium through which the signals propagate. At the risk of oversimplification, we will say that a statistician will often formulate tests

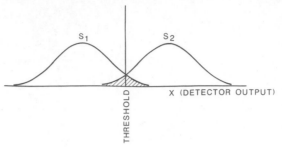

FIG. 1.   The decision regions.

to distinguish between hypotheses *before* examining the data; a signals analyst will often form hypotheses *after* analyzing data.

It has been our experience that it can be extremely helpful to look at a situation both as a statistician and as a signals analyst. It is by a dual viewing that one can perceive the delicate and the aesthetic bridges between various disciplines, the holism that abounds in mathematics.

As a simple example, let us look in two ways at a mistake that has been made over and over throughout the years. The mistake concerns the art of data reduction. In Fig. 2 there are two traces. The top trace is a trace produced by connecting sample value points produced by a white (pseudo)random process. The bottom trace was formed by connecting the points produced by the following "data reduction" routine, applied three times. First, the data were smoothed. Letting x_n be the original data sample values, a new series y_n was formed as

$$y_i = \tfrac{1}{6}(x_i + x_{i-1} + x_{i-2} + x_{i-3} + x_{i-4} + x_{i-5}) \quad (1)$$

Second, the smoothed data were differenced to form yet another series z_n according to the rule

$$z_i = 2y_i - y_{i-1} \quad (2)$$

The bottom trace appears to have a periodic component and yet it was derived from a random source. Why is this and why is it so interesting? It is interesting because this is precisely what many scientists have done to some of their data and published a false conclusion based on an apparent periodicity. In advanced statistics, the creation of a false periodicity in this manner is related to the Slutsky–Yule effect that predicts the period of the false oscillation based upon the smoothing and differencing operations. From a signal-processing viewpoint it is immediately clear what has happened. The operation (1) can be viewed as a digital filtering operation that increasingly diminishes the higher frequencies, that is, the higher the frequency the more attenuation. The operation (2), on the other hand, increasingly diminishes the lower frequencies, the lower the frequency the more it is attenuated. Thus the cascaded combination of the two operations results in an overall operation that gives preference to frequencies in a relatively narrow range. Doing the operation three times narrows or "sharpens" this range. Thus when white noise, which contains energy at all frequencies, is applied, the majority of the output signal that will be passed through will consist of energy in this narrow frequency range and the output will resemble an oscillatory signal at the "passband's" center frequency.

II. Fundamentals of Signal Processing

A. The Representation of Signals

A signal is a record or history of some physically real value over time or space. A typical signal might be the voltage over time produced by a crystal microphone exposed to the human voice or the two-dimensional array of the values of the point charges produced on a charge coupled device (CCD) scanning array as it sweeps across an illuminated scene. Signals are repre-

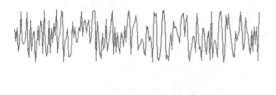

FIG. 2.   Two traces. (Courtesy of R. Kubichek.)

sented by a continuous record such as $x(t)$ for our voice voltage or $l(x, y)$ for our luminosity. This is known as the representation of signals in the time or spatial domain. Another, exceptionally important, representation of signals is in the frequency domain. The most common and significant of these representation is that provided by Fourier analysis. Fourier showed that signals could be represented as the weighted sum of sinusoids. For a periodic signal of period T, that is, a signal $x(t)$ for which $x(t) = x(t + T)$ for all t, we can write

$$x(t) = a_0 + 2 \sum_{n=1}^{\infty} \left[a_n \cos\left(\frac{2\pi nt}{T}\right) + b_n \sin\left(\frac{2\pi nt}{T}\right) \right] \qquad (3)$$

and solve for the coefficients of the sinusoids as follows:

$$a_n = \frac{1}{T} \int_{-T/2}^{T/2} x(t) \cos\left(\frac{2\pi nt}{T}\right) dt \qquad (4a)$$

$$b_n = \frac{1}{T} \int_{-T/2}^{T/2} x(t) \sin\left(\frac{2\pi nt}{T}\right) dt \qquad (4b)$$

Higher frequencies, large values of n in (4a,b), become increasingly important as the signal becomes closer to a waveform possessing discontinuities. For example, for a periodic triangular waveform that exhibits a discontinuity only in the waveform's derivatives, a_n, b_n decrease quadratically, that is, a_n, $b_n = 0(1/n^2)$; but for a function possessing a discontinuity in the waveform itself, such as a periodic square wave, a_n, b_n decrease only linearly, that is, a_n, $b_n = 0(1/n)$.

If the signal is not periodic (and of finite energy), then the continuous Fourier transform is used:

$$X(f) = \int_{-\infty}^{\infty} x(t) \, e^{-j2\pi ft} \, dt \qquad (5a)$$

$$x(t) = \int_{-\infty}^{\infty} X(f) \, e^{j2\pi ft} \, df \qquad (5b)$$

Equations (5a,b) are often spoken of as a Fourier transform pair where $X(f)$ is the Fourier transform of $x(t)$ and vice versa. [See FOURIER SERIES.]

Just why the Fourier representation is so important is probably due to a confluence of properties. First, the sinusoids are orthogonal, that is, for a periodic signal,

$$\int_{-T/2}^{T/2} \left(\begin{array}{c} \sin \\ \text{or} \\ \cos \end{array}\right) \left(\frac{2\pi mt}{T}\right) \left(\begin{array}{c} \sin \\ \text{or} \\ \cos \end{array}\right) \left(\frac{2\pi nt}{T}\right) dt$$

$$= \begin{cases} T/2, & m = n \\ 0, & m \neq n \end{cases} \qquad (6)$$

Second, the Fourier transform is a linear operation, that is, the Fourier transform of $c_1 x_1(t) + c_2 x_2(t)$ is $c_1 X_1(f) + c_2 X_2(f)$ where c_1 and c_2 are arbitrary constants. Third, the Fourier transform can be easily scaled in time or frequency. The Fourier transform of $x(ct)$ is $(1/|c|)/X(f/c)$ where c is an arbitrary scale factor. Fourth, the magnitude of the Fourier transform is invariant under time displacement, it undergoes only a phase shift, that is, the Fourier transform of $x(t - t_0)$ is $X(f) \exp(-j2\pi ft_0)$. Conversely, a frequency shift is such that the Fourier transform of $X(f - f_0)$ is $x(t) \exp(j2\pi f_0 t)$. The time (or spatial) shift property is important in pattern recognition where it is known that a particular waveform is present somewhere in time (or space). The converse frequency shift property is at the heart of many communications processes such as heterodyning. Fifth, sinusoids "occur" naturally as the modes of familiar vibrations and other natural physical processes.

Signals are often represented by a series of discrete samples $\{x(n)\}$ for a one-dimensional signal such as our voice waveform or $\{l(i, j)\}$ for an image. Making the conversion from $x(t) \rightarrow \{x(n)\}$ involves the three steps depicted in Fig. 3. The SAMPLER measures the exact value of $x(t)$ at, for our discussion, periodic intervals, $t = kT$, where T is the sampling interval and k an integer. Note that the sampling frequency is $f_s = 1/T$. The QUANTIZER maps the exact sampled value of $x(t)$ into one of L quantization levels. The ENCODER assigns a unique code word to each of the L quantization levels and accordingly to each of the sampled values of $x(t)$.

As an example of this process of converting a continuous signal to a series of discrete samples, consider that we have the time trace shown in Fig. 4. Let us now sample $x(t)$ at the 12 points

FIG. 3.   The conversion of a continuous signal to a series of discrete samples.

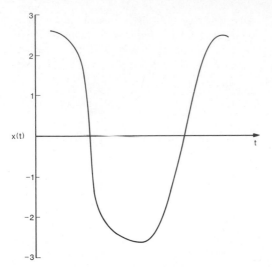

FIG. 4.   An example of $x(t)$.

TABLE I.   ENCODER's Assignment of Code Words to QUANTIZER Outputs

Quantizer output	Code word
−3	000
−2	001
−1	010
0	011
1	100
2	101
3	110
4	111

indicated by heavy dots on the time axis as shown in Fig. 5. Let the sampled values be

$$\{2.6, 2.4, 1.6, -1.9, -2.4, -2.6,$$
$$-2.6, -1.8, -0.4, 1.4, 2.4, 2.4\} \quad (7)$$

Now let us quantize the values by using the quantizer depicted in Fig. 6. After quantization, the set of values in Eq. (7) have been quantized to the set

$$\{3, 2, 2, -2, -2, -3, -3, -2, 0, 1, 2, 2\} \quad (8)$$

The ENCODER for our example is described by the operation specified in Table I. The twelve quantized values in (8) thus become the series of code words in (9).

$$\{110, 101, 101, 001, 001, 000, 000,$$
$$001, 011, 100, 101, 101\} \quad (9)$$

or the "continuous" bit stream 110101101001-00100000001011100101101.

The process we have just gone through is known in general as analog-to-digital (A/D) conversion. There are a number of important considerations in A/D conversion. Starting with the SAMPLING process we encounter the first of these and that is the question "How fast must I sample—what is the minimum acceptable value of f_s?" The answer to this basic question incor-

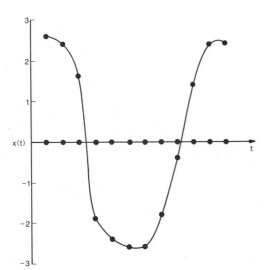

FIG. 5.   Sampling the continuous signal.

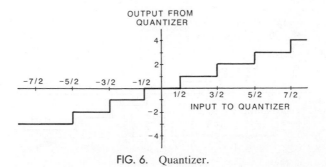

FIG. 6.   Quantizer.

FIG. 7.   Aliasing error.

porates the name of the man who answered it: Nyquist's rule. Nyquist's rule specifies that sampling must occur at a frequency greater than twice the highest significant frequency component of the waveform to be sampled. Failure to comport to Nyquist's rule can lead to "aliasing." This undesirable phenomenon "folds" high-frequency components into the lower frequencies. To see this, consider Fig. 7 where we depict a sinusoid of frequency 1 Hz sampled at 1.33 Hz, a frequency less than the required Nyquist sampling frequency of greater than 2 Hz. In Fig. 7 we have sketched a smooth waveform through the sample points. The reconstructed waveform, that is, the waveform that appears to be present from the "undersampling" is a sinusoid of period 3 sec or frequency 0.333 Hz. Thus, a relatively high-frequency component of 1 Hz has been aliased to a frequency component of one-third its frequency. For this reason, one usually prefaces the input of the SAMPLER with a low pass filter (LPF) that removes any frequencies above that allowed by Nyquist's rule.

The second item we must consider is the effect of the QUANTIZER. Clearly the QUANTIZER will be important if we wish to be able to preserve the fidelity of the sample signal. First, the QUANTIZER's dynamic range must accommodate the dynamic range of the sampled signal most, if not all, of the time. Second, the QUANTIZER must have a sufficient number of quantizing levels in order to capture the requisite nuances of the signals. The QUANTIZER of Fig. 6 of our example is an example of a multilevel mid-tread uniform quantizer. It is uniform because the spacing of the quantizaton levels is a constant. It is called mid-tread (as opposed to mid-riser) because zero and the region about zero is handled as a distinct level. Many quantization schemes have been developed and employed over the years. In digital telephony, nonuniform quantizers called companders are extremely important. For digital signal processing, however, the uniform quantizer is preeminent. A very useful "rule-of-thumb" is that the signal-to-quantizing noise ratio for a uniform quantizer increases about 6 dB for each doubling of the number of quantization levels.

The ENCODER is also important. The code word table should be matched to the signal processing data handling architecture. Note that the digital stream that results from sampling an analog signal at a rate of f_s samples per second and quantizing to L different levels and encoding those levels with a table such as Table I will be $f_s \cdot \log_2 L$ bits per second.

B. WINDOWING

Because measurements of a particular signal or image are limited to samples of finite scope in time or space, it is necessary to consider the effects of this constraint. The study of these effects falls under the general discipline of "windowing." The term is apt as what we are allowed is a "window" into our process instead of a complete time or spatial record.

The simplest window is the rectangular, sometimes called "boxcar," window. This window is the familiar rect function.

$$\text{rect}(x) = \begin{cases} 1, & |x| \leq T/2 \\ 0, & |x| > T/2 \end{cases} \quad (10)$$

What is the effect of such a function? Let us suppose that the process we wish to examine is a pure sinusoid of frequency f_0 that has existed since $t = -\infty$ and will exist to $t = +\infty$. All we see of the sinusoid is a sampling from $t = -T/2$ to $t = +T/2$.

We can calculate the magnitude of the Fourier transform of the short segment of a sinusoid that results from overlaying the rect window on a sinusoid. All we need do is take the Fourier transform of the rect function and then use the frequency shifting property of the Fourier transform. The result is a Fourier transform whose magnitude is as depicted in Fig. 8. From Fig. 8 we see that our windowing has given rise to a rather broadband signal. This may be quite undesirable in many situations and has led many researchers to look for more appropriate window functions for various problems. A triangular window, for example, deemphasizes the beginning and ending values of the sampled signal and, for our sinusoid, leads to a Fourier transform that is much "tighter" about f_0 than is the case for the rect window.

C. THE DISCRETE FOURIER TRANSFORM

In Eq. (5), we have introduced the continuous Fourier transform and its inverse transform. Analytical computations of (5) are possible for only a very few signals. In most situations $x(t)$ cannot

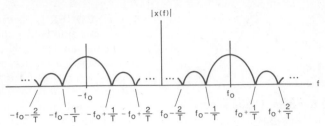

FIG. 8. The magnitude of the Fourier transform of a short segment of a sinusoid.

be expressed in an analytical form and it may even be represented by a curve or perhaps its values are known at equal intervals. These reasons point to the direction of the discrete representation of (5) and its evaluation. Before we can write the discrete representation, we should note a few aspects. First, the signal $x(t)$ needs to be represented at discrete values, say at, $t = nT$, where T is selected according to Nyquist's rule. Second, the integral of $x(t)$ over the small interval $nT \le t < (n + 1)T$ can be approximated by

$$\int_{nT}^{(n+1)T} x(t) \, e^{-j2\pi ft} \, dt \simeq Tx(nT) \, e^{-j2\pi fnT} \quad (11)$$

which is usually referred to as rectangular integration where we have assumed that the integrand does not change much in the interval. Finally, we sum over all the intervals. Now,

$$X(f) = T \sum_{n=-\infty}^{\infty} x(nT) \, e^{-j2\pi fnT} \quad (12)$$

Most signals are assumed to be nonzero only for a finite interval. Assuming $x(t) = 0$ for $t < 0$ and $t \ge NT$, where N is some integer, we can write (12) in the form

$$X(f) = T \sum_{n=0}^{N-1} x(nT) \, e^{-j2\pi fnT} \quad (13)$$

which, incidentally, is periodic with period $1/T$. Sampling $X(f)$ at intervals of $1/NT$, we can write

$$X\left(k \frac{1}{NT}\right) = T \sum_{n=0}^{N-1} x(nT) \, e^{-j(2\pi/N)kn}$$

$$k = 0, 1, ..., N - 1 \quad (14)$$

Since T and $1/NT$ are constants, they are usually omitted and the discrete Fourier transform (DFT) is expressed by

$$X(k) = \sum_{n=0}^{N-1} x(n) \, e^{-j(2\pi/N)kn}$$

$$k = 0, 1, ..., N - 1 \quad (15)$$

Using the analysis similar to the above, we can approximate (5b) by

$$x(n) = \frac{1}{N} \sum_{k=0}^{N-1} X(k) \, e^{j(2\pi/N)kn}$$

$$n = 0, 1, ..., N - 1 \quad (16)$$

which is the inverse discrete Fourier transform (IDFT).

From (15) and (16), it is clear that they can be interpreted as linear transformations and they convert one set of data from one domain to another (time to frequency and vice versa). The computation of $X(k)$ in (15) takes N^2 complex multiplications and additions, which gets large when N is large. Fast algorithms have been developed. An algorithm, that is considered to be classic now is given by Cooley and Tukey and is applicable when N is a power of 2. This algorithm and its variations are sometimes referred to as fast Fourier transforms (or FFTs). An outline of this algorithm is presented below.

The idea in this algorithm is to express an N-point DFT by two $N/2$-point DFTs. That is, if

$$A(k) = \sum_{i=0}^{(N/2)-1} x(2i) \, e^{[-j2\pi/(N/2)]ik}$$

$$k = 0, 1, 2, ..., (N/2) - 1 \quad (17)$$

$$B(k) = \sum_{i=0}^{(N/2)-1} x(2i + 1) \, e^{[-j2\pi/(N/2)]ik}$$

$$k = 0, 1, 2, ..., (N/2) - 1 \quad (18)$$

then

$$X(k) = A(k) + B(k) \, e^{-j(2\pi/N)k}$$

$$k = 0, 1, ..., N - 1 \quad (19)$$

From (17) and (18), we see that $A(k)$ and $B(k)$ are DFTs of even and odd samples, respectively. Note also that $A(k)$ and $B(k)$ are periodic with period $(N/2)$. Since an N-point DFT is reduced to two $N/2$-point DFTs, it can be seen that this will reduce the number of operations approximately by half. Successive application of this idea will reduce the number of operations to $N \log_2 N$ as compared to N^2 by direct computation. It is clear that the DFT and the IDFT have the same general forms and, therefore, the same algorithm can be used for both cases with slight modification.

Before we leave this topic, we want to point out certain facets of the FFT. First, the basic FFT algorithm is computed in $(L + 1)$ stages corresponding to 2^L data points. Let us denote the sequence of complex numbers resulting from the mth stage of computation as $X_m(i)$, $m = 0, 1, \ldots, L$. Further, let us define the input sequence in the bit reversed order and define

$$X_0(i) = x(\hat{i})$$

where i is obtained from $\hat{i}$ by bit reversion. For example, if $N = 8$, $\hat{i} = 3$, then

$$(\hat{i})_2 = 011$$

and reversing the bits, we have

$$(i)_2 = 110 = (6)_{10}$$

The heart of the FFT computation is in the so-called "butterfly operation," which simply corresponds to computing the output array for the $(m + 1)$st stage via

$$X_{m+1}(p) = X_m(p) + e^{-j(2\pi/N)r}X_m(q)$$

$$X_{m+1}(q) = X_m(p) - e^{-j(2\pi/N)r}X_m(q)$$

where r depends on the stage and indices of the variables p and q. Various other fast algorithms are available for computing the DFT when N is a power of 2, and for arbitrary N (see Oppenheim and Schafer).

The discrete Fourier transform requires multiplications that are computationally expensive compared to additions. A transform that uses only additions and subtractions is the Walsh–Hadamard transform, which is discussed below for N a power of two.

D. WALSH–HADAMARD TRANSFORM

The standard class of Walsh–Hadamard matrices of order $N = 2^L$ can be generated from the recursive relation

$$H_k = \begin{bmatrix} H_{k-1} & H_{k-1} \\ H_{k-1} & -H_{k-1} \end{bmatrix}, \qquad k = 1, 2, 3, \ldots \quad (20)$$

with $H_0 = 1$. For example, when $N = 4$, we have

$$H_2 = \begin{bmatrix} 1 & 1 & 1 & 1 \\ 1 & -1 & 1 & -1 \\ 1 & 1 & -1 & -1 \\ 1 & -1 & -1 & 1 \end{bmatrix} \quad (21)$$

The Hadamard transform can be defined by

$$\mathbf{y} = H_L\mathbf{x} \quad (22)$$

where $\mathbf{x}$ is an N-dimensional data vector and $\mathbf{y}$ is an N-dimensional transform vector. The inverse transform is given by

$$\mathbf{x} = (1/N)H_L\mathbf{y} \quad (23)$$

Various fast algorithms are available to implement (22) or (23). The algorithm that is intuitive uses the basic recursive structure

$$H_k = \begin{bmatrix} I & I \\ I & -I \end{bmatrix}\begin{bmatrix} H_{k-1} & 0 \\ 0 & H_{k-1} \end{bmatrix} \quad (24)$$

where I is an identity matrix of order 2^{k-1} and the second algorithm uses the relation

$$H_L = A^L \quad (25a)$$

where

$$A = \begin{bmatrix} 1 & 1 & 0 & 0 & \cdots & 0 & 0 \\ 0 & 0 & 1 & 1 & \cdots & 0 & 0 \\ \vdots & \vdots & \vdots & \vdots & \ddots & \vdots & \vdots \\ 0 & 0 & 0 & 0 & \cdots & 1 & 1 \\ 1 & -1 & 0 & 0 & \cdots & 0 & 0 \\ 0 & 0 & 1 & -1 & \cdots & 0 & 0 \\ \vdots & \vdots & \vdots & \vdots & \ddots & \vdots & \vdots \\ 0 & 0 & 0 & 0 & & 1 & -1 \end{bmatrix}$$

$$(25b)$$

Using either one of the structures in (24) or (25) in (22) allows for the computation of the Walsh–Hadamard transform (WHT) in $N \log_2 N$ additions and/or subtractions.

The uses of DFT and WHT are too numerous to mention. First and foremost, the DFT coefficients give the normalized spectral values of the signal. Other uses include applications in data compression, convolution, deconvolution, etc. In the data compression application, for example, we could transmit transformed data rather than raw data, as many of the DFT coefficients (or WHT coefficients) are close to zero and therefore only a portion of the coefficients need to be sent. Next let us consider convolution and correlation.

E. CONVOLUTION AND CORRELATION

Before we consider these, there are some concepts that are of interest in system theory. In our study we assume that a system is defined as a unique transformation that maps an input signal, $x(t)$, into an output signal, $y(t)$. Let us denote this by

$$y(t) = T[x(t)] \qquad (26)$$

There are three classes of systems that are of interest to us. These can be defined by putting constraints on the transformation $T[\cdot]$.

1. Linearity

Given the responses $y_1(t)$ and $y_2(t)$ corresponding to the respective inputs $x_1(t)$ and $x_2(t)$, then a system is linear if and only if

$$T[a_1x_1(t) + a_2x_2(t)] = a_1T[x_1(t)] + a_2T[x_2(t)]$$
$$= a_1y_1(t) + a_2y_2(t) \qquad (27)$$

where a_1 and a_2 are arbitrary constants. An interesting example of linearity is the delay operation

$$r(t) = T[x(t)] = x(t - t_0) \qquad (28)$$

where t_0 is some positive constant. This is a linear operation since

$$T[a_1x_1(t) + a_2x_2(t)] = a_1x_1(t - t_0) + a_2x_2(t - t_0) \qquad (29)$$

Others include those of the derivative, integration, etc. An example of a nonlinear operation is

$$r(t) = T[x(t)] = x^2(t) \qquad (30)$$

2. Time Invariance

Let $y(t)$ be the response to the input $x(t)$. Then $y(t - t_0)$ is the response to the input $x(t - t_0)$ if the system is time invariant.

Note that this requires that the response be the same anytime you apply the input except that the response is shifted according to the time of the input.

3. Causality

A causal system is a system for which the output for any time t_0 depends on the inputs for

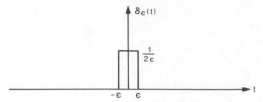

FIG. 9.   An example of an impulse function, as $\varepsilon \to 0$.

$t \leq t_0$ only. That is, the response does not depend on the future inputs and it relies only on the past and present inputs.

The above concepts allow for the characterization of a linear system. One function that is important to us is the impulse function, $\delta(t)$, which is defined in terms of the process

$$\int_{-\infty}^{\infty} x(t)\delta(t)\, dt = x(0) \qquad (31)$$

where $x(t)$ is any test function that is continuous at $t = 0$. Equation (31) is a special case of the so called sifting property

$$\int_{-\infty}^{\infty} x(t)\delta(t - t_0)\, dt = x(t_0) \qquad (32)$$

where again $x(t)$ is assumed to be continuous at $t = t_0$. An example of an impulse function is shown in Fig. 9, where we have

$$\delta(t) = \lim_{\varepsilon \to 0} \delta_\varepsilon(t) \qquad (33)$$

Intuitively we see that any function having unit area and zero width in the limit as some parameter approaches zero is a suitable representation of $\delta(t)$. A dynamite explosion is a good approximation, for example, for an impulse input, which is used in seismic exploration. In the following we will consider exclusively linear time invariant systems. The impulse response, $h(t)$, of a linear time invariant system is defined to be the response of the system to an impulse at $t = 0$. By the linear time invariant properties, we can see that the response of a linear system to the input

$$x(t) = \sum_{n=1}^{N} a_n\delta(t - t_n) \qquad (34)$$

is

$$y(t) = \sum_{n=1}^{N} a_nh(t - t_n) \qquad (35)$$

Using the sifting property in (32), we can write

$$x(t) = \int_{-\infty}^{\infty} x(t')\delta(t - t') \, dt' \qquad (36)$$

Using the rectangular integration, we can approximate (36) by

$$x(t) = \sum_{n=-N_1}^{N_2} x(n \, \Delta T)\delta(t - n \, \Delta T) \, \Delta T \qquad (37)$$

where ΔT is some small increment time. The response $y(t)$ in (35) is given by

$$y(t) = \sum_{n=-N_1}^{N_2} x(n \, \Delta T)h(t - n \, \Delta T) \, \Delta T \qquad (38)$$

As $\Delta T \to 0$, $n \, \Delta T$ approaches the continuous variable t', the sum in (38) becomes an integral, and we have

$$y(t) = \int_{-\infty}^{\infty} x(t')h(t - t') \, dt'$$

$$= \int_{-\infty}^{\infty} h(t')x(t - t') \, dt' \qquad (39)$$

which is usually referred to as a convolution integral. The second integral in (39) is obtained from the first by redefining the variable. The convolution is an important relation, which, in words, says that the response of an arbitrary input is related to the impulse response via (39). Equation (39) is symbolically written in the form

$$y(t) = x(t) * h(t) = h(t) * x(t) \qquad (40)$$

where (∗) represents convolution.

The computation of the convolution in (39) can be computed analytically only if $x(t)$ and $h(t)$ are known analytically and can be integrated. Another approach is to use transform theory. That is, if the Fourier transforms of $x(t)$ and $h(t)$ are expressed as

$$F[x(t)] = X(f) \qquad (41)$$

and

$$F[h(t)] = H(f) \qquad (42)$$

then we can show that

$$Y(f) = F[y(t)] = H(f)X(f) \qquad (43)$$

and

$$y(t) = \int_{-\infty}^{\infty} H(f)X(f) \, e^{j2\pi ft} \, df \qquad (44)$$

The transform of $h(t)$, $H(f)$, is usually referred to as a transfer function relating the input

and the output. Transfer functions play a major role in communication theory, control systems, and signal processing. We will consider the discrete convolution later. Next, let us consider the concepts of correlation for the aperiodic case.

The process of correlation is useful for comparing two signals and it measures the similarity between one signal and a time-delayed version of a second signal. The cross correlation of $x(t)$ and $g(t)$ is defined by

$$R_{gx}(\tau) = \int_{-\infty}^{\infty} g(t)x(t + \tau) \, dt \qquad (45)$$

which is a function of τ, the amount of time displacement. When $g(t) = x(t)$, $R_{gg}(\tau)$ is referred to as autocorrelation. Correlation functions are useful in many areas. For example, if $x(t) = g(t - t_0)$, then $R_{gx}(t)$ will be a maximum when $\tau = t_0$ indicating a method for measuring the delay. It is of interest that the cross correlation in (45) can be expressed in terms of convolution in (40). It can be shown that

$$R_{\bar{g}x}(\tau) = g(\tau) * x(\tau) \qquad (46)$$

where $\bar{g}(t) = g(-t)$. That is, the cross-correlation of $g(-t)$ with $x(t)$ will result in the convolution of $g(t)$ with $x(t)$.

Transforms can be used to compute correlations also. For example, if $F[x(t)] = X(f)$, $F[g(t)] = G(f)$, and $F[R_{gx}(\tau)] = S_{gx}(f)$, then

$$S_{gx} = G^*(f)X(f) \qquad (47)$$

where $G^*(f)$ is the complex conjugate of $G(f)$. Computing the inverse transfer of (47) will give the cross-correlation function. Next, let us consider the discrete computation of convolutions and correlations.

4. Computation by Discrete Methods

From (38), we can write

$$y(k \, \Delta T) = \sum_{n=-\infty}^{\infty} x(n \, \Delta T)h[(k - n) \, \Delta T] \, \Delta T$$

Suppressing ΔT for simplicity, that is, $\Delta T = 1$, we have

$$y(k) = \sum_{n=-\infty}^{\infty} x(n)h(k - n) = \sum_{n=-\infty}^{\infty} h(n)x(k - n)$$

$$(48)$$

which is usually referred to as discrete convolution and $h(n)$, the unit sample response. In a similar manner, we can write the correlation in (45) in discrete form

$$R_{gx}(k) = \sum_{n=-\infty}^{\infty} g(n)x(n + k) \qquad (49)$$

The computational procedures are very similar for both convolution and correlation and, therefore, we will concentrate on convolutions. It should be pointed out that (48) and (49) could have been derived for either a continuous or a discrete system.

For physical systems $h(n) = 0$, $n < 0$, that is, they are causal, and furthermore we will assume that for practical reasons that $h(n)$ and $x(n)$ are time limited. Let

$$x(n) = 0 \qquad \text{for } n < 0 \text{ and for } n \geq L \quad (50a)$$

$$h(n) = 0 \qquad \text{for } n < 0 \text{ and for } n \geq N \quad (50b)$$

It is clear that the lower limit on n for $x(n)$ could be different and the analysis is very similar. With (50) in (48), we can rewrite

$$y(k) = \sum_{n=0}^{L-1} x(n)h(k - n) = \sum_{n=0}^{N-1} h(n)x(k - n) \qquad (51)$$

Noting the lengths of $x(n)$ and $h(n)$ are L and N, respectively, we see that the output $y(k)$ is time limited and is nonzero for $0 \leq k \leq N + L - 2$. We stress that the convolution of two sequences will result in a sequence longer than either one. Let us illustrate these ideas by a simple example.

EXAMPLE.   Let

$$x(0) = 1, x(1) = 1 \quad \text{and} \quad x(n) = 0$$
$$\text{for} \quad n < 0 \quad \text{and} \quad n > 1$$

and

$$h(0) = 1, h(1) = -1 \quad \text{and} \quad h(n) = 0$$
$$\text{for} \quad n < 0 \quad \text{and} \quad n > 1$$

From (49), we can write

$$y(n) = 0, \qquad n < 0$$
$$y(0) = x(0)\,h(0) + x(1)\,h(-1) = x(0)\,h(0)$$
$$y(1) = x(0)\,h(1) + x(1)\,h(0) \qquad (52)$$
$$y(2) = x(0)\,h(2)$$
$$y(n) = 0, \qquad n > (N + L - 2) = 2$$

The equations in (52) can be written in a matrix form for $0 \leq n \leq 2$

$$\begin{bmatrix} y(0) \\ y(1) \\ y(2) \end{bmatrix} = \begin{bmatrix} h(0) & 0 \\ h(1) & h(0) \\ 0 & h(1) \end{bmatrix} \begin{bmatrix} x(0) \\ x(1) \end{bmatrix} \qquad (53a)$$

In symbolic form, (53a) can be written as

$$\mathbf{y} = H\mathbf{x} \qquad (53b)$$

Note the nice structure of the coefficient matrix in (53a). This structure can be generalized. Incidentally, the correlation equations have the same general structure as in (53). Direct computation of (51) is computationally intensive, especially when the number of output data, $N + L - 1$, is large. For this case, the FFT is generally used to compute convolutions and correlations. This is discussed next. For an effective use of the FFT, first, let M be a power of 2 and

$$M \geq N + L - 1 \qquad (54)$$

Second, pad $x(n)$ and $h(n)$ with 0's. That is, the new $x(n)$, $h(n)$, and $y(n)$ are

$$x'(n) = \begin{cases} x(n), & 0 \leq n \leq L - 1 \\ 0, & L \leq n < M \end{cases}$$

$$h'(n) = \begin{cases} h(n), & 0 \leq n \leq N - 1 \\ 0, & N \leq n < M \end{cases} \qquad (55)$$

$$y'(n) = \begin{cases} y(n), & 0 \leq n \leq N + L - 2 \\ 0, & N + L - 1 \leq n < M \end{cases}$$

Third, compute the DFTs of $x'(n)$ and $h'(n)$ using the FFT. That is

$$X'(k) = \text{DFT}[x'(n)]$$
$$H'(k) = \text{DFT}[h'(n)] \qquad (56)$$

Fourth, compute the products, that is,

$$Y'(k) = H'(k)X'(k) \qquad (57)$$

Fifth, compute the IDFT of $Y'(k)$, that is,

$$y'(n) = \text{IDFT}[Y'(k)] \qquad (58)$$

and delete the last $M - (N + L - 1)$ points from $y'(n)$ to obtain $y(n)$. Note the similarity between the discrete computation of $y(n)$ via the DFT and the computation of $y(t)$ via Fourier transforms. We should note the padding in (55) and the constraint in (54). Similar approaches can be given for computing the correlations.

III. Technology of Signal Processing

The technology of signal processing, like so many other technologies, got a primal boost during World War II. It was indeed quite remarkable what the scientists and engineers accomplished with what today seems bulky and crude. Figure 10 is a picture of one of the most important instruments of those days, the sound spectrograph. The sound spectrograph depicted presented a hard copy, visual display of the energy present in different frequency ranges over a time record of modest length. The vertical axis was devoted to frequency which typically spanned the range from 0 to 3+ kHz. The horizontal axis was used for time; a typical trace was of about 2 sec in duration. The sound spectrograph of the early 1940s technology thus exhibited an analysis capability whose time-bandwidth product was of the order of 10^4. [*See* ACOUSTIC SIGNAL PROCESSING.]

Figures 11–15 are typical of some research sound spectrographs produced in the early 1940s. Figure 11 is the sound spectrograph associated with thermal noise. Note the random distribution of energy over frequency and time. Figure 12 is the sound spectrograph associated

FIG. 11. Spectrograph—thermal noise. (From speech and facsimile scrambling and decoding, summary Technical Report of Division 13, NDRC Vol. 3, Washington, D.C., 1946.)

with a telephone bell. Note the significant low-frequency component and the alternating pattern in the two upper frequency components probably brought about by the clapper alternating strokes between two slightly different bells. Figure 13 is the sound spectrograph produced by crumpling paper. Notice that the crumpling action results in broadband, noise-like impulses. Figure 14 is a sound spectrograph of piano music. Note the multitone components. Figure 15 is a sound spectrograph of normal human speech of the words "one, two, three, four, five, six." Note the heavy deposits of energy in the time varying "bands." These bands are termed the formants of speech. They are the resonances excited in the oral and nasal cavities of the vocal tract by the periodic pulse train generated at the glottis for voiced sounds such as the vowels. [*See* DIGITAL SPEECH PROCESSING.]

Today, of course, most signal processing is done by high-speed electronic digital computational devices. Because so many of the signal-processing tasks share computational "modules," such as matrix-processing tasks, a class of computational engines has gained prominance in contemporary signal processing. These machines are the array processors (not "arrays" of processors that will be discussed later) that are serial, "pipelined," computational devices that are extremely efficient when dealing with certain repetitive, deterministic (nondecision, nonbranching) computational code such as is written for manipulation of matrices or numerical arrays. One such device is pictured in Figure 16. The device is capable of performing 100 million floating point computations a second when it is able to run at maximum efficiency. These array processors or, as they are sometimes called, numerical coprocessors, are very popular and widespread and also somewhat difficult to compare without actual "hands-on," problem-specific experience.

Comparisons between any competing entities are extremely difficult to make the more complex the entities are. To attack the problem, the

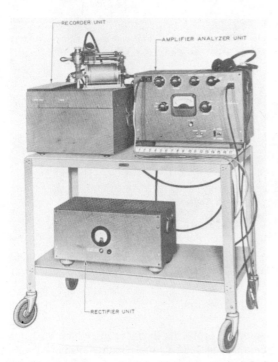

FIG. 10. Sound spectrograph. (From speech and facsimile scrambling and decoding, summary Technical Report of Division 13, NDRC Vol. 3, Washington, D.C., 1946.)

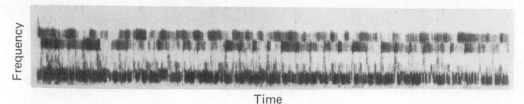

FIG. 12.   Spectrograph—telephone bell. (From speech and facsimile scrambling and decoding, summary Technical Report of Division 13, NDRC Vol. 3, Washington, D.C., 1946.)

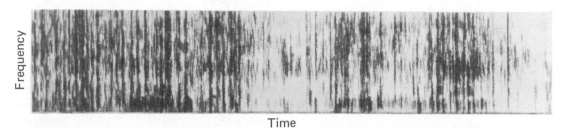

FIG. 13.   Spectrograph—crumpling paper. (From speech and facsimile scrambling and decoding, summary Technical Report of Division 13, NDRC Vol. 3, Washington, D.C., 1946.)

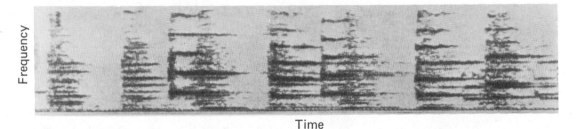

FIG. 14.   Spectrograph—piano music. (From speech and facsimile scrambling and decoding, summary Technical Report of Division 13, NDRC Vol. 3, Washington, D.C., 1946.)

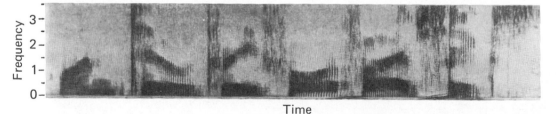

FIG. 15.   Spectrograph—normal human speech: one, two, three, four, five, six. (From speech and facsimile scrambling and decoding, summary Technical Report of Division 13, NDRC Vol. 3, Washington, D.C., 1946.)

computational community has informally and collectively devised a potpourri of comparison guidelines usually termed "benchmarks." The motivation for developing benchmarks in innate to our culture and also it is a general desire of management personnel to "dedimensionalize"

complex problems in order to reduce vast and complex issues to simple figures-of-merit.

Benchmarking has been done on (1) hardware, such as fast coprocessors, (2) software, such as operating systems, and (3) entire computational systems. The desire to concoct the per-

FIG. 16.   Array processor. (Courtesy of Star Technologies, Inc.)

fect and universal razor is an attractive intellectual pursuit. It may, however, be fruitless as the true "bottom line" is the efficacy of a piece of hardware, software, or system in solving a particular—hypothetical—problem.

One can be fully secure in the suspicion that a vendor will disseminate the performance of those benchmarks that reveal the product in the most favorable light.

The following terms are widely used when comparing processors. They have some meaning for serial machines but become inadequate, and even misleading, for highly parallel structures.

1.   MIPS: millions of instructions per second; generally the reciprocal of the cycle time.

2.   LOPS: logical operations per second; as many machines are capable of performing register-to-register logic instructions, LOPS can be thought of as the product of MIPS times the machine's word length.

3.   (M)FLOPS: (mega, or millions of) floating point operations per second; generally estimated by the inverse of the time (in microseconds) required to execute a floating point multiplication.

Looking a bit more closely at serial machines, it is often the case that a problem will be sized in time as the ratio of millions of floating point operations required to solve the problem divided by the MFLOPS capability of the machine. That the time required is often more than three times longer than predicted is surprising and disheartening but eminently reasonable and can be easily seen from the following analysis from Parkinson and Liddell.

Let t_c be the cycle, the time required to fetch an operand (from memory or write a result to

memory), t_i be the time required to perform integer arithmetic, t_L be the time required to perform a logical test, and t_{FM} be the time required to perform a floating point multiply.

Let us examine what is involved in the following program:

$$\text{DO } 1 \; I = 1, N$$

$$1 \quad Z(I) = X(I)*Y(I)$$

Each pass through the loop requires that we

1.   fetch two operands from memory: $X(I)$ and $Y(I)$,

2.   floating point multiply the operands,

3.   store the result, $Z(I)$, in memory,

4.   increment an integer counter $I \leftarrow I + 1$,

5.   logically test the counter to see if it exceeds N.

The time, T, required to perform (1)–(5) is

$$T = 3t_c + t_i + t_L + t_{FM} \qquad (59)$$

If the computation of T is limited by the floating point computation, then we might opt to invest in a high-speed floating point coprocessor, that is, we would seemingly be attracted to a device that had a high $1/t_{FM}$, a large number of megaflops. But what really happens? Let us assume that the t_{FM} before speedup is t_{FM1} and t_{FM2} after speedup. We would be tempted to believe that we had reduced our program's running time by

$$t_{FM2}/t_{FM1} \qquad (60)$$

Unfortunately this is not so. Let us designate T_1 as the value of T in (59) before floating point speedup and T_2 as the value of T after floating point speedup. The true ratio of times is then

$$\frac{T_2}{T_1} = \frac{3t_c + t_i + t_L + t_{FM2}}{3t_c + t_i + t_L + t_{FM1}} \qquad (61)$$

If, for example $t_c = 200$ nsec, $t_i = 100$ nsec, $t_L = 100$ nsec, and $t_{FM1} = 1200$ nsec, $t_{FM2} = 200$ nsec, the time is reduced by only 50%, not by one-sixth which is the ratio between t_{FM2} and t_{FM1}. It is the "overhead" that accompanies floating point operations that diminishes the ratio. This concept of overhead is crucial and will be seen again when we consider parallel architectures.

Note that even if t_{FM2} were to vanish (become zero) in (61), the results would be scarcely more dramatic. This analysis led one group of researchers to opine that "MFLOPS seems to be more popularly used as a measure for systems whose floating point hardware is so powerful that the run time is dominated by all the other terms!''

But computers are not the only way to process signals. A discovery by Lord Rayleigh, published in 1885, has great significance to contemporary signal processing—especially that processing that must be done in "real time." Rayleigh's work concerned the existence of a mode of elastic wave propagation, at most a few wavelengths deep, along the surface of a solid. These Rayleigh, or surface acoustic, waves (SAW) can be used to process analog signals over impressive bandwidths and dynamic ranges.

The usefulness of applying SAW techniques stems from the simple observation that acoustic waves travel much more slowly than electromagnetic waves in free space; typically on the order of one one-hundred thousandth the velocity. Thus, an entire pulse train, which, if sent as an electromagnetic wave through space, might stretch for a great distance, if converted by a transducer to a surface acoustic wave might be entirely captured on a small SAW device effectively serving as a delay line.

By inserting transducers along the SAW medium, it is possible to perform all sorts of signal-processing functions and this is where SAW devices have become so important. The transducer elements can be laid along the SAW to provide many useful impulse responses. Thus the SAW can perform convolution, or cross-correlation, and matched filters, both fixed and programmable, are one of the most important applications of SAW technology that is to be found fulfilling many signal-processing tasks especially in communications and radar processing.

Another separate technology that is gaining in importance for signal processing is optical processing. Optical processing will become increasingly important for those special cases that require enormous time-bandwidth products and real-time computational throughputs of relatively limited precision. Optical processing is so attractive because (1) the computations are done at the speed of light and (2) in parallel. [See OPTICAL INFORMATION PROCESSING.]

The cornerstone of optical processing is due to the remarkable property that a lens can aid in performing a Fourier transform. In a very cogent and straightforward manner, Stark has shown the relevant mathematics that we summarize. Assume that a monochromatic plane wave (such as produced by a laser) is incident

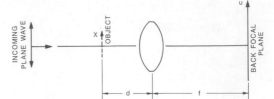

FIG. 17. A lens performing a Fourier transform.

upon an object laying in the wave's plane. The diagram in Fig. 17 depicts the object, the lens, and the position of the lens' back focal plane (f spatial units to the right of the lens' center). We need consider a one-dimensional object only. (The extension to two dimensions is straightforward.) The x-coordinate is assigned to measure spatial distance in the object's dimension and the u-coordinate for spatial frequency in the back focal plane. Assuming that the object's transmittance function is $t(x)$, the free space wavelength of the plane wave is λ, and that the lens aperture has a pupil function $P(\ )$, Stark shows that the complex amplitude of the optical radiation at the back focal plane is proportional to

$$\frac{1}{\sqrt{j\lambda f}} \exp\left[j\pi \left(1 - \frac{d}{f}\right) u^2 \lambda f \right]$$

$$\cdot \int_{-\infty}^{\infty} t(x)P(x + \lambda ud)e^{-j2\pi ux}\, dx \qquad (62)$$

Stark points out the following:

1. If the object is set at the lens' front focal plane, that is, at $d = f$, then (62) becomes the traditional Fourier transform integral. Note that the pupil function serves as a spatial "window" in the same manner as the time windows we looked at earlier.

2. If the object is set against the lens, that is, $d = 0$, then the pupil will have no effect if the object's dimensions do not exceed the lens' aperture. The phase factor (preceding the integral) will not vanish as it did for the case in which $d = f$, however, for this case the pupil will not affect the transform by attenuating the higher frequencies, a phenomenon known as "vignetting." The phase factor is of no consequence, of course, if one wishes to measure only the power spectrum, the square of the magnitude of (62).

The possibilities for signal processing implied by (62) are enormous. A simple and rather ingenious application was a device, constructed by a company for sale, that examined needles to determine if they were sufficiently sharp. As the degree of sharpness was reflected in the needle's surface shape, it was reasonable to expect that the Fourier transform of the shape could serve as a discriminant. This is exactly what the company's device did. The needles were passed by a lens that aided in taking the Fourier transform. Needles that were too dull exhibited a Fourier spectrum that had larger low spatial frequency components than needles that were sufficiently sharp.

More sophisticated and powerful concepts are also possible with relatively simple optics. For example, by placing a "Fourier transform mask"—an object whose transmittance function is itself the Fourier transform—of a scene at the back focal plane of Fig. 17 and then passing the transmitted light through another lens, it is possible to perform the convolution, or, equivalently, the cross-correlation of two scenes. An immediate application of this would be to pattern recognition in such uses as scene template matching.

We should remember that it is often through the synergy that is obtained on melding two distinct and separate technologies that we realize the true quantum leaps in any discipline. The future seems especially promising for the interplay of acoustics and optics, and much progress has already been made, but the state-of-the-art is such that electronic computers will most likely dominate as the tool of choice for signal processing for at least another decade.

IV. Contemporary Signal-Processing Topics

A. DECONVOLUTION

In an earlier section we discussed the convolution of two signals, $x(n)$, the input signal, and $h(n)$, the unit sample response. In most measuring systems, the data are recovered through a measuring device. That means what we have is not the actual data, but the data convolved with the unit sample response of the measuring device. Obviously, to get the true data we need to get rid of the measuring device's unit sample response. This process is usually referred to as deconvolution. In some applications, for example, speech processing, $x(n)$ is assumed to be known and the vocal tract transfer function needs to be determined. In either case, we need

to deconvolve the received data to find $x(n)$ or $h(n)$. This is discussed next for the case of solving for $x(n)$. The other case can be dealt with similarly.

The deconvolution problem can best be stated in terms of matrices. Given the following system of equations [see Eqn. (51)]

$$\mathbf{y} = H\mathbf{x} \qquad (63)$$

where $\mathbf{y}$ is an $M = (N + L - 1)$-dimensional vector, H a rectangular matrix of dimension $(N + L - 1) \times L$ and of rank L, and $\mathbf{x}$ an L-dimensional vector, find $\mathbf{x}$ with the assumptions that H and $\mathbf{y}$ are known. If there is a unique solution, then it is very easy to solve, as we need to solve a set of L independent equations. Unfortunately, the problem is not that simple in practice, since the received vector is not the same as $\mathbf{y}$ as it is corrupted by noise. That is, the received vector, say $\mathbf{d}$, is

$$\mathbf{d} = \mathbf{y} + \mathbf{n} \qquad (64)$$

where $\mathbf{n}$ is some noise vector. In most applications the noise is assumed to be additive. Now the problem is to minimize the difference between the desired vector $\mathbf{y}$ and the received vector $\mathbf{d}$ in some sense. The most common measure is an L_p measure, which is discussed next.

Let y_i and d_i be the ith respective components of the M-dimensional vectors $\mathbf{y}$ and $\mathbf{d}$. Furthermore, let

$$E = \sum_{i=0}^{M-1} |d_i - y_i|^p \qquad (65)$$

where $1 \le p \le 2$ is of most interest, and the error E is referred to as L_p based. In vector notation, we have

$$E = ||\mathbf{d} - \mathbf{y}||_p = ||\mathbf{d} - H\mathbf{x}||_p \qquad (66)$$

When $p = 2$, the solution is referred to as the least squares or L_2 solution, and for other ps, we simply identify them as L_p solutions. The least-squares solution or the L_2 solution is the simplest to compute and can be determined by setting

$$dE/dx_i = 0, \qquad i = 0, 1, 2, ..., L - 1 \quad (67)$$

and solving for the unknowns. The solution can be expressed in a nice compact form, and is

$$\mathbf{x} = (H^\mathrm{T} H)^{-1} H^\mathrm{T} \mathbf{d} \qquad (68)$$

The matrix $(H^\mathrm{T} H)^{-1} H^\mathrm{T}$ is sometimes referred to as the pseudo-inverse of the rectangular matrix H. Next let us consider other L_p solutions.

Unfortunately there is no simple solution as in (68) for other values of p. There are however, iterative techniques for obtaining L_p solutions. The convergence of these iterative algorithms is assured for $1 \le |p| \le 2$ and we assume that in the following.

1. Iteratively Reweighted Least-Squares (IRLS) Algorithm

The IRLS algorithm solves a set of overdetermined system of equations (Byrd and Payne)

$$H\mathbf{x} = \mathbf{d} \qquad (69)$$

by iteratively computing

$$\mathbf{x}(k + 1) = [H^T W(k) H]^{-1} H^T W(k) \mathbf{d} \qquad (70)$$

where $W(k)$ is a diagonal matrix with its diagonal entries, $W_{ii}(k)$, given by

$$W_{ii}(k) = \begin{cases} |r_i(k)|^{p-2}, & |r_i(k)| > \varepsilon \\ \varepsilon^{p-2}, & |r_i(k)| \le \varepsilon \end{cases} \qquad (71)$$

where $r_i(k)$ corresponds to the ith entry in the residual vector

$$\mathbf{r}(k) = \mathbf{d} - H\mathbf{x}(k) \qquad (72)$$

and ε is some small positive number. The IRLS algorithm is computationally intensive, as we need to find the inverse of the matrix $[H^T W(k) H]$ of order L at each stage. A simpler algorithm is the residual steepest descent algorithm (Huber and Dutter; Yarlagadda, Bednar, and Watt). This is discussed next.

2. Residual Steepest Descent (RSD) Algorithm

The RSD algorithm solves (69) by iteratively computing

$$\mathbf{x}(k + 1) = \mathbf{x}(k) - \Delta_k (H^T H)^{-1} H^T \boldsymbol{\gamma}(k) \qquad (73)$$

where

$$\boldsymbol{\gamma}(k) = \text{col}[\gamma_1(k)\gamma_2(k) \cdots \gamma_M(k)] \qquad (74)$$

with

$$\gamma_i(k) = |(Hx(k) - \mathbf{d})_i|^{p-1} \text{ sgn}(H\mathbf{x}(k) - \mathbf{d})_i \qquad (75)$$

where $\text{sgn}(t) = +1 (-1)$ if $t > 0 (t < 0)$. When $t = 0$, one can arbitrarily choose $\text{sgn}(t)$ to be either $+1$ or -1. The scale factor Δ_k in (73) is determined by minimizing

$$E(k + 1) = \| - \mathbf{d} + H\mathbf{x}(k)$$
$$- \Delta_k H(H^T H)^{-1} H^T \boldsymbol{\gamma}(k) \|_p \qquad (76)$$

with respect to Δ_k in the L_p sense. Note that Δ_k is a scalar and fast convergence can be obtained by using the IRLS algorithm to compute Δ_k.

In the case of the L_p based deconvolution, $1 \le p \le 2$, values of p close to 1 produce deconvolution filters with less sensitivity to aberrant noise than when p is close to 2. L_1 deconvolution is generally considered more robust than L_2 deconvolution.

B. Image Transforms and Their Use in Image Coding

Let P be a square matrix of dimension $2^n \times 2^n$. We will define the transform of P as

$$B = A^T P A \qquad (77)$$

where A is any nonsingular matrix. Normally the matrix describing an image, the image matrix, consists of integers with the integers identifying the image brightness of the picture elements or pixels (small regions of the image). The integers range from 0 to $2^L - 1$, where L depends on the system. For our analysis, we will assume that the entries of P, p_{ij}, are from the real field, even though the entries in P are integers. Some examples of A in (77) are the discrete Fourier transform matrix, Hadamard matrix, etc. In the following, we will use an interesting transform operator that will be useful for image coding. Consider

$$A = \alpha I + (1 - \alpha)H_n \qquad (78)$$

where H_n is the Walsh–Hadamard matrix defined earlier, $0 \le \alpha \le 1$ and $\alpha \ne 1/2$. In a way the transform in (77) with A given in (78) is a hybrid transform between the image P and the transformed image. Substituting (78) in (77) and expanding, we have

$$B = \alpha^2 P + (1 - \alpha)^2 H_n P H_n$$
$$+ \alpha(1 - \alpha)(P H_n + H_n P) \qquad (79)$$

There are three parts in this expression. The first term is the weighted original image. The second term is the weighted Hadamard transform. The third term is a cross-product term. Interestingly, the inverse of A in (78) is ($\alpha \ne 1/2$)

$$A^{-1} = \frac{1}{(2\alpha - 1)} [\alpha I - (1 - \alpha)H_n] \qquad (80)$$

indicating that the inverse transform can be obtained with almost the same structure as the forward transform.

For a significant range of α below unity, the

FIG. 18.   The original image.

FIG. 20.   The jammed NPT.

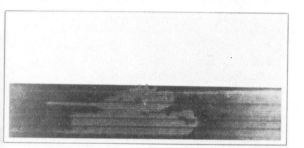

FIG. 19.   The NPT of the original image.

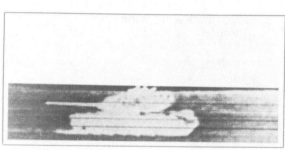

FIG. 21.   Restoration by 50 iterations.

sense of the scene is interpretable, the features are recognizable, but there is a large Hadamard component. These two properties allow for a unique approach to error correction and/or image compression. Since the scene is interpretable for a range of α, we refer to this transform as the naturalness preserving transform (NPT). Assume that a section of B in (79) is lost or discarded. We can recover P by minimizing the texture noise. Texture is hard to define. It is the feature of an image or subimage that is easy for a human to identify but difficult for a machine. We generally speak of the texture of sand, speckled tile, etc. The simplest texture is of course when all the pixel values are the same. This may be the case for a still lake or a cloudless sky.

As an example, consider the Figs. 18–21 (Yarlagadda and Hershey). Figure 18 is the original image (P). The image is 256×256 pixels with each pixel capable of 256 gray levels. Figure 19 is the transformed image [B in (17) with $\alpha = 0.9$]. Note the scene is still interpretable. Figure 20 has resulted from the opposition's jamming a high interest portion of the transmission imagery. We denote the jammed NPT by B_0. We note that the sky appears to be cloudless and therefore have a subimage that has a simple texture. If we do not know the exact pixel values, we can assume a statistical average, a constant texture for the sky. The concept is to reconstruct the image P from the jammed image B_0 with a knowledge that a constant texture image segment exists in the original image P.

The solution of the problem is obtained by iteratively computing at the ith stage (Yarlagadda and Hershey):

$$P_i = A^{-1}\hat{B}_{i-1}A^{-1}$$
$$B_i = A\hat{P}_i A \tag{81}$$

where $\hat{B}_i$ and $\hat{P}_i$ are obtained from B_i and P_i by substituting the known values from B_0 and P.

That is, the nonjammed values in B_0 will be substituted at each stage in B_i to get B_i and the known values in P (for example, the pixel values corresponding to the constant texture image segment) are substituted in P_i to get $\hat{P}_i$. The algorithm converges and the result after 50 iterations is shown in Fig. 21, which clearly indicates the reconstruction of the jammed image portion.

A few comments are in order as far as the computational complexities. First, the jammed subimage in Fig. 20 consists of 57×256 pixels. If a more common, direct approach were used, such as least squares, we would need to solve $256 \times 57 = 14{,}592$ unknowns in 32,768 equations corresponding to sky pixels. Obviously, the direct approach is computationally intensive. The proposed iterative method involves the computation of Hadamard transforms of vectors of dimension 256. It has been shown that by using a powerful array processor, the proposed method is easily implementable and can operate near real time.

V. Thoughts on the Future

It is presumptuous, at best, to imagine that one can predict, with any certainty, the future of such an encompassing discipline as signal processing. It seems safe, however, to expect continuing advances in almost all of the areas with, perhaps, the greatest gains in the computational aspects.

Looking back over the history of automatic data processing, we see that hardware dominated the cost of computation in the 1950s and early 1960s. In the latter 1960s, the 1970s, and up through the early 1980s, the dominant cost of computation was software. But, it now seems probable that the next high cost item will be new computer architectures, their control and calculational flows.

As we look through the history of signal-processing algorithms over the past two decades, we come to suspect that the truly phenomenal reductions in work realized by, say, the fast Fourier transform, and other algorithms in those halcyon early days, will probably not be repeated on anything near as significant a scale. What will be breathtaking to watch, however, will be the immense computational power and attendant signal-processing capability available through the intelligent lashing together and control and management of parallel computational structures employing thousands, and eventually millions, of individual processing nodes.

Computers developed around what has become known as the VonNeuman architecture. This simple architecture featured I/O units, a memory unit, a control unit, and an arithmetic and logic unit (ALU). The so-called "bottleneck" in the design arose not from the above modularization of functions but from the practice of passing all of the I/O operations through the ALU. This, of course, halted any computations until the I/O operations were finished. This architecture and data flow were preeminent until 1958.

In the mid-1950s, the Direct Memory Access (DMA) concept began to emerge which freed up the ALU but the controller was still held responsible for I/O control and this still kept the ALU from operating at top, theoretical efficiency. This situation was not remedied until an independent I/O channel architecture was developed and commonly integrated into computer design.

The idea of using more than one processor concurrently developed over a relatively long time period, that is, it is not at all a new idea. As early as 1958, Unger proposed the concept that computer architecture can be designed to better fit some problems:

> A general purpose digital computer can, in principle, solve any well defined problem. ... However, they are relatively inept at solving many problems where the data is arranged naturally in a spatial form. ... It appears that efficient handling of problems of the type mentioned above cannot be accomplished without some form of parallel action.

Unger's seminal concept has seen realizations from the early work done by Slotnick and others with the SOLOMON computer to today's COSMIC CUBE at the California Institute of Technology.

Some order, in the form of a hierarchical model of computer organizations, was brought about by Flynn in 1972. He made popular the terms that have stuck to the present time and have been adopted worldwide. The terms are best known by their acronyms and are as follows:

1. SISD: $single$-$instruction$ stream—$single$ $data$ stream: this is the simplest architecture—a single processor.
2. SIMD: $single$ $instruction$ stream—$multiple$ $data$ stream: there are three basic types of computer under this rubric.
   2a. Array (of) processors: typically an array of processors has one control

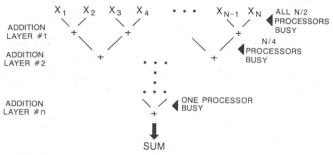

FIG. 22.   Parallel computation of the sum of $N = 2^n$ numbers.

unit and many independent and identical processors. The processors have their own local memory and are each simultaneously driven by the same instruction from the control unit.

2b. Pipelined processors: execution of operations by time interleaving the component parts of the operation among subprocessor elements tailored to the operation components.

2c. Associative processors: a type of array processor in which the processing elements are not directly addressed but rather impressed into service as needed.

3. MIMD: *M*ultiple-*i*nstruction stream—*M*ultiple *d*ata stream: this is essentially an interacting set of more than one processor. The processors are driven by different programs and they may share the same memory. They may be either "loosely" coupled, "tightly" coupled, or something in between.

The lure of multicomputer processing can be appreciated by looking at an example of what can be done with more than one CPU. Our first such example, for motivational purposes, is the conceptually simple task of adding together a list of numbers.

Let us assume, for convenience, that we have $N = 2^n$ numbers to add together, that is,

$$s = x_1 + x_2 + \cdots + x_{N-1} + x_N \qquad (82)$$

If we use a single processor, we will, of course, have to make $N - 1 = 2^n - 1$ additions. If the time required for a single processor to perform an addition of two numbers is T_A, then the single processor will take $(2^n - 1)T_A$ time units. The single processor will be kept busy for the entire computational time, that is, there will be no idle time in which computational power will be un-

used. If, on the other hand, we use $N/2 = 2^{n-1}$ processors, we can add all the numbers in $\log_2 2^n = n$ time units. How this is done is shown in Fig. 22. By using 2^{n-1} processors we have achieved a speed advantage of $(2^n - 1)/n$ over a single processor. There are a few points to make:

1. The speed advantage is $\approx 2^n/n$ which is less than 2^{n-1} for moderately sized n. The time for the parallel computation is *not* reduced in direct proportion to the number of processors.

2. When multiple processors are used on this problem, there is much idle processor time, for example, at addition layer #n only one of the 2^{n-1} processors is busy.

Being a bit more general, if we use $N_p = 2^k$ processors where $\max(N_p) = 2^{n-1}$, we can show that the time to perform the sum s of (82) is

$$T_s = 2^n/N_p + \log_2 (N_p - 1) \qquad (83)$$

and the total processor idle time, which is the sum of the number of processors idle at each step time in forming the sum, is

$$T_i = N_p[\log_2 (N_p - 1)] + 1 \qquad (84)$$

We graph T_s and T_i in Fig. 23 for the case for which $n = 10$ and $N = 2^{10} = 1024$.

How well parallel computation works—how much time it is capable of saving—depends almost exclusively on the algorithm used and its implementation. Ware has concocted a model that is very helpful in understanding and appreciating this concept. The "speedup," S, which obtains by using P processors in parallel is defined as

$$S = \frac{\text{Problem execution time for one processor}}{\text{Problem execution time for } P \text{ processors}}$$

$$(85)$$

Ware's model introduces the parameter α which is defined as the proportion of work that can be done in parallel. Ware's model assumes that at

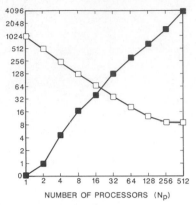

FIG. 23. T_s ($\square$) and T_i ($\blacksquare$) for $N = 2^{10} = 1024$.

any time instant all P processors are busy or only one processor is busy. Normalizing the problem execution time for one processor to unity we have

$$S = \frac{1}{(1 - \alpha) + \alpha/P} \tag{86}$$

In Fig. 24 we graph S against P for significant values of α. Note that α must be very close to unity to get anywhere near the full contribution of all P processors. Indeed, as $P \to \infty$, $S \to 1/(1 - \alpha)$. Buzbee introduced an additional term into Ware's model. This new term, $\sigma(P)$, reflects the increased "overhead" of managing parallel algorithms such as synchronization and interprocessor communications. With this new term, we have the following expression for S in the "Ware–Buzbee" model for parallel computation:

$$S = \frac{1}{(1 - \alpha) + \alpha/P + \sigma(P)} \tag{87}$$

Now, even if $\alpha = 1$, max(S) is upperbounded by $1/[1/P + \sigma(P)]$.

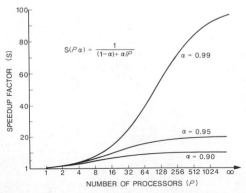

FIG. 24. S versus P for Ware's model of parallel computation.

As an example of an important process that can be paralleled in a number of ways and as an example of the importance of interprocessor communication times, let us consider the distributed computation of the radix-2 fast Fourier transform (FFT) as developed by Bhuyan and Agrawal. Consider that we wish to take the FFT of an N time-sampled waveform. For convenience, assume that N is a power of two, $N = 2^n$. Let us assume that the basic unit of FFT computation, the "butterfly," takes B time units to perform. The time to perform the $N = 2^n$ point FFT on a single processor, T_{sp}, is then

$$T_{sp} = nNB/2 \tag{88}$$

Let us now allow $P = 2^m$ processors to work on the FFT in parallel and achieve interprocessor communications by transferring data from processor to processor through a main memory as shown in Fig. 25. Let us assume that a processor to processor transfer through main memory requires 2τ time units. Then the time to perform the FFT on the architecture of Fig. 25 is

$$T_{pmm} = (nNB/2P) + mN\tau \tag{89}$$

The speedup, S, for this configuration is then

$$S = \frac{T_{sp}}{T_{pmm}} = 1 \bigg/ \left(\frac{1}{P} + \frac{2m\tau}{nB} \right) \tag{90}$$

Looking at (90) in terms of the Ware–Buzbee model we see that

$$\alpha = 1 \tag{91a}$$

and

$$\sigma(P) = (2\tau/nB) \log_2 P \tag{91b}$$

It is thus clear that the dimensionless τ/B is critical to the effective parallelism achieved. We are thus beginning to see that interprocessor communication is an extremely important issue. This leads to the consideration of other architectures such as that shown in Fig. 26. In this schema, the data are not transferred from processor to processor through a main memory but rather directly through an interconnecting network. Interconnecting networks continue to be

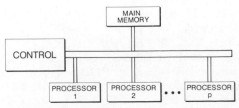

FIG. 25. Multiprocessors sharing a bus.

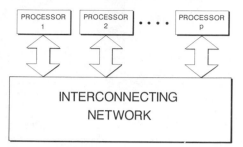

FIG. 26. Multiprocessors with interconnection network.

an extremely important topic for multiprocessor research and they bring their own complications. By necessity, an interconnecting network must, generally, be capable of executing some impressive combinatorial feats. One commonly used interconnection network is known as the "shuffle-exchange" network. It is a natural for data transfer for the FFT as it is intimately related to the butterfly structure.

However, the shuffle-exchange network takes some time and the network itself requires physical area in the realization medium. All of these tradeoffs must be recognized and appropriately weighted before settling on the final design of a signal processor.

We have caught only a glimpse of the future possibilities afforded by parallel processing. Not even all of the questions are known, and those that are, are not all yet properly phrased. But still we can sense the power that lies ahead and it promises to be as exciting a future as were the "good old days."

BIBLIOGRAPHY

Bhuyan, L., and Agrawal, D. (1982). Applications of SIMD computers in signal processing. *Nat. Comput. Conf., AFIPS Conf. Proc., Houston,* 135–142.

Byrd, R. H., and Payne, D. A. (1979). Convergence of the Iteratively Reweighted Least Squares Algorithm. Tech. Rep. 313, The John Hopkins Univ, Silver Spring, Maryland.

Flynn, M. J. (1972). Some computer organizations and their effectiveness. *IEEE Trans. Comp.* **C-21**(9), 948–960.

Hall, E. L. (1979). "Computer Image Processing." Academic Press, New York.

Huber, P. J., and Dutter, R. (1974). Numerical solution of robust regression problems. *In* COMSAT 1974, "Proc. Symp. Computational Statistics" (G. Breechmann, ed.). Physica Verlag, Wien, Germany.

Lee, S. H. (ed.) (1981). "Optical Information Processing." Springer-Verlag, Berlin and New York.

Oppenheim, A. V. (1978). "Applications of Digital Signal Processing." Prentice-Hall, Englewood Cliffs, New Jersey.

Oppenheim, A. V., and Schafer, R. W. (1975). "Digital Signal Processing." Prentice-Hall, Englewood Cliffs, New Jersey.

Papoulis, A. (1977). "Signal Analysis." McGraw-Hill, New York.

Rabiner, L. R. (1975). "Theory and Application of Digital Signal Processing." Prentice-Hall, Englewood Cliffs, New Jersey.

Robinson, E. A., and Treitel, S. (1980). "Geophysical Signal Analysis." Prentice-Hall, Englewood Cliffs, New Jersey.

Slotnick, D. L., Borck, W. C., and McReynolds, R. C. (1962). The SOLOMON computer. *AFIPS Conf. Proc.* **22**, 97–107.

Special Issue on Seismic Signal Processing. (1984). *Proc. IEEE* **72**.

Stark, J. (1982). "Applications of Optical Fourier Transforms." Academic Press, New York.

Unger, S. H. (1958). A computer oriented toward spatial problems. *Proc. I. R. E.* (*USA*) **46,** 1744–1750.

Yarlagadda, R., and Hershey, J. (1985). A naturalness preserving transform for image coding and reconstruction. *IEEE Trans. ASSP* **ASSP-33**, 1005–1012.

Yarlagadda, R., Bee Bednar, J., and Watt, T. (1985). Fast algorithms for L_p deconvolution. *IEEE Trans. ASSP* **ASSP-33,** 174–182.

SOURCE CODING, THEORY AND APPLICATIONS

N. Farvardin   *University of Maryland*

GLOSSARY

Channel: Medium through which information is transmitted.

Channel capacity: Quantity, associated with a channel, that determines the largest number of bits that can be transmitted through the channel reliably.

Quantization: Mapping (one-dimensional or multidimensional) from a continuous alphabet to a finite alphabet.

Rate-distortion function: Function $R(D)$, associated with a source, that represents the smallest amount of information necessary to represent the source with an average distortion less than or equal to D.

Source: Information that is to be transmitted to the destination.

Source alphabet: Set of possible source output values.

Data compression is the branch of communications that deals with the problem of representing a source of information with as few bits as possible while preserving a certain degree of fidelity. The foundations of data compression were laid in Shannon's original work on source coding with a fidelity criterion in which he established the relationship between the rate-distortion function of the source and the capacity of the channel in a point-to-point communication problem. Data compression schemes, also referred to as source coders or waveform coders, are now being used in various information transmission (storage) situations in which bandwidth (memory) is limited.

I. Introduction

The ever growing demand for transmission and storage of data necessitates the more efficient use of existing transmission and storage facilities. A source coding (data compression) system is a scheme that operates on source data to remove redundancies so that only those values essential to reproduction are retained. For a given source of information, the optimum performance theoretically attainable by any source coding scheme is given by the source rate-distortion function. Specifically, for a given source, the rate-distortion function $R(D)$ determines the smallest amount of information necessary to represent the source with an average distortion less than or equal to D. [*See* DATA COMMUNICATION NETWORKS.]

In any practical situation, given the source probabilistic description, one would try to design the source coding scheme in such a way as to obtain performance as close to the source rate-distortion function as possible. Indeed, practical considerations impose certain serious limitations on the system complexity. In any case, as in other real-world problems, there exists a tradeoff between complexity and performance. [*See* SIGNAL PROCESSING, GENERAL.]

In this article, we provide an overall picture of the developments on the theoretical and practical aspects of source coding. To the extent possible, we avoid mathematical details. In particular, the main theorems are stated without proof. The proofs of the theorems can be found in various text books including a few cited in the bibliography at the end of this manuscript. Particular emphasis is placed on providing motivation and intuitive feeling for the different definitions and theorems described herein. Moreover, we describe a few of the more popular practical source coding schemes and provide a performance comparison against the optimum performance theoretically attainable that is described by the source rate-distortion function.

II. Coding Theorems: Theoretical Limits

In this section, we provide a precise formulation of the data compression problem and the basic definitions necessary for subsequent developments. This is followed by the statement of several coding theorems that help shed light on the operational significance of the source entropy, the channel capacity, and the source rate-distortion function.

We start with a discussion of noiseless source coding and the operational significance of the source entropy (or entropy rate) and then extend these results to source coding with a fidelity criterion and fundamental performance limitations dictated by the source rate-distortion function. While most of the results stated herein are for discrete-alphabet sources, at the end of this section extensions to continuous-alphabet sources are also considered.

A. Source Description

Let us assume that the information source is modeled as a discrete-time discrete-alphabet random process $\{U_n\}$ taking values from the alphabet $A_U = \{a_1, a_2, ..., a_A\}$. Here, n is simply a time index. Let us assume for the present time that $\{U_n\}$ is a sequence of independent and identically distributed (i.i.d) random variables with probabilities $P_i \triangleq P_r (U = a_i)$, $i = 1, 2, ..., A$. This source is commonly referred to as a discrete memoryless source (DMS).

The self-information associated with the event $\{U = a_i\}$, $i = 1, 2, ..., A$, is defined by

$$I(a_i) = -\log_2 P_i \quad \text{bits/source symbol} \quad (1)$$

The quantity $I(a_i)$ is clearly a random variable whose expected value, given by

$$H(U) = -\sum_{i=1}^{A} P_i \log_2 P_i \quad \text{bits/source symbol} \quad (2)$$

is known as the entropy of the DMS.

The self-information associated with an event is inversely proportional to the probability of oc-

currence of that event. Thus, less likely events contain more self-information, and vice versa. In a sense, the self-information of an event can be interpreted as the amount of "surprise" associated with the occurrence of that event. The entropy of a discrete memoryless source, which is the average value of self-information, can be interpreted as the amount of uncertainty associated with the source. Or, in other words, the entropy of the source is the amount of information that needs to be provided to remove the uncertainty in the source. For data compression of DMSs the source entropy has a very important operational significance. Specifically, the entropy of the DMS determines the smallest number of bits necessary for representing the source with an arbitrarily small probability of error. This fact is more precisely described through the following theorem.

Theorem 1 (Noiseless Source Coding Theorem for the DMS). Consider a discrete memoryless source U with finite entropy $H(U)$. Consider a binary block code that takes sequences of length N of the source output and maps them into blocks of length L of binary digits such that only one source sequence can be mapped into each code sequence. Let P_e denote the probability of occurrence of a source sequence for which there is no code sequence. Then for any $\varepsilon > 0$, there is a block code such that if

$$\frac{L}{N} \geq H(U) + \varepsilon \quad (3)$$

P_e can be made arbitrarily small by making N sufficiently large. Conversely, if

$$\frac{L}{N} \leq H(U) - \varepsilon \quad (4)$$

there is no code for which P_e can be made arbitrarily close to zero.

This theorem underscores the operational significance of the entropy of the DMS. Note that in Eqs. (3) and (4) the quantity L/N is the number of binary digits (bits) per source symbol necessary for representing those source sequences for which a code sequence (code word) exists.

Roughly speaking, the theorem establishes that the entropy of a DMS is a fundamental lower limit on the number of bits per source symbol necessary for exact representation of the source. The limit has an important practical implication because it provides us with a means of measuring the performance of a particular noiseless source coding scheme against the very best possible performance.

EXAMPLE 1.   Consider a memoryless binary source U with alphabet $A_U = \{0, 1\}$. Let $p = Pr \, (U = 0)$. Then the entropy of this source $H(U)$ is given by the binary entropy function as follows:

$$H_b(p) \overset{\Delta}{=} -p \log_2 p - (1 - p) \log_2(1 - p)$$

$$\text{bits/source symbol.} \qquad (5)$$

Now consider two cases. First, let $p = 0.5$. In this case the source is called a binary symmetric source. The source entropy assumes its largest possible value equal to $H(U) = 1$, bit/source symbol, which implies that the smallest number of bits necessary to represent this source is one bit per symbol. Second, consider the case where $p = 0.1$. In this case the entropy of the source is $H(U) = H_b(0.1) = 0.469$, bits/source symbol. That is, the smallest number of bits necessary to represent this source is even smaller than half a bit per source symbol. Of course, in this case a nontrivial coding operation is needed to represent this source with an average number of bits close to the source entropy. This is interesting because the size of the source alphabet is the same in both cases.

The preceding example shows that knowledge of the probability distribution of the source plays an important role in determining the smallest number of bits necessary for representing the source. This fact can be used advantageously in many communication or storage situations in which the symbols that the source generates are highly different in probability.

Before we extend our results to more complicated sources, let us consider another version of the noiseless source coding theorem that also has some practical implications. The main idea in the following theorem is that of variable-length coding. Here, we assign a binary code word to every source symbol (or sequence of source symbols), but the code words do not necessarily possess the same length. The code words are chosen so that source outputs with large probabilities are mapped to code words with small lengths while source outputs of small probabilities are mapped to code words of larger lengths in an effort to minimize the average code-word length. In using variable-length codes one has to be careful that the code can be decoded uniquely. That is, for a received sequence of code symbols there is only one decoded sequence. Variable-length codes satisfy-

ing these conditions are said to be uniquely decodable.

Theorem 2 (Variable-length Noiseless Source Coding Theorem for the DMS).   Consider a DMS U with finite entropy $H(U)$. Suppose a variable-length source coding scheme takes sequences of length N of the source output symbol, say $\mathbf{u}$, and maps them into a binary code word of length $l_N(\mathbf{u})$ for all possible sequences $\mathbf{u}$. Then, there exists a uniquely decodable variable-length code for which the average code word length described by

$$\bar{l}_N = \sum_{\mathbf{u}} Pr(\mathbf{u}) l_N(\mathbf{u}) \qquad (6)$$

satisfies

$$\frac{\bar{l}_N}{N} \leq H(U) + \frac{1}{N} \qquad (7)$$

On the other hand, no such code exists for which

$$\frac{\bar{l}_N}{N} < H(U) \qquad (8)$$

This theorem establishes the existence of a uniquely decodable variable-length code for which the average code length per source symbol can be made arbitrarily close to the source entropy by making N large. It is important to mention that there is a major distinction between this theorem and Theorem 1 in that in Theorem 1 there is always a subset of source output sequences for which no code word exists, while in Theorem 2 there is a code word for every possible source output sequence. Therefore, the block coding of Theorem 1 always implies a vanishingly small but nonzero probability of error, while the variable-length code of Theorem 2 does not introduce any coding error.

EXAMPLE 2.   Let us consider a DMS U with alphabet $A_U = \{a_1, a_2, a_3, a_4\}$ and probabilities $P_1 = 1/2$, $P_2 = 1/4$, $P_3 = 1/8$, and $P_4 = 1/8$. The entropy of this source is easily calculated to be $H(U) = 1.75$, bits/source symbol. Consider the variable-length code that maps a_1 to 0, a_2 to 10, a_3 to 110, and a_4 to 111. First, note that this variable-length code is uniquely decodable. This is because no code word is a prefix of any other code word. Because of this condition, during the decoding operation, the end of the code words can be easily marked. Variable-length codes that satisfy this condition are called prefix condition codes. For this example the average code word length is given by $\bar{l}_1 = 1.75$ bits/source sample, which is exactly equal to the entropy of the source in this case. Therefore, as one could see from this example, variable-length codes can be used to provide very efficient source coding.

In the past 30 years various algorithms have been developed for the construction of variable-length codes. Among the most important is the Huffman algorithm. For a discrete memoryless source with given letter probabilities, this algorithm constructs a variable-length code that is optimum in the sense that it minimizes the average code word length.

So far we have considered only memoryless sources. Now we extend our results to discrete stationary sources. This class of sources, which includes stationary sources with memory, is representative of many real-world sources and therefore is more important from a practical point of view.

Consider a discrete stationary source $\{U_n\}$ and let $\mathbf{U}_N$ be the generic notation for a sequence of N consecutive source output letters, say, $(U_1, U_2, ..., U_N)$. Let $H(\mathbf{U}_N)$ denote the entropy of a source whose output is the N-vector $\mathbf{U}_N$. Then the entropy rate of the discrete stationary source is defined as

$$H_\infty(U) = \lim_{N \to \infty} \frac{1}{N} H(\mathbf{U}_N) \qquad (9)$$

The entropy rate of a discrete stationary ergodic source has the same operational significance as the entropy of a discrete memoryless source. Specifically, there are coding theorems that establish that, for a discrete stationary ergodic source with entropy rate $H_\infty(U)$, there exists a code by means of which the source can be represented with arbitrarily small probability of error so long as the average number of bits per source sample is greater than $H_\infty(U)$. Conversely, no such code exists if the average number of bits per source symbol is below the source entropy rate.

In general, the computation of the entropy rate of an arbitrary discrete stationary source is a difficult task. However, for some special sources, the entropy rate can be computed quite easily. An important class of sources that is often used in modeling real-world processes is the class of Markov sources. Such sources are characterized by a set of states, say, $A_S = \{1, 2, ..., J\}$, and an output alphabet as before, say, $A_U = \{a_1, a_2, ..., a_A\}$. The probability of entering a new state depends only on the most recent state. We denote by S_n the state of the process at time instant n. Let us assume that the conditional probability of entering state i, given that the

most recent state was j, is denoted by $\alpha_{ji} \triangleq Pr(S_n = i | S_{n-1} = j)$ and that the probability of being in state i is denoted by $\pi_i \triangleq Pr(S_n = i)$ assuming that the Markov chain is irreducible and has reached the steady state. Then the following relationship is used to compute the state probabilities.

$$\pi_j = \sum_{i=1}^{J} \pi_i \alpha_{ij} \qquad j = 1, 2, ..., J \qquad (10)$$

Upon defining the conditional entropy of the source, given a particular state i, by

$$H(U|i) = -\sum_{k=1}^{A} P_i(a_k) \log_2 P_i(a_k)$$
$$i = 1, 2, ..., J \qquad (11)$$

where $P_i(a_k)$ is the probability that the source produces the output a_k when the state is i, it is easy to show that the source entropy rate is given by

$$H_\infty(U) = \sum_{i=1}^{J} \pi_i H(U|i) \qquad (12)$$

Equation (12) can be used to compute the entropy rate of a Markov source based on the conditional entropies of each state and the corresponding state probabilities.

Now that we have characterized the source and defined the source entropy rate, we describe the communication channel and the channel capacity. From a communication theoretic point of view the interaction between the source and channel and the resulting performance limitations are determined through the source entropy rate and the channel capacity. We elaborate more on this later.

B. Channel Description

We consider only discrete memoryless channels. A discrete memoryless channel (DMC) is a channel with an input alphabet $A_X = \{b_1, b_2, ..., b_I\}$ and an output alphabet $A_Y = \{c_1, c_2, ..., c_J\}$. At time instant n, the channel maps the input variable X_n into the output variable Y_n in a random fashion. The channel is uniquely determined by specifying the conditional probabilities,

$$q(j|i) \triangleq Pr(Y_n = c_j | X_n = b_i)$$
$$i = 1, 2, ..., I, j = 1, 2, ..., J$$

Since the channel is memoryless, the conditional probability of a sequence of channel outputs, given a sequence of channel inputs, can be decomposed into the product of the corresponding conditional marginal probabilities, that is,

$$Pr(y_1, y_2, ..., y_n | x_1, x_2, ..., x_n) = \prod_{i=1}^{n} Pr(y_i | x_i)$$

(13)

Definition 1. Consider two discrete random variables X and Y with alphabets $A_X = \{b_1, b_2, ..., b_I\}$ and $A_Y = \{c_1, c_2, ..., c_J\}$, respectively. Then the average mutual information between X and Y is defined by

$$I(X; Y) = \sum_{i=1}^{I} \sum_{j=1}^{J} P_i q(j|i) \log_2 \frac{q(j|i)}{Q_j}$$

(14)

where

$$P_i = Pr(X = b_i) \qquad i = 1, 2, ..., I$$

$$Q_j = Pr(Y = c_j) \qquad j = 1, 2, ..., J$$

and

$$q(j|i) = Pr(Y = c_j | X = b_i)$$

$$i = 1, 2, ..., I, j = 1, 2, ..., J$$

The average mutual information $I(X; Y)$ is a measure of the amount of "information" that the random variables X and Y provide about one another. Notice from Definition 1 that when X and Y are statistically independent, we have $I(X; Y) = 0$, which means that X and Y do not provide any information about one another. On the other hand, when $X = Y$, we have $I(X; Y) = H(X)$. That is, the amount of information that a random variable X provides about itself is equal to its entropy $H(X)$. This is consistent with our previous interpretation of entropy as a measure of uncertainty in the source.

Definition 2. The capacity of a DMC is defined as

$$C = \max I(X; Y) \qquad \text{bits/channel use} \quad (15)$$

where the maximization is taken with respect to all possible input distributions $\{P_i, i = 1, 2, ..., I\}$.

The operational significance of the channel capacity resides in the channel coding theorem stated below.

Theorem 3 (Channel Coding Theorem). Consider a discrete memoryless channel with capacity C bits/channel use and a discrete stationary source with entropy rate $H_\infty(U)$ bits/source symbol. Let us assume that the source generates one symbol every τ_s sec and that the

channel accepts one channel symbol every τ_c sec. Then, if $H_\infty(U) \le (\tau_s/\tau_c)C$, the output of the source cannot be transmitted over the channel with arbitrarily small probability of error.

Note that the quantity $(\tau_s/\tau_c)C$ is nothing but the channel capacity measured in bits/source symbol. From now on, we refer to this quantity as $\tilde{C}$. Theorem 3 plays a fundamental role in communication theory. It establishes the operational significance of the channel capacity as the rate of transmission below which reliable communication is possible and above which reliable communication is impossible. It is important to note that the theorem establishes the existence of a code that satisfies these conditions and does not provide any insight on the construction of such codes. The construction of implementable codes that guarantee reliable communication remains an open problem.

C. SOURCE CODING WITH A FIDELITY CRITERION

While the channel coding theorem establishes the existence of a code that achieves reliable communication when $H_\infty(U) \le \tilde{C}$, it provides no information as to the performance of the best codes when $H_\infty(U) > \tilde{C}$. All it says is that arbitrarily small error probability cannot be achieved. This question was first raised by Shannon 1949 and later solved by himself in 1959. The answer to this question lies in a branch of information theory called rate-distortion theory.

Rate-distortion theory enables us to determine the smallest amount of distortion that needs to be tolerated if indeed the source entropy rate $H_\infty(U)$ is larger than the channel capacity $\tilde{C}$. To be able to define the rate-distortion function of a source, we first have to define a distortion measure.

Let us consider a discrete source with alphabet $A_U = \{a_1, a_2, ..., a_A\}$ and a reproduction alphabet denoted by $A_V = \{d_1, d_2, ..., d_B\}$. Throughout this discussion we assume that the distortion between the source output u and its reproduction v is measured by a nonnegative, real-valued function $d(u, v), u \in A_U, v \in A_V$. This distortion measure is called a single-letter distortion measure because it does not depend on the past or future values of the source or the reproduction sequence. In what follows we provide a definition of the rate-distortion function for a DMS.

Definition 3. Let a DMS with alphabet A_U and probabilities $P_i, i = 1, 2, ..., A$, be given and consider a single-letter distortion measure $d(u, v) \geq 0$. Consider a conditional probability assignment $q(j|i) = Pr(V = d_j | U = a_i)$. This probability assignment results in an average mutual information between the source U and the reproduction V, denoted by $I(\mathbf{q}) \equiv I(U; V)$ and described by

$$I(\mathbf{q}) \equiv I(U; V)$$

$$= \sum_{i=1}^{A} \sum_{j=1}^{B} P_i q(j|i) \log_2 \frac{q(j|i)}{\sum_{k=1}^{A} P_k q(j|k)} \quad (16)$$

and an average distortion given by

$$D(\mathbf{q}) = \sum_{i=1}^{A} \sum_{j=1}^{B} P_i q(j|i) d(a_i, d_j) \quad (17)$$

Then the source rate-distortion function is defined as

$$R(D) = \min I(\mathbf{q}) \quad (18)$$

where the minimization is taken with respect to the set of all conditional probabilities $\mathbf{q} = \{q(j|i), i = 1, 2, ..., A, j = 1, 2, ..., B\}$ that satisfy

$$D(\mathbf{q}) \leq D \quad (19)$$

To describe the operational significance of the rate-distortion function of the source, we have to state a coding theorem. There are various versions of the source coding theorem subject to a fidelity criterion. Let us consider the block coding theorem because it is easier to understand. First, we describe the operation of a block code.

Consider a set $\mathbf{C}$, called the code book, consisting of M vectors $\mathbf{C} = \{\mathbf{v}_1, \mathbf{v}_2, ..., \mathbf{v}_M\}$, each vector consisting of N elements in the reproduction alphabet. For each sequence of N consecutive source outputs, say $\mathbf{u} = (u_1, u_2, ..., u_N)$, the encoder selects that code word $\mathbf{v}_i \in \mathbf{C}$ that results in the smallest distortion $d(\mathbf{u}; \mathbf{C})$ given by

$$d(\mathbf{u}; \mathbf{C}) = \min_{\mathbf{v} \in C} d(\mathbf{u}, \mathbf{v}) \quad (20)$$

where $d(\mathbf{u}, \mathbf{v})$ is used to denote the distortion between the vectors $\mathbf{u} = (u_1, u_2, ..., u_N)$ and $\mathbf{v} = (v_1, v_2, ..., v_N)$, described by

$$d(\mathbf{u}, \mathbf{v}) = \sum_{n=1}^{N} d(u_n, v_n) \quad (21)$$

Therefore, the average per-letter distortion associated with this block code is given by

$$D(\mathbf{C}) = \frac{1}{N} \sum_{\mathbf{u}} Pr(\mathbf{u}) d(\mathbf{u}; \mathbf{C}) \quad (22)$$

where $Pr(\mathbf{u})$ is the N-fold probability of the source, and the summation is taken over all possible sequences of length N at the source output. Furthermore, the rate of the code is defined by

$$R = \frac{1}{N} \log_2 M \quad (23)$$

The average per-letter distortion and the rate associated with a block code are the two important characteristics of the code. The following theorem determines the operational significance of the rate-distortion function through a block coding argument.

Theorem 4 (Source Coding Theorem with a Fidelity Criterion for the DMS). Consider a DMS and a bounded distortion measure $0 \leq d(u, v) \leq d_{max} < \infty$. Let us suppose that the source rate-distortion function is given by $R(.)$. Then, for a given D and for any $\varepsilon > 0$, there exists a block code $\mathbf{C}$ of rate $R \leq R(D) + \varepsilon$ such that $D(\mathbf{C}) \leq D + \varepsilon$. Conversely, it is impossible to achieve an average distortion less than or equal to D when the code rate R is smaller than $R(D)$.

The important practical consequence of Theorem 4 can be described as follows. If we are willing to accept an average distortion equal to D, then the smallest number of bits necessary for representing the source is given by $R(D)$. In general, for $D > 0$, clearly $R(D) < H(U)$. That is, if we are willing to tolerate some nonzero distortion, we should be able to represent the source with a number of bits less than the source entropy. This is the essence of data compression with a fidelity criterion.

Note that the rate-distortion function is only an attribute of the source and has nothing to do with the channel. To further characterize the rate-distortion function, let us define D_{min} as the minimum possible value that the average distortion can assume. Also, define D_{max} as the smallest value of the average distortion for which $R(D) = 0$. Then it is possible to show that for all $D \in (D_{min}, D_{max})$, the rate-distortion function is a continuous, strictly decreasing convex $\cup$ function. A typical rate-distortion function curve is depicted in Fig. 1. It should be noted that when the distortion measure is such that small average distortion corresponds to small values of the probability of error, at zero distortion (i.e., the intersection of the $R (\cdot)$ curve with the ordinate), the value of the rate-distortion function is the source entropy, the smallest number of bits necessary to represent the source with arbitrarily small error probability.

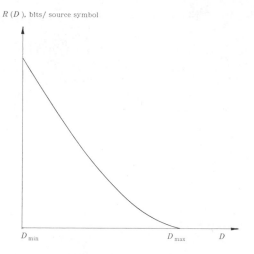

$R(D)$, bits/ source symbol

D_{min} D_{max} D

FIG. 1. Typical rate-distortion function curve.

Now we can go back to the question raised earlier as to what happens if we have a source whose entropy rate is bigger than the capacity of the channel. From the channel coding theorem it is clear that to be able to transmit reliably over a channel of capacity $\tilde{C}$, the rate R should be no bigger than $\tilde{C}$. Therefore, if the entropy of the source $H(U)$ is larger than the channel capacity, we can use a source encoder (ideal) to represent the source at a rate just below the capacity at the cost of some distortion given by the relationship $R(D) = \tilde{C}$. Now, the output of the source encoder can be considered as a new source whose entropy is below the channel capacity; hence it can be transmitted reliably over the channel. Therefore, the only source of distortion is that introduced by the source encoder. This observation is more rigorously summarized in the following theorem.

Theorem 5 (Combined Source and Channel Coding Theorem for the DMS and DMC). Consider a DMS with rate-distortion function $R(D)$, bits/source symbol, and a DMC with capacity C, bits/channel use. The source generates one symbol every τ_s sec, and the channel accepts one channel symbol every τ_c sec. Then, for every $\varepsilon > 0$, the source can be transmitted over the channel with an average distortion $D + \varepsilon$ where D is such that $R(D) < (\tau_s/\tau_c)(C - \varepsilon)$. Conversely, it is impossible to reproduce the source with an average distortion D if $R(D) > (\tau_s/\tau_c)C$.

EXAMPLE 4.   Consider a binary memoryless source U with alphabet $A_U = \{0, 1\}$ which outputs one symbol every $\tau_s = 1$ sec. The source

symbol probabilities are $Pr(X = 1) = 1 - Pr(X = 0) = \alpha$. The output of this source is to be transmitted over a binary symmetric channel with identical input and output alphabets $A_X = A_Y = \{0, 1\}$ and crossover probability $Pr(Y = 0|X = 1) = Pr(Y = 1|X = 0) = \beta$. The channel can transmit one channel symbol every $\tau_c = 0.5$ sec. We use the Hamming distance as our distortion criterion. Specifically, we assume that the reproduction alphabet is $A_V\{0, 1\}$ and that $d(u, v) = 0$ if $u = v$, and $d(u, v) = 1$ if $u \neq v$. The rate-distortion function of the source and the capacity of the channel are easily shown to be equal to

$$R(D) = H_b(\alpha) - H_b(D) \qquad \text{bits/source symbol}$$

$$\text{(24a)}$$

and

$$C = 1 - H_b(\beta) \qquad \text{bits/channel use} \quad \text{(24b)}$$

respectively. Now, let us assume that the source is symmetric (i.e., $\alpha = 0.5$). Then the first important question is whether it is possible transmit this source over this channel with arbitrarily small probability of error. The answer is yes only if the source entropy is less than the channel capacity. The source entropy, which is nothing but the value of the rate-distortion function at zero distortion, is equal to

$$H(U) = H_b(0.5) - H_b(0) = 1$$
$$\text{bit/source sample} \qquad \text{(24c)}$$

Thus, if $C > (\tau_c/\tau_s)H(U)$ or equivalently, $H_b(\beta) < 0.5$, reliable communication is possible. This corresponds to $\beta \lesssim 0.11$. The other question, which is very important, is the value of smallest distortion we should tolerate if in fact $\beta > 0.11$. This question is easily answered by equating the rate-distortion function to the channel capacity to get $H_U(D) = 2H_b(\beta) - 1$. For instance, if $\beta = 0.2$, the smallest distortion to be tolerated is $D \cong 0.092$.

We are now ready to extend the above results to more general sources, namely, stationary ergodic sources. Indeed, most real-world sources are best modeled by such sources. In general, a typical real-world source is a source that has memory, or in other words, its outputs are correlated. In such cases the advantage of source coding is much more noticeable. This is because, due to statistical dependence between the source outputs, the source can be reproduced with fewer bits than the corresponding memoryless source.

The study of source coding with a fidelity criterion for stationary ergodic sources is more complicated than that of the DMS. The first important issue is that of defining the rate-distortion function for such sources. In what follows we define the rate-distortion function of these sources.

Let us consider a discrete stationary ergodic source $\{U_n\}$ with alphabet A_U, a reproduction alphabet A_V, and a bounded distortion measure $d(u, v) \leq d_{max} < \infty$. Let $\mathbf{u} = (u_1, u_2, ..., u_N)$ and $\mathbf{v} = (v_1, v_2, ..., v_N)$ be N-vectors with elements in A_U and A_V, respectively. Then, consider a conditional transition probability assignment $q(\mathbf{v}|\mathbf{u})$. This conditional probability assignment results in an average mutual information between the source and the reproduction N-vectors denoted by

$$I(\mathbf{q}) \triangleq I(\mathbf{U}; \mathbf{V})$$

$$= \sum_{\mathbf{u}} \sum_{\mathbf{v}} Pr(\mathbf{u})q(\mathbf{v}|\mathbf{u}) \log_2 \frac{q(\mathbf{v}|\mathbf{u})}{Q(\mathbf{v})} \quad (25)$$

where $Q(\mathbf{v})$ is simply the reproduction probability induced by the transition probability $q(\mathbf{v}|\mathbf{u})$. Furthermore, an average distortion can be defined between the random N-vectors $\mathbf{U}$ and $\mathbf{V}$ by

$$D(\mathbf{q}) = \sum_{\mathbf{u}} \sum_{\mathbf{v}} Pr(\mathbf{u})q(\mathbf{v}|\mathbf{u})d(\mathbf{u}, \mathbf{v}) \quad (26)$$

Definition 4. The rate-distortion function of the stationary ergodic source described above is given by

$$R(D) = \lim_{N \to \infty} \frac{1}{N} R_N(D) \quad (27a)$$

where $R_N(D)$ is defined by

$$R_N(D) = \min I(\mathbf{q}) \quad (27b)$$

and the minimization is taken over all possible transition probability assignments that satisfy

$$\frac{1}{N} D(\mathbf{q}) \leq D \quad (27c)$$

Without repeating the source coding theorem for stationary ergodic sources, we mention here that the rate-distortion function of a stationary ergodic source has exactly the same operational significance as that of the DMS. Various source coding theorems exist to substantiate this.

At this point, we have covered all fundamental results that are relevant to source coding. Before we discuss some specific source coding schemes, there are several issues that deserve more careful consideration.

D. CONTINUOUS ALPHABET SOURCES

So far, we have considered only sources that possess a discrete alphabet for both noiseless source coding and source coding with a fidelity criterion. A large class of real-world sources are best modeled by continuous alphabet sources, that is, sources whose set of possible outputs is a continuum of values. Examples are speech and image signals. In what follows, we examine how the concepts of entropy, entropy rate, and rate-distortion function can be extended to sources with continuous alphabets.

First, consider a real-valued discrete-time continuous-alphabet memoryless source U, with alphabet $A_U \subset \mathbb{R}$. Let $f_U(\cdot)$ denote the probability density function of the random variable U. Then, if we consider an interval $(a, a + \Delta] \subset A_U$, its self-information is given by $-\log_2 Pr(a < U \leq a + \Delta)$, which approaches infinity as the length of the interval approaches zero. Roughly speaking, the amount of "surprise" associated with an event $\{U = a\}$ is infinity. Indeed, if we proceed and define the entropy of the source as the limit when Δ approaches zero of the average value of the above self-information term, we get infinity. This, in fact, makes sense. As we mentioned earlier, the entropy of the source is a measure of uncertainty or randomness in the source; for a continuous alphabet source it seems intuitively obvious that the source is "infinitely" random. In fact, for a continuous-alphabet source it seems reasonable that an infinitely large number of bits would be needed for an exact representation of the source.

Thus, for a continuous-alphabet source the entropy is not a well-defined quantity. Nevertheless, there exists a similar expression for continuous-alphabet sources, called the source differential entropy, defined by

$$h(U) = -\int_{-\infty}^{\infty} f_U(u) \log_2 f_U(u) \, du \quad (28)$$

While the differential entropy of the continuous-alphabet source carries over some of the properties of the entropy of the discrete source, there are some fundamental differences. First, the differential entropy of a source cannot be considered as a measure of uncertainty of the source. Second, the differential entropy is not necessarily finite, nor is it necessarily nonnegative. Third, in contrast with the entropy of a discrete source, the differential entropy is not invariant under coordinate transformation.

In light of the above discussion, the number of bits for an exact representation of continuous-alphabet sources is infinity. However, it is conceivable that when some average distortion can be allowed, the source can be represented by a finite number of bits per sample. Indeed, the rate-distortion function for a continuous-alphabet source can be defined in a manner similar to that for a discrete source and possesses the same operational significance. Specifically, for the above continuous-alphabet memoryless source, the rate-distortion function is defined as

$$R(D) = \inf I(q) \qquad (29a)$$

where the infinum is taken over the set of all possible conditional probability density functions from the source alphabet A_U to the reproduction alphabet A_V, denoted by $q(v|u)$ satisfying

$$D(q) \equiv \int_{-\infty}^{\infty} \int_{-\infty}^{\infty} f_U(u)q(v|u)d(u, v) \, du \, dv \leq D$$
$$(29b)$$

Here, $I(q)$ is used to denote the average mutual information between the source and the reproduction variates, defined by

$$I(q) = \int_{-\infty}^{\infty} \int_{-\infty}^{\infty} f_U(u)q(v|u)$$

$$\cdot \log_2 \frac{q(v|u)}{f_V(v)} \, du \, dv \qquad (30)$$

where $f_V(.)$ is the probability density function of the reproduction random variable V.

The rate-distortion function of the continuous-alphabet memoryless source defined above has an interpretation identical to that of the discrete source described before. Simply, for a given value of D, the corresponding $R(D)$ determines the smallest number of bits necessary to represent the source with an average distortion less than or equal to D. Extensions of the above definition to rate-distortion functions of stationary ergodic sources is essentially the same as that for memoryless sources and is omitted here.

E. Computation and Bounding of the Rate-Distortion Function

The problem of computing the rate-distortion function of a source is a formidable task. The problem is a complex nonlinear constrained optimization problem. Since 1959, when the rate-distortion function of a source was first introduced in Shannon's paper, various researchers have been involved in developing techniques for computing this function. To date, there is no known technique for exact computation of the rate-distortion function of a general source with an arbitrary distortion measure. In fact, what is known is very limited.

Let us consider memoryless sources first. Explicit expressions for the rate-distortion function of such sources is available only for some special cases, including the binary source with the probability-of-error distortion measure and the Gaussian source with the squared-error distortion measure. However, an iterative algorithm, referred to as the Blahut–Arimoto algorithm, exists which can be used for the numerical computation of the rate-distortion function. For sources with memory the problem is much more complex. The only case for which an explicit expression for the rate-distortion function is known is the stationary Gaussian source with the squared-error distortion measure. In this case the rate-distortion function is given in terms of the following parametric expression:

$$D_\theta = \frac{1}{2\pi} \int_{-\pi}^{\pi} \min(\theta, \Phi(\omega)) \, d\omega \qquad (31a)$$

$$R(D_\theta) = \frac{1}{4\pi} \int_{-\pi}^{\pi} \max\left(0, \log_2 \frac{\Phi(\omega)}{\theta}\right) d\omega \qquad (31b)$$

where $\Phi(\omega)$ is the source power spectral density.

Due to the fact that for sources with memory the rate-distortion function is defined as a limiting process, extensions of the algorithm for the numerical computation of $R(D)$ is extremely complex. To date, for sources with memory, no computationally effective algorithm for the numerical computation of $R(D)$ is available.

In view of many difficulties associated with the exact computation of the rate-distortion function, various techniques for developing upper and lower bounds on the rate-distortion function have been developed. In what follows, we describe a few of these bounds that are more commonly used in the literature.

1. Gaussian Upper Bound for Memoryless Sources

Consider a continuous-alphabet memoryless source with a given variance σ^2. For the squared-error distortion measure, the rate-distortion function of this source $R(D)$ is upper bounded by that of a memoryless Gaussian source possessing the same variance, denoted by $R_G(D)$ and described by

$$R(D) \leq R_G(D) = \frac{1}{2} \log_2 \frac{\sigma^2}{D} \qquad (32)$$

2. Shannon Lower Bound for Memoryless Sources

Consider a continuous-alphabet memoryless source with a difference distortion measure, that is, $d(u, v) = d(u - v)$. Then the following lower bound on the rate-distortion function holds.

$$R(D) \geq R_{SLB}(D) = h(U)$$

$$+ \sup_{s \leq 0} \left[sD - \log_2 \int_{-\infty}^{\infty} e^{sd(x)} \, dx \right] \qquad (33)$$

In the special case of squared-error, $d(u, v) = (u - v)^2$ the above bound reduces to

$$R(D) \geq R_{SLB}(D) = h(U)$$

$$- \frac{1}{2} \log_2(2\pi e D) \qquad \text{bits/source symbol} \quad (34)$$

3. Memoryless Upper Bound for Sources with Memory

Consider a stationary ergodic source with rate-distortion function $R(D)$. Let us denote by $R_1(D)$ the rate-distortion function of a memoryless source with the same marginal density as the original source. Then, the following bound holds.

$$R(D) \leq R_1(D) \qquad (35)$$

The interpretation of this upper bound is that for a given level of tolerable distortion D, the number of bits necessary to represent a source with memory is smaller than that for the similar source without memory, that is, memory in the source helps represent the source with smaller number of bits.

4. Gaussian Upper Bound for Sources with Memory

Consider a stationary ergodic source with a given power spectral density. Then, for the squared-error distortion measure, the rate-distortion function of this source is upper bounded by that of a Gaussian source with the same power spectral density as described in Eq. (31).

5. Wyner–Ziv Lower Bound for Sources with Memory

For a stationary ergodic source let us define the differential entropy rate of the source by

$$h_\infty(U) = \lim_{N \to \infty} \frac{1}{N} h(U_N) \qquad (36)$$

where $h(U_N)$ is the differential entropy of a source whose outputs are N-vectors of the original source. Then the following bound on the rate-distortion function of the original source holds.

$$R(D) \geq R_1(D) - [h(U) - h_\infty(U)] \qquad (37)$$

Note that the quantity $h(U) - h_\infty(U)$ can be interpreted as a measure of memory in the source. In fact, for memoryless sources it is easily shown that $h(U) - h_\infty(U) = 0$.

The above bounds enable us to develop upper and lower bounds on the rate-distortion function of a general source. While, in general, nothing can be said about the tightness of these bounds, in some special cases it is possible to prove that some of the above bounds are tight over some portion of the rate-distortion function.

Now that we are familiar with the fundamental limitations in data compression and the operational significance of the source entropy and rate-distortion function, the important question to be asked is how one should go about designing a source coding scheme whose performance coincides with that of the rate-distortion function of the source. Unfortunately, the coding theorems that establish the operational significance of the rate-distortion function do not have constructive proofs. Therefore, these theorems do not suggest direct design procedures. In general, designing a source coding scheme whose performance coincides with that of the rate-distortion function is out of reach. The best we can hope for is a coding scheme whose performance is close to the rate-distortion function.

In the past two decades a tremendous amount of research has been devoted to the development of efficient and easily instrumentable source coding algorithms. In the following section we look at some of the source coding procedures that are used in practical situations and make appropriate performance comparisons.

III. Practical Source Coding Schemes

In the previous section we described the basic theoretical limitations in source coding and showed that the source rate-distortion function determines the smallest theoretically achievable rate necessary for representing the source at a given value of distortion. The rate-distortion function of the source provides us with a lower bound on the rate versus distortion performance of practical source coding schemes. We now introduce several source coding schemes that are used in practical situations, examine the optimum performance attainable by such schemes, and provide performance comparisons against the corresponding source rate-distortion function.

We begin by zero-memory quantization of memoryless sources because it is the simplest of all schemes considered here and the easiest to understand. We then consider sources with memory and study predictive quantization techniques as well as block transform quantization. We also consider the recently developed algorithm for vector quantization for both memoryless sources and sources with memory.

A. Zero-Memory Quantization of Memoryless Sources

The simplest and most common source coding technique is zero-memory quantization of the real-valued outputs of an appropriately defined source. This technique, often referred to as analog-to-digital conversion (A/D), has been and continues to be a popular method of digitizing analog signals. It is important to emphasize that throughout this article we continue to assume that our signal is a sampled (in time) version of a waveform. If the samples belong to a continuous alphabet, the signal is called an analog signal. On the other hand, if the samples are quantized to a finite number of possible values, the signal is called a digital signal.

Let us assume that the source can be modeled as a discrete-time continuous-alphabet (analog) memoryless source represented by a sequence of (i.i.d.) random variables $\{U_n\}$ with common p.d.f. $f_U(.)$ possessing a zero mean and variance $\sigma^2 = E\{U^2\}$.

Consider a zero-memory (or one-dimensional) quantization of the above source. The adjective zero-memory is used to emphasize that the quantizer's output at a given instant depends only on the current input value and is independent of the past and future input and output values.

An N-level zero-memory quantizer is a mapping $q_N(.)$ that maps the source output $u \in \mathbb{R}$ into one of N possible values in the reproduction alphabet $A_V = \{y_1, y_2, ..., y_N\}$. This operation can be described in terms of the thresholds x_1, $x_2, ..., x_{N-1}$, partitioning the support of f_U into

N disjoint and exhaustive regions according to

$$v \triangleq q_N(u) = y_k, \quad x_{k-1} < u \le x_k,$$

$$k = 1, 2, ..., N, \tag{38}$$

where $x_0 \equiv -\infty$, $x_N \equiv \infty$, and v is the quantizer output.

Here, all x_k are called quantization thresholds, and all y_k are called representation (reconstruction) levels.

An N-level quantizer converts each source output into one of the representation levels and yields an average distortion

$$D_N(\mathbf{x}, \mathbf{y}) = \sum_{k=1}^{N} \int_{x_{k-1}}^{x_k} f_U(u) d(u, y_k) \, du \tag{39}$$

where $d(.,.)$ is a nonnegative distortion measure, and $\mathbf{x}$ and $\mathbf{y}$ are given by $\mathbf{x} = (x_1, x_2, ..., x_{N-1})$ and $\mathbf{y} = (y_1, y_2, ..., y_N)$, respectively.

Upon defining

$$P_k \triangleq Pr(x_{k-1} < U \le x_k),$$

$$k = 1, 2, ..., N \tag{40}$$

the entropy at the quantizer output is given by

$$H(V) = H_N(\mathbf{x}) = -\sum_{k=1}^{N} P_k \log_2 P_k$$

bits/source symbol   (41)

For a given vector of threshold levels $\mathbf{x}$ and a given vector of representation levels $\mathbf{y}$, the average distortion $D_N(\mathbf{x}, \mathbf{y})$ and the output entropy $H_N(\mathbf{x})$ determine the rate-distortion performance of the quantizer. Here, the quantizer's output entropy $H_N(\mathbf{x})$, in view of the noiseless source coding theorems described in Section II, is representative of the smallest number of bits necessary for representing the quantizer output.

For a given $\mathbf{x}$ and $\mathbf{y}$, we can compare the performance of the quantizer against the rate-distortion function in the following way. Let us consider the rate-distortion plane as illustrated in Fig. 1. Then the pair $(D_N(\mathbf{x}, \mathbf{y}), H_N(\mathbf{x}))$ represents a point in this plain. A lower bound to such points for any source coding scheme is given by the source rate-distortion function $R(D)$. Thus the loss in performance introduced by zero-memory quantization is equal to $\Delta R \triangleq H_N(\mathbf{x}) - R(D_N(\mathbf{x}, \mathbf{y}))$, bits per source symbol. Similarly, the loss measured in terms of additional distortion, is given by $\Delta D \triangleq D_N(\mathbf{x}, \mathbf{y}) - D(H_N(\mathbf{x}))$, where $D(R)$ is the inverse function of $R(D)$ and is called the source distortion-rate function.

Obviously, one should try to minimize the performance loss given by ΔR or ΔD. In what follows we consider the problem of choosing the vectors $\mathbf{x}$ and $\mathbf{y}$ in an effort to minimize the performance loss.

Definition 5. An N-level quantizer $q_N(.)$ is said to be an optimum entropy-constrained quantizer if it minimizes the average distortion while the output entropy is held below a given value.

According to this definition, the optimum entropy-constrained quantizer minimizes the loss ΔD for a fixed output entropy. More formally, the problem is to minimize

$$D_N(\mathbf{x}, \mathbf{y}) = \sum_{k=1}^{N} \int_{x_{k-1}}^{x_k} f_U(u) d(u, y_k)\, du \quad (42a)$$

subject to

$$H_N(\mathbf{x}) \leq R \quad (42b)$$

This problem is a nonlinear constrained optimization problem. Necessary conditions for the solution to this problem can be obtained from the celebrated Kuhn–Tucker theorem. Upon letting $(\mathbf{x}^*, \mathbf{y}^*)$ be a relative solution of Eq. (42), there must exist a scalar $\lambda \geq 0$, such that

$$\lambda \ln \left(\frac{P_{k+1}^*}{P_k^*} \right) = d(x_k^*, y_{k+1}^*) - d(x_k^*, y_k^*)$$

$$k = 1, 2, \ldots, N - 1, \quad (43)$$

and

$$\lambda[H_N(\mathbf{x}^*) - R] = 0 \quad (44)$$

where

$$P_k^* = \int_{x_{k-1}^*}^{x_k^*} f_U(u)\, du \quad (45a)$$

and all y_k^* must satisfy

$$\frac{\partial}{\partial y_k} D_N(\mathbf{x}^*, \mathbf{y}^*) = 0 \quad (45b)$$

The above equations are usually extremely difficult to solve. Let us consider the case of the squared-error distortion measure. We maintain this assumption in the rest of this article unless specifically stated otherwise. In this case it is possible to show that Eq. (43) reduces to

$$\lambda \ln \left(\frac{P_{k+1}^*}{P_k^*} \right) = (y_{k+1}^* - y_k^*)(y_{k+1}^* + y_k^* - 2x_k^*)$$

$$k = 1, 2, \ldots, N - 1 \quad (46)$$

where

$$y_k^* = \frac{1}{P_k^*} \int_{x_{k-1}^*}^{x_k^*} u f_U(u)\, du \quad (47)$$

For this special case, there are various algorithms for solving for the optimum $\mathbf{x}^*$ and $\mathbf{y}^*$. An interesting observation to be made from Eq. (47) is that for the squared-error distortion measure, the optimum representation levels are the center of probability mass of their corresponding quantization intervals.

While the above equations can be solved numerically to determine the rate-distortion performance of an N-level quantizer, it is certainly very useful to be able to come up with an explicit expression for the rate-distortion of these optimum quantizers. In general, even for very simple source distributions, it is very difficult to come up with an explicit expression for the rate-distortion performance of optimum entropy-constrained quantizers. However, for a fairly general class of sources, it is possible to show that at high bit rates and for a large number of quantization levels, a quantizer with uniformly spaced thresholds performs very close to optimal. Furthermore, in this case the average distortion of the quantizer as a function of the output entropy R is given by the simple expression

$$\hat{D}(R) = \frac{1}{12} 2^{2(h-R)} \quad (48)$$

where h is simply the source differential entropy. This result was first reported by Gish and Pierce in 1968.

Equation (48) is particularly useful in making comparisons against the source rate-distortion function. Recall from Section II that the Shannon lower bound on the rate-distortion function is given by

$$R_{\text{SLB}}(D) = h - \frac{1}{2} \log_2(2\pi e D) \quad (49)$$

It can be shown that under some mild conditions on the source distribution, the Shannon lower bound is tight in the high-bit-rate region. Therefore, the distortion-rate function of the source can be well approximated at high bit rates by

$$\hat{D}(R) = \frac{1}{2\pi e} 2^{2(h-R)} \quad (50)$$

Now, comparing Eq. (48) with Eq. (50), it is easy to see that $\hat{D}(R)/D(R) = \pi e/6$. This corresponds to 1.53 dB performance loss. Similarly, it can be shown that for a fixed value of distortion D, $\hat{R}(D) - R(D) = 0.5 \log_2(\pi e/6) \cong 0.255$ bits/symbol. Here, $\hat{R}(D)$ is the inverse function of $\hat{D}(R)$ described by Eq. (48). These results are extremely useful because they do not depend on the source distribution. While such nice explicit results do not exist for the low-bit-rate region, experimental evidence shows that for most source densities the gap between the source rate-distortion function and the optimum entropy-constrained quantizer performance narrows in the low-bit-rate region. Indeed, at $R = 0$ the two performances coincide.

These results are fundamentally important because they determine the optimal performance theoretically attainable by any zero-memory quantization scheme. However, remember that the rate of these quantizers is described in terms of the quantizer output entropy. To represent the quantizer output by a number of bits close to the entropy, some type of a variable-length coding scheme is needed. Variable-length codes suffer from a few difficulties that limit their practical value. Specifically, the decoding of variable-length codes takes place in a sequential fashion. Therefore, an error in the transmission stage may cause a run of errors in the decoded sequence. Furthermore, since most channels operate on a synchronous basis, for the transmission of variable-length codes some type of buffer between the output of the variable-length encoder and the input of the channel is needed. This buffer may overflow or underflow. The overflow and underflow are undesirable and are to be avoided.

In light of these difficulties with the implementation of entropy-constrained quantizers, we now describe a quantization scheme that transmits the representation levels by fixed-length codes and hence does not suffer from the difficulties of variable-length codes.

Definition 6. An N-level quantizer $q_N(.)$ is said to be optimum in the Lloyd–Max sense, if it merely minimizes the average distortion for the fixed number of levels N.

The rationale behind this definition is that for a fixed number of levels N (assuming that N is a power of 2), we can always represent the quantization levels by $\log_2 N$ bits. Thus, the rate is not dictated by the entropy of the quantizer output but rather by the number of quantization levels. Therefore, in this case the problem is merely to minimize

$$D_N(\mathbf{x}, \mathbf{y}) = \sum_{k=1}^{N} \int_{x_{k-1}}^{x_k} f_U(u) d(u, y_k) \, du \quad (51)$$

where the quantizer rate is given by $R = \log_2 N$. Similar necessary conditions for the optimality

of this quantizer can be derived. Without much ado over the details, let us mention that for the squared-error distortion criterion, the necessary conditions for optimality are given by

$$x_k^* = \frac{y_{k+1}^* + y_k^*}{2} \quad (52a)$$

and

$$y_k^* = \frac{1}{P_k^*} \int_{x_{k-1}^*}^{x_k^*} u f_U(u) \, du \quad (52b)$$

These conditions were apparently independently derived by Lloyd and Max, hence the term Lloyd–Max quantizer. For a certain class of sources it is possible to prove that the above conditions are also sufficient for the optimality of the quantizer. Several algorithmic procedures for the numerical solution of Eq. (52) are documented in the literature. The most important of all are two algorithms proposed by Lloyd and Max.

Similar to the previous case, in general, in the low-bit-rate region no explicit result for the rate versus distortion performance of the Lloyd–Max quantizer is available. Numerical results, however, for various sources are well-documented. There are also some suboptimal schemes for quantization that have certain implementation-related advantages. For example, uniform quantizers are widely used in practical systems. These are quantizers for which the thresholds and the representation levels are uniformly spaced. Another popular technique for quantizer design is what is known as companding. The idea here is to take the source output u and perform some type of nonlinear mapping, say $c(.)$, upon it. The mapping $c(.)$ is called a compressor. Then the output of the compressor is quantized by means of a uniform quantizer and operated upon by $c^{-1}(.)$. The mapping $c^{-1}(.)$ is called an expander. In the limiting case of a large number of quantization levels, it is possible to show that the optimum compressor is of the form

$$\frac{dc_{\text{opt}}(u)}{du} = K \frac{[f_U(u)]^{1/3}}{\int_0^\infty [f_U(x)]^{1/3} \, dx} \quad (53)$$

where $K = c(\infty)$. In this case, the average quantization distortion is easily shown to be

$$D_{\text{opt},N} = \frac{2}{3N^2} \left\{ \int_0^\infty [f_U(u)]^{1/3} \, du \right\}^3 \quad (54)$$

This is another explicit expression for the distortion of quantizers obtained by companding in the high-bit-rate region. It is also shown that quantizers designed by companding are asymptotically (high bit rate) optimal in the Lloyd–Max sense. Therefore, we can use the asymptotic result in Eq. (54) to develop performance comparisons between the Lloyd–Max quantizer and the source rate-distortion function.

EXAMPLE 5. Consider a zero-mean unit-variance Gaussian source U with probability density function

$$f_U(u) = \frac{1}{\sqrt{2\pi}} e^{-u^2/2} \qquad (55)$$

For this source, the distortion-rate function is given by $D(R) = 2^{-2R}$. The corresponding high-bit-rate Lloyd–Max performance is $D_{\mathrm{opt},N} = \pi\sqrt{3}/(2N^2)$. Noting that the rate is given by $R = \log_2 N$, it is straightforward to show that the distortion-rate performance of the quantizer is given by

$$\bar{D}_{\mathrm{opt}}(R) = \frac{\pi\sqrt{3}}{2} 2^{-2R} \qquad (56)$$

Comparing this with the source distortion-rate function, one could easily see that $\bar{D}_{\mathrm{opt}}(R)/D(R) = \pi\sqrt{3}/2$, or 4.35 dB performance loss. The performance loss is easily computed to be 0.722 bits/symbol, substantially larger than that for the entropy-constrained quantizers.

Recall that the gap between the rate-distortion function and the high rate performance of entropy-constrained quantizers is fixed (0.255 bits/symbol) independent of the source distribution. This does not hold for Lloyd–Max quantizers. In other words, the gap between the rate-distortion function and the Lloyd–Max quantizer performance depends upon the source distribution.

It can be concluded that for memoryless sources, zero-memory quantizers perform very close to the optimal performance theoretically attainable. Indeed, for most sources, the optimum entropy-constrained quantizer performs within 0.255 bits/symbol of the best possible

performance for all rates. If the source has memory, however, somehow the inherent correlation between neighboring samples must be used in the data compression operation. This cannot be done with a zero-memory quantizer because this quantizer operates on the source outputs instantaneously. Therefore, for sources with memory, to obtain performance close to the rate-distortion function, we need to use something more sophisticated than simple zero-memory quantization.

B. PREDICTIVE QUANTIZATION

Now we turn our attention to data compression of sources with memory. In sources with memory, there is correlation between adjacent samples, and hence each sample provides some information about the values of the neighboring samples. In other words, there is some redundancy in the source output samples. A "good" data compression system is that which effectively exploits the knowledge that one source sample provides about the neighboring samples, or, equivalently, effectively removes the redundancy in the source.

We now describe a data compression scheme in which instead of quantizing the source outputs directly, we quantize the difference between the source output and its predicted version. Such schemes, in general, are called predictive quantization schemes. The block diagram of a generic entropy coded predictive quantization scheme is illustrated in Fig. 2. Let us first describe the operation of this predictive quantization scheme.

Roughly speaking, the source signal can be conceived as having two parts: One part ($\hat{U}_n$) is predictable relative to the transmitted sequence Y_n and hence conveys no useful information; the other part (the prediction error E_n) is unpredictable and (since it uniquely determines the signal) therefore contains the useful information. A predictive quantization scheme attempts to discard

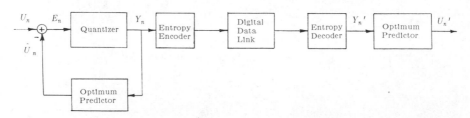

FIG. 2.   Block diagram of a generic predictive quantization scheme.

the predictable part, for it can be reproduced at the receiver, and encode only the unpredictable portion using a zero-memory quantizer.

Let us assume that the source U_n is a stationary ergodic source with zero-mean and variance σ_U^2. At time instant n, the predicted estimate of U_n based on all the past transmitted values Y_1, Y_2, ..., Y_{n-1} is given by the following conditional expectation:

$$\hat{U}_n = E\{U_n|Y_{n-1}, Y_{n-2}, ..., Y_1\} \qquad (57)$$

On the other hand, the instantaneous estimate of U_n based on observing Y_1, Y_2, ..., Y_n is described by

$$\bar{U}_n = E\{U_n|Y_n, Y_{n-1}, ..., Y_1\} \qquad (58)$$

At time instant n, the predictor in the transmitter loop makes an estimate of the value of U_n based on the previously transmitted values. Note that, assuming the transmitted sequence Y_n reaches the receiver with no error (i.e., the transmission medium is noiseless), the quantity $\hat{U}_n$ can also be produced at the receiver. Thus the transmitter computes the part of the source output that could not be predicted (the prediction error),

$$E_n = U_n - \hat{U}_n \qquad (59)$$

and quantizes it for transmission.

The analysis of the predictive quantization in such generality as just described is formidable. Even the computation of the average distortion is not a straightforward matter. So, to be able to carry the analysis further we make some specific assumptions on the structure of the source. For simplicity, we assume that the source is a first-order autoregressive process described by

$$U_n = \rho U_{n-1} + W_n \qquad n = 1, 2, ... \qquad (60)$$

where W_n is a sequence of i.i.d. random variables with zero mean and variance σ_W^2 and ρ the source correlation coefficient.

With this assumption, it is possible to show that

$$\hat{U}_n = \rho \bar{U}_n \qquad (61)$$

Furthermore, we can write

$$\bar{U}_n = \hat{U}_n + E\{E_n|Y_n, Y_{n-1}, ..., Y_1\}$$
$$n = 1, 2, ... \qquad (62)$$

If we further assume that the sequence E_n is nearly independent, we can rewrite Eq. (62) as

$$\bar{U}_n = \hat{U}_n + E\{E_n|Y_n\} \qquad n = 1, 2, ... \qquad (63)$$

Finally, if we assume that the quantization lev-

els are chosen as the center of probability mass of their respective intervals, it is easy to show that

$$E\{E_n|Y_n\} = Y_n, \qquad n = 1, 2, ... \qquad (64)$$

Therefore, with all the above assumptions, we have

$$\bar{U}_n = \hat{U}_n + Y_n \qquad (65a)$$

and

$$\hat{U}_n = \rho \bar{U}_n \qquad (65b)$$

The predictive quantization scheme described by Eq. (65) is widely used in practice and is known as the differential pulse code modulation (DPCM). Various versions of the DPCM scheme are being used in different speech and image compression situations.

In what follows we analyze the performance of the predictive coding scheme described above. To be consistent with our previous results we perform our analysis in a rate-distortion theoretic framework.

The average distortion associated with the above DPCM scheme is given by

$$D_N(\mathbf{x}, \mathbf{y}) = E\{(U_n - \bar{U}_n)^2\} \qquad (66)$$

where $\mathbf{x}$ and $\mathbf{y}$ denote the quantization thresholds and representation levels as before, and error-free transmission is assumed so that $U_n' = \bar{U}_n$. It is interesting to note that

$$U_n - \bar{U}_n = E_n - Y_n \qquad (67)$$

which means that to minimize the overall reconstruction error, it suffices to minimize the quantization error. Therefore, the problem of optimizing the performance of the DPCM scheme is reduced to that of designing the quantizer optimally.

While the optimization of the performance of DPCM may appear trivial, it is actually a formidable task. The main difficulty resides in the fact that to design the quantizer, we need to have the p.d.f. of the error sequence. However, the p.d.f. of the error sequence is not known and depends on the structure of the quantizer. So, clearly, there is a very intricate intertwining between the p.d.f. of the prediction error and the parameters of the quantizer. In fact, even for a fixed, maybe nonoptimal quantizer, the computation of the p.d.f. of the prediction error involves the solution of a complicated integral equation. Therefore, in general, it is very difficult to optimize the performance of the DPCM scheme. For first-order Gauss–Markov and Laplace–Markov sources, however, some numerical results on the performance of optimal DPCM systems are available.

In the high-bit-rate region, as in zero-memory quantization, it is possible to come up with explicit formulas for the rate versus distortion performance of optimum DPCM systems. Specifically, when the source is a first-order Gaussian autoregressive source (i.e., the source described in Eq. (60) with Gaussian W_n, it is possible to show that in the high-bit-rate region the prediction error has a Gaussian density as well, with variance

$$\sigma_E^2 = \sigma_W^2 + \rho^2 D \qquad (68)$$

where D is the average quantization distortion given by

$$D \equiv \hat{D}(R) = \frac{1}{12} 2^{2(h_E - R)} \qquad (69)$$

and h_E is the differential entropy of the prediction error sequence. Simplification of these equations implies that

$$\hat{D}(R) = \left(\frac{\pi e}{6}\right) \sigma_W^2 \, 2^{-2R} \qquad (70)$$

a result very similar to the asymptotic performance of the optimum zero-memory quantizer at high bit rates. Indeed, this is a nice way of assessing the performance improvement obtained by DPCM over zero-memory quantization. If we had used zero-memory quantization to encode this source, the average distortion at high bit rates would be exactly the same as that in Eq. (70) with σ_U^2 replacing σ_W^2. Noting that

$$\sigma_U^2 = \frac{\sigma_W^2}{1 - \rho^2} \qquad (71)$$

we would conclude that the DPCM coding gain is given by

$$G = \frac{1}{1 - \rho^2} \qquad (72)$$

Thus, for $\rho = 0$, there is no coding gain. This is to be expected as $\rho = 0$ corresponds to a memoryless source in this case, and there is nothing to be gained by using a predictive coding scheme. On the other hand, for values of ρ close to 1, corresponding to a highly correlated source, the coding gain is very large.

Another interesting comparison to be made is against the source rate-distortion function. It is possible to show that the source rate-distortion function at high bit rates is given by

$$R(D) = \frac{1}{2} \log_2 \frac{\sigma_W^2}{D} \qquad (73)$$

Comparison of Eq. (73) with Eq. (70) immediately implies that the performance loss of the optimum entropy-constrained DPCM at high bit rates is exactly 0.255 bits/symbol, as in the memoryless case. The results for the case in which a Lloyd–Max quantizer is used in the DPCM loop essentially follow the same pattern as in the zero-memory case and hence shall not be repeated here.

C. BLOCK TRANSFORM CODING

As mentioned earlier, when there is memory in the source, the data compression system should attempt to make use of the memory. The DPCM scheme utilizes the memory in the source by making a prediction of the current sample based on the previous transmitted samples. An alternative would be to quantize a block (preferably large) of the source outputs at once. This way, we can make use of the correlation in the source to reduce the transmission rate. Indeed, in the proof of the source coding theorem subject to a fidelity criterion, we did resort to a block coding argument. There are two difficulties involved in this type of quantization. First, how to choose the quantization representation vectors and, second, the complexity of encoding. We address both of these issues in Section III,D where we discuss vector quantization. However, in this subsection, we present a particular form of block quantization that is commonly referred to as block transform coding.

For the sake of consistency, let us continue to assume that our source is described by the model in Eq. (60). Let us denote by $\mathbf{U}$ a vector of L consecutive source samples, say, $\mathbf{U} = (U_1, U_2, \ldots, U_L)^T$. In a block transform coding scheme the source output vector $\mathbf{U}$ is operated upon by a nonsingular $L \times L$ transformation matrix $\mathbf{A}$ to generate an L-vector $\mathbf{Y}$ with uncorrelated components described by

$$\mathbf{Y} = (Y_1, Y_2, \ldots, Y_L)^T = \mathbf{AU} \qquad (74)$$

The intuitive idea behind this transformation is to exploit the correlation between the source samples by first generating a set of uncorrelated variables $Y_1, Y_2, \ldots, Y_L$ and then encoding these sample by sample. Thus each component of $\mathbf{Y}$ is quantized and encoded by means of a zero-memory quantizer. Therefore, we need a total of L quantizers, one for each transform coefficient. These quantizers could be entropy-constrained or Lloyd–Max quantizers. Let us denote the quantized version of Y_k by $\hat{Y}_k$.

In the receiver, an $L \times L$ transformation matrix $\mathbf{B}$ operates upon the received vector $\hat{\mathbf{Y}}$ to generate a replica of $\mathbf{U}$, say $\hat{\mathbf{U}}$, given by

$$\hat{\mathbf{U}} = \mathbf{B}\hat{\mathbf{Y}} \qquad (75)$$

For the squared-error distortion measure, the average distortion per source symbol is given by

$$D = \frac{1}{L} E\{(\mathbf{U} - \hat{\mathbf{U}})^T(\mathbf{U} - \hat{\mathbf{U}})\} \qquad (76)$$

At this point we are ready to describe the problem. Given a fixed number of bits/source symbol, our goal is to minimize the average distortion given by Eq. (76). The minimization of D is achieved by the appropriate choice of the matrices $\mathbf{A}$ and $\mathbf{B}$ and the appropriate design of the quantizers used for encoding the transform coefficients. This problem was first studied by Huang and Schultheiss in 1963; they considered the block transform quantization of stationary Gaussian sources with the assumption that Lloyd–Max quantizers are used for encoding the transform coefficients.

Huang and Schultheiss showed that for a given $\mathbf{A}$ the best matrix $\mathbf{B}$ is the inverse of $\mathbf{A}$. Furthermore, they showed that the best transformation matrix $\mathbf{A}$ is the matrix whose rows are the orthonormalized eigenvectors of the covariance matrix of $\mathbf{U}$. This optimum transformation, called the Karhunen–Loeve transformation, is a unitary transformation (i.e., $\mathbf{A}^T = \mathbf{A}^{-1}$). Thus it is easy to show that

$$D = \frac{1}{L} E\{(\mathbf{Y} - \hat{\mathbf{Y}})^T(\mathbf{Y} - \hat{\mathbf{Y}})\}$$
$$= \frac{1}{L} \sum_{i=1}^{L} \lambda_i D_i \qquad (77)$$

where λ_i is the variance of the ith transform coefficient and D_i is the variance-normalized distortion associated with the ith quantizer.

Let us assume that the rate associated with the ith quantizer is denoted by R_i. Now that the transformation matrices $\mathbf{A}$ and $\mathbf{B}$ are chosen, the problem is simply to minimize

$$D = \frac{1}{L} \sum_{i=1}^{L} \lambda_i D_i \qquad (78a)$$

subject to

$$\frac{1}{L} \sum_{i=1}^{L} R_i = R \qquad (78b)$$

Notice that at this point we are still free to choose our quantizers to be entropy-constrained or Lloyd-Max quantizers. If we choose to use entropy-constrained quantizers, R_i should denote the entropy at output of the ith quantizer. On the other hand, should we decide to use Lloyd–Max quantizers, $R_i = \log_2 N_i$, where N_i is the number of quantization levels for the ith quantizer.

In either case, the problem as stated in Eq. (78) is that of determining the optimum assignment of rates (entropies or bits) to different transform coefficients. Therefore, the problem of optimal design of block transform quantizers is reduced to that of an optimal rate assignment problem. This problem, for the case of Lloyd–Max quantizers, was solved by Huang and Schultheiss. The extension to entropy-constrained quantization has also been considered recently.

For the sake of completeness, we must mention that for Gaussian sources and in the limit of large block sizes, it is established that the block transform quantization has a performance identical to DPCM at high bit rates. This holds for both entropy-constrained and Lloyd–Max quantization. In the low-bit-rate region, explicit expressions for the performance of the block transform quantization are not available. However, numerical results indicate that block transform quantizers have performance superior to the DPCM scheme.

Block transform quantization schemes and various adaptive versions thereof are used in many practical source coding schemes. Many low-bit-rate speech coding and video conferencing schemes are based on this system. An attribute of transform coding schemes that makes them very attractive is the low complexity of these systems. Once the transformation is performed, the rest is simple, zero-memory quantization. In fact, due to the very structure of the block transform coding scheme, the system is amenable to parallel processing which is becoming very popular. To minimize the complexity of computing the matrix transformation, various suboptimal transformations have been considered in different applications. The discrete cosine transform has proven to be a very useful transformation in the sense that it provides performance very close to that obtained by the optimal Karhunen–Loeve transformation, and yet it can be implemented by means of very fast algorithms.

D. Vector Quantization

Vector quantization is simply a multidimensional extension of the zero-memory (one-dimensional) quantization scheme. An L-dimensional, N-level vector quantizer q_N (.) is a mapping from $\mathbb{R}^L$ to the set of reproduction vectors $A_Y = \{\mathbf{y}_1, \mathbf{y}_2, ..., \mathbf{y}_N\}$. Here, each $\mathbf{y}_i$ is a vector in $\mathbb{R}^L$. Associated with the vector quantizer is a partition of $\mathbb{R}^L$, say $A_S = \{S_1, S_2, ..., S_N\}$, where each $S_i \subset \mathbb{R}^L$. Here, the A_S is a collection of disjoint and exhaustive regions in $\mathbb{R}^L$. The operation of the vector quantizer is then described by

$$q_N(\mathbf{u}) = \mathbf{y}_i \quad if \ \mathbf{u} \in S_i \qquad (79)$$

Here, $\mathbf{u} = (u_1, u_2, ..., u_L)$ is an L-dimensional vector of source outputs.

This operation is identical to the one we used in the proof of the source coding theorem subject to a fidelity criterion. Thus we know that in the limit of large L, there exists a vector quantizer with appropriate A_Y and A_S, whose performance is very close to the rate-distortion function of the source.

Ever since Shannon established the usefulness of block source encoding techniques, there has been considerable interest in finding good block encoders or vector quantizers. Until 1980, the developments on this subject were limited to the developments of uniform vector quantizers for which some asymptotic properties were also developed. A break-through in the area of design of optimal vector quantizers was made in 1980 by Linde, Buzo, and Gray. They provided an iterative algorithm, referred to hereafter as the LBG algorithm, for the design of optimal error minimizing vector quantizers based on a training sequence of data. Essentially, the LBG algorithm is a multidimensional extension of an algorithm for one-dimensional quantizer design suggested by Lloyd. Since the invention of the LBG algorithm, a tremendous amount of research activity has been expended on various applications and refinements of vector quantization. We briefly describe the LBG algorithm.

Let us suppose that the distortion is given by a nonnegative function $d(\mathbf{u}, \mathbf{v})$. For a fixed number of representation vectors N, we wish to minimize the average distortion

$$D = \frac{1}{L} E\{d(\mathbf{U}, \mathbf{V})\} \qquad (80)$$

where $\mathbf{V}$ is used to denote the quantizer output vector. Let us suppose that the partition A_S is fixed. Then, it is possible to show that the best set of reproduction vectors A_Y is such that $\mathbf{y}_i$ satisfies

$$E\{d(\mathbf{U}, \mathbf{y}_i)|\mathbf{U} \in S_i\} = \min_{\mathbf{x}} E\{d(\mathbf{U}, \mathbf{x})|\mathbf{U} \in S_i\}.$$

$$(81)$$

Conversely, assume that the set of reproduction vectors A_Y is fixed. Then it is easy to show that the best partition A_S is such that

$$\mathbf{u} \in S_i \quad \text{only if}$$

$$d(\mathbf{u}, \mathbf{y}_i) \leq d(\mathbf{u}, \mathbf{u}_j) \qquad \text{all } j \qquad (82)$$

The LBG algorithm is simply a successive application of the conditions described by Eqs. (81) and (82). In practice, the actual implementation of the algorithm is not based on computing the expectation in Eq. (81). In fact, computing this expectation in high dimensions involves multidimensional integration over regions of possibly irregular shapes and, hence, is computationally very difficult. Linde, Buzo, and Gray, in their original paper, suggested a remedy to this problem by using a training sequence of data which, in turn, replaces multidimensional integration, by simple summation.

An additional advantage of this algorithm is that the average distortion decreases at each step of the algorithm, and hence the LBG algorithm guarantees convergence to a locally optimum quantizer.

A few comments about the performance and complexity of the LBG-based vector quantization are in order. First, it should be noted that in the absence of an entropy constraint, for a given block size L, the vector quantizer provides the best possible performance. Of course, there is no guarantee that the locally optimum quantizer that the algorithm has converged to is also globally optimum. The choice of the initial conditions dictates the locally optimum quantizer the algorithm converges to. Unfortunately, there is no way a priori of determining the initial choice that results in the globally optimum solution.

To provide a complete comparison, in Tables I and II we consider the performance of optimum DPCM, optimum block transform quantization (both entropy-constrained and Lloyd–Max), and optimum vector quantization, for a first-order Gauss–Markov source as described in Eq. (60) with $\rho = 0.5$ and $\rho = 0.8$, respectively. It should be noted that the results for the vector quantizer are only available at $R = 1$ bit/sample. At higher bit rates the computational requirement of the vector quantizer design is beyond current practical limits. In Tables I and II, the results corresponding to $L = 1$ describe the performance of the zero-memory quantizer. The performance gains of the DPCM, block trans-

TABLE I.   Signal-to-Noise Ratios (in dB) of Optimum Entropy-constrained Block Transform Quantization and Other Schemes for a First-order Gauss–Markov Source with $\rho = 0.5$.

R (bits/sample)	Consecutive source samples (L)				DPCM[a]	$R(D)$[b]
	1	2	4	8		
1	4.58	5.43	5.78	5.96	5.22	7.27
	(4.40)[c]	(4.71)	(5.30)	(5.41)		
	[4.40][d]	[5.18]	[5.69]			
2	10.51	11.14	11.45	11.60	11.58	13.29
	(9.30)	(9.33)	(10.12)	(10.24)		
3	16.53	17.15	17.47	17.62	17.77	19.31
	(14.62)	(14.62)	(15.34)	(15.46)		

[a] Optimum differential pulse code modulation.
[b] Rate-distortion function.
[c] Huang–Schultheiss scheme, values in parentheses.
[d] Vector quantization scheme, values in brackets.

form coding, and vector quantization compared with zero-memory quantization are obvious. Notice that these gains are more substantial for the larger value of ρ.

Note that the rate associated with the vector quantizer is given by $R \triangleq (\log_2 N)/L$, or $N = 2^{LR}$. Therefore, the number of representation vectors grows exponentially with the rate R and the block size L. This exponential growth limits both the design and the implementation of vector quantizers at high rates and for large block sizes. In an actual implementation of the vector quantizer, the distortion between the source output vector and each representation vector is computed, and the vector with smallest distortion is used to represent the source vector. Therefore, the number of distortion computations is directly proportional to the number of representation vectors. To overcome these difficulties, various suboptimal schemes for the implementation of the vector quantizers are under investigation.

Vector quantization has found practical applications in many source coding situations. It has made a considerable effect on the performance of speech encoders based on linear predictive coding (LPC). It has also had an impact on speech recognition research. More recently, some researchers have studied the performance of vector quantizers in image coding situations.

E. RELEVANT ISSUES

Before closing this section, we wish briefly to describe two issues that are inherently important in any data compression situation.

TABLE II.   Signal-to-Noise Ratio (in dB) of Optimum Entropy-constrained Block Transform Quantization and Other Schemes for a First-order Gauss–Markov Source with $\rho = 0.8$.

R (bits/sample)	Consecutive source samples (L)				DPCM[a]	$R(D)$[b]
	1	2	4	8		
1	4.58	7.45	8.47	9.02	7.56	10.46
	(4.40)[c]	(6.87)	(7.90)	(8.39)		
	[4.40][d]	[6.85]	[8.18]			
2	10.51	12.74	13.86	14.41	14.44	16.48
	(9.50)	(11.29)	(12.64)	(13.20)		
3	16.53	18.75	19.84	20.31	20.96	22.50
	(14.62)	(16.93)	(17.78)	(18.44)		

[a] Optimum differential pulse code modulation.
[b] Rate-distortion function.
[c] Huang–Schultheiss scheme, values in parentheses.
[d] Vector quantization scheme, values in brackets.

1. Sensitivity to Transmission Errors

The function of a source encoder is to remove redundancy in the source. This removal of redundancy results in additional sensitivity to transmission errors. This is an inherent problem with data compression schemes. However, some data compression schemes, due to their very structure, are more sensitive to this problem. For instance, variable-length codes could be extremely sensitive to channel errors due to the sequential nature of the decoding operation. Currently most data compression schemes are designed based on the assumption that the channel is noiseless. It is felt that if the data compression scheme is designed based on the assumption that the channel is actually noisy, considerable performance improvements could be obtained.

2. Adaptivity and Robustness

In most data compression schemes it is assumed that the source statistical knowledge is available. In practical cases, however, either a complete probabilistic description of the source is not available or the source statistics undergoes temporal and/or spatial variations. There are two ways of combating this problem. First, one could design the source encoder in such a way that it performs well for a class of possible source statistics, in which case the design is called robust. Second, the system could be designed to be adaptive in the sense that it senses variations in the source statistics and adapts itself to the current statistics by changing some parameters. Both of these issues are fundamentally important in the design of any data compression system for real-world applications.

BIBLIOGRAPHY

Berger, T. (1971). "Rate-Distortion Theory: A Mathematical Basis for Data Compression." Prentice-Hall, Englewood Cliffs, New Jersey.

Davisson, L. D. (1972). Rate-distortion theory and applications. *Proc. IEEE* **60**, 800–808.

Farvardin, N., and Modestino, J. W. (1984). Optimum quantizer performance for a class of non-Gaussian memoryless sources. *IEEE Trans. Inform. Theory* **IT-30**, 485–497.

Farvardin, N., and Modestino, J. W. (1985). Rate-distortion performance of DPCM schemes for autoregressive sources. *IEEE Trans. Inform. Theory* **IT-31**, 402–418.

Gallager, R. G. (1968). "Information Theory and Reliable Communication." Wiley, New York.

Gray, R. M., ed. (1982). Special issue on quantization. *IEEE Trans. Inform. Theory* **IT-28**, 127–262.

Jayant, N. S., and Noll, P. (1984). "Digital Coding of Waveforms—Principles and Applications to Speech and Video." Prentice-Hall, Englewood Cliffs, New Jersey.

Linde, Y., Buzo, A., and Gray, R. M., (1980). An algorithm for vector quantizer design. *IEEE Trans. Commun.* **COM-28**, 84–95.

Lynch, T. J. (1985). "Data Compression—Techniques and Applications." Lifetime Learning Publications, Belmont, Calif.

Makhoul, J., Roucos, S., and Gish, H. (1985). Vector quantization in speech coding. *Proc. IEEE* **73**, 1551–1588.

TELECOMMUNICATION SWITCHING

F. F. Taylor   *AT&T Bell Laboratories*

GLOSSARY

Bell operating companies (BOCs): Major telephone operating companies that were a part of the Bell system prior to divestiture and were grouped into seven holding companies known as regional Bell operating companies (RBOCs). The BOCs comprise those companies for which AT&T was a major shareholder prior to divestiture. These 23 companies account for approximately 80% of the telephone lines served in the United States.

Central office switching system: Automatic switching system that provides the connection of telephones in a local exchange area such as a suburb or a segment of a metropolitan area. Such systems may serve as few as 100 lines in rural areas or 50,000 lines or more in a metropolitan exchange. In large cities, more than one such system can be found in a single telephone operating company building.

Centrex service: Business customer service in which the switching equipment is part or all of a central office switching system, usually on telephone operating company premises and owned and operated by that operating company.

Common channel signaling (CCS): Interoffice signaling method wherein the signals for the traffic for a number of interoffice trunks are transmitted between switching systems via a separate channel, common to the group of trunks being served.

Integrated services digital network (ISDN): Network in which customers can access voice, data, and video services over a single communications link between a customer's communications terminal and the telecommunications network.

Interoffice signaling: Signals transmitted between switching systems to direct the establishment, supervision, and disconnection of communications links for voice or data calls.

Interoffice trunks: Transmission facilities for interconnecting switching system nodes within a telecommunications network.

Key system: Small telephone switching system to serve business customers having an associated group of telephones, from less than ten to several hundred. The name derives from the illuminated pushbuttons or keys on the telephone sets that control feature capabilities. The keys can be used to select among outside calls coming into the key system, to put incoming calls into a "hold" state, to bridge more than one key telephone on an incoming call, and so on.

Local access and transport area (LATA): In the United States, the exchange area served by telephone operating companies on an exclusive or monopoly basis. The continental United States is divided into approximately 160 LATAs.

Local area network (LAN): Network confined to a relatively small area such as an office building or a university campus to provide a community of data communications, such as terminals to some and minicomputers to others so as to share data bases and other software. This approach segregates voice and data communications into separate networks because of the differing nature of the traffic generated for voice and for data.

Modem: Abbreviation for modulator–demodulator. A device to convert the direct-current binary one and zero signals into alternating-current signals to operate over voice-grade telephone loops and switching system circuit connections. A modem facilitates communication between keyboard terminals and computers, usually over voice-grade switched circuit lines. Transmission rates most commonly in use are 1200 bits per second. More sophisticated modems are available that operate at rates up to 9600 bits per second. Higher bit rates require special treatment of the loops and switching circuits involved in the connections.

Network control point (NCP): Node in the common channel signal network comprising a computer with associated data bases. The information is accessed by switching systems to store and retrieve data to provide service and feature capabilities to network users.

Private branch exchange (PBX): Switching system to serve medium to large business customers with telephone extensions ranging from a few hundred to thousands. Initially providing voice connection capability, PBXs are now providing voice and data capabilities to meet growing needs for business customers, for computers, for word processing, and for other office functions such as electronic mail between terminals attached to the PBX.

Signal transfer point (STP): Node in a common channel signaling network that links the signal transmission channels and directs the packets of information between other STPs, switching systems, and network control points to carry the information across the network to set up, supervise, and take down calls and to provide for service capabilities to the network users.

Switching network: Also switching connection network. That part of a telecommunications switching system that provides the means for interconnecting terminals connected to it.

Telephone loop: Traditionally, the pair of wires extending from a central office switching system to individual telephones. Copper loops are increasingly being replaced, in part, by electronic carrier systems and fiberoptic multiplex systems to carry many conversations over many fewer pairs of wires or optical fibers.

Time slot interchanger: Switching element to switch connections in a time-division multiplexed stream of calls by interchanging the order in which the information travels in the stream of signals being transmitted.

A telecommunications network serves to interconnect terminals selectively to allow the flow of information among them, the flow traveling across some transmission media. Telecommunication switching provides selective interconnection and reduces the amount of transmission media required. Telecommunication switching first served telephone and telegraph networks but are now found in networks connecting computers and remote terminals and for the interconnection of one-way and two-way video networks. The networks connecting voice, data, and video have been traditionally separate because of the differing demands placed on the transmission and switching elements, but the trend is toward integrating these networks into a single network. New telecommunication switching systems are being developed to interconnect simultaneously these multiple modes of information flow.

I. Network Elements

Telecommunications networks include three principal elements: terminal nodes equipped with terminals (telephone instruments, data terminals, computers, and the like), links of transmission facilities to interconnect the terminal nodes and internal nodes for communication, and switching systems at the internal nodes to provide selective interconnection of two or more terminal nodes. These internal nodes are, in telephone switching networks, sometimes referred to as central offices or exchanges and are known as switching systems.

Figure 1 shows an example of a public telephone network such as that owned and operated by telephone administrations, either government organizations or private corporations. In addition to public networks, businesses may own and operate their own private internal switching system. Systems designed for use by large business firms are called private branch exchanges (PBXs). Small business systems, once known as key systems, are designed for customers with less than several hundred telephones.

The link or telephone line connecting a terminal to a switching system is called a loop, because it is usually a pair of copper wires that, with the terminal and a battery at the central office, form an electrical loop for the transmission of voice and data.

The transmission links interconnecting the switching systems are referred to as trunks. Usually a group of trunks interconnect the switching systems since each trunk carries one call at a time, and there may be a number of telephone calls simultaneously in progress and switched between two central offices, A and B in Fig. 1, for example. Calls traversing two or more central offices are known as interoffice calls. If, however, the telephone conversation takes place between telephones 1 and 2, then switching system B makes that connection internally, and interoffice trunks are not used. Such connections are referred to as intraoffice calls.

Signaling is the fourth major element of a telecommunication network. Terminals and switching systems send and receive signals to convey the intelligence throughout a communication switching network to direct the establishment of circuit connections and other services. A prearranged plan of signaling codes and sequences enables the use of telephones to access the network. Electromagnetic signals among switching systems further communicate the intelligence to make connections through one or more switching systems to communicate locally, between two cities, across the country, and throughout the world.

International standards bodies on telecommunications include the Consultative Committee on International Telephony and Telegraphy (CCITT). It functions under the auspices of the United Nations to establish the necessary worldwide recommendations for telecommunications signaling and related practices to permit most telephones in the world to be reached from any other telephone. Technical representatives from most countries around the world attend regular working meetings of various technical committees and develop new recommendations for standards to improve the technology and capability of the global telecommunications network.

The fifth major element of a telecommunications network is its structure. The topology or interconnection of the terminals and the switching system nodes via the web of transmission facilities or trunks determines the routes available to making the connections for calls. Among the myriad of possible choices, the number of connecting links and the number of calls over each link is limited by the cost of these connections. Alternate routes from point to point in these networks allow for the efficient use of the total network to carry traffic by use of all available resources. To do otherwise renders the service too expensive or too undependable and unresponsive.

This redundancy among the switching system and transmission facilities also provides for the survival of communications capabilities in the face of the loss of switching and transmission elements by the use of alternate routes to avoid the faulty elements. This robustness of operation can be crucial in times of regional or national emergencies, to aid in recovery from natural catastrophes or in support of military defense.

The engineering of the placement and size of switching systems and the choice of routes, alternate routes, and sizes of interconnecting transmission facilities are the keys to providing a good, efficient structure for the telecommunications network. [See COMMUNICATION SYSTEMS, CIVILIAN.]

Although the emphasis of the descriptions that follow are on voice communications using traditional telephones, it should be understood that data communications have an ever increasing role in telecommunication switching networks. In the past, voice and data have tended to occupy separate networks. This is changing rapidly; integration of terminals, transmission, and switching to handle both voice and data together, simultaneously or alternatively, has be-

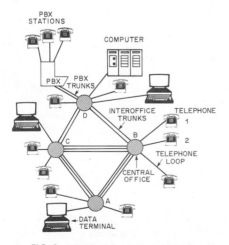

FIG. 1.   Public telephone network.

gun during the mid- to late-1980s. [*See* DATA TRANSMISSION MEDIA; TELECOMMUNICA-TIONS; TELECOMMUNICA-TIONS; VOICEBAND DATA COMMUNICA-TIONS.]

II. Telecommunications Terminals

Figure 1 shows the most common types of telecommunications terminals to be found in to-day's networks: telephones, data terminals and computers, key systems, and PBXs.

A. THE TELEPHONE

To place a call with a telephone equipped for pushbutton signaling, the customer lifts the handset from the cradle, which causes the dc resistance of the set, monitored by the central office, to change from a high value to a low value. This causes the loop current, provided by a central office battery, to increase. A sensor in the central office alerts the switching system to prepare to receive a telephone number, in this case, a sequence of digits represented by different combinations of pairs of audible tones. [*See* TELEPHONE SIGNALING SYSTEMS, TOUCH-TONE.]

The rotary dial, still in widespread use in most public networks, sends the same information via pulsations in the loop resistance. The pulse rate, which can be reliably generated and detected with traditional copper loops and rotary dials, ranges from 10–20 pulses/sec. This method of signaling is slower, and the repertoire of digits and symbols that can be transmitted has been limited.

The telephone set normally includes a ringer or other acoustic sounder to alert the user to incoming telephone calls. Additionally there is transmission circuitry to separate the incoming and outgoing voice signals that appear simultaneously on the loop, so that the speakers do not hear too much of their own voices. These and other functions are included to provide clear transmission and reception over a wide range of loop lengths and operating temperatures.

B. DATA TERMINALS AND COMPUTERS

Although the amount of data traffic has been relatively small through the mid-1980s, there is a growing use of the public telephone networks to send data between computers and remote entry and display terminals. Keyboard characters and numbers are encoded into sequences of binary signals, and these binary one or zero elements of information are converted into voice-band alternating-current signals suitable for transmission over the public voice networks. Special circuits known as modems (abbreviation for modulator–demodulators) are interposed between the loop and the data terminals to handle the signaling and conversion.

Modems in the United States public network can be used to communicate between most terminal nodes over the public network at standard transmission rates up to 9600 bits per second (bps) over circuits designed to carry signals limited to 4 kHz for voice. (The maximum rate for data has moved steadily upward over time. Increasingly sophisticated techniques have been devised and economically implemented using complex integrated circuits.)

Higher bit rates for data transmission have been limited to special networks or special switched circuits designed to handle this traffic. The higher rates are needed, for example, for rapid transfer of large amounts of data among computer systems.

C. PBXS, SMALL BUSINESS SYSTEMS, AND LOCAL AREA NETWORKS

Many places of business have a number of telephones for internal communications and also to connect with the public network. Small business systems and PBXs (known as PABXs or private automatic branch exchanges outside of the United States) are private switching systems, usually located on the customer premises to provide the internal connections required within the business and to interface the public network for communications links to its clientele.

Small business systems are designed for small businesses that require anywhere from less than ten to several hundred telephones or stations. The former name, key system, refers to the keys or illuminated pushbuttons on the telephone set that are used to select among incoming calls or to connect one telephone or station to another in the group of telephones served.

Key systems were originally based on the wired logic or electromechanical circuits such as relays to provide a basic business service. More recently, microprocessor chips with sophisticated programs stored on VLSI memory chips provide the control of the small business system operation and include many more features with much less physical space and power required.

Microprocessors have migrated downward from PBXs to small business systems and to the telephone set itself. The limited array of illuminated keys have been replaced by larger groups of pushbuttons, multicolored light-emitting diodes, and alphanumeric displays to provide much more versatility. The functional differences and feature capabilities between the large and small business systems have rapidly dwindled with the ubiquitous application of sophisticated computer control.

For larger businesses, PBXs serve as switching systems for internal calls and are connected to the public telephone network by a group of external lines or trunks. These are engineered to handle the traffic entering and leaving the PBX. The number of these external lines is usually much less than the number of stations served by the PBX. These systems include attendant positions for the screening of certain calls entering or leaving the PBX.

PBXs serve anywhere from less than 100 telephone stations to large installations with more than 10,000 stations. The equipment is located on customer premises and may be leased or purchased. Also, like small business systems, PBXs have become much more versatile and smaller in size with the use of integrated circuit technology.

An alternative approach to placing business systems on the customers' premises is to have all the telephone and data terminals connect directly to a central office switching system in the public network. Centrex service thus provides the functional equivalent to the traditional PBX services. No special switching equipment is required on customer premises, and all the maintenance and administrative activities are done by telephone company maintenance personnel in the central office. Depending on the situation, these and other advantages may outweigh the costs of having a line per terminal connecting to the central office.

More recently, local area networks (LANs) have come into being that allow switched interconnections among a local community of computers and computer terminals. Such communities of communications can be found in large business firms, universities, and hospitals. These switching systems use a different switching technique known as packet switching for efficiently handling data, a technique that is relatively inefficient for switching voice communications.

Packet switching evolved from the need to in-terconnect networks of computers and has proven to be economical for data networks covering much larger distances and areas than the LANs. Data traffic, particularly that generated by interactive sessions between video-display keyboard terminals and computers, often consists of short bursts for queries and responses followed by longer intervals of silence.

Packet switching, described in more detail below, does not require the use of a dedicated circuit for the duration of an interactive session and uses network facilities more efficiently. Accordingly, packet switching is considered a non-circuit switching technique differing from space-division and time-division switching techniques that dedicate circuits and/or fixed timing channels for the duration of the call or interactive session.

PBXs using digital time-division circuit switching techniques can provide circuit switched connections for voice or data at rates up to 64 kilobits per second (Kbps). Digital transmission and time-division switching techniques became the dominant modes for new circuit-switched telecommunication products and networks beginning in the late 1970s (see next section).

III. Modes of Transmission and Switching

Transmission of voice and data can take place in either analog or digital format. Switching systems use either space-division or time-division circuit switching network arrangements. Until recently, telecommunication networks were based predominantly on analog transmission and space-division switching. Today, the trend is moving rapidly toward digital transmission and digital time-division switching techniques.

There are circuit compatibilities and circuit synergies with these latter techniques. VLSI circuits and fiber-optic transmission systems are much easier to design and build using digital techniques. Also, digital-based telecommunications networks offer more flexibility in the features and services that can be provided.

A. ANALOG AND DIGITAL TRANSMISSION

Transmission of communications signals has been largely analog in nature, that is, an electrical signal or a radio wave is a replica in intensity or frequency of the intensity of the acoustical waves of speech. The major exception to this has been the use of the dots and dashes (short and long pulses) of codes such as the Morse code for transmitting keyboard characters in wire and radio telegraph systems.

Beginning in the 1960s, time-division multiplexing (TDM) for digital transmission was introduced to use interoffice trunk facilities more efficiently. A number of schemes have been introduced, using both analog and digital formats for TDM transmission.

The digital transmission technique most widely used in the world today is known as pulse-code modulation or PCM. Two formats were developed and used in the 1960s, one in North America and a second, slightly different format, in Europe. Both formats were accepted and standardized by CCITT and are in worldwide use.

PCM uses binary encoding of periodic samples of speech waveforms for the transmission of information. The binary sequence of pulses is compressed to a few microseconds in width, so the samples for a number of simultaneous conversations can be interleaved and merged (multiplexed) and transmitted over a facility that formerly carried only one analog voice channel.

At the receiving end of the transmission link the signal may be demultiplexed to a number of individual channels and reconverted to analog format for switching or retransmission in the analog format. This step is usually necessary to switch the channels to separate destinations in a space-division switch and to transmit the signals over ordinary loops to traditional terminals that were designed to work only with the analog format.

Initially, relatively expensive electronics were required for the terminal equipment for transmitting and receiving PCM signals. In addition to the terminal equipment, electronic reshaping and amplification using circuits known as repeaters are required to extend digital transmission for longer distances. These repeaters were periodically spaced every few kilometers to guard against excessive signal degradation.

Initial applications were relatively limited. First applications were to congested copper cables carrying interoffice trunks in ducts buried under busy streets in major metropolitan areas. The advent of VLSI and fiber-optic transmission systems has significantly lowered the costs of PCM transmission, making it the choice in much more widespread applications.

Today, PCM digital transmission dominates the growth of transmission circuits, especially

among interexchange trunks. This form of transmission is also beginning to appear in portions of the loop between the central office and the terminal equipment.

The introduction of fiber optics is hastening the deployment of PCM digital transmission in loop and interoffice trunk applications. The advantages include increased distance required between repeaters and much greater capacities for additional channels and bandwidth.

Digital subscriber loop carrier systems are used to serve new housing subdivisions and multistory residence buildings to save on loop transmission facilities. Similarly, central office switching systems may have portions of their switching and control elements distributed in the loop plant for larger but remote areas that are more economically served in this fashion. The planning for growth of new telephone serving areas must include all of these alternatives to arrive at the best solution.

Figure 2 depicts the pulse-code modulation scheme widely used with TDM transmission. By sampling speech signals 8000 times per second and encoding the samples into an 8-bit binary equivalent, voice signals can be represented by a digital stream of 64 Kbps. It can be shown that this is sufficient (and until recently necessary) to represent the voice signal with fidelity equivalent to that of analog transmission techniques.

By compression of the 8-bit samples, 24 voice channels can be transmitted at slightly more than 1.5 megabits per second (Mbps). (The European standard compresses 30 channels into

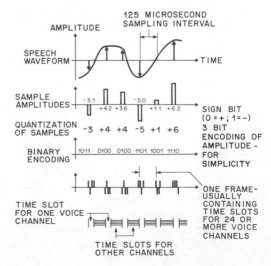

FIG. 2.   PCM encoding and multiplexing.

slightly more than 2.0 Mbps.) This transmission is one-way, so two pairs of wires are required for 24 conversations. Here 22 pairs have been regained for new circuits for growth in traffic. For this reason, TDM transmission systems are sometimes referred to as pair-gain systems.

B. SPACE-DIVISION AND TIME-DIVISION SWITCHING

Figure 3a shows an elementary example of space-division switching for establishing circuit connections between telecommunications terminals. The two-dimensional space-division switch consists of horizontal input buses and vertical output buses that carry telephone signals. Inputs to the switch can be connected to outputs by operating individual crosspoints, as shown in the figure. Input A is connected to output 2, and B to 1, with spatial separation of the transmission paths (hence, the term space-division switching).

Note that the crosspoint is operated to set up the connection and then remains operated for the duration of the call, which lasts anywhere

from a few seconds to hours or even days. Electromechanical relays and switches were originally employed in the earliest automatic switching systems. To operate and release the connections in such switches typically requires several hundred milliseconds, but this amount of time is inconsequential in a typical telephone call lasting 2–3 min.

The number of crosspoints grows very quickly; if M and N are the same, then the number grows as the square of the number of inputs. Relatively fast and reliable connections with this type of switch limits N to 10 to 20 inputs in practice. For larger numbers of connections, arrays of switches in numbers of stages provide access between more inputs and outputs. If properly designed the number of crosspoints grows much less rapidly than the square of N. This also helps control the cost of a space-division network.

Figure 3b shows two stages of arrays with three switches in each array to connect nine inputs to nine outputs. Only 54 crosspoints are required as opposed to the 81 of a single-stage switch; however, the network cannot always have every input connected to an output, depending on the choice of connections to be made.

Such a network is said to be "blocking": Certain connections block attempts to set up other connections. In this example the frequency of occurrence of blocking is not acceptable. For larger numbers of inputs and larger values of N for the individual space-division switches, designs exist that have better chances of not blocking and require much less than N^2 crosspoints.

A simplified example of a time-division switch is shown in Fig. 4. Assume that three channels (A, B, and C) of time-division multiplexed signals are presented to the inputs of the switch. They are carried in time slots 1, 2, and 3, respectively, in the multiplexed stream. The switch has a time-slot memory and two control memories. The input and output crosspoints are operated sequentially at times t_1, t_2, t_3, t_1, t_2, ... in synchronism with the incoming stream A_1, B_1, C_1, A_2, B_2, ...

The crosspoints to operate are indicated by the contents of the control memories. The input crosspoints are operated in the sequence 1, 2, 3, 1, 2, ... in synchronism with the output crosspoints operating 3, 2, 1, 3, 2, ... The net effect is to change the time-slot sequence of signals leaving the switch to B_1, A_1, C_1, B_2, A_2, ... or switch the signals in time. By providing different

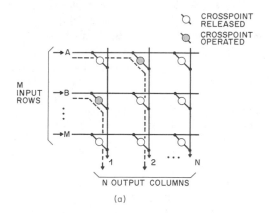

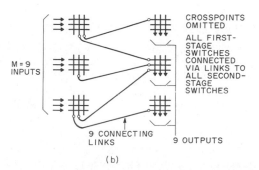

FIG. 3.   Space-division switching. (a) Space division matrix switch and (b) two-stage switching networks.

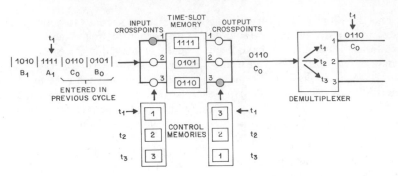

FIG. 4.   Time-division switching using a time-slot interchanger.

sequences to the control memories, any arbitrary interchanging of incoming time slots can be achieved. The demultiplexed signals can be switched to any one of the output crosspoints 1, 2, or 3, depending on the contents of the control memories.

Note that the crosspoints in this example are required to operate at much higher speeds than for space-division switching, previously described. In actual practice, switching speeds of a few nanoseconds or less are required. The faster the switches can operate, the more channels can be switched. Increasing the speed of operation is very desirable to make time-division switches much more efficient than space-division switches in terms of equipment, floor space required, and costs.

The example shown here is a particular form of time-division switch called a time-slot interchanger (TSI). There are other techniques, and switching system connection networks can be made of arrays and stages of several kinds of time-division switches, similar to the techniques used with space-division networks.

Space-division switches require the demultiplexing of time-division transmission signals before performing the switching. If the outputs from the switch are to be transmitted to another switch by digital time-division techniques, then the signals have to be multiplexed again. These demultiplexing/remultiplexing steps are not required of time-division switching systems. Consequently, as Fig. 5 shows, a synergy exists where digital transmission and digital time-division switching have to interface.

Another advantage of time-division switching networks over their space-division counterparts is the ability to expand the switching system connecting networks over a wider range of terminal sizes, from very small to very large, without suffering diseconomies. A single design of switching system network can conveniently serve small central offices having a few hundred lines to large metropolitan applications having 100,000 lines or more. The single design advantages include commonality of hardware and software design and common parts to manufacture and to keep as spares over a wider range of applications. The operating administrations enjoy the advantage of having to operate and maintain a lesser diversity of equipment with time-division digital switching systems than with the earlier space-division designs.

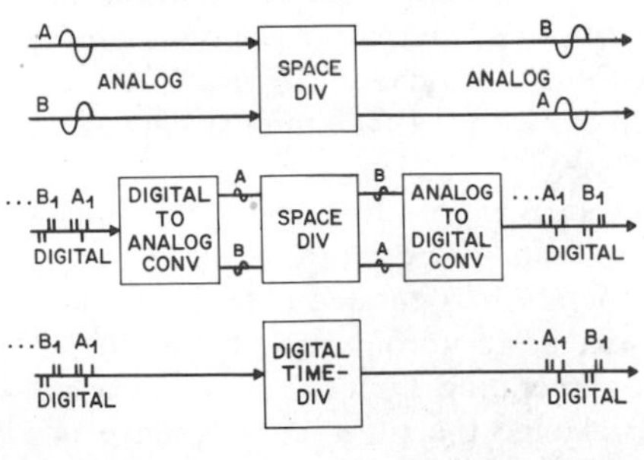

C. NONCIRCUIT SWITCHING TECHNIQUES

Telecommunications networks that establish a path for voice or data communication for the duration of the call (circuit switching systems) have been most efficient for voice communication but may not be the best for data communications; for example, a number of computer terminals operating interactively with a host computer offer relatively short bursts of communication followed by longer periods of inactivity. This kind of traffic can be handled relatively efficiently with packet switching. As each

FIG. 5.   Synergy of digital transmission and switching.

message is presented to the network, the terminal gear segments the message into a sequence of packets. Along with the data to be transmitted are appended additional bits of information. These include an address to indicate the intended destination and a sequence number to indicate the order of the segment in the overall message.

As each successive packet is transmitted to a packet switching node, it is directed over a transmission facility to the next node toward the destination. However, at each node the packet can be held or delayed if all links are busy or can be directed over several alternate routes leading to the destination.

Successive packets may encounter different delays, and the arrival time of the segments may vary or even be reversed. By appending sequence numbers to the packets, the reversals can be detected and remedied.

Packet switching systems are efficient for some modes of data communication, but part of their efficiency derives from the acceptability of variable delays in the delivery of the packets of information. If one works on a computer terminal, and if occasionally the response takes more than one or two seconds, then the short pauses are acceptable, particularly since the computer appears to be willing to wait for terminal users to provide their responses.

Speech can also be packetized by first going through the encoding process as previously described for TDM transmission; then the successive coded samples, with appropriate manipulation, can be processed as a stream of data. Lightly loaded packet switches transmit this information with delays undetectable to the ear and provide an acceptable quality of transmission. However, the bit rate for voice is relatively high compared with most interactive terminal operations and is ongoing instead of appearing as occasional bursts; also, under relatively moderate loads for processing the bursty traffic, the quality of speech degrades due to variable delays of the packet transmission.

With newer circuit technologies coming into being, the ability to manufacture extremely fast, high-capacity packet switching systems will eventually make such a switch practical. Techniques to reduce the size of the bit stream representing speech will also tend to make such an approach more attractive. For example, encoding techniques have improved enough to compress TDM transmission from 64 Kbps per voice channel to 32 Kbps at a quality level acceptable for standardization in the telecommunications industry.

Another technique, long in use in transoceanic submarine cables, can also be used to reduce the bit rate of voice being presented to a packet switch. Time assignment speech interpolation (TASI) takes advantage of the large number of silent intervals encountered in two-way voice conversations. The omitted intervals are used to inject syllables or phrases from other conversations.

The adaptation of this technique to packet switching uses two special packets to denote the time of the beginning of the silent interval and of its end. The two packets replace hundreds or thousands of packets that would otherwise be generated to describe 8000 samples of silence per second.

In summary, sophisticated digital processing techniques, large-scale integrated circuits, and optical transmission systems have vastly extended the capabilities of telecommunications transmission systems and at the same time lowered their capital and operating costs.

IV. Telecommunication Switching Systems

A. ELEMENTS OF A SWITCHING SYSTEM

Figure 6 shows the basic elements of a communication switching system based on circuit switching techniques. The four basic elements are the switching connection network, the control elements, the signaling elements, and the human–machine interface.

The switching connection network is used to set up connections between one telephone loop and others that are connected to that same switch, to service circuits for signaling operations, and to trunk circuits for interoffice calls. It is assumed that the switching connection network provides connections between any telephone on the left and any trunk, service circuit, or junctor on the right.

The junctor circuit in this system (other network arrangements, not shown, may not require this special circuit) is used to provide a loop for connecting two telephones in the same network: one to each port of the junctor.

The control elements may be centralized as in Fig. 6 or distributed in the signaling elements and the network. The controls of a central office switching systems are predominantly a central-

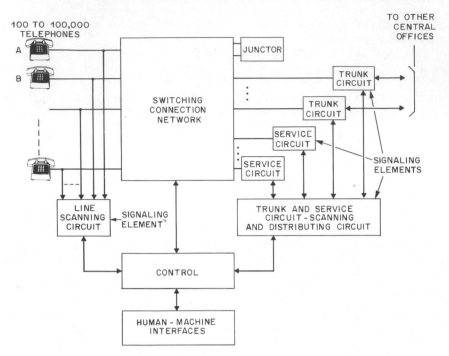

FIG. 6.   Basic elements of a switching system.

ized mainframe computer, a minicomputer, or a group of microprocessors operating as distributed controls or as a centralized multiprocessing array. The major portion of the complexity and sophistication lies in the stored programs that operate the central office switching system.

The advantage of architectures that use distributed microprocessors is a common design to serve economically either a few hundred lines in rural villages or the largest central offices in metropolitan cities. A single design to serve many applications conserves software development, which is the major component of designing such systems for architecture. This is very important since the work to define what the programs must do and to write and fully test such large software systems requires hundreds of millions of dollars, more than the development of sophisticated hardware. This approach is to be found in most second-generation digital switching systems.

A centralized mainframe processor best serves the larger offices, and the system designers do not have to deal with interprocessor communication. Before the advent of microprocessors and large semiconductor memory chips, the choice leaned very heavily toward a single mainframe approach.

The signaling elements of the hardware are the scanners, distributors, and service circuits. Scanning and distributing elements in the trunk circuits and junctors are used to transmit and receive signaling information from loops and trunks to and from other switching systems.

The human–machine interface combines elements of hardware and software to allow operating, administrative, and maintenance personnel to communicate with the system. The hardware usually includes displays of system elements and keys or buttons to allow reconfiguration of the system in case of difficulties that cannot be overcome automatically. Also, there are video display units with keyboards for entering data to change and add services for individuals or groups of customers. These terminals also may be used to aid maintenance personnel to locate faulty hardware in the system quickly and safely. Extensive software to support these links and human factors considerations are important since these interfaces can make a big difference in the effectiveness and efficiency of the craftsperson. A very large part of the annual operating expenses is the salaries for these people; inefficient designs put an otherwise good switching system at an economic disadvantage compared with its competitors.

B. SOFTWARE FOR A SWITCHING SYSTEM

A modern central office switching system is controlled by a large collection of integrated program modules that provide the equivalent of a real-time time-shared computer serving a large number of interactive terminals. It is a real-time system in that it must respond to varying demands for telephone connections without noticeable delay in sending signals such as dial tone and ringing and setting up talking connections. It is time-shared in that it must serve, simultaneously, 100,000 telephones or more without delay except in all but the heaviest of traffic.

To provide a software design that can be developed and managed by large teams of computer scientists and telephone engineers, modularity is the key to success. Often teams of hundreds of skilled individuals are required to complete within a few years a successful design of a large body of programs of a million lines or more of program steps.

Over time, telecommunications networks and the switching systems that provide much of the control functionality of these networks have been evolving as technology permits and customer needs have become more sophisticated. One consequence of this evolution is that the software design must change and grow to meet the expanding requirements for new capability. Without modularity, and the flexibility and clarity it can bring, the desired software changes may become too costly to make even if they are realistically possible.

An operating system (OS) is used to stitch together the programs that process calls and carry out actions in system hardware. Some of these programs are operating continuously on loose or tight timing schedules; others operate only as calls come and go in the system. Different programs carry out different steps of the call. The selection of the call processing programs depends on the type of call being made.

The system action programs are segregated from the call processing programs to isolate the latter from the details of the switching system hardware. The designer of the call processing programs is free to think about the capabilities and features at a higher level.

This partitioning also permits the use of higher level languages to increase the productivity of this programming activity and to enhance understandability of the software. Properly designed, this approach also permits portability: the ability to run the same software on different processors with a minimum of changes. Newer, faster processors can be adapted much more economically with portability, and the amount of damage to functional equivalence of the two versions of the programs is reduced.

C. SWITCHING SYSTEM MAINTENANCE CONSIDERATIONS

Telephone switching down-time objectives (where service is denied to any group of 100 or more customers) is stated to be no more than 1 hour in 20 years. Maintenance programs provide preventive and remedial support to keep the system running in the face of hardware, software, or craftsperson-related problems. Other programs are used to exercise faulty equipment to pinpoint faulty components, usually to within a very few suspected plug-in circuit packs.

In addition to the complex real-time, time-shared requirements for telephone switching applications, switching systems are required to provide extensive automatic maintenance capabilities to ensure highly dependable operation. Typically, a central office switching system has a dependability objective of not denying service to more than 100 customers at a time for more than 3 minutes a year over a 20-year expected life of the system. In addition, not more than one call in a thousand should be mishandled. A much lower failure rate for creating errors in billing data generated by the switching system is required to be acceptable by most operating administrations.

To attain these objectives, equipment is redundantly provided, and extensive error checking of program instructions and data is done as they flow through the equipment. High-speed circuits are included to automatically switch suspected faulty equipment units out of service and switch into service the redundant replica that had been standing by.

Robust, fault-tolerant software is also an essential part of the design. To provide dependable operation of a telecommunications switching system, far more than bug-free operating systems and call-processing software are required. For example, special programs can periodically audit data gathered by the call-processing programs for internal consistency. Inconsistencies due to transient errors can be detected and eradicated before they spread and destroy the ability to continue operating in a rational manner. This is just one example of a maintenance program, in this case for preventive maintenance. Frequently, almost as much programming is required to provide maintenance software. These programs provide much of the control of preventive and remedial actions.

In addition to supporting the detection and elimination of hard faults and transient errors that affect the ability to process and bill telephone calls, there are large diagnostic test programs that exercise suspected faulty equipment to localize the problem to one or two plug-in circuit packages.

These capabilities assure dependable operation and at the same time simplify and reduce the cost of maintaining these systems. With today's cost-effective technologies, and even with the high degree of automation through system software, the cost of operating and maintaining these systems dominates the total cumulative capital and expense expenditures. A total view of the life cycle costs should be understood in selecting a modern communications switching system.

D. OVERVIEW OF SOFTWARE STRUCTURE

The operating system (OS) directs the routine operations that are run periodically, one example being the scanning of all the lines served every few hundred milliseconds. The OS controls the timing of these programs and schedules other programs to carry out all the call-processing and maintenance functions.

In addition, the OS monitors the real time consumed as the traffic of telephone calls change in the office. Built into the OS, or associated with it, are overload controls that reschedule priorities and frequencies of activities to ensure that the system continues to process as many calls as practical whenever the offered traffic exceeds the capacity of the system. One way to do this is to restrict the number of call originations that can be generated per unit time. This has the effect of delaying the dial tone for an extra few seconds for some calls, but this is far better than having the number of calls handled drop to a very low percentage of the peak capacity.

The OS must also handle the interleaving of unscheduled maintenance programs to recover the system from faults so as to minimize the mutilation of calls in progress at the time of occurrence of the fault. The OS also must schedule segmented diagnostic programs to allow the craftsperson to be able to run tests and get results in a timely fashion, but not at the expense of deferring call-processing programs so much that calls are mutilated or lost.

During normal periods of traffic handling there are preventive maintenance programs that take advantage of unused real time to run test programs of hardware and to audit system memory contents for correctness and consistency. In periods of high traffic these programs must be deferred, and it is the responsibility of the OS to schedule them so as not to interfere with peak demands on call processing.

There are many ways to organize data structures in memory, for permanent and transient information and for the call-processing software that operates with real-time time-shared inputs and the data. Stored-program control of telecommunication switching systems has evolved over a period of more than 30 years. Faster and cheaper processors, distributed microprocessors, higher-level software languages and compilers, relational data bases, and data base managers have all been employed or have influenced the evolution of the large software packages required for switching systems. The result has been improved modularity and flexibility to absorb the evolution of hardware technology and feature capabilities without unduly damaging the original software.

In the description of tracing a simple call, the methodology assumed resembles that of the 5ESS® switching system, which has a software architecture designed in the 1980s and uses the technologies and techniques noted above. However, to simplify the description, many details are omitted and others changed. The description (that follows) is written as if there were a single centralized control. In fact, the control and data structures of the 5ESS® system are distributed, and a number of different processors are involved in different parts of the call setup process.

Tracing a Call in a Switching System

The control (the processor executing the programs) is traced through its thread of actions for one call. In reality the control is time-sharing its many activities among all calls as well on other administrative and maintenance tasks. The OS directs the control from task to task to create the time-sharing of functions. It activates processes, (programs and associated blocks of transient memory) as each process is needed. Once the process is completed, it is deactivated, and the blocks of memory are released to become resources for other processes.

The steps of handling an individual call begin with the customer lifting the handset of the telephone, going off-hook in telephone parlance.

Among the periodically scheduled activities is a system process (which is ongoing and does not require activation) that scans all of the lines served every few hundred milliseconds looking for originations. The line scanner sensors detect the increase in loop current associated with a telephone going off-hook.

When customer A picks up the telephone to make a call, the next scan of that loop results in a report to the control. This triggers another process that causes the control to consult its memory records to determine whether loop A is equipped with a rotary dial or pushbuttons with Touch-Tone dialing capability. The control also arranges for the line scanner not to send additional reports of the same origination in subsequent scans.

As part of the modularity of the software design, there are terminal processes associated with the terminals of the network: the lines and trunks. The call processing for the initial steps is handled by an originating terminal process, that is, it handles the steps associated with the actions of the originating telephone line or network terminal.

The line being called in this example (the terminating line) has a separate process associated with it: the terminating terminal process. It handles the completion of the call in this example. Different originating and terminating processes are activated depending on the type of terminal. A different process is required for the origination of a call from a residential telephone and a public coin telephone.

The control activates the proper originating terminal process for this call; associated with this step is the assignment of a block of memory for the process to store transient information associated with the call.

This same program portion of the originating process is activated for other calls and run concurrently as needed. The OS directs the same program to process different calls as required in sequence to carry out the actions for concurrent call originations. The distinction among these processes includes different blocks of transient memory for each process.

Assume that the memory records indicate that loop A is equipped for Touch-Tone service. The originating process calls another process to obtain the line-associated data for this line. Other information required, for example, would be the directory number of the originating line for billing purposes.

The control then sets up a connection between loop A and a service circuit equipped with a dial-tone generator and a Touch-Tone dialing

receiver that detects the pushbutton signals. The outputs of the detector are scan points connected to the trunk and service circuit scanner. These scan points indicate the values of detected digits. A separate process handles the details of finding and establishing the desired paths in the switching control network; complex details of handling the network are shielded from the many terminal processes.

The originating process for this line then causes the control to examine periodically the Touch-Tone dialing receiver scan points for digits. This scan takes place at higher speeds than the line scanning to ensure that no digits are missed. The OS and hardware-generated timing signals control the frequency of the actions according to the timing needs. If the scanner reports a digit, the originating process stores that value in the locations reserved for digits in its memory block.

Upon detection of the first digit, the service circuit is directed to disconnect the dial-tone generator but to continue to keep the connection to the Touch-Tone dialing receiver.

As digits are collected, they are also analyzed to see if all of the digits have been collected to begin the next steps of setting up the call. For example, if the first three digits dialed are 911, then no more digits are expected because this is the number for emergency service in the United States dialing plan.

Timing may also have to be initiated. If the first digit dialed is 0 then there may or may not be additional digits. When the initial 0 is detected, a timing counter associated with the originating process's block of memory is started. Upon a completion of a fixed timing interval, say 15 seconds after the first 0, if no additional digits are received, and if the loop is still in the dialing state, then the call is recognized as one requesting connection to an operator for assistance.

Once all of the digits have been collected for a call, which for some calls include those to designate an inter-LATA carrier, the control can take the next step in call setup. Before processing the collected digits, the control must disconnect loop A from the Touch-Tone dialing receiver to make it available for other calls. There is normally a pool of these receivers, which are engineered for the level of expected traffic in the office, and must be treated by the switching system control as a limited resource.

The originating terminal process then asks other processes to determine the location of the line being called and to activate the process to complete the call setup. Assume the customer on loop A dialed the telephone number associated with loop B. The physical location of loop B's connection to the switching has no relation to the dialing number found in the telephone directory. For each directory number associated with the switching system, there is stored in memory the relationship of the directory number to the equipment number corresponding to the called line.

At this point the terminating terminal process is associated with the called line and carries out the rest of the steps of call setup. The equipment number is used by the terminating process to operate a line scanner to examine the state of the line being called.

A scan is directed to loop B to determine if B's telephone is idle and available for the call. If that is not the case, A's loop is connected to a service circuit that provides a busy-tone signal to indicate that B's line is busy.

If loop B is idle, the control arranges to connect a service circuit to both loop A and loop B. Loop B is connected to a ringing-tone generator to alert B to an incoming call. An audible replica of the ringing tone is connected to loop A to inform A that B's loop is being rung.

Once these connections have been established, the control advises the trunk-scanning programs to examine periodically the scan points associated with the selected ringing-tone and audible ringing-tone service circuits. In this manner the system monitors loop B for off-hook indicating the call has been answered. It also monitors loop A since B may not answer; if that is so, A eventually abandons the attempt and goes on-hook.

Assume A remains on the line, and B answers. The control responds to this change in call status by removing the connections to ringing and establishing the talking connection through the network to loops A and B. In the example of the switching system portrayed in Fig. 6, there is a special service circuit known as a junctor that serves to complete this connection. The control directs loop A to be connected to one port of the junctor and B to the other so conversation can begin.

The terminating process also directs a supervisory scanning program to examine periodically the two lines for the end of the call by detecting the on-hook condition occurring on loop A or loop B.

Once loop A or loop B hangs up, the connection must be taken down to complete the call.

This step frees the loops for other calls and releases the network paths for other connections.

Note that the hanging up of the telephones on loops A and B is not usually a simultaneous event with respect to the scanning of those loops as seen at the junctor scan points. Special timing blocks in memory and disconnect timing scans are initiated by the terminating process. This avoids the possibility of having the second line to go on-hook being falsely detected as an origination unless that line stays off-hook for some period of time, say 10 seconds. A longer time is taken as a request to place another call.

If a time out occurs, a new originating terminal process starts the processing of the next call. In either case, the terminating terminal process is now finished with the call and is deactivated, and its call memory block is released for use on other calls.

Associated with each call is a call record maintained in transient memory. The call-processing and maintenance programs have, among their tasks, the reporting of call events to an administrative process that updates the record as the call proceeds. Other administrative programs can access these data as required to generate traffic reports, uncover traffic sensitive network or terminal problems, and gather the information for billing.

Depending on whether the calling number has a fixed billing rate or some form of measured service in terms of call distance and duration, additional processes are instigated. If the call is to be billed, the necessary data can be retrieved from the call record upon completion of the call. The data can either be stored on disk for retrieval during off-hours or sent immediately via a data link to a billing accounting center that handles the conversion of data into paper records to be mailed to the customer.

In this example a simple intraoffice call has been traced. At the same time that this call is being handled, hundreds or thousands of other calls are occurring at random intervals and must be handled correctly in real time. The description given here is for only one type of simple call. Calls to and from other central offices must also be handled. Long-distance calls, calls to emergency numbers, directory assistance, and calls to special operators and to overseas destinations all require different call processing sequences. Special services, such as toll-free calling and customized services, require special handling and take additional real-time resources.

In addition, the programs must interleave in a noninterfering fashion all the maintenance and administrative processes to keep the switching system providing dependable service and having its data memories frequently updated to meet customer requests for changes or added specialized services. Many other complications were omitted or only alluded to.

When all these requirements are taken into account, and when all the time-sharing and real-time concerns are noted, the detailed requirements and the resulting software are very large, on the order of 1 million lines of program steps. The undertaking of such a project requires a substantial development effort, on the order of 1000 person-years, and the capital backing of hundreds of millions of dollars to support the designing, writing, and debugging of the integrated hardware–software system. It is for these reasons that modularity, flexibility, and comprehensibility are essential.

V. The Telecommunications Network in the United States

Figure 7 shows the hierarchical structure of the U.S. telephone network as it appeared before divestiture in 1983. There were some 20,000 local switching systems serving approximately 120 million telephones. In addition to the interconnecting of small numbers of local switching systems, overall connection in the telephone network is accomplished with local tandem switching systems and a four-level hierarchy of toll switching systems. These last form the long-distance network operated by AT&T and other inter-exchange carriers.

The hierarchical structure is one way of orga-

nizing redundancy in the toll network to provide the continuation of long-distance service in the face of occasional failures of a particular switching system or transmission link. In this way the network can work around traffic congestion to provide good service in periods of very high traffic and in the event of natural disasters.

Figure 7 demonstrates the redundancy of this network. A call from telephone A to telephone B is first attempted to be completed via the intertoll trunk in the group of trunks numbered 1. From there it proceeds to the distant local central office, either directly via the toll-connecting trunk marked i or via the local tandem over the route ii. If all trunks in group 1 were busy, the toll center would look for alternate routes, trying the routes marked 2, then 3, and then 4. If all the links shown in the figure were equipped (only an ideal case), there would be up to eight possible ways through the network. In this way, as trunk routes are used up on other calls, the network attempts to use the remaining facilities to assume efficiency and good call-completion rates during the busy hours.

It should be noted that this hierarchical structure, established by AT&T for its long-distance network in 1930, is now being changed to an alternate routing scheme that adapts dynamically with the time of day and traffic to provide even greater efficiencies than the 5-level hierarchical structure.

When the proper sequence of digits has been dialed, a call can be automatically placed by the network to almost any of the telephones in the country. Service at the local level for calls completed locally and the portion of access to the toll network provided in these local switching systems or local tandem switching systems is provided by local operating companies. There are some 1200 operating companies ranging in size from companies with a single switching system serving a community of several hundred lines to companies such as one of the Bell operating companies (BOCs), General Telephone and Electric, United Telephone Company, and Contel. These large companies serve millions of customers, some in many states and spanning thousands of miles.

As part of the settlement of an antitrust suit between AT&T and the Department of Justice, AT&T agreed to divest itself of its local Bell operating companies. The BOCs are no longer able to provide intrastate long-distance service across certain boundaries. This service is now called interexchange service, and the long-dis-

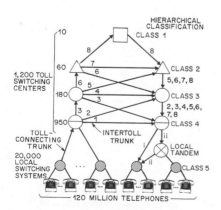

FIG. 7.   United States hierarchical network.

tance telecommunications companies are sometimes referred to as interexchange carriers. Similarly, the local areas are referred to as local exchanges and the local operating companies as local exchange carriers.

These local areas are sometimes called local access and transport areas (LATA). The 22 Bell operating companies were reorganized into 7 regional holding companies, and more than 160 LATAs were formed to define the extent of local exchange and interexchange service.

Competition exists among the interexchange carriers for service; the local exchange companies provide exclusive monopoly services, subject to regulation by local and state regulatory bodies. AT&T is also subject to state and federal regulation to a lesser degree because of the existence of competition.

Certain elements of local exchange products and services were opened to competition at the time of divestiture. These included the lease or sale of customer premises equipment including telephones, key systems, and PBXs. In addition, the FCC, in allowing mobile cellular telephone service to be introduced, required that these services would also be open to competition by nonwireline carriers.

The Department of Justice also stipulated that the BOCs provide equal access so that a person dialing a long-distance call would be able to handle the call in the same fashion for different long-distance companies. The consequences of that stipulation has had a significant impact on the design of switching systems.

VI. Loop and Interoffice Signaling

Signaling begins with the act of lifting the telephone handset. An increase in loop current is detected at the central office as a request for origination. Dial tone, dialing, ringing, and answer are all familiar signaling steps. Less apparent to customers are the steps of signaling between switching systems for calls that require additional transmission links and switching systems for their completion. Also less apparent are the signals to sustain the connection for the duration of the call and the signals at the end of the call to stimulate the taking down of the connection.

Automatic switching systems operating under the control of rotary dials have been in operation for nearly 100 years. With its pulse rate of a nominal 10 pulses/sec to represent decimal digits, the dial telephone transmits intent of call

destination at a modest few bits per second. Pushbutton telephones using Touch-Tone service have speeded up that element of signaling by a factor of two or more. Automatic Touch-Tone dialers can dial a local call in less than two seconds, about five times as fast as a rotary dial.

The rate of transmission of information among switching systems within the network has risen more dramatically from a few tens of bits per second with voice-band signaling techniques to the present speeds of 56 Kbps with common channel signaling (CCS).

A. SIGNALING TYPES AND THE U.S. NUMBERING PLAN

Several types of signaling have been described in the tracing-a-call example. These can be defined as alerting, addressing, and supervisory.

Alerting signals provide stimuli for customer actions and convey information as to the state of a call. Dial-tone informs the customer that the switching system is ready to accept digits. Ringing and audible-ringing tones indicate that the call has been placed and that the terminating party should answer. Busy tone indicates that the call cannot be completed, and the caller should hang up.

Address signals refer to the dialed digits, starting with the originating telephone, as they are translated and conveyed in varying formats across the network in setting up the call. The structure of these addresses depends on a national numbering plan in the public network. A somewhat simplified description of the numbering plan used in the United States follows.

Seven-digit codes for the form NNX-XXXX define the format within a numbering plan area. N can be any of the digits 2 through 9 and X any digit from 0 through 9. The 7-digit codes provide all the numbers required for all but the largest metropolitan areas or entire states in many instances.

To reach outside such areas requires a 3-digit prefix of the form NBN, where B is either a 0 or a 1; as before, N is limited to 2 through 9. In the United States, this 3-digit prefix is often referred to as the numbering plan area code.

More recently the long-distance prefix has to be prefixed with a 0 or a 1 to provide additional direction of such calls. The 1 indicates direct completion of the call automatically; the 0 is used to request operator assistance or to trigger additional steps for billing via credit card authorization. Additional digits defined in an international numbering plan may be prefixed to extend automatic placement of calls world-wide to most telephones in hundreds of countries.

The structures of numbering plans are designed to make the most efficient use of the numbers available to provide consistency of dialing procedures for telephone customers wherever they may be. They are intended also to be efficient in requiring the shortest dialing sequences to reach all telephones in an area, a nation, and the world.

Telephone switching systems must be designed to handle the complexity of this numbering plan structure. Any changes in the numbering plan have significant impact on the switching equipment and, of course, on the ease of the customer's ability to use the network.

A major change that took place in the United States in the 1980's was to allow customers to use the numbering plan to select the long-distance carrier of their choice. Three more digits had to be added to the dialing plan, and extensive changes and additions are required to implement this capability in tens of thousands of telecommunications switching systems.

Supervisory signals are ongoing signals in the loop circuits and network of which the telephone customer is normally unaware. Supervision is necessary for locating idle switching and transmission equipment and for maintaining a connection, once set up, across all of the required equipment. The same supervisory signals provide the stimulus for taking down a call connection once a call is finished. In the example of tracing an intraoffice call, the scanning of the junctor circuits for disconnect represents monitoring of the supervisory signals for the talking state of that call.

B. Loop Signaling

Loop signaling has been discussed briefly in Section IV,E, and Touch-Tone service is described elsewhere. [See TELEPHONE SIGNALING SYSTEMS, TOUCH-TONE.]

The rotary dial operates at 10–12 pulses/sec. Modems for data terminals include dialers that can operate with either signaling method. Personal computers and more sophisticated modems work as auto-dialers so that telephone numbers can be entered directly from keyboards or from repertoires of frequently used numbers stored in memory. With the advent of integrated services digital networks (ISDNs), the signaling

capability from telecommunications terminals is expected to expand.

C. Signaling between Telecommunication Switching Systems

Figure 8 shows a simplified comparison of conventional interoffice signaling and common channel signaling (CCS). Both signaling examples show a single toll or interchange trunk connection, but there may be two or more transmission links and additional switching systems required depending on the routing of the call.

The conventional signaling methods operate by selecting or seizing an idle trunk to establish the link between switching systems. Either a dc voltage condition or a voice-frequency tone indicates the idle supervisory state of the trunk. The act of seizing the trunk causes the supervisory signal to change to the busy condition. Then the address signals, ringing, and answer supervision pass back and forth over this link, which will eventually be used for the call itself.

Associated with each trunk circuit are senders and receivers to transmit and receive this information. The signals may be single-frequency or dual-frequency tones in the voice band and travel the same transmission path as the subsequent conversation. (The dual-frequency tones operate at different frequencies than that for Touch-Tone service.)

These signals operate at moderate speeds (a few tens of bits per second) and result in call

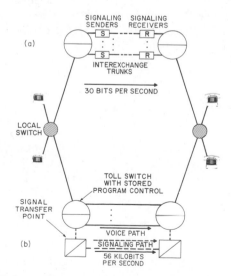

FIG. 8. (a) Conventional (call-completion time up to 10 sec) and (b) common channel signaling (call-completion time 1–2 sec) between toll centers.

setup times that take as long as 10 seconds to route calls across four or five links. Call setup time is defined to be the span between the completion of dialing the last digit and the beginning of ringing for that call.

The 4ESS™ switching systems have been rapidly deployed in the AT&T long-distance network beginning in 1976. They now carry more than 80% of the traffic in that network. With them came the new common channel signaling (CCS). With its use in these and most of the other AT&T toll switching systems, virtually all calls in the network are handled by CCS.

The configuration shown is based on the CCITT recommendations and was designed for the U.S. long-distance network. The recommendations include other configurations than the one depicted. Some of these may be developed for other networks with different structures or demands.

Separate links are used to carry the CCS data across the network in packet switching networks that improve the speed of operation. The special packet switching systems in this network are known as signal transfer points (STP).

This arrangement uses a relatively few STPs, only 14 are presently anticipated to carry the signaling traffic, compared with the 175 AT&T-Communications toll switching systems. This network carried some 33 million calls on the average business day in 1985. For most of these calls the completion time has dropped to less than two seconds.

Figure 9 shows the scheme implemented in the AT&T-C interexchange network. The United States is divided into seven geographical regions, and a pair of STPs serves each region. Toll switching systems in a given region are con-

nected to both STPs, and redundant links interconnect all 14 STPs. The redundancy of switching and transmission equipment protects against failure due to individual failures of STPs or their connecting transmission links.

Because the conventional interoffice signaling allowed the customer access to its voice-frequency tones, the toll network was vulnerable to fraudulent use to avoid billing. Since CCS does not share the trunk circuit transmission path for signaling, this vulnerability is being removed from the AT&T-C network.

The most important advantage of this new signaling configuration is the provision of features and services on a network basis. CCS originated on the AT&T long-distance network and has provided a pervasive base of capabilities. This method of signaling is beginning to reach into the local networks to provide services at this level of the network. The deployment of computerized switching systems in terms of total lines and span of coverage is still unequaled elsewhere in the world.

VII. Evolving Telecommunication Services

In additional to the basis capabilities of providing interconnections between residence and business telephones and to private networks, many optimal services are available to customers today. With the advent of stored program control in these systems the variety and convenience of such services have grown substantially. Starting in the late 1960s, as the deployment of the systems increased to millions of customers in the United States, operating companies began to offer such services on a commercial basis.

A. Customized Calling Services

The ability to provide residential or business service from individual telephones is sometimes referred to as plain old telephone service (POTS). With the advent of trials with stored-program control switching systems by AT&T beginning in 1960, experimental added service capabilities were tried with telephone customers to allow them more flexibility in placing and completing their telephone calls. The trials were met with enthusiastic response, and widespread introduction followed the introduction of stored program control central office switching systems in the United States.

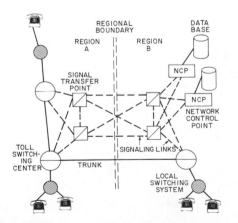

FIG. 9.   Common channel signaling (CCS) network.

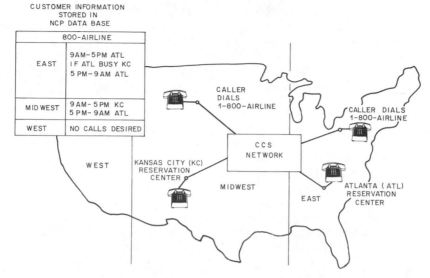

FIG. 10. Advanced 800 service.

The first of these, custom calling services, was introduced commercially by the Bell operating companies in 1969. Custom calling allowed telephone users to have, without special telephones, new control over the use of their telephones. This includes the ability to have calls automatically transferred to another telephone number, the ability to be informed that another party is trying to reach their telephone when it is already in use on another call, and the ability to use short 2- or 3-digit codes to dial a repertoire of up to 30 different numbers. The numbers stored could include 7-digit numbers for local calls, 10 or 11 digits for long-distance calls, and 14 or more digits for international calls. Furthermore, customers could add to the list or make changes in the repertoire by using their telephones to signal the central office switch of the desired changes.

Centrex is a collection of service capabilities to provide a group of business telephones with service equivalent to that provided with a PBX. Centrex capabilities were extended to include customer features and control far beyond those practical with the earlier switching systems, which lacked stored-program controls.

B. DATA-BASE SERVICES

Referring again to Fig. 9, note the network control points (NCPs) and the associated data bases. These processors and data bases can be accessed from toll switching systems in the CCS

connected network. This access is under control of programs written to provide new network services that are based on the information in the data bases. Figure 10 shows an example of a network service operating in this manner.

In the United States, 800-service refers to the ability to place an 11-digit call of the form 1-800-NNX-XXXX and not be billed for the call even though it may be an interexchange call or an overseas call. The business telephone customer pays for this service capability and is billed for the call usage. In this way, airlines, hotels, department stores, and other businesses attract their customers to call for reservations and purchases without the deterrent of paying for a long-distance call.

The business has the choice of coverage of such toll-free calling—international, nationwide, or limited to smaller zones according to the market reached. As it was initially implemented, prior to the introduction of the CCS network, extensive call coverage usually required more than one 800 number depending on where the call originated. The different numbers identified the source of the call to allow the billing charges to be accumulated according to the zone that corresponded to the distance of the call.

In addition, a large business customer would have several offices for receiving 800 calls, strategically located to reduce the total distance of all calls. With the distinctive 800 numbers, the calls could be routed to the closest available office. In the example in Fig. 10, two reservation offices are shown for the airline. In actual practice there could be as many offices as there are 800 numbers assigned. However, the customer must choose the correct 800 number or the call is not completed.

With CCS, it is relatively easy to add the software control to the signaling to transmit not only destination address information but also the originating address. In the example, the originating numbering plan area (NPA) is provided to the NCP along with the 800 number. Given this information, an array of 800 numbers for a single business is no longer needed. Nationwide business coverage with 800 service requires only a single 800 number rather than a long list of such numbers. The location of the origination of the call is included in the address signaling transmitted with CCS. This information is used to select the proper destination in the data base according to the desires of the business customer.

With the signal number, the business customer can also select an 800 number that represents the corporate name, product, or line of business. The mnemonic *airline* used in Fig. 10 corresponds to the digits 247-5463 on the dial. The advertising for the reservation service would read 1-800-AIRLINE, which is easy for the traveler to remember.

To help the business customer further, the selection of the destination of the call could depend on the time of day. In the example in the figure, if one office is closed, all calls are directed to the remaining office until the first office re-opens the following morning. Again, the number of offices and alternates can be extensive and can be easily modified by changing the contents of the table in the data base. Not shown, but also implemented, is the ability to redirect calls depending on the day of the week.

With the approximately 150 computerized switching systems controlling most of the AT&T interexchange network, new software to provide new features or to elaborate existing ones can be efficiently designed and rapidly deployed to improve service to all customers of the telecommunications network in the United States. Similar networks have started their introduction in other national networks as well.

C. INTEGRATED SERVICES DIGITAL NETWORKS (ISDN)

With the digital encoding of voice for transmission and switching, the fundamental differ-

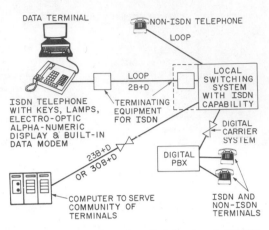

FIG. 11.   Integrated services digital network (ISDN).

ence between voice and data has been eliminated. The stream of bits in digital networks can be speech, streams of characters of data, or encoded graphics or other pictorial information.

ISDN provides the diverse capability for transmitting voice, data, and graphics on a single telephone line rather than separate facilities. Its early beginnings included activities in the CCITT to ensure the timely standardization activities to telephone administrations and manufacturers.

The first steps toward ISDN are indicated in Fig. 11. Special circuitry added to the loop is required to extend the bandwidth capabilities from the 64 Kbps required for one voice channel to 192 Kbps. This allows two channels of 64 Kbps for voice or data; the remaining 16 Kbps are available for signaling and additional information flow. The 64-Kbps channels are known as B channels, the 16-Kbps channel in this example is referred to as the D channel; hence the term 2B + D.

The single telephone line could simultaneously support a telephone conversation and data communication to a different distant destination. The remaining D channel can also simultaneously support messages on an optoelectronic display, informing the customer of another message without disturbing the voice and data calls already in progress.

A computer could also use a 2B + D channel if the data transfer rates were low enough. However, it can also use a 23B + D channel and transmit data to other computers or to a large number of data terminals at rates up to nearly 1.5 Mbps with the aid of a digital carrier system. To accommodate the larger number of B channels, the D channel capacity is expanded to 64 Kbps. (Note that the CCITT recommendations include 30B + D to match the European PCM carrier format, which operates at 2.0 Mbps.)

If the local digital switch has full ISDN capability, it will be able to handle packet switching operations as well as circuit switched connections. Then the computer could transmit its data in packetized format, and the local switch would redirect the packets to individual data terminals for interactive sessions. With the use of statistical multiplexers on a 2B + D channel, it would be possible to support a cluster of interactive work stations via a single ISDN loop.

Note the dotted line in Fig. 11, which indicates that the loop terminating equipment at the central office end may be an integral part of the ISDN switching system. For example, the 5ESS® digital switching system includes both internal terminations and the capability to do both circuit and packet switching for ISDN.

ISDN represents an entirely new format of signaling. Innovation in circuits, software, and services have been actively pursued in the 1980s. Standardization activities have also been substantial during this period in national and international standards organizations to permit different networks and equipment to work together to provide a new form of universal service where voice, data, and other forms of communication are readily available to all telecommunications customers who need them.

BIBLIOGRAPHY

Bohacek, P. K., ed. (1982). "Stored Program Control Network." Special issue, *Bell System Technical Journal* **61**(7), Part 3, September. AT&T Customer Information Center, Indianapolis.

Hayward, W. S., Jr., ed. (1985). "5ESS® Switch." Special issue, *AT&T Technical Journal* **64**(6), Part 2, July–August. AT&T Customer Information Center, Indianapolis.

Joel, A. E., Jr., et al. (1982). "A History of Engineering and Science in the Bell System. Switching Technology (1925–1975)," Vol. III. AT&T Customer Information Center, Indianapolis.

McDonald, J. C., ed. (1983). "Fundamentals of Digital Switching." Plenum, New York.

Pokress, R. L., ed. (1984). "Integrated Services Digital Network." Special issue, *IEEE Communications Magazine* **22**(1), January.

Rey, R. F., ed. (1983). "Engineering and Operations in the Bell System. Telecommunications in the Bell System 1982–1983," 2nd ed. AT&T Customer Information Center, Indianapolis.

TELECOMMUNICATIONS

Herbert S. Dordick   *Temple University*

GLOSSARY

Analog: Representation of information by continuous wave forms that vary as the source varies; method of transmission by this method. Representations that bear some physical relationship to the original quantity, usually a continuous representation.

Bandwidth: Difference in hertz (cycles per second) between the highest and lowest frequencies in a signal or that are needed for transmission of a given signal; width of an electronic transmission path or circuit in terms of the range of frequencies it can pass.

Broadband communication: Communications system with a bandwidth greater than voice bandwidth. Cable television is a broadband communication system with a bandwidth usually from 5 to 450 MHz.

Carrier: Signal with known characteristics (frequency, amplitude, and phase) that is altered or modulated to carry information. Changes in the carrier are interpreted as information.

Channel: Segment of bandwidth used to establish a communications link. A channel is defined by its bandwidth; a television channel, for example, has a bandwidth of 6 megahertz (MHz), a voice channel about 4000 hertz (Hz).

Circuit switching: Process by which a physical interconnection is made between two circuits or channels.

Coaxial cable: Metal cable consisting of a conductor surrounded by another conductor, in the form of a tube, which can carry broadband signals by guiding high-frequency electromagnetic radiation.

Common carrier: Entity that provides transmission services to the public at nondiscriminatory rates and exercises no control of the message content; organization licensed by the Federal Communications Commission (FCC) and/or by various state public utility commissions to provide communication services to all users at established and stated prices.

Demodulate: Process by which information is recovered from a carrier.

Digital: Function that operates in discrete steps as contrasted to a continuous or analog function. Digital computers manipulate numbers encoded into binary (on–off) forms, while analog computers sum continuously varying forms. Digital communications is the transmission of information using discontinuous, discrete electrical or electromagnetic signals that change in frequency, polarity, or amplitude. Analog intelligence may be encoded for transmission on digital communications systems.

Final mile: Communication system required to get the information or program from an earth station to where the information or program is to be received and used. Terrestrial broadcasting from local stations, the public switched telephone network, and/or cable television systems provide the final mile for today's satellite networks.

Frequency spectrum: Range of frequencies of electromagnetic waves (in radio terms). The range of frequencies useful for radio communications is from about 10 kilohertz (kHz) to 3000 gigahertz (GHz).

Geostationary satellite: Satellite with a circular orbit 22,300 miles in space that lies in the plane of the earth's equator and has the same period as that of the earth's rotation; thus the satellite appears to be stationary when viewed from the earth.

Local loop: Wire pair that extends from a telephone central office to a telephone instrument. The coaxial cable in a broadband or CATV system that passes by each building or residence on a street and connects with the trunk cable at a neighborhood node is often called the subscriber loop.

Message switching: Computer-based switching technique that transfers messages between points not directly connected. The system receives messages, stores them in queues or waiting lines for each destination point, and retransmits them when a receiving facility becomes available.

Modulation: Process of modifying the characteristics of a propagating signal such as a carrier so that it represents the instantaneous changes of another signal. The carrier wave can change its amplitude (AM), its frequency (FM), its phase, its duration (pulse code modulation), or combinations of these.

Multiplexing: Process of combining two or more signals from separate sources into a single signal for sending on a transmission system, from which the original signals may be recovered.

Pulse code modulation (PCM): Technique by which a signal is sampled periodically, each sample quantified by means of a digital binary code.

Slow-scan television: Technique for transmitting video signals on a narrow-band circuit such as a telephone line that results in a picture that changes every few seconds.

Subcarrier: Additional signal imposed on the broadcaster's carrier. This subchannel carries information, sound, or data that are different from those carried by the carrier and require a special receiver.

Teleprocessing: Data processing wherein the data manipulation is performed at a processor electrically connected to but remote from the place where the data are entered or used.

Twisted pair: Two wires that connect local telephone circuits to the telephone central office.

Throughout history man has communicated to man and to animal in many ways. By means of sign language and primitively verbalized sounds, as we have seen in imaginative films, humans have sought ways to transmit their feelings and thoughts to others. These means of communication are still used today, enhanced by the invention of language. Communication takes place face-to-face, across mountains by yodeling, and over two tin cans connected by a string; when communication takes place with the aid of electricity and electromagnetism, it is telecommunications.

I. Telecommunications and the Information Society

A. VIEWS OF THE INFORMATION SOCIETY

The idea of an information society may have originated in Japan. Sparked by Fritz Machlup's 1962 book, "The Production and Distribution of Knowledge," a Japanese science, technology, and economics study group that was formed by the government to provide guidance to economic planners coined the term *johoka shakai,* or informationalized society. Clearly this phrase was chosen to be analogous to the term industrialized society, which had emerged out of the industrial revolution of the 18th century.

For Japan, this new society would be one in which there is an abundance in quantity and quality of information with all the necessary facilities for its distribution. This information would be easily, quickly, and efficiently distributed and converted into whatever form and for whatever purpose the user might want. The information society would be universal and affordable.

Many versions of information societies have surfaced since its introduction by the Japanese planners. In the United States, Porat showed in 1977 that the number of information workers had exceeded workers in agriculture, industry, and in the provision of services. A society in which information work dominates can, certainly, be said to be an information society.

In the 1980s, American and Japanese researchers undertook the long and tedious task of counting words produced by all media and words consumed in their respective countries and concluded that the rates of production and consumption of words in newspapers and books, on radio and television, via the telephone, telegraph, telex, and facsimile, and words in the form of data via data communications networks, and so forth, were growing rapidly in both countries. Japan and the United States were both recognized as being information societies. Consequently, the number of words produced and consumed and the rate of this production and consumption could be seen as indicators of the informationization of society.

B. THE NETWORK MARKETPLACE

In the United States and increasingly in many of the highly industrialized nations of the world, a new industry is emerging from the marriage of computers and telecommunications. This industry permits users to interact directly with one or more computers and with distributed information systems which may be in banks, shops, airline terminals, libraries, or schools.

Some analysts have suggested that this potential for delivering services on a network, or network information services, could virtually transform not only the information and communications activities in a society but the very nature of the society itself. They have suggested that new services will be offered, such as remote shopping, news on demand, electronic message delivery or mail, remote medical consultation and diagnosis, electronic banking, remote and interactive education and training, and conceivably, remote working.

Network information services will connect the needs and resources of users to the capabilities and services of producers and facilitate transactions between them. All of the usual services of a marketplace can be offered in a large information network. Products and services can be advertised; buyers and sellers located; and ordering, billing, and delivery of services facilitated. All manner of transactions can be consummated, including wholesale, retail, brokering, and mass distribution. Indeed, the entire range of the products and services of business, industry, consumers, and government can be viewed as a as a marketplace—a marketplace on a communication network, or the network marketplace.

Communicating and computing technologies, the information technologies, have not only reached into the office but also have created an electronic environment in the home, which one could call the networked home. The telephone reaches into more than 96% of all of the nation's households and has become a sophisticated information instrument, its simple keypad expanded to typewriter format and married to a cathode-ray terminal or to the television set.

It can be argued that a society capable of accessing information for work and play on telecommunications networks, whether they be cable or the electromagnetic spectrum of broadcasting are information societies. These networks are prerequisites for any information society. [See COMMUNICATION SYSTEMS, CIVILIAN.]

C. TELECOMMUNICATIONS AND THE INFORMATION SOCIETY

Just as the hammer and chisel are the tools of the stone cutter and the plane and saw those of the carpenter, so are the telecommunications technologies the tools of the information worker in the information society. The telecommunications technologies make information societies possible. Telecommunications provides the means by which information is communicated and thereby becomes useful knowledge and of value to a society.

The decades following the Second World War have seen a veritable explosion in communications technology. Telephone subscribership in the United States almost doubled during the period from 1945 to 1975, and oceans were crossed with more transoceanic cables than had been laid in the previous century. POTS (plain old telephone services) learned the new language of bits, bytes, and bauds and to leap across continents through outer space as well as through wires.

The rapid growth in telecommunications technologies is not solely a U.S. phenomenon or even a phenomenon of the most developed nations of the world. It is occurring around the globe. Consequently, telecommunications has moved to the top of the political agendas of nations. New technology has facilitated the breaking up of the monopoly control over telecommunications; AT&T, the British Post and Telegraph, and the French Telecommunications monopolies are under competitive challenges. The powerful Ministry of Posts and Telecommunications in Japan is under fire from their trade ministry to open their doors more rapidly to competition. This growing importance of telecommunications throughout the world is the result of three important factors:

1. Many nations have realized that they cannot compete in the development of the informa-

tion industries without good telecommunications technology

2. Because of the convergence of the computer and communications technologies these industries see the possibility of growth lying in each other's territory.

3. Multinational corporate operations require worldwide, round-the-clock communications, global networks for banks, air traffic control, travel reservations, news, and trade.

II. The Telecommunication System

A. COMMUNICATION SYSTEMS AND NETWORKS

All communication systems, whether face-to-face, via tin cans and string, or shouting and yodeling, have the same elements. These are the following:

1. An information source, such as man.

2. An information encoder, such as the vocal tract, which takes ideas and encodes them into vibrations that are coded into language.

3. A communicating channel, such as the string between the tin cans, the copper wires between the telephones, or the space that has been acted upon by electromagnetic changes for broadcasting.

4. A receiver, such as the telephone, radio, or tin can

5. A decoder, such as the ear and brain, which translate sound into words, music, and ideas.

6. A recipient, such as the person (or animal) to whom the communications is directed.

There are many different technologies and systems for each of these functions, and the study of telecommunications is concerned with what they are, how they work, what they are good for, and how to choose among them for the task at hand. For example, the source might be an orchestra on the stage at Carnegie Hall, the encoder one or more television cameras encoding the scene for transmission over a broadcast channel to television antennas on rooftops, which are linked to the receivers, the television sets in living rooms where the signals are decoded by the picture tube and its electronics for delivery to the intelligent recipient, the viewer–listener.

Some forms of communications are virtually instantaneous, the only delay being the physical limitations of the speeds of sound or light. Electrons travel at the speed of light, delayed only by the nature of the material through which they travel. Telephone conversations, live television, either over-the-air (broadcast) or over-the-wire (cablecast), and semaphores, still used by passing ships at sea, all provide instantaneous (real-time) communications.

In other forms of communications, the message is recorded and then stored for some time before the recipient gets it. Books, films, paintings, recordings, Telex messages, messages by computer to computer, and the garbled notes on your telephone answering machines are all examples of delayed communications.

This discussion has dealt with two persons communicating and an orchestra performing for many people. These types of communications are called point-to-point and point-to-many points communications, respectively. Point-to-many points communication is broadcasting.

When many people wish to communicate with each other point-to-point simultaneously, networks must be provided. One kind of network has a separate channel between each information source and recipient. Each person has a switch for selecting among all possible persons on the network with whom to connect. Each person must have as many switches as there are people on the network, less one for connecting to oneself. Consequently, in a network of 10 terminals there are 90 switches; and there will be 45 transmission paths,

$$L = \frac{N(N-1)}{2} \qquad (1)$$

where L is the number of transmission paths and N the number of terminals to be interconnected.

An alternative that was adopted very early in telecommunications is a centrally switched network. A switched network uses fewer and shorter wires or cables and requires only one switch, which must be very reliable and is often quite complex. [See DATA TRANSMISSION MEDIA].

B. TELECOMMUNICATIONS SYSTEMS

Telecommunication systems use electricity and electromagnetism to transmit messages or information. The telegraph, for example, works by sending electric current over a wire. The telegrapher's sending key alternately opens and closes the connection between the sender and the receiver, the circuit, thereby allowing current to flow. At the receiving end, the current flows though an electromagnet that pulls on the receiver key causing it to click.

Because the electrical impulses are either on, off, long, or short, a code is required. The Morse code uses dots and dashes (a dash is three times longer than a dot) to represent letters.

In practice, when a voltage is applied to a transmission line, the current does not begin to flow immediately. It builds up over time and at a rate that depends on the nature of the material through which it is passing, a wire or a cable. Consequently, there is a difference between the transmitted signal and the received signal. The information transmitted is not received as it is transmitted because of distortion.

Signal distortion due to the characteristics of the transmission system is but one of the problems facing designers of telecommunication systems. Electrical noise is also present in the form of noise such as that generated by fluorescent lights and the sparking in automobile engines. There is also the noise caused by nature, both in the atmosphere (lightning, for example) and in space when the sun is very actively discharging flares. Finally, there is cosmic noise emitted from suns or stars light-years away.

The time required for any electrical operation such as turning a current on or off depends on certain physical characteristics of the cable or wire, capacitance, and resistance. When a wire or a cable is conducting a current, it builds up a capacitance between itself and any neighboring conductors or electromagnetic fields. The electrical capacitance of any transmission system is defined as the ratio of the charge to the difference in potential between the wires carrying the current or between the wires and a surrounding electromagnetic field. The longer the path over which the current is being carried or the larger the conductive surfaces facing each other, the greater the capacitance. Also, the capacitance is inversely proportional to the distance between the conductors or surfaces. The larger the capacitance the greater the interference with the current-carrying ability of the wire.

The resistance of a conductor, which also slows the rate at which the charge is carried in the wire or the signal is being transmitted along the wire, is proportional to the length of the wire and inversely proportional to the cross-section area of the wire. When a signal is put on a wire at one end, it does not immediately appear at the other end even though electricity travels at the speed of light. Because of capacitance and resistance, the signal increases from zero to the maximum value slowly.

If the signal at the transmitting end is impressed and taken away in too short a time, it may not be detected at all. For a signal to be detected requires that it be above a certain threshold value. This threshold value depends on both the sensitivity of the receiving equipment and the magnitude of the noise in the system. We live in a very electrically noisy world. If the signal is too small with respect to the noise, the signal transmitted will not be detected.

Noise is always present and unwanted. A measure of noise as it affects the detectability of a signal is given by the signal-to-noise ratio, which is measured in decibels. A decibel (dB) is used to express the magnitude of a change in the level of power, voltage, current, or sound intensity and is defined as 1/10 bel. A signal-to-noise ratio of 10 (dB) means that the signal power is 10 times the noise power. A signal-to-noise ratio of 20 dB means that the signal power is 100 times the noise power. A ratio of −3 dB means that the signal power is half the noise power. The number of decibels is ten times the logarithm to the base 10 of the ratio.

$$N = 10 \log_{10} \frac{S_p}{N_p} \qquad (2)$$

C. The Telegraph as a Digital Communications System

The telegraph, one of the earliest telecommunications systems, going back more than 150 yr, is a digital system. If one modifies the three-symbol Morse code by eliminating the dot and using only the dash and calls the dash a one (1) and the time space a zero (0), one gets the digital language of modern computers. In digital communication systems two symbols, a one and a zero, are used; the presence of a tone or a voltage is represented by a 1 and the absence of a tone or voltage is represented by a 0, if one assigns that meaning to the two digits.

D. The Telephone as an Analog System

Unlike the telegraph, the telephone is an analog system; the human voice is converted to an electrical signal that is similar to or analogous to the voice signal. The information transmitted over the telephone is represented by a continuous and smoothly varying signal amplitude or frequency within a given range of frequencies.

It is interesting that the transmission of digital signals via the telegraph preceded the development of the analog telephone, despite the fact that the transmission of analog voice signals with reasonable fidelity is quite difficult. This is especially true for long-distance telephone transmission, in which amplification of the relatively complex voice signal is required. Amplification of digital signals is made much simpler by the ease with which it is possible to regenerate (rather than amplify) digital signals that are sent over very long distances, without the encumbrance of excessive noise. This is but one reason why modern telephone networks are returning to the digital system of the early telegraph. It required the development of electronic components capable of very high switching rates to make digital voice transmission a reality.

E. INFORMATION AND CHANNEL CAPACITY

To most people, information is what is obtained by asking for directions, talking to friends, scanning and reading newspapers, listening to the radio or attending to television, and dialing directory assistance. The sort of information and its content and meaning are what is important to the user of a telecommunications system. But when engineers design systems, they are concerned with how much information. This is the critical distinction between how social scientists deal with information in their study of human communications and how engineers must define information to design telecommunication systems for human communications. The mathematical theory of communications bridges these two concepts of information. [*See* RADIO SPECTRUM UTILIZATION.]

When using a flashlight, a semaphore, a television picture, or a telephone for transmitting information, if the flashlight does not blink, the semaphore does not move, the television picture does not change, or the voice tones do not vary then no information is conveyed. Information is conveyed only when there is doubt and uncertainty present and the information can remove the doubt and resolve the uncertainty. Information that resolves greater uncertainties or assists in making better choices among many alternatives is valued more highly than information that does not do so.

In 1948, Claude Shannon of the Bell Laboratories provided a mathematical theory that corresponds to human experience and common sense. He pointed out that

1. for a signal to carry information, the signal must be changing; and
2. to convey information, the signal must resolve uncertainty; the greater the uncertainty

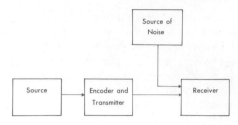

FIG. 1. Shannon's model of the communications process.

the greater the information contained in its resolution.

Shannon's model of the information process includes the information source that generates the message, the transmitter that encodes the message from the source to the signal, the communication channel that conveys or transmits the signal, the noise that affects the signal in unpredictable ways, and the receiver that decodes the signal and reproduces the message as generated by the information source (Fig. 1).

The measure of the amount of information is a probabilistic one; it depends on the how many different messages the message source can produce. If the source can produce only one message, and there is no uncertainty about that message, then zero information is transmitted as noted in above. If the source produces n messages, each with equal probability, there is a great deal of uncertainty that must be reduced. Shannon defined the uncertainty of the outcome between two equally likely message possibilities as the unit of information; thereby he could use as his measure the binary digit or bit.

Suppose that an event has occurred and that a message is transmitted telling about this occurrence. The amount of information received in that message is defined as

information received

$$= \log \frac{\text{probability at receiver of event after message is received}}{\text{probability at receiver of event before message is received}}$$

(3)

If there is no noise in the communications system, then the information received is

information received

$$= -\log \left[\begin{array}{c} \text{probability at receiver of event} \\ \text{before message is received} \end{array} \right]$$

(4)

In binary digit terms, the quantity of information is $-log$ 2. However, not all of the events or messages are equally probable. It is necessary, therefore, to consider the average information for each event or message in a stream. Therefore,

average information received

$$= - \sum_j p_j \log 2p_j \qquad (5)$$

where j is the probability of the jth event or message symbol.

An interesting finding of Shannon's work is that the above equation has been familiar to physicists ever since the classical researches of Maxwell, Boltzman, and Gibbs' classical mechanics and thermodynamics. Boltzman used the symbol H for this summation and showed that in ideal systems it is proportional to the thermodynamic quantity of entropy. While these classical physicists recognized that there was some connection between information and entropy, they did not develop a theory of information because they were not transmitting information. Shannon, however, used H to define the average language entropy per message,

$$H = - \sum_j p_j \log p_j \qquad (6)$$

Of primary interest to designers of communications systems is how to transmit information in a noisy environment and how to make optimum use of the transmission channel. Shannon showed that even a noisy channel has a capacity to transmit information accurately. For a channel with bandwidth B in which noise of power N is added to a signal of power S, Shannon showed that the channel capacity C is expressed as

$$C = B \log_2 \left(1 + \frac{S}{N}\right) \qquad (7)$$

This is an important design limitation toward which engineers must work to achieve reliable information transmission. This equation shows that it is better to use more rather than less bandwidth in transmitting a signal of a given power in a noisy environment.

Yet engineers have a limited bandwidth with which to work, and they must find ways to use that limited bandwidth in the most efficient matter, because bandwidth costs money, whether in spectrum space or copper cable. (With the advent of optical fiber transmission systems, the cost is significantly lessened.) Shannon recognized that not all symbols in a language were equally probable and that there was a considerable amount of redundancy in any message. Hence, by judicious coding, a limited bandwidth can be used to transmit information reliably, even in the presence of noise.

F. CONSERVING BANDWIDTH: THE DESIGNER'S CHALLENGE

The use of a central switch rather than point-to-point interconnection is one means of conserving bandwidth: there then are fewer copper wire interconnections. More efficient use of bandwidth can also be achieved by sending messages more rapidly on the transmission path, thereby sending more messages per hour. However, the bandwidth required would be greater, and it is not clear if the additional bandwidth would be worth the additional cost.

Multiplexing is a more efficient way of conserving bandwidth. Transmitting more than one signal on a channel at the same time can be achieved by shifting the frequency of the signal, thereby giving each signal a different place in the channel, or by sharing the time on the channel. Multiplexing is the process of putting several signals on one channel or sharing the frequency spectrum.

G. FREQUENCY DIVISION MULTIPLEXING

Telephone systems use multiplexing to conserve copper and duct space. Telephone carriers multiplex 12 telephone channels on a single cable. Each signal requires 4000 Hz of bandwidth; so by using a carrier in the range of 60 to 108 kHz, or a bandwidth of 48 kHz, the system assigns 4 kHz of space to each of the 12 channels, thereby filling the 48-kHz single cable with what would normally have required 12 cables. Frequency separations or guard bands are required between each of the 12 channels to prevent interference between the transmitted signals.

H. TIME DIVISION MULTIPLEXING

Transmitting a message requires the allocation of both time and frequency. Frequency division multiplexing divides frequency or bandwidth. Time division multiplexing divides time. Frequency multiplexing occupies a certain bandwidth throughout all of the time of transmission. Time division multiplexing (TDM), on the other hand, uses the time of transmission sporadically, taking up small portions of the time with one signal or many signals. In a time division multiplexing, signals are intermixed in time. To reconstruct the transmitted message requires the ability to keep track of what signal is being sent at what time. Time division multiplexing in its many forms is the most promising means for low-cost transmission of information (i.e., for the cost-effective use of bandwidth).

To time division multiplex analog signals requires that selected portions of the signal be transmitted without destroying the meaning of the information being communicated. This requires sampling of the analog signal. The rate at which the analog signal is sampled is critical to the reliability with which the transmitted signal or message is reproduced. Sampling at a very high rate is expensive and adds little to the intelligibility of the signal. Sampling at a very low rate risks missing some of the variations in the signal, the information being transmitted. There is a minimum sampling rate that insures that the sampled signal is an accurate reproduction of the original message. The sampling theorem of Hartley and Shannon specifies that to insure reliable information transmission the sampling rate should be at least twice the bandwidth of the signal to be sampled. Thus, for a voice signal in the assigned channel capacity of 4000 Hz, the sampling rate should be at least 8000 Hz.

III. The Telephone and the Telephone Network

A. THE TELEPHONE AS A SOCIAL SYSTEM

The telephone is probably society's most significant telecommunications technology, yet it has all but been overlooked by social researchers. While the social impact of broadcasting, the steam engine, the printing press, and the automobile has been traced by historians, the telephone has for the most part escaped scrutiny.

The desire to communicate at a distance is as old as legend. It is no surprise that at least five telephone inventors were racing each other to the patent office, with Bell winning by 30 min over Gray. But the consequences of the telephone have always appeared to affect society in counterintuitive ways. The telephone was thought to save physicians from making house calls, but physicians initially found that it increased them; patients could summon the doctor rather than travel to the doctor. The phone in-

vades privacy with its strident, insistent ring but protects privacy by allowing the transaction of business away from homes. It allows the dispersal of centers of authority, but it also allows tight and continuous supervision of field offices away from these centers. The telephone makes information available but reduces or even eliminates written records that document often important information.

Today there are some 400 million telephones in more than 200 countries operating on what is essentially a single global network. This network carries annually about 400,000 million conversations, a load that is increasing at the rate of 20% per year. By the end of this century there will be 1500 million telephones and more than a million million calls each year. [See VOICE-BAND DATA COMMUNICATIONS.]

B. TELEPHONE TECHNOLOGY

Three subsystems constitute the telephone system: the subscriber loop or, as it is often called, the local loop; the switching system where so many of the new "chip" developments are reshaping the telephone; and the long-distance or line-haul subsystem.

1. The Subscriber Loop

The subscriber or local loop consists of handsets and the cable and twisted copper pair in the local distribution facilities. The local loop has four functions.

1. Voice transmission by means of signals that vary the amplitude of a direct current in the transmission line in accordance with the sound pressure that is produced at the handset microphone.
2. Dialing by interrupting the dialing signal either by tone dialing or by rotary dialing; this is a form of pulse modulation.
3. Ringing by applying alternating current at the switch end of the line, thereby ringing the telephone instrument bell.
4. Off-hook. When the handset at the receiving phone (the called number) is picked up, a switch closes, causing the current in the line to activate the microphone in the handset. At the receiving or called phone, the amplitude modulated carrier with the message signal is heard; at the calling party telephone the dial tone is heard.

The local loop provides access to the worldwide telephone network. Without the local loop, access to the information world would not be possible. Until recently the local loop has not been susceptible to new technology. The monopoly structure of the telephone industry in the United States and throughout the world classified the local loop as a natural monopoly with an exclusive franchisee assigned to an area. However, new transmission technologies have been developed that can compete with the traditional two-wire twisted pair of the local loop. In particular, cable television using coaxial cable can provide point-to-point wide bandwidth transmission services to the local area, including some limited provisions for switched broadband services. Similarly, high-frequency microwave in the gigahertz range can deliver broadband transmission services over limited ranges suitable for data and video transmission across the city and between cities. Finally, it is believed that cellular radio technology can deliver radio telephone services at costs that could be competitive with the traditional telephone local loop if market demand increases significantly.

The AT&T divestiture of 1984 has loosened constraints on these competitors for the local loop. Firms offering several technologies including cable television, cellular radio, and digital termination services (gighertz microwave across limited distances) are profitably providing local loop services for special applications. These technologies, often referred to as by-pass technologies because they by-pass the traditional local loop are segmenting the local loop market.

2. Switching

A switched network, as noted previously, required shorter cables and thus conserved bandwidth by conserving copper. While a non-switched network can use many relatively low-cost switches, the switched network requires large, complex, and often expensive switches.

The switching functions are the following:

1. To detect the telephone off-hook and to inform the system when the telephone is lifted off the hook either to make or to receive a call.
2. To connect the pulse receiver at the central office so as to be able to determine the number being dialed.
3. To apply the dial tone to the line to signal the caller that the caller can now place the call.
4. To make the connection to the line specified by the dialing signals. How this is done depends on the type of call made, which determines the type of switch to be used. If the call is a local call, that is, the parties are in the same exchange, the connection is applied directly to the called subscriber's line. If the call is to someone in another exchange, there may be intermediate switching centers through which the call must pass. The local exchange does what was done for the local call: connect the calling to the called parties.
5. To ring the called party if the line is not being used or busy. If the line is busy, the busy signal is generated.
6. To detect that the call has been completed. When the telephone handset is placed back on the hook, or on-hook, the switch breaks the connection. Modern switches also record information about the call for billing purposes.

Telephone progress can be measured in terms of switching progress, for as new developments in switching technology were made available to the system, new services were offered. The history of switching from the step-by-step switch that replaced the manual operators who were the human controllers of switching during the early part of this century to the digital switches that were first installed in the mid-1960s is the story of the development and introduction of microelectronics and computers in telecommunications.

There are two types of switching, sequential and common control switching. In a sequential switching system, connections are made progressively through switch paths as each digit is dialed. Sequential switching occupies valuable switch and line capacity while competing the connection. Because switch capacity is expensive as are trunks, switching resources are conserved. This is accomplished by the second mode of switching, common control switching.

In common control switching, a portion of the switch, called the control segment, records the called party's number and by means of a separate signaling path, the common control path, checks to see if the called number is busy before making any connections. If the called party is not busy, the switch looks for the best switching path at the time and makes the connection through the path found. Common control signals can be information rich—information to instruct other parts of the switch and the system to perform an increasing number of functions, such as call-forwarding, call-waiting, call storing, and arranging teleconferences.

a. Step-by-Step Switch Technology. Step-by-step switching systems were installed in the 1920s and are still operating in many areas of the nation. These systems perform sequential switching as described above. Dial pulses con-

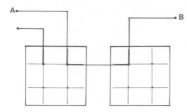

FIG. 2. Crossbar matrix relay switch. A and B are end points or terminals in the transmission; A "calls" B.

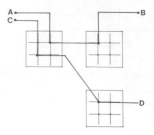

FIG. 3. Reduce blocking by adding a matrix switch. A, B, C, and D are end points or terminals in the transmission; A "calls" B, and C "calls" D.

trol the relay that steps the switches along, selecting one of 100 connections per switch step. From each step there are pairs of wires that connect to local telephones. These switches are in use primarily in local exchanges, or exchanges into which the telephone instrument connects directly, and in the telephone system architecture are called Level 1 Exchanges.

b. Crossbar Switching Technology. Level 1 Exchanges connect callers within the same exchange. However, if a call is made to another exchange or requires the long-distance network, the call must go through several exchanges. The step-by-step switch cannot carry the increased loads from several local exchange areas. To meet these needs the crossbar switch was introduced in the 1940s. The crossbar switch is a matrix-type relay with four wires connected as shown in Fig. 2. With four wires arranged in the matrix form, there are four possible connecting points rather than only the one that would be available with two wires intersecting.

Horizontal and vertical bars select contacts, or crosspoints, thereby connecting the signal path through which the parties communicate when an open line is found. When the telephone is returned on-hook, the crossbar circuit is opened, and the crossbars are released to look for other open lines to interconnect. While the crossbars are searching for available connections for a call, the called numbers are stored. To reduce blocking or waiting for switch pats to open, the capacity of matrix switches is increased by adding matrix relays as shown in Fig. 3.

c. Electronic Switching Technology. In 1965, not long after the invention of the transistor, electronic switches were introduced. While the switching was not performed electronically, it was controlled by computers or equipment that performed computerlike functions. The early electronic switches were computer-controlled crossbar switching systems. Electrome-

chanical relays were later replaced by solid-state devices. [*See* TELEPHONE SIGNALING SYSTEMS, TOUCH-TONE.]

d. Common Control Switching. The crossbar switch with its matrix format is the basis for today's modern common control exchange. The switches need not be electronic; as we noted previously, they can be and often are electromechanical crossbar switches. Often the switch is simply several banks or stages of crossbar switches; however, the switch is controlled by microprocessors performing computer functions. These computer functions are performed by concentrators, common control processors, links, registers, the sender, and the scanner. When a call is made, the scanner detects the signal and sends it to the common control. The common control searches for a free register, a place to store numbers, and when it finds one, it transmits the dial tone and records the called number in the register. After determining that the called number is not busy, the common control searches for open switch paths. If one is found, the register releases the stored request into the link, and the conversation can begin. The register is released to perform future searches.

In this manner, more efficient use of expensive switching equipment and copper cables is made while decreasing the wait time for a connection. Computer functions in telephone switching systems that can remember and store information for later use are the bases for Stored Program Control Systems. [*See* TELECOMMUNICATION SWITCHING.]

3. Line Haul or Long-Distance Transmission

Line haul systems provide transmission paths to carry subscriber-to-subscriber communications between local exchanges. There are considerable design differences between line haul and local loop transmission systems. These differences are the result, primarily, of the characteristics of local loop and long-distance calling patterns. The nation's long-distance network was responsible for the rapid growth of the telephone industry in the United States. Planners of the system recognized that the more subscribers on a network, the more valuable that network is to the user as well as the provider. As the number of subscribers increased, especially in the years after World War II, and as the number of telephone companies throughout the country sought to interconnect, demand for the long-distance pairs of wires soon resulted in intolerably long waiting times. Consequently, much of the research at telephone laboratories was devoted to increasing the capacity of these long-distance circuits. Many of today's multiplexing techniques were initially applied to the line haul networks, greatly increasing the number of signal channels carried by the network. To increase the efficiency and lower the cost of long-distance calling, new transmission technologies including microwave, satellites, and optical fiber systems have been introduced. A local loop circuit is used by a single pair of subscribers at a time, but long-distance trunks are time-shared among many subscribers.

Because line haul circuits are interexchange networks, and because of their special uses, they differ from local loop circuits in two major respects:

1.   There is no need to maintain compatibility with the handsets, and a variety of input and output signals may be used. The long-distance network is, therefore, a "transparent" network.
2.   The long transmission requires that the signals be amplified. Amplification requires that long-distance circuits be four-wire circuits so that amplification can be in one direction at a time. This also leads to echoes in the system, which must be eliminated.

Over long distances—transcontinental, transoceanic, and through space—losses not only can make the original message unintelligible but they can make it disappear altogether. To ensure that the signal is greater than the noise requires the insertion of amplifiers. Unfortunately, the noise is amplified as well as the signal. Consequently, the signal-to-noise ratio of the amplifiers must be sufficiently high to guarantee that the signal is received in the presence of the noise.

When amplifiers are used in long-distance transmission lines, it is necessary that the signals travel in only one direction at a time. Otherwise, the amplified signal would interfere with the original signals and lead to cancellation. Transmission lines with amplifiers must have two circuits—one in each direction—for two-way communications, or two wires in each direction. It is necessary to match the two-wire local loop circuit to the four-wire long-distance circuits by means of hybrids. Hybrids separate the sending signals from the receiving signals. Often this separation is incomplete, and coupling or echoes result. Voice-activated echo suppressors are added at each end of a four-wire circuit. These operate on the principle that when one party is talking, the other party is listening. Consequently, one loop can be closed while the other is open. This often led to unacceptable voice clipping, so the echo cancelors now being used are tuned to eliminate only the reflected signals rather than another party's speech. These cancelors store the transmitted speech for the short time it takes for the speech segment to come back to the talker and play back what has been stored in opposition to the incoming echo, thereby canceling it out.

4. Transmission Economics

The design objectives of long-distance transmission are to provide the greatest amount of transmission capacity on the network for the least cost. Consequently, the designer uses the multiplexing techniques discussed previously. The choice of which to use depends on the relative cost per circuit mile for each of the options available. The cost of the channel and terminal equipment including handsets and multiplexing must be considered in the analysis.

Since line costs are proportional to the distance covered, and terminal costs are fixed, multiplexing becomes more economic than individual circuits only beyond a certain distance. For a particular multiplex system, the minimum distance at which there is an economic advantage is called the prove-in-point. In Fig. 4 two multiplexing methods are compared with the individual circuit option. Multiplex system A uses frequency multiplexing and multiplex system B time division multiplexing. System A's cost per circuit mile becomes lower than the individual circuit cost per circuit mile at a distance of L_1 miles. System B's cost per circuit mile does not prove-in until a distance of L_2 miles. System A has an economic advantage over both system B and the individual circuits up to a distance of L_3 miles. At that point it proves-out in favor of system B.

Dramatic reductions in circuit-mile costs have been made during the past 50 yr, sometimes by as much as 100 : 1. This has been achieved by building larger and larger capacity systems. But these lower costs can be achieved only if there is a demand for that many circuits on a particular route. In planning a voice or data communications network that includes long distance as well as local loop transmission, the choice of the best technology for each part of the network is essentially based on traffic estimates for the various portions of the network. Because there have been more choices of technology in long-distance transmission than in local loop transmission, more cost savings have been achieved there than in local loops. Long-distance costs have been steadily falling while local loop costs have remained essentially constant. While this had a great deal to do with regulatory practices of the state public utility commissions, who sought to ensure universal telephone services in their state for humane as well as political reasons, line haul transmission technologies have had a great deal to do with this cost trend. Several choices are now available to system planners for line haul systems.

1. Cable wire pairs, the old fashioned local loop
2. Coaxial cables, which were developed over 40 yr ago and are used today not only for long-distance telephone but also for cable television
3. Microwave radio relay systems, which have for several years been replacing the coaxial cable for cross-country transmission, but which are now being supported by satellites and optical fiber transmission systems.

It is a truism of almost all technology, and especially so of the telecommunications technologies, that new technologies do not entirely dis-place older technologies. For many years to come all of these techniques for providing long-distance communications will be economic in specific situations. Thus, for example, microwave guide systems with very large channel capacities in the neighborhood of 230,000 to 460,000 voice circuits are being developed in parallel with ever higher-frequency satellites and optical fiber transmission systems.

5. Optical Transmission

Using light as a means of communicating goes far back in history. Mirrors were used by armies to send reflected-sun signals from mountaintop to mountaintop. More than a century ago, Alexander Graham Bell demonstrated the photophone, a device consisting of mirrors and selenium detectors that carried speech no more than a few city blocks; nevertheless, he showed that intelligence could be transmitted on a light beam. [See OPTICAL FIBER COMMUNICATIONS.]

Modern optical systems were made possible by the invention of the laser as a light source. Optical transmission systems are being installed in several places in the United States and throughout the world, and the reasons for the great interest in them are twofold: Optical transmission provides almost unlimited bandwidth and can deliver unlimited channel capacity, and the glass fibers that carry these channels take up little space and are very inexpensive. It is expected that in time optical transmission systems will replace not only certain portions of today's long-distance transmission systems but also will find their way into the local loops and into homes.

An optical fiber is thinner than a human hair. But because glass is a very strong material in tension but is brittle and easily cracked and nicked when bent too much or handled too roughly, the fiber optical cable is made up mostly of material to cushion the glass and to provide mechanical support.

Optical transmission is not very different from transmission by copper wire. The functions required to generate a signal and to detect and make intelligible the information transmitted are the same as those required to send signals along a copper wire. If the fiber is manufactured correctly, its refractive index (ratio of the velocity of light in a vacuum to its velocity in the glass fiber) in concert with its surrounding material prevents light from escaping from the sides of the fiber, thereby keeping the losses down.

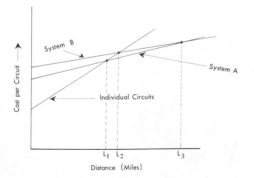

FIG. 4.   Prove-in and prove-out.

Optical fiber transmission is attractive for several reasons:

1. Glass is more plentiful than copper.
2. The small size of the cables is attractive for increasing the capacity of ducts that are now crowded with conventional cables.
3. All forms of electrical interference are eliminated.
4. The bandwidth potential far exceeds that of electrical cables.
5. The low resistance or attenuation even at very high frequencies reduces the number of amplifiers required to deliver signals over very long distances.

6. Satellite Transmission

In his famous "Wireless World" paper of October 1945, Arthur C. Clarke suggested that "a true broadcast service giving constant field strength at all times over the whole of the globe would be invaluable, not to say, indispensable, in a world society." He proposed that "rockets" in suitable extraterrestrial positions could overcome the difficulties with high-frequency radio transmissions and the "peculiarities of the ionosphere" and make feasible not only worldwide radio and telephone but also television. Thus was born satellite communications.

INTELSAT, the worldwide satellite carrier, is used primarily for the delivery of telephone voice service and for data transmission. As interest in digital transmission grows, so does the use of satellites for sending information in the form of data. This is because the satellite offers very wide bandwidth at costs that are insensitive to distance.

Sir Isaac Newton hypothesized that if he could climb a high enough mountain and shoot a cannonball into the air, the ball would fall toward the earth at the same rate as the earth curves away from Isaac. In this way, the cannonball would continue to circle the earth without ever falling to the earth; it would be an artificial moon around the earth, or a satellite.

Newton could not find his mountain; even if he had, the earth's atmosphere would have slowed the cannonball so much that it would soon have crashed. But if he could have shot a cannonball out of the earth's atmosphere at the speed of 8 km/sec (5 miles/sec), it would never have returned; it would have circled the earth indefinitely with no expenditure of power.

When a satellite is circling the earth in this way, it is in a stable orbit. There are an infinite number of possible stable orbits. An escape speed of 5 miles/sec can lift a satellite to a stable orbit close to the earth, just outside the atmosphere, where it will circle the globe once every 90 min. It will be an elliptical orbit with an apogee (the highest point) of 300 miles and a perigee (lowest point) of 100 miles. However, a satellite in this orbit would be of limited communications value for radio transmission since this satellite would be in view intermittently, i.e., for only about a quarter of an hour in each orbit.

Increasing the velocity at which a satellite leaves the earth places the satellite in a more distant orbit, and the velocity at which the satellite circles the earth decreases since gravity is less and the vehicle thus needs less centrifugal force to balance the pull of the earth. With orbital radii of 6000 (perigee) and 12,000 (apogee) miles from the earth, a satellite circles the earth every 5–12 hr and can be seen from any point for about 2 hr. In a geosynchronous orbit of 22,300 miles, the satellite would appear stationary since it is rotating at the same speed as the earth. This orbit is admirably suited for communications.

Satellites as repeaters in space are well suited for long-distance telephone transmission. Most satellite transponders (repeaters) have a bandwidth of 36 MHz or some multiple thereof such as 72 MHz, although there are some variations in satellites now in orbit. Transponder bandwidth depends on the amount of traffic the buyer or renter wants to transmit. If, for example, the communications system is used as a common carrier, that is, used by a great many firms, there is a great deal of voice, video, image, and data traffic; so larger bandwidth transponders are required.

There are usually 12 of these transponders on a vehicle, reused by means of polarization and therefore providing 24 wideband communications. Satellite bandwidth is costly bandwidth, hence it is used for high-valued services. These include television, data, and long-distance voice transmissions. A typical satellite provides, for example, up to two color-television channels, 1200 voice channels, 16 channels for data transmission at the rate of 1.544 megabits/sec (MBs), 400 channels for data at 64,000 bits/sec (or 64 kilobits/sec, KBs), and 600 channels for data at 40 KBs. The first INTELSAT satellite launched in 1971 provided 4000 voice circuits and two color-television channels; INTELSAT V launched in 1981 has a capacity to provide 12,000 two-way voice circuits plus two color-television channels.

Accessing satellite communications requires traversing of the final mile. After a 44,600 mile journey to and from space, messages must get to and from the earth station to be transmitted and received. The telephone system, through its long-distance operations, provides both the master control and the television receive-only for long-distance telephone services.

The economic benefits of the satellite rest not only on its ability to link distant parts of the earth with a great deal less difficulty and, of course, lower cost than is possible with earth-based technology, but also on the increased amount of effective bandwidth that can be obtained from the satellite transponders through frequency reuse, digital transmission, and time division multiple access technologies. The lease cost of satellite circuits have been falling rapidly and can be expected to continue to fall as satellite lifetimes are increased from 7 to 10 yr. Competition from optical transmission, so well suited for line haul systems, will keep satellite costs down.

In 1981 there were 47 million business telephones in use and 135 million residence telephones in 97% of all households. These telephones, both business and residential, were used to send more than 900 million messages. Operating income from the domestic use of telephones was greater than $55 billion, with another $1.5 billion coming from international services. About 30% of this telephone traffic was long-distance traffic, the kinds of traffic likely to be delivered by satellite.

Telephone traffic is expected to grow at about 8%/yr. With transnational businesses growing and domestic firms expanding their facilities throughout the United States, a significant proportion of this growth will be for long-distance services, which are candidates for satellite delivery. [*See* SATELLITE COMMUNICATIONS.]

IV. Telephony in the Next Century

Control signaling provides the in-channel signals heard by the subscriber:

1. The dial tone that informs the subscriber that the system is ready to receive the number dialed
2. The signals that transmit the number dialed to the appropriate switching office
3. The busy signal that informs the subscriber to try again

4. The ringing that informs the subscriber that there is a call on the line

A more comprehensive signaling system is required to integrate the entire network, one that can impart intelligence—memory and decision making—to the telephone system.

In the early days of the telegraph and the telephone, it was often necessary for operators and engineers to communicate with each other to inform connecting networks of scheduled downtimes, to discuss and solve transmission problems, and as the communications on these networks became crowded, to manage transmissions by planning ahead. A separate channel called an order wire was often used. Today this order wire has reappeared as the CCIS systems in modern telephone plants throughout the world.

The CCIS signals are out-of-channel signals, not heard by the subscriber, that manage the network:

1. Informing a switching center that for one reason or another the call cannot be completed
2. Telling the center that the call has been completed and to disconnect and start a new call
3. Managing the flow of messages in the network by keeping track of and determining the status of various transmission paths in order to avoid trouble spots and to route calls around congested routes
4. Trouble-shooting the network and isolating trouble spots for dispatching maintenance and repair services
5. Collecting and processing billing information for the use of over 1500 telephone systems in the nation and for international systems as well

A. NEW TELEPHONE SERVICES

Signaling converts the plain old telephone system into a multipurpose information instrument. Additional uses for the traditional plant, without the need for major system upgrading, creates additional revenues with relatively small new plant investment. The local loop network has remained essentially unchanged for well over 75 yr and is where the major telephone plant costs have been "buried"; it consists of twisted copper pairs of wires running into every subscriber's household and every office building and local switching center. This accounts for about 70% of the $162 billion dollar telephone plant investment in the nation.

Over the years the long-distance plant had been upgraded as noted previously. Over this period, switching emerged from the mechanical Strowger technology to the stored-program computer-controlled switches we know to be among the largest computers in operation today. The introduction of control signaling and the common control signaling systems created the opportunities for the telephone companies to use their existing analog system in ways that could attract new revenues without requiring replacement of a major portion of their plant. Consequently, with a relatively small investment, primarily in semiconductor devices, new revenues are being found from innovative uses of the traditional telephone plant. Some of these new services include audioteleconferencing, enhanced with visuals and documents; calling and managing conferences worldwide; call forwarding; automatic dialing of frequently called numbers; store and forward messages and eliminating "telephone tag"; personal message answering and message delivery service; call waiting, and speed dialing.

B. TELECOMMUNICATIONS–TRANSPORTATION TRADEOFF

In 1902, H. G. Wells anticipated that in the 20th century the growing use of new means of communications to conduct business at a distance would reduce land transportation. In 1914, *Scientific American* predicted a decline in transit congestion as new communications services grew. When the Arab embargo of 1972 led to an almost tenfold increase in oil prices, it was widely assumed that auto travel would be significantly replaced by the telephone, and several researchers showed just how much energy could be saved by trading off transportation for telecommunications. But travel and congestion, in the air and on the highways, seems to have grown as rapidly as the new communications technologies arrived on the scene.

The latest estimate of the cost of executive and staff travel in U.S. firms was in the neighborhood of $70 billion per year and rising. Transportation accounts for about 25% of the nation's total energy use with the automobile consuming almost 42% of this amount. The latest census showed that household budgets for commuting to work, shopping trips, and entertainment, and visiting are growing more rapidly than other expenditures. If only 10% of the travel of executives in the nation could be reduced, for example, by eliminating 2–3 person trips per month between New York and Chicago, or between Los Angeles and San Francisco, a savings of as much as $4 billion in time and travel costs could be realized, taking into account the cost of the telecommunications substitute.

There are many theories as to just why the convenience of electronic communications has had so much difficulty turning commuters into telecommuters. Past experience with audioconferencing systems generally supported these theories. Organizing a teleconference was difficult, and voice quality often poor. Researchers argued that audioteleconferencing was not a good way to meet people for the first time and that the conference would be more successful if the conferees were well known to each other. Further, telephone conferencing was not suitable for bargaining and negotiating. Researchers were concerned about social presence, a synthesized measure of several factors that affects the quality of human communications: the ability to transmit facial expressions, posture, the direction of looking, dress, and other nonverbal cues. All of these factors contribute to the social presence of a medium, and audioconferencing using the telephone did not rank well when compared with a multimedia system. More recent research has indicated that despite the relatively poor social presence of telephone conferencing, it was, nevertheless, quite good for giving orders, making quick decisions in a crisis, settling differences of opinion, giving briefings, and seeking information.

There is growing evidence that as people do more conferencing by telephone, these disadvantages disappear. Telephone conference meetings are shorter than in-person meetings, and participants believe they are more persuasive in their arguments and that their fellow participants are more trustworthy. Despite the early fears that disembodied voices would make conferencing so impersonal as to be uncomfortable, frequent users of audioconferences are pleased with the results. The intelligent telephone, capable of delivering "custom calling" options, has gone a long way toward making telephone conferencing easy to arrange and almost as effective as the in-person meeting.

C. THE VISUAL IN AUDIOVISUAL COMMUNICATIONS

Visuals can also be introduced to the telephone conference. Remote writing or telewriting using electronic blackboards can be dialed up on a separate line and controlled through the conference telephone. Information written with a small location indicator attached to the writing instrument on special conductive surfaces (the electronic blackboard) is translated into signals of varying frequencies, a form of frequency modulation. These signals are transmitted, stored, and reproduced as continuous writing at the receiving end.

If participants wish to use slides in their presentations, projectors can be activated at each location by sending coded signals very similar to the kinds of multiple-frequency tones used for pushbutton dialing. The remote projectors decode these tones, converting the signals into instructions for the projector to step ahead or return to a previous slide.

To transmit a given quantity of information, specific amounts of time and bandwidth are required. It is possible to trade-off time and bandwidth to send information. Thus, it is possible to send the information on a 78 rpm record $2\frac{1}{3}$ times faster than at $33\frac{1}{3}$ rpm by increasing the bandwidth or frequency range by $2\frac{1}{3}$. The same practice can be used to transmit video images. To send a black and white television picture requires about 4 MHz (4,000,000 cycles per second). But with only 4 kHz of bandwidth available on the voice channel we need to increase the transmission time. So instead of sending the video in real time we record the images at the transmitting end or freeze the pictures or frames and send one frame every 30 sec. The viewer is scanning the video image at a slower rate than it is being recorded; hence the term slow-scan television.

D. TELECOMMUNICATIONS–TIME TRADEOFF

When the central office consisted of a human operator at a switchboard, it was possible to leave messages for later delivery or instruct the operator to intercept calls for delivery at later and perhaps more appropriate times.

Digitizing analog signals makes possible the transmission of many messages in the same bandwidth space by means of time division multiplexing. The same technique can be used for storing digitized voice on magnetic disks similar to those used by computers for storing large amounts of information. This mass storage facility is accessed by the electronic switch, either in the telephone central office or in the firm's private automatic branch exchange. Voice store

and forward services enable us to trade telecommunications for time. [*See* DIGITAL SPEECH PROCESSING.]

E. DISTRIBUTING INTELLIGENCE IN THE TELEPHONE SYSTEM

When businesses learned how valuable and important good communications were and increased the number of telephones in their offices, they installed a switchboard and an operator on their own facilities. This was the first private branch exchange (PBX), and it was as intelligent as the persons operating it.

When mechanical switching was introduced into the central offices, it was also introduced into the private automatic branch exchanges (PABX) in offices. These were large and noisy devices hidden behind heavy doors, often near receptionists desks in outer offices. But as computing became an integral part of modern switching, semiconductor microprocessors reduced the PABXs to small desk-top equipments, attractive and intelligent.

Today's telephone system is a dual communications network; one network performs the task for which the telephone was invented, the transmission of people-to-people messages, and a second network controls, manages, and ensures that the people-to-people network works efficiently. Control signaling travels along this second network, interconnecting seats of intelligence along the network. These are locations where we collect the information necessary for network management and control, which we store and use in ways that enhance the services the message network can deliver. This intelligence is distributed at the various switching centers and at PABXs located in offices. Today, intelligence is also located in the telephone instrument itself.

F. DATA COMMUNICATIONS ON THE TELEPHONE NETWORK

Business activities, whether carried out from the home or from the office, require the use of many media, often simultaneously, including voice, data, one or more forms of image distribution, full motion or slow-motion video, and fasimile. This need for diversity in communications emerged in the United States in the mid-60s. The post-World War II industrial and business explosion had resulted in the distribution of plants and offices throughout almost every corner of the nation and leaped across oceans as national firms went multinational. Communications and transportation costs grew rapidly, and before long, computers wanted to talk to one another even more frequently and for longer periods of time than did the human workers in the plants and offices.

It was during this period that pressures increased on the telephone carriers for providing more diverse forms of communication services at rates more fairly reflecting their actual costs. The high costs of corporate telecommunications and growing demand for data and visual communications encouraged the telephone carriers to find ways to lower costs and deliver more varied communications services on the telephone without having completely to replace their multibillion dollar capital investments.

The telephone system was designed for voice transmission. Once on the line, the connection is held until the conversation or silences are completed.

If data-carrying pulses are transmitted along telephone cables, there is distortion. A 500-microsecond (μsec) transmitted pulse is distorted to one requiring almost 1500 μsec when received. The consequence of many pulses or bits being sent along the ordinary telephone line is interference that distorts the information.

While it is possible to send pulses in this manner on the subscriber loop, perhaps over distances as much as 10 miles, it is not possible to send such pulses along the trunks between switching centers because frequency multiplexing at high carrier frequencies is used to minimize the copper and duct space. To send pulses or digital on–off signals along a telephone line as tone signals requires modems, devices that can translate the pulses into modulations of the carrier at the transmitting end and that can demodulate the resulting modulated carrier signals into pulses or digital information at the receiving end.

The traditional telephone can transmit digital signals at surprisingly high speeds using tone-modulated carriers. Teleprinters or automatic telegraphs, the most familiar of which is the Teletype©, have been widely used in the United States, while in Europe switched teleprinter networks or telex systems have been used extensively and for many years served as the basic communication network in the absence of the telephone for most businesses.

Most teleprinters transmit 60–100 words/min.

An average English word consists of five letters and one space, or six characters. These letters must be translated into bits, and a 5-bit code known as a Baudot code has been made standard for most teleprinters throughout the world. Each letter of the word consists of 5 bits. To know when a word begins and ends, additional bits must be added. Except for the bit that indicates the end of the word, which is half again as long as the other bits, all of the bits are equal in time. For each character of a word, 7.5 pulses (i.e., $1 + 5 + 1.5$) per character are transmitted. A group of bits that are to be sent or processed together is called a byte. A teleprinter system operating on a telephone line sends 50 pulses/sec. One pulse/sec is one baud. At 50 baud the teleprinter transmits about 67 words/min. At this rate, one can multiplex many such teleprinters on a voice channel. Because there is interpulse distortion, the number of digital transmissions on a voice channel is usually limited to 20. In a single voice channel of 300–3400 Hz, digital transmissions are assigned bandwidths according to the speed at which the data are transmitted. If a transmission is to take place at 110 baud, 170 Hz is assigned, for 220 baud, 340 Hz, and for 440 Baud, 680 Hz.

When spoken over a telephone, words and sentences are readily recognized. However, when voice transmissions take place over the satellite, the quarter-second delay due to the 44,600 miles traversed can often cause difficulties. Speakers often feel they must resort to the techniques used by radio operators who end their sentences with "over" or "over and out." These are the rules of the road, or protocols, required for communicating in special situations.

Communicating information as sequences of ones and zeros is one such special situation, and rules of the road are required to make the messages intelligible. Decision rules are built into the sending and receiving devices to permit one to understand what the other is doing. For example, if an 8-bit per second (bps) stream of bits such as 0 0 1 1 1 0 1 0 is transmitted, and the receiver rate is adjusted to receiving 4 rather than 8 bps, it will read that word as 0 1 1 1.

The receiver must be synchronized with the transmitter. In both the receiving and transmitting equipment there are devices, event clocks, that generate accurate internal pulses or events. They are activated when a transmission is begun. When a string of bits representing a message is transmitted, a certain number of bits are used as a code to tell the receiver that a message is coming and that the clock is running, before the receiver begins to read the message.

Computers generally communicate asynchronously; indeed, most telex and data communications are asynchronous communications. In asynchronous communications the message is broken up into short groups corresponding to the code for one character. Thus, the telex Baudot message is a 5-bit group, most personal computers use an 8- or 16-bit group, and some new ones are opting for a 32-bit group. There is no event clock, so synchronization between the transmitter and the receiver must be reestablished for each group. This is done by adding extra events such as a start and a stop bit. Eliminating the need for event clocks in modems reduces their cost; hence the use of asynchronous communications for most data communications networks using the telephone system.

Using the entire voice channel to send data is not very efficient since the transmission of data is a bursty transmission; there is a great deal of time between bursts of data even in a string that represents a single word. And while these bursty or discontinuous streams of data are being transmitted, both transmission paths and switches are occupied. Designers assume that the average line and switch are used only about 5–10% of the time even during the busiest hours. When computers communicate with one another, they are likely to be on the line for hours on end.

The amount of data on the voice network is increasing. Data communications speeds on the telephone network vary from high-speed computer-to-computer traffic to slow data transmissions for monitoring fire alarm and security systems in firms and households and for monitoring and controlling energy consumption. These latter signal bursts may only travel at 100 baud or at most 1000 baud, which is much less than the 9600 baud engineers have found to be the practical limit for a 4000-Hz voice channel connected directly from transmitter to receiver, without any switching in between or with leased private lines.

The voice channel can be used to send both data and voice at the same time on the channel. This is done by using only a portion of the channel for voice, for example, up to 2040 Hz. This leaves the bandwidth from 2040 to 3400 Hz for data. Some telephone carriers have found that they can transmit up to about 960 baud as data

over voice in that space. At the central office serving the subscriber, the data are separated from the voice signals and transmitted from that point onward along separate networks.

The system serves the subscriber in two very different ways. One circuit allows the subscriber to use both voice and data up to about 1.2 KBs, quite suitable for sending computer messages, banking, shopping on the network, and other services, which we discuss later in considerable detail. On the other line, a subscriber can use higher speed business data services up to about 4.8 Kbs as well as voice. In this way the local loop is essentially untouched; the additional equipment required is at the subscriber's home or office and in the central office. The cost of the central office equipment is shared by all of the subscribers, those using voice only as well as those using voice and data. The additional subscriber equipment is paid for by the subscribers who order these new services.

Another means for using the telephone for data transmission is by transmitting data under voice. In this manner, very high-speed data, in the range of 1.5 Mbs, are delivered under the master group frequency where voice transmissions begin. Special coding techniques are used to squeeze 1.5 MBs into a 500-kHz space.

End-to-end digital telephone transmission paths are gradually replacing these mixed modes for transmitting data on the telephone network. To capture all the nuances of the human voice, the signal must be sampled at least twice its bandwidth or 8000 times per second and quantized using an 8-bit code. Thus, if we want to send voice over the telephone, the rate should be 64,000 bps (8 bits × 8000 samples per second).

Time division multiplexing several signals requires switching designed to ensure that signals sampled from several sources are correctly reconstructed at the receiver. This is achieved by switching in time and space. For example, assume there are three separate voice conversations, A, B, and C, which are sampled 8000 times per second, coded with the 8-bit number, and ready for transmission. The sampling system takes a sample from A and starts it down the transmission line, then one from B and sends that one down the line, and then one from C and sends that one down the transmission line. At the other end the signals are decoded, and the sampling is repeated. The messages are reconstructed but without any missing segments. However, instead of delivering all of the A samples to channel A they are delivered to channel B. Switching has enabled channel A to communicate with channel B.

G. INTEGRATED SERVICES DIGITAL NETWORK

Today's telephone system is essentially a dual system, operating in both analog and digital fashion. Messages are transmitted primarily in analog throughout the system, and control or supervisory signals are transmitted digitally in the message system as well as along a separate pathway. The telephone system has been modernized by overlaying computer techniques on traditional technologies to meet the demands of business and industry while still providing universal service at affordable costs for residential subscribers.

This has been achieved by modernizing those portions of the system that are used in common by all users, the switching and line haul or long-distance systems of the network. Indeed, without these developments, intelligent telephones with revenue-producing services would be impossible. Today's business users require access to a variety of media: voice, image, video, and data. The household on the network marketplace also requires the ability to call up the most appropriate communication mode for the task at hand. To do so requires an end-to-end digital pathway or transmission system, from office to office, from home to home, and from home to office, in which video images, data, and still pictures are all transmitted as bits, indistiguishable from one another except in the information they encode.

There are many reasons for having an end-to-end digital system. Channel capacity can be more efficiently used with time division multiplexing than with frequency division multiplexing. Switching is integrated, that is, control and messages move along a single path, thereby monitoring performance more easily, accurately, and reliably. Signal-to-noise ratios of digital transmission systems are higher than those for analog systems, and there is likely to be less interference from external high-frequency sources. Finally, pulse regeneration, rather than signal amplification, greatly reduces the cost of transmission while ensuring highly reliable transmission.

However, there are additional costs for the integrated digital transmission system. Voice is the heaviest traffic on telephone systems today and will still account for more than 90% of all traffic on the U.S. network even in the year 2000. Transmitting digital voice requires analog-to-digital and digital-to-analog conversion at every telephone in the system. With almost 300 million telephones in the United States, all designed for analog transmission, the conversion cost would be astronomical. Where telephones are just now being provided in large numbers, as in many developing countries, an end-to-end digital system is seen as the preferred route for the future. Thailand and Brazil, for example, are just now constructing their telephone infrastructure and have chosen to go all digital. France is greatly expanding its telephone infrastructure, going from about 20% penetration a decade or so ago to 80% penetration by the end of the century and is doing so with an all-digital system.

The end-to-end digital system is the integrated services digital network (ISDN). This network can provide voice, video, image, and data services and can provide transmission rates from 10 baud for alarm services to 96 Mbaud for full motion television. A subscriber can order whatever service (and transmission speed) the subscriber wants from a network terminal. The ISDN is presumed to be intelligent, and consequently there is no need for intelligent telephones and terminals.

V. Politics of the Telephone

Theodore Vail, the organizational genius who made Bell a household word, had a grand design for a universal service: one communication network that would interconnect a nation. The nation is now interconnected by over 1500 independent telephone companies that act as a single system and provide telephone service at affordable rates, that is to say, at rates that can be met by more than 98% of the households in the nation.

Vail's dream was given reality by the Communication Act of 1934; the Bell System was, essentially, made a national monopoly to provide this universal service at affordable costs. To merit this special grace, the Bell System agreed, in a 1956 consent decree, not to engage in any services that could not be classified as transmission. The convergence of the communicating and computing technologies blurs the distinction between computing and communicating. Information transmitted can be manipulated, translated, stored, even modified before being received and the processes performed to achieve efficiencies in communications.

The consequences of almost 10 years of debate has been the Greene Decision of 1983 (effective 1984) which deregulated AT&T, thereby allowing a reconstructed AT&T to enter into nontransmission businesses. In return for lifting the consent decree of 1956, AT&T agreed to dispose of its 22 operating companies (the local Bell operating companies) and to set up a separate subsidiary for its nontransmission (computer) ventures. AT&T's manufacturing arm, the Western Electric Company and its Long Lines Division, will now be competing even more aggressively to provide computerized and other specialized terminals, with information hardware, systems, and software, as well as firms offering long-distance services by satellite and other modes.

Local loop service remains the responsibility of the local Bell operating companies, albeit in a new configuration of seven large holding companies into which the individual operating companies have been folded. These operations remain regulated by the state utility commissions, who now have the sole responsibility to see to it that telephone rates remain affordable, if they deem this to be a social responsibility or a political necessity.

Some observers have suggested that there will be a single, all-digital network for all forms of communications including voice, data, video, and images. While technologically this is quite feasible, it is more likely that different services will continue to seek different transmission means. Banks will look to the high-speed data services cable might offer, while the small business will continue to use the telephone to reach the bank. Airlines will want to use special radio equipment and frequencies quite differently from those used by the radio listener in the automobile. Satellites will continue to be used for long-distance transmission while the twisted pair or the cable, and perhaps the optical fiber, will be used for voice and personal computer transmissions in the home. A variety of transmission modes will be used side by side because it is economical and useful to do so. Users will pay for the bandwidth they need.

The United States has chosen a competitive telephone infrastructure, one that provides the consumer, business and residential, a rich choice of terminal devices or instruments and

services. With the divestiture of AT&T the United States has added to an already diversified telephone carrier industry; the seven holding companies, in which the former Bell operating companies are now housed, are independent of one another and from their parent. They join the almost 1500 independent telephone companies that have been operating these many years, but these new entities deliver the telephone services to the vast majority of households and businesses in the United States. The United States has strayed far from a monolithic system characterized by the ISDN as conceived by the national postal telephone and telegraph (PT&T) monopolies in many nations throughout the world.

The United States has opted for a marketplace of networks delivering a variety of services to clients who are willing to pay for these services. These services range from the traditional telephone services most residences require, basic services, to the special services required by business and industry such as digital transmissions, facsimile, teleconferencing, and voice store-and-forward, the enhanced services made possible by intelligent telephones and networks, and the ability to send digital transmissions over the telephone network.

The United States has achieved universal telephone service at affordable costs by means of cross-subsidization from higher-valued and higher-cost long-distance services and business uses.

After divestiture, this form of cross subsidy is no longer possible; business and industry users of enhanced services now pay only for the services they purchase. They do not wish to support low-cost universal services, arguing that these social goals have already been met. Further, they argue, residential subscribers should pay the real cost of their basic telephone services.

Unlike countries that have nationalized PT&T monopolies where tax revenues underwrite system developments, costs for advanced developments for the U.S. telephone system will eventually come out of the subscriber's pockets. Consequently, whether it is necessary for all subscribers to have access to all of the capacity potentially to be offered by an ISDN is an oft-raised question. Why should a subscriber pay for 6.312 MBps capacity when the subscriber only uses in 56 KBps for voice transmission? With cable television systems providing wired

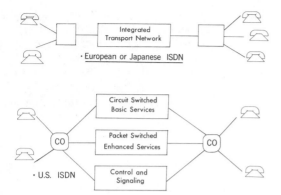

FIG. 5.   Competing versions of the integrated services digital network (ISDN).

video in the United States, why should a telephone subscriber pay for 96 MBps for video over the ISDN? Even for the personal computer users wishing to access remote databases, 1200 baud is more suitable than 1.544 MBps.

Specialized carriers in the emerging marketplace of networks in the United States are offering enhanced services to their business customers and to those households with sophisticated personal computer users. But it is not likely that there will be a single monolithic ISDN in the United States with the architecture envisioned by several European nations and by Japan. Figure 5 compares the U.S. ISDN concept with those of several European nations and Japan.

In the European and Japanese systems, the integrated transport network, the all-digital transmission network, reaches from household to household, from business establishment to business establishment. Every node on the network has available to it the bandwidth necessary for delivering a range of services with bandwidths from 10 bps to 96 Mbps.

On the other hand, the U.S. model of the ISDN makes a distinction between basic and enhanced services. There are three transport paths: circuit-switched analog for voice, packet-switched for enhanced data services, and the all-digital system, extending from central office to central office instead of subscriber to subscriber. While the subscriber loop will, in time, become all-digital, there will likely continue to be a competitive market for customer equipment for both basic and enhanced services in the United States.

The decision of the United States to go this competitive route is one that has created some

considerable concern throughout the world. The telephone has been an international instrument; over the last decade or so, international direct dialing has made it convenient to place calls from New York to Cape Town, from Los Angeles to Bangkok, and were it not for cold war politics, from Chicago to Moscow. Worldwide standards created by the CCITT (the International Telegraph and Telephone Consultative Committee) has made this possible. When complex communication systems have common or similar architectures, it is relatively easy to set up standards so that systems can talk to one another. But now that there are significant differences between the U.S. telephone industry structure and that which appears to be emerging throughout the world, it is feared that this ability to communicate across nations will become more difficult. Certainly, creative standards will be required to ensure that the world is not disconnected.

Of greater concern to the United States and other nations with high-technology industries is that telecommunications hardware and systems developed for the U.S. market will be unable to operate on foreign networks and vice versa, thereby creating serious trade barriers in the very industry that is considered to be one of the world's great growth industries.

The international politics of the telephone are the politics of trade. For if the PT&Ts of the world provide their form of ISDN, there will continue to be little opportunity for trade in telephones and terminal equipment. However, many nations, particularly the United States, seek freer trade and the opportunity to provide terminal equipment to what is probably the most lucrative business of the future, telecommunications equipment and information services.

While there may be good technical and economic reasons for embarking on the all-digital path, many believe that the PT&Ts also have political motives for their drive to the ISDN. High-valued telecommunications services in many nations help defray the operating costs of postal services, provide jobs, and serve as a reminder to a nation of its national sovereignty. For these reasons, domestic and international telephone charges tend to be higher overseas compared with rates imposed by U.S. carriers. There is every good reason, then, why a nation wishes to maintain its PT&T as a monopoly. One way of doing so is to ensure that all telephone hardware, from office to office and from home to home, is provided by the telephone monopoly. Providing an ISDN with all of the intelligence built into the network, where there is no differentiation between basic telephone services and such enhanced services as those requiring the transmission digital signals, preserves the carrier monopoly and helps keep out foreign competition.

VI. Social Uses of the Telephone

The uses to which the telephone is put are more varied and more numerous than many have believed. Clearly the telephone is important for making appointments, obtaining information-on-the-spot, and managing personal and business affairs. It is probably the executive's most important management tool, extending management reach far beyond the geographical limits of the office, second only to face-to-face interaction for effective direction and motivation. The telephone is important for the student and researcher, assisting in keeping in touch with assignments missed and following up on recommended readings and sources of information.

These are the instrumental uses of the telephone, uses for which one can usually rationalize an economic payoff; it saved time and transportation costs, or performed some task that otherwise would not have been performed. There are also the intrinsic uses of the telephone, for which there may not be an economic value but rather a return that has an affective value. Gossiping on the telephone is sometimes necessary for relieving personal stress, calling up a loved one to determine if all is well, just to keep in touch, coordinating a community affair on behalf of a volunteer organization, and polling voters on issues. These are the intrinsic uses of the telephone that are often overlooked when pricing telephone services and determining if nations should receive loans for telephone system development.

Telephone services are differentiable products; not every use of the instrument is of the same value to the user. The telephone belongs to that unusual class of instruments that affect society in ways determined by society. The telephone is a facilitating device; the consequences of its use depend in great part on purposive human calculation as do economic behavior and games. If one wants to substitute communications for transportation, it can be done. If one wishes to construct skyscrapers, the telephone is necessary for interfloor and office communications, otherwise space would be devoted to elevators and stairs rather than to offices.

Understanding the social uses of telecommunications is necessary for the making of intelligent telecommunications policy.

BIBLIOGRAPHY

Bell System (1982). 1A voice storage system. *Bell System Technical Journal* **61**(5), 811–911.

Brock, G. W. (1981). "The Telecommunications Industry." Harvard Univ. Press, Cambridge, Massachusetts.

Crane, R. J. (1979). "The Politics of International Standards; France and Color TV War." Ablex Publishing, Norwood, New Jersey.

de Sola Pool, I. (1983). "Forecasting the Telephone." Ablex Publishing, Norwood, New Jersey.

de Sola Pool, I., Inose, H., Takasaki, N., and Hurwitz, R. (1984). "Communication Flows; A Census in the United States and Japan." University of Tokyo Press, Tokyo, pp. 109–120.

Dordick, H. S. (1986). "Understanding Modern Telecommunications." McGraw-Hill, New York.

Dordick, H. S., and Williams, F. (1986). "Innovative Management Using Telecommunications: A Guide to Opportunities, Strategies, and Applications." Wiley, New York.

Dordick, H. S., et al. (1986). "The Emerging Network Marketplace," 2nd ed. Ablex Publishing, Norwood, New Jersey.

Fike, J. J., and Friend, G. E. (1983). "Understanding Telephone Electronics." Radio Shack, Ft. Worth, Texas.

Johansen, R., Vallee, J., and Spangle, K. (1979). "Electronic Meetings: Technical Alternatives and Social Choices." Addison-Wesley, Reading, Massachusetts.

Miya, K. (1981). "Satellite Communications Technology." KDD Engineering and Consulting, Tokyo.

Pelton, J. W. (1981). "Global Talk." Sijthoff & Noordhoff, Rockville, Maryland.

Pierce, J. R. (1981). "Signals: The Telephone and Beyond." Freeman, San Francisco.

Shannon, C., and Weaver, W. (1949). "The Mathematical Theory of Communications." Univ. of Illinois Press, Champaign-Urbana.

Williams, F., and Dordick, H. S. (1983). "The Executive's Guide to Information Technology; How to Increase Your Competitive Edge." Wiley, New York.

TELEPHONE SIGNALING SYSTEMS, TOUCH-TONE

Leo Schenker *AT&T Bell Laboratories*

GLOSSARY

Common control: Central office switch type of system indicating that switching connections are set up by common equipment once the complete dial information has been received.

Guard action: Means for reducing the probability of talk-off.

Loop: That part of the telephone plant that connects subscribers' premises to the local central office.

Multifrequency key pulsing (MFKP): Method used for signaling between telephone switches.

Progressive control: *See* Step-by-Step.

Slope: Change with frequency in the attenuation of transmission.

Station set: Telephone set.

Step-by-Step: Central office switching equipment of the type where the incoming dial pulses directly effect the position of switches and thus set up a talking path.

Talk-off: Imitation of control signals by speech.

Twist: *See* Slope.

Touch-Tone dialing is a voice frequency means of signaling from telephone sets to central office switches to indicate the destination of a call. The user interface is a set of pushbuttons. Compared to the older dialing method with the rotary dial, Touch-Tone signals have such frequency and level characteristics that they can be transmitted beyond central offices to any point in a telephone network. Hence, Touch-Tone signals can be used for the transmission of data after a connection has been set up. Touch-Tone dialing is now widely used in North American and is beginning to be introduced in other countries.

I. Introduction

Touch-Tone dialing is a method of sending signals from telephone customers' premises to central offices and beyond. It was first introduced in 1964. Compared to rotary dialing, its principal advantages are

1. All the signaling energy is in the voice frequency band, making it possible to transmit signaling information to any point in the telephone network to which voice can be transmitted, that is, "end to end."

2. Touch-Tone dialing is faster, reducing the dialing time for users and, equally important, reducing the holding time for central office common equipment.

3. It provides a means for transmitting more than ten distinct signals: twelve in all standard implementations (and the scheme is capable of sixteen).

4. It is a more convenient signaling method, as attested by users.

II. History of Development and Introduction

The development work that led to the introduction of Touch-Tone dialing began at Bell

Laboratories in the mid-1950s. This was not the first occasion on which the merits of a "pushbutton" dial (as it was then called) were studied. Following the introduction of operator multifrequency key pulsing (MFKP), a system of ac signaling for station equipment was designed, built, and tested in the laboratory in 1941. World War II interrupted further work, and seven years elapsed before equipment was installed and used on an experimental basis in a small trial at Media, Pennsylvania, in 1948. This system was based on the transmission of damped oscillatory signals at two out of an available six frequencies and required special receivers at the central office to detect and convert the signals. The station set used in the Media trial is illustrated in Fig. 1. The dialing unit had mechanical linkages that plucked two out of six metal reeds, each of which was resonant at a specified frequency. When a customer pushed any one of the ten buttons, two reeds were plucked to form a signal coded to the corresponding digit. The energy so generated was transferred inductively to coils in the station set network and so transmitted to the receiver at the central office. Although this mechanism was cumbersome, the performance of the equipment and the reaction of the customers pointed the way to an ultimately feasible system and indicated a favorable public response to pushbutton signaling. The technical and economic aspects of the system tested at Media were not attractive and further work was deferred.

The year of the Media trial, 1948, is also noteworthy because in June of that year the discovery of the transistor was announced. Concepts that formerly were considered exotic became practical. The activity in electronic device development resulted in an entirely new array of circuit components. Miniature capacitors, ferrite inductors, precision resistors, and printed wiring became available, and electric power consumption ceased to be a formidable obstacle.

Using this new technology, a compact multifrequency oscillator, equipped with pushbuttons for selecting and controlling voice frequency signals, was developed in the second half of the 1950s. The oscillator was particularly adapted to the low and variable power available from the central office battery over the range of existing loops to the station set. The oscillator design provided for control of the signal at a level high enough to be reliably detectable at the central office in the presence of the noise from numerous sources that is always present on telephone circuits, but not so high as to exceed crosstalk requirements. The concept of a four-by-four frequency code resulted in a relatively simple mechanical system at the station set.

Concurrently with these electrical and mechanical developments, human factors engineers studied pertinent psychophysical factors and performance ratings of button arrangement were made. The optimum size, spacing, travel, and operating force of the buttons were determined. It was also established that feedback of the signal tones through the telephone receiver was desirable. These studies indicated that customer dialing could be speeded up and generally facilitated by the use of a pushbutton operation, probably without increasing dialing irregularities seriously.

On the basis of satisfactory results in the areas of electromechanical design and psychophysical studies, a moderate-sized field trial of pushbutton dialing was designed and carried out in 1959. Two more trials were to follow before the system was introduced for general use. The chief objective of the first trial was to determine the effectiveness and ease of use of a modern button-operated mechanism with then current types of switching systems, when placed in the hands of a typical segment of the public. Two locations were selected. One of these, at Elgin, Illinois, was equipped with fast-operating common control equipment; the other, at Hamden, Connecticut, used the slower step-by-step switching equipment. The necessary central office receiver and converter equipment was provided on a "black box" basis. At Hamden, this equipment included a digit storage facility and a dc

FIG. 1.   Pushbutton telephone used in 1948 trial test. [Courtesy of AT&T Bell Laboratories.]

outpulser geared to the speed of the step-by-step system. Approximately 120 customers in each central office area were provided with Touch-Tone sets of a preliminary design.

As in the Media trial, customers in both areas were enthusiastic about this experimental service, basing their reaction on the increased speed and general ease of use. In a relatively short period, the customers achieved a dialing speed with the Touch-Tone sets that was almost twice their established rotary dial speed. It was anticipated from laboratory tests that customer signaling irregularities would tend to occur at a higher rate with pushbuttons than with the slower and more familiar rotary dial, and this was confirmed in the field trial. Within a relatively short period, the Touch-Tone error rate of the trial customers showed a trend that led to the projection that with longer experience, accuracy equal to that of rotary dialing would be achieved.

Concurrent development work on the electromechanical design of Touch-Tone dials and receiver-converter equipment resulted in designs proposed for manufacture, and a second set of trials, technical trials of this equipment, were planned and carried out in 1960. As in the Hamden and Elgin trials, two locations were selected. One of these was at Hagerstown, Maryland, served by common control switching equipment; the other was the Cave Spring office at Roanoke, Virginia, served by step-by-step equipment. Figure 2 shows the telephone sets used in the technical trials.

The purpose of the technical trials was to test the operational capabilities of the system and its components. These included

1. The dialing and supervisory capabilities of the system over typical and limiting plant conditions. A sample of adequate size and diversity, including various gauges of loaded and non-loaded cable and open wire and exposure to various environmental conditions, was needed. This objective was achieved by examining the outside plant and selecting stations that represented both typical and extreme conditions.

2. The effectiveness of the protection provided in the system against false signals. It was anticipated that signal-like impulses would be generated by speech, line noise, room noise, atmospheric disturbances, test supervisory tones, etc., that exist on telephone circuits.

3. The reliability, stability, and maintenance requirements of the equipment and its components under typical environmental conditions.

FIG. 2.   Touch-Tone telephone used in 1960 technical trials. [Courtesy of AT&T Bell Laboratories.]

4. Customer usage characteristics such as dialing speed, learning rate, signal pulse duration and intersignal interval duration, error types, and rates.

The results of this second set of trials led to the conclusion that a practical new dialing scheme could be developed based on the technical concepts of the system and components used in the technical trials. The technical performance of station sets and central office equipment brought no surprises. Signal imitations from speech and other noises were not a problem. Customers continued to be highly supportive.

Touch-Tone calling was introduced as a premium service on a market trial basis at Findlay, Ohio, and at Greensburg, Pennsylvania, in 1961. The equipment supplied for the station installations in these areas included some minor refinements relative to that used in the technical trials, and some modifications of the dial were made to adapt the design to the full line of station sets, that is, to desk, wall, "Princess®," keyset, Call Director®, and coin station sets, all of which were equipped with Touch-Tone dials.[1] Figure 3 is a photograph of the Touch-Tone desk set used in the market trials. The appearance design is by Henry Dreyfuss and is familiar because of the widespread use of sets of this design. The cen-

[1] Princess® and Call Director® are registered trademarks of AT&T Inc.

FIG. 3.   Touch-Tone desk telephone. [Courtesy of AT&T Bell Laboratories.]

tral office equipment also included minor refinements, but was essentially unchanged.

Central offices beyond those in the market trial were converted for Touch-Tone dialing beginning in 1963. The new method of dialing was available at the New York Worlds Fair in 1964.

III. Touch-Tone Dialing Scheme

In the previous section, several advantages of Touch-Tone dialing over rotary dialing were listed. When the development program that led to Touch-Tone dialing began in 1953, not all of these were considered equally important. Initially, the only objective was to find a way to reduce dialing time, probably through a scheme that would involve a pushbutton person–machine interface. It was known from the Media trial that customers would be favorably disposed toward pushbutton dialing.

The development environment was governed by the following factors. The rotary dial had been around for several decades and was very inexpensive. The transistor had been invented a few years earlier; there existed no mass applications of this new device. Power dissipation, variation of gain among devices, reliability, and cost were all issues. Thus, it was risky to consider a scheme that would involve an oscillator in millions of telephones and in millions of diverse locations. Some signaling method not requiring an active device was likely to be more practical. For example, one scheme that was studied involved the generation of damped oscillatory waves by interrupting the direct current through the coil of an inductor-capacitor tuned circuit. By switching the capacitor to one of ten taps on the inductors, ten different digit-identifying frequencies could be generated, all in the telephone voice frequency range. But such signals were highly likely to be imitated by the user's (inadvertent) speech in the dialing interval. Hence, in this scheme, each pushbutton operation also caused a stepwise reduction in the direct current drawn by the telephone from the central office. The damped oscillations were ignored by the central office receiver unless accompanied by a dc step.

Very soon it became clear that it was essential to be able to transmit customers' signals "end to end," that is, from one customer's premises, through one or more central offices, to the premises of another customer, anywhere that voice can be sent. Two requirements resulted from the end-to-end signaling objective:

1.   The signals must not contain an out-of-band component such as a dc step.
2.   Sustained rather than damped signals must be used to maintain adequate signal-to-noise margins for the wider range of transmission losses when two customer loops are involved.

The first of these two requirements—the need for the signals to be wholly contained within the voice frequency band—also brings with it the problem of vulnerability to talk-off. The second reintroduced the uncharted domain of active devices. [*See* VOICEBAND DATA COMMUNICATIONS.]

A. Choice of Code

When only voice frequencies are employed, protection against talk-off must rely heavily on statistical tools. This protection is required only during interdigital intervals; speech interference with valid signals can be avoided by transmitter disablement when a pushbutton is operated. Since signals with a simple structure are prone to frequent imitation by speech and music, some form of multifrequency code particularly difficult of imitation is indicated. If the signal frequencies are restricted in binary fashion to being either present or absent, the greatest economy in frequency space results from the use of all combinations of N frequencies, yielding $n = 2^N$ different signals. However, some of these are no more than single frequencies and are therefore undesirable from the standpoint of talk-off. Another drawback is that as many as N frequencies must be transmitted simultaneously; these involve an N-fold sharing of a restricted amplitude range and may also be costly to generate. If $n = 10$, N would need to be at least four. At the expense of using up more frequency range, one is led to a P-out-of-N code, yielding $n = N!/P!(N - P)!$ combinations. There is statistical advantage in knowing that there are always P components in all valid signals, no more and no less.

To minimize the number of circuit elements, as well as to reduce the sharing of amplitude range, P should be as small as possible, yet be larger than unity for the sake of talk-off protection. Let us then examine codes in which $P = 2$. If one can be found that is not readily imitated by speech or music, there is no merit in choosing P higher than two, provided that the total number of frequencies N needed for the required number of combinations can be accommodated in the available frequency spectrum. With $P = 2$, N must be at least five to provide ten combinations (for the ten digits). With $N = 6$, fifteen combinations are available. The MFKP signaling scheme makes use of a two-out-of-six code. There were reasons to assume that MFKP would not have adequate talk-off performance. As discussed below, a code was chosen with attributes that provided protection against talk-off by several means and with which fewer than one talk-off in 5000 calls was achieved, a degree of talk-off performance that was deemed minimal for customer acceptance.

There are advantages, as we shall see, in imposing the further restriction that, with $P = 2$, the frequencies for each combination fall respectively into two mutually exclusive frequency bands. If, for example, fifteen or more combinations are required, N must be at least eight. In the four-by-four code, eight signal frequencies are divided into two groups: group A, the lower four frequencies, and group B, the upper four. Each signal is composed of one frequency from group A and one from group B, resulting in sixteen signal combinations.

B. Band Separation and Limiter Action

With a two-group arrangement, it is possible at the receiver to separate the two frequencies of a valid signal by band filtering before attempting to determine the two specific components. This separation of the two components of a signal renders reliable discrimination between valid signals and speech or noise simpler for two rea-

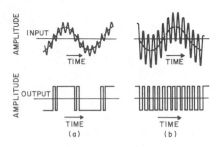

FIG. 4. Instantaneous output for two-frequency input.

sons: (1) each component can be amplitude-regulated separately, thus compensating for "slope" (or "twist"), and (2) an instantaneous extreme "limiting" can be applied to each component after band separation, thus providing a substantial guard action, as shown below.

It is a characteristic of extreme instantaneous limiters that they accentuate differences in levels between the components of an incoming multifrequency signal. This property is used with frequency modulation radios and is referred to as limiter capture. This may be used to provide guard action, that is, action to reduce the probability of false response to speech or other unwanted signals.

Figure 4a shows the input and output of such a limiter with an input signal composed of two frequencies. The lower frequency (dashed line) has the larger input amplitude. Assuming infinite gain, the limiter output may be constructed from the axis crossings of the input. The higher frequency produces some interference, but the low frequency dominates in the output. In this example, noise could be substituted for the higher frequency. In Fig. 4b the higher frequency has the larger amplitude and dominates in the output. Quantitative relationships are shown in Fig. 5, where the abscissa is the ratio at the limiter input of the power in the interference to the power at the wanted frequency, and the ordinate is the power at the wanted frequency at the limiter output relative to its value in the absence of any interference at the input.

Curve 1 in Fig. 5 is for the case where the unwanted component is random noise; curve 2 applies where the unwanted component is a single frequency, not a harmonic of the wanted frequency. Experimental results are also shown for the case of a wanted frequency of 900 Hz and an unwanted frequency of 500 Hz.

What is the significance of the data shown in Fig. 5 with respect to guard action? Speech may contain components that simulate proper sig-

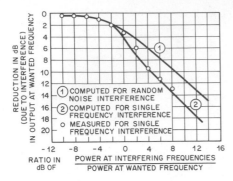

FIG. 5.   Effect of interference at limiter input on its output.

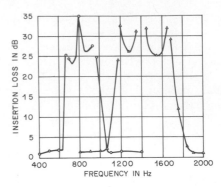

FIG. 7.   Band elimination filter insertion loss.

nals, but it is likely to include energy at other frequencies also. The selective circuitry that follows the limiter is designed to recognize a signal as bona fide, not only when it falls within a rather narrow passband, but also when it appears at an amplitude within about 2.5 dB of the full output that the limiter is capable of delivering. Thus, when a burst of speech contains components at more than just the two signal frequencies, this fact is used to inhibit detection of the signal frequencies in the burst. Inspection of Fig. 6, a simplified block diagram of the receiver, may be helpful at this point.

Aside from considerations of guard action, it is interesting to note that, in divesting signal tones of amplitude variations, a limiter endows a selective circuit with an appreciably higher degree of selectivity than would normally be associated with the selective circuit's response characterisics. Thus, rather simple tuned circuits can adequately perform the function of frequency discrimination.

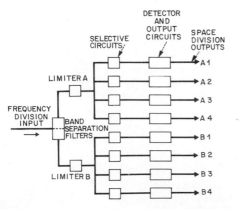

FIG. 6.   Simplified diagram of the Touch-Tone receiver.

Guard action of the type that has been discussed requires that only one of the two tones making up a valid signal be admitted to each limiter. To derive the full benefit of limiter guard action, as much of the speech spectrum as possible should be given access to the limiter. Clearly, a bandpass filter preceding the limiter, to separate the two components of a valid signal, would defeat this objective, since it would permit competition for limiter capture between a signal frequency and only that portion of the speech burst lying in the same band (A or B). However, a filter attenuating merely the other group of frequencies allows competition with the whole speech spectrum except with the attenuated band (B or A). Suitable characteristics for the band elimination filters are shown in Fig. 7. The loss in the elimination band is more than 17 dB. It will be shown that the two components of a bona fide signal differ at most by 5 dB; consequently, the interfering effect at the limiter input of the complementary component is never greater than that of an unwanted signal 12 dB below that at the wanted frequency. It can be seen from Fig. 5 that the effect of noise 10 dB or more below the signal is insignificant.

C. Choice of Frequencies

Attenuation and delay distortion characteristics of typical combinations of transmission circuits were such that it is desirable to keep the frequencies of a telephone signaling system within the 700–1700 Hz range.

The choice of frequency spacing depends in part on the accuracies of the signal frequencies. It was expected that signals generated at the station set could be held within 1.5% of their nominal frequency values and that the pass bands of the receiver selective circuits could be maintained within 0.5% of their nominal ranges. On the basis of these numbers, the selective circuits of central office receivers need to have recognition bands of at least ±2% about the nominal frequencies.

The standardization of amplitude at the output of the limiters permits an accurate definition of recognition bands in the receiver, independently of levels at the receiver input. As a result, frequencies may be spaced closely, approaching the recognition bandwidth of 4%. If the lowest frequency is chosen as 700 Hz, the next frequency must then be more than 728 Hz. Wider spacing makes the precise maintenance of the bandwidth less critical.

Another factor can profitably be taken into account in the selection of a frequency spacing. To reduce the probability of talk-off, the combinations of frequencies representing bona fide signals should be such that they are not readily imitated by the output from the speech transmitter. In a receiver with the guard action described, no sound composed of a multiplicity of frequencies at comparable levels is likely to produce talk-off. Thus, consonants present no problem. Vowels do, however, present a problem, as do single frequency sounds such as whistles that are loud enough to encounter some harmonic distortion in a carbon transmitter. It was shown by H. Fletcher that an electrical analog of the mechanism involved in the articulation of vowels is a buzzer (the vocal cords producing a fundamental and a long series of harmonics) followed by a selective network that shapes the harmonics into formants of the vowel sound. As a result, the spectrum of any sustained vowel contains a number of frequencies bearing harmonic relationships to each other. Hence, it is desirable that the pairs of frequencies representing valid signals avoid as many of these harmonic relationships as possible.

A family of frequencies that avoids a large proportion of troublesome combinations and also meets all the other requirements discussed so far is as follows, the adjacent frequencies in each group being in the fixed ratio of 21:19, with 2.5 times this interval between the groups:

Group A (Hz)	Group B (Hz)
697	1209
770	1336
852	1477
941	1633

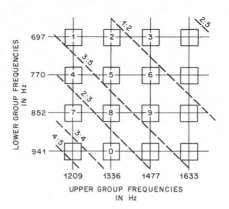

FIG. 8.   Window diagram for signal frequencies.

All frequencies are essentially within the 700–1700 Hz range, and the spacing is adequate to accommodate the recognition bands.

The 16 pairs of frequencies representing valid signals avoid low-order ratios. This is illustrated in Fig. 8 where frequencies are plotted on two logarithmic scales: those below 1000 Hz as ordinates and those above 1000 Hz as abscissae. Any valid signal is represented by a pair of coordinates, namely, its two component frequencies. The 16 square windows represent the ±2% recognition bands required.

The diagonal lines in Fig. 8 are the loci of pairs of frequencies having simple harmonic relationships, that is, 1:2, 2:3, etc. The avoidance of these particular diagonals is beneficial because they represent the effects of harmonics not only of the corresponding order but also of higher order, for example, the third and sixth, the ninth and fifteenth, and so on. Not all applicable diagonals are shown in Fig. 8, and 18 of the 65 potentially troublesome combinations are not avoided. However, if two frequencies in speech representing two fairly high-order harmonics (e.g., 4:7 and 7:12) happen to fall at proper signal frequencies, their fundamental will be low in frequency. The odds are then good that other harmonics within the telephone speech band will together have sufficient relative intensity to bring guard action into play.

Comparable benefits with respect to talk-off can be derived from other sets of frequencies with the same geometric spacing but displaced up or down on the frequency scale. The effect of such displacement is to shift all the windows in a direction parallel to the diagonals.

D. CHOICE OF AMPLITUDES

Since signaling information does not bear the redundancy of spoken words and sentences, and yet must be transmitted with a high degree of reliability, it is advantageous for the signal power to be as large as permitted by the environment. A nominal combined signal power for the two frequencies of 1 dB above 1 mW at the telephone set terminals was adopted as a realistic value, a ± 3 dB tolerance about this nominal value being acceptable.

For subscriber loops, the maximum slope between 697 and 1633 Hz is about 4 dB, the attenuation increasing with frequency. In two subscriber loops (as in end-to-end signaling), the slope may be 8 dB, although, statistically, this is unlikely. A reduction in the maximum level difference at the receiver in the two signal components can be achieved by transmitting the group B frequencies at a level 3 dB higher than that of the group A frequencies. In this way, the nominal amplitude difference at the receiver input between the two components of a valid signal is never more than 5 dB (end to end) and 1 dB at the central office. The nominal output powers were chosen as -3.5 dBm and -0.5 dBm for groups A and B, respectively, adding up to 1.3 dBm. In more than 99% of station-to-central office connections, the 1000 Hz loop loss was expected to be less than 10 dB (during the early 1960s). Similarly, the station-to-station loss at 1000 Hz was estimated to be less than 27 dB in more than 99% of all connections. Making some allowance for slope and variations in the generated power, the minimum signal power at a central office receiver was estimated to be 15.5 dBm. At a receiver involved in end-to-end signaling, the minimum power was estimated to be 32.5 dBm. Higher minimum levels can probably be expected today.

E. Signal Duration

In the development of the Touch-Tone dialing system, two important parameters had to be fixed before adequate data were available: the duration of a valid signal and the signaling rate. The probability of talk-off can be reduced by increasing the duration of the test applied to a signal by the receiver before accepting the signal as valid. It was clearly undesirable, however, to require even the most adept users to extend a pushbutton operation beyond an interval determined by their innate or acquired mental and physical acuity. A frame of reference for setting requirements for these intervals can be formed in two ways. First, competent typists achieve a speed of 60 words per min, and, at five letters per word, the corresponding time interval be-

tween letters is 200 msec. Second, musical compositions occasionally include 64th notes; although the speed of performance is at the discretion of the performer or conductor, the duration of a 64th note on a piano is estimated to be rarely as short as 45 msec. A minimum of 40 msec was set as the objective for both signal and intersignal intervals, allowing for a dialing rate over 10 signals per sec.

The allocation of a minimum of 40 msec to these intervals proved to be reasonably sound. The median tone time found in the field trials was approximately 160 msec; durations of 30 to 40 msec were observed but only approximately 0.15% of dialing starts failed because of short tones. The median interdigital time was approximately 350 msec; no case of failure because of short interdigital intervals was observed. Fortunately, consideration never had to be given to increasing the minimum signal time to improve talk-off performance.

IV. Touch-Tone Dial Telephone

The scheme described in the previous section would have remained just another scheme if it had not been feasible to develop a dial that could be manufactured and maintained at a reasonable cost. The Touch-Tone dial represented a bold departure from past practices in a product line that today would be called consumer products. A proven mechanical device, honed to near-perfection as a result of extensive use and several redesigns, was to be replaced by a new mechanism and circuitry, including a relatively immature electronic active component, the germanium transistor, to be installed in millions of homes and offices.

A. Touch-Tone Dial Circuit

The basic circuit is a transformer-coupled version of the Meacham bridge stabilized oscillator and is shown in Fig. 9. Its operation is as follows. A feedback path is provided in the emit-

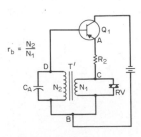

FIG. 9.   Basic oscillator.

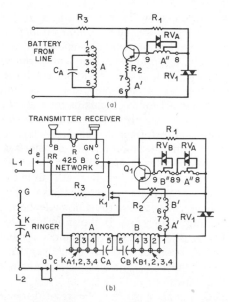

FIG. 10.   Touch-Tone dial circuit. (a) Basic oscillator circuit. (b) Shown as an applique to the Western Electric 500 type telephone set circuit.

ter–follower amplifier by a resistor R_2 and a transformer T coupling the emitter output to the base. The transformer is tuned by capacitor C_A and shunted by an amplitude-sensitive resistive element RV. Since the voltage gain from base to emitter is essentially unity, the voltage ratio of T from emitter circuit to base must exceed unity to provide enough positive feedback to maintain oscillation. The net gain around the feedback path is regulated at exactly unity by the action of RV.

Figure 10a shows a more practical form of the oscillator. Here, the coupling transformer has a separate tuned winding, which provides freedom both to select its impedance level and to locate it conveniently in the overall circuit. Also shown are taps on winding A for connection to C_A to provide frequency selection, and a tap for varistor RV_A on winding A'' to determine, as will be explained, the amplitude at which unity feedback is established.

Direct current from the telephone line flowing through R_1 and the silicon junction varistor RV_1 makes the relatively constant forward conduction voltage of the silicon junction available for biasing the transistor to operate in the class A mode. Since this voltage is essentially constant, the same direct emitter current will flow for all practical values of loop length. R_2 is selected to fix this current at a value only slightly in excess

of the maximum peak alternating current in order to minimize collector dissipation. The low ac impedance of RV_1 when conducting avoids the need for a bypass capacitor around the bias source.

The function of RV_A in limiting the amplitude of oscillation may conveniently be described by reference to Fig. 10a. This device is a silicon junction varistor that has the well-known property of falling sharply in forward resistance when a characteristic forward voltage (about 0.6 V) is exceeded. Being bridged across a winding of the tuned transformer, RV_A will rapidly degrade Q when this voltage is reached. Component values and turns ratios can be chosen so as to result in positive feedback and, hence, growing amplitude of oscillation until the peak amplitude across RV_A reaches 0.6 V. Beyond this amplitude, the varistor resistance falls so rapidly that the peak ac voltage may be considered fixed at 0.6 V across that portion of winding A'' bridged by RV_A. It will be noted that this method of amplitude regulation permits the transistor to operate linearly.

While the foregoing analysis applies to a single-frequency design, the actual design being described is a dual-frequency arrangement of the circuit. This is shown in Fig. 10b, arranged as an applique to the 500 type telephone set. Dual-frequency operation is accomplished by employing two completely independently tuned transformers, A and B, with their respective windings in series, using the basic circuit of Fig. 10a. Simultaneous oscillation at two frequencies is possible because the active element is permitted to operate linearly.

In Fig. 10b, the circled cross-points on the tuned windings represent frequency selecting contacts, one of which on each winding is closed early in the operating stroke of each pushbutton. The tuned windings are shown subjected to the dc voltage drop across RV_1 through contacts on K_1, thus storing magnetic energy in the transformer cores. Opening these contacts late in the pushbutton stroke interrupts the direct current in the windings, shock-exciting the oscillator into immediate oscillation at full amplitude. The pushbutton mechanism is described in a later section.

The question arises whether the single-frequency analysis holds for dual-frequency operation; it does, provided that the frequency of the other tuned circuit is sufficiently remote to make its impedance at the emitter winding negligibly low compared with that of the corresponding winding being considered in the analysis. This requirement is not entirely met in this case, especially when the highest frequency in the low group and the lowest frequency in the high group are being generated. Under such conditions, a significant reactive impedance is seen in the other emitter winding. This impedance is automatically corrected (to make the loop gain $\mu\beta = 1/0°$) by a small frequency shift that produces a counter reactance in the emitter winding under analysis. Such a correction is known as frequency "pulling." Analysis of this behavior, confirmed by experimentation, has led to the following expression for the frequency change due to pulling:

$$\Delta f_A = \frac{f_A f_B}{4 Q_A Q_B (f_B - f_A)}$$

where f_A and f_B are the nominal frequencies being generated by transformers A and B, and Q_A and Q_B are the circuit quality factors (henceforth denoted as Q) obtained at f_A and f_B. For typical values of Q at the closest frequencies of the 4 × 4 code (941 and 1209 Hz), the pulling is approximately 3 Hz for each frequency. By adjusting the location of the tuning tap, this error is reduced by roughly one-half, so that for the various frequency combinations, the resulting error is approximately ±0.15%. Frequency pulling is another consideration why the two-group 4 × 4 code was selected over the MFKP code. The expression shown above results in a three times greater frequency pulling for MFKP if a single transistor dual-frequency oscillator is used, an important economic consideration at the time the code was selected. As will be seen, a few years later, when Touch-Tone dialing was implemented in the Trimline® telephone, the single transistor solution was abandoned.[2]

A wide range of operating conditions must be considered. For example, the dc resistance of customer loops (including the source resistance at the central office) ranges from approximately 400 to 2000 Ω while supervisory requirements limit the current drawn by the set to a minimum of approximately 0.02 A. Also, the lower ac loss in short loops as compared with long ones results in a wide range of signal levels at the central office receiver terminals.

The circuit of Fig. 10b, although used in early field trials, does not perform satisfactorily when subjected to combinations of extreme conditions. The outstanding faults are excessive

[2] Trimline® is a registered trademark of AT&T Inc.

power dissipation in the transistor on short loops, marginal available voltage on long loops for producing full-amplitude signals, and lack of amplitude regulation with loop length. Since the speech network of the 500 type telephone set limits the power dissipated in the set components on short loops, regulates the transmitted amplitudes (lower levels on short loops), and contains a transformer that enables signals to be generated at low amplitude for transmission at higher amplitudes, a circuit was devised for integrating the Touch-Tone caller with this network. The form of this integrated circuit, as used in the technical trials and in production telephone sets for a number of years, is shown in Fig. 11.

This article would not be complete if mention were not made of two key components in the earlier designs: the tuned transformers and the tuning capacitors.

The requirements for the transformers included high Q (about 30), good inductance stability with temperature and time, adjustability of inductance, small size, and low cost.

Early in the development, it became apparent that a transformer employing a manganese-zinc ferrite core held the greatest promise of meeting these requirements. Production of special shapes by molding reduced cost, and very low losses at speech frequencies led to small size. While the temperature variation of permeability is high for this material, provision of an air gap sufficient to dilute this variability to tolerable values still affords a core of small size and satisfactory Q. A cup type core was chosen because it promised all the desired features, including adjustability.

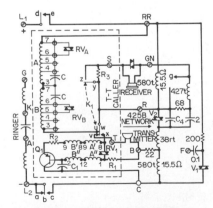

⊘ = SCREW TERMINAL ON 425B NETWORK
◎ = SCREW TERMINAL ON "TOUCH-TONE" CALLER

FIG. 11.   Integrated Touch-Tone caller circuit used in trials.

Polystyrene capacitors met the requirements for the mating tuning elements, including a negative temperature coefficient that canceled part of the positive temperature coefficient of the tuned transformers. With this combination of components, it was possible to hold a frequency variation over the temperature range of $-30°C$ to $55°C$ of about $+0.5\%$.

B. HUMAN FACTORS DESIGN

Extensive tests were performed to determine customer preference, dialing speed, and error rate as various physical design parameters were changed. These parameters included button size and spacing, stroke, force displacement characteristics and many others. One important parameter, although included in the test programs, ended up being determined by mechanical design considerations: the button arrangement. A rectangular array of three rows of three buttons, plus one button in the center of the fourth row $(3 \times 3 + 1)$ was selected because it resulted in an arrangement that required fewer electrical contacts. Of course, later, with the addition of the * and # buttons, this became a 3×4 array. Fortunately, this button arrangement turned out to be close to optimal, both in performance and preference.

Consideration was given to numbering the button in the same sequence as an adding machine, that is, 7, 8, and 9 in the top row, the "Sundstrand" arrangement. User preference was the reverse order of row numbers, that is, 1, 2, and 3 in the top row. Not enough people were users of adding machines in the late fifties; handheld calculators did not exist.

User preference and performance studies, coupled with design considerations, eventually resulted in the now familiar attributes for desk type Touch-Tone telephones: $\frac{3}{8}$-in. square buttons, separated by $\frac{1}{4}$ in.; $\frac{1}{8}$-in. stroke without snap action; about 100 g force at the bottom of the stroke.

C. MECHANICAL DESIGN

The pushbutton mechanism is the means by which the customer controls the operation of the dual-frequency oscillator. In the early part of the downstroke of a pushbutton, each tuning capacitor is connected to a tap on its associated tuning coil to select the two frequencies of the signal for that button. Later in the downstroke, a

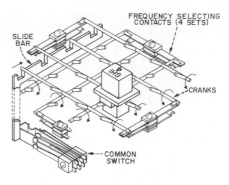

FIG. 12.   Mechanism schematic for the Touch-Tone caller.

switch (K_1, Fig. 11) common to all buttons is operated. The common switch has the following four functions:

1.   Attenuation is inserted into the receiver circuit to reduce the level of side tone during signaling to a comfortable value. This must occur first.

2.   The transmitter path is opened to prevent speech or background noise from interfering with the signal.

3.   Power is applied to the transistor.

4.   The direct current through the tuning coils is interrupted to initiate the signal at full amplitude by shock excitation. This should occur last so that none of the energy stored in the tuning coils is wasted prior to enabling the transistor.

In the original design a mechanism was conceived that kept the number of frequency selection contacts to a minimum (7), rather than to have each button operate two frequency selection contacts that could have required 20 and, later, 24 contacts. This type of mechanism is probably still in the majority of Touch-Tone sets in use today.

A schematic of the mechanism is shown in Fig. 12. Two sets of cranks operate the coil-tap switches. The cranks in the column direction operate switches connected to the high-band coil and the cranks in the row direction operate the low-band switches. When a button is pushed, the pair of cranks crossing below that button are rotated, and one coil tap contact is made for each tuned circuit, thus selecting the two appropriate frequencies. An arm on each of the "row" cranks moves a slide bar when any one of them is rotated. Movement of this slide bar

FIG. 13.   Touch-Tone dial assembly. [Courtesy of AT&T Bell Laboratories.]

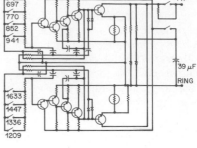

FIG. 14.   Second generation Touch-Tone dial circuit.

operates the common switch. Figure 13 shows the front and back of a Touch-Tone dial.

D. EVOLUTION OF TOUCH-TONE DIAL

The circuitry and mechanism of the Touch-Tone dial have evolved significantly from the early versions. The design had to be modified for the Trimline (dial-in-handset) telephone. Because of space limitations, the ferrite core tuning transformers had to be made smaller, resulting in smaller Q. This increased the tendency to frequency pulling in the dual frequency, single transistor version; hence, two separate oscillators had to be used. Otherwise, the original design endured virtually unchanged for more than a decade.

The availability of silicon and tantalum thin film integrated circuits led to a second generation of designs. Figure 14 is the schematic of this version. Physically, this consists of two glass substrates glued together, with about 20 leads to the crank mechanism and the speech network. Most transistors, diodes, and some resistors are implemented in a silicon integrated circuit chip, with another smaller chip for amplitude-regulating diodes. The tuning elements are capacitors and resistors implemented in tantalum thin film form. Because of the elimination of the ferrite cores, this is much lighter and more robust. The dimensions of the substrates are about 1×0.5 in. As technology evolved, it became possible to eliminate the separate diode chip and to combine the two glass substrates into a single ceramic substrate.

The third and current generation is based on very large scale integration silicon technology—hundreds of devices on a chip—and is produced by a variety of manufacturers. In some cases, the very large scale integration chips provide functions beyond the generation of the Touch-Tone signals, for example, storage of telephone numbers that can then be transmitted by pushing one or two buttons. Some of the chip designs can also generate the rotary dial type direct current pulses. Figure 15 shows a photograph of one of these new types of telephones.

The mechanical design of Touch-Tone dials has also changed. There are no more crank type actuating mechanisms. Rubber dome keypads or membrane keypads are used in the latest designs. The frequency selection and common switch functions are effected by the closing of a

FIG. 15.   Touch-Tone telephone, 1985. [Courtesy of AT&T Bell Laboratories and Panasonic.]

single contact under a pushbutton and the use of logic circuitry.

V. Touch-Tone Receiver

Prior to the introduction of automatic switching, the means of "signaling" the destination of a call involved the spoken language or, more accurately, languages. In areas where more than one language was in widespread use, multilingual operators were required at switchboards. The advent of dialing simplified this situation because dialing became a universal people machine "signaling language." With the introduction of a second customer signaling scheme, the earlier multilingual requirement on operators was now imposed on the central office signal reception equipment.

Historically, automatic switching systems can be divided into progressive control and common control systems. In progressive (step-by-step) systems, the dc pulses resulting from the operation of a rotary dial directly control a series of switches, one for each digit of the dialing sequence, to establish the talking path between calling and called customers. In common control systems the information transmitted by the dialing signals is stored in a register and action to establish a talking path is taken only after the information for the complete dialing sequence has been received. Since progressive switching systems are obsolete, only the arrangement for a common control system, No. 5 crossbar, will be described, which is also much simpler. The same approach applies to modern electronic systems.

Figure 16 shows the necessary modification of a No. 5 crossbar central office for Touch-Tone calling. The receiver takes the 4 × 3 tones and the translator transforms the information into a

2-out-of-5 code that can be recorded directly in the existing originating registers of the No. 5 crossbar office. The receiver is bridged across the line and if dial pulses are received instead of Touch-Tone signals, they continue on to the originating register as before.

A. RECEIVER DESIGN

Figure 17 is a schematic diagram of the receiver. A Touch-Tone signal first passes through a high input impedance amplifier. This amplifier also includes a high pass filter to attenuate the 60- and 180-Hz components of induced power line noise and to attenuate dial tone, standardized as a combination of 350 and 440 Hz. This "standard" dial tone replaced a variety of tones used in older offices and was required for compatibility with the guard action described earlier.

After amplification, the two signaling frequencies are separated into their respective groups by the band elimination filters.

The response of each limiter to the single Touch-Tone frequency at the band elimination filter output is a symmetrical square wave of fixed amplitude containing the fundamental and odd harmonics of the signal frequency. Each limiter drives a group of circuits used for signal frequency recognition. In each group, the tuned circuit corresponding to the incoming signaling frequency responds to the fundamental of the square wave with an amplitude sufficient to operate a detector. However, the receiver does not deliver output signals the instant the detectors

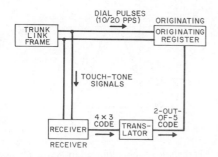

FIG. 16.   Touch-Tone arrangement for No. 5 crossbar.

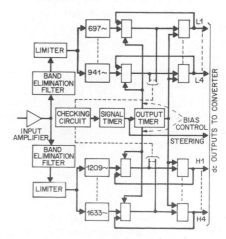

FIG. 17.   Schematic diagram of the Touch-Tone receiver.

operate. Instead, it makes a check to determine the presence of one and only one valid tone in each of the two-frequency groups. This check must continue uninterrupted for a prescribed time interval before an output signal duration test is performed by the checking circuit and signal timer that are shown in the block diagram of Fig. 17.

The check for one and only one frequency in each group is achieved by a combination of simple OR and AND logic circuitry and a characteristic of the limiter permitting the operation of only one detector in a group at a time. When two frequencies of comparable amplitude are applied to a limiter input, neither frequency has sufficient energy at the limiter output to operate a detector.

Once the signal duration test is completed, the receiver is ready to deliver output signals. On minimum-length input signals, there is very little time left after tuned circuit buildup and the signal duration test for delivery of outputs signals. Clearly, some sort of memory is required to prolong the output signals. In the Touch-Tone receiver, this memory is achieved by providing an output timer and output gates with positive feedback to the associated detector circuits.

At the end of the signal duration test, the output timer is turned on. This timer is a monopulser that remains in its quasi-stable (on) state for the required duration of the output pulse. The timer engages the output gates so that the output leads corresponding to the two operated detectors are energized. A portion of each output signal is fed back to the associated detector and holds the detector in the operated condition, even though the input signal may have ceased before the output timer returns to its normal (off) state. The output gate and detector combination remains locked in as long as the output timer is on. The output timer also increases the threshold level on all detectors during its on period so that if the input signal ends, noise or speech will not operate another detector while the outputs are enabled. When the output timer turns off, it disables the output gates, terminating the channel output signals, and restores the detector threshold levels to normal.

In addition to the channel outputs (one per signal frequency), there is a steering output that controls the advance of steering logic in the attached digit registering equipment from one digit storage position to the next. The steering lead remains energized as long as output signals are present or signal tones are present, whichever is longer. On short signals, the locked-in detectors hold the checking circuit and signal timer on so that the steering lead remains energized until after the output timer goes off. On long input signals, when the output timer goes off, the detectors remain operated from the tuned circuit outputs and hold the steering output in the operated condition until the end of the Touch-Tone pulse. This avoids double registration of long signals.

The recognition bandwidth of each channel circuit is dependent on the detector threshold level, the amplitude of the fundamental component of the limiter output, and the Q of the tuned circuit. A recognition bandwidth of approximately 5% of the nominal channel frequency is required to allow for variations in received frequencies and receiver frequency response. Knowing the required recognition bandwidth, the most important of those variables to be determined is the Q of the tuned circuit.

The points where the bounds of recognition bandwidth intersect the tuned circuit response curve mark the required threshold level (see Fig. 18). For a high Q circuit, the threshold level would be far below the peak of the tuned circuit response curve. This would be poor from a limiter guard action point of view (consider the limiter guard action as pushing the response curve down with respect to the threshold level). Furthermore, if the threshold level is more than 3

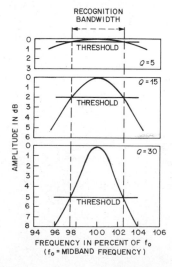

FIG. 18.   Threshold levels required for various values of tuned circuit Q.

dB below the peak of the response curve, it is possible to operate two detectors in a group simultaneously. A very low Q tuned circuit enhances limiter guard action, but the correspondingly small slopes of the response curve at the band edges makes the recognition bandwidth very sensitive to changes in Q, limiter output and detector threshold levels. A tuned circuit Q of approximately 15 was chosen. This results in a threshold level approximately 2 dB below the peak of the tuned circuit response curve.

Receivers designed according to these principles were used in the various trials, and performance was good, including digit simulation performance, which was one talk-off in 10,000 calls. (The rate of digit simulation was poorer when the simulation of the six unused signals is also taken into account.)

B. Evolution of Touch-Tone Receiver

The architecture of the receiver has essentially remained unchanged over the years. However, the physical design has changed radically, as would be expected in light of technological advances since the 1960s. The receiver used in the various trials and in production through about 1975 used all discrete components, 50–60 transistors, coils, and so on. The very early receivers took up about 6 in. of height on 23-in.-wide racks. Soon a redesign resulted in two central office receivers being accommodated in the same space.

During the initial phase, a receiver intended for use in PBXs was developed with a simplified architecture to save space and cost. This was justified on the basis that amplitude sensitivity and talk-off requirements could be less stringent. However, in many PBX applications this turned out to be unsatisfactory and the receivers were replaced by central office receivers with special mounting arrangements.

Receivers introduced in 1975 had the passive LC filters replaced by active RC filters. Integrated circuits on thin film substrates incorporating resistors were used, as well as operational amplifier chips and chip capacitors. Typical physical designs involved three 5×7 in. printed circuit boards.

Receivers in the most current applications are essentially on one silicon chip. For example, for a digital central office, where advantage is taken of the fact that the analog signals offered to the receiver have been encoded into pulse code modulation format by common input circuitry, a Touch-Tone receiver is accommodated on a 40-pin digital signal processor chip that has 45,000 transistors. The program contained in the chip emulates all the signal processing functions described earlier.

VI. Conclusions

Touch-Tone dialing, conceived in the 1950s and introduced in the 1960s, has been widely accepted by the public in the United States. It is estimated that between 40 and 60 million Touch-Tone telephones are in use in the United States. The rate of penetration is probably governed by the higher cost of the Touch-Tone telephone sets and the premium charge for Touch-Tone access lines. In some telephone company territories, an additional factor is the extent to which central offices have been converted to Touch-Tone service.

Touch-Tone dialing is also widely used in Canada and has been introduced in Japan. It is beginning to be available in some European countries. In Europe and elsewhere, pushbutton dial phones are used that emulate the direct current signals produced by rotary dials. Such instruments are usually capable of storing the information input by the actuation of the pushbuttons and transmitting the dc pulses at the rate at which the central office can accept them. In this way, non-tone-generating pushbutton schemes satisfy the needs of users for more rapid and convenient dialing, but they do not significantly reduce the interval between the start of dialing and the setting up of a connection. In fact, they may increase the interval between the end of dialing and the setting up of a connection. The availability of inexpensive non-tone-generating pushbutton dials may slow down the use of Touch-Tone when Touch-Tone capable central offices are eventually introduced in these countries.

A major advantage of Touch-Tone is the potential for data transmission and remote control. A powerful application gaining increasing use is DIVA (data in voice answer). An example is AT&T's Advanced 800 Service. A customer calling an airline may receive a voice announcement: dial "1" for reservation and "2" for flight information. Other services requiring interaction between the telephone user and service providers such as banks and investment companies are also coming into use. [*See* DATA COMMUNI-

CATION NETWORKS; DIGITAL SPEECH PRO-CESSING.]

BIBLIOGRAPHY

Battista, N., Morrison, C. G., and Nash, D. H. (1963). Signaling system and receiver for Touch-Tone calling. *IEEE Trans. Commun. Electron.* **82**.

Benson, L., Crutchfield, F. L., and Hopkins, H. F. (1963). Application of Touch-Tone calling in the Bell system. *IEEE Trans. Commun. Electron.* **82**.

Berry, R. W., Miller, P., and Rickert, R. M. (1966). A tone-generating integrated circuit. *Bell Lab. Rec.* **44**.

Boddie, J. R., Sachs, N., and Tow, J. (1981). Digital signal processor: Receiver for Touch-Tone service. *Bell Syst. Tech. J.* **60**.

Gagne, D. J., and Schulz, C. J. (1963). Central office and PBX arrangement for Touch-Tone calling. *IEEE Trans. Commun. Electron.* **82**.

Ham, J. H., and West, F. (1963). A Touch-Tone caller for station sets. *IEEE Trans. Commun. Electron.* **82**.

Schenker, L. (1960). Pushbutton calling with a two-group voice frequency code. *Bell Syst. Tech. J.* **39**.

TELEVISION IMAGE SENSORS

Paul K. Weimer *Consultant, formerly RCA Laboratories*

GLOSSARY

Charge-coupled device (CCD): Electronic device in which charge is transferred across a semiconductor substrate by means of clock pulses applied to gate electrodes; used for scanning in image sensors and for signal delay and semiconductor memories.

Image intensifier: Nonscanning device that produces an intensified image of the radiation pattern focused on its photosensitive area.

Integrated circuit: Circuit whose components and internal connections are fabricated together in a series of processing steps, usually on a silicon chip.

Interlaced scanning: Scanning raster in which all the odd numbered lines are scanned in field 1 and then the even lines are scanned in field 2.

Photocathode: Cathode, typically of a cesium compound, having the property of emitting electrons when activated by light or other radiation.

Photoconductor: Material whose electrical conductivity is increased by absorption of light or other radiation.

Raster: Pattern of television scanning lines used in sampling the image.

Target: Area in a camera tube where the light-induced charge pattern is stored during integration prior to scanning.

Vidicon: Generic name for a type of camera tube in which the optical image is projected on to a photoconductive target scanned by an electron beam.

An image sensor is the photosensitive component of a television camera analogous to the retina in the human eye or the film in a photographic camera. Its function is to scan electronically an optical image focused upon its sensitive area, thereby generating a video signal from which the image can be reconstructed. Two forms of image sensors currently in use are (1) camera tubes employing electron beam scanning and (2) solid-state image sensors, which are scanned by integrated circuits. Camera tubes are sometimes called television pickup tubes. Solid-state image sensors are also called solid-state imagers or self-scanned sensors.

I. Role of Image Sensors Today

The development of sensors for television cameras was started during the 1920s specifically for use in broadcasting. Today television cameras are found in many applications, in the home as well as in science and industry. The earliest image sensors were vacuum tubes that generated the video signal by scanning an optical image of the scene with an electron beam. Such camera tubes have been vastly improved by many years of development and are still widely used in broadcasting and home video cameras. Although tubes are gradually being displaced by solid-state sensors for many applications, some of their performance features are unique and have not yet been duplicated.

Research on solid-state image sensors was begun in the 1960s, concurrently with the development of integrated circuits. These compact and

FIG. 1. A studio camera for television broadcasting, and a home video camera, shown in insert. (Courtesy of RCA New Products Division, Lancaster, Pennsylvania.)

versatile devices are more rugged than tubes and are being rapidly improved in performance. Their compatibility with the digital world of computers offers significant advantages for scientific measurements and robotic machine vision. As shown in the inset to Fig. 1 and in Fig. 26, they are already found in electronic movie and still cameras, leading to an entirely new concept in amateur photography.

II. Design Considerations

The remarkable performance of the human eye in sensitivity, resolution, and freedom from spurious signals has set a standard for image-pickup devices that is not easily met. It soon became apparent that high sensitivity could not be achieved by an imaging device unless virtually all of the photons comprising the optical image were used in generating the video signal. The first requirement was to increase the responsivity of the photoelectric process to yield as close one electron per photon as possible. The early demands of television provided a strong motivation for many years of industrial research on photoemission and photoconductivity.

A second requirement for a sensitive camera was that none of the photogenerated carriers be wasted in the scanning process. To avoid this, the pickup device could provide a capacitor at each picture element (pixel) of the image area to store all the charges generated at that pixel during the period between successive scans. Without such means for integrating the total charge

developed during the entire scanning period (usually 1/30 sec), the incident light would have to be increased by a factor equal to the total number of pixels in the image (~250,000). An example of the extreme insensitivity resulting from lack of storage was an otherwise excellent early camera tube known as the Farnsworth image dissector, which used only the charges released during the instant of scan.

The large number of pixels required to achieve adequate resolution of picture detail has increased the difficulties of building a satisfactory pickup device. Defective or nonuniform sensor elements will show in the transmitted picture as fixed spots or background patterns. Other spurious signals include random "noise" and conspicuous streaks or blobs resulting from light overload in certain areas. All of these problems were met with both camera tubes and solid-state image sensors and were solved in different ways for the two technologies. [See SIGNAL PROCESSING, GENERAL.]

The image sensors to be reviewed here are for use with scenes illuminated by normal light sources independent of the scanning process. Alternative pickup systems requiring light-spot scanning of the scene itself will not be discussed.

III. Photoemissive Camera Tube Development

A. The Iconoscope (First Use of Storage)

The importance of storage was recognized very early by V. K. Zworykin, who incorporated the idea into a camera tube that he called the iconoscope. Figure 2 compares this tube with a more modern camera tube (a vidicon) and a solid-state image sensor. All of them have a photosensitive image area that contains an array of capacitors for integrating the signal charge developed at each pixel. In the iconoscope, photoemissive silver oxide cesium islands were formed on a large sheet of insulator having a conducting coating known as the "signal plate" on the reverse side. The photosensitive surface was sufficiently nonconducting that this so-called target structure behaved as an array of separate capacitors with the signal plate as a common electrode. An optical image focused on the photosensitive area caused electrons to be emitted from each island in proportion to the

FIG. 2.   Television image sensors, showing an early iconoscope, a vidicon, and a solid-state image sensor. (Courtesy Thompson-CSF, Michel Mathieu.)

image brightness and a positive charge pattern to be formed in the capacitor array.

The video signal was generated in the iconoscope by scanning the charge pattern in a raster of horizontal lines by means of a beam of electrons from the electron gun. As the charges at each pixel were discharged in turn by the beam, a time-varying video signal was induced in the signal plate output lead.

Although revolutionary in its concept, the iconoscope did not bring as large an increase in sensitivity over the nonstorage image dissector as expected. Its storage efficiency was only about 5% because its high velocity scanning beam struck the target with sufficient energy to knock out secondary electrons rather than simply depositing electrons as required for reading and erasing the stored pattern. The iconoscope was used successfully in early broadcasting for pickup from movie film, but it was too insensitive for satisfactory live pickup. It was soon replaced by an improved camera tube called the "orthicon."

The orthicon had a photoemissive storage target similar to the iconoscope, but achieved 100% efficiency in conversion of the stored charge to video signal. This improvement was achieved by the use of a low velocity scanning beam that struck the target with a secondary emission ratio less than unity. The beam charged the target negatively until each pixel was returned to a true reference potential determined by the gun cathode potential. The development of low energy scanning was an important advance, which is

now incorporated into all modern camera tubes. The orthicon, however, had problems of instability in the presence of bright lights, and it was replaced by the "image orthicon" before commercial broadcasting was started in the United States following World War II.

B. THE IMAGE ORTHICON AND THE IMAGE ISOCON

The image orthicon achieved a sensitivity 100 times greater than the orthicon by incorporating two important features: an image intensification section and an electron multiplier for the output signal. The tube structure was relatively complicated as shown by the cross-sectional drawing in Fig. 3, but it served for practically all television broadcasting until the mid-1960s. In operation, electrons ejected from the photocathode by the optical image were focused electron optically to form a charge pattern on the thin semiconducting target within the tube. The resistivity and thickness of the thin target were chosen so that the lateral leakage did not degrade resolution of the charge pattern while keeping the conductance through the target sufficient for the low velocity electron beam to neutralize the charge from the reverse side. The video signal was derived from the unused portion of the beam returning to the electron multiplier surrounding the gun.

The unique storage target was responsible for the remarkable tolerance of the image orthicon to large differences of light levels in the scene. A fine mesh screen mounted close to the thin semi-

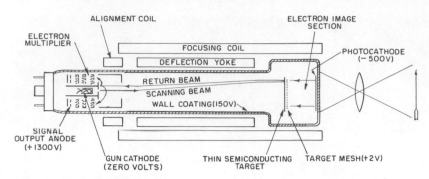

FIG. 3.   The image orthicon camera tube, used for U.S. television broadcasting from 1945 to 1965. [From Rose, Weimer, and Law, (1946). The image orthicon: A sensitive television pickup tube. *Proc. IRE.* **34**, 424. © 1946 IRE (now IEEE).]

conductor served to collect the secondary electrons ejected by the electron image and to preserve contrast with difficult lighting.

The electron multiplier provided a major increase in sensitivity by raising the output signal level far above the amplifier noise, which had been a limiting factor for the iconoscope and orthicon. However, the sensitivity was still limited by the shot noise in the return beam. Unfortunately, the percentage modulation of the return beam was small and, due to the inverted polarity of the signal, was a maximum in the darker areas where the beam noise was most conspicuous.

The image isocon is a camera tube derived from the image orthicon having improved noise characteristics. Instead of obtaining the signal from the total return beam, as in the image orthicon, the signal was derived from only those electrons that had been "scattered" by striking the target in the brightest (most positive) areas. The "reflected" electrons whose lateral paths were unaffected by their encounter with the target were prevented electron optically from entering the output multiplier. The remaining fraction of the incident beam was deposited on the target for neutralization of the charge pattern, as in the image orthicon. The result was a tube whose output current was a maximum in the light areas and whose beam noise was a minimum in the dark areas. Because of its reduced noise, the image isocon, in spite of its complexity, is still sold commercially for low light level applications, while the image orthicon is now obsolete.

C. Image Intensifier Tubes

Although the quantum efficiency of photoemitters is less than for photoconductors and

the other solid-state devices to be discussed later, photoemissive tubes offer such a variety of methods for image intensification that they are often preferred for applications requiring extremely low light levels. The intensifier can either be incorporated into the same bottle with an image sensor or built into a separate bottle and coupled to the image sensor by use of fiber-optic face plates.

An example of a high-gain intensifier built into a camera tube is the silicon intensifier tube, or so-called SIT tube. Photoelectrons ejected from the photocathode are focused at high energy onto a thin silicon diode target of a type to be described in the next section. Bombardment-induced conductivity in the target yields a gain of roughly 1000 times at 7000 V acceleration. Such a tube does not need a signal multiplier for high sensitivity, so the video output lead can be connected directly to the target.

An example of a separate image intensifier that can be coupled optically to a camera tube is shown in Fig. 4. This tube contains only a photo-

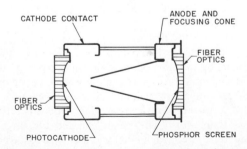

FIG. 4.   An image intensifier tube, with fiber-optic face plates. (From RCA Electro Optics Handbook EOH-11, 1974. RCA New Products Division, Lancaster, Pennsylvania.)

cathode at one end and a light-emissive phosphor at the other, with an electron lens structure in between. Fiber-optic face plates permit more than one intensifier to be cascaded, if desired. The optical gain of an intensifier tube would typically be 30–40 to 1. The addition of such an intensifier to the SIT tube produces the I-SIT combination, which will transmit useful images under starlight illumination (see Fig. 16). However, such intensifiers will limit the resolution to 400–500 lines.

IV. Photoconductive Camera Tubes

A. PHOTOCONDUCTOR REQUIREMENTS FOR TELEVISION PICKUP

The increasing complexity of the photoemissive camera tubes stimulated a parallel development of the vidicon type of photoconductive camera tube beginning in the late 1940s. An advantage of photoconductivity for this application was its increased quantum yield, which approached 1.0 electron per photon, in contrast to about 0.1 for photoemitters. Good sensitivity was therefore possible in a simple tube consisting mainly of an electron gun and a photoconductive target, upon which the image is focused.

The cross section of a vidicon is shown in Fig. 5. The target consists of a thin layer of photoconductor deposited on a transparent conducting coating on the inside face of the tube. The surface of the photoconductor is scanned directly by the electron beam, generating the video signal in the transparent conducting signal plate.

In operation, the signal plate is biased 10–20 V positive with respect to the scanned surface of the photoconductor, which is continually reset to the gun cathode potential at 1/30-sec intervals by the low-velocity scanning beam. The condition for charge integration at each pixel is met by choice of a photoconductor having an effective dark resistivity of 10^{12} Ω cm or more. Charge transport in the photoconductor during the interval between scans causes the potential of the scanned surface at each pixel to rise in proportion to the local brightness of the image. Neutralization of this charge pattern by the beam induces small voltage changes in the signal plate, providing the output signal.

The thickness and dielectric constant of the photoconductor must be chosen to produce a total capacitance at each pixel adequate to hold the maximum signal and not so large as to exceed the charging capability of the low-velocity beam. The latter condition would cause "lag" or smearing of the image for rapid motion. Lag also results when the photoconductor contains traps that produce a slow response to changes in light intensity. Since both photoconductive lag and capacitive lag become worse at low signal levels, there has been little incentive to improve the sensitivity of the vidicon by use of a signal multiplier, such as used in the image orthicon or isocon. However, the sensitivity of the vidicon without a multiplier has been adequate for many applications.

A significant advantage of the photoconductive vidicon target is that it introduces no geometrical structure into the image other than the lines associated with the scanning raster. Such vidicons are therefore particularly suitable for high-definition television systems. The structureless feature does not apply to silicon vidicon

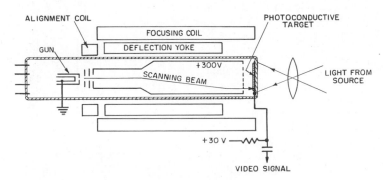

FIG. 5.   The vidicon photoconductive camera tube. [From Weimer, Forgue, and Goodrich (1951). *RCA Review* **12**, 306.]

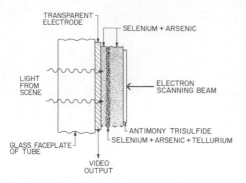

FIG. 6. Cross-sectional view of the deposited layers in the Saticon photoconductive target. (Courtesy of RCA New Products Division, Lancaster, Pennsylvania.)

targets or to the solid-state image sensors discussed below.

B. Modern Vidicon Type Camera Tubes

Modern vidicons range in diameter from 1.5 to less than 0.5 in. Their beams can be focused and deflected either magnetically or electrostatically. The following tubes differ primarily in the materials and construction of their targets.

1. Antimony Sulfide Vidicon

The first vidicon to be introduced commercially contained a porous form of antimony sulfide deposited by evaporation in the presence of a residual gas. It can be produced at relatively low cost, and it has the unique feature that it has a "gamma" of approximately 0.6, where the signal current is proportional to the light intensity raised to the power of gamma. It also has the feature that its sensitivity can be controlled by adjusting the bias potential of the signal plate, thus providing an electronic "iris" for the camera.

2. Plumbicon

The lead oxide photoconductor and excellent gun of the Plumbicon raised its performance level to the point where it had displaced the image orthicon for color broadcasting by 1965. The lead oxide was processed in such a way as to produce a heterojunction that was reverse biased in operation. The blocking contact to the signal plate resulted in low lag and a linear response particularly suitable for use in color cameras.

3. Saticon

The Saticon target, whose cross section is shown in Fig. 6, consists primarily of amorphous selenium that is doped with arsenic to prevent recrystallization and with tellurium near the signal plate to produce a heterojunction. It is capable of excellent resolution and is widely used for single-tube color cameras for home use.

4. Newvicon

An improved version of this target contains CdSe, ZnTe, and CdTe, with a small amount of In_2Te_3. It was designed particularly for resistance against damage due to strong overexposure. It also incorporates a heterojunction and is mass produced in a small $\frac{1}{2}$-in. size for color cameras.

5. Ultricon (or Silicon Vidicon)

The silicon vidicon development in 1967 was important because it demonstrated that reverse-biased diodes in silicon were sufficiently insulating to provide full charge storage in solid-state image sensors. When the photodiode array is used as the target in a vidicon, it provides a highly stable camera that can be focused directly at the sun without damage to the target. As shown by the cross-sectional drawing in Fig. 7, the silicon sheet containing the array must be made relatively thin for light to penetrate from the reverse side to the photosensitive depletion regions surrounding each diode. The same target can also be activated with high-energy electrons from a photocathode as is done in the SIT tube (described in Section III,C).

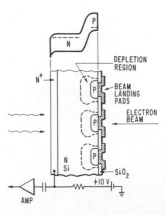

FIG. 7. Cross-sectional view of a silicon vidicon photodiode target. (Courtesy of R. Rodgers, RCA.)

V. Solid-State Image Sensors: Multiplexed X-Y Scanning

The first experimental approach to an integrated self-scanned solid-state image sensor was to scan an array of photoconductive elements by means of X-Y address buses connected to peripheral scanning circuits. This early work employed thin film techniques for fabricating the peripheral scanning circuits as well as the photosensitive array. Figure 8 shows a photograph and circuit of an experimental 180 × 180 pixel photoconductive array scanned by two 180-stage shift registers, which was fabricated in 1966 by RCA. The entire integrated circuit, including thin-film transistors, resistors, photoconductors, and Schottky diodes, was deposited onto four 1-in.-square glass substrates. The sensor shown was incorporated into an experimental hand-held camera that transmitted its signal directly to a nearby TV receiver. Although this sensor did not have charge storage, it provided a measure of light integration by virtue of "excitation storage" in the high-gain photoconductor.

The development of silicon metal oxide semiconductor (MOS) transistors and their use in integrated circuits led to an improved sensor ele-

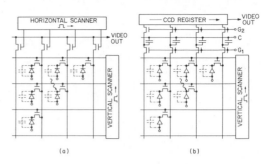

FIG. 9.   A silicon MOS photodiode sensor, showing two methods of horizontal readout: (a) Multiplexed readout, and (b) CCD (charge-coupled device) readout. [From Terakawa et al. (1980). *IEEE Electron Device Lett.* **EDL-1,** 86. © 1980 IEEE.]

ment that had full charge storage and could readily be scanned by conventional silicon circuits. As shown in Fig. 9, each pixel consists of a reverse-biased photodiode, shunted by its depletion-layer capacitance, connected in series with an MOS transistor. Each time the vertical scan pulse turns on the MOS transistors along one row, the photodiode potentials along that row are reset positively with respect to the substrate. During the integration period between scans, each transistor is turned off while the capacitor shunting the reverse-biased photodiode is gradually discharged in proportion to the light intensity. Readout of each pixel occurs when its transistor is gated on again by the vertical register, resetting the potential of each photodiode to its reference potential and transferring a temporary charge deficiency to the adjacent column bus. Reset of the column buses in sequence by the horizontal scanner delivers the video signal to an external load resistance. The charge–discharge cycle of the photodiode is closely analogous to that of a silicon vidicon.

A problem with the multiplexed method of readout of an X-Y array as shown in Fig. 9a is the large capacitance of the output bus that increases the noise level of the output amplifier. Figure 9b shows an alternative method of readout employing a charge-coupled device (CCD) output register that reduces the noise and permits operation at lower light levels.

More complex versions of the MOS photodiode sensor are now used in single-chip solid-state color cameras such as the one shown in the inset of Fig. 1. (For additional details on the structure and operation of the MOS photodiode sensors, see Figs. 20b and 25.)

X-Y multiplexed scanning is used with an-

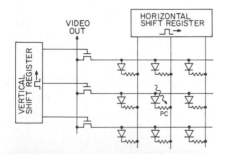

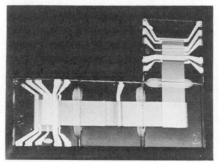

FIG. 8.   Diagram and photograph of an early X-Y photoconductive image sensor, having 180 × 180 pixels and integrated thin film scanning circuits, built in 1966. [From Weimer P., et al. (1967). *Proc. IEEE* **55,** 1591. © 1967 IEEE.]

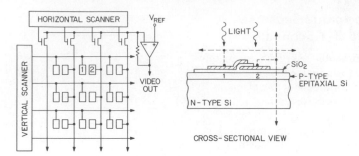

FIG. 10.   A CID (charge-injection device) sensor, showing the sensor array and the pixel cross section [From Weimer, P., and Cope, A. (1983). Image sensors for television and related applications. *In* "Advances in Image Pickup," Vol. 6 (Kazan, B., ed.). Academic Press, Orlando. © 1973 IEEE.]

other type of silicon sensor, known as the charge-injection device (CID). Instead of having a photodiode at each pixel, the photocharge is collected in potential wells under two capacitive electrodes, as shown in Fig. 10. One electrode is connected to the adjacent row bus and the other is connected to the column bus. The size of the charge is measured by transferring it from the row bus to the column bus or vice versa. After each pixel is read out, its charge is injected back into the substrate by simultaneously pulsing both gates into accumulation. An interesting feature of the CID is that the signal charge can be read out repeatedly without destroying the charge.

The charge-injection sensor is reported to be particularly resistant to spurious "blooming" or spreading of charge caused by light overload in a local area. It is currently used in industrial equipment requiring robotic vision.

Although sensors with X-Y address buses offer significant advantages for some applications, the signal-to-noise limitations introduced by the buses have led to an entirely different method of scanning (discussed in the next section).

VI. Solid-State Image Sensors: Scanning by Charge Transfer

A. APPLICATION OF THE CCD TO SOLID-STATE SCANNING

The development of analog charge transfer registers such as the CCD by Bell Laboratories opened the way in 1970 to a new system for scanning of solid-state sensors. Figure 11 shows a cross-sectional view of a CCD register which, when activated by clock voltages on the gate electrodes, causes charges to be transferred along a channel in the silicon under the gates. By introducing a pattern of charges into the register (by direct illumination or otherwise) and then transferring all charges toward the on-chip output amplifier, a video signal is obtained by simply measuring the size of the charge packets as they arrive sequentially. Transfer efficiency in CCD registers can be very high (>0.9999 per stage), permitting hundreds of transfers with small transfer noise or degradation in pattern definition. Since the input capacitance of the on-chip amplifier need be only slightly larger than

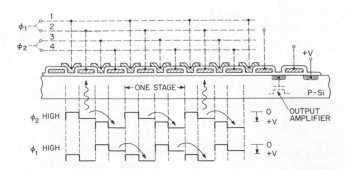

FIG. 11.   Cross section of a CCD (charge-coupled device) sensor.

the storage capacitance of a single pixel, the amplifier input noise can be kept much lower than any previous type of sensor. Because of this feature, CCD sensors have exceeded the sensitivity of all sensors reported to date except those incorporating electron multipliers or high-gain image intensifiers (see Fig. 16).

The CCD structure shown in Fig. 11 has four conducting gate electrodes per stage and can be operated as a four-phase or two-phase register. The lower drawings show the channel potentials in the silicon for each phase when operated in the two-phase mode. Each time the clock potentials are reversed, the charge under electrode 2 or 4 is advanced one-half stage toward the right because of the built-in barriers under electrodes 1 and 3.

Charge-coupled device registers can be used as an inexpensive delay line for analog (or digital) signals that are introduced sequentially at the input end. When used as a scanner for an image sensor, the charge pattern is introduced into all stages in parallel while the clock voltages are held fixed. In some sensors, the register is illuminated directly for this step, or alternatively, the register can be filled by charges from adjacent photodiodes. For readout, the clock voltages move the entire pattern past the floating diffused electrode that is connected to the gate electrode of the on-chip output amplifier.

Figure 12 illustrates three types of CCD registers that are used in two-dimensional sensors. The three-phase register shown in Fig. 12a uses a separate layer of polycrystalline silicon as the conducting gate for each phase. The direction of transfer is controlled by the phase relationship between the clocks. In the two-phase register shown in Fig. 12b the direction of transfer is determined by channel implants that provide an asymmetrical barrier at the rear of the potential well under each electrode.

The register shown in Fig. 12c, which requires only one clock electrode per stage, is called a virtual-phase CCD. By use of a more complex implantation schedule in the channel, the potential at the surface of the semiconductor in the virtual-phase region remains pinned to the potential of the surrounding p-type channel stops. The register then behaves similarly to a two-phase register operating in a single-clock mode with one gate held at a fixed voltage while the other is driven above and below that voltage. The virtual-phase register offers a simpler gate structure and the advantage that the register can be illuminated more effectively from the gate side than the registers requiring complete coverage of the channel by the gates.

B. Transfer Architecture for Area-Type CCD Sensors

Although television systems normally employ horizontal scanning, the principal direction of charge transfer in an area type sensor can be either vertical or horizontal. Three types of CCD scanning architecture are shown in Fig. 13.

1. Vertical Frame Transfer CCD Sensors

A typical frame transfer sensor has an array of approximately 400 vertically oriented CCD registers, each having 512 stages operated in parallel. The lower half of each register (labeled "A" in Fig. 13a) is illuminated by the optical image. During the first (odd line) field period, the clock voltages in this area are held fixed while a charge

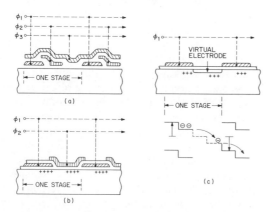

FIG. 12. Charge-coupled device registers currently used for image sensors: (a) three-phase CCD; (b) two-phase CCD; and (c) a virtual-phase CCD.

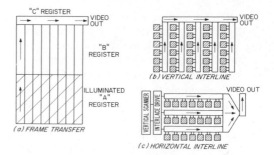

FIG. 13. Three types of CCD scanning architecture for solid-state area arrays: (a) frame transfer sensor; (b) vertical interline photodiode sensor; and (c) horizontal transfer interline photodiode sensor.

pattern is collected in potential wells under the positive electrodes. The clock voltages are reactivated during the vertical retrace to transfer the entire pattern quickly to the upper storage area, labeled "B." During the next (even line) field period, the odd line charge pattern continues to advance slowly toward the output register labeled "C" while the A register clocks are reset in the opposite phase to collect electrons along the interlaced even lines. As each row of charge in the B register reaches the C register, the entire row is transferred to it in parallel during the horizontal retrace period. The video signal is generated at the output amplifier as the charges arrive in sequence during the horizontal scan period. This simple method of generating a standard interlaced scan raster is effective, but it results in a double-width scanning spot that overlaps each of the adjacent scan lines by one-half.

The frame transfer architecture was used by Bell Laboratories in the earliest CCD cameras and is still being used by some manufacturers. Although it requires more silicon area for a given image size, its complete separation of the sensing and scanning areas permits higher sensitivity and very effective suppression of some types of spurious signals. It is now used in some three-sensor color cameras for broadcasting.

2. Vertical Transfer Interline Photodiode Sensors

The interline sensor shown in Fig. 13b differs from the frame transfer sensor in that the sensing elements are photodiodes interleaved between the vertical registers. The charge pattern is integrated on the photodiodes for 1/60 or 1/30 sec and then transferred directly to the adjacent register during the vertical retrace period. These registers advance the entire charge pattern toward the horizontal output register during each field period, just as described for the B and C registers in the frame transfer system.

The use of two-phase CCDs in the vertical registers with a separate photodiode for each phase is convenient for interlace, and it permits the sensor to be operated either with or without vertically overlapped pixels. The latter mode gives higher vertical resolution, which is advantageous for single-chip color cameras using checkerboard color filter patterns. Since the vertical readout registers are within the image area, they must be shielded from the light. Much effort has been spent on this type of sensor trying

to minimize spurious signals caused by the direct collection of photoelectrons by these registers.

The vertical interline sensors have received intensive development by engineers in Japan who have preferred this approach for single-chip color cameras for home video applications.

3. CCD Sensors with Charge Transfer Parallel to the Scan Lines

The horizontal transfer sensor shown in Fig. 13c illustrates an entirely different format that offers potential advantages for some applications. Since each register serves for one scan line, digital on-chip scanning circuits are required to switch on the clock drive for each register in turn. Interline photodiodes can be used or the CCD registers can be illuminated directly. The output signals from each line can be merged in a charge-funnel register, with extra CCD stages added prior to the output amplifier to make the total readout delay from each line equal to one line time. The reduced delay is to be contrasted with the vertical transfer sensors in which the added delay extends up to a 1/60 sec field time. The horizontal transfer mode of operation has received less development than either of the vertical transfer sensors.

VII. Performance Characteristics of Image Sensors

A. RESPONSIVITY

The radiant responsivity for various image sensors is plotted in Fig. 14 in milliamperes per watt of incident radiation as a function of wavelength. All curves represent effective responsivity, including the effect of sensor losses such as the reduced responsive area in each pixel of the interline photodiode sensor (curve 5). The dotted arcs (labeled QE = 100 to 10%) provide a scale for estimating quantum efficiency in electrons per photon at each wavelength. Curve 4 in Fig. 14 is for an illuminated register frame transfer CCD sensor, which has been thinned to approximately 10 μm and illuminated through the substrate. Curves 2 and 3 are for the Saticon and Newvicon camera tubes, with deposited photoconductors. Curve 1 represents the sensitive photocathode used in image intensifiers and in the image isocon camera tube. No curve is shown for the thinned photodiode target of the silicon vidicon, whose responsivity is comparable to that of the CCD wells in curve 4.

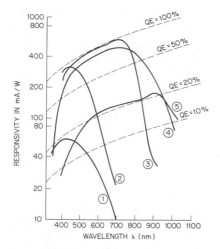

FIG. 14. Effective responsivity and quantum efficiency for various types of sensors plotted as a function of the wavelength of the incident light. Numbered curves as follows: (1) S-20 photocathode used in sensors and intensifiers; (2) Se-As-Te photoconductor used in Saticons; (3) Zn-Cd-In-Se-Te photoconductor used in Newvicons; (4) rear-illuminated silicon CCD register with thinned substrate; and (5) front-illuminated silicon interline photodiode CCD sensor.

It should be noted that television sensors are not limited to the visible spectrum. Both camera tubes and solid-state sensors have been built for the near infrared and for the ultraviolet, extending down into the x-ray range.

The curves of Fig. 14 show that both camera tubes and solid-state sensors have been developed whose responsivity approaches one electron per photon, the maximum to be expected from a directly illuminated storage target. That this quantum efficiency is more than an order of magnitude higher than for photographic film is very useful for scientific applications. However, the actual operating sensitivity of a camera is also a function of its noise background and other factors. The fact that cameras with photoemissive image intensifiers can function at much lower light levels than the current vidicons and CCD sensors shows the importance of noise considerations.

B. SIGNAL TO NOISE RATIO

The output signal from an image sensor can be expressed either by an output current I_s or by the number of carriers N_s that are integrated at each pixel during the integration period T. The latter designation is more commonly used in discussing solid-state image sensors. The quantity

N_s can be calculated from the image irradiance H (in watts per square centimeter) and the responsivity S (in amperes per watt) by means of the following expression:

$$N_s = HSA_pT/q \qquad (1)$$

where A_p is the area of one pixel (in square centimeters) and q the electronic charge (in coulombs). The maximum value of N_s is limited by the storage capability of the pixel given by

$$N_s\,(\text{max}) = C_pV_{\text{max}}/q \qquad (2)$$

where C_p is the pixel capacitance and V_{max} is the maximum voltage change of the capacitor during integration. The output signal current I_s is related to N_s by the expression

$$I_s = N_sqf \qquad (3)$$

where f is the pixel scan rate.

It is well known that the number of photons comprising the image and the number of photoelectrons generated are random variables whose standard deviation gives rise to "noise" in the transmitted picture. If N_s electrons are collected at each pixel during the integration period, the root-mean-square noise is equal to the square root of N_s. A fundamental limitation on the signal to noise ratio (S/N) is therefore given by

$$\text{S/N} = N_s/\sqrt{N_s} = \sqrt{N_s} \qquad (4)$$

An image sensor whose signal is limited only by this so-called shot noise would be considered ideal, provided that the signal represents full charge storage.

A second important component of noise in the television signal can be generated by the input stage of the video amplifier. It has been shown that noise due to this source is a function of the capacitance shunting the output terminal of the sensor, as well as of the video bandwidth and noise properties of the input circuit. Figure 15 illustrates how a large amplifier noise background (N_A) can exceed the inherent shot noise in the signal, thereby degrading performance. The resultant signal to noise ratio is derived from the root-mean-square sum of all sources of noise including the shot noise in the signal.

For best performance at low illumination levels, it is necessary to either reduce the amplifier noise (as was done in the CCD sensors) or find an appropriate way of increasing the signal level before it enters the external amplifier (as was done in camera tubes with image intensifiers or signal multipliers).

Figure 16 shows a normalized plot of signal

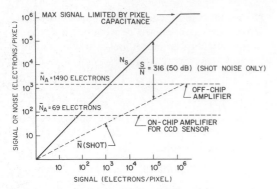

FIG. 15.   Signal N_s, shot noise $\bar{N}$, and amplifier noise N_A for sensors, expressed in rms electrons per pixel. S/N, signal to noise ratio. [From Weimer, P., and Cope, A. (1983). Image sensors for television and related applications. *In* "Advances in Image Pickup," Vol. 6 (Kazan, B., ed.) Academic Press, Orlando.]

output versus image illuminance (measured in lux on the sensor face) for various types of image sensors. These curves are useful primarily in showing the range of light levels at which each device can function. The image isocon signal level is higher because it is the only device shown using an electron multiplier in the output signal. The vidicon and solid-state sensors (CCD and MOS) are all capable of more than 60 dB (1000 : 1) signal to noise ratio at higher light levels. Comparisons between sensor types are only approximate since not all test conditions have been accurately specified.

The remarkable ability of the photoemissive SIT and I-SIT tubes to function at 10^{-4} lux illuminance on the sensor (the sensor illuminance

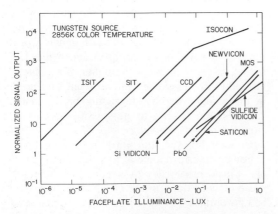

FIG. 16.   Signal output of various types of image sensors plotted as a function of sensor illuminance in lux. [From Flory, R. (1985). Image acquisition technology. *Proc. IEEE* **73.** © 1985 IEEE.]

for a scene illuminated by starlight) results from their use of one or two stages of image intensification prior to storage. At very low light levels, the output S/N from these tubes approaches very close to the limit set by the shot noise in the photocurrent. Image intensifiers have also been combined with other sensitive tubes such as the isocon and with thinned solid-state sensors that are bombarded directly by high-voltage electrons from a photocathode. Such devices exceed the low-light capability of the current directly illuminated CCDs having an amplifier noise of 50 to 60 rms electrons.

Charge-coupled device sensors with on-chip amplifiers offer the best prospects of any of the solid-state sensors for achieving signal to noise ratios limited only by shot noise in the signal. The input capacitance of the on-chip amplifier and the lead coupling it to a floating diode in the CCD register can be kept very small compared to the shunt capacitance of an external amplifier. Typical on-chip amplifiers consist of one or two MOS transistors with their output connected to an off-chip "correlated double-sampling" circuit to minimize reset noise and $1/f$ noise generated by the amplifier.

Recent studies have identified five kinds of noise in the CCD that contribute to a read noise floor such as shown in Fig. 15. These are as follows:

1.   Output amplifier noise (discussed above).
2.   Background noise resulting from variable amounts of charge being added to the signal charge in each pixel from sources such as the following: (a) thermally generated charge (dark current); (b) optical or electrical bias charge ("fat-zero") sometimes added to aid the process of charge transfer; and (c) internal luminescence generating photons that are reabsorbed in neighboring pixels. Such luminescence may occur at structural blemishes or at $n - p$ diffusions used in the input or output structure of the CCD.
3.   Charge transfer noise caused by a low charge transfer efficiency (CTE) or by the generation of a transfer-related spurious charge. The first is relatively unimportant because of the high CTEs of modern CCDs. The second is unique to the virtual-phase CCD and can be minimized by use of three-level clocking.
4.   Reset noise caused by the uncertainty in voltage to which the output node can be reset after the charge on each pixel is read. A correlated double-sampling circuit is effective in removing this noise source.

5. Trapping state noise caused by the slow release of trapped charge by surface states in the CCD channel. This problem has been greatly alleviated by the current use of buried channel CCDs that prevent such interactions.

A typical value for total read noise for CCD sensors operated with a video bandwidth of 4 MHz is presently in the range of 50 to 60 rms electrons. For scientific applications, where the sensor is cooled and a slow scan rate of about 50 kHz is used, a CCD read noise of 4 rms electrons can be obtained. Further reduction is believed possible.

The above discussion has been concerned only with temporal noise and has ignored the fact that fixed pattern noise can be generated at high light levels by a sensor that has appreciable variation in sensitivity from one pixel to the next. Patterns of this kind would increase linearly with signal as opposed to shot noise which increases as the square root of the signal. Solid-state sensors are more prone to such variations than modern camera tubes whose image areas involve far less geometric structure. Fixed pattern noise has been a difficult manufacturing problem that is gradually being brought under control.

C. RESOLUTION: MODULATION TRANSFER FUNCTION

The resolution of an imaging system is preferably characterized by its modulation transfer function (MTF) or by its contrast transfer function (CTF), rather than by its limiting resolution as observed visually with a test chart. MTF is the sine wave spatial frequency amplitude response, and it is sufficiently general to apply to practically all components of an imaging system, including the optics. An advantage of the MTF designation is that the MTF of an entire system can be derived from the product of the MTFs of its components. CTF response curves are sometimes plotted instead of MTF curves because they can be measured using "square wave" bar patterns rather than sine wave patterns. CTFs cannot be cascaded to evaluate overall response, but the MTF curve can be readily calculated from the measured CTF, and vice versa.

The fundamental limitations on MTF for camera tubes and solid-state sensors have much in common, but their practical limits are different and will be discussed separately.

1. Resolution Limitations of Camera Tubes

Vidicon type photoconductive tubes, having only an electron gun and a structureless target,

are presently capable of the highest resolution of conventional camera tubes. They have been used in experimental cameras by Japanese workers for evaluating a proposed new high-definition television system having 1125 lines instead of 525, for picture quality comparable to that in a movie theater. To satisfy these requirements, the vidicon gun has been redesigned for better control of the electron velocity distribution in the low-velocity beam. It is now possible to reduce the beam spot size without destroying its effectiveness in discharging the target. The deposited photoconductive target of the Saticon is particularly suitable for high-resolution application because of its freedom from internal light scatter and diffusion of the signal charges.

Figure 17 shows CTF response curves for three Saticon tubes. The 2-in.-diameter tube with the return beam multiplier has excellent resolution and a high signal to noise ratio, but because of its large target capacitance, it has too much discharge lag for live pickup. This problem was eliminated by the development of a new diode gun and by the use of smaller size targets. The 1-in.-diameter tube is considered adequate for the 1125 line system, and the $\frac{2}{3}$-in.-diameter tubes are currently used for broadcast cameras and single-tube color cameras. Still smaller photoconductive tubes are now available.

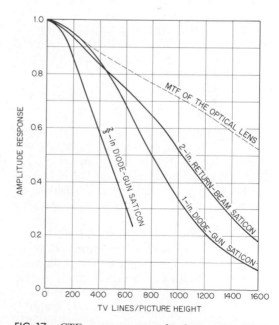

FIG. 17. CTF response curves for three types of Saticon sensors. [From Isozaki, Y., et al. (1981). 1-inch Saticon for high-definition television cameras. *IEEE Trans. Electron Devices* **ED-28,** 1500. © 1981 IEEE.]

The high resolution of the photoconductive camera tubes cannot be matched by the image intensifier tubes, such as the I-SIT and SIT tubes. MTF degradation in the intensifier section can be caused either by electron optical limitations or by the electrode structure in a secondary emission type intensifier.

2. Resolution Limitations of Solid-State Imagers

Generation of the video signal by the camera is a three-dimensional (x, y, t) sampling process, which is followed by time-divisional multiplexing of the samples. In solid-state sensors, the taking of discrete samples in the x dimension replaces the continuous horizontal sampling of a scanning beam. Discrete sampling in all three dimensions is capable of good results provided the number of pixels is adequate for the required MTF. However, in some solid-state sensors a portion of each pixel area is required for scanning purposes, thus preventing this area from being responsive to light. As will be shown below, the shape of the MTF curve is affected by the geometric arrangement of the responsive area within the pixel. Other factors limiting MTF include lateral diffusion of charge between pixels and resolution losses in the scanning process. These factors will be discussed separately.

a. Geometric Limitations on MTF. The geometric limit on MTF is given by the Fourier transform of the responsive area of each pixel. For a responsive cell of dimension Δx repeated with a center-to-center pixel spacing p_x, the geometrical MTF in the x direction is

$$\text{MTF}_x = \frac{\sin[(\pi/2)(f/f_\text{n})(\Delta x/p_x)]}{(\pi/2)(f/f_\text{n})(\Delta x/p_x)} \quad (5)$$

where f is the spatial frequency of the test pattern and f_n the Nyquist limit, or the highest frequency that can be resolved (given by $f_\text{n} = 1/2p_x$). A similar expression could be written for the MTF in the y direction which would be different only if the pixel spacing p_y or Δy were different.

Figure 18 illustrates the use of Eq. (5) to determine the effect on MTF of the sensing cell geometries for the frame transfer CCD and the vertical interline CCD. For ease of comparison, both sensors are assumed to have a total of 480 (vertical) × 640 (horizontal) pixels equally spaced in both directions, with $p_x = p_y = p$. The MTF response in Fig. 18 is plotted against the spatial frequency of a sine wave pattern, measured in TV lines per picture height (PH). The 480 active

scan lines and the 3 : 4 aspect ratio are in agreement with the 525 line U.S. system. The Nyquist limit in both the vertical and horizontal directions is 480 TV lines/PH, and the horizontal clock frequency would have to be approximately 12 MHz. The shaded portion under each curve indicates that the picture frequencies in excess of the Nyquist limit are invalid and subject to aliasing with the pixel structure.

The difference in the horizontal responses of the two sensors arises because the width of the photodiodes Δx in the interline sensor is equal to $p/2$, while in the frame transfer sensor $\Delta x = p$. The MTF response of the interline sensor is therefore higher at the Nyquist limit, but the aliasing would be worse for patterns having spatial frequencies beyond this limit.

The vertical responses of the two sensors are different because $\Delta y = 2p$ for the vertically interlaced frame transfer sensor, while $\Delta y = p$ for the interline sensor. The vertical overlap of the responsive areas in successive fields of the

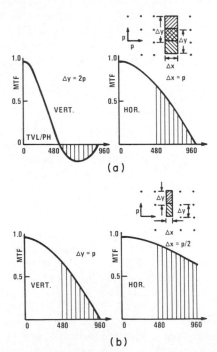

FIG. 18.   Geometric limitations on horizontal and vertical MTF for two types of interlaced solid-state sensors, as calculated from Eq. (5) for various sensing cell dimensions (x, y) and pixel spacing p: (a) Frame transfer CCD sensor with overlapped lines in successive fields, and (b) vertical interline photodiode CCD sensor with contiguous lines. [From Weimer, P., and Cope, A. (1983). Image sensors for television and related applications. *In* "Advances in Image Pickup," Vol. 6 (Kazan, B., ed.). Academic Press, Orlando. © 1973 IEEE.]

frame transfer sensor causes the MTF response to pass through zero at the Nyquist limit, degrading the MTF but also reducing the aliasing for higher frequencies.

The above analysis relates only to the effect of the geometry of the responsive area on MTF. It can be applied equally well to other sensors such as the horizontal transfer CCD sensor or the MOS photodiode sensor.

b. MTF Degradation Caused by Charge Diffusion. Lateral diffusion of photo-generated carriers is more likely to degrade MTF in a silicon solid-state sensor than in a vidicon having a deposited photoconductor, in which the diffusion lengths are much shorter. In the preceding section, it was assumed that in the frame transfer CCD carriers generated in the channel stop regions between registers would migrate by diffusion to the nearest potential well, so that horizontally adjacent responsive areas would be contiguous. In practice, carriers can also be generated in the nondepleted silicon substrate at a sufficient distance from the storage well to degrade MTF—an effect that is strongly dependent on the wavelength of the light. Such effects become progressively more pronounced as the pixel spacings are reduced relative to the image size.

For illumination from the gate side, the degradation for visible light entering the silicon substrate is small, but for infrared light it is substantial. Illumination from the substrate side requires thinning of the substrate for blue response, but the thinning also reduces the MTF degradation (at the expense of red response).

c. MTF Degradation Caused by Solid-State Scanning. MTF losses due to scanning in a CCD sensor can arise from charge transfer inefficiency in the registers. The major portion of the charge that fails to be transferred during a given clock cycle will remain in the well to combine with subsequent charge packets, thereby degrading signal contrast and lowering MTF. The loss in MTF at each location in the sensor is a function of the total number of transfers from that location to the output diode and the loss per transfer.

In practice, the transfer efficiency of well-made buried channel CCD registers is remarkably high: 0.99998 per transfer or greater. For example, in a large two-phase sensor having 1000 pixels on a side, 92% of the charge collected in the pixel farthest from the output will remain in the same charge packet after it has been transferred to the output (4000 transfers). The remainder is lost through recombination or it dribbles out in later packets.

In an X-Y sensor MTF losses due to scanning can be caused by the distributed resistive and capacitive properties of the address buses and by inadequate switching speed of the MOS gates. The output amplifier must provide a low-input impedance or else compensate for high-frequency losses caused by the large capacitance of the output bus.

D. Spurious Signals Generated by the Image Sensor

A necessary criterion of sensor performance is that it should not generate false signals that would degrade the quality of the final picture. Typical situations that can result in an inaccurate response from the camera are (1) light overload in certain areas of the image, (2) design weaknesses in the sensor, and (3) fabrication defects in the sensor.

1. Spurious Signals Caused by Light Overload

The necessity for obtaining charge integration at low-light levels, without having charge spreading in the bright areas, has presented a problem for most types of sensors. The image orthicon was the first of the camera tubes to combine high sensitivity with freedom from charge spreading in overloaded areas. However, the high-light stability of the image orthicon was achieved by the return of secondary electrons to the target in the image section, which also produced an objectionable black halo around any excessively bright object.

Light overload in a vidicon target results in excess charge that the beam may fail to neutralize in one scan, causing lag or trailing behind moving objects. These effects can be minimized by improved guns or by dynamic feedback to the gun to increase beam current where needed. The spreading of the signal charge around a very bright area, known as "blooming," is not a major problem with deposited photoconductors, in which the diffusion range is relatively small. Blooming was initially a serious problem with silicon vidicon targets, but it can be substantially eliminated by incorporating a "sink" electrode between the diodes to capture excess charge in the bright areas.

Blooming in silicon buried channel CCD sensors can be particularly objectionable because

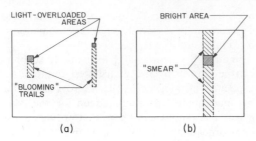

FIG. 19. Two kinds of spurious signals surrounding a bright area in a vertical transfer CCD sensor: (a) blooming and (b) smear. [From Weimer, P., and Cope, A. (1983). Image sensors for television and related applications. *In* "Advances in Image Pickup," Vol. 6 (Kazan, B., ed.). Academic Press, Orlando. © 1973 IEEE.]

the excess charge tends to spill down the register, producing bright streaks in the picture. As shown in Fig 19a, the streaks are vertical for the frame transfer and vertical interline sensors, and the length of each streak depends on the amount by which the illumination on the overloaded pixel exceeds the light required to fill its normal storage capacity.

Several methods are available for minimizing the blooming phenomenon in buried channel CCD sensors. The method shown in Fig. 20a for frame transfer sensors is to provide an antiblooming drain adjacent to each register to collect excess photoelectrons before they can spread down the register. For this to be effective, the antiblooming drain must be separated from the register by a barrier that is slightly lower than the voltage barrier separating successive stages during the charge integration period.

An alternative antiblooming structure, shown in Fig. 20b for an MOS photodiode sensor, uses the silicon substrate itself for the overflow drain. In this arrangement, the n-type photodiode is placed in a lightly doped shallow p-well in the n-type substrate. If the diode is overloaded by light, the excess charge is drained off vertically into the positively biased substrate rather than spilling across the MOS transistor or down the register in an interline CCD sensor. This system has the advantage that none of the active area of the pixel needs to be sacrificed to make room for laterally spaced drain lines, thereby permitting the pixels to be smaller.

2. Smear Signals from Solid-State Sensors

The appearance of spurious streaks above and below a white area in the picture is called "smear." This effect, shown in Fig. 19b, often occurs simultaneously with blooming, but it does not require light overload to be present. Camera tubes are free of this problem, but most solid-state sensors seem to have it and for different reasons.

Smear occurs in a frame transfer CCD sensor because its registers normally remain illuminated by light from the scene while the accumulated charge pattern is being transferred out of the sensing area. A completely effective solution is to use a light shutter to cut off the light from the scene during each vertical retrace period while this transfer is being made. However, a

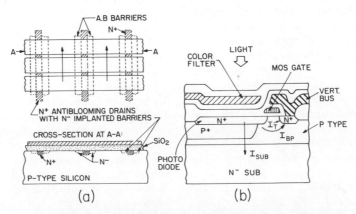

FIG. 20. Two methods of controlling blooming caused by light overload in local areas of the image: (a) laterally spaced antiblooming drains in a frame transfer sensor, and (b) N-P-N (p-well) structure in an X-Y MOS photodiode sensor. [From Aoki et al. (1982). *IEEE Trans. Electron Devices* **ED-29,** 745. © 1982 IEEE.]

mechanical shutter may not be fully acceptable for all applications.

A vertical interline photodiode sensor is particularly vulnerable to smear because the CCD registers must be kept from capturing any photoelectrons during the relatively long field period in which they are transferring charges to the output register and the adjacent photodiodes are collecting charges for the next field. Not only must the registers be optically shielded from the incident light, but they must be prevented from capturing photoelectrons excited deep within the substrate. The p-well construction described in the preceding section for MOS sensors will reduce the smear in interline sensors by more than 20 times compared to that of earlier interline sensors having a p-type substrate. Other interline structures, which permit more rapid transfer of charge from the illuminated area, have been found to reduce the smear by an additional 60 times. These structures can be combined with the p-well construction for even more effective smear reduction.

3. Spurious Signals Caused by Sensor Defects

Throughout the entire period of television development, the technical demands on the camera have always required sensors to be designed at the limit of the available technology. The unique characteristic of imaging devices to exhibit their internal imperfections for all to see has set severe standards for quality in manufacture.

Early photoemissive camera tubes required mechanical perfection in fine mesh screens, storage targets, and electron gun structures to avoid spurious glitches or nonuniform shading in the transmitted picture. Local variations in photoemission, secondary emission, or contact potentials were also visible to any viewer.

The vidicon photoconductive tubes, consisting of a gun and a structureless target, represented a design less prone to defects and was accordingly less costly to manufacture. For many important applications, these tubes are still preferred to solid-state sensors.

Solid-state imagers, relying on the mature silicon integrated circuit technology, have reversed the trend toward simplicity and are now among the most complex circuits manufactured. The increasing demands on resolution and stability have constantly required more pixels with smaller structural dimensions. The need for ob-

taining an analog signal with a high signal to noise ratio from every pixel introduces fabrication constraints beyond those required for large memory chips. The use of redundancy in image sensors to replace defective pixels does not appear feasible. External circuits for substituting a signal from adjacent pixels for that from a defective pixel has found only limited usage to date. Signal nonuniformity, or "fixed pattern noise" (of a level no more than a few decibels) is often sufficiently visible to require rejection in manufacture. Nevertheless, solid-state sensors in 1985 were beginning to compete with photoconductive camera tubes for both the home video and broadcasting markets.

VIII. Special Applications

A. SENSORS FOR COLOR TELEVISION

Modern television systems require the camera to generate three simultaneous signals corresponding to the primary color components red, green, and blue. The current practice in broadcasting is to use a camera having three sensors and an optical lens system, such as shown in Fig. 21, which forms a primary colored image of the scene on each sensor. Efficient color separation is obtained by the use of dichroic filters in

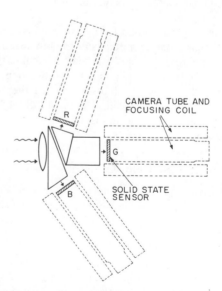

FIG. 21. Optical system for a three-sensor color camera (having three camera tubes or three solid-state sensors). [From Weimer, P. (1975). Image sensors for solid state cameras. *In* "Advances in Electronics and Electron Physics," Vol. 37 (Marton, L., ed.). Academic Press, New York.]

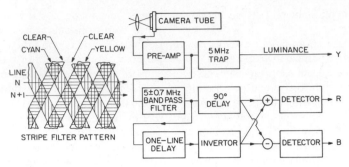

FIG. 22. Striped color filters and color separation circuit used in a popular single-tube home video color camera. [From Flory, R. (1985). Image acquisition technology. *Proc. IEEE* **73**. © 1985 IEEE.]

combination with internally reflecting prisms. The sensors are of conventional design and can be either camera tubes or solid-state sensors.

Although three-sensor cameras are capable of excellent performance, their extra complexity, size, weight, and cost are unacceptable for many applications (see Fig. 1). A major disadvantage of three-tube cameras is the requirement for registration of the three transmitted images, where the adjustment of as many as 80 knobs may be necessary for satisfactory coincidence. Portable broadcast cameras employing three solid-state sensors have their sensors permanently attached to their optical systems and provide much more stable registry.

The long-term need for a simple reliable single-sensor color camera has been recognized by television researchers for the last 35 years. This work led to the development of the popular home video cameras, all of which use either one camera tube or one solid-state sensor chip. The principles behind these special sensors will be discussed separately.

1. Single-Tube Color Cameras

For a single camera tube to generate three simultaneous color signals, the image area must be subdivided by a fine-patterned, three-color filter preferably built into the tube in close proximity to the photoconductive target. The type of color pattern required depends on the method by which the three independent signals are generated by the single scanning beam. Two general methods of signal separation have been used in color tubes marketed to date:

1. signal separation based on the filter pattern being resolved well enough by the beam to

generate a high-frequency color signal carrier by scanning the pattern, and

2. signal separation based on having three output leads from the target providing separate color signals independent of beam resolution.

a. Color Separation Requiring the Beam to Resolve a Patterned Color Filter. Figure 22 shows a color filter pattern and camera circuit currently used in many home video cameras for separating the color components. The filter pattern consists of two superimposed sets of color stripes, one having alternating cyan and clear stripes, and the other having yellow and clear stripes. Since a cyan filter passes green and blue light and absorbs only red, the scanning of the electron beam across this pattern will generate a carrier frequency that is modulated by the red component of the video signal. Similarly, the yellow-striped pattern will generate a carrier signal modulated by the blue component of the signal, since the yellow filter rejects only the blue light. The green light is not absorbed by either of the striped filters and therefore produces no color carrier.

With the filter patterns symmetrically oriented as shown in Fig. 22, the red and blue signal components can be separated into separate channels by combining the total signals from each successive pair of scan lines using a circuit containing the one-line CCD delay as shown. Although the beam intercepts the red and blue patterns at essentially the same frequency, the difference in phase of the two carriers on successive lines permits this circuit to direct the red and blue signals into separate channels. The green component is derived in another circuit by subtracting the sum of the red and blue signals from the total luminance signal.

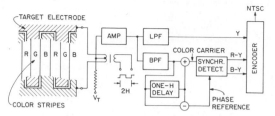

FIG. 23. The Trinicon color camera tube that encodes three primary colors on a single carrier generated by the scanning beam. [From Flory, R. (1985). Image acquisition technology. *Proc. IEEE* **73**. © 1985 IEEE.]

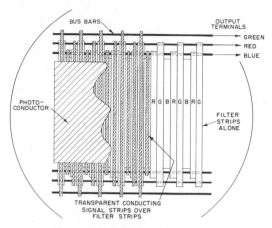

FIG. 24. Multiple electrode target and filter for a tricolor vidicon that permits the stripe pattern to be finer than can be resolved by the beam. [From Weimer, P., et al. (1960). *IRE Trans. Electron Devices* **ED-7**, 147. © 1960 IRE (now IEEE).]

Typical home video cameras of this type in use in 1985 employ a single ⅔-in. Saticon or Newvicon photoconductive tube with built-in stripes of sufficient fineness to generate red and blue carriers of about 4 MHz frequency. The resulting limiting resolution in the horizontal direction is about 250 TV lines, adequate for home use but not for broadcasting.

Another single-tube color camera uses a special camera tube called the "Trinicon," which encodes all three color primaries on a single carrier frequency. The color pattern consists of red, green, and blue stripes, as shown in Fig. 23. The modulated color signal behaves very much like the phase- and amplitude-modulated suppressed carrier system used in color broadcasting. For white light the carrier amplitude drops to zero, but when the color shifts from one primary to another, the carrier shifts in phase.

To be able to measure the phase shift in the presence of variations in scan velocity, the Trinicon signal plate is made in two interleaved sections so that the tube itself provides a reference frequency. The color carrier and reference frequency are separated from the combined output by means of a band-pass filter and, after separation from each other by means of a one-line delay, are used to generate the color difference signals (red minus Y and blue minus Y) with a pair of synchronous detectors. The luminance signal (Y) is obtained from the output of the low-pass filter. Although the target structure of the Trinicon tube is more complex than that of the two-carrier system described above, its encoding system permits higher resolution and more stable color for the same number of stripes.

b. Color Separation by Means of a Multiple Electrode Target. The necessity for subdividing the image area of a single-tube camera by

color filter stripes can cause spurious color signals to be generated when the optical image contains detailed patterns having a comparable spatial frequency. This problem can be minimized either by restricting the image detail by means of an optical low-pass filter (as done in the home video cameras) or by increasing the number of stripes in the image area. Figure 24 shows an alternative type of single-tube color vidicon target, which generates three color signals directly without requiring the filter pattern to be coarse enough to be resolvable by the beam. The target of this early vidicon contained 870 red, green, and blue multilayer interference filter stripes, with a separate transparent conductive strip over each filter. All conductive strips for each color were connected to provide a separate output lead for each color. The target structure was covered with a standard vidicon photoconductor and scanned with an electron beam whose scanning spot size had no effect on the efficiency of color separation.

A single-Saticon camera incorporating this type of target having 1059 red, green, and blue stripes was built recently and reported to yield 500 lines limiting resolution, the highest reported to date for a single-tube camera.

2. Single-Chip Solid-State Color Cameras

An extensive research program has been carried out in Japan during the 1980–1985 period to develop a single-chip color camera for the home video market. Although many approaches were

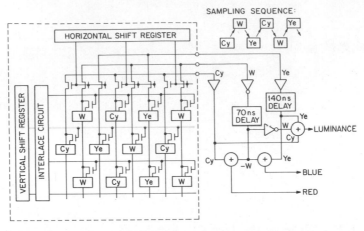

FIG. 25.   An MOS photodiode color sensor providing separate output electrodes for three color channels. [From Flory, R. (1985). Image acquisition technology. *Proc. IEEE* **73.** © 1985 IEEE.]

investigated simultaneously, the principal effort appears to have been based on the vertical interline CCD sensors and the X-Y MOS photodiode sensors. Typical sizes had the $\frac{2}{3}$-in. format with approximately 500 (vertical) × 400 (horizontal) pixels for use with the 525-line NTSC system or 580 (vertical) × 475 (horizontal) for the 625-line European system. The MOS photodiode color cameras were introduced for home use in the United States in 1984, but the interline CCD camera may have been delayed by the smear and blooming problem.

The design of the single-chip color cameras has followed the earlier development of the single-tube cameras, but the problem is simpler in some respects. The filter materials are not required to withstand rigorous tube-processing temperatures, and checkerboard filter patterns can be permanently aligned with the pixel structure. Excellent color performance has been obtained using checkerboard patterns formed from the primary colors red, green, blue and also from the subtractive primaries cyan and yellow combined with white. The filter patterns and associated camera circuits can be chosen to minimize the generation of false colors by the picture detail. Special integrated camera circuits containing a one-line delay and other components for separating the colors have been developed. Vertical interline sensors having 580 (vertical) × 475 (horizontal) pixels have been reported to yield as many as 340 lines horizontal resolution and more than 400 lines vertical resolution with no erroneous colors.

The need for a one-line delay in the camera can be eliminated as shown in Fig. 25 using an X-Y MOS photodiode sensor having a separate output lead for each of the three color components. A 485 (vertical) × 384 (horizontal) pixel sensor of this type yielded 350 lines resolution vertically and 280 lines horizontally.

Single-chip color cameras are expected to replace the current single-tube cameras for the home video market, as soon as their production is cost competitive. The CCD cameras will be capable of higher sensitivity, have a smaller size and weight, and have the dependability and long life expected of all silicon devices. Figure 26 shows an electronic snapshot camera containing

FIG. 26.   An electronic camera for still pictures employing a CCD color sensor and a built-in "still video floppy" disk for recording. (Courtesy of Matsushita Electric.)

a single CCD color sensor and a "still video" floppy disc for recording as many as 50 pictures.

B. Special Sensors for Science and Industry

Although low-cost vidicon cameras have been used largely for surveillance purposes for many years, the much greater geometric and photometric accuracy of solid-state sensors will permit many new applications. Equally important is the ease with which the design of solid-state sensors can be modified for added functions. A few of the applications for which special purpose sensors have been built are listed below.

1. Single-Line Sensors

Earth mapping from a satellite or robotic inspection of parts in a fast-moving assembly line represent two instances where a single row of self-scanned sensors may be preferable to an area type sensor. The relative motion of the scene perpendicular to the line of sensing elements permits a two-dimensional picture to be acquired, with more rapid feedback of information than from an area type sensor. Sensors with higher resolution are more easily obtained since only a single row of pixels needs to be fabricated.

Figure 27 shows a typical arrangement for a line sensor having about 2000 pixels on 12-μm centers. The simultaneous use of two CCD scanning registers for readout from a single row of photodiodes permits the diodes to be more closely spaced. Four CCD registers have been used in sensors having 3456 pixels in a row. An alternate design of line sensor for page-reading applications, which does not require a lens for imaging, involves the use of a very long sensor

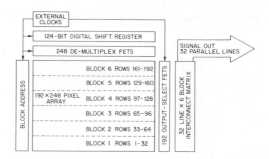

FIG. 28. Scanning organization of a sensor designed for high-speed motion analysis at 2000 frames per second. [From Lee, et al. (1982). *IEEE Trans. Electron Devices* **ED-29,** 1469. © 1982 IEEE.]

that is in direct contact with the page. Large contact type sensors have been built consisting of photoconductive elements scanned by thin-film transistors. Alternatively, shorter silicon sensors can be butted together for a single long sensor.

2. Sensor for High-Speed Imaging

A special sensor, shown in Fig. 28, was developed at Eastman Kodak Laboratories for motion analysis studies. By scanning 32 lines in parallel, the entire array can be scanned in less than 1/2000 sec while requiring a video bandwidth in each channel only 1/32 as wide as if all pixels had been scanned in sequence in 1/2000 sec. To attempt such parallel operation with a camera tube would require 32 carefully aligned scanning beams.

3. Sensors for Astronomical Observations

Photoconductors and photoemitters have a quantum efficiency one to two orders of magnitude higher than that of photographic film. Astronomers have long been aware of this advantage of electronic imaging, but the geometric and photometric instability of camera tubes has prevented their widespread use for scientific observations. Now, improved solid-state sensors with associated equipment are available, and they are rapidly becoming a part of every observatory. Sensors are selected for their high responsivity, large number of pixels, and low readout noise. In operation, they are cooled to liquid nitrogen temperatures, reducing the dark current sufficiently to allow exposures of many minutes. The readout rate can be reduced to a fraction of that required for broadcast, thereby reducing still further the readout noise. Sensors for this pur-

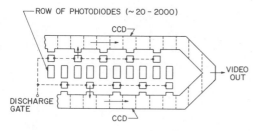

FIG. 27. A single-line sensor for page reading applications employing two CCD registers for readout. [From Weimer, P., and Cope, A. (1983). Image sensors for television and related applications. *In* "Advances in Image Pickup," Vol. 6 (Kazan, B., ed.). Academic Press, Orlando. © 1973 IEEE.]

pose will soon be available having 2000 (vertical) × 2000 (horizontal) pixels.

For operation at extremely low photon flux, where very high resolution is not required, an image intensifier stage can be introduced ahead of the sensor, while operating the sensor in a "photon-counting" mode, with integration occurring in an external digital memory.

BIBLIOGRAPHY

Barbe, D., and Campana, S. (1977). Imaging arrays using charge-coupled concept. *In* "Advances in Image Pickup and Display" (Kazan, B., ed.), vol. 3, pp. 171–296. Academic Press, New York.

Flory, R. E. (1985). Image Acquisition Technology. *Proceedings of the IEEE* **73**, 613–637.

Janesick, J. R., Elliott, T., Collins, S., Marsh, H., Blouke, M., and Freeman, J. (1984). The future scientific CCD. *In* "State of the Art Imaging Arrays and their Applications," *SPIE*, **501.**

Rose, A. (1973). "Vision, Human and Electronic." Plenum, New York.

Weimer, P. K. (1960). Television camera tubes: A research review. *In* "Advances in Electronics and Electron Physics" (Marton, L., ed.), vol. XIII, pp. 387–437. Academic Press, New York.

Weimer, P. K., and Cope, A. D. (1983). Image sensors for television and related applications. *In* "Advances in Image Pickup and Display" (Kazan, B., ed.), vol. 6, pp. 177–252. Academic Press, New York.

VIDEOTEX AND TELETEXT

Antone F. Alber *Bradley University*

GLOSSARY

Access time: Elapsed time between when a frame is called and the instant it begins to appear on the screen.

Alphageometric: Creation of graphic displays from primitive geometric shapes in the form of points, lines, arcs, rectangles, and polygons.

Alphamosaic: Creation of graphic displays by filling in points on the screen to form mosaic images.

Frame: Physical unit of information equivalent to what would be seen on a display screen.

Gateway: Hardware and software that form the interface between two videotex systems.

Information provider: Entity that prepares information for presentation on a videotex or teletext system.

North American Basic Teletext Specification: Interim standard for teletext transmission.

North American Presentation Level Protocol Syntax: Standard that specifies the coding scheme for videotex and teletext services.

Page: Logical unit of information in a videotex system consisting of one or more frames.

Scan line: One of the horizontal lines that makes up a TV picture. The National Television System Committee standard for TV broadcasting in the United States has 525 lines.

Teletext: Broadcasting system that displays selected frames of information as they are being continuously recycled by the originator of the signal.

Vertical blanking interval: Twenty-one scan lines in a TV signal not seen by the viewer, some of which may be used to carry a teletext signal. The vertical blanking interval is seen as a black bar when the picture rolls.

Videotex: Easy-to-use interactive electronic service.

Videotex and teletext are easy-to-use electronic services. Videotex is interactive and permits a user to access a computer from home or work and retrieve information for display on a specially equipped TV set or display terminal. Teletext is one-way and permits a user to select frames of information as they are being continuously recycled by the originator of the signal.

Three features distinguish videotex from conventional computer time sharing. Videotex deals in information, not data. Data are facts, concepts, and instructions suitable for communication, interpretation, or processing by human or automatic means. Information is data that have been transformed into meaningful and useful forms. In a videotex system information is often stored, accessed, and displayed in a convenient unit called a frame.

A second distinguishing feature is the ease of accessing the information. Videotex systems, particularly those designed for the consumer marketplace, can be operated by a preschooler. The third feature is the use of color and graphics. This last feature is not absolutely necessary and is frequently not found on videotex system devoted to disseminating news and financial information.

I. Interactive Videotex Systems

A videotex system can be divided into three major segments: the service provider, the communication network, and the user.

A. SERVICE PROVIDER

The service provider operates the computer system and either maintains the information base and services that are accessed by the users or supports a gateway to other computer systems where the information and services reside.

Service providers are often telephone companies, cable television operators, public data networks, and companies already in the business of providing information such as newspapers. Telephone companies are interested in videotex because it offers a chance to increase revenue from higher levels of telephone traffic and the sale of electronic yellow pages. Cable television operators have used the promise of interactive services to acquire new franchises and view it in the long run as an enhanced service that can be sold to cable subscribers.

Public data networks provide long-distance communications and regard videotex as a way to expand the amount of carriage and open new markets. Publishers, especially newspapers, have been actively involved with videotex since its inception. Publishers have been faced with increasing costs to produce and distribute paper-based products and continuing encroachment by media competitors such as broadcasters. Videotex is regarded as a long-run answer to the economic problems of the publishing industry and a method to enter new markets.

1. Hardware

The major hardware components needed to provide a videotex service are a computer, a magnetic tape unit, disk drives, information provider terminals, and a line printer (Fig. 1).

The size computer that has been used runs the gamut from microcomputers to large general-purpose business machines, but most videotex systems use a computer that is generally classified as a minicomputer. Popular examples are

Minicomputer

Tape Drive

Disk Drives

Line Printer

IP System

    Full Editing Keyboard

    Input Devices

FIG. 1.   Service provider hardware.

the PDP 11/44 manufactured by the Digital Equipment Corporation and the Series/1 made by IBM.

There are four performance criteria considered when selecting a computer for a videotex service. It should be able to store and retrieve information efficiently. A videotex computer does very little data processing in the traditional sense. Instead it retrieves information that is requested; supports some transactional services such as electronic banking and shopping; and adds, deletes, and updates information within the system.

It must be able to support enough simultaneous users to make the system cost-effective. The greater the number of simultaneous users possible, the larger the total population that can be served by one computer. The size of the population that can be served is important in a commercial service because it determines how much each user must be charged.

A rule of thumb for determining the theoretical number of simultaneous users is to multiply the number of requests that can be serviced per second by the amount of time it takes a user to interact with the system. For example, if the computer can satisfy 25 requests per second and a user takes 40 sec to read a frame and transmit another request, 1000 simultaneous users could be served. However, from a practical standpoint, the number of simultaneous users permitted is less than this due to physical constraints on the computer and communication services such as telephone lines that are available.

The total population that will be served by a computer is a business decision. The loading rates are calculated by dividing the population to be served by the number of simultaneous users. For example, a population of 10,000 using a computer that permitted 1000 simultaneous users would have a loading ratio of 10 to 1. During peak periods some users will not be served. Management must make a conscious decision between reducing the loading ratio by adding additional computer resources and foregoing potential revenue from lost sales.

The third performance criterion is the ability to operate within a data communications environment. Ideally, the system should support different forms of communications services (high speed, low speed), allow different communications protocols (asynchronous, synchronous), and permit different types of terminals to be used (text, graphic).

In applications where it is important that the

service is always available for access, a dyadic or fault-tolerant computer may be justified. Both of these approaches involve the duplication of critical parts of the machine so that it can continue operating if individual components fail. This added capability may be justified for financially based transactions such as electronic banking and shopping.

In addition to the computer, a magnetic tape unit is often included to permit bulk updating of the information base and to enable the information base to be archived inexpensively. The information base is stored on disk to allow direct access. The amount of disk storage needed depends on the coding scheme used and the number of frames in the information base. Generally, the time to access the information on the disk drive and not the amount of storage is the constraining factor in a videotex service.

The information provider (IP) system prepares the frames that are stored in the information base and later retrieved and displayed. The choice of equipment is very important because it determines the graphic quality of the display and the cost to create the frame.

The IP system may be either on-line or off-line. On-line systems are less expensive and there is a wider variety of equipment available. In a text-only videotex service, data and word processing terminals can be used.

Off-line IP equipment must have its own processing and storage capability and can usually interface electronically with the videotex computer to transfer the frames that are created. Several advantages result from operating off line. The videotex computer does not have to manage a network of frame creation terminals and expend the overhead resources that this would entail. Communication costs are generally lower because the IP equipment can store frames and transmit them all at one time. Frames can be transferred physically on floppy disks or magnetic tape. And perhaps most important, security of the information base is improved because access is easier to control.

An IP system may contain a variety of devices to facilitate the frame creation and modification process. A full editing keyboard is generally found in every system. This consists of an alphanumeric keyboard, a set of function keys for color and graphics control, and a numeric cluster. Input devices that are sometimes included are digitizer boards, joysticks, trackballs, mouse devices, light pens, video cameras, and spatial digitizers. The last-mentioned device resembles

a micrometer and is moved over and around the surface of a solid object. This enables the x, y, and z coordinates of the object to be calculated and stored digitally in the system.

Practically all commercially available IP systems possess the following minimum features: eight colors consisting of red, blue, green, yellow, magenta, cyan, white, and black; variable character height; upper- and lowercase text; flashing; drawing and fill commands; cursor control; and a variable data rate to accommodate the most economical transmission service.

2. Software

A videotex system requires the control and processing software found in most computer systems. Control software consists of the executive routines that coordinate the operation of the hardware, input–output programs that move information to and from memory, and communication programs that manage the flow of information between the system and the users and information providers. Processing software consists of utility routines for transferring information among storage devices, language translators to convert programs written in high-level languages such as COBOL and C to a form the machine can process, and software to manage the information base.

In addition, a system must possess a large number of application-dependent programs to support videotex functions. These programs may be grouped into the five categories shown in Fig. 2.

Information base management software incorporates the functions needed to create, maintain, and archive the frames stored in the system. Creation involves defining the information structure as well as creating and storing the frames. After the information base is stored, it must be possible to edit the contents of frames, modify the structure of the information base by changing the relationship among frames, and in-

Information Base Management

Service Administration

Information Retrieval

User Services

Telesoftware

FIG. 2.   Videotex software functions.

sert and delete selected frames. For security reasons, it is necessary to periodically copy and archive the entire contents of the information base.

Administration software manages user involvement by handling logging on and off activities, sending bulletins to users logging on, and disabling unauthorized users. It also collects data for billing purposes and to permit system tuning and modifications to reflect usage patterns.

The information retrieval software establishes the connections between the user and the information base. Common activities permitted are the selection of a frame from a menu or by direct entry of a frame address; return to a previously accessed frame; return to the main index frame; scrolling; and the ability to correct a key entry. A feature found on virtually all systems is a help function to assist users who are having difficulty.

User services vary widely among systems and include activities such as electronic messaging, distributed processing, opinion polling, interest matching, telemetering, and bill paying. These types of videotex applications are discussed more fully later.

Telesoftware are computer programs transmitted from the videotex computer to a microcomputer where they are executed or stored. To make telesoftware possible the videotex computer must have software to convert the program to be transmitted into frame-size units with the appropriate headers and error-detecting blocks added. At the receiving end, the microcomputer must have an emulator or driver program that simulates the videotex computer operation. The function of the driver program is to reassemble the frame-size units to form the original program.

The concept behind telesoftware is similar to retrieving books from a library. It provides to the user an ability to perform a large variety of tasks on demand without making a significant financial investment and is a revolutionary step in the distribution of computing power to the user.

B. COMMUNICATION NETWORK

The backbone of the videotex system is its communication network. There are several features a network must possess to make it suitable for videotex.

It must be easily expandable. Videotex ser-

vices are new and relatively unknown. Consequently, the subscriber base in most countries is small. To make the services as affordable as possible, it is necessary to minimize the investment in hardware. Therefore, the technique usually followed is to use a modular approach and add hardware components as the customer base expands. If the initial system was configured for 10,000 subscribers when only 1000 existed, the cost per subscriber might be so high as to make it impossible to retain the 1000.

It must be possible in many cases to access a variety of computers through what are referred to as gateways. A gateway is simply the combination of hardware and software that forms the interface between two videotex systems and enables a user of one system to communicate with a second system. One approach for implementing a gateway on a public videotex service is to have a special gateway frame that is retrieved by the subscriber. The selection of the frame automatically connects the user to the other system. The process may take slightly longer than the normal response time to retrieve a frame, but the user may not even be aware that the process has occurred.

The system should be able to accommodate a variety of terminals and multiple-transmission schemes. Ideally, unprogrammable terminals as well as microcomputers should be operable on the system from a variety of manufacturers. Depending on the application involved and the communication services available, it should be possible to access the videotex computer on narrowband or wideband circuits.

Last, the service must be reliable and secure. Like the lights or water, the user should be able to turn on a receiving unit and gain access to the system. Any applications that involve financial transactions or confidential information should be secure from unauthorized access and manipulation. [See DATA COMMUNICATION NETWORKS.]

1. Phone-Based Networks

Phone-based networks may involve either or both circuit-switching and packet-switching services. Circuit-switched service is normally used for voice transmission and is what most people would be thinking of when discussing the telephone network. Circuit switching connects two or more points and permits the exclusive use of the circuit until the connection is released. Packet switching involves the breaking up of a message into small blocks called packets. The packets are introduced to the network and travel by the most expeditious routes to their destination, where they are reassembled to form the original message. An analogy would be mailing each page of a 10-page letter in separate envelopes. When the 10 envelopes reach their destination, the letter would be reassembled and read. In many cases a circuit-switched network is the link between a user and a packet-switched network.

There are two major advantages of using circuit switching for videotex. It is ubiquitous. Ninety-three percent of all American households have a telephone. Also, it is relatively inexpensive, especially when there is a flat rate charge for local service regardless of how many calls are made. The most frequently cited disadvantage is that videotex ties up the phone unless a second circuit is installed. Another problem is related to the paint time on the display device. Paint time refers to how long it takes for a frame to be displayed. The typical phone line is capable only of low-speed transmission and a frame containing a lot of graphic content can require an annoying amount of time to be displayed.

Packet switching is ideal for videotex applications because it is very cost-effective for transmitting short messages over long distances. Also, most packet networks can interface with equipment from different manufacturers. Its major shortcoming is that it is not widely available to carry transmissions locally, such as within the same city. [*See* TELECOMMUNICATIONS.]

2. Cable TV Networks

Cable-based networks have entered their third stages since being introduced shortly after World War II. Community Antenna TV (CATV) was originally intended to provide television reception in mountainous areas where the terrain interfered with signal reception. In the 1960s cable TV became a national medium for delivering a broad spectrum of viewing choices. In the 1980s cable is emerging as a two-day medium for delivering videotex and other forms of enhanced services.

Cable TV has two advantages over a phone-based network. The most important is its much wider bandwidth. A cable may have 50 or more channels and each channel can carry as much information as 1200 telephone lines in the same time period. A second advantage is the cable industry is much less regulated. In most in-

stances, cable systems operate under local franchises and owners can negotiate with the authorities in each community served for specific services and the best terms possible. Unfortunately, only about 3% of the systems in the United States can support the two-way flow of information needed to sustain a videotex service and to upgrade existing one-way systems can easily cost in excess of $20,000 per mile. The major difficulty encountered in cable systems that offer videotex services has been shielding the signal from outside electrical interference. Television signals can withstand spurious electrical interference without a noticeable impact on the picture, but a videotex signal consisting partly of text becomes meaningless gibberish if electrical interference is introduced because the cable passes too close to a high-voltage power line or a mobile two-way radio operates near the cable.

For purposes of discussion, a two-way cable system can be divided into the headend and the transmission network. The headend is an operations center for the system. It is where the videotex signals are received from another source or produced and then amplified and converted for transmission over the cable network.

The two-way flow of information is made possible by either installing two cables, one for each direction of flow, or a single bidirectional cable. The latter approach is the more complex of the two options possible. To provide the bidirectional flow, the frequency spectrum of the cable is divided into a downstream path originating at the headend and an upstream path originating at the subscriber. The bandwidth of the cable is normally in the 400 MHz (megahertz) range. The downstream cable path usually employs frequencies above 50 MHz and the upstream frequencies below 50 MHz. The range and user applications are shown in Fig. 3. The range between 32 MHz and 50 MHz is highly susceptible to electrical disturbances and is not used. Am-

Downstream	50 MH$_z$ and Above
Upstream	
Video	20 MH$_z$–32 MH$_z$
Data	12 MH$_z$–20 MH$_z$
Teleservices	5 MH$_z$–12 MH$_z$

FIG. 3.   Transmission frequencies for two-way cable.

plifiers are distributed throughout the network to boost the signals. At points where the cable divides, signal splitters are installed. A directional coupler is inserted where drop cables from the subscribers' residence are attached to the distribution lines on the street.

C. USER COMPONENTS

In order to view a frame, a decoder is necessary to receive and translate the videotex signal, an access device is needed to select the particular frame desired, and a terminal is needed to present the frame. These items may be consolidated into a single piece of equipment or operate as three separate but interconnected components.

1. Decoder

Videotex decoders may be designed to operate with a display device such as a television set, with a terminal of the type used in data processing, or with a microcomputer. Decoders for television sets and data-processing types of terminals are classified as either set-tops or built-ins. Set-top units are generally cheaper and permit the owner to upgrade any display unit in a few minutes by connecting the decoder to the antenna jacks. Built-in units provide a better picture because they are connected to the RGB (red, green, blue) color guns rather than to the antenna jacks. A microcomputer can function as a decoder by inserting an interface card in one of its expansion slots or by using a software program that converts the microcomputer into a decoder.

2. Access Devices

Devices for interacting with the decoder can be divided into hand-held keypads, alphanumeric keyboards, and touch-sensitive screens.

The simplest hand-held keypads contain 12 or 16 keys and resemble a television channel selector. Early units were attached to the decoder by wire, but today most units can be operated remotely by using either infrared or ultrasonic transmission. Keypads are suitable for retrieving frames through a menu screen and when the specific address of the frame is known, but they are not convenient for electronic messaging and applications that require a large number of letters and digits such as some commercial transactions.

An alphanumeric keyboard resembles a typewriter. In contains a full complement of letters, digits, special symbols, and function keys such as a "next" key that facilitates access to the system. Most business applications require a full keyboard. Touch-sensitive screens are commonly found on public access terminals (PATS) located in motel lobbies and shopping centers. Touch-sensitive screens eliminate the need for a keyboard and thus reduce the terminals susceptibility to vandalism and stuck keys. They are most appropriate when the information is displayed in a highly structured manner and a choice can be selected by touching the appropriate spot on the screen of the terminal.

3. Terminals

Most videotex terminals are visual display units. In many cases the units have a built in decoder and are suitable only for receiving and displaying videotex transmissions. However, practically any television set with an attached decoder can serve as a videotex terminal. A television set is well suited for videotex because it is reliable, well accepted, and relatively inexpensive. Its only serious shortcomings are its association in the user's mind with entertainment (which makes it more difficult to sell consumers on the need for videotex) and the bandwidth (which can limit the graphics possible).

Printers, microcomputers, and a variety of special-purpose devices are also used as terminals. Hard copy produced by a printer is sometimes needed for confirmation of commercial transactions and sometimes simply to produce a more permanent record of what appeared on the screen of a display device. A variety of special-purpose devices such as Braille printers and speech synthesizers for the physically handicapped are also available.

D. THE INFORMATION BASE

Information in a videotex system is stored, modified, and retrieved in a physical unit called a frame. The frames are often linked together in such a way that it is possible to pass from frame to frame as each is displayed. The information contained in the frame may be presented in a graph. It may have variable-size lettering and specially created characters, and it may appear in a rainbow of colors.

1. Information Providers

The person or organization that prepares information for presentation through a videotex medium is known as an information provider (IP). Any person or organization can be an IP, but for purposes of discussion the three major categories are publishers, advertisers, and government units include public agencies.

Publishers are the principal source of information for videotex services today. This includes publishers of print-based products. Newspapers, magazines, newsletters, advisory services, and reference services are often prepared electronically by using computer-driven text composition processes. Only the final product is distributed on paper. For some information providers the transition to electronic delivery is a relatively effortless and inexpensive process. For example, today hundreds of newsletters are available electronically to owners of microcomputers and computer terminals. Publishers also include the news wire services and the nearly 2000 computerized public data bases available in the United States.

Advertisers are another major source of information. Often the information is directed solely at marketing a product, but it may also be sponsored. An analogy is the sponsorship by advertisers of television shows that are underwritten by paid commercials. In some videotex systems a frame or set of frames is sponsored in much the same way.

Government units and public agencies constitute a third major source of information. In Europe and Canada there is a considerable body of information sponsored by these two groups. Government-sponsored information has usually focused on telling the constituency about how to contact certain officials and the different types of services available. Public agencies have dealt with issues such as consumer rights, health care, and charitable contributions.

2. Organization

The conceptual design of the information base is considerably different that the actual design. Conceptually, the information base is organized like an inverted tree with its branches radiating outward from the trunk (Fig. 4).

There are two important advantages obtained from viewing the information organized in this way. It is extremely easy to train someone how to find a frame. There is a logical numbering of the frames. For example, the frames in level one can be numbered from left to right 1, 2, 3, and so forth. The frames in level two can be numbered from left to right 11, 12, 13, 14, . . . , 21, 22,

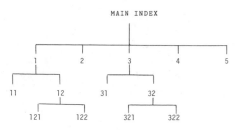

FIG. 4.   Conceptual view as an inverted tree.

23, . . . , 31, 32, ... The logical numbering makes it possible to group frames and sell them much like a radio uses its call numbers. In Great Britain, many of the information providers market their frames by advertising the three-digit numbers issued to them by the service provider.

The actual design of the information base is more complex than is portrayed by the tree structure. Frames are linked together so that it is possible to jump from branch to branch and from one level to another. Because of these linkages, the information base is really a network rather than the tree structure imagined by the user.

3. Access

There are many different kinds of frames. The first frame that is often seen is called the main index. It provides an overview of what is in the information base and where it is located. At each level there are routing frames that are similar to the main index but only list the contents for that particular branch. The sought-after information is contained on information frames. Response frames allow the user to input information such as a credit card number and quantity wanted when placing an order. A gateway enables a user to access a different computer.

There are three principal ways to access frames and several hybrid techniques based on modifications and combinations of the three. All videotex systems support treeing. This involves stepping through the information base from routing frame to routing frame by using the prompt numbers on each frame until the desired information frame is reached. Treeing is the simplest technique to use but may involve accessing a large number of frames to reach the desired information.

Direct access involves entering the unique number on each frame. This is fast but requires prior knowledge of the correct number. If the user has accessed the frame before, the number may be remembered. Paper directories exist for some systems so that a user can look the number up and enter it.

The most convenient is a keyword access. However, it is the most costly of the three access forms to maintain because rigid controls must exist to prevent duplicate keywords and someone must be responsible for assigning and publicizing the keywords.

E. APPLICATIONS

Videotex applications may be divided into the two general categories of consumer and corporate applications. In the United States, consumer applications have received most of the publicity, but corporate applications have been the most successful.

1. Consumer Applications

The six classes of consumer applications are listed in Fig. 5. Information retrieval, or electronic publishing as it is sometimes called, is the most popular application. It may take the form of news, weather, airline schedules, and financial data such as stock prices.

Commercial transactions consist of electronic banking and shopping. Electronic banking allows a user to check statement balances, review recent checks that have cleared, transfer funds between accounts, perform personal budgeting, and pay bills. Also referred to as home banking and telebanking, it is an extension of the banking industry's automated teller machines to the home environment.

Electronic shopping encompasses the retrieval of information and the payment for goods and services. The advantages of shopping electronically are that it is convenient since the consumer can do it from home and that it can save money since comparison shopping is much easier. From the merchants' standpoint it is cost-efficient since money does not have to be invested in retail floor space.

Electronic messaging is a system designed for originating, transmitting, storing, and receiving messages without the intervention of third parties. It can eliminate time-consuming and annoying telephone tag when two people are trying to reach one another by phone, and it is a convenient medium for reaching more than one person because messages can be broadcast to multiple addresses. Text, graphic, and oral messaging could be supported by the same system.

Educational services are found on most if not all consumer videotex services. These services usually consist of information about education activities in the area, self-paced reading and mathematics exercises, conference information, and meeting schedules. Videotex has been successfully used in regions where students cannot conveniently travel back and forth to school due to terrain and for handicapped students.

Personal transactions such as interest matching, information storage, opinion polling, and electronic voting are also possible. Interest matching operates much like the personal column of a newspaper. It unites people with a common interest or objective and supports an exchange of ideas or goods and services. Opinion polling has been a feature on several videotex services. Voting for people seeking a political office is very feasible but has yet been carried out through the medium of videotex.

Upstream and downstream computing exists on many systems. Upstream computing involves sending data to the videotex host computer where it is processed and then the results are returned for display on the terminal. Downstream computing is the reverse process. In this application computer software called telesoftware is retrieved by the user from the videotex host computer and downloaded to a microcomputer serving as a videotex terminal where it is processed.

Gaming is another popular application. It may involve playing against the videotex computer or playing against someone else with the computer acting as the referee. If the videotex service supports telesoftware-based applications, it may involve downloading the game to a microcomputer where it is played.

Teleservicing is the least-developed consumer application. Teleservicing may be divided into

Information Retrieval

Commercial Transactions

   Electronic Banking

   Electronic Shopping

Electronic Messaging

Educational Services and Personal Transactions

Computations and Gaming

Teleservices

FIG. 5.   Consumer applications.

telecontrol, telemetering, and telemonitoring. Telecontrol is the ability for the utility company to make physical adjustments to appliances in the home such as furnaces and air conditioners. Telemetering is the remote reading of meters, and telemonitoring is remote surveillance for the detection of fire and burglars. Teleservicing has been tested in a consumer videotex environment, but it is not commonly available on commercial service.

2. Corporate Applications

There is an overlap between consumer and corporate applications, especially those applications directed at information retrieval, electronic banking, and electronic messaging that are acquired through a public service that consumers also use.

Some companies operate in-house videotex services for conveying company news, distributing product announcements and price lists, order taking, and staying in touch with a widely disbursed sales force. Some companies have made extensive use of their system for special education and training programs. The graphics capability found on some systems has enabled videotex terminals to be interfaced with special cameras for producing slides and for supporting computer-aided design and manufacturing applications. And the communications network, messaging features, and ease of use built into most videotex systems has enabled companies to let selected employees telecommute by working at home.

F. GRAPHICS CREATION

There are two widely supported standards for graphics creation and a large set of variations. The two principal standards are the North American Presentation Level Protocol Syntax (NAPLPS) and the Videotex Presentation Layer Protocol (VPLP). Both standards create textual displays from a character set identified as ISO 646, which allows each country to change a couple characters in the set to reflect the needs of that country. An example would be to use a dollar sign in the United States and a pound sign in Great Britain for currency transactions.

1. Alphageometric

The NAPLPS is an alphageometric standard. Pictorial displays are created primarily with pic-

ture description instructions (PDIs), but the standard also includes a mosaic set, a macroset, and a dynamically redefinable character set (DRCS).

A PDI is a geometric primitive. There are seven PDIs defined: point, line, arc, rectangle, polygon, incremental, and control. The incremental PDI draws a point, line, or polygon in a piecewise manner. Control provides control over the modes of the drawing commands such as the color of an image.

A picture is created by connecting the various PDIs. Depending on the patience and skill of the artist, these geometric primitives can be connected on the screen of a display and stored for access later just as well as the same image could be rendered on a canvas with paints and brushes.

The mosaic approach to graphics is discussed in the next section. The macroset feature incorporated in the NAPLPS works in the following way. A string of characters is stored in the terminal with a code. When the code is received and invoked, the string is displayed. A DRCS involves downloading a pattern of data such as a corporate logo and storing it in the terminal until it is displayed.

The NAPLPS has four advantages when compared to the VPLP. The quality of the image and its resolution are better. The number of characters stored and transmitted to produce a comparable display is less. Last, there is complete independence between the information base and the terminal that is used.

2. Alphamosaic

The VPLP is an alphamosaic standard developed in Europe by the Conference of European Post and Telecommunications Administrations (CEPT). The standard is commonly referred to as CEPT after the organization that proposed it. Mosaic, geometric, and photographic displays are encompassed in the standard, but in the initial version of the standard only a mosaic capability was included.

An image is created by joining together mosaic characters. In the default version of the standard, a display screen is 40 characters wide and 24 characters high. Each character is formed from a character cell that is two elements wide and three elements high. By filling in one or more of the elements with colors a character can be formed. By joining together the characters an image can be formed. The principle is the same as an artist would use to create a picture on the wall of a building with small ceramic tiles.

The major advantage the mosaic approach has over the geometric approach is price. A mosaic-based decoder requires 2000 or fewer bytes of storage. Very little intelligence is needed by the unit, and the computer at which the information base is stored can use simple fixed-frame-size storage strategies that minimize the computing overhead that is required to support the videotex service.

G. History

Sam Fedida has been given credit for conceiving the idea of videotex in 1970. At the time he was a research engineer with the British Post Office. Earlier in his career, Fedida had been involved in the development of a computerized system to track vacancies in European hotels. One of the discoveries made during the systems development was that 80% of the cost of the system was for salaries and overhead of the clerks who interrogated the computer for customers. It was clear that allowing customers to look up the information about vacancies would reduce these overhead costs considerably. Later, when working on efforts to enhance the appeal of the Viewphone, a device that would allow persons speaking on the telephone to see one another, it became apparent that enhancements such as information retrieval could be accomplished by using the telephone network and a modified television set rather than the special communications facilities that would have to be built to support the Viewphone.

Coupling the two ideas of letting people without training operate a computer system with some of the enhancements proposed for the Viewphone gave birth to a working model of videotex in July 1974. The British phone company endorsed the concept as a way to increase the number of phone lines and the volume of traffic, particularly during periods of low demand in the evening and on weekends.

In Canada and France the ground work was being laid for similar systems, and the pace quickened as developments in Great Britain continued. In August 1978 the Canadian government demonstrated a technology called Telidon, which was based on alphageometric graphics. In France, the government began a program called Telematique in the mid-1970s that was to combine the advantages of telecommunications and data processing. By the late 1970s videotex-related efforts were under way throughout Europe, North America, Japan, and many other areas of the world.

In 1981 the alphamosaic approach was adopted by the CEPT for Europe. This was followed in November 1983 by the approval of the NAPLPS in North America. With the adoption of standards public services began to proliferate.

H. Operating Services

By the end of 1985 there were more than two dozen public videotex services operating around the globe. Representative examples are described in this section.

1. Prestel

The Prestel public service was launched in September 1979 in Great Britain. The service is run by British Telecom and had over 70,000 subscribers by early 1987. The information base is open to anyone who wishes to pay a nominal charge to store information. There are over 300,000 frames on the system that are maintained by approximately 1000 information providers. Many of the information providers are hired by other firms to develop and store frames, and so the number of companies actively involved is considerably higher than 1000. The service has several interconnected computer centers where the information is stored and most of the population can access the system from anywhere in Great Britain by a local phone call.

Prestel was the first public videotex service and as such it has been a model for many other systems. Former employees of Prestel have gone on to jobs throughout the world.

Prestel has not yet lived up to the expectations of its founders. The problem has been termed the "chicken or the egg problem." It has been very difficult to market the service because of its cost and the information available on it, and it has not been possible to reduce the cost because there are so few users and growth has been slow.

2. Teletel

The impetus for the French videotex service was a massive effort to modernize the telephone system beginning in 1975. These efforts were organized under the Telematique program and consisted of four product groups: videotex, electronic directory service, mass-fax, and telewriter. Teletel, the videotex service, and the electronic directory service were designed around an inexpensive black-and-white terminal that was distributed to telephone subscribers to replace the paper telephone directory. By early 1987 over 2.5 million of these terminals had been placed in user hands, and by 1990 it is anticipated every telephone subscriber in France will have one.

Mass-fax and Telewriter require new and special-purpose terminals that use the telephone network. Mass-fax is a facsimile service for transmitting fixed images such as charts, drawings, and legal documents. Telewriting transmits handwritten messages and graphics.

The "chicken or the egg problem" has been resolved by establishing an auxiliary use for videotex. In this case it is telephone directory information. However, once the terminals are in place, they may be used for information retrieval, electronic banking and shopping, and so forth. The electronic directory service was launched in February 1983 and almost overnight created a videotex industry.

3. United States

More than three dozen videotex experiments have been carried out in the United States. One of the largest and most complete was conducted in Florida and resulted in the public launching in October 1983 of the first NAPLPS-based service in the United States. Viewtron operated in the Miami area until March 17, 1986, when it was closed for financial reasons.

The most successful videotex services operating in this country have been text-only services. The largest is CompuServe, which has over 270,000 subscribers and supports a full range of applications including news, weather, and sports information; shopping; messaging; financial services; travel information; and educational exercises. Two other popular services are Dow Jones News/Retrieval and The Source.

Unlike in other countries, the federal government has provided very little direct financial support and encouragement for the development of a videotex industry. Because of the large capital outlay required and the synergism that results, companies have found it advisable to form partnerships. Three of the larger partnerships are Trintex (IBM, Sears), COVIDEA (AT&T, Chemical Bank, Bank of America), and CNR Partners (CITICORP, Nynex Corporation, RCA).

II. Teletext

Teletext is a broadcasting system that displays selected frames of information as they are being continuously recycled by the originator of the signal. The frames are combined into a block called a magazine. The size of the magazine and the form of transmission determine how long the cycle is before a frame is repeated and thus the average access time to a sought-after frame. For example, a 100-frame magazine transmitted on four scan lines of the vertical blanking interval (VBI) has an average access time of 6.92 sec. A magazine with 200 frames takes approximately twice as long. Changing the bandwidth of the transmission will directly affect the access time. A 100-frame magazine that uses an entire TV channel rather than just a few scan lines will reduce the access time for a specific frame to less than a second.

A. System Components

The major components of a teletext system are the service provider's equipment, the information base, the transmission network, and the equipment that receives the teletext signal.

The computer at the service provider's site can be a micro- or minicomputer. To it are attached one or more display terminals, a printer, and special terminals for creating and storing frames. A teletext system operates in the following way. Frames are created and stored in a temporary memory called the editing store. When they are to be included in the broadcast cycle, they are transferred to the main working store. From the working store the frames are continuously cycled through the encoder, where the information is converted into a serial teletext format and broadcast.

The information in a teletext service is more temporal and general in nature than the information in a two-way service. The first frame displayed when the teletext option is selected is the main index of the magazine chosen. The items in the index have prompt numbers that are entered by the user with a keypad or keyboard to "grab" that frame as it cycles past. Popular frames are often repeated at several locations in the magazine to reduce the access time. If there is a companion two-way service, it may be accessed for more information about a subject first viewed on the teletext service.

A variety of options are available for transmitting information for a teletext service (Fig. 6). The vertical blanking interval (VBI) is used more than all the others combined. The VBI is a portion of the television signal that contains information for the correct operation of the set such as equalizing and synchronization pulses. It is sometimes visible as a black line when the television picture rolls. Only half of the VBI is needed for synchronization and control, leaving the balance for teletext. Full-channel teletext can carry over 100 times the amount of information as the VBI, but it requires that an entire channel be dedicated to teletext.

There are many less commonly used ways to transmit a teletext signal. The subsidiary communications authorization (SCA) transmits teletext on an unused portion of the same signal that an FM radio station broadcasts its music. Subscription TV (STV) is the broadcast of a television signal over the air to subscribers with specially designed addressable decoders. The addressability feature enables the broadcaster to control and charge for special services such as teletext. Multipoint distribution services (MDS) can support VBI and full-channel teletext transmissions over the air by using microwaves. An MDS service requires that subscribers have an antenna and a special decoder. Direct broadcast satellite (DBS) transmission involves the transmission of television and enhanced services from a new form of high-power satellite to small and relatively inexpensive antennas. The antennas are approximately 1m in diameter and can be fastened to windowsills and rooftops. Direct broadcast satellite services are expected to become popular in the latter half of the 1980s, especially in areas of the country where cable is not available.

There are two quasi standards that are being promoted in the United States. The North American Basic Teletext Specification (NABTS) incorporates NAPLPS for graphics. World System Teletext is based on the British teletext specification and uses alphamosaics.

Teletext and videotex decoders are often combined in the same unit. This enables them to share circuitry and reduce the total cost compared to what it would be if each decoder were built separately.

B. APPLICATIONS

Teletext is suited for information retrieval, education-based applications, and limited forms of computations and gaming. Information of general interest with a short life is most appropriate. The first requirement exists because the broadcaster has to limit the size of the magazine to keep the access time low, and so information that will appeal to the widest audience possible must be selected. Since the magazine can be read in just a few minutes, it must be modified continuously. In some public services, editors make changes and add frames on the fly as frames cycle past. Most services offer news headlines, weather, financial information, television and radio listings, and sports. In a few instances, educational material and simple games such as puzzles and quizzes may be broadcast. In Great Britain, the two teletext services, CEEFAX and ORACLE, support telesoftware applications, and programs can be downloaded to microcomputers for execution.

C. HISTORY

Teletext was conceived by the British Broadcasting Corporation. The concept was first mentioned in internal documents in December 1970. By October 1972 work had progressed to the point that a public announcement was made describing teletext and a public demonstration was held in January 1973. During the next two years, a series of meetings were held with the British

Vertical Blanking Interval (VBI)

   Over the Air

   Cable

Full Channel TV

   Over the Air

   Cable

Subsidiary Communications Authorization

Subscription Television

Multipoint Distribution Services

Direct Broadcast Satellite

FIG. 6.   Transmission options for teletext.

Broadcasting Corporation, the Independent Broadcasting Authority, and developers of Prestel that resulted in the publication of a unified standard. After a test service and a pilot trial were conducted, the first public teletext service, CEEFAX, was launched in November 1976. The French public service called Antiope was started the next year.

Elsewhere around the world teletext trials were being conducted and events in Great Britain and France were being monitored closely. In March of 1983, the Federal Communications Commission made its long-awaited decision that has commonly been called the open market ruling. This ruling allows licensees of commercial television stations, public television, and low-power television stations to operate teletext services and choose the type of service and the technical system for broadcasting the signal.

The only restriction is that the teletext signal not interfere with existing broadcasting services. In April 1983 CBS began operating its teletext service, and NBC followed with its service a month later.

BIBLIOGRAPHY

Alber, A. F. (1985). "Videotex/Teletext: Principles and Practices." McGraw-Hill, New York.

Gecsei, J. (1983). "The Architecture of Videotex Systems." Prentice-Hall, Englewood Cliffs, New Jersey.

Hurly, P., Laucht, M. and Hlynka, D. (1985). "The Videotex and Teletext Handbook." Harper & Row, New York.

Lipis, A. H., Marschall, T. R., and Linker, J. H. (1985). "Electronic Banking." Wiley, New York.

Strauss, L. (1983). "Electronic Marketing." Knowledge Industry Publications, New York.

VOICEBAND DATA COMMUNICATIONS

Stephen B. Weinstein   *Bell Communications Research*

GLOSSARY

Analog: Signaling proportional to the original information waveform.

Asynchronous: Signaling bursts not synchronized with one another or a reference clock.

Baseband: Signals in their unmodulated form, usually including spectral components near zero frequency.

Baud: Pulsing rate (symbols per second) of a data signal.

Bit rate: Information rate of a data signal, not necessarily equal to the baud. Measured in bits per second.

Digital: Signaling with elements from a discrete finite set.

Equalization: Correction of channel transmission characteristics.

Full duplex: Simultaneous two-way communication.

Half-duplex: One direction at a time communication.

Intersymbol interference: Interference arising in pulse detection from the "tails" of neighboring pulses.

Modem: Device, including *mo*dulation–*dem*odulation operations, for communicating digital data over an analog channel.

Nyquist pulse: Signaling pulse that requires minimal bandwidth for a specified spacing between noninterfering pulses.

Packet communication: Data communication by separate packets of data individually conveyed through a network.

Passband: Transmission band, clearly excluding frequencies near zero, that carries a modulated communication waveform.

Private line: Point-to-point communication circuit leased from a communications carrier for the exclusive use of the leasing party.

Pulse: Elementary signaling waveform whose level or other parameter is, for digital communication, selected from a discrete set.

Symbol: Idealized digital representation, within an elementary signaling interval, selected from a discrete set.

Synchronous: Signaling at uniform clocked intervals.

Transversal filter: Linear filter consisting of a tapped delay line with the output the weighted sum of the voltages at the taps.

Trellis coding: Convolutional coding to introduce correlation among successive levels, as in channel trellis coding to improve bit error rate performance.

Voiceband: Having transmission characteristics typical of a telephone channel, especially a frequency bandwidth of roughly 300 to 3300 Hz.

Electrical data communication goes back to the Morse code telegraph systems that were built along the railroads in the mid-nineteenth century. Although many data are still carried over special facilities, the demand for broader access has stimulated the rapid development of voiceband data communication, which is the transmission of a digital data stream through a channel designed for a single voice signal. A telephone channel is the most common voice channel. Modems, which translate digital data streams into analog signals appropriate for the voiceband channel and correct for some channel deficiencies, are at the heart of modem data communication. Sophisticated modulation and adaptive signal processing techniques, together with large-scale integration in semiconductor chips, have made it possible to transmit data through long-distance telephone channels at rates as high as 19,200 bits/sec. The basic principles are of bandwidth-efficient modulation and adaptive compensation of channel distortions. The same principles have found new applications in access to digital networks, in digital microwave carrier systems, and in fiber optic communication.

I. Applications and Networks

Data communication has long been with us, represented by written languages, semaphore signaling, and many other formats. Electrical data communication, including "modern" concepts of multiplexing and efficient coding, was developed in the telegraph systems of the nineteenth century pioneered by Samuel F. B. Morse. In the first half of the twentieth century, teletype and telephoto services were implemented and became essential to the publishing industry. User-to-user data communication systems such as telex also became available and remain important commercial services. Only since the beginning of the computer era in the 1960s, however, has electrical data communication become important to society generally.

In an information society, data are what carry information among people and machines. Word processing and document transmission, airline reservation and credit card authorization systems, automatic teller machines, and personal computers all represent and process information as data. In some cases these data are produced and used in one isolated location, but increasingly data are shuttled from one machine to another through a variety of communication facilities. Even if everything is taking place in the confines of one building, as it might in a company office, a factory, or a local computing environment, a local area network (LAN) is likely to connect equipments together. The interconnection of LANs by metropolitan area networks (MANs) is a rapidly developing area of computer communications that illustrates the growing role of data communication in removing geographic constraints on the dissemination and use of electronic information. [See COMMUNICATION SYSTEMS, CIVILIAN.]

Although many kinds of communication channels are used for data communication, the importance of the voiceband channel lies in the ubiquity of the switched telephone network. When the need for communication from widely dispersed terminals to host computers began to be felt in the 1960s, no specialized data networks with this kind of reach were available. Whether communicating data between nodes of an airline reservations network or calling in a credit authorization from a point of sale or transferring data from a bank branch to the central computer, the only—or at least the most economical—facilities available were private or dialed telephone lines. [See DATA COMMUNICATIONS, NETWORKS.]

Unfortunately, communicating equipments such as terminals and computers produce digital signals designed to travel only a short distance to other equipments nearby. A need was born for telephone line modems to convert these digital signals to waveforms capable of passing through analog passband telephone channels (Fig. 1) with bandwidth, distortion, and noise characteristics much more accommodating to voice than to data. Telephone channels are also, in general, line switched, maintaining a fixed connection for the duration of a call. This is not efficient for the "bursty" activity of a typical terminal. Despite the later proliferation of specialized data networks with packet-switched communication for efficient sharing of communication links by bursty users, the need for access from locations not directly served by such networks calls for voiceband data communications through the telephone network. Millions of modems are sold annually for use with computer terminals, professional workstations, and personal computers that must communicate through telephone lines for at least part of the connection to distant host computers. Even if the entry node to a specialized data network is relatively close by at the network end of the subscriber line (Fig. 2), a data modem or transceiver is needed. [See ACOUSTIC SIGNAL PROCESSING.]

Users' equipments vary greatly in data format and speed, so that many types of modems are

FIG. 1.   Data transmission through a regular telephone channel. M, Modem; flag, terminal.

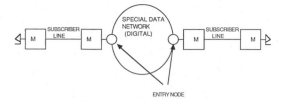

FIG. 2.   Limited-distance data transmission over copper wires for data network access. M, Modem; flag, terminal.

produced, as described in Section III. A fundamental distinction is made between asynchronous and synchronous communication. An asynchronous data stream does not have uniformly spaced pulse transitions, although, as Fig. 3 suggests, it often consists of bursts of uniformly spaced pulses that begin at arbitrary times. A burst may correspond to a particular character key that the user has just pressed. The receiving equipment is synchronized to each character burst, aided by "start" and "stop" indicators. A parity check bit (modulo 2 sum of the bits in a character representation) can be added to the character burst to detect some transmission errors. Lower-cost (and lower-speed) terminals are often asynchronous, whereas more expensive (and higher-speed) equipments are synchronous. The standardization of specialized data networks to synchronous data formats is driving a trend toward broader use of synchronous terminals.

Data communication in commercial applications is often characterized by a network of connections, not just a single point-to-point circuit. Figure 4 illustrates a private inquiry-type data communication network that could be found in a transactional application, such as a reservations system. The links in this example are all voicegrade private lines, permanently connected, leased from local telephone companies and interexchange carriers. Where there are several terminals in one location, a cluster controller coordinates their operation so that they share the communication link appropriately. Concentrators combine lower-rate data streams into higher-rate streams for further savings on communication lines. Modems at each end of each communication link convert between the baseband digital data streams created or absorbed by terminating equipments and the continuous passband signals required on the communication circuits. A front-end processor at the host computer manages the network, leaving the host computer free for data processing functions. Although the links in this network are permanently connected, the traffic can be packet-switched if appropriate packet switches and formatters are installed by the network operators.

The same network could well include dial-in terminals calling through ordinary telephone circuits from locations with insufficient traffic to justify a dedicated private line. This has become especially common for card-reading credit authorization terminals in stores and for access from personal computers to office computers, electronic mail services, and electronic information providers. A public packet-switched network, with dial-in ports closer to the user, is often a more economical way to access the private network or the host computer directly. [See TELECOMMUNICATION SWITCHING; TELEPHONE SIGNALING SYSTEMS, TOUCH-TONE.]

Data traffic is not necessarily digital in origin, such as characters sent from a keyboard or a bulk file transfer between computers. Digital data streams can be derived from analog signals, transmitted to distant locations with no degradation in quality, and converted back to analog signals. A device called a "codec" converts the analog signal to a digital data stream. Different coding techniques result in different data rates and conversion distortions, with 32 kbit/sec adaptive differential pulse code modulation (ADPCM) becoming the standard for high-quality telephone speech. Other encoding schemes can "compress" speech into data streams at rates from 1200 to 16,000 bits/sec, of lower recovered quality but appropriate for secure digital transmission through voiceband channels. Once an analog-to-digital conversion penalty has been accepted, the advantages of digital

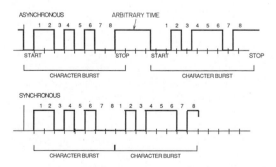

FIG. 3.   Asynchronous and synchronous data streams, illustrated for character burst traffic. In a synchronous data stream there is no need for start and stop indicators to define timing, but rather uniform clocking of all pulse transitions.

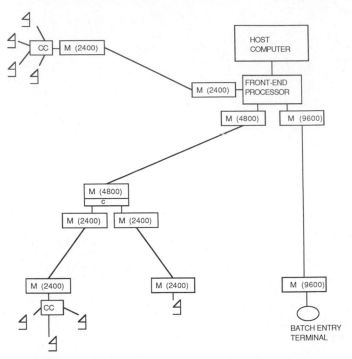

FIG. 4. Inquiry-type data communications network utilizing modems, cluster controllers, concentrators, and a front-end processor. CC, Cluster controller; M (R), modem operating at a rate of R bits/sec; C, concentrator; flag, terminal.

transmission include nearly perfect regeneration of digital symbols regardless of the length of the circuit, flexibility in mixing traffic of different media and service types, and more reliable and easily maintained transmission equipment.

Digital circuits carrying analog traffic are usually internal to communication neteworks and not apparent to the end user. Digital carrier systems, consolidating the outputs of a number of codecs, are widely used to multiplex voice signals in the telephone network. A variety of copper wire, microwave radio, and optical fiber transmission systems, operating at data rates of 1.544 Mbit/sec (in the United States) and up, carry interleaved streams of digitized voice and other signals. [See DATA TRANSMISSION MEDIA; MICROWAVE COMMUNICATIONS; OPTICAL FIBER COMMUNICATIONS.]

Although the high-speed links used in these carrier systems require much greater bandwidth than that of a voice-grade channel, they use many of the same signal modulation and channel equilization techniques originally derived for the humble telephone circuit. It is ironic, and a reminder of the progress that has yet to be made toward end-to-end digital communication, that a signal may originate in digital form at a terminal, be converted to analog form for transmission through a telephone circuit, be converted to a much higher rate digital signal within the network for transmission on a digital carrier system, be converted back to analog form for delivery to the other end of the circuit, and be converted there to a replica of the original data signal.

The telephone networks of the major industrialized countries are evolving toward all-digital, multimedia networks that will eventually provide end-to-end carriage of digitized voice, data, and video traffic. Some new services will involve integrated handling of different media and extensive user signaling to and through the network to set up a variety of calling and receiving options. The integrated services digital network (ISDN), a concept for a family of standard digital interfaces and services, is receiving growing acceptance. Voiceband data communication may eventually be superceded by end-to-end digital communication via ISDN, but the principles of data communication developed for analog voice circuits will continue to find applications on the links and subscriber lines internal to the digital network. [*See* DIGITAL SPEECH PROCESSING.]

II. Voiceband Channels

The focus of this article is on the telephone voice channel, the one the telephone subscriber sees and the channel that is often used for links in private data networks. This channel is also the one for which most of the advances in data communications have been made. It is not a single or unique entity, but rather a family of types and characteristics, as this section will describe as a prelude to the introduction to modems in Section III.

The telephone voice channel is rarely, if ever, a physically separate wire or dedicated radio system. If it is a long-distance circuit, it is usually a tandem connection of a number of different transmission facilities. As the last section suggested, some of these may be carrier facilities (Fig. 5). A carrier system is a bulk transmis-

sion system between switching offices of telephone companies or other network operators. A voice channel entering a carrier system passes through a line signal unit (a codec in a digital carrier system), which processes the signal in some appropriate way and multiplexes it into one of many trunk circuits in the carrier system. The analog to 32 kbit/sec digital coding of the new ADPCM codecs doubles the capacity of the old 64 kbit/sec pulse code modulation (PCM) based carrier systems but can introduce new nonlinear and other distortions into data signals.

A trunk circuit may be connected in tandem with other network trunks and finally with a destination subscriber circuit, usually a copper wire pair. It is useful to know how a voiceband circuit is put together when investigating the sources of channel impairments, but the user will usually be satisfied to think of the circuit as a "black box" with specified characteristics. These characteristics will be described later in this section.

The telephone voiceband channel comes in a

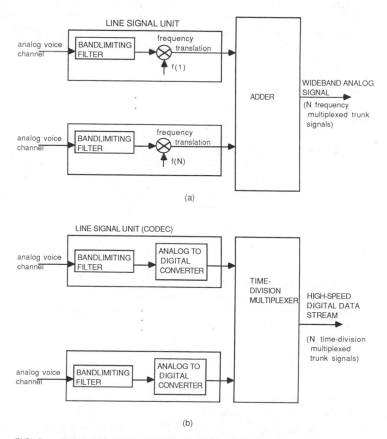

FIG. 5.   Channel banks in (a) frequency division multiplexed and (b) time-division multiplexed carrier systems.

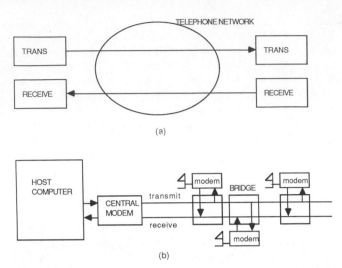

(a)

(b)

FIG. 6.   Private line types. (a) A four-wire private line is a dedicated connection through the network with a separate channel for each direction of transmission. (b) A multipoint circuit supports geographically separated terminals polled by a host computer.

number of forms. Reference has already been made to the private line, usually a four-wire circuit (Fig. 6a) in which separate two-wire transmission channels are provided for the two directions. Data communication can be full duplex (i.e., in both directions at the same time). Because transmission *within* the telephone network is, in fact, almost entirely over separate channels in each direction, all that is required to complete a four-wire circuit to a subscriber is two two-wire subscriber lines. Dedicated assignments of carrier system trunks can be made, but "virtual" private lines, which act as hard wired circuits but may actually consist of different facilities at different times, have become feasible through the developing "intelligence" of a network controlled by software.

An extension to multiple locations is offered in the multipoint circuit (Fig. 6b). This is a four-wire private line used in polling applications, where a central processor interrogates in turn a number of terminals. By connecting up to 20 terminals to one line, the transmission costs are considerably lower than what they would be if each terminal had an individual connection to the central processor. The connection points are bridge circuits that match impedances and may or may not have control and diagnostic capabilities such as the capability to switch a particular terminal in or out. Much attention has been given in data communications research to efficient operation in polling applications, espe-

cially quick adaptation of the channel equalizer in the receiver of the central (polling) modem as it jumps from listening to one distant transmitter to another.

Private lines are attractive to large corporations, data network operators, and others who can keep them busy enough to justify the monthly leasing rates. They have the important advantage of immediate connectivity, without the setup time of a direct connection.

A dialed line is two wire (communication in both directions on the same pair of wires), despite the fact that most of the circuit may be on four-wire carrier facilities. The advantages of a dialed line are its availability from any subscriber location, and toll charges (if any) that are paid for the actual time of use only. It is appropriate for individuals at home and any location with relatively low traffic.

Unless the dialed line is operated simplex (one way only) or half-duplex (one way at a time), the signals traveling in the two directions must be separated by the modems to prevent self-interference of a receiver by its own transmitter. Full-duplex data transmission (Fig. 7) is achieved either by frequency separation, in which the two directions of transmission are in separate frequency bands, or by echo cancellation, described in more detail later, in which the same frequency band is used for both directions and self-interference is adaptively canceled.

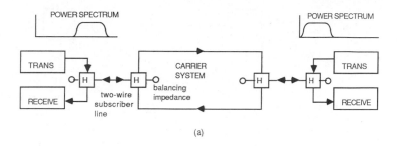

(a)

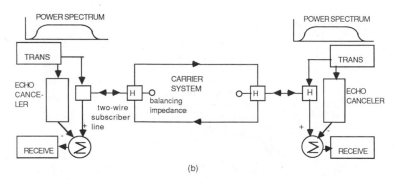

(b)

FIG. 7. Full-duplex data communication over dialed telephone lines. (a) Separate frequency band (a little less than half the channel) in each transmission direction. (b) Use of same full-channel frequency band in both directions, with echo cancellation to suppress self-interference.

The conversion from two-wire to four-wire circuit is carried out by a network element called a hybrid coupler, shown in Fig. 7. This is a passive four-port device that functions as a directional coupler, theoretically preventing west-going signals from looping around to the eastbound carrier leg, and east-going signals from looping around to the westbound leg. In practice, impedance unbalances lead to some leakage and hence to returning echo. These are not harmful in a transmission scheme using separate frequency bands for the two transmission directions (Fig. 7a) but must be effectively canceled when both directions use the same bandwidth (Fig. 7b).

The actual routing of a full-duplex long-distance call between local access transport areas (LATAs) is suggested in Fig. 8. There is a handoff from local exchange to interexchange carrier in each LATA and a hierarchy of switching offices. This is transparent to the user, although the delay and distortion characteristics of the connection may change with the number and type of carrier systems involved.

TELEPHONE CHANNEL CHARACTERISTICS AND IMPAIRMENTS

The most prominent characteristic of a telephone channel is its linear transmission characteristic (Fig. 9). This is a complex function, the Fourier transform of the channel impulse response. Note that it is a passband characteristic, with little transmission below 300 Hz or above 3300 Hz.

Although the transmission characteristic can be represented by its real and imaginary parts, it is more convenient to use the equivalent representation of magnitude and envelope delay distortion. The delay distortion is related to the phase $\phi(\omega)$ as the derivative

$$D(\omega) = -d\phi(\omega)/d\omega$$

where $\omega = 2\pi f$ is the angular frequency, measured in radians per second, and f is frequency in hertz. The delay distortion expresses the dispersion of the channel through relative differences between the transmission delays of different spectral components. It is conventional to

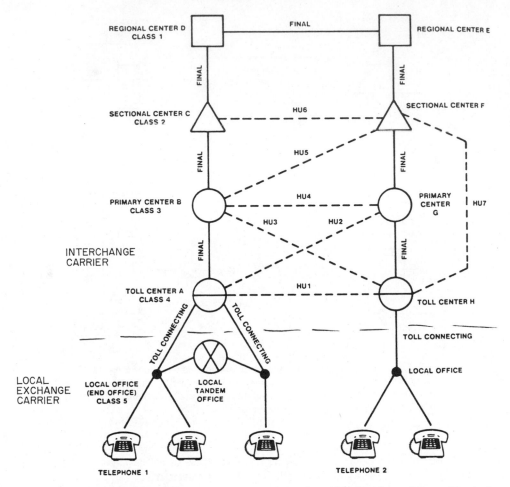

FIG. 8. Switching hierarchy. [Reproduced with permission from "Engineering and Operations in the Bell System." (1984). Copyright 1984 AT&T Bell Laboratory, Murray Hill, New Jersey.]

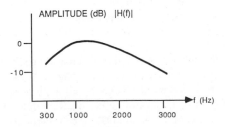

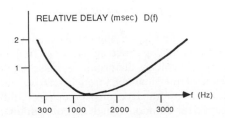

neglect the absolute delay, which would be the constant part of $D(\omega)$, and picture just the relative delay of different spectral components. Both the amplitude and the delay distortions contribute to the smearing of signaling pulses and the consequent intersymbol interference that must be corrected in higher-speed modems.

Linear distortion arises from the practical limitations of network elements, particularly codec filters (Fig. 5). These filters are designed to limit the spectra of voice signals to avoid cross talk

FIG. 9. Typical amplitude and delay distortion of a telephone channel. [Reproduced with permission from Gitlin, R. D., and Weinstein, S. B. (1981). *Proc. IEEE Int. Conf. Commun.* © 1981 IEEE.]

among channels in carrier systems. They were not, in the past, intended to minimize phase distortion in the passband, since it does not degrade voice signals. [*See* SIGNAL PROCESSING, GENERAL.]

The second most prominent characteristic of a telephone channel is noise. It is the sum of many small contributions of thermal noises, power-supply ripples, radio noises, atmospheric electrical discharges, cross talk interference, and other products of natural and man-made processes. Although there can be occasional large noise spikes, the sum noise is usually well modeled as Gaussian noise, which is analytically convenient. The noise is usually at a low level compared with signal strength. The important criterion is the ratio (SNR) of the power of the data signal at the input of a modem receiver to the power of the noise present at that point. [*See* NOISE IN ELECTRONIC CIRCUITS.]

For the telephone voice channel, this ratio is usually high enough—perhaps 20–30 dB—to correspond to a maximum possible error-free information transfer rate (channel capacity) of 20,000 to 25,000 bits/sec. This value is determined from the well-known formula of information theory

$$C = W \log_2(1 + P/WN_0) \qquad (1)$$

where, for a linear channel of bandwidth W hertz perturbed by white (uniform-spectrum) Gaussian noise of spectral power density N_0 watts per hertz, C is the channel capacity in bits per second. The derivation of this formula provides additional insight into the signaling problem, in particular the result that power should not be applied uniformly across the frequency spectrum passed by the channel, but instead more heavily at those frequencies that the channel attenuates the least. Specifically, a "water-pouring" analogy is found (Fig. 10), in which power is poured into a trough formed by the inverse of the ratio of the channel transfer function squared to the (one-sided) noise spectral power density. This distribution of signal power is not necessarily the first priority in signal design, which has other important criteria to meet.

Although the channel capacity is as high as 25,000 bits/sec, it may take an inordinate coding or decoding effort to approach it in practice. It is important, in the design of suboptimum data receivers, to avoid or minimize noise enhancement in the process of compensating other impairments. In the end, it can turn out that the other significant channel impairments described

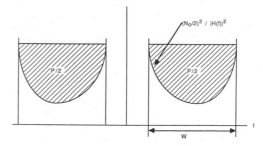

FIG. 10. "Water-pouring" analogy for distribution of signal power in a linear channel with Gaussian noise of uniform spectral density. In order to achieve channel capacity, signal power must be concentrated where the channel has the best transmission properties.

below can be the most serious constraints on performance at very high speeds.

One of these additional impairments is perturbation of carrier phase. Phase jitter and frequency offset (or phase roll) are associated with the oscillators used for frequency translation of voice channels into frequency slots in frequency division multiplexed (FDM) carrier systems (Fig. 5a). The spectral magnitude of random phase jitter tends to concentrate (Fig. 11) at harmonics of the power line frequency, suggesting pickup or inadequate filtering in power supplies used within the network. Some of the lower-frequency component of phase jitter can be traced to the 20-Hz telephone ringing tone. Frequency offset, a uniform shift in all spectral components of as much as several hertz, can result from slight differences in the "up" and "down" frequency translations in FDM channel banks. It is becoming a less intractable problem as digital carrier systems begin to predominate.

A further perturbation, very difficult to compensate for, is nonlinear distortion. It can sometimes be modeled (Fig. 12) as a memoryless third-degree polynomial function imbedded between linear distortions. The levels of second- and third-degree spectral components are typically 30 or 40 dB below the fundamental.

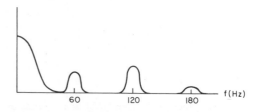

FIG. 11.   Spectral magnitude of typical random phase jitter process.

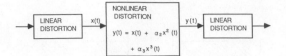

FIG. 12. Rough model for channel with both linear and nonlinear distortions (noise and phase perturbations not shown).

Nonlinear distortion can be the limiting factor in very high speed voiceband data communication. Considerable data have been accumulated characterizing telephone channels and the performance of modems through them.

When private lines are leased, they are usually conditioned to meet specified limits on the expected degradations. Figure 13 shows how delimiting boxes are defined for the linear characteristics, and maximum noise, phase jitter, and harmonic distortion levels stated. Linear characteristics, such as those of Fig. 9, must lie

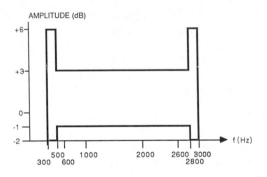

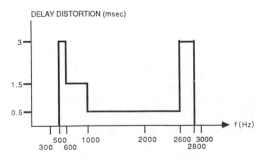

FIG. 13. Specifications for a C2-conditioned private line. Noise: at least 24 dB below signal. Phase jitter: maximum 10° peak to peak. Frequency offset: ±0.5 Hz. Nonlinear distortion: second-harmonic inband components more than 25 dB below fundamental; third-harmonic inband components more than 30 dB below fundamental. [Reproduced with permission from "Transmission Systems for Communications." Copyright Western Electric Company.]

within the boxes to meet the conditioning specifications. Test methods are included in some of the specifications.

There are additional abrupt disturbances—impulse noise, phase jumps, total outages—on telephone circuits stemming from the peculiarities and operating environments of various transmission and switching systems. These are usually neglected in the design of data transmission equipment for general telephone circuit use.

To communicate data full duplex, steps must be taken to disable certain equipments that aid voice communication but interfere with data communication. Long-distance telephone channels have been equipped since the 1920s with echo suppressors to prevent the return of disturbing echoes to a speaking party. On the assumption that only one party at a time is likely to be talking, an echo suppressor blocks a signal from the other end of the line in addition to the undesired echoes. Simultaneously two-way conversation, be it voice or data, is prevented. These units are gradually being replaced by echo cancelers, which selectively eliminate the most distant echoes without blocking the signal from the other party but still, because of adaptation requirements, cause trouble for full-duplex data conversations. Both kinds of equipment can be disabled by specified signaling and kept disabled through maintenance of a minimum signal power in a specified frequency interval.

III. Modems

Modems have already been defined as the translators (Fig. 1) between digital terminal equipments and analog (as the user sees them) transmission circuits. Voiceband modems may be separate units or built into equipments. They may be one-way or full duplex, for private or dialed lines, and operate at data rates from 300 bits/sec to as high, at the time of writing, as 19,200 bits/sec.

For a given data rate, the modem must simultaneously achieve several conflicting objectives. The first is the necessary spectral efficiency, measured in bits/sec Hz. For example, with a nominal channel bandwidth of 2400 Hz, the spectral efficiency varies from 1 bit/sec Hz for a 2400 bit/sec modem to 6 bits/sec for a 14,400 bit/sec modem. High spectral efficiency calls for modulation techniques that pay a price either in lowered noise immunity from larger signal sets or in increased channel distortion from placing signal energy in the distorted edges of the transmission band (or both). From a performance point of view, the price is paid in the second objective, low bit error rate. Bit error rate is the statistical expectation of the proportion of data bits at the receiver output that are incorrect, and in many applications is considered high if it exceeds 10. Data communications technology has made great strides in increasing spectral efficiency while maintaining an acceptably low bit error rate. A third objective, for higher-speed modems that must adaptively equalize, or correct, the transmission channel, is quick adaptation to changes in the channel or switches to different channels. Rapid adaptation, like spectral efficiency, conflicts with good steady-state error rate performance. A fourth objective, perhaps the most important of all, is low cost for a given level of performance. Very large scale integrated (VLSI) circuits have made possible dramatic reductions in modem costs and increases in their computational sophistication.

Figure 14 illustrates the structure of a synchronous modem, which is a transmitting and receiving device that creates data pulses on a strictly periodic schedule and clocks data in and out accordingly. Mnay lower-speed modems are asynchronous, with pulse transitions occurring whenever the data terminal equipment generates them. A pulse is generated each symbol interval, which does not restrict the pulse tails from extending over several symbol intervals. Pulses are modulated, usually amplitude modulated, by a level that may derive from several bits in the incoming data stream. Section IV describes these and other concepts of baseband signaling.

The information-modulated data pulse train in turn modulates a carrier signal to produce the inband line signal. The receiver's demodulator recreates the baseband signal. Frequency, phase, and amplitude modulation are commonly used (Section VIII). Line signal level codings and modulation techniques extending over several symbol intervals can be exploited to achieve better performance for a given signal-to-noise ratio. Phase perturbations in the received signal can be tracked by an adaptive local oscillator in the receiver. The correct sampling time can also be selected by the adaptive control circuitry.

The heart of a higher-speed modem is its channel equalizer (Section V), an adaptive linear filter that largely corrects for the linear distortion of the transmission channel. In many channels, the linear distortion causes severe intersymbol interference among pulses. At rates of 4800 bits/sec and higher, channels that yield totally unacceptable error-rate performance without equalization are conventionally equalized to

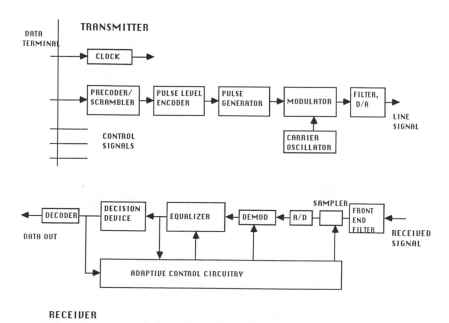

FIG. 14.   Full-duplex, four-wire, synchronous, and digital modem architecture. A/D, Analog-to-digital converter; D/A, digital-to-analog converter.

TABLE I. CCITT Recommendations for Modems[a,b]

Designation	Speed	Signaling format	Application
V.16	(Analog)	FM, half-duplex, subcarrier frequency 950 Hz	ECG transmission on dialed circuits
V.19	10 char/sec	Asynchronous, half-duplex parallel tones from "A" set (697, 770, 852, 941 Hz) and "B" set (1209, 1336, 1477, 1633 Hz)	Parallel (telephone pushbutton tones) signaling on dialed circuits
V.20	20–40 char/sec	Asynchronous, half-duplex parallel tones (as in V.19, but faster transmission and a possible third signal set	As in V.19
V.21	Up to 300 band (asynchronous)	FSK, full duplex, carrier frequencies in two directions, 1080 and 1750 Hz, frequency deviation +100 Hz	Dialed network
V.22	1200 bits/sec (optional 600) (synchronous)	Four-phase DPSK, full duplex, carrier frequencies in two directions, 1200 and 2400 Hz	Dialed network and leased circuits
V.22 bis	2400 bits/sec (synchronous)	QAM, 16 signal points, carrier frequencies in two directions, 1200 and 2400 Hz; fixed compromise equalization	Dialed network and leased circuits
V.23	Up to 1200 baud (plus 75-baud reverse channel) (asynchronous, or synchronous)	FSK, half-duplex, carrier 1700 Hz, frequency deviation +400 Hz	Dialed network
V.26	2400 bits/sec (plus 75-baud supervisory channel (synchronous)	Four-phase DCPSK, full duplex, carrier frequency 1800 Hz	Four-wire leased circuits
V.26 bis	2400 bits/sec plus optional 75-baud reverse channel (1200 bits/sec reduced rate capability (synchronous)	As in V.26, but half-duplex	Dialed network
V.26 ter	2400 bits/sec (1200 bits/sec reduced rate capability)	Four-phase DCPSK, full duplex, compromise or adaptive equalization, echo cancellation to separate directions, carrier frequency 1800 Hz	Dialed network and leased circuits
V.27	4800 bits/sec (plus 75-baud reverse channel) (synchronous)	Eight-phase DCPSK, full duplex or half-duplex, manual equalizer, carrier 1800 Hz	Four-wire or two-wire leased circuits
V.27 bis	—	As above, with automatic equalizer	—
V.27 ter	—	As above, half-duplex	—
V.29	9600 bits/sec (synchronous)	QAM, 16 signal points, full duplex, automatic equalization, carrier 1700 Hz	Four-wire leased circuits
V.32	Up to 9600 bits/sec (synchronous)	QAM, full duplex, 32 signal points with channel trellis coding (16 otherwise), echo cancellation, 2400-baud pulsing, carrier frequency 1800 Hz	Dialed network and leased circuits
V.33	14,400 bits/sec	AM–PM, 128 signal points	Four-wire leased circuits

[a] From "Data Communication over the Telephone Network." (1985). CCITT Red Book, Vol. III, Fascicle VIII. 1. International Telecommunications Union, Geneva.

[b] Abbreviations: FSK, frequency-shift keying; DCPSK, differentially coherent phase-shift keying; QAM, quadrature amplitude modulation (rectangular array of AM–PM signal points); AM–PM, simultaneous amplitude-phase carrier modulation, equivalent to pairwise amplitude modulation on quadrature cosine and sine carrier waveforms.

the point where bit error rates of 10^{-6} and lower are observed. Slow changes in the linear channel characteristics are easily tracked. Advances in VLSI technology have made possible very high speed digital signal processing chips of moderate cost, which can carry out the millions of multiplications per second required for the digital signal processing operations of equalization, filtering, phase tracking, and other modem functions.

A number of equalizer structures have been invented and used, but the most commonly used are the tapped delay line, or transversal, filter, and the decision-feedback equalizer. In both cases, decisions made on the equalized data stream are used to drive adaptation of the tap weights, which are the variable elements. An alternative approach, which can also be combined with an equalizer, is to compute the most likely sequence of data symbols transmitted in a data communication session, based on observation of the noisy and mutually interfering data pulses received from the channel. The maximum likelihood sequence estimation (MLSE) receiver structure (Section VI) has been too computationally complex to implement in most general use modems.

The CCITT, a world standards-making organization under the International Telecommunications Union, has issued a long series of recommendations for modems at various speeds and for different applications, all meant to facilitate interworking between units made by different manufacturers. A number of these are summarized in Table I. Explanations of terms under "Signaling format" are given in later sections. The V.32 standard specifies, in addition to automatic adaptive channel equalization as provided in V.29, channel trellis coding to improve the signal-to-noise ratio, and echo cancellation to make possible dialed-line operation with full-duplex data rates of 4800 bits/sec and higher. These techniques are described in Sections VII and VIII, respectively.

Single-chip frequency-shift keyed (FSK) modems are commercially available, and very large scale integration is making the higher-speed modems smaller and cheaper. The 1200 bit/sec full duplex modems commonly used today for microcomputer and personal terminal communications via dialed telephone lines, provided as plug-in boards or small external units, will probably be yielding to 4800 bit/sec units by the time this appears in print.

IV. Baseband Signaling

Baseband (nonmodulated) signaling has its own applications and is a fundamental element of passband data communications. The modulation function in a modem effectively transforms a baseband pulse train into a passband line signal whose bandwidth and other spectral characteristics may be identical to those of the baseband signal.

In the usual case, a baseband pulse train conveys information in a stream of pulses that are identical except for amplitude. The signal designer, faced with a noisy channel of finite bandwidth, has a number of opportunities and tradeoff options:

1. Desirable spectral characteristics can be realized through shaping the frequency spectrum of the pulse or the frequency spectrum of the correlated amplitude sequence.

2. Bandwidth can be reduced at the expense of noise immunity, a possibility implied by the capacity formula [Eq. (1)]. A typical way to do this is to use multilevel signals. For the same average signal power, use of more signal levels reduces immunity to noise.

3. Although pulses may overlap, they can be designed not to interfere with one another or to interfere in controlled ways. Intersymbol interference caused by linear channel distortion can be largely compensated for by further pulse shaping in the transmitter and receiver.

The designer must also see that timing information is present, that is, that there are frequent enough transitions between signal levels so that the receiver can sample the pulse amplitude in the strongest part of the pulse.

Assume, for the moment, binary pulses transmitted at symbol intervals of T seconds. The usual system model (Fig. 15) has a transmitter, a receiver, and a channel with noise and linear distortion. Figure 16 illustrates three popular line codes, or codings of input data into amplitude-modulated pulse trains. Rectangular-shaped pulses are shown for clarity. Binary antipodal signaling is perhaps the most straightforward but has energy at low frequencies, which is undesirable in many systems, and, in long sequences of data "0" or data "1," has a dc component and no signal transitions. Alternate mark inversion (AMI) coding eliminates the energy at low frequencies but still has the problems of no signal transitions, and in fact no en-

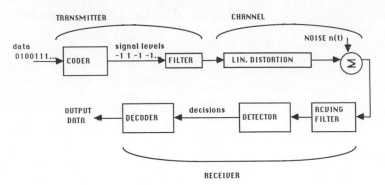

FIG. 15.   Baseband binary data communication system model.

ergy at all, during long sequences of data zeros. This problem is sometimes handled by replacing long sequences of zeros by special sequences with sufficient transitions. Manchester coding keeps energy on the line at all times and has at least one transition per symbol interval but pays a price in bandwidth.

A rectangular pulse—or any pulse wholly contained within one symbol interval of T seconds—has the virtue that in a pulse train, no pulse interferes with any other. A pulse constrained to a T-second interval, however, consumes more bandwidth than is necessary. The only requirement a pulse has to satisfy, to avoid intersymbol interference, is that all pulse samples at intervals T, except the one that carries the information, be zero. The Nyquist pulse (Fig. 17) can easily be proved to satisfy this condition with the minimum possible bandwidth of

$1/2T$ Hz, but since it is difficult to generate and work with a pulse with the abrupt *frequency* spectrum that this one has, a modem will compromise by using a pulse of slightly greater bandwidth and smooth frequency characteristic that retains the property of zero samples at symbol intervals. A frequency characteristic that is often used is the "raised cosine" of Fig. 17, for which the edge of the frequency characteristic is a cosine function rolling off in an additional fraction (50 or 100% in Fig. 17) of the minimum $1/2T$ Hz.

The detection of a signal level (for a modem, in the "decision device" of Fig. 14) is realized as a sampler followed by a circuit that decides which decision region the sample is in. If a binary antipodal signal has been perturbed by symmetrically random noise and distortion, the optimum (minimum probability of error) deci-

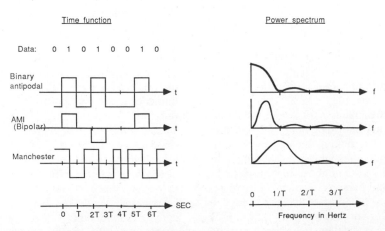

FIG. 16.   Binary antipodal, alternate mark inversion (AMI), and Manchester line signal codings (T is the symbol interval).

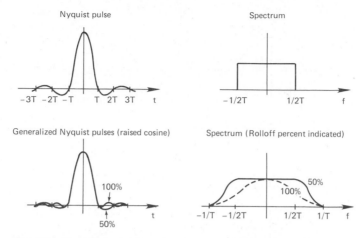

FIG. 17.   Band-limited pulses that do not contribute intersymbol interference.

sion regions are simply the sets of positive and negative numbers, as indicated in Fig. 18.

Pulses that do not quite satisfy the condition for no intersymbol interference can also be used. By allowing some intersymbol interference, the frequency spectrum can be further smoothed, with energy moved away from band edges. With a properly chosen coding of input data into pulse amplitudes, the original data stream can be recovered from received signal samples, which contain contributions from two or more pulses. Several of these pulses, called partial response because only part of a received pulse is detected in a single symbol interval, are illustrated in Fig. 19. The class 1 pulse in this figure was named a duobinary signal by its discoverer, and the class 4 pulse a modified duobinary signal. These pulse shapings are realized as the combination of filter shapings in transmitter and receiver.

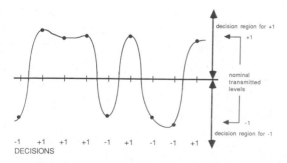

DECISIONS

FIG. 18.   Decision regions on symbol-interval samples of a binary antipodal pulse train received through a noisy, distorted channel.

Samples taken, at T-second (symbol) intervals, of a pulse train of binary partial response pulses will have more than two levels. Figure 20 illustrates a pulse train of class 4 partial response pulses. The samples taken at symbol intervals have three possible levels. For a fixed average power and symbol-by-symbol detection, there is less margin against noise here than in a pulse train using noninterfering pulses, and the bit error rate is higher. The degradation for the class 4 pulse is equivalent to a reduction in signal-to-noise ratio of 2.1 dB. This loss does not occur if detection is carried out as one large operation on the entire transmitted pulse sequence, rather than symbol-by-symbol, but sequence estimation techniques are computationally complex.

The built-in redundancy of partial response line signals does, however, help to detect errors, because some sample sequences are impossible unless noise causes an error. In Fig. 20, for example, a transmitted line signal value of +2 or of −2 will not be repeated two symbol intervals later in any sequence of pulse amplitudes. An error at the receiver in the decision for a given sample value may appear to propagate in later decisions, since the sample values are heavily correlated, but appropriate precoding operations on the input data stream will prevent this.

The selection of a pulse shape depends on the relative noise and bandwidth constraints in a given application. For high-speed modems, raised cosine pulses with rolloffs of 10 to 15% are often chosen. Furthermore, the pulse trains may be multileveled rather than just binary. In going to more levels, the data rate of a pulse–

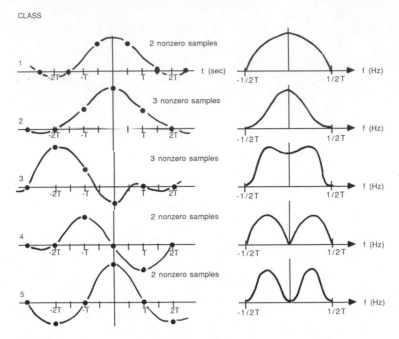

FIG. 19.   Partial response pulses, shown sampled at T-second intervals and their Fourier transforms.

amplitude-modulated (PAM) baseband line signal is increased with no change in the signal bandwidth, thus increasing spectral efficiency. The factor increase in data rate (or equivalently spectral efficiency) is log (number of amplitude levels). Figure 21 illustrates a four-leveled PAM data train. Of course, for a given average power, the spacing between levels decreases as the number of levels increases, so that immunity to noise is reduced and the error rate, for a given

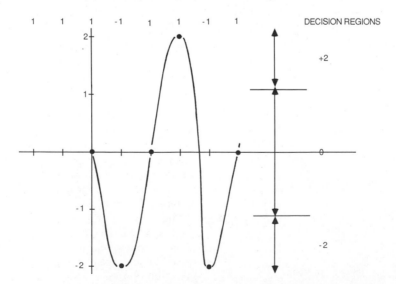

FIG. 20.   Typical transmitted line signal levels with use of class 4 partial response pulses.

33

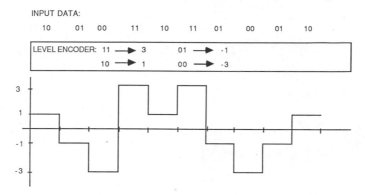

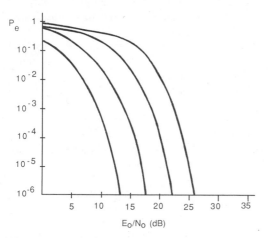

FIG. 21.   Coding of a binary data stream into a four-level PAM data train.

signal-to-noise ratio, is higher, as Fig. 22 shows.

The combined effects of noise and distortion can be represented visually on an oscilloscope screen in the eye pattern (Fig. 23). This pattern is the superposition of signals representing all possible pulse sequences just before the detector in the data receiver. For a binary system, the eye pattern indeed looks like an eye, which closes if noise and distortion are too large. The more open the eye, the greater is the margin of safety against errors. The best sampling time is where the eye is widely open and a small error in

FIG. 22.   Probability of bit error versus the ratio of energy per bit to noise power per hertz for PAM signals with different numbers L of levels. [Reproduced with permission from Lucky, R., Salz, J., and Weldon, N. (1968). "Principles of Data Communication." Copyright 1968 McGraw-Hill, New York.]

the sampling line does not lead to a significantly smaller margin of safety. Automatic adjustment circuits can be devised to adust the sampling time, sometimes in conjunction with a data scrambling system to guarantee an adequate number of data level transitions.

Eye closure at the sampling time can be quantified by an analytical expression such as mean-squared error (MSE), which is the expected value of the squared difference between a sample value and the signal level that should have been received if the channel were ideal. Most channel equalizers (described in Section V), which reduce intersymbol interference and thereby improve the eye pattern, are adapted by algorithms seeking to minimize the MSE.

Estimation of the probability of error when the channel characteristics are known is useful for system design but difficult because of the effects of intersymbol interference. Useful bounds on the probability of error have been derived.

V. Channel Equalization

For the baseband PAM communication model described in Section IV, and assuming that the channel characteristics are exactly known, a receiver can be designed that decides for that entire transmitted sequence which is most likely to have produced the observed received signal. The maximum likelihood sequence estimation (MLSE) receiver is impractical to implement directly, since the number of computations increases exponentially with the length of the sequence, which could contain millions of

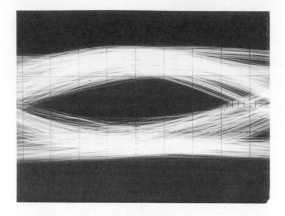

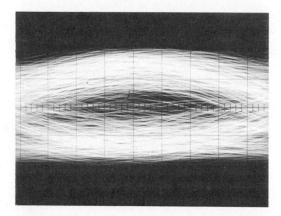

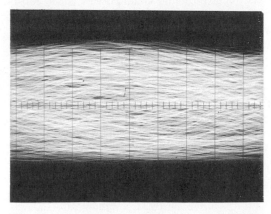

FIG. 23. Eye patterns of binary antipodal pulse trains, over one symbol interval. These illustrate an open eye, a barely open eye, and an eye closed from severe interference.

symbols. A dynamic programming technique known as the Viterbi algorithm, applicable to decoding a convolutionally encoded level sequence, reduces the complexity to (sometimes) manageable proportions by converting the de-

tection problem to that of optimizing individual transitions from one system "state" to the next.

The computational complexity is relatively low if the dispersion, or "memory," of the channel is small. The performance loss described for partial response signaling using symbol-by-symbol detection can be fully recovered by an MLSE detector. The Viterbi algorithm is a nonlinear receiver structure because it cannot be described by a linear filtering operation alone.

A suboptimal linear receiver structure is more commonly used, sometimes in conjunction with the Viterbi algorithm when channel trellis coding (Section VI) is included. It can be shown that the optimum linear receiver is a filter matched to the received signal pulse, followed by a transversal filter that forms the weighted sum of the voltages at symbol-interval taps on a delay line (Fig. 24). For a high-speed telephone channel modem, this "synchronous" equalizer typically has 32 taps.

The values of the tap weights are selected to reduce intersymbol interference, that is, shape the overall system impulse response so that all but one of its symbol-interval samples are very small, through some performance criterion. As with the Viterbi algorithm, the channel must be known exactly to make these assignments. The zero-forcing equalizer with $N + 1$ taps reshapes the overall signaling pulse so that the N samples closest to the central information-bearing sample are zero. The more robust minimum mean-squared error (MMSE) equalizer minimizes the expected squared error, at the equalizer output, between the sample value and the ideal signal level that it should be.

A channel equalizer can dramatically improve the performance of a communication system, compensating for both amplitude and delay distortion in the channel. The structure of Fig. 24 does, however, have a serious limitation. If there is a severe null in the channel transfer function (frequency spectrum), the equalizer's attempt to beef up the received pulse spectrum in that frequency range will amplify noise. In realizations based on processing of symbol-interval samples, an incorrect matched filter for the chosen sampling time (inevitable with the compromise bandpass filter used in practice) can actually cause such spectral nulls.

The partial solution is to use one structure, the fractionally spaced equalizer (FSE, Fig. 25), which is equivalent to the combination of matched filter and symbol-interval (synchronous) equalizer. The tap spacing must be small

36

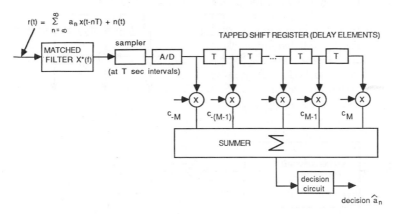

FIG. 24.   Optimum linear receiver (with properly selected tap weights) for the PAM baseband communication system of Fig. 16. The tapped delay line spans the dispersion of the channel. $r(t)$, Received signal; $X(f)$, transfer function from combined transmitter and channel shapings; T, symbol interval (sec); $\{c_M\}$, tap weights.

enough so that there is no aliasing distortion (frequency foldover) over the entire signal spectrum. For the telephone channel and symbol rates of the order of 2400/sec, a $T/2$ (1/4800-sec) spacing is adequate, with perhaps 64 to 128 taps. On severely phase distorted channels, the FSE may outperform the synchronous equalizer with correctly chosen timing by 2 or 3 dB. Furthermore, by removing the problem of matched filter variation with sampling time, the FSE significantly eases the timing recovery function of a modem.

A third type of equalizer, useful when postcursors, the nonzero symbol-interval samples *following* the desired information-bearing sample, are significant, is the decision-feedback equalizer (DFE, Fig. 26). The design presumption, as in the other equalizers, is that the channel pulse is known exactly, so that the postcursor values can be used as tap weights in the DFE to cancel the postcursor intersymbol interference entering the subtractor. Also, the signal improvement depends on the assumption that the decision data levels fed back through the delay line are almost always correct. When a decision error is made, performance can degrade for more than one symbol interval through error propagation.

Other equalizer structures exist as variations of these, some making preliminary data decisions that are used to improve the final decisions. Entirely different structures, such as

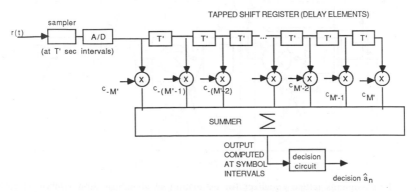

FIG. 25.   Fractionally spaced equalizer; T' is typically $T/2$ for telephone channel applications. For symbols, see Fig. 24.

which "tentative" decisions are made, fed through a nonlinear canceler analogous to a DFE, and the output of the nonlinear canceler subtracted from the filtered received signal can gain 1–3 dB in some channels.

VI. Passband Data Communication

To go through a telephone channel with the passband characteristics suggested in Fig. 9, a baseband information signal must be modulated onto a carrier waveform. The telephone channel is narrowband, compared with the information rates that users want to send through it, so that spectrally efficient modulation formats are needed. Linear modulation techniques, which code data streams into two PAM pulse trains impressed on sine and cosine carriers, respectively, achieve a frequency translation from baseband to passband that does not expand bandwidth. The spectral efficiency of multileveled PAM carries through to these "two-dimensional" modulated signals, as do the structures of channel equalizers and much of the analysis of their performance. The main additional feature is adaptive phase tracking circuitry that must function simultaneously with the adaptive channel equalizer.

Before describing the structure of a two-dimensional modem, already introduced in Fig. 14, it is worth noting that at lower data rates, for which spectral efficiency is not so critical, frequency-modulation techniques work well. FSK modems operating full duplex at 1200 bits/sec on dialed lines are widely used for personal computer communications. Separate half-channels, as shown in Fig. 7a, are used for each direction of transmission. In multileveled FSK, each level selects a different frequency, with the deviation ratio selected to achieve a desirable spectral shape (e.g., one without spikes). Unlike linear modulation, the bandwidth increases with the number of levels. FSK has the advantage of a constant signal (envelope) amplitude, which makes it robust against the nonlinearities of power-limiting devices in the channel, and if care is taken to make it phase continuous at the boundaries between symbols, the spectrum for binary FSK is only slightly wider than that for binary amplitude modulation.

Phase modulation also produces a constant-amplitude passband signal but is, like amplitude modulation, a linear format. Figure 29 shows the general forms of a modulator and demodulator for M-phase phase-shift keying (PSK), with the assumption that the carrier phase is known ex-

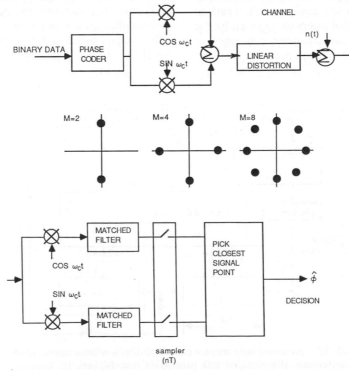

FIG. 29. PSK transmitter and receiver and signal constellations for two-, four-, and eight-phase systems (with rectangular pulse shape).

actly to the receiver, and the two-dimensional signal constellations for two-, four-, and eight-phase PSK. A two-dimensional signal constellation is the set of possible amplitude pairs, or signal points, for the cosine and sine carrier waveforms or equivalently, the set of magnitude and phase pairs. The constant-magnitude signal points of PSK shown in Fig. 29 correspond to codings of 1, 2, and 3 bits, respectively, per symbol. Gray coding, for which the codings for adjacent signal points on the circle differ in only one bit position, can minimize the cost of a symbol (phase decision) error.

The bit error probability increases rapidly with the number of phases above 4 and is especially vulnerable to phase jitter. Phase modulation is usually used with no more than eight signal points. In order to avoid the carrier recovery (local oscillator reference phase determination) problem, differential phase-shift keying (DPSK) may be used, in which data bits are coded into a change in phase rather than an absolute phase,

but at a price in error rate performance. Figure 30 shows bit error rates for PSK and DPSK.

Phase modulation is a two-dimensional signal format with the signal points constrained to lie on a circle. High spectral efficiency and good error-rate performance have been obtained with 16 or more signal points distributed more uniformly in the two-dimensional signal space. A more uniform distribution increases the minimum distance between signal points and thus the immunity to error caused by noise, phase jitter, and other impairments. The commercially significant quadrature amplitude modulation (QAM) and CCITT V.29 constellations are illustrated in Fig. 31.

The transmitter for data communications with two-dimensional signals is essentially the same as in Fig. 29, with the addition of pulse-shaping filters. The receiver structure, including a pass-band equalizer, is sketched in Fig. 32, an expansion of Fig. 14. It can be shown that it is equivalent to have the equalizer either precede or

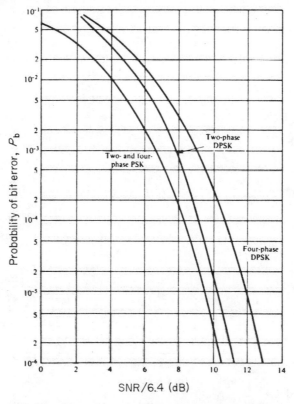

SNR/6.4 (dB)

FIG. 30. Bit error probabilities for binary- and four-phase PSK and DPSK. [Reproduced with permission from Proakis, J. G. (1983). "Digital Communications." Copyright 1983 McGraw-Hill, New York.]

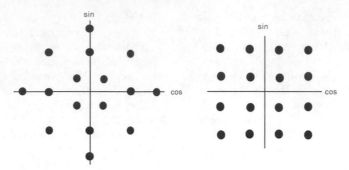

FIG. 31.   CCITT V.29 and QAM two-dimensional signal constellations.

follow the demodulator, but there is an advantage, for the phase-tracking demodulation system, to take the delay introduced by the equalizer out of that adjustment loop. Whether in passband or baseband format, the equalizer must be designed to process a complex analytical pair of signals differing by a 90° phase in each frequency component and in no other way, and has been shown to have the cross-coupled structure illustrated in Fig. 33, where the equalizer is shown following the demodulator for analytical convenience. The incoming signal is converted to a complex analytical signal by creating a new "imaginary" signal in a Hilbert filter, which is a 90° phase rotator. A complex variable notation is useful for analysis, and the filtering and updating operations of the cross-coupled equalizer are simply represented by

$$\tilde{\mathbf{u}}_n = \tilde{c}_n' \tilde{q}_n$$
$$\tilde{\mathbf{c}}_{n+1} = \tilde{\mathbf{c}}_n - \beta \tilde{e}_n \tilde{q}_n^* \tag{5}$$

where $\tilde{\mathbf{c}}_n$ is a complex tap weight vector, $\tilde{q}_n$ a complex tap voltage vector, $\tilde{u}_n$ the complex equalizer output voltage, and $\tilde{e}_n$ the equalizer output error. Figure 33 suggests that there is four times as much computational complexity in a cross-coupled equalizer as in a real baseband equalizer. More than a million 16-bit times 16-bit multiplications per second are required for a 64-tap FSE, and the decoder and phase tracker require additional capacity. Only in recent years have VLSI signal processing devices made this level of complexity possible in a moderate-cost modem.

Three major innovations have been developed for high-speed voiceband modems since the FSE. One is channel trellis coding, which convolutionally encodes the transmitted signal points so as to increase the minimum distance between the transmitted *sequences* of signal points, even though the minimum distance between signal points in the signal constellation is decreased in order to obtain the required coding redundancy. Increasing the minimum distance between signal point sequences reduces the probability of error, assuming that MLSE detection such as the Viterbi algorithm is used. Figure

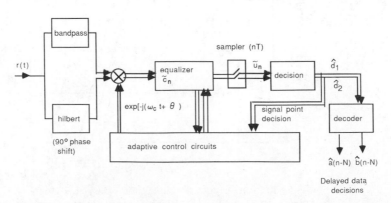

FIG. 32.   Receiver structure for two-dimensional signals. Complex variables simplify description and analysis.

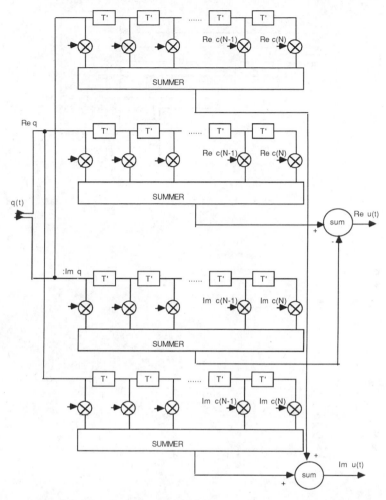

FIG. 33.   Cross-coupled equalizer structure of Fig. 32.

34 shows the convolutional encoder and the 32-point signal constellation of draft recommendation V.32 of the CCITT. This 32-point constellation is an alternative to the 16-point constellation used for 9600 bit/sec transmission in the absence of channel trellis coding. Another important element of channel trellis coding, the partitioning procedure for assignment of encoder outputs to signal points, is indicated by the bits-to-signal-point encodings under the signal points in the figure. An improvement in signal-to-noise ratio of several decibels has been demonstrated by using this technique on severely distorted channels.

The second innovation is echo cancellation, illustrated in Fig. 7. The hybrid coupler does not succeed in keeping all of the transmitted signal energy (near echo) out of the receiver, nor can it block the long-distance echoes that return from the distant end of the circuit. An adaptive transversal filter very much like a channel equalizer can be used to generate replicas of the near and distant echoes, and it can be driven by the data levels rather than the transmitted signal in order to minimize the computational complexity. Figure 35 illustrates the modem architecture. The bulk delay preceding the distant echo canceler has to be determined from a sounding of the channel or some equivalent adaptation procedure. An echo canceler can suppress as much as 60 dB of echo and make it possible to operate at rates up to 9600 bits/sec full duplex on dialed telephone lines.

As VLSI implementations come down in

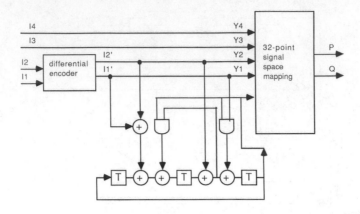

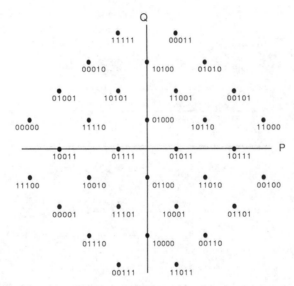

FIG. 34. Convolutional encoder and 32-point signal constellation for 9600 bit/sec data communication in CCITT recommendation V.32. Binary number under each signal point is data sequence coded into that point. © 1984 IEEE. [Reproduced with permission from Wei, L. F. (1984). *IEEE J. Selected Areas Commun.*]

price, 4800 bit/sec modems may supplant 1200 bit/sec modems for personal computer communications over telephone circuits. Echo cancellation has a second important application in the access line to the integrated services digital network. In combination with multilevel or partial-response line codes chosen to minimize bandwidth occupancy, echo cancellation transceivers operating at rates of the order of 150 kbit/sec full duplex can realize the two 64 kbit/sec and one 16 kbit/sec digital channels constituting the "basic" ISDN subscriber interface

on all but a few existing subscriber lines.

The third innovation is multisymbol line signal coding. By coding an input data train into pairs, or even triplets, of successive symbols, for example, coding 8 bits into the 256 possible signal points in a four-dimensional signal space corresponding to two symbol intervals in a 16-point QAM system, the minimum separation of signal points is increased and the immunity to noise is increased. A 1- to 3-dB signal-to-noise ratio advantage can be gained in this way.

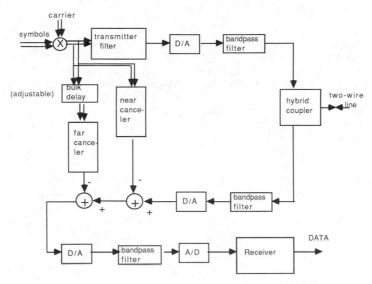

FIG. 35.   Structure for a modem using echo cancellation to operate full duplex on dialed telephone lines. [Reproduced with permission from Werner, J. J. (1984). *IEEE J. Selected Areas Commun.* **SAC-2**(5). © 1984 IEEE.]

VII. Future of Data Communication

The world has been moving to digital formats for many reasons. Equipment is more robust and easier to maintain, signal quality does not degrade with distance or number of tandem connections, and different media can be integrated in one transport, switching, and processing system. However, a vast investment has been made in the subscriber plant for voice communications, mostly in the copper twisted pair lines connecting subscribers with telephone offices. This plant will not be abandoned for a very long time, nor will every telephone office at the other end of a subscriber line be an entry node to the digital network. Voiceband data communications will continue to be of great importance, with user demand for higher data rates at lower cost to approach the service available to those large users or fortunate small users who *are* on digital circuits.

Even for digital subscribers line, that is, those with subscriber interfaces accepting digital data signals directly, data communications technology is necessary. A transceiver sends data to an receives data from the digital network node at the other end of the subscriber line. The 144 kbit/sec "basic" ISDN interface utilizes baseband pulse transmission and echo cancellation for simultaneous two-way transmission in the same bandwidth. Here, too, a demand can be expected for higher data rates at lower costs and

for reliable operation on marginal as well as on good-quality lines. Digital microwave carrier systems have already adopted many of the voiceband data communication techniques, and some are using QAM with a remarkable 256 signal points. Finally, even the fiber optic links of the evolving broadband network call for sophisticated modulation, equalization, and detection techniques to obtain the best possible performance for the lightwave channel. These are the reasons that voiceband data communication has had an impact beyond the conventional telephone circuit and has become a foundation technology of the information age.

BIBLIOGRAPHY

Biglieri, E., Gersho, A., Gitlin, R., and Lim, T. L. (1984). Adaptive Cancellation of Nonlinear Intersymbol Interference for Voiceband Data Transmission. *IEEE J. Selected Areas Commun.* **SAC-2**(5)

"Data Communication over the Telephone Network." (1985). CCITT Red Book, Vol. VIII, Fascicle VIII. 1." International Telecommunications Union, Geneva.

"Engineering and Operations in the Bell System." (1984). AT&T Bell Labs, Murray Hill, New Jersey.

Falconer, D. D. (1976). Jointly Adaptive Equalization and Carrier Recovery in Two-Dimensional Digital Communication Systems. *Bell Sys. Tech. J.* **55**(3).

Forney, G. D. (1973). The Viterbi Algorithm. *Proc. IEEE* **61**(3), 268–278.

Gallager, R. (1968). "Information Theory and Reliable Communication." Wiley, New York.

Gersho, A., and Lawrence, V. B. (1984). Multisymbol Signal Coding. *IEEE J. Selected Areas Commun.*

Gitlin, R., and Weinstein, S. (1981). Fractionally-Spaced Equalization: An Improved Digital Transversal Equalizer. *Bell Sys. Tech. J.* **60**(2).

Hayes, J., and Weinstein, S. (in press). "Data Communications," Plenum, New York.

Lender, A. (1966). The Duobinary Technique for High-Speed Data Transmission. *IEEE Trans. Commun. Electronics* March, 214–218.

Lucky, R., Salz, J., and Weldon, N. (1968). "Principles of Data Communication." McGraw-Hill, New York.

Proakis, J. G. (1983). "Digital Communications," McGraw-Hill, New York.

Ungerboeck, G. (1982). Channel Coding with Multilevel/Phase Signals. *IEEE Trans. Inf. Theor.* **IT28**(1),

Werner, J. J. (1984). An Echo-Cancellation Based 4800 bits/s Full-Duplex DDD Modem. *IEEE J. Selected Areas Commun.* **SAC-2**(5).

Widrow, B., and Stearns, S. D. (1985). "Adaptive Signal Processing." Prentice Hall, Englewood Cliffs, New Jersey.